de Gruyter Expositions in Mathematics 46

Groups of Prime Power Order

Volume 1

by

Yakov Berkovich

Walter de Gruyter · Berlin · New York

Author
Yakov Berkovich
Jerusalem str. 53, apt. 15
Afula 18251
Israel
E-Mail: berkov@math.haifa.ac.il

Mathematics Subject Classification 2000: 20-02, 20D15, 20E07

Key words: Finite p-group theory, counting of subgroups, regular p-groups, p-groups of maximal class, characterizations of p-groups, characters of p-groups, p-groups with large Schur multiplier and commutator subgroups, $(p-1)$-admissible Hall chains in normal subgroups, powerful p-groups, automorphisms of p-groups, p-groups having nonnormal subgroups only, Alperin's problem on abelian subgroups of small index.

∞ Printed on acid-free paper which falls within the guidelines
of the ANSI to ensure permanence and durability.

ISSN 0938-6572
ISBN 978-3-11-020418-6

Bibliographic information published by the Deutsche Nationalbibliothek

The Deutsche Nationalbibliothek lists this publication in the Deutsche Nationalbibliografie; detailed bibliographic data are available in the Internet at http://dnb.d-nb.de.

Typeset using the author's TeX files: Kay Dimler, Müncheberg.
Printing and binding: Hubert & Co. GmbH & Co. KG, Göttingen.
Cover design: Thomas Bonnie, Hamburg.

Contents

List of definitions and notations . . . ix

Foreword . . . xiv

Preface . . . xvii

Introduction . . . 1

§1 Groups with a cyclic subgroup of index p. Frattini subgroup. Varia . . . 22

§2 The class number, character degrees . . . 58

§3 Minimal classes . . . 69

§4 p-groups with cyclic Frattini subgroup . . . 73

§5 Hall's enumeration principle . . . 81

§6 q'-automorphisms of q-groups . . . 91

§7 Regular p-groups . . . 98

§8 Pyramidal p-groups . . . 109

§9 On p-groups of maximal class . . . 114

§10 On abelian subgroups of p-groups . . . 128

§11 On the power structure of a p-group . . . 146

§12 Counting theorems for p-groups of maximal class . . . 151

§13 Further counting theorems . . . 161

§14 Thompson's critical subgroup . . . 185

§15 Generators of p-groups . . . 189

§16 Classification of finite p-groups all of whose noncyclic subgroups are normal . . . 192

§17 Counting theorems for regular p-groups . . . 198

§18 Counting theorems for irregular p-groups . . . 202

§19 Some additional counting theorems . . . 215

§20 Groups with small abelian subgroups and partitions . . . 219

§21 On the Schur multiplier and the commutator subgroup . . . 222

§22 On characters of p-groups . . . 229

§23 On subgroups of given exponent . . . 242

§24 Hall's theorem on normal subgroups of given exponent . . . 246

§25 On the lattice of subgroups of a group . . . 256

§26 Powerful p-groups . . . 262

§27 p-groups with normal centralizers of all elements . . . 275

§28 p-groups with a uniqueness condition for nonnormal subgroups . . . 279

§29 On isoclinism . . . 285

§30 On p-groups with few nonabelian subgroups of order p^p and exponent p 289

§31 On p-groups with small p'-groups of operators . . . 301

§32 W. Gaschütz's and P. Schmid's theorems on p-automorphisms of p-groups 309

§33 Groups of order p^m with automorphisms of order p^{m-1}, p^{m-2} or p^{m-3} 314

§34 Nilpotent groups of automorphisms . . . 318

§35 Maximal abelian subgroups of p-groups . . . 326

§36 Short proofs of some basic characterization theorems of finite p-group theory . . . 333

§37 MacWilliams' theorem . . . 345

§38 p-groups with exactly two conjugate classes of subgroups of small orders and exponent $p > 2$ . . . 348

§39 Alperin's problem on abelian subgroups of small index . . . 351

§40 On breadth and class number of p-groups . . . 355

§41 Groups in which every two noncyclic subgroups of the same order have the same rank . . . 358

§42 On intersections of some subgroups . . . 362

§43 On 2-groups with few cyclic subgroups of given order . . . 365

§44 Some characterizations of metacyclic p-groups . . . 372

§45 A counting theorem for p-groups of odd order . . . 377

Appendix

A.1 The Hall–Petrescu formula . . . 379

A.2 Mann's proof of monomiality of p-groups . . . 383

A.3 Theorems of Isaacs on actions of groups . . . 385

A.4 Freiman's number-theoretical theorems . . . 393

A.5 Another proof of Theorem 5.4 . . . 399

A.6 On the order of p-groups of given derived length . . . 401

A.7 Relative indices of elements of p-groups . . . 405

A.8 p-groups with absolutely regular Frattini subgroup . . . 409

A.9 On characteristic subgroups of metacyclic groups . . . 412

A.10 On minimal characters of p-groups . . . 417

A.11 On sums of degrees of irreducible characters . . . 419

A.12 2-groups whose maximal cyclic subgroups of order > 2 are self-centralizing 422

A.13 Normalizers of Sylow p-subgroups of symmetric groups . . . 425

A.14 2-groups with an involution contained in only one subgroup of order 4 . 431

A.15 A criterion for a group to be nilpotent 433

Research problems and themes I 437

Author index 505

Subject index 507

List of definitions and notations

Set theory

$|M|$ is the cardinality of the set M (if G is a finite group, then $|G|$ is called its order).

$x \in M$ ($x \notin M$) means that x an element (not an element) of the set M. $N \subseteq M$ ($N \nsubseteq M$) means that N is (is not) a subset of the set M; moreover, if $M \neq N \subseteq M$ we write $N \subset M$.

$\varnothing$ is the empty set.

N is called a nontrivial subset of M, if $N \neq \varnothing$ and $N \subset M$. If $N \subset M$ we say that N is a proper subset of M.

$M \cap N$ is the intersection and $M \cup N$ the union of sets M and N. If M, N are sets, then $N - M = \{x \in N \mid x \notin M\}$ is the difference of N and M.

$\mathbb{Z}$ is the set (ring) of integers: $\mathbb{Z} = \{0, \pm 1, \pm 2, \dots\}$.

$\mathbb{N} = \{1, 2, 3, \dots\}$ is the set of natural numbers.

$\mathbb{Q}$ is the set (field) of rational numbers.

$\mathbb{R}$ is the set (field) of real numbers.

$\mathbb{C}$ is the set (field) of complex numbers.

Number theory and general algebra

p is always a prime number.

π is a set of primes; π' is the set of all primes not contained in π.

m, n, k, r, s are, as a rule, natural numbers.

$\pi(m)$ is the set of prime divisors of m; then m is a π-number if $\pi(m) \subseteq \pi$.

n_p is the p-part of n, n_π is the π-part of n.

(m, n) is the greatest common divisor of m and n.

$m \mid n$ should be read as: m divides n.

$m \nmid n$ should be read as: m does not divide n.

GF(p^m) is the finite field containing p^m elements.

$\mathbb{F}^*$ is the multiplicative group of a field $\mathbb{F}$.

$\mathcal{L}(G)$ is the lattice of all subgroups of a group G.

If $n = p_1^{\alpha_1} \dots p_k^{\alpha_k}$ is the standard prime decomposition of n, then $\lambda(n) = \sum_{i=1}^{k} \alpha_i$.

Groups

We consider only finite groups which are denoted, with a pair exceptions, by upper case Latin letters.

If G is a group, then $\pi(G) = \pi(|G|)$.

G is a p-group if $|G|$ is a power of p; G is a π-group if $\pi(G) \subseteq \pi$.

G is, as a rule, a finite p-group.

$H \leq G$ means that H is a subgroup of G.

$H < G$ means that $H \leq G$ and $H \neq G$ (in that case H is called a *proper* subgroup of G). $\{1\}$ denotes the group containing only one element.

H is a nontrivial subgroup of G if $\{1\} < H < G$.

H is a maximal subgroup of G if $H < G$ and it follows from $H \leq M < G$ that $H = M$.

$H \trianglelefteq G$ means that H is a normal subgroup of G; moreover, if, in addition, $H \neq G$ we write $H \lhd G$ and say that H is a proper normal subgroup of G. Expressions 'normal subgroup of G' and 'G-invariant subgroup' are synonyms.

$H \lhd G$ is called a nontrivial normal subgroup of G provided $H > \{1\}$.

H is a minimal normal subgroup of G if (a) $H \trianglelefteq G$; (b) $H > \{1\}$; (c) $N \lhd G$ and $N < H$ implies $N = \{1\}$. Thus, the group $\{1\}$ has no minimal normal subgroup.

G is simple if it is a minimal normal subgroup of G (so $|G| > 1$).

H is a maximal normal subgroup of G if $H < G$ and G/H is simple.

The subgroup generated by all minimal normal subgroups of G is called the *socle* of G and denoted by $\mathrm{Sc}(G)$. We put, by definition, $\mathrm{Sc}(\{1\}) = \{1\}$.

$\mathrm{N}_G(M) = \{x \in G \mid x^{-1}Mx = M\}$ is the normalizer of a subset M in G.

$\mathrm{C}_G(x)$ is the centralizer of an element x in G : $\mathrm{C}_G(x) = \{z \in G \mid zx = xz\}$.

$\mathrm{C}_G(M) = \bigcap_{x \in M} \mathrm{C}_G(x)$ is the centralizer of a subset M in G.

If $A \leq B$ and $A, B \trianglelefteq G$, then $\mathrm{C}_G(B/A) = H$, where $H/A = \mathrm{C}_{G/A}(B/A)$.

$A \operatorname{wr} B$ is the wreath product of the 'passive' group A and the transitive permutation group B (in what follows we assume that B is regular); B is called the active factor of the wreath product). Then the order of that group is $|A|^{|B|}|B|$.

$\operatorname{Aut}(G)$ is the group of automorphisms of G (the automorphism group of G).

$\operatorname{Inn}(G)$ is the group of all inner automorphisms of G.

$\operatorname{Out}(G) = \operatorname{Aut}(G)/\operatorname{Inn}(G)$, the outer automorphism group of G.

If $a, b \in G$, then $a^b = b^{-1}ab$.

An element $x \in G$ inverts a subgroup $H \leq G$ if $h^x = h^{-1}$ for all $h \in H$.

If $M \subseteq G$, then $\langle M \rangle = \langle x \mid x \in M \rangle$ is the subgroup of G generated by M.

$M^x = x^{-1}Mx = \{y^x \mid y \in M\}$ for $x \in G$ and $M \subseteq G$.

$[x, y] = x^{-1}y^{-1}xy = x^{-1}x^y$ is the *commutator* of elements x, y of G. If $M, N \subseteq G$ then $[M, N] = \langle [x, y] \mid x \in M, y \in N \rangle$ is a subgroup of G.

$o(x)$ is the order of an element x of G.

An element $x \in G$ is a π-element if $\pi(o(x)) \subseteq \pi$.

G is a π-group, if $\pi(G) \subseteq \pi$. Obviously, G is a π-group if and only if all of its elements are π-elements.

G' is the subgroup generated by all commutators $[x, y]$, $x, y \in G$ (i.e., $G' = [G, G]$), $G^{(2)} = [G', G'] = G'' = (G')'$, $G^{(3)} = [G'', G''] = (G'')'$ and so on. G' is called the *commutator* (or *derived*) subgroup of G.

$\mathrm{Z}(G) = \bigcap_{x \in G} \mathrm{C}_G(x)$ is the center of G.

$\mathrm{Z}_i(G)$ is the i-th member of the upper central series of G; in particular, $\mathrm{Z}_0(G) = \{1\}$, $\mathrm{Z}_1(G) = \mathrm{Z}(G)$.

$\mathrm{K}_i(G)$ is the i-th member of the lower central series of G; in particular, $\mathrm{K}_2(G) = G'$. We have $\mathrm{K}_i(G) = [G, \dots, G]$ ($i \geq 1$ times). We set $\mathrm{K}_1(G) = G$.

If G is nonabelian, then $\eta(G)/\mathrm{K}_3(G) = \mathrm{Z}(G/\mathrm{K}_3(G))$.

$\mathcal{M}(G) = \langle x \in G \mid \mathrm{C}_G(x) = \mathrm{C}_G(x^p) \rangle$ is the Mann subgroup of a p-group G.

$\mathrm{Syl}_p(G)$ is the set of p-Sylow subgroups of an arbitrary finite group G.

S_n is the *symmetric* group of degree n.

A_n is the *alternating* group of degree n

Σ_{p^n} is a Sylow p-subgroup of S_{p^n}.

$\mathrm{GL}(n, F)$ is the set of all nonsingular $n \times n$ matrices with entries in a field F, the n-dimensional *general linear* group over F, $\mathrm{SL}(n, F) = \{A \in \mathrm{GL}(n, F) \mid \det(A) = 1 \in F\}$, the n-dimensional *special linear* group over F.

If $H \leq G$, then $H_G = \bigcap_{x \in G} x^{-1}Hx$ is the *core* of the subgroup H in G and $H^G = \bigcap_{H \leq N \trianglelefteq G} N$ is the *normal closure* or *normal hull* of H in G. Obviously, $H_G \trianglelefteq G$.

If G is a p-group, then $p^{b(x)} = |G : C_G(x)|$; $b(x)$ is said to be the *breadth* of $x \in G$, where G is a p-group; $b(G) = \max\{b(x) \mid x \in G\}$ is the *breadth* of G.

$\Phi(G)$ is the Frattini subgroup of G (= the intersection of all maximal subgroups of G), $\Phi(\{1\}) = \{1\}$, $p^{d(G)} = |G : \Phi(G)|$.

$\Gamma_i = \{H < G \mid \Phi(G) \leq H, |G : H| = p^i\}, i = 1, \dots, d(G)$, where $G > \{1\}$.

If $H < G$, then $\Gamma_1(H)$ is the set of all maximal subgroups of H.

$\exp(G)$ is the exponent of G (the least common multiple of the orders of elements of G). If G is a p-group, then $\exp(G) = \max\{o(x) \mid x \in G\}$.

$k(G)$ is the number of conjugacy classes of G (= G-classes), the class number of G.

K_x is the G-class containing an element x (sometimes we also write $ccl_G(x)$).

C_m is the cyclic group of order m.

G^m is the direct product of m copies of a group G.

$A \times B$ is the direct product of groups A and B.

$A * B$ is a central product of groups A and B, i.e., $A * B = AB$ with $[A, B] = \{1\}$.

$E_{p^m} = C_p^m$ is the elementary abelian group of order p^m. G is an elementary abelian p-group if and only if it is a p-group $> \{1\}$ and G coincides with its socle. Next, $\{1\}$ is elementary abelian for each prime p.

A group G is said to be *homocyclic* if it is a direct product of isomorphic cyclic subgroups (obviously, elementary abelian p-groups are homocyclic).

$ES(m, p)$ is an *extraspecial* group of order p^{1+2m} (a p-group G is said to be extraspecial if $G' = \Phi(G) = Z(G)$ is of order p). Note that for each $m \in \mathbb{N}$, there are exactly two nonisomorphic extraspecial groups of order p^{2m+1}.

$S(p^3)$ is a nonabelian group of order p^3 and exponent $p > 2$.

A *special* p-group is a nonabelian p-group G such that $G' = \Phi(G) = Z(G)$ is elementary abelian. Direct products of extraspecial p-groups are special.

D_{2m} is the *dihedral* group of order $2m$, $m > 2$. Some authors consider E_{2^2} as the dihedral group D_4.

Q_{2^m} is the *generalized quaternion* group of order $2^m \geq 2^3$.

SD_{2^m} is the *semidihedral group* of order $2^m \geq 2^4$.

M_{p^m} is a nonabelian p-group containing exactly p cyclic subgroups of index p.

$\mathrm{cl}(G)$ is the *nilpotence class* of a p-group G.

$\mathrm{dl}(G)$ is the *derived length* of a p-group G.

$\mathrm{CL}(G)$ is the set of all G-classes.

A p-group of *maximal class* is a nonabelian group G of order p^m with $\mathrm{cl}(G) = m-1$.

$\Omega_m(G) = \langle x \in G \mid o(x) \leq p^m \rangle$, $\Omega_m^*(G) = \langle x \in G \mid o(x) = p^m \rangle$ and $\mho_m(G) = \langle x^{p^m} \mid x \in G \rangle$.

A p-group is absolutely regular if $|G/\mho_1(G)| < p^p$.

A p-group is *thin* if it is either absolutely regular or of maximal class.

$G = A \cdot B$ is a *semidirect product* with *kernel* B and *complement* A.

A group G is an extension of $N \trianglelefteq G$ by a group H if $G/N \cong H$. A group G splits over N if $G = H \cdot N$ with $H \leq G$ and $H \cap N = \{1\}$ (in that case, G is a semidirect product of H and N with kernel N).

$H^\# = H - \{e_H\}$, where e_H is the identity element of the group H. If $M \subseteq G$, then $M^\# = M - \{e_G\}$.

An automorphism α of G is *regular* (= *fixed-point-free*) if it induces a regular permutation on $G^\#$ (a permutation is said to be *regular* if it has no fixed points).

An *involution* is an element of order 2 in a group.

A *section* of a group G is an epimorphic image of some subgroup of G.

If $F = \mathrm{GF}(p^n)$, then we write $\mathrm{GL}(m, p^n), \mathrm{SL}(m, p^n), \dots$ instead of $\mathrm{GL}(m, F)$, $\mathrm{SL}(m, F), \dots$.

$c_n(G)$ is the number of cyclic subgroups of order p^n in a p-group G.

$s_n(G)$ is the number of subgroups of order p^n in a p-group G.

$e_n(G)$ is the number of subgroups of order p^n and exponent p in G.

$\mathcal{A}_n$-group is a p-group G all of whose subgroups of index p^n are abelian but G contains a nonabelian subgroup of index p^{n-1}. In particular, $\mathcal{A}_1$-group is a minimal nonabelian p-group for some p.

$\alpha_n(G)$ is the number of $\mathcal{A}_n$-subgroups in a p-group G.

Characters and representations

$\mathrm{Irr}(G)$ is the set of all *irreducible* characters of G over $\mathbb{C}$.

A character of degree 1 is said to be *linear*.

$\mathrm{Lin}(G)$ is the set of all *linear* characters of G (obviously, $\mathrm{Lin}(G) \subseteq \mathrm{Irr}(G)$).

$\mathrm{Irr}_1(G) = \mathrm{Irr}(G) - \mathrm{Lin}(G)$ is the set of all *nonlinear* irreducible characters of G; $\mathrm{n}(G) = |\mathrm{Irr}_1(G)|$.

$\chi(1)$ is the *degree* of a character χ of G,

χ_H is the *restriction* of a character χ of G to $H \leq G$.

χ^G is the character of G induced from the character χ of some subgroup of G.

$\bar{\chi}$ is a character of G defined as follows: $\bar{\chi}(x) = \overline{\chi(x)}$ (here $\bar{w}$ is the complex conjugate of $w \in \mathbb{C}$).

$\mathrm{Irr}(\chi)$ is the set of irreducible constituents of a character χ of G.

If χ is a character of G, then $\ker(\chi) = \{x \in G \mid \chi(x) = \chi(1)\}$ is the *kernel* of a character χ.

$\mathrm{Z}(\chi) = \{x \in G \mid |\chi(x)| = \chi(1)\}$ is the *quasikernel* of χ.

If $N \trianglelefteq G$, then $\mathrm{Irr}(G \mid N) = \{\chi \in \mathrm{Irr}(G) \mid N \not\leq \ker(\chi)\}$.

$\langle \chi, \tau \rangle = |G|^{-1} \sum_{x \in G} \chi(x)\tau(x^{-1})$ is the *inner product* of characters χ and τ of G.

$\mathrm{I}_G(\phi) = \langle x \in G \mid \phi^x = \phi \rangle$ is the *inertia subgroup* of $\phi \in \mathrm{Irr}(H)$ in G, where $H \lhd G$.

1_G is the *principal character* of G ($1_G(x) = 1$ for all $x \in G$).

$\mathrm{M}(G)$ is the *Schur multiplier* of G.

$\mathrm{cd}(G) = \{\chi(1) \mid \chi \in \mathrm{Irr}(G)\}$.

$\mathrm{mc}(G) = \mathrm{k}(G)/|G|$ is the *measure of commutativity* of G.

$\mathrm{T}(G) = \sum_{\chi \in \mathrm{Irr}(G)} \chi(1)$, $\mathrm{f}(G) = \mathrm{T}(G)/|G|$.

Foreword

After the successful classification of finite "quasithin" simple groups (M. Aschbacher and S. Smith, 2003), the classification of finite simple groups is finally (after 22 years of delay!) completed. Since the theory of finite solvable groups is developed to a satisfactory level, there remains to classify finite p-groups.

But the classification of finite p-groups in the classical sense is not even possible. The reason is that a finite p-group has "too many" normal subgroups and consequently there is an extremely large number of nonisomorphic p-groups of a given fixed order. For example, there are 267 nonisomorphic groups of order 2^6. P. Hall has replaced the concept of "isomorphism" with "isoclinism" in order to get larger families of p-groups and then he has tried to classify p-groups up to isoclinism. But this idea did not have much success (apart from p-groups of small order).

In this book it is proposed another approach to finite p-groups. We classify p-groups with various properties in such a way that sufficiently large sets of p-groups are covered. In addition, we try to choose these properties so that they cover in some way all finite p-groups. For example, we study "regular p-groups" and "irregular p-groups", p-groups with "small" abelian subgroups and p-groups with "large" abelian subgroups, modular p-groups and nonmodular p-groups, etc. Of course, the class of nonmodular p-groups is too large and so we determine (at least) all minimal nonmodular p-groups. A specialty of this book are many "counting theorems" as an important application of the enumeration principle of P. Hall.

Of special interest are finite 2-groups. In fact, if G is a nonabelian finite simple group and the structure of its Sylow 2-subgroup P is known, then the structure of G is almost uniquely determined. Consequently, various structural results on 2-groups simplify considerably the classification process of finite simple groups. Therefore, we devote a large part of the book to finite 2-groups.

This book can be used as a textbook for graduate students and as a reference book for research workers in finite p-group theory and to my knowledge it is the first book of that kind.

There are numerous exercises in this book with various degrees of difficulty. For difficult exercises the solutions are also given. In the section "Research problems and themes" there are 1400 open questions concerning p-groups. A large number of open questions are also formulated in the main text. Hence in the book are presented many problems which wait for the future research workers in this very rich and interesting field.

The book splits in two parts, an elementary one (Volume 1) and more advanced (Volume 2). This makes the book suitable for students, teachers and researchers.

A tremendous effort was taken to make the exposition as simple as possible with numerous new proofs of famous theorems which are much easier than their original proofs and this makes the book well readable. An enormous amount of original literature was analyzed in order to select for the book the most significant contributions from recent research. Most of the book content consists of results originally appeared during recent 15 years. For the first time many p-groups have been classified with given structures of its minimal nonabelian subgroups. This shows that the elementary methods have still a great potential.

The book is self-contained and apart from some applications of the character theory, it is completely elementary. To the reader with some knowledge of the rudiments of group theory and elementary algebra, this book should present no difficulty.

Zvonimir Janko (Mathematical Institute, Heidelberg University)

Preface

Let G be a finite group, let a prime p divide $|G|$, $|G| = p^\alpha \cdot n$, where $p \nmid n$. By Sylow's theorem, G has a subgroup P of order p^α; P is called a *Sylow p-subgroup* of G. By the same theorem, all maximal p-subgroups are conjugate in G (and so they are Sylow subgroups) and their number is congruent to $1 \pmod{p}$. The structure of finite groups is closely related to the properties and the embedding their Sylow p-subgroups, and there are many problems which can be solved only by a detailed examination of the relevant p-groups. The history of finite group theory shows that the investigation of p-subgroups is one of its most powerful method. Finite p-groups are ideal objects for combinatorial and cohomological investigations.

In the study of finite p-groups the main difficulty lies in the fact that the number of such groups is very large; for example, there are exactly 267 nonisomorphic groups of order 2^6 (P. Hall and J. Senior; see [HS]), 2328 groups of order 2^7, 56092 groups of order 2^8 (E. A. O'Brien), 10494213 groups of order 2^9 (B. Eick and E. A. O'Brien), 49487365422 groups of order 2^{10} (H. U. Besche, B. Eick and E. A. O'Brien), 504 groups of order 3^6. Therefore, it is very difficult to find nontrivial properties of almost all p-groups. (The following results are most general properties of p-groups: nilpotence, monomiality, Burnside's basis theorem, counting theorems of Sylow, Miller and Kulakoff.) So it is natural to seek common properties for sufficiently large sets of p-groups.

In this elementary book we will prove a number of deep theorems on finite p-groups.

Some basic properties of finite p-groups were proved by Sylow, Frobenius and Burnside. But namely Philip Hall (1904–1982) laid the foundations of modern p-group theory in three fundamental papers [Hal1, Hal2, Hal3] (note that these are the only his papers devoted to finite p-groups; at the time of publication of the last of these papers he was 36). I consider these papers as No's 1, 2 and 3 in finite p-group theory. The first of these papers presents the detailed investigation of regular p-groups, a wide subclass of p-groups in the sense that for given n there are a finite number of primes p such that there exist irregular p-groups of order p^n. In the second paper Hall proves a number of deep properties of irregular p-groups and establishes some strong regularity criteria. All modern approaches to classification of finite p-groups are based on the third Hall's paper. Thus, Hall transformed p-group theory from a collection of miscellaneous facts and results into an organized field and the very essential part of finite group theory.

Undoubtedly, Norman Blackburn's papers [Bla3] and [Bla5] are the most outstanding achievements in p-group theory after Hall. In those papers Blackburn studied very important p-groups of maximal class, which, as a rule, are irregular, and deduced from obtained results a number of deep properties of irregular p-groups. He also investigated and characterized a number of important classes of p-groups. Most of investigations in post Hall era of finite p-group theory develop Blackburn's ideas.

Avinoam Mann in his important paper [Man6] simplified the proofs of a number of basic theorems concerning p-groups of maximal class and also contributed significantly in the theory of regular p-groups in his subsequent papers (see [Man2–Man5]) and obtained a lot of important results about characters of p-groups (see [Man12]). Next, in the seminal paper [LubM], he and Lubotzky introduced and studied so called powerful p-groups – the class groups which obtained wide applications in finite p-group theory and in the theory of analytic pro-p-groups.

Recently, after 2000, Zvonimir Janko wrote a long series important papers (see Bibliography) devoted, in most cases, to 2-groups which I consider as highest achievements in the theory of 2-groups, the most prosperous part of p-group theory. Exposition of main results of these authors is the main aim of the present book.

We prove a number of important properties of regular p-groups and p-groups of maximal class; our exposition is based on mentioned above papers of Hall, Blackburn, and Mann. These results are central. Next, we prove almost all significant counting theorems that are known up to date; see §§1, 5, 10, 12, 13, 17–19, 36, 37 (the book contains all the material which is necessary to understand the proofs of those theorems). In some places we use §§4.3, 4.4 of Suzuki's brilliant book [Suz] essentially (see §§7, 29, and Appendix 1 and Appendix 18 in Volume 2). Most proofs are new. We assume that the reader is familiar with the basic facts of finite group theory and character theory. To do the book more or less intelligible even to non-specialists, we omitted the proof of such difficult result as Zassenhaus commutator identity; for its proof see [Hup, §3.9]. Apart of this, the book is self-contained. The Hall–Petrescu identity is presented in Appendix 1 and its proof is taken from [Suz, §4.3]. Since our book is entirely elementary, such notions as varieties and Lie algebras do not appear in what follows. Hence, we have to omit a number of important topics. We did not intend to give an encyclopedic exposition of the subject.

We omit or consider briefly a number of important topics such as Sylow p-subgroups of important groups, coclass and breadth of p-groups, the Burnside problem and so on. The forthcoming book "Finite p-Groups" by A. Mann must fill some of these gaps. Our consideration of the Schur multiplier is fairly fragmental (we recommend to the interested reader Karpilovsky's book [Kar], giving the encyclopedic presentation of this matter; for more elementary exposition see [BZ, Chapter 6]). There is an excellent exposition of p-group theory in Chapters 3 and 8 of three volume book by Huppert and Blackburn. However all mentioned books have small intersections with our one and not so elementary.

To the reader who has some knowledge of the rudiments of combinatorics, finite groups, elementary algebra and number theory, this book should present no difficulty.

All groups in this book are finite and, as a rule, have prime power order.

Introduction and Sec. 1 contain some preparatory material. The starting point for many our considerations is fundamental Frobenius' theorem on the number of solutions to $x^n = 1$ in a group (the nice proof of this theorem, due to I. M. Isaacs and G. R. Robinson [IsR], is presented). The important Fitting's lemma on the class of product of two normal nilpotent subgroups is proved. In Sec. 1 we prove a number of results which important in what follows (Lemma 1.1, Theorems 1.2, 1.10, 1.17 and so on). In Sec. 2, the nice proof of Hall's expression for the class number of a p-group, due to Mann, is presented. In Sec. 4 the description of p-groups with cyclic Frattini subgroup is given. The proof is based essentially on counting theorems from Sec. 1.

In Sec. 5 a number of elementary counting theorems for p-groups is proved. As a rule, the proofs are based on Hall's enumeration principle – Theorem 5.2 or his variants. We present enumeration principle free proofs of some strong counting theorems. We also present the new enumeration principle and prove with its help nice Y. Fan's result [Fan] (see Theorem 5.17). Most of the material of this section appears in the book form at the first time. In Sec. 6, we prove theorems like Maschke's. Namely, we show that if a p'-group X acts on the abelian p-group P and R is an X-invariant subgroup of exponent p in P, then $P = S \times S_1$, where S, S_1 are X-invariant and $\Omega_1(S) = R$. In Sec. 7 some basic facts on regular p-groups are proven. As we have noticed, our exposition follows closely to [Suz, Chapter 4] and Mann [Man2–5].

In Sec. 9 the basic properties of p-groups of maximal class are proved. A number of proofs is due to Mann. The main result of Sec. 10 is Theorem 10.1, a natural generalization of Alperin's theorem on centralizers of normal abelian subgroups [Alp3]. As a consequence, we prove assertions on the number of elementary abelian subgroups of orders p^3 and p^4 in p-groups of odd order; for further results and another approach, see [KonJ, JonK]. The main results of Sec. 11 are due to Mann; the power structure of p-groups is investigated there in detail. The p-groups satisfying certain of the basic properties of regular groups, are studied. On this way, some interesting new criteria of regularity are proven. These results we do not use in what follows (however, groups introduced in that section, play important role in §88).

In Sec. 12 we prove a number of counting theorems for p-groups of maximal class. However, the most important result of that section is Blackburn's fundamental Theorem 12.1 on p-groups without normal subgroups of order p^p and exponent p. In Sec. 13 most important counting theorems are presented (Theorems 13.2, 13.5 and 13.6). In Sec. 13 we prove a number of counting theorems of new type (see, for example, Theorem 13.18).

In Sec. 15 we prove nice theorems of Thompson and Mann. For example, Thompson's theorem asserts, that if G is a p-group, $p > 2$, such that $\Omega_1(G) \leq \mathrm{Z}(G)$, then $\mathrm{d}(G) \leq \mathrm{d}(\Omega_1(G))$. According to theorem of Mann, if G is a 2-group such that $\Omega_2(G) \leq \mathrm{Z}(G)$, then $\mathrm{d}(G) \leq \mathrm{d}(\Omega_2(G))$.

In Sec. 16 the p-groups all of whose nonnormal subgroups are cyclic, are classified. A number of counting theorems is proved in Sec. 17, 18.

In Sec. 24 we offer a new proof of Hall's theorem on Hall chains in normal subgroups (further development of this theme see in Sec. 88 from Volume 2).

In Sec. 26 we prove basic properties of powerful p-groups.

In Sec. 31 we consider the p-groups with small p'-groups of operators. Sec. 32–34 are devoted to automorphisms of p-groups.

In Sec. 36 we offer short proofs of a number of classical characterization theorems (most of them were proved in previous sections). Part of these proofs is based on Blackburns characterization of metacyclic p-groups (see Theorem 36.1).

In Appendix 1, we prove the result of fundamental importance — the Hall–Petrescu formula. Other appendices supplement the main text.

More detailed information about topics considered in the book, see in Contents and Subject Index.

We inserted in the text numerous exercises of varied degrees of difficulty, and they constitute its essential part. Many of them are given with hints or full solutions. Some exercises are, in fact, theorems or open questions.

Both parts of the book are concluded by comprehensive lists of problems which I began to write more than 25 years ago. At least 50 problems from these lists were solved mainly, by Janko (almost all his solutions are presented in Volume 2). There are two comprehensive lists of unsolved problems with interesting comments published by Mann [Man20] and Shalev [Sha5]. These lists and our one do not overlap.

The bibliography includes a number of the most important papers devoted to topics considered in this book. For more comprehensive bibliography, see Internet (in particular, MathSciNet).

The list of most important notations and definitions follows Contents. Our definitions and notations, as a rule, are standard.

I am indebted to Avinoam Mann for numerous useful discussions and help. The correspondence with Martin Isaacs allowed me to acquaint the reader with a number of his old and new results. Moreover, Mann and Isaacs familiarized me with a number of their papers prior of publication. I am especially indebted to Zvonimir Janko and Noboru Ito. Janko helped me generously in checking the whole text (some places he read many times); he also wrote the Foreword, Sections 16, 27, 28, 35, subsection 2° in Section 26, Theorem 34.8 and Appendices 12, 14. Note that Janko is a coauthor of Volume 2 of this book. Ito carefully read the whole this part and all appendices from Volume 2 and made a lot of corrections and useful suggestions. Lev Kazarin helped me with Sec. 22 (and also with Sec. 46, 63 and 65 from Volume 2). The first hundred pages of this Volume were read by M. Y. Xu (Beijing University), and he sent me the lists of misprints and suggestions. I am also indebted to Gregory Freiman, Marcel Herzog (both at Tel-Aviv University), Moshe Roitman, who also wrote Appendix 7, and Izu Vaisman (both at University of Haifa) for help and support.

I dedicate this volume to the memory of my parents Sarah (1916–1983) and Gilya (1911–1999) and my friends Grisha Karpilovsky (1940–1997) and Emanuel Zhmud (1918–2007).

Introduction

In this section we prove or formulate some basic results of p-group theory and also consider some main questions which we shall treat in what follows in greater detail. It is fairly difficult to establish the exact authorship of some basic results.

Lemma 1. *Let $a \in G$ be of order $g = mn$, where $(m,n) = 1$. Then $a = a_m a_n$, where $o(a_m) = m$, $o(a_n) = n$ and a_m, a_n are powers of a. If $a = bc = cb$, where $o(b) = m$, $o(c) = n$, then $b = a_m$, $c = a_n$.*

Proof. It follows from $(m,n) = 1$ that there exist $x, y \in \mathbb{Z}$ such that $mx + ny = 1$. Set $a_m = a^{ny}$, $a_n = a^{mx}$; then $a_m a_n = a_n a_m = a$. We have $(x, y) = 1 = (x, n) = (y, m)$. If $o(a_m) = m_1$, then $1 = (a_m)^{m_1} = a^{nm_1 y}$ so mn divides $m_1 ny$ and hence m divides m_1 since $(m, ny) = 1$. As $(a_m)^m = a^{mny} = 1$, $o(a_m) = m_1$ divides m. Thus, $m_1 = m$ so $o(a_m) = m$ and, similarly, $o(a_n) = n$.

Assume that $a = bc = cb$, where $b, c \in G$ and $o(b) = m$, $o(c) = n$. We claim that $b = a_m$ and $c = a_n$. We have $(a_n)^m = (a_m a_n)^m = a^m = (bc)^m = b^m c^m = c^m$. Let $x \in \mathbb{Z}$ be such that $mx \equiv 1 \pmod n$; then $a_n = (a_n)^{mx} = ((a_n)^m)^x = (c^m)^x = c^{mx} = c^{mx+ny} = c$ since $o(c) = n$. It follows that $a_m a_n = a = bc = ba_n$ so $b = a_m$. □

Lemma 2 (Cauchy). *Let G be an abelian group. If a prime p divides $|G|$, then G has an element of order p.*

Proof. Suppose that the lemma is true for all proper subgroups of G. One may assume that G is not cyclic. Then G has two different maximal subgroups A and B; $G = AB$ and $|G| = \frac{|A|\cdot|B|}{|A\cap B|}$, by the product formula (see below), so p divides either $|A|$ or $|B|$, and now the result follows by induction. □

Lemma 2 is a corollary of the following basic result — Sylow's theorem: If G is a group of order $p^a m$, where p is a prime, $a, m \in \mathbb{N}$ and $p \nmid m$, then G has a subgroup of order p^a. Below Wielandt's proof of this theorem follows. Let $\mathcal{M}$ be the set of p^a-element subsets of G; then $|\mathcal{M}| = \binom{p^a m}{p^a}$ and $p \nmid |\mathcal{M}|$. Let $\mathcal{M} \times G \to \mathcal{M}$ be the action of G on $\mathcal{M}$: $(M, g) \mapsto Mg$. Then $\mathcal{M} = \mathcal{M}_1 \cup \dots \cup \mathcal{M}_r$ is a partition of $\mathcal{M}$ in G-orbits. Since $p \nmid |\mathcal{M}|$, one may assume that $p \nmid |\mathcal{M}_1|$. Take $M \in \mathcal{M}_1$ and let H be the G-stabilizer of the "point" M. Then $|\mathcal{M}_1| = |G : H|$, and so p^a divides $|H|$, $|H| \geq p^a$. Since $MH = M$, M is a union of left cosets of H. However, $|M| = p^a \geq |H|$ so we get $M = H$, as desired.

Let G be abelian and $\pi(G) = \{p_1, \dots, p_s\}$. For $p \in \pi(G)$, let $G_p = \{x \in G \mid \pi(o(x)) = \{p\}\}$. Then $G_p \leq G$ and $|G_p| = |G|_p$, i.e., $G_p \in \mathrm{Syl}_p(G)$. Thus,

Lemma 3. *If G is abelian and $\pi(G) = \{p_1, \dots, p_s\}$, then $G = G_{p_1} \times \dots \times G_{p_s}$.*

If all $\neq 1$ elements of an abelian group G have order p; then $|G| = p^m$ for some $m \in \mathbb{N}$. In this case, G is a direct product of m groups of order p (check!); we say that G is an *elementary abelian* p-group. Two elementary abelian p-groups of the same order p^m are isomorphic; we denote such group by E_{p^m} and call it *elementary abelian p-group*. It is easy to show that every subgroup is complemented in G and $|\mathrm{Aut}(G)| = (p^m - 1) \dots (p^m - p^{m-1})$ (check!). Next, one can consider G as an m-dimensional vector space over the finite field $\mathrm{GF}(p)$, the Galois field with p elements. It follows that $\mathrm{Aut}(G) \cong \mathrm{GL}(m, p)$, the *$m$-dimensional general linear group* over $\mathrm{GF}(p)$.

Given an abelian p-group $G > \{1\}$ and $n \in \mathbb{N}$, write

$$\Omega_n(G) = \{x \in G \mid x^{p^n} = 1\}, \ \mho_n(G) = \{x^{p^n} \mid x \in G\}.$$

Then $\Omega_n(G), \mho_n(G)$ are characteristic subgroups of G and

$$\exp(\Omega_n(G)) = \exp(G/\mho_n(G)) \le p^n, \ \Omega_n(G) \cong G/\mho_n(G)$$

(this follows from Lemma 4(c), below), and $\exp(G/\Omega_1(G)) < \exp(G)$.

Lemma 4. *Let G be an abelian p-group.*

(a) If G has only one subgroup of order p, it is cyclic.

(b) If Z is a cyclic subgroup of G of maximal order, it is a direct factor of G so G is a direct product of cyclic subgroups.

(c) If $G = Z_1 \times \dots \times Z_k$, where $Z_1, \dots, Z_k$ are cyclic, then $\Omega_n(G) = \Omega_n(Z_1) \times \dots \times \Omega_n(Z_k)$, $\mho_n(G) = \mho_n(Z_1) \times \dots \times \mho_n(Z_k)$, $G/\mho_n(G) \cong Z_1/\mho_n(Z_1) \times \dots \times Z_k/\mho_n(Z_k)$.

Proof. We proceed by induction of $|G|$. The first assertion in (c) follows from (b), other assertions in (c) are obvious.

(a) Suppose that G is noncyclic. If $A < G$ is maximal, then A is cyclic. Take $b \in G - A$. By assumption, $o(b) > p$. Let $b^p = a^{p^s}$, where $\langle a \rangle = A$. Then $s \ge 1$ since G is noncyclic. Set $c = b^{-1}a^{p^{s-1}}$; then $c \in G - A$ is of order p, a contradiction.

(b) Assume that $A < G$ is of order p such that $A \not\le Z$ (see (a)). Then $ZA/A \le G/A$ is cyclic of maximal order (note that $|ZA/A| = |Z|$ since $A \not\le Z$). Indeed, if $\langle z_1 A \rangle \le G/A$ and $o(z_1 A) > |Z|$, then $o(z_1) > |Z|$, a contradiction. Therefore, by induction, $G/A = (ZA/A) \times (D/A)$, where $D/A < G/A$. We have $G = ZAD = ZD$. Since $Z \cap D = (Z \cap ZA) \cap D = Z \cap (ZA \cap D) = Z \cap A = \{1\}$, we get $G = ZD = Z \times D$. □

Let $\mathrm{CL}(G) = \{K_1, \dots, K_r\}$ be the set of G-classes, where $r = \mathrm{k}(G)$ is the *class number* of G. Set $|K_i| = h_i$, $1 = h_1 = \dots = h_z < h_{z+1} \le \dots \le h_r$, $z = |\mathrm{Z}(G)|$.

Then

$$|G| = z + h_{z+1} + \cdots + h_r$$

(this is the *class equation* for G) so

Lemma 5. *If $p \in \pi(G)$ and p divides h_i for $i = z+1, \dots, r$, then p divides $|Z(G)|$. In particular, the center of p-group $> \{1\}$ has an element of order p.*

Note that the length of a chain $G = G_1 \geq \cdots \geq G_n \geq G_{n+1} = \{1\}$ equals n.

Let G be a group. Set $Z_0(G) = \{1\}$, $Z_1(G) = Z(G)$. Suppose that $Z_i(G)$ has been defined for $i \leq k$. Define $Z_{k+1}(G)$ as follows: $Z_{k+1}(G)/Z_k(G) = Z(G/Z_k(G))$. The chain $\{1\} = Z_0(G) \leq Z_1(G) \leq \cdots \leq Z_k(G) \leq \cdots$ is said to be the *upper central series* of G. All members of that series are characteristic in G.

Definition 1. For elements $x, y \in G$, their *commutator* $x^{-1}y^{-1}xy$ is written as $[x, y]$. If $X, Y \subseteq G$, then $[X, Y]$ is the subgroup generated by all commutators $[x, y]$ with $x \in X, y \in Y$.

The *lower central series* $G = K_1(G) \geq K_2(G) \geq \cdots$ of G is defined as follows: $K_1(G) = G$, $K_{i+1}(G) = [K_i(G), G]$, $i > 0$. All members of that series are characteristic in G. We have $K_i(G)/K_{i+1}(G) \leq Z(G/K_i(G))$. If $H \leq G$, then $K_i(H) \leq K_i(G)$ for all i.

Since $[y, x] = [x, y]^{-1}$, we have $[Y, X] = [X, Y]$. We write $[G, G] = G'$; G' is called the *commutator subgroup* (or *derived subgroup*) of G. We also write $G^{(0)} = G$, $G' = G^{(1)}$. Then the subgroup $G^{(i+1)} = [G^{(i)}, G^{(i)}]$ is called the $(i+1)$-th derived subgroup of G, $i \geq 0$. The chain $G = G^{(0)} \geq G^{(1)} \geq \cdots \geq G^{(n)} \geq \cdots$ is called the *derived series* of G; all members of this series are characteristic in G and all factors $G^{(i)}/G^{(i+1)}$ are abelian. The group G is said to be *solvable* if $G^{(n)} = \{1\}$ for some n. The length of the derived series of a solvable group G is said to be the *derived length* of G and we denote it by $\mathrm{dl}(G)$. In particular, the derived length of a nonidentity abelian group equals 1, and $\mathrm{dl}(\{1\}) = 0$. If S is a section of a solvable group G (i.e., $S = H/K$, where $K \trianglelefteq H \leq G$), then S is also solvable and $\mathrm{dl}(S) \leq \mathrm{dl}(G)$ so sections of solvable groups are solvable.

By definition, $[x_1, \dots, x_n] = [[x_1, \dots, x_{n-1}], x_n]$ $(n > 2)$, and similarly for subgroups. Then $K_i(G) = [G, \dots, G]$ (i times); see below. If $K \leq H$ are normal in G, then H/K is a *central factor* of G if $[H, G] \leq K$. Obviously, $K_2(G) = G'$.

Definition 2. A group G is said to be *nilpotent* if the upper central series of G contains G. If, in addition, the length of the upper central series of G is c, then G is said to be of *class* c; we write $c = \mathrm{cl}(G)$ (in other words, $G > \{1\}$ is nilpotent of class c, if $Z_c(G) = G$ but $Z_{c-1}(G) < G$). In particular, the class of the identity group is 0 and the class of a nonidentity abelian group is 1.

Clearly, $\mathrm{cl}(Z_i(G)) \leq i$. Using Lemma 5 and induction, we get

Lemma 6. *A p-group is nilpotent.*

Proof. One may assume that $G > \{1\}$. Then $\mathrm{Z}(G) > \{1\}$ (Lemma 5). Applying induction to $G/\mathrm{Z}(G)$, we complete the proof. □

Lemma 7. *Let $H < G$, where G is nilpotent. Then $H < \mathrm{N}_G(H)$.*

Proof. Suppose that $k \geq 0$ is such that $\mathrm{Z}_k(G) \leq H$ but $\mathrm{Z}_{k+1}(G) \not\leq H$. One may assume that $\mathrm{Z}_k(G) = \{1\}$; then $k = 0$ and $\mathrm{Z}_{k+1}(G) = \mathrm{Z}(G) \leq \mathrm{N}_G(H)$. □

Exercise 1. Let G be a nilpotent group.

(a) G is the direct product of its Sylow subgroups.

(b) Every maximal subgroup is normal in G and has prime index in G.

(c) If $\{1\} < N \triangleleft G$, then $N \cap \mathrm{Z}(G) > \{1\}$.

(d) Sections of G are nilpotent.

(e) If $\{1\} < N \triangleleft G$, then N contains a G-invariant subgroup of prime index.

Hint. (a) If $P \in \mathrm{Syl}(G)$ and $\mathrm{N}_G(P) \leq H \leq G$, then $\mathrm{N}_G(H) = H$ (Sylow and Frattini) so $H = G$ (Lemma 7). Then $\mathrm{N}_G(P) = G$, i.e., $P \triangleleft G$, and (a) is clear. (b) follows from Lemma 7. To prove (c), we may assume that G is a p-group and apply the class equation. Clearly, (e) follows from (c).

Exercise 2. Suppose that G is a noncyclic nilpotent group. Then (a) If $a \in G$, then $\langle a^x \mid x \in G \rangle < G$ and (b) $G/\mathrm{Z}(G)$ is noncyclic.

Exercise 3. If G is a noncyclic nilpotent group, then G/G' is noncyclic. (*Hint.* Use Exercise 1(a,e).)

Part (a) of Exercise 4 is called the *basic theorem on abelian groups*.

Exercise 4. (a) (Kronecker; see also [FS]) An abelian group G is a direct product $G = Z_1 \times \cdots \times Z_n$ of cyclic subgroups such that $|Z_1| \mid |Z_2| \mid \cdots \mid |Z_n|$. The numbers $|Z_1|, \dots, |Z_n|$ are determined uniquely. The group G is called the abelian group of type $(|Z_1|, \dots, |Z_n|)$; all abelian groups of the same type are isomorphic. Every cyclic subgroup of order $|Z_n|$ is a direct factor of G (= Lemma 4.)

(b) If G, H are abelian and such that $\mathrm{c}_n(G) = \mathrm{c}_n(H)$ for all n, then $G \cong H$.

Exercise 5. (a) If $\mathrm{N}_G(H) > H$ for all $H < G$, then G is nilpotent. (*Hint.* Take $H = \mathrm{N}_G(P)$, where $P \in \mathrm{Syl}(G)$.)

(b) A direct product of nilpotent groups is nilpotent.

(c) G is nilpotent if and only if any two elements of G of coprime orders commute.

Recall that $\Phi(G)$, the intersection of all maximal subgroups of a group G, is said to be the *Frattini subgroup* of G. If $a \in \Phi(G)$, $X \subseteq G$ and $G = \langle a, X \rangle$, then $G = \langle X \rangle$. This is true for arbitrary (infinite) groups (B. H. Neumann). It is easy to check, using Sylow's theorem and Frattini's lemma (if $H \trianglelefteq G$ and $P \in \mathrm{Syl}_p(H)$, then $G = H\mathrm{N}_G(P)$), that $\Phi(G)$ is nilpotent.

Exercise 6. (a) (Wielandt) A group G is nilpotent if and only if $G' \le \Phi(G)$. (*Hint.* If $P \in \mathrm{Syl}_p(G)$ is not normal in G and $N_G(P) \le M$, where M is maximal in G, then $N_G(M) = M$; however, $(G' \le \Phi(G) <)M \triangleleft G$, a contradiction.)

(b) If all maximal subgroups H of G are such that $p \nmid |G : H|$, are normal in G, then $P \in \mathrm{Syl}_p(G)$ is normal in G (= G is p-closed) and G/P is nilpotent.

(c) Suppose that $H < G$ and $\Phi(H) \triangleleft G$. Prove that $\Phi(H) \le \Phi(G)$. In particular, if $H \triangleleft G$, then $\Phi(H) \le \Phi(G)$.

(d) If H is subnormal in G, then $\Phi(H) \le \Phi(G)$.

Exercise 7 (Ito–Szep). There is a nilpotent $N \le G$ such that $N^G = G$. (*Hint.* If $P \in \mathrm{Syl}(G)$ is not normal in G, then $(N_G(P))^G = G$. Use induction on $|G|$.)

Exercise 8 (Zhmud; see [BZ, Ch.6]). Let G be an abelian p-group. Prove that there exists a cyclic subgroup $C < G$ such that G/C is a direct product of two isomorphic groups and the order of C depends only on G.

Exercise 9 (Matsuyama [Mat]). If P is a p-subgroup of G such that p divides $|G : N_G(P)|$, then there exists $x \in G - N_G(P)$ such that $P^x \le N_G(P)$.

Solution. Let $P < S \in \mathrm{Syl}_p(G)$; then P is not normal in S so one may assume that $S = G$ is a p-group. Set $N = N_G(P)$; then $N < G$. Let $N < H \le G$, where $|H : N| = p$. Then $N \triangleleft H$ so for $x \in H - N$ we get $P^x \neq P$ and $P^x < N$.

Exercise 10. If $A, B \le G$, then by the *product formula*, $|AB| = \frac{|A|\cdot|B|}{|A\cap B|}$. If, in addition, $G = AB$ and $A \le C \le G$, then $C = A(C \cap B)$ (the *modular law*).

Solution. (i) Let $A = \bigcup_{i=1}^{r} t_i(A \cap B)$ be a partition. Then $AB = \bigcup_{i=1}^{r} t_i B$, and this is also a partition. We are done since $r = |A : (A \cap B)|$.

(ii) If $c = ab \in C$, where $a \in A, b \in B$, then $b = a^{-1}c \in B \cap C$ since $a \in A \le C$. It follows that $c = ab \in A(C \cap B)$. We get $C = A(B \cap C)$ since $A(B \cap C) \le C$.

Exercise 11. (a) If $H \triangleleft G$, then $(G/H)' = G'H/H$.

(b) If G is nilpotent of class $c > 1$ and $x \in G$, then $H = \langle x, G' \rangle$ is of class $< c$.

Solution. (a) Set $F/H = (G/H)'$. Since $G/F \cong (G/H)/(F/H)$ is abelian, we get $G'H \le F$. If $G'H < F$, then $F/H > (G/H)'$, which is not the case. (b) We have $H < G$ since $G' \le \Phi(G)$. Next, $H/K_3(G)$ is abelian (Exercise 2(b)). It follows that $K_3(G) \ge H' = K_2(H)$. Then $K_4(G) \ge [H', G] \ge [H', H] = K_3(H)$, and so on. If $c = \mathrm{cl}(G)$, then $K_c(H) \le K_{c+1}(G) = \{1\}$.

Exercise 12. The lengths of upper and lower central series of a nilpotent group G are equal (see Appendix 6). Next, $[K_i(G), Z_i(G)] = \{1\}$. (*Hint.* The last assertion is true for $i = 1$. For $i > 1$ use the Three Subgroups Lemma; see Exercise 13(b).)

Exercise 13. Let G be an arbitrary (even infinite) group and $x, y, z \in G$.

(a) Prove the following commutator identities:

$$[x, yz] = [x, z][x, y]^z, \qquad [xy, z] = [x, z]^y[y, z],$$
$$[x, y^{-1}, z]^y[y, z^{-1}, x]^z[z, x^{-1}, y]^x = 1 \quad \text{(Hall–Witt identity)}.$$

(b) (Three Subgroups Lemma (Hall)) Let $X, Y, Z \le G$ and set $X^* = [Y, Z, X]$, $Y^* = [Z, X, Y]$, $Z^* = [X, Y, Z]$. If $N \trianglelefteq G$ and $X^*, Y^* \le N$, then $Z^* \le N$.

(c) If $X, Y, Z \trianglelefteq G$, then $Z^* \le X^*Y^*$ (we retain the notation of (b)). In particular, if $X^* = \{1\} = Y^*$, then $Z^* = \{1\}$.

Hint (Hall). (a) We have $(xy)^z = xy[xy, z]$, $(xy)^z = x^z y^z = x[x, z]y[y, z]$; then $xy[xy, z] = x[x, z]y[y, z]$, and the second identity follows. The first identity is proved similarly. We recommend to the reader to check the third identity. (b) Suppose that $X^*, Y^* \le N$ and $N \trianglelefteq G$. Suppose that in the third identity of (a), $x \in X$, $y \in Y$, $z \in Z$. The last two factors on the left both lie in N since $N \trianglelefteq G$. Hence, $[x, y^{-1}, z] \in N$; so every element z of Z commutes (mod N) with every generator $[x, y^{-1}]$ of $[X, Y]$. Thus, $Z^* = [X, Y, Z] \equiv \{1\} \pmod N$, or, what is the same, $Z^* \le N$. To prove (c), put $N = X^*Y^*$ and apply (b).

It follows from Exercise 13(c) that $[G', Z_2(G)] = \{1\}$ (Grün's theorem). Indeed, $[G, Z_2(G), G] \le [Z(G), G] = \{1\}$ and $[Z_2(G), G, G] \le [Z(G), G] = \{1\}$ so $\{1\} = [G, G, Z_2(G)] = [G', Z_2(G)]$.

Here is a variant of the Hall–Witt identity: $[x, y, z^x][z, x, y^z][y, z, x^y] = 1$ for all $x, y, z \in G$. It follows from Exercise 13(a) that, if $\mathrm{cl}(G) \le 2$ and $x, y, z \in G$, then $[x, y]^n = [x, y^n]$ for $n \in \mathbb{N}$, $[x, yz] = [x, y][x, z]$ and $[xy, z] = [x, z][y, z]$.

Remarks. 1. We use Exercise 13(c) to prove the following Kaloujnine's result. Let G be a group and let $H = H_0 \ge H_1 \ge H_2 \ge \cdots \ge H_i \ge \dots$ be a series of normal subgroups of G. If $L \trianglelefteq G$ is such that $[L, H_{i-1}] \le H_i$ $(i = 1, 2, \dots)$, then $[K_j(L), H_i] \le H_{i+j}$ $(i \ge 0;\ j \ge 1)$. We use induction on j. If $j = 1$, then the inclusion holds since $K_1(L) = L$. We have, by Exercise 13(c),

$$\begin{aligned}[K_j(L), H_i] &= [K_{j-1}(L), L, H_i] \le [L, H_i, K_{j-1}(L)][H_i, K_{j-1}(L), L]\\ &\le [H_{i+1}, K_{j-1}(L)][H_{j-1+i}, L] \le H_{i+1+j-1}H_{j-1+i+1} = H_{i+j}.\end{aligned}$$

2. For any group G, we define [Bla3, page 47] $\eta_i(G) \le G$ as follows:

$$\eta_i(G)/K_i(G) = Z(G/K_i(G)), \quad i \ge 2.$$

Then $\eta_2(G) = G$ and $[\eta_i(G), G] \le K_i(G)$ for $i \ge 2$. Putting, in Remark 1, $L = G$, $H = \eta_i(G)$, and define the subgroups H_k inductively by the rules $H_0 = H$ and $H_{k+1} = [H_k, G]$ for $k \ge 0$, we see that $[K_j(G), H_k] \le H_{j+k}$. Then $H_1 = [H, G] = [\eta_i(G), G] \le K_i(G)$; hence, $H_j \le K_{i+j-1}(G)$, for $j \ge 1$. Thus, by Remark 1, $[K_j(G), \eta_i(G)] = [K_j(G), H] \le H_j \le K_{i+j-1}(G)$. In what follows we write $\eta_3(G) = \eta(G)$. We use $\eta(G)$ in §9 (see also Theorem 1.40).

Exercise 14. Let a group A act on a group G, let $U \trianglelefteq G$ be A-invariant abelian. Suppose that elements of A induce identical automorphisms on G/U and U. Prove that A' centralizes G. (*Hint.* For $g \in G$ and $a, b \in A$, compute $g^{[a,b]}$.)

A group is *homocyclic* if it is a direct product of isomorphic cyclic groups.

Exercise 15. Let G be an abelian p-group.

(i) G is a direct product of homocyclic groups of pairwise distinct exponents.

(ii) If $\Phi(G)$ is homocyclic and $\mathrm{d}(\Phi(G)) = \mathrm{d}(G)$ then G is also homocyclic.

(iii) Let $\exp(G) = p^e$. Then G is homocyclic if and only if $\mho_{e-1}(G) = \Omega_1(G)$.

Now we prove the celebrated theorem of Frobenius on the number of solutions of equation $x^n = 1$ in a finite group. This theorem together with Sylow's theorem is the first and most fundamental counting theorem in finite group theory. The new proof, presented below, is due to I. M. Isaacs and G. R. Robinson [IsR]. For other proofs, see [BZ, Chapters 4 and 5]. Let $f_n(G)$ be the number of solutions of equation $x^n = 1$ in G and let n_p denote the largest p-power dividing n.

Theorem 8 (Frobenius). *If n divides $|G|$, then n divides $f_n(G)$.*

Lemma 9. *Given a group G, $n \in \mathbb{N}$ and prime p, write $q = n_p$ and let T be a transversal for those conjugacy classes of elements $y \in G$ for which $y^{n/q} = 1$. Then*

$$f_n(G) = \sum_{t \in T} |G : C_G(t)| \cdot f_q(C_G(t)). \tag{1}$$

Proof. By Lemma 1, if $g \in G$ with $g^n = 1$, then $g = xy$, where $xy = yx$, $o(x) = o(g)_p$ divides q, $o(y) = o(g)/o(g)_p$ divides n/q, and such expression is unique. It follows that $f_n(G) = \sum_{y \in G,\, y^{n/q}=1} f_q(\mathrm{C}_G(y))$. Indeed, if g is as above, its contribution in $f_q(\mathrm{C}_G(y))$ equals 1 if y is the p'-part of g and 0 if y is not a p'-part of g. The contribution of g in $f_n(G)$ is also 1. Since $f_q(\mathrm{C}_G(y))$ remains constant as y runs over the $|G : \mathrm{C}_G(t)|$ elements in the conjugacy class represented by $t \in T$, formula (1) follows. □

Definition 3. A group G has the *p-Frobenius* property if p^α divides $f_{p^\alpha}(G)$ whenever p^α divides $|G|$.

Lemma 10. *Let q be a power of p such that q divides $|G|$. Suppose that $H \le G$ is a subgroup having the p-Frobenius property. Then q divides $|G : H| \cdot f_q(H)$.*

Proof. One may assume that $q_0 = |H|_p < q$ (otherwise, the lemma is trivial). Then $f_q(H) = f_{q_0}(H)$ is divisible by q_0, by hypothesis, and $|G : H|$ is divisible by $|G|_p/q_0$. It follows that $|G|_p$ divides $|G : H| f_q(H)$. Since q divides $|G|_p$, all is done. □

Lemma 11. *G has the p-Frobenius property.*

Proof. We use induction on $|G|$. Let q be a p-power such that q divides $|G|$. First suppose that $q = |G|_p$. Applying Lemma 9 with $n = |G|$, we obtain

$$|G| = n = f_n(G) = |T \cap Z(G)| \cdot f_q(G) + \sum_{t \in T - Z(G)} |G : C_G(t)| \cdot f_q(C_G(t)).$$

By induction and Lemma 10, q divides $|G : H| f_q(H)$ for $t \in T - Z(G)$, where $H = C_G(t)$. Since q divides $|G|$, we get q divides $|T \cap Z(G)| \cdot f_q(G)$, by the displayed formula, and so it is enough to show that $p \nmid |T \cap Z(G)|$. We have $T \cap Z(G) = \{y \in Z(G) \mid y^{n/q} = 1\}$, i.e., $|T \cap Z(G)| = f_{n/q}(Z(G))$, and so $p \nmid |T \cap Z(G)|$, by Lemma 2, since $p \nmid (n/q)$.

Now suppose that $q < |G|_p$. As q divides $|G|_p$ divides $f_{|G|_p}(G)$, by the previous paragraph, it is enough to show that q divides $f_{|G|_p}(G) - f_q(G)$. The last number is the cardinality of the set $\mathfrak{M}$ of elements of G having p-power order exceeding q. If a is one of such elements and $o(a) = p^s (> q)$, then the number of elements of order $> q$ in $\langle a \rangle$ is $p^s - q$, and q divides $p^s - q$ since, by assumption, the p-power $q < p^s$. Then the set $\mathfrak{M}$ is partitioned in subsets of cardinalities that are multiples of q. □

Proof of Theorem 8. It suffices to show, for each prime p, that n_p divides $f_n(G)$. But this follows from Lemmas 9–11. □

Exercise 16 ([IsR]). Let G be a group, $a \in G$, $n \in \mathbb{N}$. Set $f_n(G, a) = |\{x \in G \mid x^n = a\}|$. By the general Frobenius theorem, $f_n(G, a)$ is divisible by $(n, |C_G(a)|)$. The proof of the last result follows closely to the proof of Theorem 8 using the identity $f_n(G, a) = \sum_{t \in T} |G : C_G(t)| \cdot f_q(C_G(t), a)$. where $q = n_p$ and T is a transversal for the conjugacy classes of elements $y \in G$ such that $y^{n/q} = a$.

Let $G \cong S_p$. If $x \in G$ with $x^p = 1$, then either $x = 1$ or x is a p-cycle. The number of p-cycles in G equals $(p-1)!$ so $f_p(G) = (p-1)! + 1$ and, by Theorem 8, $(p-1)! + 1 \equiv \pmod{p}$ (this is the theorem of Wilson).

Proposition 12 (N/C-Theorem). *If $H \leq G$, then $N_G(H)/C_G(H)$ is isomorphic to a subgroup of* Aut(H).

Proof. One may assume that $H \trianglelefteq G$. Then, for $x \in G$, a mapping $\tau_x : h \mapsto xhx^{-1}$ $(h \in H)$ is an automorphism of H. Thus, $x \mapsto \tau_x$ is a homomorphism of G into Aut(H) with kernel $C_G(H)$. □

Exercise 17 (Ito). Let $|G| = p^n$. Then G is called *one-stepped* (Ito) if there exists an ordered sequence $x_1, \dots, x_n \in G$ of elements of order p such that $|\langle x_1, \dots, x_i \rangle| = p^i$ for $i = 1, \dots, n$. We see that one-stepped p-groups are generated by elements of order p. Prove that a p-group is one-stepped if and only if it is generated by elements of order p. Any p-group is isomorphic to a subgroup of a one-stepped p-group. (*Hint.* (b) A Sylow p-subgroup of the symmetric group S_{p^n} is one-stepped, by (a).)

A p-group G is called *k-stepped* if it has an ordered sequence of elements $x_1, \dots, x_t$ of orders p^k such that $G = \langle x_1, \dots, x_t \rangle$ and $|\langle x_1, \dots, x_i \rangle| \le p^{ki}$ for all $i = 1, \dots, t$. Set $\Omega_k^*(G) = \langle x \in G \mid o(x) = p^k \rangle$. If G is k-stepped, then $\Omega_k^*(G) = G$. Now let $\Omega_k^*(G) = G$. We prove by induction on $|G|$ that G is k-stepped. Let H be a maximal k-stepped subgroup of G and assume that $H < G$; then $H^G < G$ and $\Omega_k^*(H^G) = H^G$ so H^G is k-stepped, by induction, and so $H^G = H$, i.e., $H \lhd G$. Take $x \in G - H$ with $o(x) = p^k$. Then $T = \langle H, x \rangle > H$ and $\Omega_k^*(T) = T$ with $|T| \le p^k |H|$ so T is k-stepped, a contradiction. Thus, $G = H$.

Let $G = G_1 \times \dots \times G_n$, $A \le G$ and $a \in A$. Then $a = (a_1, \dots, a_n)$, where $a_i \in G_i$ for all i. Define a *projection* $\pi_i : A \to G_i$, setting $\pi_i(a) = a_i$, all $a \in A$. Then π_i is a homomorphism and $A_i = \pi_i(A)$ is an epimorphic image of A. Obviously, $A \le A_1 \times \dots \times A_n$.

Exercise 18. Let A be a maximal abelian (nilpotent) subgroup of the direct product $G = F \times H$ of groups F and H. Then $A = (A \cap F) \times (A \cap H)$.

Exercise 19. Let $x \in G$ with $o(x) = p$, where G is an abelian p-group. Then $x \in C$, where C is a cyclic direct factor of G, and $|C|$ depends only on x.

Exercise 20. Let R be an elementary abelian subgroup of an abelian p-group G. Prove that $G = S \times S_1$, where $S, S_1 \le G$ and $\Omega_1(S) = R$.

Exercise 21. If G is a group such that $\mathrm{Aut}(G) = \{1\}$, then $|G| \le 2$. Classify the groups G such that $|\mathrm{Aut}(G)| = 2$. If $\mathrm{Aut}(G) > \{1\}$ is cyclic, then 2 divides $|\mathrm{Aut}(G)|$.

Exercise 22. Let $G = \langle x \rangle \cong \mathrm{C}_n$. Then the mapping $\sigma_m : x \mapsto x^m$ is an automorphism of G if and only if $(m, n) = 1$, and every automorphism of G has such form. Show that $\sigma_m = \sigma_t$ if and only if $m \equiv t \pmod{n}$, and so $\mathrm{Aut}(G) \cong (\mathbb{Z}/n\mathbb{Z})^*$, the group of units of the n-element ring $\mathbb{Z}/n\mathbb{Z}$ of residues modulo n. Thus, $\mathrm{Aut}(G)$ is abelian of order $\varphi(n)$, where $\varphi(*)$ is the Euler totient function.

Exercise 23. If $G = A \times B$, $(|A|, |B|) = 1$, then $\mathrm{Aut}(G) \cong \mathrm{Aut}(A) \times \mathrm{Aut}(B)$.

Exercise 24. $\mathrm{Aut}(\mathrm{C}_p) \cong \mathrm{C}_{p-1}$.

Exercise 25. Let $G \cong \mathrm{C}_{p^n}$, $n \in \mathbb{N}$. (a) If $p > 2$, then $\mathrm{Aut}(G)$ is cyclic. (b) If $p = 2$, $n > 2$, then $\mathrm{Aut}(G) \cong \mathrm{C}_2 \times \mathrm{C}_{2^{n-2}}$. In that case the subgroup C_2 is generated by the automorphism $x \mapsto x^{-1}$ and the subgroup $\mathrm{C}_{2^{n-2}}$ is generated by the automorphism $x \mapsto x^5$. (*Hint.* See the text following Theorem 1.2.)

Exercise 26. $\mathrm{Aut}(\mathrm{C}_n)$ is cyclic if and only if $n = 2$ or 4, or $n = 2^\mu p^k$, where $\mu \in \{0, 1\}$ and $p > 2$, $k \in \mathbb{N}$.

Exercise 27. If $M, N \lhd G$, then $G/(M \cap N)$ is isomorphic to a subgroup of $(G/M) \times (G/N)$. (*Hint.* Consider the mapping $\alpha : G \to (G/M) \times (G/N)$ defined as follows: $\alpha(x) = (xM, xN)$. Check that α is an epimorphism and $\ker(\alpha) = M \cap N$.) It follows that if $G > \{1\}$ is a p-group, then $G/\Phi(G)$ is elementary abelian.

Exercise 28. Prove that, for any group G, $G/\mathrm{Z}(G) \not\cong \mathrm{Q}_{2^n}$.

Exercise 29. Suppose that $H = AB$, where A, B are distinct cyclic such that $A \cap B > \{1\}$. Prove that, for any group G, $G/\mathrm{Z}(G) \not\cong H$.

Solution. Assume that $G/\mathrm{Z}(G) = AB = H$. Let $U/\mathrm{Z}(G) = A$, $V/\mathrm{Z}(G) = B$; then U, V are abelian and $UV = G$. We have $\mathrm{Z}(G) < U \cap V \leq \mathrm{Z}(G)$, a contradiction.

Exercise 30. A p-group G of order p^n has exactly $n + p - 1$ nontrivial normal subgroups if and only if it is of class $n - 1$.

Exercise 31 (G. Freiman). Let G be a nonabelian group. Prove that the following conditions are equivalent: (a) Whenever a, b are two noncommuting elements of G, then $a^2 = b^2$. (b) All subgroups of G are normal.

Solution. We will prove only that (a) $\Longrightarrow$ (b). Let $x, y \in G$ be such that $xy \neq yx$. Then also $x \cdot xy \neq xy \cdot x$. It follows from $x^2 = y^2 = (xy)^2$ that $x = yxy$ and $y = xyx$. Then $x = yxy = xyx \cdot xy = xy \cdot x^2 \cdot y = xy^4$ so $y^4 = 1$. Similarly, $x^4 = 1$. We have $y^{-1}xy = y^{-1} \cdot yxy \cdot y = xy^2 = x^3$ so $\langle x \rangle \trianglelefteq G$ for all $x \in G - \mathrm{Z}(G)$. It follows that all cyclic subgroups so all subgroups are normal in G.

Exercise 32. Let G be a p-group of exponent $p^e > p$. Then every two elements of G of distinct orders are permutable if and only if $\Omega_{e-1}(G) = \langle x \in G \mid o(x) \leq p^{e-1} \rangle \leq \mathrm{Z}(G)$.

Exercise 33 ([Ito8]). A proper subgroup of an arbitrary group G, which is the centralizer of some element in G, is called a *fundamental* subgroup of G. A fundamental subgroup F of G is said to be *free* if F has no proper fundamental subgroups of G and is not contained in a fundamental subgroup of G properly. Prove that a free fundamental subgroup of a nonabelian group G is a direct product of a p-subgroup and an abelian p'-subgroup for some prime p. (See Proposition A.23.2.)

Exercise 34. Suppose that a Sylow p-subgroup of an arbitrary group G has a cyclic subgroup of index p. Improve, in that case, the part of Sylow's theorem on the number of p-subgroups of given order in G.

Exercise 35. If a group G is noncyclic so $\mathrm{Aut}(G)$ is noncyclic.

Exercise 36. Show that C_{2^3} is the only 2-group G such that $|\mathrm{Aut}(G)| = 2^2$.

Exercise 37. Does there exist a 2-group G such that $\mathrm{Aut}(G) \cong \mathrm{E}_{2^3}$? Classify the groups G such that $\mathrm{Aut}(G) \cong \mathrm{D}_8$. Note that $\mathrm{Aut}(\mathrm{D}_8) \cong \mathrm{D}_8 \cong \mathrm{Aut}(\mathrm{C}_4 \times \mathrm{C}_2)$. (Mann) Does there exist other p-groups G satisfying $\mathrm{Aut}(G) \cong G$?

Exercise 38. Study the normal structure of $\mathrm{Aut}(G)$, where $G \cong \mathrm{C}_{p^m} \times \mathrm{C}_{p^n}$.

Exercise 39. Classify the cyclic p-groups G such that the sequence of automorphism groups $\mathrm{Aut}(G)$, $\mathrm{Aut}(\mathrm{Aut}(G)), \ldots$ contains $\{1\}$.

Exercise 40. If G is a nonabelian p-group and $G/\mathrm{Z}(G)$ is abelian of type $(p^{a_1}, p^{a_2}, \dots, p^{a_d})$, where $a_1 \geq a_2 \geq \dots \geq a_d$, then $a_1 = a_2$.

Solution. Assume that $a_1 > a_2$. Then $L/\mathrm{Z}(G) = \mho_{a_1-1}(G/\mathrm{Z}(G))$ has order p. If $C/\mathrm{Z}(G)$ is cyclic of order p^{a_1}, then C is abelian and contains L. Since all such C generate G, we get $\mathrm{Z}(G) < L \leq \mathrm{Z}(G)$, a contradiction.

Theorem 13. *(a) If $H, K \leq G$, where G is arbitrary, then $[H, K] \trianglelefteq \langle H, K\rangle$.*

(b) Let $X, Y, Z \leq G$ and let Y normalize X and Z. Then $[XY, Z] = [X, Z][Y, Z]$.

Proof. (a) We will prove that $H \leq \mathrm{N}_G([H, K])$. It suffices to show that for any $u \in H$ and for any generator $[h, k]$ $(h \in H, k \in K)$ of $[H, K]$, we have $[h, k]^u \in [H, K]$. By Exercise 13(a), we have $[h, k]^u = [hu, k][u, k]^{-1} \in [H, K]$ so that $H \leq \mathrm{N}_G([H, K])$. By symmetry, $K \leq \mathrm{N}_G([K, H]) = \mathrm{N}_G([H, K])$.

(b) We have $XY \leq G$ so $[XY, Z] = \langle [xy, z] \mid x \in X, y \in Y, z \in Z\rangle$. By (a), $[X, Z] \trianglelefteq \langle X, Z\rangle$. Next, $[Y, Z] \leq Z \leq \mathrm{N}_G([X, Z])$ so $[X, Z][Y, Z] \leq G$. By Exercise 13(a), we get $[xy, z] = [x, z]^y[y, z] = [x^y, z^y][y, z] \in [X, Z][Y, Z]$ so $[XY, Z] \leq [X, Z][Y, Z]$. The reverse inclusion is obvious. □

Exercise 41. In the notation of Theorem 13(a), $[H, K]K = K^{\langle H,K\rangle}$.

If a central product $G = A * B$ (i.e., $G = AB$ and $[A, B] = \{1\}$), then $G \cong (A \times B)/Z_0$, where $Z_0 \leq \mathrm{Z}(G)$ is isomorphic to some subgroups of $\mathrm{Z}(A)$ and $\mathrm{Z}(B)$.

Exercise 42. Suppose that $G = A * B$ is of order 16, where $A \cong \mathrm{D}_8$ and $B \cong \mathrm{C}_4$. Prove that $G = Q * B$, where $Q \cong \mathrm{Q}_8$.

Exercise 43. Let L be a p-group with cyclic $\mathrm{Z}(L)$ and let $G = A \times B$ be a direct product of two p-groups that have no subgroups isomorphic to L. Prove that G has no subgroups isomorphic to L.

Exercise 44. Let $G = A * B$ and $H = C * D$ be groups of order 32, where $A, B, D \cong \mathrm{D}_8$, $C \cong \mathrm{Q}_8$. Prove that $G \not\cong H$. (*Hint.* Find $\mathrm{c}_1(G)$ and $\mathrm{c}_1(H)$.)

Exercise 45. Let $G = A * B$, $H = C * D$ be groups of order 32, where A, B are dihedral of order 8, C, D are ordinary quaternion. Prove that $G \cong H$.

Exercise 46. Let G, H be groups of order 2^{2n+1}, $n > 1$, G the central product of n dihedral groups of order 8, H the central product of $n - 1$ dihedral groups of order 8 and the ordinary quaternion group. Prove that $G \not\cong H$.

Exercise 47 (Mann). Let G be a p-group such that $|G' : \mathrm{K}_3(G)| = p$ and $\mathrm{K}_3(G) > \{1\}$. Then $|G/\mathrm{K}_3(G) : \mathrm{Z}(G/\mathrm{K}_3(G))| = p^2$.

Solution. One may assume that $|\mathrm{K}_3(G)| = p$; then $|G'| = p^2$ and $\mathrm{Z}(G) \cap G' = \mathrm{K}_3(G)$. In that case, $C = \mathrm{C}_G(G')$ has index p in G. By the Three Subgroups

Lemma, $[C', G] = [C, C, G] \leq [C, G, C][G, C, C] \leq [G', C] = \{1\}$ so $C' \leq Z(G) \cap G' = K_3(G)$. Then $C/K_3(G)$ is an abelian maximal subgroup of $G/K_3(G)$ and $|(G/K_3(G))'| = |G'/K_3(G)| = p$. Now the result follows from Lemma 1.1.

Exercise 48 (Mann). Let G be a p-group such that $|G' : K_3(G)| = p$ and $K_3(G) > \{1\}$. Let $H < G$ be maximal. Then $H' \leq K_3(G)$ if and only if $Z(G/K_3(G)) \leq H/K_3(G)$.

Solution. Suppose that $H' \leq K_3(G)$ so $H/K_3(G)$ is an abelian maximal subgroup of a nonabelian group $G/K_3(G)$. It follows that $Z(G/K_3(G)) < H/K_3(G)$. Now assume that $Z(G/K_3(G)) \leq H/K_3(G)$. Then $|(H/K_3(G)) : Z(G/K_3(G))| = p$ (Exercise 47) hence $H/K_3(G)$ is abelian so $H' \leq K_3(G)$.

In what follows, we assume that the reader is familiar with basic facts of character theory. Isaacs's book [Isa1] is the best standard textbook in character theory for beginners (see also [BZ, Part 1]). We recall some definitions and basic results.

Let G be a finite group, $n \in \mathbb{N}$ and $\mathbb{C}$ the field of complex numbers, $V = V(n, \mathbb{C})$ a vector space of dimension n over $\mathbb{C}$ and $GL(V)$ the group of all nonsingular linear transformations of V. A homomorphism $T : G \to GL(V)$ is said to be a *representation* of G of degree n. A *character* of T is a function $\chi = \chi^T : G \to \mathbb{C}$ defined by $\chi(g) = \mathrm{tr}(T(g))$ $(g \in G)$, where $\mathrm{tr}(\phi)$ is the trace of the linear transformation $\phi : V \to V$. It is known that $\mathrm{tr}(\phi)$ is independent of a basis of V. We also express this fact saying that the representation T *affords* the character $\chi^T = \chi$. The number $\deg(T) = \chi(1)$, where $1 \in G$, is said to be the *degree* of the character χ. Since $T(g)^{|G|} = T(g^{|G|}) = T(1) = I_{\deg(T)}$, it follows that eigenvalues of the matrix of $T(g)$ are $|G|$-th roots of unity. Since $\mathrm{tr}(T(g))$ is the sum of eigenvalues of $T(g)$, it follows that $\chi(g)$ is an algebraic integer and $|\chi(g)| \leq \deg(T) = \chi(1)$. By definition, $\ker(\chi) = \ker(T)$. It is easy to prove that $\ker(\chi) = \{g \in G \mid \chi(g) = \chi(1)\}$. If $V = \mathbb{C}^n$, we can identify $GL(V)$ with $GL(n, \mathbb{C})$. In that case, we obtain matrix representations.

Given a representation $T : G \to GL(V)$, one can define an action of G on V: $g \cdot v = T(g)(v)$, $g \in G$, $v \in V$. A representation $T : G \to GL(V)$ is said to be *irreducible* if $\{0\}$ and V are the only G-invariant subspaces of V. A character χ of G is said to be *irreducible* if it is afforded by an irreducible representation of G.

Two representations $T : G \to GL(V)$ and $T_1 : G \to GL(V_1)$ are said to be *equivalent* if there exists an isomorphism $U : V \to V_1$ such that $T(g)(v) = (U^{-1}T_1(g)U)(v)$, all $g \in G$, $v \in V$, or, what is the same, $T(g) = U^{-1}T_1(g)U$ for all $g \in G$. Characters of equivalent representations of G are equal and such representations have equal kernels (check!).

A representation T is said to be *faithful* if $\ker(T) = \{1\}$. A character χ of G is said to be *faithful*, if it is afforded by a faithful representation of G. In other words, χ is faithful if $\ker(\chi) = \{1\}$.

Given representations $T_i : G \to GL(V_i)$, $i = 1, 2$, we define their sum $T_1 + T_2$ by $((T_1 + T_2)(g))(v_1 + v_2) = T_1(g)(v_1) + T_2(g)(v_2)$, $g \in G$, $v_1 \in V_1$, $v_2 \in V_2$. Then

$T_1 + T_2 : G \to \mathrm{GL}(V_1 \oplus V_2)$ is also a representation of G. The sum of any finite number of representations is defined similarly.

Exercise 49. $\ker(T_1 + T_2) = \ker(T_1) \cap \ker(T_2)$.

Exercise 50. $\chi^{T_1+T_2} = \chi^{T_1} + \chi^{T_2}$. So if χ^1, χ^2 are characters of G then $\ker(\chi^1 + \chi^2) = \ker(\chi^1) \cap \ker(\chi^2)$.

Let $\mathrm{Irr}(G)$ be the set of all irreducible characters of G. It is known that $|\mathrm{Irr}(G)| = \mathrm{k}(G)$, where $\mathrm{k}(G)$ is the class number of G. Let $\mathrm{Irr}(G) = \{\chi^1, \dots, \chi^r\}$, $r = \mathrm{k}(G)$. We have

$$\chi^1(1)^2 + \cdots + \chi^r(1)^2 = |G|. \tag{2}$$

Given characters χ, τ of G, their scalar product $\langle \chi, \tau \rangle$ is defined as follows:

$$\langle \chi, \tau \rangle = |G|^{-1} \sum_{g \in G} \chi(g)\tau(g^{-1}). \tag{3}$$

If χ is a character of G and $g \in G$, then $\chi(g^{-1}) = \overline{\chi(g)}$, where $\bar{c}$ is the complex conjugate of $c \in \mathbb{C}$. Therefore, one can rewrite (3) as follows:

$$\langle \chi, \tau \rangle = |G|^{-1} \sum_{g \in G} \chi(g)\overline{\tau(g)}. \tag{4}$$

Suppose that $H \trianglelefteq G$ and θ is a character of G/H. Define a function $\chi_\theta : G \to \mathbb{C}$ by $\chi_\theta(g) = \theta(gH)$. Then χ_θ is a character of G such that if $\ker(\theta) = K/H$ then $\ker(\chi_\theta) = K$. χ_θ is said to be the *inflation* of θ onto G. If θ is irreducible, then χ_θ is also irreducible.

Let χ be a character of G, $H \trianglelefteq G$ and $H \le \ker(\chi)$. We define a function $\chi_0 : G/H \to \mathbb{C}$ as follows: $\chi_0(xH) = \chi(x)$, all $x \in G$. It is easy to check that χ_0 is a character of G/H and, if χ is irreducible, then χ_0 is also irreducible. Then $\ker(\chi_0) = \ker(\chi)/H$. Clearly, χ is the inflation of χ_0 onto G. Usually we identify χ_0 with χ.

Theorem 14. *If g, h are conjugate elements of G, then $\chi(g) = \chi(h)$ for all characters χ of G, i.e., characters of G are class functions.*

Proof. If A and B are $n \times n$ matrices, then $\mathrm{tr}(AB) = \mathrm{tr}(BA)$ so, if $g = h^x$ for $g, h, x \in G$ and χ is a character afforded by a representation T of G, then

$$\begin{aligned} \chi(g) &= \mathrm{tr}(T(g)) = \mathrm{tr}(T(x^{-1}hx)) = \mathrm{tr}(T(x)^{-1}T(h)T(x)) \\ &= \mathrm{tr}(T(x)^{-1}T(x)T(h)) = \mathrm{tr}((T(h)) = \chi(h). \end{aligned}$$ □

Let M be a set. Define a function $\delta : M \times M \to \mathbb{R}$ as follows: $\delta_{a,b} = 0$ if $a \ne b$ and $\delta_{a,a} = 1$. So defined function is said to be *Kronecker delta* on the set M.

Theorem 15. *(a)* The First Orthogonality Relation. *If $\chi, \tau \in \mathrm{Irr}(G)$, then $\langle \chi, \tau \rangle = \delta_{\chi,\tau}$, where δ is the Kronecker delta on the set* $\mathrm{Irr}(G)$.

(b) The Second Orthogonality Relation. *If $g_i \in K_i$, $g_j \in K_j$, where $\mathrm{CL}(G) = \{K_1, \dots, K_r\}$ is the set of conjugacy classes of G, then $\sum_{\chi \in \mathrm{Irr}(G)} \chi(g_i)\overline{\chi(g_j)} = \delta_{i,j}|C_G(g_i)|$, where δ is the Kronecker delta on the set $\{1, \dots, r\}$.*

By Maschke's theorem (see [BZ, Chapter 1]), a representation T of G is equivalent to a sum of irreducible representations of G. The last sum can be written so: $a_1T_1 + a_2T_2 + \cdots + a_kT_k$, where $T_1, \dots, T_k$ are pairwise nonequivalent and $a_1, \dots, a_k$ are positive integers. Then $\chi^T = \sum_{i=1}^{k} a_i \chi^{T_i}$. By Theorem 15(a), $\langle \chi^T, \chi^T \rangle = a_1^2 + \cdots + a_k^2$ and $\langle \chi^T, \chi_i \rangle = a_i$, all i.

If χ is a character of G and $\langle \chi, \chi \rangle = 1$, then χ is irreducible.

Theorem 16. *If $g, h \in G$ are such that $\chi(g) = \chi(h)$ for all $\chi \in \mathrm{Irr}(G)$, then g, h are conjugate in G.*

This follows from Theorem 15(b).

Let χ be a character of G. We define the character $\bar{\chi}$ as follows: $\bar{\chi}(g) = \overline{\chi(g)}$, where $\bar{z}$ is the complex conjugate of $z \in \mathbb{C}$. If a matrix representation $T = (\alpha_{ij})$ of G affords the character χ and $\bar{T} = (\bar{\alpha}_{ij})$, then $\bar{T}$ is also a representation of G and $\chi^{\bar{T}} = \bar{\chi}$. Let us find $\bar{\chi}(g)$ for $g \in G$. Suppose as above that a representation T of G affords χ. If $\deg(T) = n$ and $\epsilon_1, \dots, \epsilon_n$ are eigenvalues of $T(g)$, then $\chi(g) = \epsilon_1 + \cdots + \epsilon_n$ since $T(g)$ is equivalent to the diagonal matrix $\mathrm{diag}(\epsilon_1, \dots, \epsilon_n)$, by Schur's lemma. It follows that $\mathrm{diag}(\epsilon_1^{o(g)}, \dots, \epsilon_n^{o(g)}) = T(g)^{o(g)} = T(g^{o(g)}) = \mathrm{I}_{o(g)}$ so $\epsilon_i^{o(g)} = 1$ for all i, i.e., $\epsilon_1, \dots, \epsilon_n$ are $o(g)$-th roots of unity. We have

$$T(g^{-1}) = \mathrm{diag}(\epsilon_1^{-1}, \dots, \epsilon_n^{-1}) = \mathrm{diag}(\bar{\epsilon}_1, \dots, \bar{\epsilon}_n) = \bar{T}(g),$$

so $\chi(g^{-1}) = \overline{\chi(g)}$ which we write as $\bar{\chi}(g)$ (this justifies (4)). If $\chi \in \mathrm{Irr}(G)$, then $\bar{\chi} \in \mathrm{Irr}(G)$. Indeed, $\langle \bar{\chi}, \bar{\chi} \rangle = |G|^{-1} \sum_{g \in G} \bar{\chi}(g)\overline{\bar{\chi}(g)} = \langle \chi, \chi \rangle = 1$.

Theorem 17 (Ito). *If $\chi \in \mathrm{Irr}(G)$, then $\chi(1)$ divides the index of any abelian normal subgroup of G.*

Let $H \le G$ and λ a class function on H. Let $G = \bigcup_{i=1}^{n} x_i H$ be a partition. Define a function $\lambda^G : G \to \mathbb{C}$ as follows: $\lambda^G(g) = |H|^{-1} \sum_{x \in G} \dot{\lambda}(x^{-1}gx) = \sum_{i=1}^{n} \dot{\lambda}(x_i^{-1}gx_i)$ for all $g \in G$, where $\dot{\lambda}(g) = \lambda(g)$ if $g \in H$ and 0 if $g \in G - H$. Clearly, $\dot{\lambda}$ is a class function on G. If λ is a character of H, then

(a) λ^G is a character of G; λ^G is said to be the character *induced* from H onto G or an *induced character* (indeed, λ^G is a character, by (c), below).

(b) $\ker(\lambda^G) = \ker(\lambda)_G$, where $H_G = \bigcap_{x \in G} H^x$ for $H \le G$.

(c) (Frobenius Reciprocity) If $H \le G$, $\chi \in \mathrm{Irr}(G)$, $\lambda \in \mathrm{Irr}(H)$, then $\langle \lambda^G, \chi \rangle = \langle \lambda, \chi_H \rangle$. Here χ_H is the restriction of χ to H.

(d) $\lambda^G(1) = |G : H|\lambda(1)$.

Set $\mathrm{Lin}(G) = \{\chi \in \mathrm{Irr}(G) \mid \chi(1) = 1\}$. Elements of the set $\mathrm{Lin}(G)$ are called *linear characters* of G; all other irreducible characters of G are called *nonlinear*. If $\lambda \in \mathrm{Lin}(G)$ is afforded by a representation T of G and $x, y \in G$, then $\lambda(xy) = T(xy) = T(x)T(y) = \lambda(x)\lambda(y)$ since $\lambda(x)$ can be identified with $T(x)$. It follows that a linear character of G is a homomorphism of G in $\mathbb{C}^*$, the multiplicative group of the field $\mathbb{C}$. It is clear that a function $1_G : G \to \mathbb{C}$ such that $1_G(x) = 1$ for all $x \in G$, is a linear character of G. This character is called *principal*. If V is a one-dimensional vector space, then a representation $T : G \to \mathrm{GL}(V)$ such that $(T(g))(v) = v$ for all $g \in G$ and $v \in V$, affords the principal character. If V is n-dimensional, then the representation T defined as above, affords the character $n \cdot 1_G$, a multiple of the principal character of G. If a representation T of G affords the character χ, then the function $\det(\chi) : G \to \mathbb{C}$, defined by $(\det(\chi))(g) = \det(T(g))$ $(g \in G)$, is a linear character of G, the *determinant* of χ. If $\lambda \in \mathrm{Lin}(G)$, then, obviously, $G' \le \ker(\lambda)$ (indeed, if $x, y \in G$, then $\lambda([x, y]) = \lambda(x)^{-1}\lambda(y)^{-1}\lambda(x)\lambda(y) = 1 \in \mathbb{C}^*$). We have $|\mathrm{Lin}(G)| = |G : G'|$. It follows that $G = G'$ if and only if $\mathrm{Lin}(G) = \{1_G\}$; then $\det(\chi) = 1_G$ for all $\chi \in \mathrm{Irr}(G)$.

A character χ of a group G is said to be *monomial* if there exist $H \le G$ and $\lambda \in \mathrm{Lin}(H)$ such that $\chi = \lambda^G$. Obviously, linear characters are monomial. A group G is said to be an M-*group* if all its irreducible characters are monomial. If $\chi \in \mathrm{Irr}(G)$ is monomial, then G has a subgroup of index $\chi(1)$.

Theorem 18 (Blichfeldt). *All p-groups are M-groups.*

Let χ_1, χ_2 be characters of G, $g \in G$, $\chi_i = \chi^{T_i}$, $i = 1, 2$, where T_1, T_2 are matrix representations of G. Then $(T_1 \otimes T_2)(g)$ is defined as the Kronecker product $T_1(g) \otimes T_2(g)$ of the matrices $T_1(g)$ and $T_2(g)$. Since

$$\begin{aligned}(T_1 \otimes T_2)(gh) &= T_1(gh) \otimes T_2(gh) = T_1(g)T_1(h) \otimes T_2(g)T_2(h)\\ &= (T_1 \otimes T_2)(g)(T_1 \otimes T_2)(h),\end{aligned}$$

it follows that $T_1 \otimes T_2$ is a representation of G, the *tensor product* of T_1 and T_2; its degree is $\deg(T_1) \cdot \deg(T_2)$. The product $\chi_1\chi_2$ of characters χ_1, χ_2 of G is defined as follows: $(\chi_1\chi_2)(g) = \chi_1(g)\chi_2(g)$, $g \in G$. We have $\chi^{T_1 \otimes T_2} = \chi^{T_1}\chi^{T_2}$ so the product of two characters of G is also a character of G. If $\chi_1, \chi_2 \in \mathrm{Irr}(G)$, then $\ker(\chi_1) \cap \ker(\chi_2) \le \ker(\chi_1\chi_2)$ and inequality is possible.

Let χ be a character of G and $\lambda \in \mathrm{Lin}(G)$. If T is a representation of G with character χ, then $T_1(x) = \lambda(x)T(x)$ defines the representation T_1 of G with character $\chi_1 = \lambda\chi$. We have $\langle \lambda\chi, \lambda\chi \rangle = \langle \chi, \bar{\lambda}\lambda\chi \rangle = \langle \chi, \chi \rangle$ so, if χ is irreducible then $\lambda\chi$ is also irreducible. It follows from what has just been said, that $\mathrm{Lin}(G)$ is a multiplicative abelian group acting on the set $\mathrm{Irr}(G)$ via multiplication.

Let $H \trianglelefteq G$. If λ is a character of H and $g \in G$, we define $\lambda^g : H \to \mathbb{C}$ by $\lambda^g(h) = \lambda(ghg^{-1})$. The function λ^g is a character of H (check!; to this end, build the representation of H affording λ^g). We say that λ^g is *conjugate* to λ in G. It is easy to see that $(\lambda, g) \mapsto \lambda^g$ defines an action of G on the set of characters of H. This action is isometric, i.e., $\langle \lambda^x, \mu^x \rangle = \langle \lambda, \mu \rangle$ for all characters λ, μ of H and $x \in G$. In particular, if $\lambda \in \mathrm{Irr}(H)$, then $\lambda^x \in \mathrm{Irr}(H)$. The stabilizer $\mathrm{I}_G(\lambda) = \{g \in G \mid \lambda^g = \lambda\}$ of a 'point' λ in G is called the *inertia subgroup* of $\lambda \in \mathrm{Irr}(H)$ in G. Since λ is a class function on H, we get $H \leq \mathrm{I}_G(\lambda)$.

Theorem 19 (Clifford). *Let $H \trianglelefteq G$ and $\chi \in \mathrm{Irr}(G)$. Then the following Clifford decomposition holds: $\chi_H = e(\lambda_1 + \cdots + \lambda_t)$, where $e \in \mathbb{N}$ is the ramification of χ with respect to H, $\lambda_1 \in \mathrm{Irr}(H)$, $\{\lambda_1, \dots, \lambda_t\}$ is the G-orbit of λ_1 (in particular, $t = |G : \mathrm{I}_G(\lambda_1)|$). Next, there exists $\theta \in \mathrm{Irr}(\mathrm{I}_G(\lambda_1))$ such that $\theta^G = \chi$ and $\theta_H = e\lambda_1$. The number e divides $|\mathrm{I}_G(\lambda_1) : H|$.*

Theorem 20. *If $\chi \in \mathrm{Irr}(G)$, then $\chi(1)^2 \leq |G : \mathrm{Z}(G)|$.*

Proof. By Theorem 19, $\chi_{\mathrm{Z}(G)} = \chi(1)\lambda$, where $\lambda \in \mathrm{Lin}(\mathrm{Z}(G))$. By reciprocity, $\langle \lambda^G, \chi \rangle = \langle \lambda, \chi_{\mathrm{Z}(G)} \rangle = \langle \lambda, \chi(1)\lambda \rangle = \chi(1)\langle \lambda, \lambda \rangle = \chi(1)$ so $\chi(1)\chi$ is a constituent of λ^G. In that case, $|G : \mathrm{Z}(G)| = \lambda^G(1) \geq (\chi(1)\chi)(1) = \chi(1)^2$. □

Thus, if G is a p-group then $\chi(1)^2$ divides $|G : \mathrm{Z}(G)|$ for all $\chi \in \mathrm{Irr}(G)$.

Exercise 51. Let $H \trianglelefteq G$, $\lambda \in \mathrm{Irr}(H)$. Then λ^G is irreducible if and only if $\mathrm{I}_G(\lambda) = H$.

Exercise 52. Using Theorem 19, prove Theorem 17.

Solution. Let $A \trianglelefteq G$ be abelian and $\chi \in \mathrm{Irr}(G)$. By Theorem 19, $\chi_A = e(\lambda_1 + \cdots + \lambda_t)$ (the Clifford decomposition). We have $\chi(1) = et$, since λ_i are linear for all i. By Theorem 19, $(\chi(1) =)\, et$ divides $|\mathrm{I}_G(\lambda) : A||G : \mathrm{I}_G(\lambda)| = |G : A|$.

Consider a function $\rho_G : g \mapsto \delta_{1,g}|G|$ $(g \in G)$, where δ is the Kronecker delta on the set G. Clearly, ρ_G is a class function vanishing on $G^\# = G - \{1\}$. If $\chi \in \mathrm{Irr}(G)$, then $\langle \chi, \rho_G \rangle = |G|^{-1}\chi(1)\delta_{1,1}|G| = \chi(1)$ so $\rho_G = \sum_{\chi \in \mathrm{Irr}(G)} \chi(1)\chi$ is a character of G, the *regular* character of G. We have $\rho_G(1) = |G|$. On the other hand, $\rho_G(1) = \sum_{\chi \in \mathrm{Irr}(G)} \chi(1)^2$, and we obtain (2).

Exercise 53. ρ_G is faithful. In particular, the following equality holds:

$$\bigcap_{\chi \in \mathrm{Irr}(G)} \ker(\chi) = \{1\}.$$

Next, if $\{1\} = H \leq G$, then $\rho_G = (1_H)^G$.

Solution. Let us prove the last assertion. If $\chi \in \mathrm{Irr}(G)$, then, by reciprocity, $\langle \chi, (1_H)^G \rangle) = \langle \chi_H, 1_H \rangle = \langle \chi(1)1_H, 1_H \rangle = \chi(1)$, so $(1_H)^G = \sum_{\chi \in \mathrm{Irr}(G)} \chi(1)\chi = \rho_G$.

The set of nonlinear irreducible characters of G equals $\mathrm{Irr}_1(G) = \mathrm{Irr}(G) - \mathrm{Lin}(G)$. We claim that $D = \bigcap_{\chi \in \mathrm{Irr}_1(G)} \ker(\chi) = \{1\}$. Take $x \in D$. In view of Exercise 53, it suffices to show that $x \in \ker(\lambda)$ for all $\lambda \in \mathrm{Lin}(G)$. Indeed, let $\chi \in \mathrm{Irr}_1(G)$ and $\lambda \in \mathrm{Lin}(G)$. Then $\lambda\chi \in \mathrm{Irr}_1(G)$ so $\chi(1) = (\lambda\chi)(1) = (\lambda\chi)(x) = \lambda(x)\chi(x) = \lambda(x)\chi(1)$ so $\lambda(x) = 1$.

Exercise 54. A p-group G has a faithful irreducible character if and only if $\mathrm{Z}(G)$ is cyclic. (See [BZ], §1.7; see also Exercise 55(a).)

Let χ be a character of G. The *quasikernel* $\mathrm{Z}(\chi)$ of χ is the set of all $g \in G$ such that $|\chi(g)| = \chi(1)$. If $\chi = \chi^T$, then $\mathrm{Z}(\chi) = \{x \in G \mid T(x) \text{ is a scalar matrix }\}$.

Exercise 55. Let T be a representation of a p-group G and $\chi = \chi^T$.

(a) $g \in \mathrm{Z}(\chi)$ if and only if $T(g)$ is a scalar matrix. In particular, $\mathrm{Z}(\chi) \trianglelefteq G$ and $\mathrm{Z}(G) \le \mathrm{Z}(\chi)$. (*Hint.* If $g \in \mathrm{Z}(\chi)$, then $|\chi(G)| = \chi(1)$ so all eigenvalues of $T(g)$ are equal.)

(b) If χ is irreducible, then $\mathrm{Z}(\chi)/\ker(\chi)$ is cyclic. If, in addition, χ is faithful, then $\mathrm{Z}(\chi) = \mathrm{Z}(G)$ is cyclic. (*Hint.* Clearly, $\ker(\chi) \le \mathrm{Z}(\chi)$. Use Exercise 54.)

(c) $\bigcap_{\chi\in\mathrm{Irr}(G)} \mathrm{Z}(\chi) = \mathrm{Z}(G)$.

(d) $\lambda \in \mathrm{Irr}(G)$ is linear if and only if $\mathrm{Z}(\lambda) = G$.

Solution. (c) If $x \in D = \bigcap_{\chi\in\mathrm{Irr}(G)} \mathrm{Z}(\chi)$, then by the Second Orthogonality Relation, $|\mathrm{C}_G(x)| = \sum_{\chi\in\mathrm{Irr}(G)} \chi(x)\overline{\chi}(x) = \sum_{\chi\in\mathrm{Irr}(G)} \chi(1)^2 = |G|$ so $x \in \mathrm{Z}(G)$. Since $\mathrm{Z}(G) \le D$, our claim follows.

Exercise 56. A group G is abelian if and only if $\mathrm{Irr}(G) = \mathrm{Lin}(G)$.

Exercise 57. Prove that $\bigcap_{\lambda\in\mathrm{Lin}(G)} \ker(\lambda) = G'$. (*Hint.* $\rho_{G/G'}$ is a faithful character of G/G'.)

Exercise 58. Let $H \trianglelefteq G$ and $x \in G$. Then $|\mathrm{C}_{G/H}(xH)| \le |\mathrm{C}_G(x)|$ with equality if and only if $\chi(x) = 0$ for every $\chi \in \mathrm{Irr}(G)$ not containing H in its kernel.

Let $g \in G$ be such that $\chi(g) = 0$ for all $\chi \in \mathrm{Irr}_1(G)$. By the Second Orthogonality Relation, we have $|\mathrm{C}_G(g)| = \sum_{\chi\in\mathrm{Irr}(G)} |\chi(g)|^2 = |G : G'|$ so all elements of the coset xG' are conjugate in G.

Exercise 59. Using the Second Orthogonality Relation, prove Theorem 16.

Exercise 60. If a p-group G has a faithful irreducible character χ of degree p, then it has an abelian subgroup of index p. (See [BZ, Chapter 18].)

Exercise 61. Let G be an abelian p-group and Z a cyclic direct factor of G. Find the number of direct complements to Z in G.

Exercise 62. Let $G = Z_1 \times \cdots \times Z_n$ be the direct product of cyclic p-groups Z_i, $i = 1, \dots, n$, $|Z_1| \ge \cdots \ge |Z_n|$. Find the number of direct complements M to Z_1 in G such that $M \cap Z_i = \{1\}$ for all i.

Exercise 63. Let $G = A \times B$ be an abelian p-group. Find the number of direct complements to A in G.

Exercise 64. Let a p-group $G = E \times A$, where E is nonabelian with $|\mathrm{Z}(E)| = p$ and A is abelian.

(a) Find the number of direct complements to E in G. (*Hint.* (a) Let T be a direct complement to E in G. Then $T \leq \mathrm{Z}(G) = \mathrm{Z}(E) \times A$ and T is a complement to $\mathrm{Z}(E)$ in $\mathrm{Z}(G)$, and conversely.)

(b) Find the number of direct complements to A in G.

Exercise 65. Let $G = M_1 \times M_2$, where M_1 and M_2 are nonabelian of order p^3. Show that the number of direct complements to M_1 in G equals p^2. (*Hint.* All direct complements to M_1 in G lie in $\mathrm{C}_G(M_1) = M_2 \times \mathrm{Z}(M_1)$.)

Exercise 66. Let $G = E_1 \times E_2$, where $|\mathrm{Z}(E_1)| = |\mathrm{Z}(E_2)| = p$. Find the number of direct complements to E_1 in G.

Exercise 67. Let a p-group $G = M \times E$, where M is of maximal class (see §1) and E is extraspecial (see §4). Find the numbers of direct complements to M and E in G.

Exercise 68. Let $P \in \mathrm{Syl}_p(G)$ be of order p^m and $n \leq m$. Denote by $s'_n(G)$ the number of subgroups of order p^n in G that are not contained in P. Show that $s'_n(G) \equiv 0 \pmod{p}$.

Exercise 69. Let $A, B \leq G$ be abelian and suppose that they normalize one another. Prove that AB is nilpotent of class at most two.

Theorem 21 (Fitting). *If $A, B \lhd G$ are nilpotent of classes a, b, respectively, then $AB \trianglelefteq G$ is nilpotent and* $\mathrm{cl}(AB) \leq a + b$.

Proof. One may assume that $G = AB$. Set $Z = \mathrm{Z}(A) \cap \mathrm{Z}(B)$; then $Z \leq \mathrm{Z}(G)$. Assume that $a, b > 0$. Set $e = a + b$ and use induction on e. Since $\{1\} < \mathrm{Z}(A) \lhd G$, one can consider the quotient group $\bar{G} = G/\mathrm{Z}(A)$. Then $\bar{G} = \bar{A}\bar{B}$, where $\bar{A} = A/\mathrm{Z}(A)$, $\bar{B} = B\mathrm{Z}(A)/\mathrm{Z}(A) \cong B/(B \cap \mathrm{Z}(A))$. Therefore $\bar{G}$ is a product of two normal subgroups: $\bar{A}$ of class $a - 1$ and $\bar{B}$ of class at most b. By induction, $\bar{G}$ is nilpotent of class at most $e - 1$. Similarly, $G/\mathrm{Z}(B)$ is nilpotent of class at most $e - 1$. Then G/Z, as a subgroup of $(G/\mathrm{Z}(A)) \times (G/\mathrm{Z}(B))$ (Exercise 27), is nilpotent of class at most $e - 1$. Since $Z \leq \mathrm{Z}(G)$, we get $\mathrm{cl}(G) \leq e = a + b$. □

Exercise 70. Let G be a p-group. Suppose that the class of every maximal subgroup of G is at most k. Then $\mathrm{cl}(G) \leq 2k$. Let, in addition, $k > 1$ and $\mathrm{cl}(G) \geq k + 2$. Let K be the subgroup generated by centers of all maximal subgroups of G. Then $K \leq \Phi(G)$ and K is abelian.

Solution. If $K \leq \Phi(G)$, it is abelian. Assume that $K \not\leq \Phi(G)$. Then there are distinct maximal A, B such that $\mathrm{Z}(B) \not\leq A$. In that case, $G = A\mathrm{Z}(B)$ so $\mathrm{cl}(G) \leq k + 1$, by Theorem 21, a contradiction.

Exercise 71 ([Bert2]). Let G be a nonabelian p-group. Then there exists $x \in G - \mathrm{Z}(G)$ such that $|\mathrm{C}_G(x)|^2 > |G|$.

Exercise 72. If $H < G$ and $\chi \in \mathrm{Irr}(G)$ is such that $\chi = \lambda^G$ for $\lambda \in \mathrm{Lin}(H)$, then $H' < G'$.

Exercise 73. Let G be a group such that $\phi : x \mapsto x^3$ is an endomorphism of G. Show that if $3 \nmid |G|$, then G is abelian.

Solution (M. Roitman). Prove that ϕ is a bijection. Deduce that $(yx)^2 = x^2y^2$ for $x, y \in G$. It follows that $(yx)^4 = (x^2y^2)^2 = y^4x^4$, and so $(xy)^3 = y^3x^3$. Thus, $y^3x^3 = (xy)^2(xy) = y^2x^2xy$ so $y\phi(x) = yx^3 = x^3y = \phi(x)y$. It follows that $y \in \mathrm{Z}(G)$ since ϕ is a surjection. Since y is arbitrary, G is abelian.

Exercise 74. Let G be an abelian p-group of exponent p^e and $V < G$. If G/V is homocyclic of exponent p^e, then $G = V \times W$. (*Hint.* Let $W \leq G$ be as small as possible such that $VW = G$. Show that $|W| = |G : V|$.)

Exercise 75. Let G be a p-group. For $p^n \leq \exp(G)$, let σ_n, c_n denote the numbers of solutions of $x^n = 1$ and cyclic subgroups of order p^n in G, respectively. (a) Suppose that we know numbers c_n for all divisors p^n of $\exp(G)$. Is it possible to find all numbers σ_n? (b) Suppose that we know numbers σ_n for all divisors p^n of $\exp(G)$. Is it possible to find all numbers c_n? (The answers are 'yes'. Indeed according to Miller, in that case, $\sigma_{p^n} = \sum_{i=0}^{n} \varphi(p^i)\mathrm{c}_i$, and this solves (a). Step by step, this also solves (b).)

Exercise 76. Let $H < G$, where G is a (finite) group. Suppose that, for every $a \in G$, there exists $x \in G$ such that $H^x \leq \mathrm{C}_G(a)$. Prove that $H \leq \mathrm{Z}(G)$. In particular, if a prime $p \nmid |G : \mathrm{C}_G(a)|$ for all $a \in G$, then $p \nmid |G : \mathrm{Z}(G)|$. (*Hint.* $\bigcup_{x \in G} \mathrm{C}_G(H)^x = G$ so $\mathrm{C}_G(H) = G$.)

Exercise 77. Let A_1, A_2 be nonincident subgroups of exponent $\leq p^e$ of a p-group G. If $A_1 \cup A_2$ is the set of all elements of order $\leq p^e$ in G, then A_1, A_2 are the only maximal subgroups of exponent $\leq p^e$ in G and $\exp(G) > p^e$. At least one of subgroups A_1, A_2 has exponent p^e.

Solution. By hypothesis, $A_1 \cup A_2 \neq G$ so $\exp(G) > p^e$. Assume that $A_1 < B < G$, where $\exp(B) \leq p^e$. Then $B - A_1 \subseteq A_2$ so $A_1 < B = \langle B - A_1 \rangle \leq A_2$, a contradiction. Thus, A_1 is a maximal subgroup of exponent $\leq p^e$ in G, and the same is true for A_2. Let U be a maximal subgroup of exponent $\leq p^e$ in G such that $A_1 \neq U \neq A_2$. Then the nonempty set $U - A_1$ is contained in A_2 so $U = \langle U - A_1 \rangle \leq A_2$, which is not the case. Thus, A_1 and A_2 are the only maximal subgroups of exponent $\leq p^e$ in G. The last assertion now follows since $\exp(G) > p^e$.

Exercise 78 ([PR, Lemma 2.29]). Let a 2-group $G = A_1A_2$, where $A_1, A_2 < G$ are elementary abelian. If for each $a \in A_2 - A_1$ we have $\mathrm{C}_{A_1}(a) = A_1 \cap A_2$, then $A_1 \cup A_2 = \{x \in G \mid x^2 = 1\}$.

Remark 3. Let G, A_1 and A_2 be such as in Exercise 78. If $a_1 \in A_1 - A_2$ and $a_2 \in A_2 - A_1$, then $\langle a_1, a_2 \rangle \cong \mathrm{D}_8$. Indeed, $K = \langle a_1, a_2 \rangle \cong \mathrm{D}_{2^n}$ for some $n > 2$ since a_1 and a_2 do not commute. Any two distinct four-subgroups of K are contained in distinct A_i's so K has exactly two four-subgroups hence $n = 3$.

Exercise 79 ([PR, Lemma 2.30]). Suppose that G is a 2-group and $G_0 \trianglelefteq G$. Also suppose that A_1 and A_2 are elementary abelian subgroups of G_0 with $G_0 = A_1 A_2$ and that for each $a \in A_2 - A_1$ we have $\mathrm{C}_{A_1}(a) = A_1 \cap A_2$. Then every normal elementary abelian subgroup of G normalizes both A_1 and A_2.

Solution. Let $A \lhd G$ be elementary abelian. By Exercise 78, all involutions of G_0 lie in $A_1 \cup A_2$. Set $B = A \cap G_0 (\trianglelefteq G)$. Assume that $B \not\leq A_1$. Then $\langle B - A_1 \rangle = B$ so $B \leq A_2$ since $B - A_1 \subseteq A_2$. Therefore, $[A, A_2] \leq [A, G_0] \leq A \cap G_0 = B \leq A_2$ and so A normalizes A_2. The claim now follows as A_1 and A_2 are all maximal elementary abelian subgroups of G_0 (Exercise 77).

Exercise 80 ([PR, Lemma 2.21]). Suppose that G is an operator group of the p-group Q. Let $\{1\} = Q_0 \trianglelefteq Q_1 \trianglelefteq \cdots \trianglelefteq Q_{n-1} \trianglelefteq Q_n = Q$ be a G-invariant series of Q. For $i = 1, \ldots, n$ set $\bar{Q}_i = Q_i / Q_{i-1}$. Then $|Q : \mathrm{C}_Q(G)| \geq \prod_{i=1}^n |\bar{Q}_i : \mathrm{C}_{\bar{Q}_i}(G)|$.

Exercise 81. Suppose that Q is a p-group and $A \leq \mathrm{Aut}(Q)$. If $[Q, A] \leq \mathrm{Z}(Q)$ and $[Q, A, A] \leq \Omega_k(\mathrm{Z}(Q))$, then $[\mho_k(Q), A, A] = \{1\}$.

Solution. By Exercise 13(a), for $x \in Q$ and $a \in A$ we have $[x^2, a] = [x, a]^x [x, a] = [x, a]^2$ since $[x, a] \in \mathrm{Z}(Q)$. It follows, by induction, that $[x^n, a] = [x, a]^n$ for $n \in \mathbb{N}$. Then, for $x \in Q, a, a_1 \in A$ we have $[x^{p^k}, a, a_1] = [[x^{p^k}, a], a_1] = [[x, a]^{p^k}, a_1] = [x, a, a_1]^{p^k} = 1$ since $[x, a, a_1] \in \Omega_k(\mathrm{Z}(Q))$. Moreover, if $x_1, \ldots, x_s \in Q$ and $a, a_1 \in A$, then $[\prod_{i=1}^s x_i^{p^k}, a, a_1] = \prod_{i=1}^s [x_i, a, a_1]^{p^k} = 1$, and we are done.

Exercise 82. Suppose that Q is a p-group and $x \in Q$. Then $|Q : \mathrm{C}_Q(x)| \leq |[Q, x]|$. If, in addition, $[Q, x] \leq \mathrm{Z}(Q)$, then $|Q : \mathrm{C}_Q(x)| = |[Q, x]|$.

Solution. We have $|Q : \mathrm{C}_Q(x)| = |\{[w, x] \mid w \in Q\}| \leq |[Q, x]|$. Indeed, the number of Q-conjugates of x equals the number of Q-conjugates of x^{-1}. The set of Q-conjugates of x^{-1} is $\{w^{-1} x^{-1} w \mid w \in Q\}$, and our formula holds since $|\{[w^{-1} x^{-1} w x \mid w \in Q\}| = |\{w^{-1} x^{-1} w \mid w \in Q\}|$. In the event that $[Q, x] \leq \mathrm{Z}(Q)$, for $w_1, w_2 \in Q$, we have $[w_1, x][w_2, x] = [w_1 w_2, x]$, and so $\{[w, x] \mid w \in Q\} = [Q, x]$.

Exercise 83. Let $U_1, \ldots, U_r$ be proper subgroups of a group G. If $G = U_1 \cup \cdots \cup U_r$, then $r \geq p + 1$, where p is the minimal prime divisor of $|G|$.

Exercise 84 (see Appendix 7). Let G be a p-group, $N \lhd G$, and let $g \in G$ be such that $[N : \mathrm{C}_N(g)| \leq p$. Setting $H = \langle g, N \rangle$, prove that $H' \leq \mathrm{C}_N(g)$.

Exercise 85. Let $G = \langle a, b \mid a^{p^n} = b^{p^2} = 1, a^b = a^{1+p^{n-1}} \rangle$. Prove that G contains exactly p normal cyclic subgroups of order p^n.

Exercise 86 ([Dol]). Let M be a maximal subgroup of an abelian p-group G. Then there exists a subgroup of M that has a cyclic (direct) complement in G.

Solution. We use induction on $|G|$. Let $\exp(G) = p^e$ and suppose that $\exp(M) = p^e$; then M contains an element y of order p^e. Set $T = \langle y \rangle$; then $G = G_1 \times T$, where $G_1 < G$ (Lemma 4(b)). We have $M = T \times (M \cap G_1)$, by the modular law. Since $M_1 = M \cap G_1$ is maximal in G_1, we get, by induction, that $G_1 = N_1 \times C$ with $N_1 \leq M_1$ and C is cyclic. Then $G = T \times G_1 = T \times N_1 \times C = N \times C$, where $N = T \times N_1 \leq M$. Now suppose that $\exp(M) < p^e$ and take again $y \in G$ with $o(y) = p^e$. We have $\langle y \rangle \cap M = \langle y^p \rangle$ and $M = N \times \langle y^p \rangle$ for some $N \leq M$ (Lemma 4(b)) since $o(y^p) = \exp(M)$. Then $G = N \times \langle y \rangle$.

Exercise 87 (Hall). Suppose that a p-group S acts faithfully on an abelian p-group A. Then $\mathrm{C}_A(S) < \mathrm{C}_A(S')$.

Solution. Let $Z = \mathrm{C}_A(S)$; then Z is the center of the natural semidirect product $S \cdot A$ and let $Z_2/Z = \mathrm{C}_{A/Z}(S)$. Then $[S, Z_2, S] \leq [Z, S] = \{1\}$ and $[Z_2, S, S] \leq [Z, S] = \{1\}$. Therefore, by Three Subgroups Lemma, $[S', Z_2] = [S, S, Z_2] = \{1\}$. It follows that $Z_2 \leq \mathrm{C}_A(S')$. Since $Z_2 > Z$, we are done.

Exercise 88. If a group G is solvable, then $\mathrm{C}_G(\mathrm{F}(G)) \leq \mathrm{F}(G))$, where $\mathrm{F}(G)$ is the Fitting subgroup of G.

Solution. Assume that $\mathrm{C}_G(\mathrm{F}(G)) \not\leq \mathrm{F}(G)$. Let L be the least G-invariant subgroup of $\mathrm{C}_G(\mathrm{F}(G))$ not contained in $\mathrm{F}(G)$. Then $L\mathrm{F}(G)/\mathrm{F}(G) \cong L/(L \cap \mathrm{F}(G))$ is a prime power group. Since $L \cap \mathrm{F}(G)) \leq \mathrm{Z}(\mathrm{F}(G))$, the subgroup L is nilpotent. However, by Theorem 21, $L \leq \mathrm{F}(G)$, contrary to the choice of L.

Exercise 89. If G is a group, $x, y \in G$ and $n \in \mathbb{N}$, then $y^{-n}x^{-n}(xy)^n \in G'$.

Exercise 90. Let G be a nonabelian group. Then all $\chi \in \mathrm{Irr}_1(G)$ are faithful if and only if G' is a unique minimal normal subgroup of G.

§1

Groups with a cyclic subgroup of index p. Frattini subgroup. Varia

In this long section we first classify the p-groups with a cyclic subgroup of index p and deduce a number of consequences of that basic result. Another our intention is to familiarize the reader with some main themes of this book. An essential part of the material of this section is presented in book form for the first time.

The following result is basic since it is used in our book hundreds times.

Lemma 1.1. *Let A be an abelian subgroup of index p of a nonabelian p-group G. Then $|G| = p \cdot |G'| \cdot |Z(G)|$.*

Proof. Note that $A \lhd G$. Let $g \in G - A$ and let $\phi : A \to A$ be a mapping defined as follows: $\phi(a) = [a, g]$ $(a \in A)$. If $a, b \in A$, then

$$\phi(ab) = [ab, g] = [a, g]^b \cdot [b, g] = [a, g] \cdot [b, g] = \phi(a)\phi(b),$$

since $[a, g], b \in A$ and A is abelian, i.e., ϕ is a homomorphism. Next, $\ker(\phi) = \{a \in A \mid [a, g] = 1\} = C_A(g) = Z(G)$ since $G = \langle g, A\rangle$ is nonabelian. Let $K = \mathrm{im}(\phi)$; then $K \le G' < A$. Since $[a, g]^g = [a^g, g] \in K \cap A$ and A is abelian, it follows that $K \lhd G$. Since $a^g = a[a, g] \in aK$, g centralizes A/K so G/K is abelian and $K = G'$. Since $A/\ker(\phi) \cong \mathrm{im}(\phi)$, we get $|G| = p \cdot |A| = p \cdot |\ker(\phi)| \cdot |\mathrm{im}(\phi)| = p \cdot |Z(G)| \cdot |G'|$. □

According to [Tua1], if a p-group G has a normal abelian subgroup A with cyclic quotient group G/A, then $A/(A \cap Z(G)) \cong G'$ (see also [Isa1, Lemma 12.12]).

A group G is said to be *minimal nonabelian* if it is nonabelian but all its proper subgroups are abelian (in §65 such G is called an $\mathcal{A}_1$-group).

Exercise 1. If G is a minimal nonabelian p-group, then $|G'| = p$ and $G = \langle x, y\rangle$ for some $x, y \in G$. (*Hint.* If $A, B < G$ are distinct maximal, then $A \cap B = Z(G)$ has index p^2 in G. By Lemma 1.1, $|G'| = p$. Let x, y be noncommuting elements of G; then a nonabelian subgroup $\langle x, y\rangle = G$.)

Groups described in the following theorem, play important role in the book.

Theorem 1.2 (Burnside, 1897, in the first edition of his book). *Let G be a nonabelian p-group of order p^{n+1} with cyclic subgroup $A = \langle a\rangle$ of index p. Then G is isomor-*

phic to one of the following groups:

(a) $\mathrm{M}_{p^{n+1}} = \langle a, b \mid a^{p^n} = b^p = 1,\ b^{-1}ab = a^{1+p^{n-1}} \rangle$, *where* $n \geq 3$ *if* $p = 2$. *In that case,* $|G'| = p$, $\mathrm{Z}(G) = \Phi(G)$, $|\Omega_1(G)| = p^2$.

(b) $p = 2$, $\mathrm{D}_{2^{n+1}} = \langle a, b \mid a^{2^n} = b^2 = 1,\ bab = a^{-1} \rangle$, *the dihedral group. All elements in* $G - \langle a \rangle$ *are involutions.*

(c) $p = 2$, $\mathrm{Q}_{2^{n+1}} = \langle a, b \mid a^{2^n} = 1,\ b^2 = a^{2^{n-1}},\ b^{-1}ab = a^{-1} \rangle$, *the generalized quaternion group. The group* G *contains exactly one involution, all elements in* $G - \langle a \rangle$ *have the same order* 4, $G/\mathrm{Z}(G)$ *is dihedral if* $n > 2$.

(d) $p = 2$, $\mathrm{SD}_{2^{n+1}} = \langle a, b \mid a^{2^n} = b^2 = 1,\ bab = a^{-1+2^{n-1}} \rangle$, $n > 2$, *the semidihedral group. We have* $\Omega_1(G) = \langle a^2, b \rangle \cong \mathrm{D}_{2^n}$, $\langle a^2, ab \rangle \cong Q_{2^n}$ *so maximal subgroups are characteristic in* G, $G/\mathrm{Z}(G)$ *is dihedral.*

In cases (b)–(d), we have $|G : G'| = 4$, $|\mathrm{Z}(G)| = 2$, $\mathrm{cl}(G) = n$.

Proof. Since G is nonabelian, $n > 1$. Let $b \in G - A$; then $G = \langle b, A \rangle$. Since $A \triangleleft G$ and G is nonabelian, we have $b^{-1}ab = a^r$, where $r \in \mathbb{N}$ and $1 < r < p^n$. We have $a^{r-1} = a^{-1}a^b = [a, b] \in G' \leq \langle a^p \rangle$ so p divides $r - 1$. Since $b^p \in \mathrm{Z}(G)$, we get $a = b^{-p}ab^p = a^{r^p}$, $a^{r^p-1} = 1$, and so $o(a) = p^n$ divides $r^p - 1$. Let b and a be chosen so that $b^p = a^{p^s}$, where $s \leq n$ is as large as possible; then $s > 0$ since G is not cyclic. This choice yields: $o(b) \leq o(x)$ for all $x \in G - A$. We have $[b, a^{p^s}] = 1$, $[a, b] = a^{r-1}$ so $o(a) = p^n \nmid r - 1$ since $\langle [a, b] \rangle = G' < \langle a \rangle$.

(i) Let $p = 2$; then r is odd and $o(a) = 2^n$ divides $r^2 - 1 = (r-1)(r+1)$, and so, if $n > 2$, 2^{n-1} divides exactly one of the numbers $r-1, r+1$ since $\mathrm{GCD}(r-1, r+1) = 2 < 2^{n-1}$.

(i1) Suppose that $n = 2$. Then $|G| = 2^3$. In any case, $b^{-1}ab = a^{-1}$ since $a \to a^{-1}$ is a unique nonidentity automorphism of A. If $b^2 = a^2$, then a^2 is a unique involution of G since, by the choice of b, $G - A$ has no involutions. Then $G = \langle a, b \mid a^4 = 1,\ a^2 = b^2,\ a^b = a^{-1} \rangle \cong \mathrm{Q}_{2^3}$, the ordinary quaternion group. If $b^2 = 1$, then any element of $G - A$ is an involution since it has the form ba^i, $i = 0, 1, 2, 3$, and $(ba^i)^2 = (ba^ib)a^i = a^{-i}a^i = 1$. In that case, G is generated by two involutions b and ba so $G = \langle a, b \mid a^4 = b^2 = 1,\ a^b = a^{-1} \rangle \cong \mathrm{D}_{2^3}$, the dihedral group of order 2^3. Now let $n > 2$.

(i2) Suppose that 2^{n-1} divides $r - 1$. Then $r = 1 + 2^{n-1}$.

Let $o(b) = 2$; then $bab = a^{1+2^{n-1}}$ and $G = \langle a, b \mid a^{2^n} = b^2 = 1,\ bab = a^{1+2^{n-1}} \rangle \cong \mathrm{M}_{2^{n+1}}$. We have $(a^2)^b = a^{2(1+2^{n-1})} = a^2$ so $\mathrm{Z}(G) = \langle a^2 \rangle$. Next, $G' = \langle a^{-1}bab \rangle = \langle a^{2^{n-1}} \rangle$ is of order 2. Thus, G is as in (a).

Now let $o(b) > 2$. Set $b_1 = b^{-1}a^{2^{s-1}}$; then $b_1 \in G - A$, and so $G = \langle a, b_1 \rangle$. We have $b_1^2 = b^{-1}a^{2^{s-1}}b \cdot b^{-2}a^{2^{s-1}} = a^{2^{s-1}r} \cdot a^{-2^s}a^{2^{s-1}} = a^{2^{n-1+s-1}}$. By the choice of b, $n - 1 + s - 1 \leq s$ so $n \leq 2$, which is not the case.

(i3) Suppose that 2^{n-1} divides $r + 1$. Then $r = -1 + \epsilon \cdot 2^{n-1}$, where $\epsilon \in \{0, 1\}$.

Let $o(b) > 2$ for all $b \in G - A$. Then G has only one involution $a^{2^{n-1}}$.

Let $\epsilon = 0$; then $r = -1$, $b^{-1}ab = a^{-1}$ and $a^{2^s} = b^{-1}a^{2^s}b = a^{-2^s}$. Therefore, $a^{2^{s+1}} = 1$ so $2^{s+1} = 2^n$, $s = n-1$ and $b^2 = a^{2^{n-1}}$; then G is as in (c).

Let $\epsilon = 1$. Then $r = -1 + 2^{n-1}$. If $o(b) = 2$, then G is as in (d). Now let $o(b) > 2$; then $s < n$. We get $b^{-1}ab = a^{-1+2^{n-1}}$, $a^{2^s} = b^{-1}a^{2^s}b = a^{2^s(-1+2^{n-1})}$, and so $o(a) = 2^n$ divides $2^s(-1+2^{n-1}) - 2^s = 2^{s+1}(-1+2^{n-2})$. Thus, $s = n-1$, $b^2 = a^{2^{n-1}}$. We have $(ab)^2 = ab^2 \cdot b^{-1}ab = a^{1+2^{n-1}-1+2^{n-1}} = a^{2^n} = 1$, contrary to the choice of b since $ab \in G - A$ and $o(ab) = 2 < o(b)$.

(ii) Let $p > 2$. Then $o(a) = p^n$ divides $r^p - 1$. Set $r = t + 1$. Then

$$\frac{r^p - 1}{r - 1} = \frac{(t+1)^p - 1}{t} = p + \binom{p}{2}t + \cdots + \binom{p}{p-1}t^{p-2} + t^{p-1} =: x.$$

We have $p^n \nmid r - 1$ since $1 < r < p^n$ so p divides x. Since p divides t, we get p^2 divides $x - p$ so $p^2 \nmid x$. Then p^{n-1} divides $r - 1$ so $r = 1 + kp^{n-1}$ for some $k \in \{1, \dots, p-1\}$. It follows from $b^{-1}a^pb = a^{p(1+kp^{n-1})} = a^p$ that $a^p \in \mathrm{Z}(G)$ so $\mathrm{cl}(G) = 2$. Set $b_1 = b^i$, where i satisfies the congruence $ik \equiv 1 \pmod p$. Then $r^i = (1+kp^{n-1})^i \equiv 1+ikp^{n-1} \equiv 1+p^{n-1} \pmod{p^n}$. Therefore, $b_1^{-1}ab_1 = b^{-i}ab^i = a^{r^i} = a^{1+p^{n-1}}$. Since $|G'| = p$, we get $(a^jb_1^{-1})^p = a^{jp}b_1^{-p}$. We have $b_1^p = a^u \in \langle a \rangle$, where $u = pv$ for some integer v since G is not cyclic. Setting $b_2 = a^vb_1^{-1}$, we get $b_2^p = (a^vb_1^{-1})^p = a^{vp}b_1^{-p} = 1$ so $o(b_2) = p$. Since $a^{b_2^{-1}} = a^{b_1a^{-v}} = (a^{b_1})^{a^{-v}} = b_1^{-1}ab_1 = a^{1+p^{n-1}}$, we get $G = \langle a, b_2 \mid a^{p^n} = b_2^p = 1,\ b_2ab_2^{-1} = a^{1+p^{n-1}} \rangle$, and we obtain group (a) with odd p. We have $G' = \langle a^{p^{n-1}} \rangle$ so $|G : \mathrm{Z}(G)| = p|G'| = p^2$ (Lemma 1.1) and G is minimal nonabelian.

Let us prove that groups (b), (c), (d) are all of class n. In cases (b) and (c) we have $[a, b] = a^{-2}$, in case (d) we have $[a, b] = a^{-2+2^{n-1}}$, and these commutators have order 2^{n-1}. It follows that $|G : G'| = 4$ so $|\mathrm{Z}(G)| = 2$ (Lemma 1.1). Set $\bar{G} = G/\mathrm{Z}(G)$. If $n = 2$, our claim is trivial since then $|G| = 8$. Let $n > 2$. Since $\mathrm{Z}(G) < G'$, we get $|\bar{G} : \bar{G}'| = 4$ so $|\mathrm{Z}(\bar{G})| = 2$. Working by induction on $|G|$, we conclude that $\mathrm{cl}(\bar{G}) = n - 1$ so $\mathrm{cl}(G) = n$ since $|\mathrm{Z}(G)| = 2$.

We suggest to the reader to prove all remaining assertions. □

Remark 1 (general remark on cyclic extensions [Scot, Theorem 9.7.1]). Consider the determination of all extensions of A by B in case of cyclic B. Let A be a group and B cyclic of order n. Let $A \lhd G$ and $G/A \cong B$; then $G = \langle A, x \rangle$, $x^n = y \in A$, $y^x = y$ and x^n and y give the same automorphism on A. Let $t \in \mathrm{Aut}(A)$, $y \in A$, $t(y) = y$ and $t^n : a \mapsto a^y$ $(a \in A)$. Then there are G and x such that $A \lhd G$, $G = \langle A, x \rangle$, $G/A \cong B$, $x^n = y$ and $a^t = a^x$ for all $a \in A$. It follows from this that groups (a-d) of Theorem 1.2 exist. Since each finite p-group is a sequence of cyclic extensions, one can apply the above result in that case. We use this fact freely in what follows.

Theorem 1.2 describes all automorphisms of order p of cyclic p-groups.

Let $\mathfrak{H}_n = \mathrm{Hol}(p^n)$ be the holomorph of the cyclic group $\mathrm{C}_{p^n} = \langle a \rangle$.

Let $p > 2$. We have $\mathfrak{H}_1 = \langle a, \sigma \rangle$, where $a^\sigma = a^r$ with $o(r \pmod p)) = p - 1$. If $n > 1$, then $\mathfrak{H}_n = \langle a, \sigma, \tau \rangle$, where σ is as above, $a^\tau = a^{1+p}$ and $\langle \sigma, \tau \rangle$ is cyclic.

Let $p = 2$. We have $\mathfrak{H}_1 = \langle a \rangle$, $\mathfrak{H}_2 = \langle a, \sigma \rangle$, where $a^\sigma = a^{-1}$ so $\mathfrak{H}_2 \cong \mathrm{D}_8$. Now let $n > 2$. Then $\mathfrak{H}_n = \langle a, \sigma, \tau \rangle$, where $a^\sigma = a^{-1}$ and $a^\tau = a^5$. It follows that $\langle a, \sigma \rangle \cong \mathrm{D}_{2^{n+1}}$, $\langle a, \tau^{2^{n-3}} \rangle = \mathrm{M}_{2^{n+1}}$. $\langle a, \sigma\tau^{2^{n-3}} \rangle \cong \mathrm{SD}_{2^{n+1}}$, $\langle \sigma, \tau \rangle \cong \mathrm{C}_{2^{n-2}} \times \mathrm{C}_2$.

Definition 1. Let Γ_1 be the set of all maximal subgroups of an arbitrary finite group $G > \{1\}$. The subgroup $\Phi(G) = \bigcap_{H \in \Gamma_1} H$ is called the *Frattini subgroup* (or Φ-subgroup) of G. By definition, $\Phi(\{1\}) = \{1\}$.

Let $G \cong \mathrm{SD}_{2^{n+1}}$. Then $(ab)^2 = abab = aa^b = a^{2^{n-1}}$ and $(a^2)^{ab} = (a^2)^b = a^{2(-1+2^{n-1})} = a^{-2}$ so that $H_1 = \langle a^2, ab \rangle \cong \mathrm{Q}_{2^n}$. Similarly, $H_2 = \langle a^2, b \rangle \cong \mathrm{D}_{2^n}$. Thus, $\Gamma_1 = \{H_1, H_2, \langle a \rangle\}$.

Let $M < G$ and let $A < G$ be as small as possible and such that $G = \langle M, A \rangle$; then $M \cap A \leq \Phi(A)$. Indeed, assume that this is false. Then A has a maximal subgroup B such that $A \cap M \not\leq B$ so $A = \langle B, A \cap M \rangle$ and $G = \langle A, M \rangle = \langle B, A \cap M, M \rangle = \langle B, M \rangle$, contrary to the choice of A.

Proposition 1.3. *Suppose that a p-group G of order p^m has only one nontrivial subgroup S of order p^n, $0 < n < m$. Then G is cyclic, unless $n = 1$ and $G \cong \mathrm{Q}_{2^m}$.*

Proof. We use induction on $|G|$. If $m = n + 1$, then $|\Gamma_1| = 1$ so G is cyclic. Let $m > n + 1$. If $H \in \Gamma_1$, then H is either cyclic or generalized quaternion, by induction. Let $R \lhd G$ be of order p^2. Then $|G : \mathrm{C}_G(R)| \leq p$ since $|\mathrm{Aut}(R)|_p = p$. Let $R \leq M \leq \mathrm{C}_G(R)$, where $M \in \Gamma_1$. Then M is cyclic since $M \not\cong \mathrm{Q}_{2^{m-1}}$, and the result follows from Theorem 1.2. □

We consider the expressions 'G-invariant subgroup' and 'normal subgroup of G' as synonyms.

Lemma 1.4. *Let G be a p-group and $N \trianglelefteq G$. If N has no abelian G-invariant subgroups of type (p, p), then N is either cyclic or one of the groups (b), (c), (d) of Theorem 1.2. If, in addition, $N \leq \Phi(G)$, then N is cyclic.*

Proof. In view of Proposition 1.3 and Theorem 1.2, one may assume that N is nonabelian (then $|N| > p^2$) and N has no cyclic subgroups of index p. Let $A < N$ be maximal G-invariant abelian; then A is cyclic of order $> p$ since $\mathrm{C}_N(A) = A$ (see Lemma 1.9, below) so $|N : A| > p$. Let $R \leq A$ be of order p^2; then $R \lhd G$. Let B/A be a G-invariant subgroup of order p in N/A such that $B \leq \mathrm{C}_N(R)$. By the choice and hypothesis, the nonabelian subgroup B has no characteristic subgroups of type (p, p) and B is not a group of Theorem 1.2(b,c,d). Then B is cyclic (Theorem 1.2), contrary to the choice of A. Now let $N \leq \Phi(G)$ and R be as above. Since $|G : \mathrm{C}_G(R)| \leq p$, we get $N \leq \Phi(G) < \mathrm{C}_G(R)$ so N is cyclic, by the above. □

Let G be a p-group and let $\chi \in \mathrm{Irr}(G)$ be faithful. If G is neither cyclic nor as in Theorem 1.2(b,c,d), then there exists $N \in \Gamma_1$ such that χ_N is reducible. Indeed, by Lemma 1.4, there is $R \trianglelefteq G$ of type (p, p). Let N be a maximal subgroup of G such that $R \leq N \leq \mathrm{C}_G(R)$. Then χ_N is reducible since $\mathrm{Z}(N)$ is noncyclic and χ_N is a faithful character of N.

Recall that $\mathrm{Z}_i(G)$ is the i-th member of the upper central series of G.

Corollary 1.5. *Let N be a normal noncyclic subgroup of a p-group G. If $N \cap \mathrm{Z}_2(G)$ is cyclic, then N is as in Theorem 1.2(b,c,d).*

Proof. Let R be a G-invariant subgroup of order p^2 in N. Since $R \leq N \cap \mathrm{Z}_2(G)$, R is cyclic, and the result follows from Lemma 1.4. □

Proposition 1.6 (O. Taussky). *Let G be a nonabelian 2-group and $|G : G'| = 4$. Then G is as in Theorem 1.2(b,c,d).*

Proof. Suppose that G is a counterexample of minimal order. In view of Theorem 1.2, one may assume that G has no cyclic subgroups of index 2; then $|G| > 2^3$. Let $R \leq \mathrm{Z}(G) \cap G'$ be of order 2. By induction, G/R is one of the groups (b), (c), (d) of Theorem 1.2 since it is nonabelian. Let T/R be a cyclic subgroup of index 2 in G/R. Then $T = R \times Z$ is abelian of type $(2^n, 2)$ for some $n > 1$, $|Z| = 2^n$. By Lemma 1.1, $|\mathrm{Z}(G)| = \frac{1}{2}|G : G'| = 2$ which is not the case since $\mathrm{Z}(G) \geq R \times \Omega_1(Z) \cong \mathrm{E}_4$. □

A nonabelian p-group G has only one normal subgroup of index p^2 if and only if $G' = \Phi(G)$ is of index p^2 in G. In particular, a nonabelian 2-group has only one normal subgroup of index 4 if and only if it is as in Theorem 1.2(b,c,d).

Definition 2. A group G of order p^m is said to be of *maximal class* if $m > 2$ and $\mathrm{cl}(G) = m - 1$.

Corollary 1.7. *Let G be a 2-group of maximal class. Then it is as in Theorem 1.2(b,c,d) since $|G : G'| = 4$.*

Below we use the following fact. If $B < G$ is of order p, then $\mathrm{C}_G(B) = \mathrm{N}_G(B)$.

Proposition 1.8 (M. Suzuki). *Let G be a nonabelian p-group. If $A < G$ of order p^2 is such that $\mathrm{C}_G(A) = A$, then G is of maximal class.*

Proof. We use induction on $|G|$. Since $p^2 \nmid |\mathrm{Aut}(A)|$, $\mathrm{N}_G(A)$ is nonabelian of order p^3, by N/C-theorem; then $\mathrm{Z}(G) < A$ so $|\mathrm{Z}(G)| = p$. Set $\bar{G} = G/\mathrm{Z}(G)$. Since $\mathrm{C}_{\bar{G}}(\overline{\mathrm{N}_G(A)}) \leq \mathrm{C}_{\bar{G}}(\bar{A}) = \overline{\mathrm{N}_G(A)}$ is of order p^2, $\bar{G}$ is of maximal class so G is also of maximal class since $|\mathrm{Z}(G)| = p$. □

Remark 2. Let $Z < G$ be maximal cyclic, $|Z| > p$. We claim that if $Z \cap Z_1 \in \{\{1\}, Z\}$ for every cyclic $Z_1 < G$, $|Z_1| = |Z|$, then $p = 2$ and G is of maximal class. If $Z < M \leq G$ with $|M : Z| = p$, then $p = 2$ and M is of maximal class (Theorem

1.2) so $C_G(Z) = Z$ and $Z(G) < Z < M$; then $Z(G) = \Omega_1(Z)$ is of order 2. Assume that G is not of maximal class. Then $|Z| > 4$, by Proposition 1.8, and G has a normal four-subgroup R, by Lemma 1.4. We have $|Z \cap R| = 2 = |RZ : Z|$ and RZ is not of maximal class since $|RZ| > 2^3$, contrary to what has just been proved.

Exercise 2. If nonabelian p-group G is minimal noncyclic, then $G \cong Q_{2^3}$.

Exercise 3. If any two distinct maximal cyclic subgroups of a nonabelian p-group G, $p > 2$, have trivial intersection, then $\exp(G) = p$.

Exercise 4. Let a nonabelian p-group G have an abelian subgroup of index p. Then the following assertions are equivalent: (a) $|Z(G)| = p$, (b) $|G : G'| = p^2$, (c) G is of maximal class, (d) nonabelian subgroups of G are of maximal class and the number of nonabelian subgroups of index p^n in G equals p^n provided $|G| \geq p^{n+3}$.

Exercise 5. Suppose that G is a nonabelian p-group, $|Z(G)| = p$ and let $H \in \Gamma_1$ be such that $H = Z(G) \times F$ for some $F < H$. Then G is a subgroup of $\Sigma_{p^2} \in \mathrm{Syl}_p(S_{p^2})$. In particular, G is of maximal class and H is elementary abelian. (*Hint.* $|G : F| = p^2$ and $F_G = \{1\}$. Use Exercise 4.)

Exercise 6. Let G be a nonabelian p-group.

(a) (Miller) The number of abelian subgroups of index p in G is 0, 1 or $p + 1$. (*Solution.* If $A, B \in \Gamma_1$ are distinct abelian, then $A \cap B = Z(G)$. All abelian members of the set Γ_1 contain $Z(G)$.)

(b) G is generated by its minimal nonabelian subgroups. (See Theorem 10.28.)

(c) If $d(G) = 2$ and $\mho_1(G) = G'$, then either G is of maximal class or $p > 2$ and $G \cong M_{p^3}$. (*Hint.* If $p = 2$, use Proposition 1.6. Let $p > 2$. Assuming that $|G'| > p$, take $R \leq G' \cap Z(G)$ of order p. By induction on $|G|$, $G/R \cong M_{p^3}$. Then G is minimal nonabelian, contrary to Exercise 1.)

(d) Let $H \lhd G$ and $H \not\leq \Phi(G)$. Suppose that all maximal subgroups of G not containing H, are abelian. Prove that H is abelian.

Exercise 7 (Miller). If $N \lhd G$ has an abelian subgroup of index p, then N has a G-invariant abelian subgroup of index p. In particular, if G has an abelian subgroup of index p^2, it has a normal abelian subgroup of the same index. (*Hint.* Use Exercise 6(a). It is proved in [Kon1] that if $p > 2$ and $N \lhd G$ has an abelian subgroup of index p^2, then it has a G-invariant abelian subgroup of index p^2.)

Lemma 1.9. *Let G be a p-group and $N \lhd G$. If A is a maximal G-invariant abelian subgroup of N, then $C_N(A) = A$.*

Proof. Assume that $C = C_N(A) > A$. Then $C = N \cap C_G(A) \lhd G$. Let B/A be a G/A-invariant subgroup of order p in C/A. Then abelian $B \lhd G$ and $A < B \leq N$, contrary to the choice of A. □

Exercise 8. Let G be a p-group, $N \trianglelefteq G$, $|N| > p^3$. Then N has a G-invariant abelian subgroup of order p^3. (*Hint.* Use Exercise 7.)

Let G be a p-group. Set $\Omega_n(G) = \langle x \in G \mid x^{p^n} = 1 \rangle$, $\mho_n(G) = \langle x^{p^n} \mid x \in G \rangle$.

Let G be a p-group of class 2 and $\mho_1(G) \le \mathrm{Z}(G)$. We claim that then $\exp(G') = p$. Since G' is abelian, it suffices to show that all commutators have orders $\le p$. Take $x, y \in G$. Then $[x, y]^p = [x, y^p] = 1$, and we are done. In this case, for $p > 2$ and $x, y \in G$, we have $(xy)^p = x^p y^p$ (such groups G are called *p-abelian*).

Suppose that G is a p-group of class 2, $p > 2$ and $\mho_1(G) \le \mathrm{Z}(G)$. If a group A acts on G in such a way that $[A, \mathrm{Z}(G)] = \{1\}$ and $\langle g^{-1}g^a \mid g \in G, a \in A \rangle = G$, then $\exp(G) = p$. Indeed, G is p-abelian and $\exp(G') = p$, by the previous paragraph, so $\phi : x \to x^p$ is a homomorphism of G to $\mathrm{Z}(G)$. Let $x \in G$ and $a \in A$. Then $(x^{-1}x^a)^\phi = (x^p)^{-1}(x^p)^a = (x^p)^{-1}x^p = 1$. Since $G = \langle x^{-1}x^a \mid x \in G, a \in A \rangle$, we get $g^\phi = 1$ for all $g \in G$, and this is the same as to say that $\exp(G) = p$.

Assertion (a) of Theorem 1.10 is a part of celebrated Kulakoff's Theorem 5.3. Let $\mathrm{c}_n(G)$ be the number of cyclic subgroups of order p^n in G.

Theorem 1.10. *Let G be a noncyclic p-group, $p > 2$, $n > 0$.*

(a) (G. A. Miller [Mil1], A. Kulakoff [Kul]) $\mathrm{c}_1(G) \equiv 1 + p \pmod{p^2}$.

(b) (G. A. Miller [Mil1]) If $n > 1$, then p divides $\mathrm{c}_n(G)$.

Proof. Since G is noncyclic, there is $T \triangleleft G$ such that G/T is abelian of type (p, p). Let $F_1/T, \dots, F_{p+1}/T$ be all subgroups of order p in G/T. We claim that

$$\mathrm{c}_n(G) = \sum_{i=1}^{p+1} \mathrm{c}_n(F_i) - p\mathrm{c}_n(T). \tag{1}$$

Indeed, let $C < G$ be cyclic of order p^n. Then $CT < G$ since G/T is noncyclic. Let $b = |\{i \mid C \le F_i\}|$. It is clear that $b \in \{1, p+1\}$. If $b = 1$, then $C \not\le T$ and the contribution of C in the right-hand side of (1) is 1 since CT is a unique element of the set $\{F_1, \dots, F_{p+1}\}$ containing C. If $b = p + 1$, then $C \le T$ and the contribution of C in the right-hand side of (1) is $(p+1) \cdot 1 - p \cdot 1 = 1$ again, proving (1).

Suppose that the theorem is proved for all proper subgroups of G.

(a) Let $n = 1$. If one of F_i's is cyclic, then $\mathrm{c}_1(G) = p + 1$. Indeed, $|\Omega_1(G)| = p^2$, by Theorem 1.2 (if G is abelian, this is also true); then $\mathrm{c}_1(G) = \mathrm{c}_1(\Omega_1(G)) = p + 1$. Suppose that all F_i are not cyclic. Then, by induction, $\mathrm{c}_1(F_i) \equiv 1 + p \pmod{p^2}$, all i. By Sylow (see Exercise 9(a), below), $p\mathrm{c}_1(T) \equiv p \pmod{p^2}$. Therefore, by (1), $\mathrm{c}_1(G) \equiv (p+1)(1+p) - p \equiv 1 + p \pmod{p^2}$, completing the case $n = 1$.

(b) Let $n > 1$. If one of F_i's is cyclic, the result follows by Theorem 1.2: in that case, $\mathrm{cl}(G) \le 2$ and $\mathrm{c}_n(G) = \frac{|\Omega_n(G)| - |\Omega_{n-1}(G)|}{\varphi(p^n)} = \frac{p^{n+1} - p^n}{p^{n-1}(p-1)} = p$. Now let all F_i are noncyclic. Then, by induction, p divides $\mathrm{c}_n(F_i)$ for all i, and we are done, by (1). □

Lemma 1.11 (see Introduction, Proposition 12). *Let $M_1, \dots, M_n$ be normal subgroups of G and $D = \bigcap_{i=1}^n M_i$. Then G/D is isomorphic to a subgroup of $(G/M_1) \times \dots \times (G/M_n)$.*

Theorem 1.12 (Burnside's basis theorem). *Let G be a p-group and $|G : \Phi(G)| = p^d$.*

(a) $G/\Phi(G) \cong \mathrm{E}_{p^d}$. Moreover, if $N \trianglelefteq G$ and G/N is elementary abelian, then $\Phi(G) \le N$.

(b) Every minimal system of generators of G contains exactly d elements.

(c) $\Phi(G) = \mho_1(G)G'$. In particular, if $p = 2$, then $\Phi(G) = \mho_1(G)$.

Proof. (a) The first assertion follows from Lemma 1.11 since G/M is of order p for all $M \in \Gamma_1$. If G/N is elementary abelian, then $\Phi(G/N) = \{1\}$ so N is the intersection of some members of the set Γ_1.

(b) Let $G = \langle x_1, \dots, x_k \rangle$. Then $G/\Phi(G) = \langle x_1\Phi(G), \dots, x_d\Phi(G) \rangle \cong \mathrm{E}_{p^d}$, where $\{x_1, \dots, x_d\} \subseteq \{x_1, \dots, x_k\}$. Because $G = \langle x_1, \dots, x_d, \Phi(G) \rangle$, we get $G = \langle x_1, \dots, x_d \rangle$ so that $d \le k$ always. The number $d = \log_p(|G : \Phi(G)|)$.

(c) Since $G/\mho_1(G)G'$ is elementary abelian, we get $\Phi(G) \le \mho_1(G)G'$, by (a). Next, $\mho_1(G)G' \le \Phi(G)$ since $x^p \in \Phi(\langle x \rangle) \le \Phi(G)$ for all $x \in G$ and $G' \le \Phi(G)$ (Introduction, Exercise 6(a)). If $p = 2$, then $G' \le \mho_1(G)$ since $G/\mho_1(G)$ is abelian as a group of exponent 2. □

Let G be a p-group. If $H \le G$, then $\Phi(H) = \mho_1(H)H' \le \mho_1(G)G' \le \Phi(G)$. If $L \triangleleft G$ and $L \cap \Phi(G) = \{1\}$, then G splits over L. Indeed, let $A < G$ be such that $G = AL$ and A is as small as possible. Then $L \cap A \le \Phi(A) \cap L \le \Phi(G) \cap L = \{1\}$.

By Schur–Zassenhaus' theorem, $\pi(G/\Phi(G)) = \pi(G)$ (G is arbitrary finite).

A group G is said to be *metacyclic* if it contains a cyclic normal subgroup Z such that G/Z is cyclic. Sections of metacyclic groups are metacyclic.

Exercise 8a ([Red]). Let G be a minimal nonabelian p-group. Then $|G'| = p$ and G/G' is abelian of rank two and G is one of the following groups:

(a) $G = \langle a, b \mid a^{p^m} = b^{p^n} = 1,\ a^b = a^{1+p^{m-1}} \rangle$, $m \ge 2$, $n \ge 1$, $|G| = p^{m+n}$ (G is metacyclic).

(b) $G = \langle a, b \mid a^{p^m} = b^{p^n} = c^p = 1,\ [a,b] = c,\ [a,c] = [b,c] = 1 \rangle$ is nonmetacyclic of order p^{m+n+1} and if $p = 2$, then $m + n > 2$. Next, G' is a maximal cyclic subgroup of G.

(c) $G \cong \mathrm{Q}_{2^3}$.

Solution. We have $|G'| = p$ and $G = \langle a, b \rangle$, by Exercise 1(a). Since $\mathrm{d}(G) = 2$ and $G/\mathrm{Z}(G) \cong \mathrm{E}_{p^2}$, we get $\mathrm{Z}(G) = \Phi(G)$. Let $G' = \langle c \rangle$. Write $U = \Omega_1(G)$; then $|U| \le p^3$ since $\Omega_1(G/G') \cong \mathrm{E}_{p^2}$. (i) Suppose that $|U| = p^3$. If U is abelian, then G is as in (b) with $m + n > 2$. If U is nonabelian, then $U = G$ so $\exp(G) = p$ for $p > 2$ (then G is as in (b) with $m = n = 1$) and $G \cong \mathrm{D}_8$ for $p = 2$ (then

G is as in (a) with $m = 2, n = 1$). (ii) Now let $|U| = p^2$. Then, by Theorem 6.1, $G/G' = (C/G')\times(D/G')$, where $U \le C$. In that case, C is not cyclic so $C = G'\times Z$ so $G = Z \cdot D$, a semidirect product. Then D must be cyclic. and so G is as in (a). (iii) If $|U| = p$, then $p = 2$ and G is generalized quaternion (Proposition 1.3). In that case, $G \cong Q_8$ since $|G'| = 2$.

Exercise 9 (L. Sylow (1872)). (a) Let G be a group of order p^m and $n < m$. Then the number of subgroups of order p^n in G is $\equiv 1 \pmod p$.

(b) Let G be a group of order $p^m s$, $p \nmid s$, $n \le m$. Then the number of subgroups of order p^n in G is $\equiv 1 \pmod p$.

Solution. (a) We use induction on $|G|$. Let $\mathfrak{A} = \{A_1, \dots, A_r\}$ be the set of all subgroups of G of order p^n, $\Gamma_1 = \{M_1, \dots, M_t\}$ the set of all maximal subgroups of G. Let α_i be the number of elements of the set Γ_1 containing A_i and β_j the number of elements of the set $\mathfrak{A}$ contained in M_j, $i = 1, \dots, r$, $j = 1, \dots, t$. By induction, $\beta_j \equiv 1 \pmod p$, all j. Since α_i equals the number of all maximal subgroups in $G/A_i\Phi(G)$, we get $\alpha_i \equiv 1 \pmod p$. Since $t = |\Gamma_1| = 1 + p + \dots + p^{\mathrm{d}(G)-1} \equiv 1 \pmod p$, reading the equality $\alpha_1 + \dots + \alpha_r = \beta_1 + \dots + \beta_t$ modulo p yields $r \equiv 1 \pmod p$. (b) Let $P \in \mathrm{Syl}_p(G)$ and let $\mathcal{M} = \{A_1, \dots, A_r\}$ be the set of all subgroups of order p^n in G not contained in P. Considering the action of P on $\mathcal{M}$ via conjugation, we get $|\mathcal{M}| = r \equiv 0 \pmod p$. Now the result follows from (a).

Exercise 10. Let H be a p-group. Then $H \cong L < \Phi(G)$ for some p-group G.

Solution. Set $G = H \,\mathrm{wr}\, C_p$, the standard wreath product with active factor $C_p = \langle c \rangle$. Let h be an element of the first coordinate subgroup H_1 of the base of G (see Appendix 13). Then $(hc^{-1})^p = h \cdot h^c \dots h^{c^{p-1}}$ is an element of the diagonal subgroup $D \cong H_1 \cong H$ of the base. If h runs over H_1, then $(hc^{-1})^p$ runs over D, and so the latter is a subgroup of $\mho_1(G) \le \mho_1(G)G' = \Phi(G)$.

Problem. Is it true that, in Exercise 10, a p-group G can be chosen so that $\exp(G) = \exp(H)$?

Proposition 1.13 (Compare with [Hob1]; see Lemma 1.4). *Let G be a p-group and let $N \le \Phi(G)$ be G-invariant. If $\mathrm{Z}(N)$ is cyclic, then N is also cyclic.*

Proof. Assume that N is nonabelian. Let $A \le N$ be G-invariant of order p^2. Then $A \le \mathrm{Z}(\Phi(G)) \cap N$ so A is cyclic. By Lemma 1.4, N is cyclic. □

Theorem 1.14. *Let G be a nonabelian p-group, $p > 2$. If $\Phi(G) > \{1\}$ is cyclic, then $\Phi(G) \le \mathrm{Z}(G)$. Let $C < G$ be cyclic of maximal order such that $\Phi(G) \le C$. Then $G = C\Omega_1(G)$ and $G' \le \Omega_1(\Phi(G))$ is of order p. In particular, if $\Phi(G)$ is a maximal cyclic subgroup of G, then $|\Phi(G)| = p$ and $\exp(G) = p$.*

Proof. Assume that G is not as in Theorem 1.2(a). Let C be a maximal cyclic subgroup of G containing $\Phi(G)$ and let $A/C < G/C$ be of order p. Since $p > 2$, we have $\Omega_1(A) \cong E_{p^2}$ so $A = C\Omega_1(A)$ (Theorem 1.2). Since $\Omega_1(A) \triangleleft G$ centralizes $\Phi(G)$, it follows that $A = C\Omega_1(A) \leq C_G(\Phi(G))$. All such A generate G so $\Phi(G) \leq Z(G)$. Since $G/Z(G)$ is elementary abelian and G' is cyclic, it follows that $|G'| = p$ (see Exercise 19, below). By the above, $G = C\Omega_1(G)$. Therefore, if $C = \Phi(G)$, then $G = \Omega_1(G)$ so $\exp(G) = p$ (Exercise 19). □

Theorem 1.15. *Let G be a group, $\alpha \in \mathrm{Aut}(G)$. If $(o(\alpha), |\Phi(G)|) = 1$ and α induces the identity automorphism on $G/\Phi(G)$, then $\alpha = \mathrm{id}_G$.*

Proof. Assume that $\alpha \neq \mathrm{id}_G$. One may assume that $o(\alpha) = p$. Let $W = \langle\alpha\rangle \cdot G$ be the natural semidirect product with kernel G. By assumption, $T = \langle\alpha\rangle \cdot \Phi(G) \triangleleft W$ and $\langle\alpha\rangle \in \mathrm{Syl}_p(T)$. Therefore, by Frattini's lemma, $W = N_W(\langle\alpha\rangle)T = N_W(\langle\alpha\rangle)\langle\alpha\rangle\Phi(G) = N_W(\langle\alpha\rangle)\Phi(G)$. Since $\Phi(G) \leq \Phi(W)$, we get $N_W(\langle\alpha\rangle) = W$, i.e., $\langle\alpha\rangle \triangleleft W$; then $W = \langle\alpha\rangle \times G$, contrary to the assumption. □

Remark 3. We offer another proof of Theorem 1.15. As above, we assume that $o(\alpha) = p$ is a prime. Let $x \in G - \Phi(G)$. Then α has a fixed point, say x_0, on the coset $x\Phi(G)$ since $o(\alpha) = p \nmid |x\Phi(G)| = |\Phi(G)|$. Take, in each coset of $\Phi(G)$, a fixed point of α, and let F be the set of these fixed points. It follows from $\langle F, \Phi(G)\rangle = G$ that $G = \langle F\rangle$ so $\alpha_G = \mathrm{id}_G$, a contradiction.

Theorem 1.16 ([Hal1]). *Let G be a p-group and $|G : \Phi(G)| = p^d$. Then $|\mathrm{Aut}(G)|$ divides the number $(p^d-1)(p^d-p)\dots(p^d-p^{d-1})|\Phi(G)|^d = |\Phi(G)|^d|\mathrm{Aut}(E_{p^d})|$.*

Proof. Let $\alpha \in \mathrm{Aut}(G)$ and let $B = \{x_1, \dots, x_d\}$ be a minimal basis of G. Then $\alpha(B)$ is a minimal basis of G of the same type (i.e., in both these bases G has the same defining relations). Let $\mathfrak{B} = \mathrm{Aut}(G)(B)$ be the $\mathrm{Aut}(G)$-orbit of B. Since $\mathrm{Aut}(G)$ acts transitively and regularly on $\mathfrak{B}$ (check!), it follows that $|\mathrm{Aut}(G)| = |\mathfrak{B}|$. Obviously, the set of all minimal bases of G is the union of $\mathrm{Aut}(G)$-orbits; every $\mathrm{Aut}(G)$-orbit is the set of bases of the same type and contains exactly $|\mathrm{Aut}(G)|$ members. This means that $|\mathrm{Aut}(G)|$ divides the number of minimal bases of G which equals $(p^d-1)(p^d-p)\dots(p^d-p^{d-1})|\Phi(G)|^d = |\Phi(G)|^d|\mathrm{Aut}(E_{p^d})|$. □

Exercise 11. Suppose that $G = D_{2^{n+1}}$. All (exactly two) noncyclic maximal subgroups of G are dihedral if $n > 2$ and abelian of type $(2, 2)$ if $n = 2$. Next, $|\mathrm{Aut}(G)| = 2^{2n-1}$ (see Proposition 34.8 for description of $\mathrm{Aut}(G)$) and $c_1(G) = 2^n + 1$. Next, $\mathrm{Aut}(D_{2^3}) \cong D_{2^3}$. Every subgroup of composite order in G contains $Z(G)$. Show that $c_k(G) = 1$ for $1 < k \leq n$. Check that G has exactly three conjugacy classes of involutions, $k(G) = 2^{n-1} + 3$, G has exactly two conjugate classes of four-subgroups.

Exercise 12. Let $G = Q_{2^{n+1}}$ and let T be a cyclic subgroup of index 2 in G. Two noncyclic maximal subgroups of G are generalized quaternion if $n > 2$. Show that all elements in $G - T$ have order 4 so $c_2(G) = 1 + 2^{n-1}$. Next, $k(G) = 2^{n-1} + 3$.

Exercise 13. Let $G = \mathrm{SD}_{2^{n+1}}$. Then $\Gamma_1 = \{\mathrm{C}_{2^n}, \mathrm{D}_{2^n}, \mathrm{Q}_{2^n}\}$, i.e., all maximal subgroups of G are characteristic. Show that $c_1(G) = c_1(\mathrm{D}_{2^n}) = 1 + 2^{n-1}$, $c_2(G) = c_2(\mathrm{Q}_{2^n}) = 1 + 2^{n-2}$, $\Omega_1(G) \cong \mathrm{D}_{2^n}$, $\langle x \in G \mid o(x) = 4\rangle \cong \mathrm{Q}_{2^n}$.

Exercise 14. Let G be a 2-group of maximal class. Prove that the number of nonabelian subgroups of index 2^k in G is 2^k.

Exercise 15. Find $c_1(\mathrm{E}_{2^m} \operatorname{wr} \mathrm{E}_{2^n})$.

Exercise 16. Find the number of involutions in $W = G \operatorname{wr} \mathrm{C}_2$, where G is one of 2-groups of Theorem 1.2, $n > 2$. Find $c_k(W)$ for $k > 1$. Prove that W has no normal elementary abelian subgroups of order 8, unless $G \cong \mathrm{M}_{2^{n+1}}$.

Exercise 17 (Burnside). Let G be a p-group, N a nonabelian G-invariant subgroup in $\Phi(G)$. Then $|N : N'| > p^2$. (*Hint.* Use Proposition 1.13.)

Exercise 18. Let G be a nilpotent group of class 2, $x, y, z \in G$, $n \in \mathbb{N}$. Then

$$[x, yz] = [x, y][x, z], \quad [xy, z] = [x, z][y, z], \quad [x, y]^n = [x^n, y] = [x, y^n],$$
$$(xy)^n = x^n y^n [y, x]^{\binom{n}{2}}.$$

Exercise 19. Let G be a nilpotent group of class 2. Then $\exp(G')$ divides $\exp(G/\mathrm{Z}(G))$. Let n divides $|G|$. If n is odd, then $\exp(\langle x \in G \mid x^n = 1\rangle)$ divides n. If n is even, then $\exp(\langle x \in G \mid x^n = 1\rangle)$ divides $2n$. If, in addition, G is a p-group, $p > 2$, $n = p^k$, then $\exp(\Omega_k(G)) \le p^k$.

As we saw (see Exercises 11–13), $c_1(G) \equiv 1 \pmod 4$ if G is either cyclic or a 2-group of maximal class. The following theorem shows that all other 2-groups G satisfy $c_1(G) \equiv 3 \pmod 4$.[1]

Theorem 1.17. *Suppose that a 2-group G is neither cyclic nor of maximal class.*

(a) ([Isa1, Theorem 4.9] and [Ber1, §5].) $c_1(G) \equiv 3 \pmod 4$.

(b) ([Ber1, §5].) If $n > 1$, then $c_n(G)$ is even.

Proof. One may assume that G is not abelian. In view of Theorem 1.2 and Exercises 11–13, one may assume that G has no cyclic subgroups of index 2. Then $|G| > 2^3$. We use induction on $|G|$.

Let $|G| = 2^4$. Since G is not of maximal class. we have, by Proposition 1.6, $|G'| = 2$ so $|\mathrm{Z}(G)| = 4$ (Lemma 1.1) since G has an abelian subgroup of index 2.

[1] I proved Theorem 1.17 in 1966. Recently (13/05/02) I learned that Miller [Mil7] proved the same theorem many years ago. His proof is based on different ideas and fairly long. As Miller noticed, Theorem 1.17(a) follows from Theorem 1.17(b); below his argument is reproduced. Suppose that a 2-group G is neither cyclic nor a group of maximal class and $\exp(G) = 2^e > 2$, $|G| = 2^m$. Then $|G| = 2^m = 1 + c_1(G) + \sum_{i=2}^{e} \varphi(2^i) c_i(G) \equiv 1 + c_1(G) \pmod 4$ so $c_1(G) \equiv 2^m - 1 \equiv 3 \pmod 4$. Similarly, Theorem 1.10(a) follows from Theorem 1.10(b).

Then $G/\mathrm{Z}(G) \cong \mathrm{E}_4$ so $G/\mathrm{Z}(G)$ contains three distinct subgroups $F_i/\mathrm{Z}(G)$ of order 2, $i = 1, 2, 3$, and F_i are abelian. We have (see the proof of Theorem 1.10)

(2) $$\mathrm{c}_k(G) = \mathrm{c}_k(F_1) + \mathrm{c}_k(F_2) + \mathrm{c}_k(F_3) - 2\mathrm{c}_k(\mathrm{Z}(G)).$$

Since F_i are noncyclic, we get $\mathrm{c}_1(F_i) \equiv 3 \pmod 4$ so, by (2), $\mathrm{c}_1(G) \equiv 3 \cdot 3 - 2 \equiv 3 \pmod 4$. Next, if $k = 2$, then $\mathrm{c}_2(F_i)$ is even for all i so, by (2), $\mathrm{c}_2(G)$ is also even.

Next let $|G| > 2^4$. By Lemma 1.4, there is $R \lhd G$ of type $(2, 2)$.

Suppose that G/R is cyclic. Then $|G : \mathrm{C}_G(R)| = 2$, since G is nonabelian. Next, $\mathrm{C}_G(R)$ is abelian and contains all involutions in G. Therefore, (a) is true since in that case $\Omega_1(\mathrm{C}_G(R)) \in \{\mathrm{E}_4, \mathrm{E}_8\}$ so $\mathrm{c}_1(G) = \mathrm{c}_1(\Omega_1(G)) \in \{3, 7\}$. Let $n > 1$; then $\Omega_n(G)$ and $\Omega_{n-1}(G)$ are subgroups of G of exponent 2^n and 2^{n-1}, respectively. In that case, $\mathrm{c}_n(G) = \frac{|\Omega_n(G) - \Omega_{n-1}(G)|}{\varphi(2^n)}$. We get $|\Omega_n(G)| > |\Omega_{n-1}(G)| \geq 2^n$ so $\mathrm{c}_n(G)$ is even. Next assume that G/R is not cyclic.

Then G/R has a normal subgroup T/R such that $G/T \cong \mathrm{E}_4$. Let $F_1/T, F_2/T$ and F_3/T be all subgroups of order 2 in G/T. Then again we have equality (2). Since F_1, F_2 and F_3 are not of maximal class since $R < F_i$ and $|F_i| \geq 2^4$, we get $\mathrm{c}_1(F_i) \equiv 3 \pmod 4$ and $\mathrm{c}_n(F_i)$ is even for $n > 1$ and all i, by induction, $2\mathrm{c}_1(T) \equiv 2 \pmod 4$, by Exercise 9. Thus, (a) and (b) follow from (2). □

M. Herzog [Her1] has extended Theorem 1.17(a) to arbitrary finite groups.

Remark 4. Let $H \in \Gamma_1$ be of maximal class, where G is a 2-group. Suppose that G does not split over H. In that case, $\mathrm{c}_1(G) = \mathrm{c}_1(H) \equiv 1 \pmod 4$. Then by Theorem 1.17(a), G is of maximal class so one of the following holds: (i) $G = \mathrm{Q}_{2^{n+1}}$; (ii) $G = \mathrm{SD}_{2^{n+1}}$ and H is dihedral (Theorem 1.2).

Exercise 20. If a p-group G is nonabelian but all its proper sections are abelian, then either $G \in \{\mathrm{D}_{2^3}, \mathrm{Q}_{2^3}, \mathrm{M}_{p^n}\}$ or G is nonabelian of order p^3 and exponent $p > 2$. (*Hint.* G is minimal nonabelian with cyclic center.)

Definition 3. A group is said to be *Dedekindian* if all its subgroups are normal.

Dedekindian groups are nilpotent so it suffices to classify prime power Dedekindian groups.

Lemma 1.18. *If G is a Dedekindian p-group, $p > 2$, then G is abelian.*

Proof. Suppose that G be a counterexample of minimal order. Let $A, B < G$ be distinct of order p (Proposition 1.3); then G/A and G/B are abelian so $G' \leq A \cap B = \{1\}$. □

Lemma 1.19. *If a minimal nonabelian p-group G is Dedekindian, then $G \cong \mathrm{Q}_{2^3}$.*

Proof. By Lemma 1.18, $p = 2$. Assume that the lemma is false. Then $Z(G) = \Phi(G) = \mho_1(G)$ has index 4 in G (Exercise 1.8a). If $Z(G)$ is cyclic, G has a cyclic subgroup of index 2 so it is isomorphic to Q_8 (Theorem 1.2). If $Z(G)$ is noncyclic, it contains a subgroup $L \neq G'$ of order 2. Then, by induction, $G/L \cong Q_8$. We have $\Phi(G) = L \times G'$. Since $\Phi(G) = \mho_1(G)$, G has elements x and y of order 4 such that $x^2 \neq y^2$. Then $\langle x\rangle \cap \langle y\rangle = \{1\}$ so $G = \langle x\rangle \times \langle y\rangle$ is abelian, a contradiction. □

Theorem 1.20. *Suppose that a 2-group G satisfies the following conditions: (i) G contains a subgroup $Q \cong Q_8$, (ii) all subgroups of order ≤ 4 are normal in G. Then $G = Q_8 \times E_{2^s}$, $s \geq 0$, is Dedekindian.*

In particular, if a nonabelian group G is Dedekindian, then $G = Q \times E \times A$, where $Q \cong Q_8$, $\exp(E) \leq 2$ and A is abelian of odd order. Indeed, a minimal nonabelian subgroup of G is $\cong Q_8$ (Lemma 1.18).

Proof of Theorem 1.20. $Q \trianglelefteq G$ since Q is generated by (normal in G) cyclic subgroups of order 4. Assume that $C = C_G(Q)$ contains a cyclic subgroup $Z = \langle z\rangle$ of order 4. If $Q \cap Z = \{1\}$, then $QZ = Q \times Z$ contains a nonnormal cyclic subgroup $\langle az\rangle$ of order 4, where $a \in Q - Z(Q)$, and this is a contradiction. If $Q \cap Z > \{1\}$, then $QZ = Q * Z = D * Z$, where $D \cong D_8$ (Appendix 16), a contradiction since D_8 is non-Dedekindian. Thus, $\exp(C) = 2$. Next, G/C is isomorphic to a subgroup of $\mathrm{Aut}(Q) \cong S_4$ containing a subgroup isomorphic to $Q/(Q \cap C) \cong E_4$ so $G/C \in \{E_4, D_8\}$. Assume that $G/C \cong D_8$. Then G/C contains a nonnormal subgroup $\langle xC\rangle$ of order 2 so $\langle x\rangle$ of order ≤ 4 is not normal in G, a contradiction. Thus, $G/C \cong E_4$ so $G = Q * C$. If $C = Z(Q) \times E$, then $G = Q \times E$, where $\exp(E) \leq 2$, and we are done. □

Theorem 1.21 ([Pas]). *Suppose that all cyclic subgroups of a nonabelian p-group G of order $\leq p^s$ are normal. Then either $\Omega_s(G) \leq Z(G)$ or else $p = 2$, $s = 2$ and G is nonabelian Dedekindian.*

Proof. Since $\Omega_1(G) \leq Z(G)$, we assume that $s > 1$. If $\exp(G) \leq p^s$, then G is Dedekindian. Therefore, let $\exp(G) > p^s$. If G has a subgroup isomorphic to Q_8, then, by Theorem 1.20, G is Dedekindian. Next we assume that G has no subgroups isomorphic to Q_8. To prove that $\Omega_s(G) \leq Z(G)$, we use induction on $|G|$ and s. Then $\Omega_s(H) \leq Z(H)$ for all $H < G$ and $\Omega_{s-1}(G) \leq Z(G)$.

Assume that $\Omega_s(G) \not\leq Z(G)$. Then there is a cyclic $U < G$ of order p^s such that $U \not\leq Z(G)$. If $U \leq H \in \Gamma_1$, then $U \leq Z(H)$. It follows that H is the unique member of the set Γ_1 containing U so G/U is cyclic hence G is metacyclic. By Theorem 1.2, $|G : U| > p$. Then $\mho_1(U) \leq Z(G)$, by induction, and $G/\mho_1(U)$ is abelian with two distinct cyclic subgroups $A/\mho_1(U)$ and $B/\mho_1(U)$ of index p so A and B are abelian. In that case, $Z(G) = A \cap B$ is of index p^2 in G so G is minimal nonabelian. Then $G = S \cdot T$ is a semidirect product with cyclic kernel $T = \langle x\rangle \cong C_{p^s}$ and cyclic S

with $|S| > p^s$ (Exercise 8a). Let $\Omega_s(S) = \langle y \rangle$. Then the cyclic subgroup $\langle xy \rangle$ of order p^s is not S-invariant, a final contradiction. □

Lemma 1.22 ([Pas]). *Let H be maximal among nonnormal subgroups of a p-group G. Then $\mathrm{N}_G(H)/H$ and $\mathrm{Z}(G)/(H \cap \mathrm{Z}(G))$ have at most one subgroup of order p.*

Proof. If $\mathrm{N}_G(H)/H$ has two distinct subgroups A/H and B/H of order p, then $H = A \cap B \lhd G$, a contradiction. Since $H\mathrm{Z}(G) \le \mathrm{N}_G(H)$, it follows that $H\mathrm{Z}(G)/H \cong \mathrm{Z}(G)/(H \cap \mathrm{Z}(G))$ is cyclic. □

Proposition 1.23 ([Pas]). *If a p-group G is not Dedekindian, then there exists a G-invariant subgroup K of index p in G' such that G/K is not Dedekindian.*

Proof. Let G be a counterexample of minimal order and take $K \le G' \cap \mathrm{Z}(G)$ of order p. Then G/K is nonabelian Dedekindian so $p = 2$ and $|G'/K| = 2$, by induction and Theorem 1.20, hence $|G'| = 4$. By Theorem 1.20, G/K has a normal subgroup $W/K \cong \mathrm{Q}_8$. By Theorem 1.2, W is not of maximal class. By Taussky's theorem, $|W'| = 2$ so $W' \le \mathrm{Z}(G)$ and $W' \ne K$; then $G' = W'K = W' \times K \le \mathrm{Z}(G)$. Let $H < G$ be nonnormal of maximal order; then $G' \not\le H$. By Lemma 1.22, $\mathrm{Z}(G)/(H \cap \mathrm{Z}(G))$ is cyclic so $H \cap G' > \{1\}$. Now, $G/(H \cap G')$ is not Dedekindian, since $H/(H \cap G') \ntriangleleft G/(H \cap G')$, a contradiction since $|G' : (H \cap G')| = 2$. □

Corollary 1.24 ([Pas]). *If all nonnormal subgroups of a non-Dedekindian p-group G have order p, then $|G'| = p$.*

Proof. Assume that $|G'| > p$. Then G is non-Dedekindian so G' has a G-invariant subgroup K of index p such that G/K is non-Dedekindian (Proposition 1.23), hence G/K has a nonnormal subgroup H/K. Then $|H| > |K| \ge p$, a contradiction. □

Theorem 1.25 ([Pas]). *Suppose that all nonnormal subgroups of a non-Dedekindian p-group G have the same order p. Then one of the following holds:*

(a) $G \cong \mathrm{M}_{p^n}$.

(b) $G = Z * G_0$, *where Z is cyclic and G_0 nonabelian of order p^3. If $p = 2$, then $G_0 = \mathrm{D}_8$ and if $p > 2$ then $\exp(G_0) = p$.*

(c) $p = 2$ *and* $G = \mathrm{D}_8 * \mathrm{Q}_8$ *is extraspecial of order* 2^5.

Proof. By Corollary 1.24, $|G'| = p$. If $x, y \in G$, then $[x^p, y] = [x, y]^p = 1$ so $\Phi(G) = G'\mho_1(G) \le \mathrm{Z}(G)$; in particular, $G/\mathrm{Z}(G)$ is elementary abelian (by Lemma 4.2, below, it is of even rank). Let H be a nonnormal subgroup of G of order p.

(i) First suppose that $\mathrm{d}(G) \le 3$. We may assume that $|G| > p^3$ and G has no cyclic subgroups of index p. By Lemma 1.22, $\mathrm{Z}(G) \cong H\mathrm{Z}(G)/H$ is cyclic. As we have noticed, $G/\mathrm{Z}(G)$ is of type (p, p). Let $\mathrm{Z}(G) < M_i < G, i = 1, 2, M_1 \ne M_2$. Because M_1, M_2 are noncyclic abelian and $\mathrm{Z}(G)$ is cyclic, we have $M_i = \mathrm{Z}(G) \times J_i$, where $|J_i| = p, i = 1, 2$. Set $G_0 = \langle J_1, J_2 \rangle$; then $G = \mathrm{Z}(G)G_0$ so G_0 is nonabelian.

Since $\mathrm{cl}(G_0) = 2$, it follows that $G_0 \cong \mathrm{D}_8$ for $p = 2$ and $\exp(G_0) = p$ for $p > 2$ (Exercise 19). Let $p > 2$. Since $|G_0 \cap \mathrm{Z}(G)| = p$, we get $|G_0| = p^3$, by the product formula. The same is true for $p = 2$. Thus G is a group from (b).

(ii) Suppose that $\mathrm{d}(G) > 3$. Set $N = \mathrm{N}_G(H)$; then $N \in \Gamma_1$ since H is a subgroup of G-invariant subgroup of order p^2 so $H^G \le N$. By Lemma 1.22, N/H is either cyclic or ordinary quaternion. Since $\mathrm{d}(G) \ge 4$, we see that $\mathrm{d}(N) \ge 3$ so N/H is noncyclic; then $N/H \cong \mathrm{Q}_8$, $p = 2$ and $|G| = 32$. Moreover, $\mathrm{Z}(G) < N$, and since N has exactly three subgroups of order 2 (one of them is $H \not\le \mathrm{Z}(G)$), we get $|\mathrm{Z}(G)| = 2$. Since $|G'| = 2$, we get $G' = \mathrm{Z}(G)$, and so G is extraspecial. Then G has no elementary abelian subgroups of order 8 (otherwise, that subgroup were central), and we get $G = \mathrm{D}_8 * \mathrm{Q}_8$. □

Supplement to Theorem 1.25 ([Pas]). *Let $n > 1$ and $p > 2$. All nonnormal subgroups of a nonabelian p-group G of order $> p^3$ have the same order p^n if and only if $G = \langle a, b \mid a^{p^m} = b^{p^n} = 1, a^b = a^{1+p^{m-1}}\rangle$, $m \ge n$.*

Proof. Let $H \ntrianglelefteq G$. Since all maximal subgroups of H are G-invariant, H is cyclic. By hypothesis, $\Omega_1(G) \le \mathrm{Z}(G)$ is elementary abelian. Then $\Omega_1(G) \cong \mathrm{E}_{p^2}$ (otherwise, $H\Omega_1(\mathrm{Z}(G))/H$ is elementary abelian noncyclic, contrary to Lemma 1.22). By Theorem 13.7, G is metacyclic. Since all subgroups of $G/\Omega_1(G)$ are normal (indeed, their inverse images in G are noncyclic), it is abelian (Lemma 1.18). Since the cyclic subgroup $G' \le \Omega_1(G) \cong \mathrm{E}_{p^2}$, we get $|G'| = p$, and it follows from Lemma 65.2(a) that G is minimal nonabelian so, by Exercise 8a, $G = \langle a, b \mid a^{p^m} = b^{p^n} = 1, a^b = a^{1+p^{m-1}}\rangle$. If $n > m$, then $\langle ab^{p^{n-m}}\rangle$ is not b-invariant. Thus, $n \le m$. Let us show that G satisfies the hypothesis. The subgroup $\langle b\rangle < G$ is nonnormal of order p^n. Let $H < G$ be nonnormal (cyclic). Then $G' \not\le H$ so $\langle a\rangle \cap H = \{1\}$. It follows that $|H| \le |G : \langle a\rangle| = p^n$. Assume that $|H| < p^n$. Then $H < \Omega_{n-1}(G) \le \mathrm{Z}(G)$, a contradiction. Thus, $|H| = p^n$. □

Now we present some results of Blackburn [Bla7]. Let $R(G)$ denote the intersection of all nonnormal subgroups of a non-Dedekindian group G. Clearly, $R(G)$ is cyclic of prime power order. Blackburn [Bla7] classified the groups G with $R(G) > \{1\}$.

Proposition 1.26 ([Bla7]). *(a) Suppose that the intersection of any two nonnormal cyclic subgroups of minimal nonabelian p-group $G \not\cong \mathrm{Q}_8$ is $> \{1\}$. Then $p = 2$ and $G \cong \mathcal{H}_2 = \langle a, b \mid a^4 = b^4 = 1, a^b = a^{-1}\rangle$. (In that case, $R(G) = \langle b^2\rangle > \{1\}$.)*

(b) Let G be a nonabelian p-group and $H \le G$ minimal nonabelian. If the intersection of any two nonnormal cyclic subgroups of G is $> \{1\}$, then $p = 2$ and $H \in \{Q_8, \mathcal{H}_2\}$.

Proof. (a) Let $G = \langle a, b\rangle$. If G is not metacyclic, then it is possible to choose a and b so that $\langle a\rangle \cap \langle b\rangle = \{1\}$, a contradiction since $\langle a\rangle$ and $\langle b\rangle$ are not normal in G. Thus, $G = \langle a, b \mid a^{p^m} = b^{p^n} = 1, a^b = a^{1+p^{m-1}}\rangle$ is metacyclic; then

$n > 1$ so $\Omega_1(G) \le Z(G)$. Set $B = \langle b \rangle$. Let $L < \Omega_1(G)(\cong E_{p^2})$ of order p is different from $\Omega_1(B)$ and G'. Assume that G/L has a nonnormal cyclic subgroup U/L. Then $G' \not\le U$ so $\Omega_1(G) \not\le U$. It follows that U is cyclic. Then $B \cap U = \{1\}$, a contradiction. Thus, a nonabelian group G/L has no nonnormal cyclic subgroups so $G/U \cong Q_8$. We have $\exp(G) = 4$ (Theorem 1.2). Since $\mathcal{H}_2$ is the unique nonabelian metacyclic group of exponent 4, (a) follows.

(b) follows from (a) since, if $H \le G$ is minimal nonabelian, then either $H \cong Q_8$ or H satisfies hypothesis in (a). □

Corollary 1.27 ([Bla7]). *Let G be a non-Dedekindian p-group. If $R(G) > \{1\}$, then $p = 2$ and $|R(G)| = 2$.*

Proof. It suffices to show that $|R(G)| = 2$. If all minimal nonabelian subgroups of G are isomorphic to Q_8, then $G = Q \times E$, where Q is generalized quaternion and $\exp(E) \le 2$ (see Corollary A.17.3). In that case, obviously, $|R(G)| = 2$. If there is in G a minimal nonabelian subgroup $H \cong \mathcal{H}_2$, then $|R(H)| = 2$ so $|R(G)| = 2$. □

Blackburn [Bla7] classified the non-Dedekindian groups G with $R(G) > \{1\}$. For another proof of his result, see Corollary 92.7.

Exercise 21. Let G be a 2-group. If for every $x, y \in G$, there exists $z \in \mho_1(\langle x, y\rangle')$ such that $(xy)^2 = x^2y^2z^2$, then G is abelian. (*Hint.* If G is a minimal counterexample, it is minimal nonabelian so $|G'| = 2$.)

Exercise 22. A nonabelian 2-group, which is generated by involutions, generated by dihedral subgroups of order 8.

Exercise 23. Let $G = A \times B$, where A, B are arbitrary groups. Then $\Phi(G) = \Phi(A) \times \Phi(B)$.

Exercise 24. If a 2-group G has an odd number of irreducible characters of degree 2, then it is of maximal class. (*Hint.* Use Taussky's theorem.)

Exercise 25 (Miller). Classify the minimal non-Dedekindian 2-groups G. (*Hint.* G is either minimal nonabelian or isomorphic to Q_{16}; see Appendix 17.)

Exercise 26. Let $G \cong SD_{2^{n+1}}$. Is it true that $G = \langle a, b \mid a^{2^n} = 1, b^2 = a^{2^{n-1}}, a^b = a^{-1+2^{n-1}} \rangle$?

Exercise 27. Let G be a minimal nonabelian p-group. If $Z(G)$ is cyclic, then G is either of order p^3 and exponent $p > 2$ or $G \in \{D_{2^3}, Q_{2^3}, M_{p^n}\}$.

Let K be a group. A group G is said to be K-free if all sections of G are not isomorphic with K. Let $S(p^3)$ be nonabelian of order p^3 and exponent $p > 2$.

Exercise 28. If a 2-group G is D_{2^3}-, Q_{2^3}- and M_{2^n}-free for all $n > 3$, it is abelian.

Exercise 29. If a p-group G, $p > 2$, is $S(p^3)$- and M_{p^n}-free for all $n > 2$, it is abelian.

Exercise 30. Suppose that a noncyclic p-group G is not generated by subgroups of index p^2. Then $G \in \{Q_8, M_{p^m}, C_{p^{m-1}} \times C_p\}$.

Exercise 31 ([Hal2, Lemma 4.61]). Let $S \in \mathrm{Syl}_p(G)$ be cyclic of order p^m and let $1 \le k < m$. Then $c_k(G) \equiv 1 \pmod{p^{m-k+1}}$. (*Hint.* Let S act on the set of all subgroups of G of order p^k, not contained in S, via conjugation.)

Exercise 32. Let $P \in \mathrm{Syl}_2(G)$ be of maximal class and order 2^n. Find (a) $c_2(G)$ (mod 2^n) if P is generalized quaternion, (b) $c_1(G)$ (mod 2^{n-1}) if P is either dihedral, or semidihedral.

Let p divides $|G|$. Set $H_p(G) = \langle x \in G \mid o(x) \neq p\rangle$ ($H_p(G)$ is said to be the H_p-subgroup of G). Obviously, $\exp(G/H_p(G))$ divides p.

Exercise 33. Suppose that $\{1\} < H_2(G) < G$. Then $|G : H_2(G)| = 2$ and $H_2(G)$ is abelian.

Solution. Assume that $|G : H_2(G)| > 2$. Let $xH_2(G)$, $yH_2(G)$ be two distinct cosets of $H_2(G)$, $x, y \in G - H_2(G)$. Then $xy \in G - H_2(G)$, hence $o(x) = o(y) = o(xy) = 2$, and so $xy = yx$. This means that $C_G(x)$ contains all cosets $yH_2(G)$, where $H_2(G) \neq yH_2(G) \neq xH_2(G)$, and so $C_G(x) = G$. Thus, $G = \langle G - H_2(G)\rangle \le Z(G)$ so G is abelian of composite exponent; then $G = H_2(G)$, a contradiction.

If $\{1\} < H_3(G) < G$, then $|G : H_3(G)| = 3$ [SS], but the similar result is not true for $p = 5$ [Wal].

Exercise 34 ([SS]). Let $\{1\} < H_3(G) < G$, $x \in G - H_3(G)$, $h \in H_3(G)$. Then $[h, h^x] = 1$.

Solution. It follows from $(hx^{-1})^3 = 1$ that $hh^xh^{x^2} = 1$. Therefore, $[h, h^x] = h^{-1}(h^{-1})^x \cdot hh^x = h^{-1}(h^{-1})^x(h^{-1})^{x^2} = 1$.

Theorem 1.28 (Compare with [Man18, Lemma 7]). *Let G be a non-Dedekindian p-group, $|G| > p^3$.*

(a) *Let $\exp(G) > p$. Suppose that every nonnormal subgroup of G is of exponent p. Then either $|G'| = p$ or $p = 2$ and $G \cong D_{24}$. Next, if B is a minimal nonabelian subgroup of G, then B is either of order p^3 or isomorphic to M_{p^n}, $n > 3$.*

(b) *Let $\exp(G) = p$. Suppose that every minimal nonabelian subgroup of G is normal. Then either $|G'| = p$ or G is of order p^4.*

Proof. (a) For more general result, see §63.

(b) Assume that G is a counterexample, i.e., $|G'| > p$ and $|G| > p^4$. Let A be a minimal nonabelian subgroup of G. Then $|A| = p^3$ (Exercise 8a) and $A \lhd G$. If $A < B < G$, then B is generated by its minimal nonabelian subgroups (Exercise

6(b)) so $B \lhd G$. It follows that G/A is elementary abelian so $G' \leq \Phi(G) < A$ ($<$ since $\Phi(G)$ of order p^3 must be abelian, by Proposition 1.13). Thus, G' is contained in all minimal nonabelian subgroups of G. Since $|\mathrm{Aut}(A)|_p = p^3$ and $|G| > p^4$, we get $\mathrm{C}_G(A) \not\leq A$. Let $L < \mathrm{C}_G(A)$ of order p be such that $L \neq \mathrm{Z}(A)$. Set $H = AL = A \times L$; H has exactly p^2 minimal nonabelian subgroups. Since a subgroup of index p^2 in H is contained in at most $p+1$ maximal subgroups of H, it follows that $D = \bigcap A_i$, where A_i runs over all minimal nonabelian subgroups of H, has order at most p. Since $G' \leq D$, we get $|G'| = p$. □

Lemma 1.29. *Let $\{1\} < A < G$ be cyclic and G a noncyclic p-group. If $\mathrm{C}_G(A) > A$ is cyclic then $p = 2$ and G is of maximal class.*

Proof. Obviously, $\mathrm{C}_G(A) = \mathrm{N}_G(A) = G$ if $|A| = p$, a contradiction. Thus, $|A| > p$. Assume, in addition, that G is not a 2-group of maximal class. We have $\mathrm{Z}(G) < A$. By Lemma 1.4, there is $\mathrm{E}_{p^2} \cong R \lhd G$. Since $|R \cap \mathrm{Z}(G)| = p$, it follows that $|R\mathrm{C}_G(A) : \mathrm{C}_G(A)| = p$ and $M = R\mathrm{C}_G(A)$ is nonabelian so $M \cong \mathrm{M}_{p^n}$ for some $n > 3$ (Theorem 1.2 and Proposition 1.8). However, $A \not\leq \mathrm{Z}(M)$ and $A \leq \Phi(\mathrm{C}_G(A)) \leq \Phi(M) = \mathrm{Z}(M)$, a contradiction. □

Theorem 1.30. *Suppose that $\{1\} < A$ is a cyclic subgroup of a noncyclic p-group G. Let $k > 0$ and let $\mathfrak{M}$ be the set of all cyclic subgroups of G containing A as a subgroup of index p^k. If G is not a 2-group of maximal class, then p divides $|\mathfrak{M}|$.*

Proof. We use induction on $|G|$. For $D \leq G$, let $\alpha(D)$ be the set of all elements of the set $\mathfrak{M}$ contained in D. One may assume that $\mathfrak{M} \neq \varnothing$. Obviously, all members of the set $\mathfrak{M}$ are contained in $\mathrm{C}_G(A)$, $\mathrm{C}_G(A) > A$ and $\mathrm{C}_G(A)$ is noncyclic (Lemma 1.29). Clearly, $\mathrm{C}_G(A)$ is not a 2-group of maximal class. If $\mathrm{C}_G(A) < G$, then p divides $\alpha(\mathrm{C}_G(A)) = \alpha(G)$, by induction. Now we assume that $\mathrm{C}_G(A) = G$, i.e., $A \leq \mathrm{Z}(G)$. If $B \in \mathfrak{M}$, then $A \leq \Phi(B) \leq \Phi(G)$. As in proof of Theorem 1.10,

$$(3) \qquad |\mathfrak{M}| = \alpha(G) \equiv \sum_{H \in \Gamma_1} \alpha(H) \pmod{p}.$$

If one of $H \in \Gamma_1$ is cyclic, the result follows from Theorem 1.2, so let all members of the set Γ_1 are noncyclic. Suppose that one of members of the set Γ_1, say H, is a 2-group of maximal class. Since $A \leq \mathrm{Z}(G)$, we get $|A| = 2$. By Theorem 5.4, Γ_1 has exactly 4 members that are of maximal class (say H_1, H_2, H_3, H_4). Then $\alpha(H_i)$ is odd for $i = 1, 2, 3, 4$. For other $H \in \Gamma_1$, $\alpha(H)$ is even, by induction. In that case, by (3), $\alpha(G)$ is even. If all $H \in \Gamma_1$ are neither cyclic nor 2-groups of maximal class, then, by induction, p divides $\alpha(H)$ for all $H \in \Gamma_1$, and so p divides $|\mathfrak{M}|$, by (3). □

Lemma 1.31 ([Isa7, Lemma C]). *Set $H = [G, \sigma] = \langle g^{-1}g^{\sigma} \mid g \in G \rangle$, where G is a group and $\sigma \in \mathrm{Aut}(G)$. Suppose that σ fixes all elements of some normal subset X of G. Then $X \subseteq \mathrm{C}_G(H)$.*

Proof. By Introduction, Theorem 13(a), $H = [G,\sigma] \trianglelefteq G$. Since $K = \langle X\rangle \trianglelefteq G$ and $[K,\langle\sigma\rangle] = \{1\}$, we have $[G,K,\langle\sigma\rangle] \le [K,\langle\sigma\rangle] = \{1\} = [K,G,\langle\sigma\rangle]$. Therefore, by Three Subgroups Lemma, $[H,K] = [\langle\sigma\rangle,G,K] = \{1\}$, so $X \subseteq K \le C_G(H)$. □

Proposition 1.32 ([Isa7]). *Let $\sigma \in \mathrm{Aut}(G)$ be of order p and $[G,\sigma]$ a p-group. If σ fixes all elements of order p and 4 in $[G,\sigma]$, then $[G,\sigma]$ is elementary abelian.*

Proof. Assume that $o([g,\sigma])$ divides p for all $g \in G$. Then $[G,\sigma] = \langle [g,\sigma] \mid g \in G\rangle \le \mathrm{Z}([G,\sigma])$ (Lemma 1.31) so $[G,\sigma]$ is abelian; then $[G,\sigma]$ is elementary abelian, as required. Hence, we have to prove that $o([g,\sigma])$ divides p. Since $\langle\sigma\rangle \cdot [G,\sigma]$ is a p-group, we get $[G,\sigma] < G$. Working by induction on $|G|$, therefore, one may assume that $[G,\sigma,\sigma]$ is an elementary abelian p-group.

Let $g \in G$ and write $x = [g,\sigma]$ and $y = [x,\sigma]$ so $g^\sigma = gx$ and $x^\sigma = xy$. Then $y \in [G,\sigma,\sigma] < [G,\sigma]$ has order dividing p, and hence $y^\sigma = y$, by hypothesis. Also, by Lemma 1.31, $y \in \mathrm{Z}([G,\sigma])$, and so $xy = yx$. It remains to prove that $x^p = 1$. For each $r \in \mathbb{N}$, we have

$$g^{\sigma^r} = gx^r y^{\binom{r}{2}}. \tag{4}$$

Formula (4) holds when $r = 1$ since $g^\sigma = gx$, and it can be proved for $r > 1$ by induction on r using that $xy = yx$. Let us apply (4) for $r = p = o(\sigma)$ to deduce that $g = gx^p y^{\binom{p}{2}}$, and thus $x^p y^{\binom{p}{2}} = 1$. In that case, we have $1 = (x^p y^{\binom{p}{2}})^2 = x^{2p} y^{2\binom{p}{2}} = x^{2p}$ since $o(y)$ divides p and p divides $2\binom{p}{2}$. If $p > 2$, the equality $x^{2p} = 1$ forces $x^p = 1$. Let $p = 2$. Then $x^4 = 1$ so σ fixes x. In that case $y = 1$ since $x = x^\sigma = xy$. Then $1 = x^2 y^{\binom{2}{2}} = x^2$, as required. □

For an arbitrary p-group G and integer $i \ge 0$, write $W_i = W_i(G) = \langle x \in G \mid o(x) \le p^i\rangle$. In general, $W_i(G)$ need not be a subgroup. Note, however, that $\langle W_i\rangle = \Omega_i(G)$.

Proposition 1.33 ([Isa7, Corollary 2.2]). *Let a p-subgroup $P \trianglelefteq G$. Assume that all elements of order p in P and also, if $p = 2$, all elements of order 4 in $[G,P]$ are central in G. Then each subset $W_i = W_i(P)$ is a subgroup. Furthermore, if we write $\bar{G} = G/W_1$, then the image $\bar{P}$ of P in $\bar{G}$ inherits the hypotheses for each subscript $i \ge 0$. In particular, G acts trivially on each of factors W_{i+1}/W_i. In addition, if all elements of order 4 in P are central in G, then that hypothesis is also inherited by $\bar{P}$ in $\bar{G}$.*

Proof. By hypothesis, $W_1 \le \mathrm{Z}(G)$. Write $\bar{G} = G/W_1$ and observe that W_{i+1} is the preimage of $W_i(\bar{P})$ in G. Everything will follow by induction on $\exp(P)$ when we show that $\bar{P}$ inherits the original hypotheses on P.

Suppose that $\bar{u} \in \bar{P}$ is of order p. We have $u^p \in W_1 \le \mathrm{Z}(G)$, and hence u induces an automorphism σ of G with order dividing p. We see that $[G,\sigma] = [G,u] \le [G,P] \le P$, and so $[G,\sigma]$ is a p-group. Also, since σ is an inner automorphism of G, it fixes all elements in W_1 and hence all elements of order p in $[G,\sigma] \le P$. If $p > 2$, therefore, Lemma 1.32 applies and we deduce that $[G,\sigma] = [G,u]$ is elementary abelian. Thus, $[G,u] \le W_1$, and we conclude that $uW_1 = \bar{u} \in Z(\bar{G})$.

If $p = 2$ and $\bar{u}$ is as in the previous paragraph, then the inner automorphism σ fixes all elements of order 4 in $[G, P]$ since by hypothesis, these elements are central in G. Since $[G,\sigma] = [G,u] \le [G, P]$, it follows that σ fixes all elements of order 4 in $[G,u]$, and thus Proposition 1.32 applies in this case too. Again we deduce that $[G,u] \le W_1$ and $\bar{u} \in Z(\bar{G})$, as required.

In the case where $p = 2$, we must also consider elements $\bar{x}$ of order 4 in $\bar{P}$. (Note, that if $x \in \bar{x}$ and $o(\bar{x}) = 4$, then $o(x) = 8$.) We must show that if $\bar{x} \in [\bar{G}, \bar{P}]$, then $\bar{x} \in Z(\bar{G})$ and in the case where we are assuming that every element of order 4 of P is central in G, we must show that $\bar{x} \in Z(\bar{G})$ without assuming that $\bar{x} \in [\bar{G}, \bar{P}]$.

Note that $[\bar{G}, \bar{P}]$ is the image of $[G, P]$ under the natural map $G \to \bar{G}(= G/W_1)$, and thus in the case where we are taking $\bar{x} \in [\bar{G}, \bar{P}]$, we can assume that $x \in [G, P]$. Since $o(x) = 8$, we see that $o(x^2) = 4$, and so, in any case, $x^2 \in Z(G)$. It follows that the inner automorphism τ of G induced by x has order dividing 2. Also, the elements of orders 2 and 4 in $[G,\tau]$ are central in G (since they lie in $[G, P]$), and thus are fixed by τ. By Lemma 1.31, $[G,x] = [G,\tau] \le W_1$ and $\bar{x} \in Z(\bar{G})$. □

Corollary 1.34 ([Isa7]). *Let P be a p-group. Assume that $\Omega_1(P) \le Z(P)$ if $p > 2$ and $\Omega_2(P') \le Z(P)$ if $p = 2$. Then $\exp(\Omega_i(P)) \le p^i$ for all i and $G/\Omega_1(G)$ satisfies the same properties.*

Exercise 35. Classify the p-groups which are not generated by noncyclic subgroups of index p.

Exercise 36. Classify the metacyclic groups (not necessarily primary) without characteristic subgroups of prime index.

Exercise 37. Classify the p-groups which are not generated by subgroups of index p^3.

Exercise 38. Classify the p-groups whose proper subgroups of order $p^n > p^2$, but one, are cyclic.

Exercise 39. Classify the 2-groups all of whose proper subgroups of order $2^n > 2^3$, but one, are cyclic or of maximal class.

Exercise 40. (a) If any two distinct maximal subgroups of a nonabelian p-group G have abelian intersection, then $d(G) < 4$. (*Hint.* Assume that $d(G) > 3$. Let A be a minimal nonabelian subgroup of G. Then $G/A\Phi(G)$ is noncyclic.)

(b) If a p-group G has $\le p$ maximal subgroups of exponent $= \exp(G)$, then $d(G) = 2$. (*Hint.* For $x \in G$ with $o(x) = \exp(G)$, consider $G/\langle x\rangle\Phi(G)$.)

Exercise 41. Let $A \in \{D_{2^m} \times C_2, Q_{2^m} \times C_2\}$. Find all 2-groups G such that $c_n(G) = c_n(A)$ for all n.

The group $\mathrm{UT}(r, p)$ of upper unitriangular $r \times r$ matrices over the field $\mathrm{GF}(p)$ is a Sylow p-subgroup of $\mathrm{GL}(r, p)$ (compare the orders!). It is easy to check that $\exp(\mathrm{UT}(p^n, p)) = p^n$. Next, $\mathrm{UT}(r, p)$ is isomorphic to a subgroup of $\mathrm{UT}(r+1, p)$.

Exercise 42 ([Kin1, Lemma 1]). Let $\mathrm{E}_{p^r} \cong N \triangleleft G$, G is a p-group and $r \le p^n$. Then $[N, \mho_n(G)] = \{1\}$.

Solution. By N/C-Theorem, $G/\mathrm{C}_G(N)$ is isomorphic to a p-subgroup of $\mathrm{GL}(r, p) \le \mathrm{GL}(p^n, p)$. Since the exponent of a Sylow p-subgroup of $\mathrm{GL}(p^n, p)$ is p^n, we get $\exp(G/\mathrm{C}_G(N)) \le p^n$ so $\mho_n(G) \le \mathrm{C}_G(N)$.

Exercise 43 ([Kin1, Corollary 2]). Let N be a normal abelian subgroup of a p-group G, $\mathrm{d}(N) = r \le p^n$, $\exp(N) = p^{k+1}$, $k \ge 0$. Then $[N, \mho_{n+k}(G)] = \{1\}$.

Solution. We use induction on k. In view of Exercise 42, one may assume that $k \ge 1$. By induction, applied to $N/\mho_k(N)$ (of exponent p^k), we have $[\mho_{n+k-1}(G), N] \le \mho_k(N)$. By Exercise 42, $[\mho_n(G), \mho_k(N)] = \{1\}$ since $\exp(\mho_k(N)) = p$ (N is abelian!). Let $x \in N$, $g \in G$. From the first of the above relations we get $x^{g^{p^{n+k-1}}} = xs$ for some $s \in \mho_k(N)$. From the second of the above relations we get $s^{g^{p^n}} = s$. Therefore, setting $g^{p^{n+k-1}} = u$, we get $x^u = xs$ so $x^{g^{p^{n+k}}} = x^{(g^{p^{n+k-1}})^p} = x^{u^p} = xs^p = x$ since $s^p = 1$.

Theorem 1.35 ([Kin1, Theorem 3]). *Let $N \le \mho_n(G)$ be a normal subgroup of a p-group G, $n \ge 1$, $\mathrm{d}(N) = r$. Then (a) If $p > 2$ and $r \le p^n$, then $N' \le \mho_1(N)$. (b) If $p = 2$ and $r \le 2^{n-1}$, then $N' \le \mho_2(N)$.*

Exercise 44 (see Theorem 44.13). Let $N \le \Phi(G)$ be a two-generator normal subgroup of a p-group G, $p > 2$. Then N is metacyclic. (*Hint.* Use Theorem 9.11.)

Exercise 45. Let $G = H \cdot F$ be a metacyclic p-group, H and $F \triangleleft G$ are cyclic. Then $\Phi(G) = \Phi(H) \cdot \Phi(F)$.

Exercise 46 (Reported by Mann). We describe a method of construction of automorphisms. Let G be a group (not necessarily finite) and ϕ a homomorphism of G into $\mathrm{Z}(G)$ such that $\phi(\phi(x)) = 1$ for all $x \in G$. Then the map $\sigma : x \mapsto x\phi(x)$ is an automorphism of G. For every $x \in G$ and $n \in \mathbb{N}$, we have $\sigma^n(x) = x\phi(x)^n$.

Solution. By definition, $\mathrm{im}(\phi) \le \ker(\phi)$. Since $\mathrm{im}(\phi) \le \mathrm{Z}(G)$, we have $\sigma(xy) = xy\phi(xy) = x\phi(x) \cdot y\phi(y) = \sigma(x)\sigma(y)$, and so σ is an endomorphism of G that is identity on $\mathrm{im}(\phi)$ since $\sigma(\phi(x)) = \phi(x)\phi(\phi(x)) = \phi(x) \cdot 1 = \phi(x)$. Thus, $\mathrm{im}(\phi) \le \mathrm{im}(\sigma)$, so $x = \sigma(x)\phi(x)^{-1} \le \mathrm{im}(\sigma)$, and hence $\sigma(G) = G$. Since we do not assume that G is finite, we have to show that $\ker(\sigma) = \{1\}$. Let $y \in \ker(\sigma)$, i.e., $y\phi(y) = 1$. Then $\phi(y) = y^{-1}$ and $1 = \phi(\phi(y)) = \phi(y^{-1}) = \phi(y)^{-1}$, i.e., $\phi(y) = 1$. Therefore, $1 = \sigma(y) = y\phi(y) = y$. It follows that $\ker(\sigma) = \{1\}$ so $\sigma \in \mathrm{Aut}(G)$. Since $\sigma_{\mathrm{im}(\phi)} = \mathrm{id}_{\mathrm{im}(\phi)}$, the last assertion follows by induction on n.

Exercise 47. Let $P = \mathrm{E}_{2^3}$, $Q \in \mathrm{Syl}_2(\mathrm{Aut}(P))$. Then $Q \cong \mathrm{D}_8$. Let $R = Q \cdot P$ be the natural semidirect product with kernel P. Show that $|\mathrm{Z}(R)| = 2$.

Exercise 48. Let G be a p-group.

(a) Let $L \lhd G$ be such that $L \not\leq \mathrm{Z}_s(G)$. Then $|L \cap \mathrm{Z}_s(G)| \geq p^s$ so $|L| > p^s$.

(b) Suppose that $s > 2$ and $\mathrm{cl}(G) \geq s + 1$. Then, if $\mathrm{Z}_s(G)$ is of maximal class, its order is equal to p^{s+1}.

Solution. (b) By (a), $|\mathrm{Z}_s(G)| \geq p^s$. Assume that (b) is false, i.e., $|\mathrm{Z}_s(G)| = p^s$. Let $T = \mathrm{Z}_2(G)$; then T is of order p^2. Set $L = \mathrm{C}_G(T)$. In that case, $|G : L| = p$ and $\mathrm{Z}_s(G) \not\leq L$ since $|\mathrm{Z}(L)| \geq p^2$ and $|\mathrm{Z}(\mathrm{Z}_s(G))| = p$: $\mathrm{Z}_s(G)$ is of maximal class. By (a), however, $|L \cap \mathrm{Z}_s(G)| \geq p^s = |\mathrm{Z}_s(G)|$, and this is a contradiction.

Given a p-group G, let $\mathcal{A}_1(G)$ be the set of all elementary abelian subgroups of G of maximal order, $\mathcal{A}_2(G)$ the set of all abelian subgroups of G of maximal order, and $\mathcal{A}_3(G)$ the set of all maximal abelian subgroups of G of maximal exponent. Set $J_i(G) = \langle A \mid A \in \mathcal{A}_i(G)\rangle$, $i = 1, 2, 3$.

Exercise 49. Let G be a p-group.

(a) If $Q < G$ and $Q \cong J_i(G)$, then $Q = J_i(G)$, $i = 1, 2, 3$.

(b) If $G = G_1 \times G_2$, then $J_i(G) = J_i(G_1) \times J_i(G_2)$, $i = 1, 2, 3$.

(c) $\Omega_1(\mathrm{Z}(J_1(G))) = \bigcap_{A \in \mathcal{A}_1(G)} A$, $\mathrm{Z}(J_i(G)) = \bigcap_{A \in \mathcal{A}_i(G)} A$, $i = 2, 3$.

Exercise 50. If G is a p-group and $x \in G$ of order p is such that $|G : \mathrm{C}_G(x)| = p$. Then (a) [GLS, Lemma 10.20] $\mathcal{A}_1(\mathrm{C}_G(x)) \subseteq \mathcal{A}_1(G)$. (b) $\mathcal{A}_2(\mathrm{C}_G(x)) \subseteq \mathcal{A}_2(G)$.

Solution of (a). Let $A \in \mathcal{A}_1(G)$, $C = \mathrm{C}_G(x)$ and $B = A \cap C$. We may assume that $B < A$; then $|A : B| = p$ and $x \notin A$. In that case, $\langle x, B\rangle \leq C$ is elementary abelian of order $p|B| = |A|$ so $\langle x, B\rangle \in \mathcal{A}_1(C) \cap \mathcal{A}_1(G)$.

Exercise 51. If G is a p-group with $\Omega_1(G)$ of order p^e and exponent $p > 2$ and $n \in \mathbb{N}$ is such that $p < p^n \leq \exp(G)$, then p^{e-1} divides $\mathrm{c}_n(G)$.

Exercise 52. Let S and N be normal subgroups of a p-group G such that $S \leq N \leq \Phi(G)$. If $\mathrm{C}_S(N)$ is cyclic, then N is cyclic.

Exercise 53. Let G be a p-group and $H < G$ a subgroup of maximal class. Suppose that $H \cap Z = \{1\}$ for every cyclic subgroup Z of G not contained in H. Then $p > 2$ and $|H| \leq p^{p+1}$. (Use Theorems 9.6 and 13.9.)

Exercise 54. Let $H \cong \mathrm{M}_{p^n}$ be a proper subgroup of a p-group G, $p > 2$. Study the structure of G provided $H \cap L = \{1\}$ for every cyclic $L < G$ such that $L \not\leq H$.

Exercise 55. The same question as in Exercise 54 in the case where H is abelian of type (p^n, p), $n > 1$.

Exercise 56. Classify the p-groups in which every maximal cyclic subgroup coincides with its centralizer.

Exercise 57. Let G be a homocyclic group of exponent p^e. Then every maximal cyclic subgroup of G has order p^e and is a direct factor of G.

Exercise 58. If all maximal cyclic subgroups of an abelian p-group G have the same order, then G is homocyclic.

Exercise 59. Let G be a nonabelian p-group. Suppose that $Z(G)$ is a maximal cyclic subgroup of G. Then $|G : C_G(P)| = p$ for some $P < G$ such that $Z(G) < P \leq Z_2(G)$ and $|P/Z(G)| = p$.

Exercise 60. Let $P \in \mathrm{Syl}_2(G)$ be of maximal class and order 2^m. Prove that $c_k(G) \equiv 1 \pmod{2^{m-k}}$ for $m > k > 2$.

Exercise 61. Let G be a nonabelian group. Suppose that $C_G(x)$ is abelian for all $x \in G - Z(G)$. Then, if $a, b \in G - Z(G)$ are such that $C_G(a) \neq C_G(b)$, then $C_G(a) \cap C_G(b) = Z(G)$. In particular, $G/Z(G)$ has a nontrivial partition all of whose components are abelian.

Exercise 62 (M. Morigi). Let $G = AB$ be a finite group with abelian A and $Z(G) > \{1\}$. Then $A_G B_G > \{1\}$.

Solution. Consider $1 \neq ab \in Z(G)$ with $a \in A$ and $b \in B$. Suppose that $A_G B_G = \{1\}$; then $A \cap Z(G) = \{1\} = B \cap Z(G)$ so that $a \neq 1 \neq b$. Then for each $x \in A$ we have $1 = [ab, x] = [a, x]^b [b, x] = [b, x]$, so $[b, A] = \{1\}$. Therefore, $\langle b \rangle^G = \langle b \rangle^{AB} = \langle b \rangle^B \leq B$, and $\{1\} \neq \langle b \rangle^G \leq B_G$, a contradiction.

Exercise 63 (M. Morigi; see [Ito7]). Let $G = AB$ be a group and $V \leq Z(A)$, $W \leq Z(B)$ are such that $AW = WA$ and $BV = VB$; then $[V, W] \leq Z([A, B])$.

Solution. Let $a \in A$, $b \in B$, $v_1 \in V$, $w_1 \in W$ and put $b^{v_1} = v_2 b_2$, $a^{w_1} = w_3 a_3$, where $v_2 \in V$, $b_2 \in B$, $w_3 \in W$, $a_3 \in A$. Then

$$\begin{aligned} [a,b]^{v_1 w_1} &= [a, b^{v_1}]^{w_1} = [a, v_2 b_2]^{w_1} = [a, b_2]^{w_1} \\ &= [a^{w_1}, b_2] = [w_3 a_3, b_2] = [a_3, b_2], \\ [a,b]^{w_1 v_1} &= [a^{w_1}, b]^{v_1} = [w_3 a_3, b]^{v_1} = [a_3, b]^{v_1} \\ &= [a_3, b^{v_1}] = [a_3, v_2 b_2] = [a_3, b_2]. \end{aligned}$$

Then $[[A, B], [V, W]] = \{1\}$, and, since $[A, B] \geq [V, W]$, we get $[V, W] \leq Z([A, B])$. (In particular, if $V = A$ and $W = B$, then $[A, B] = G'$ is abelian [Ito7].)

Exercise 64 (M. Morigi). Let $G = AB$ be a p-group, where A is abelian and $|B'| = p^n$. Then $\mathrm{dl}(G) \leq n + 2$.

Exercise 65 (M. Morigi). Let $G = AB$ be a p-group such that $|A'| = p^m$ and $|B'| = p^n$, with $m \geq n$. Then $\mathrm{dl}(G) \leq m + 2n + 2$.

Example. Let $G = C_2 \operatorname{wr} E_4$ be the standard wreath product. Let $E_4 = \{1, (1,2)(3,4), (1,3)(2,4), (1,4)(2,3)\}$ and $\{a_1, a_2, a_3, a_4\}$ generators of the base B of the wreath product G. Then $U = \langle a_1, (1,2)(3,4)\rangle \cong D_8$, $Z(U) = \langle a_1a_2\rangle$. Set $V = \langle a_2a_3a_4, (1,3)(2,4)\rangle \cong D_8$ with $Z(V) = \langle a_2a_3a_4 \cdot a_4a_1a_2\rangle = \langle a_1a_3\rangle$. We claim that $UV = G$. Indeed, it is easy to check that $U \cap V = \{1\}$ so $|UV| = 2^6$. Since $|G| = 2^4 \cdot 2^2 = 2^6 = |UV|$, our claim follows. As $Z(G) = \langle a_1a_2a_3a_4\rangle$, and $a_1a_2a_3a_4$ is not contained in $Z(U) = \langle a_1a_2\rangle$ and $Z(V) = \langle a_1a_3\rangle$, we get $U_G V_G = \{1\}$. Thus, $G = UV$, where $U \cong D_8 \cong V$ and $U_G V_G = \{1\}$.

Exercise 66. Let $p > 2$, $n > 1$ and H be the Sylow p-subgroup of the holomorph of the cyclic group $C \cong C_{p^n}$. Prove that $|H'| = p^{n-1}$ and $|Z(H)| = p$.

Exercise 67. If a p-group G has exactly p proper nonabelian subgroups, then $Z(G) < \Phi(G)$ and $G/Z(G)$ is nonabelian of order p^3 and $G/Z(G) = \Omega_1(G/Z(G))$.

Exercise 68. If $P \cong M_{p^n}$, then $\pi(\operatorname{Aut}(P)) \subseteq \pi(p(p-1))$. (See Theorem 34.8.)

Exercise 69 ([Man12]). Let G be a p-group.

(a) (See also Kazarin's thesis (1971).) If distinct $A, B \in \Gamma_1$, then $|G' : A'B'| \le p$.

(b) Let $M \in \Gamma_1$. If $M' < G'$, then there exists $L \in \Gamma - \{M\}$ such that $M'L' < G'$.

Solution. (a) Since $A'B' \le \Phi(G) \le A \cap B$, there are in $G/A'B'$ two distinct abelian maximal subgroups $A/A'B'$ and $B/A'B'$ so $|G/A'B' : Z(G/A/B)| \le p^2$. By Lemma 1.1, $|G'/A'B'| \le p$.

(b) One may assume that $M' = \{1\}$; then $G' > \{1\}$. Let $N < G'$ be G-invariant of index p. By Lemma 1.1, $|G/N : Z(G/N)| = p|G'/N| = p^2$. As $(G/N)/Z(G/N)$ is noncyclic, it contains a subgroup $(L/N)/Z(G/N)$ of order p such that $L \ne M$. Obviously, L/N is abelian so $L' \le N$. Then $M'L' = L' \le N < G'$.

Exercise 70. Let a 2-group N of maximal class be a normal subgroup of an arbitrary finite group G. Then $N \not\le \Phi(G)$.

Now we prove, using [Hal4], some properties of central series and commutator subgroups of groups.

Definition 4. A word $\theta(x)$ in the variables $x_1, x_2, \dots$ is a formal expression $\theta(x) = x_{i_1}^{r_1} x_{i_2}^{r_2} \dots x_{i_n}^{r_n}$ with integers $r_1, r_2, \dots, r_n$. Given a group G, $\theta(G)$ is the subgroup generated by all elements of the form $\theta(a)$ in G for any choices of $a_1, a_2, \dots$. $\theta(G)$ is fully invariant in G, i.e., $\theta(G)^\epsilon \le \theta(G)$ for all $\epsilon \in \operatorname{End}(G)$.

Lemma 1.36. *Let $\theta(x)$, $\phi(y)$ be any two words involving the variables $x_1, \dots, x_m$ and $x_{m+1}, \dots, x_{m+n}$ at most, respectively. Define*

$$\psi(x) = [\theta(x_1, \dots, x_m), \phi(x_{m+1}, \dots, x_{m+n})].$$

Then for any group G, $\psi(G) = [\theta(G), \phi(G)]$.

Proof. Clearly, $\psi(G) \leq [\theta(G), \phi(G)]$. But (mod $\psi(G)$), every θ-value commutes with every ϕ-value so the converse inclusion holds, proving the lemma. □

In particular, $\mathrm{K}_n(G) = \phi(G)$, where $\phi(x) = [x_1, \ldots, x_n]$.

Given $C \subseteq G$, we write $C^G = \{c^g \mid c \in C, g \in G\}$; then C^G is a G-invariant subset. (In other places of this book C^G is the normal closure of C in G.)

Lemma 1.37. *Let $G = \langle A \rangle$ and let $H = \langle B^G \rangle \trianglelefteq G$, where $A, B \subseteq G$. Then $[H, G] = \langle C^G \rangle$, where $C = \{[a, b] \mid a \in A, b \in B\}$.*

Proof. Note that $C \subseteq [G, H] \leq H$ since $B \subseteq H \trianglelefteq G$. Write $K = \langle C^G \rangle$. Since C^G is G-invariant, $K \trianglelefteq G$ and $K \leq [H, G]$ since $C \subseteq [H, G] \trianglelefteq G$. But (mod K), every $a \in A$ commutes with every $b \in B$. Hence each Kb belongs to the center of G/K for all $b \in B$ since $G = \langle A \rangle$. Thus $H/K \leq \mathrm{Z}(G/K)$ and this is the same as to say $[H, G] \leq K$. Therefore, $K = [H, G]$. □

Exercise 71. Suppose that a group $G = \langle A \rangle$, where $A \subseteq G$. Then $\mathrm{K}_2(G) = G' = \langle [a_1, a_2]^g \mid a_1, a_2 \in A, g \in G \rangle$. In particular, if $G = \langle a, b \rangle$, then $\mathrm{K}_2(G) = G' = [G, G] = \langle [a, b]^G \rangle = \langle [a, b]^g \mid g \in G \rangle$.

Lemma 1.38. *Let $G = \langle A \rangle$, where $A \subseteq G$. Then $\mathrm{K}_n(G)$ is generated by $\mathrm{K}_{n+1}(G)$ together with all the commutators $[a_1, \ldots, a_n]$ (all $a_i \in A$).*

Proof. By Lemma 1.37, $\mathrm{K}_n(G)$ is generated by the commutators $u = [a_1, \ldots, a_n]$ together with all their G-conjugates, $a_i \in A$ for all i (see Exercise 71). But $u^x = u[u, x]$ and $[u, x] \in \mathrm{K}_{n+1}(G)$ for all $x \in G$. (Hence, if $G = \langle a, b \rangle$, then $G'/\mathrm{K}_3(G)$ is cyclic.) □

Exercise 72. (a) If $H, K, L \trianglelefteq G$, then $[HK, L] = [H, L][K, L]$.

(b) If G is a group, $x \in G$ and $y \in \mathrm{K}_r(G)$, then for any integer n, we have $[x^n, y] \equiv [x, y^n] \equiv [x, y]^n \pmod{\mathrm{K}_{r+2}(G)}$ and $(xy)^n \equiv x^n y^n [y, x]^{\binom{n}{2}} \pmod{\mathrm{K}_{r+2}(G)}$.

Exercise 73. Let G be a p-group.

(a) If G/G' is the abelian group of type $(p^{n_1}, \ldots, p^{n_r})$ and $n_1 \geq \cdots \geq n_r$, then $\exp(G'/\mathrm{K}_3(G)) \leq p^{n_2}$.

(b) If $\exp(\mathrm{K}_i(G)/\mathrm{K}_{i+1}(G)) = p^m$ for some $i \geq 2$, then $\exp(\mathrm{K}_{i+1}(G)/\mathrm{K}_{i+2}(G)) \leq p^m$.

(c) If G/G' is an abelian group of type (p^λ, p^μ), where $\lambda \geq \mu$, then $G'/\mathrm{K}_3(G)$ is cyclic of order at most p^μ.

It follows from Exercise 73(a,b), that if, for a p-group G, $\exp(G/G') = p^m$, then $\exp(\mathrm{K}_i(G)/\mathrm{K}_{i+1}(G)) \leq p^m$ for all $i \geq 2$. (See §26).

Theorem 1.39. *Let $X = \langle A \rangle$, $Y = \langle B \rangle$ be subgroups of a group G. Let C be the set of all commutators $[a, b]$ with $a \in A$, $b \in B$. Then $[X, Y] = \langle (C^X)^Y \rangle$.*

Proof. We have $[X, Y] \trianglelefteq \langle X, Y \rangle$. Hence $H = \langle (C^X)^Y \rangle \leq [X, Y]$ since $C \subseteq [X, Y]$. Conversely, let $x = a_{i_1}^{r_1} \dots a_{i_n}^{r_n}$ be any element of X, where each $a_{i_\alpha} \in A$. Then for $b \in B$ we have $b^{-1}xb = \prod_{\alpha=1}^{n}(a_{i_\alpha}[a_{i_\alpha}, b])^{r_\alpha}$. Each $[a_{i_\alpha}, b] \in C$. Hence $b^{-1}xb \equiv x \pmod{\langle C^X \rangle}$ so that $[x, b] \in H$ for all $x \in X$, $b \in B$. Let $y = b_{j_1}^{s_1} \dots b_{j_m}^{s_m}$ be any element of Y, where all $b_{i_\beta} \in B$. Then $x^{-1}yx = \prod_{\beta=1}^{m}(b_{j_\beta}[b_{j_\beta}, x])^{s_\beta}$. Since each $[b_{j_\beta}, x] \in H = H^Y$ and each $b_{j_\beta} \in Y$, we have $x^{-1}yx \equiv y \pmod{H}$ or $[y, x] \in H$. It follows that $[X, Y] = [Y, X] \leq H$, completing the proof. □

Exercise 74. If G is a group and $N \triangleleft G$ with cyclic G/N, then $G' = [G, N]$.

Recall (see Introduction, Remark 2) that $\eta(G)/\mathrm{K}_3(G) = \mathrm{Z}(G/\mathrm{K}_3(G))$ (in that remark, instead of $\eta(G)$ we wrote $\eta_3(G)$). Part (a) of the following theorem is important in the study of subgroups of maximal class in p-groups (see §12). Part (b) of that theorem allows us to control the class of a p-group G.

Theorem 1.40. *Let G be nilpotent and* $H \leq G$.

(a) *[Bla3, Theorem 1.3] If* $H\eta(G) = G$, *then* $H \trianglelefteq G$ *and* $\mathrm{K}_i(H) = \mathrm{K}_i(G)$ *for* $i \geq 2$ *so* $\mathrm{cl}(H) = \mathrm{cl}(G)$.

(b) *[Schn2, Lemma 2.1] If* $G' = H'\mathrm{K}_3(G)$, *then* $\mathrm{K}_i(G) = \mathrm{K}_i(H)$ *for all* $i \geq 2$. *Moreover,* $H \trianglelefteq G$.

Proof. We prove only the first assertion.

(a) Let $\mathrm{cl}(G) = m - 1$. We prove that $\mathrm{K}_i(H) = \mathrm{K}_i(G)$ by induction on $m - i$. For $i = m$ this is trivial. For $i < m$, by induction, $\mathrm{K}_{i+1}(H) = \mathrm{K}_{i+1}(G)$. Now by hypothesis, $G = \langle \eta(G) \cup H \rangle$; then $\mathrm{K}_i(G)$ is generated by $\mathrm{K}_{i+1}(G)$ and all commutators $y = [y_1, \dots, y_i]$, where every component y_j of y is either an element of $\eta(G)$ or H (Lemma 1.38). But by Introduction, Remark 2, $[\eta(G), \mathrm{K}_j(G)] \leq \mathrm{K}_{j+2}(G)$ ($j = 2, 3, \dots, m - 2$), and so any commutator y, one of whose components is an element of $\eta(G)$, belongs to $\mathrm{K}_{i+1}(G)$. Hence $\mathrm{K}_i(G)$ is generated by $\mathrm{K}_{i+1}(H) = \mathrm{K}_{i+1}(G)$ and all commutators of the form $y = [y_1, \dots, y_i]$, where $y_j \in H$. Since all these commutators belong to $\mathrm{K}_i(H)$, it follows that $\mathrm{K}_i(G) \leq \mathrm{K}_i(H)$, and therefore $\mathrm{K}_i(H) = \mathrm{K}_i(G)$, as required. We have $H \geq \mathrm{K}_2(H) = \mathrm{K}_2(G)$ so $H \trianglelefteq G$. □

Let G be nilpotent. Part (a) of Theorem 1.40 follows from (b) since $\mathrm{K}_3(G) \leq \eta(G)$. If $H \leq G$ be such that $H' = G'$, then $\mathrm{K}_i(H) = \mathrm{K}_i(G)$ for all $i \geq 2$ (Theorem 1.40(b)).

Supplement to Theorem 1.40(b) ([Schn2, Corollary 2.2]). *Let G be a p-group.*

(a) *If* $|G'/\mathrm{K}_3(G)| = p$, *then G has a two-generator normal subgroup H such that* $\mathrm{K}_i(H) = \mathrm{K}_i(G)$ *for all* $i \geq 2$.

(b) *If* $G'/\mathrm{K}_3(G) \cong \mathrm{E}_{p^2}$, *then G has a three-generator subgroup H such that* $\mathrm{K}_i(G) = \mathrm{K}_i(H)$ *for all* $i \geq 2$.

Let G be a 2-group such that $G' = \Phi(G)$ and $|G'/\mathrm{K}_3(G)| = 2$. Then G has a subgroup of maximal class H such that $\mathrm{K}_i(H) = \mathrm{K}_i(G)$ for all $i \geq 2$.

Remark 5. Let $A, B \in \Gamma_1$ be two distinct abelian, where G is a nonabelian p-group. Then $G = U\mathrm{Z}(G)$, where U is minimal nonabelian. If, in addition, $\exp(B) = p$, then $G = U \times V$, where V is elementary abelian and $|U| \leq p^4$. Indeed, $A \cap B = \mathrm{Z}(G)$ is of index p^2 in G and, if $U \leq G$ is minimal nonabelian, then $G = U\mathrm{Z}(G)$. If, in addition, $\exp(B) = p$, then $\mathrm{Z}(G)$ is elementary abelian so $\mathrm{Z}(G) = \mathrm{Z}(U) \times V$ for some $V < \mathrm{Z}(G)$. Then $G = U \times V$. Since $U \cap B$ is elementary abelian of index p in U, it follows that $|U| \leq p^4$ (by Exercise 8a, all subgroups of U are three-generator).

Exercise 75. Let p-groups G and H be such that, for each $i \geq 0$, they contain the same number of classes of size p^i (irreducible characters of degree p^i). Prove that if G is extraspecial then H is also extraspecial.

Exercise 76. Let G be a 2-group. Suppose that $\mathrm{d}(G) = 2$ and $\Phi(G)$ is homocyclic of rank two. Then $G = AB$, where A and B are cyclic with $A \cap B = \{1\}$.

Exercise 77. Let $Z \leq \mathrm{Z}(G)$ be maximal cyclic in a p-group G, $|Z| = p$. Prove, for every $n > 1$, that $\mathrm{c}_n(G) = p\mathrm{c}_n(G/Z)$.

Exercise 78. Classify the 2-groups G with $H \in \Gamma_1$ of maximal class.

Exercise 79. Let $A \cong \mathrm{M}_{2^n}$. Prove that if G is a 2-group such that $\mathrm{s}_m(G) = \mathrm{s}_m(A)$ for all $m \in \mathbb{N}$, then G is either abelian of type (p^{n-1}, p) or isomorphic to A.

Exercise 80. Let $G = HL$ be a 2-group, where $H \in \Gamma_1$ and $L \triangleleft G$ is cyclic. If H is of maximal class with cyclic subgroup Z of index 2, $|L \cap H| = 2$ and $[Z, L] = \{1\}$, then $|\mathrm{Z}(G)| = 4$. (*Hint.* Use Lemma 1.1.)

Exercise 81. Let G be a nonabelian p-group such that G/G' has a cyclic subgroup of index p. Prove that $|G' : \mathrm{K}_3(G)| = p$.

Theorem 1.41. *If a nilpotent group G is generated by the set M, then* $\mathrm{cl}(G) \leq c$ *if and only if any commutator of weight $c + 1$ in elements of M is trivial.*

Proof. For each $n > 0$ define the subgroup $M_n = \langle [m_1, \dots, m_n] \mid m_i \in M \rangle^G$. It suffices to show that $M_n = \mathrm{K}_n(G)$ for all n. Clearly, $M_n \leq \mathrm{K}_n(G)$. To prove the reverse inclusion, we use induction on n. For $n = 1$, $M_1 = \langle M \rangle = G = \mathrm{K}_1(G)$ so let $n > 1$. Using the commutator identities and the fact that every element of G is a product of elements from M, we conclude that every coset $[m_1, \dots, m_{n-1}, g]M_n \in \mathrm{Z}(G/M_n)$ for $m_1, \dots, m_{n-1} \in M$ and $g \in G$. Thus, $[M_{n-1}, G] \leq M_n$. By induction, $\mathrm{K}_n(G) = [\mathrm{K}_{n-1}(G), G] = [M_{n-1}, G] \leq M_n$. □

Exercise 82. For each $a \in G$ there is one-to-one correspondence between the set of commutators of the form $[g, a]$ ($g \in G$) and the set of right cosets of $\mathrm{C}_G(a)$.

Solution. Let $\psi(C_G(a)g) = [g,a]$, $g \in G$. The mapping ψ is well defined and surjective since, if $g' = cg$, $c \in C_G(a)$, then $[g',a] = [c,a]^g[g,a] = [g,a]$. If $[x,a] = [y,a]$, then $[xy^{-1},a] = [x,a]^{y^{-1}}[y^{-1},a] = [y,a]^{y^{-1}}[y^{-1},a] = [yy^{-1},a] = 1$ so $xy^{-1} \in C_G(a)$, i.e., $C_G(a)x = C_G(a)y$. It follows that ψ is injective so bijective.

Remark 6. Let $A, B \le G$ and $|G : A| \le m$, $|G : B| \le n$. Then $|G : (A \cap B)| \le mn$. Indeed, $|AB| = \frac{|A||B|}{|A \cap B|} \le |G|$ so $|A : (A \cap B)| \le |G : B| \le n$ and, similarly, $|B : (A \cap B)| \le |G : A| \le m$. Then $|G : (A \cap B)| = |G : A||A : (A \cap B)| \le mn$.

Corollary 1.42. *Suppose that G is an s-generator group, $s < \infty$, and that the set of commutators of G consists of n elements. Then $|G : Z(G)| \le n^s$.*

Proof. By Exercise 82, $|G : C_G(g)| \le n$ for each $g \in G$. Let $G = \langle a_1, \dots, a_s \rangle$. Then $C_G(a_1) \cap \dots \cap C_G(a_s) \le Z(G)$ so $|G : Z(G)| \le n^s$ (Remark 6). □

Theorem 1.43. *Suppose that G is a group with $|K_s(G)| = n$. Then G has a characteristic nilpotent subgroup of class $\le s$ and index $\le (n-1)!$.*

Proof. Set $T = C_G(K_s(G))$; then T is characteristic in G and $|G : T| \le |\mathrm{Aut}((K_s(G))| \le (n-1)!$. It remains to show that $[a_1, \dots, a_s, a_{s+1}] = 1$ for any $a_1, \dots, a_{s+1} \in T$. Indeed, $[a_1, \dots, a_s] \in K_s(G)$ so $[a_1, \dots, a_s, a_{s+1}] = 1$ since $a_{s+1} (\in T)$ centralizes $K_s(G)$. Thus, T is the desired subgroup. □

Let G be a p-group and let nonabelian M have index p in $\Phi(G)$. If $Z(M)$ is cyclic, then M is a direct factor of $\Phi(G)$ since $Z(\Phi(G))$ has an abelian subgroup of type (p, p).

Exercise 83. Let G be a p-group and let M of maximal class be of index p in $\Phi(G)$. Then $|M| = p^3$ and $\Phi(G) = M \times C$, where $|C| = p$. (*Hint.* Use Lemma 1.4.)

Exercise 84. Let G be a p-group and let a nonabelian M have index p in $\Phi(G)$. If $Z(M)$ is cyclic, then $\Phi(G) = M \times C$, where $|C| = p$.

Exercise 85. Let a noncyclic p-group G be metacyclic. If G is not a 2-group of maximal class, then $\Omega_1(G) \cong E_{p^2}$.

Exercise 86. Let G be a noncyclic p-group of exponent $> p$. Prove that the following assertions are equivalent: (a) Every cyclic subgroup of order p^2 is contained in a unique maximal subgroup of G. (b) $\exp(\Phi(G)) = p$ and $d(G) = 2$.

Exercise 87. Let H be a normal subgroup of a p-group G, $|H| > p$, $|G : H| > p$ and $Z(G/H)$ is cyclic. If $F/H \le Z(G/H)$ is of order p, then F is not of maximal class. (See Lemma 9.2.)

Exercise 88. If a nonabelian 2-group $G = \langle a \rangle \cdot E$, where $\exp(E) = 2$ and $o(a) = 2$, then $G = E_1 \times U$, where $\exp(E_1) \le 2$ and U is special of order $2|Z(U)|^2$.

Solution. We have $G' = [a, E]$ and $\exp(G) = 4$. If $x \in E$ and $xa \neq ax$, then $\langle a, x\rangle \cong D_8$ so $[a, x] \in E$ hence $C_G([a, x]) \geq \langle E, a\rangle = G$. Thus, $G' \leq Z(G) < E$. Set $Z(G) = G' \times E_1$. Since $\Omega_1(G) = G$, we get $G' = \Phi(G)$. Therefore, $G = E_1 \times U$ for some $U \leq G$. Clearly, $U' = G'$, $Z(U) = U \cap Z(G) = G' = U'$ and $U' = \Phi(U)$ so U is special. Since U has the abelian subgroup $E \cap U$ of index 2, we get, by Lemma 1.1, $|U| = 2|U'||Z(U)| = 2|Z(U)|^2$.

Exercise 89. Let G be a p-group. (a) If G is generated by G-invariant subgroups of order p^n, then $\mathrm{cl}(G) \leq n$. (ii) If G is generated by normal cyclic subgroups, then $\mathrm{cl}(G) \leq 2$. (*Hint.* (a) $Z_n(G)$ contains every G-invariant subgroup of G of order p^n. (b) Every G-invariant cyclic subgroup centralizes G'.)

Exercise 90. Let $G = M_{p^3} * C_{p^2}$ be of order p^4, $p > 2$. Show that then $G = S(p^3) * C_{p^2}$, where $S(p^3)$ is nonabelian of order p^3 and exponent p.

Exercise 91. Classify the p-groups containing only one abelian subgroup of order p^3 and exponent p^2.

Exercise 92. Let $A < G$ be abelian 2-groups and assume that $\exp(G) > 2$ and all elements of the set $G - A$ are involutions. Then (a) A is abelian and normal in G, $C_G(A) = A$ and $\exp(Z(G)) = 2$, (b) $|G : A| = 2$ and (c) A is characteristic in G.

Exercise 93. Let G be a 2-group and let a nonabelian $N < G$ be such that all subgroups of N are normal in G. If $x \in G$ has order 8, then $x^2 \notin N$.

Solution. Assume $x^2 \in N$. By Theorem 1.20, $\exp(N) = 4$ so there is $Q \leq N$ with $Q \cong Q_8$. Then there is $y \in Q$ of order 4 such that $x^2 y \neq y x^2$ since $\exp(C_N(Q)) = 2$. Since $\langle y\rangle \triangleleft G$, we get $[x, y] \in \langle y^2\rangle$ so $\langle x, y\rangle/\langle y^2\rangle$ is abelian hence $\mathrm{cl}(\langle x, y\rangle) = 2$. Then $[x^2, y] = [x, y]^2 = 1$, contrary to what has just been said.

Exercise 94. Let G be a nonabelian p-group. Suppose that, for every nonabelian subgroup H of G, we have $N_G(H) = HZ(G)$. Then $G = AZ(G)$, where A is minimal nonabelian and $|G : Z(G)| = p^2$.

Solution. Let A be a nonabelian normal subgroup of G of minimal order; then $G = AZ(G)$. Let $B < A$ be maximal; then $B \triangleleft G = AZ(G)$. By the choice of A, B is abelian so A is minimal nonabelian. Since $Z(A) \leq Z(G)$, we are done.

Exercise 95. If G is a group, $m, n \in \mathbb{N}$, then (a) $[K_m(G), Z_n(G)] = Z_{|m-n|}(G)$ and (b) $[K_m(G), K_n(G)] \leq K_{m+n}(G)$. (*Hint.* Use Three Subgroups Lemma.)

Exercise 96. Let G be a p-group. If $N \triangleleft G$ is such that $N \leq K_n(G)$ and $|N| \leq p^n$, then $N \leq Z(K_n(G)) \cap Z(Z_n(G))$.

Exercise 97 (P. Hall). Let G be a p-group. If $N \triangleleft G$ is such that $N \leq K_n(G)$ and $\mathrm{cl}(N) > t$, then $|K_t(N) : K_{t+1}(N)| \geq p^n$.

Solution. By Exercise 96 we have $|\mathrm{K}_t(N)| > p^n$ since $\mathrm{K}_t(N) \not\leq \mathrm{Z}(N)$ so $\mathrm{K}_t(N) \not\leq \mathrm{Z}(\mathrm{K}_n(G))$. Let $R < \mathrm{K}_t(N)$ be G-invariant of index p^n. By Exercise 96, $\mathrm{K}_t(N)/R \leq \mathrm{Z}(\mathrm{Z}_n(G/R))$ so $R \geq \mathrm{K}_{t+1}(N)$. It follows that $p^n = |\mathrm{K}_t(N) : R| \leq |\mathrm{K}_t(N) : \mathrm{K}_{t+1}(N)|$, as required.

Exercise 98. Let G be a p-group with $\mathrm{dl}(G) = n + 1 > 3$. Using $G^{(i)} \leq \mathrm{K}_{2^i}(G)$ and Exercise 97, prove that $|G/G^{(n)}| \geq p^k$, where $k = 2 + 3 + 2^2 + 2^3 + \cdots + 2^{n-1} = 2^n + 1$. In particular, if $n = 5$, we get $|G/G^{(5)}| \geq p^{33}$. (See also Appendix 6.)

Exercise 99. Let $G = Q * C$, where Q is a 2-group of maximal class and order 2^m and C is cyclic of order 2^n, $Q \cap C = \mathrm{Z}(Q)$. Find $\mathrm{c}_k(G)$ for all $k \in \mathbb{N}$. (See Appendix 16.)

Exercise 100. Let $G = Q * M$ be a central product of 2-groups Q and M of maximal class, $|Q| = 2^n$, $|M| = 2^m$, $Q \cap M = \mathrm{Z}(Q) = \mathrm{Z}(M)$. Find $\mathrm{c}_k(G)$.

Exercise 101 (Miller). Let G be a group of exponent p^e. Show that $|G| = 1 + \sum_{i=1}^{e} \varphi(p^i)\mathrm{c}_i(G)$. Next, if $k < e$, then $\mathrm{sol}_k(G) = 1 + \sum_{i=1}^{k} \varphi(p^i)\mathrm{c}_i(G)$, where $\mathrm{sol}_k(G) = |\{x \in G \mid o(x) \leq p^k\}|$. Hence, if we know $|G|$ and $\mathrm{c}_i(G)$ for all $i < e$, then we can find $\mathrm{c}_e(G)$. Use this to deduce Theorem 1.17(a) from Theorem 1.17(b).

Exercise 102. Let H be a nonnormal subgroup of index 4 in a 2-group G. Then G has a normal subgroup N of index 4 such that $N \cap H = H_G$.

Exercise 103. Prove that there does not exist a 2-group G of order $> 2^{n+1}$ such that $G/\mho_n(G) \in \{\mathrm{Q}_{2^{n+1}}, \mathrm{SD}_{2^{n+1}}, \mathrm{M}_{2^{n+1}}\}$. If $p > 2$, there does not exist a p-group G of order $> p^{n+1}$ such that $G/\mho_n(G) \cong \mathrm{M}_{p^{n+1}}$.

Exercise 104. Give the proof, independent of Theorem 1.17(b), of the following assertion: If a noncyclic 2-group G contains exactly one cyclic subgroup, say Z, of order $2^n > 2$, then G is of maximal class.

Exercise 105. Let G be a nonabelian 2-group. Prove that a Sylow 2-subgroup of $\mathrm{Aut}(G)$ is not isomorphic to Q_{2^n}. (*Hint.* $G/\mathrm{Z}(G) \cong \mathrm{Inn}(G) \leq \mathrm{Aut}(G)$ so $G/\mathrm{Z}(G) \cong \mathrm{Q}_{2^m}$, where $m \leq n$. Show that this is impossible.)

Exercise 106. Let G be a noncyclic 2-group of exponent > 2. If all members of the set Γ_1 but one have exponent 2, then $G \cong \mathrm{D}_8$. (*Hint.* We have $G = A\mathrm{Z}(G)$, where A is minimal nonabelian. Show that $A \cong \mathrm{D}_8$.)

Exercise 107. Let $G = U \times V$, where U and V are 2-groups of maximal class. Describe all minimal nonabelian subgroups of G.

Exercise 108 ([Zha]). Suppose that a group G is nonabelian with $\mathrm{Z}(G) > \{1\}$. Then there are two distinct noncentral G-classes of the same size.

Solution [Bal]. Assume that this is false. Let $z \in \mathrm{Z}(G)$ be of prime order, say, p, and $g \in G - \mathrm{Z}(G)$ be of prime power order, say, q^k. Since $\mathrm{C}_G(g) = \mathrm{C}_G(gz)$, g and gz are conjugate in G, by assumption, so $o(g) = o(gz)$. It follows that $q = p$ so G is a p-group; let $|G| = p^m$. Every noncentral conjugacy class of G has size $\leq p^{m-2}$ and $|\mathrm{Z}(G)| < p^{m-1}$. Therefore, $|G| = p^m \leq p + \cdots + p^{m-2} + |\mathrm{Z}(G)| < 1 + p + \cdots + p^{m-1} = \frac{p^m - 1}{p-1} < p^m$, a contradiction.

Exercise 109 ([Li1]). Let G be a nonabelian p-group. If, whenever $A < G$ is abelian, then $\mathrm{C}_G(A) \in \{A, \mathrm{N}_G(A)\}$. Prove that then $|G| = p^3$.

Solution (Berkovich). We use induction on $|G|$. Assume that we have $|G| > p^3$. If G is Dedekindian, then $p = 2$ and there is $\mathrm{Q}_8 \cong Q < G$. If $A < Q$ is of order 4, then $\mathrm{C}_G(A) \notin \{A, \mathrm{N}_G(A) = G\}$, a contradiction. Suppose that G is minimal nonabelian. First let G be metacyclic. Then it has a normal cyclic subgroup L such that G/L is cyclic and $\mathrm{C}_G(L)$ has index p in $G = \mathrm{N}_G(L) > \mathrm{C}_G(L)$. It follows that $\mathrm{C}_G(L) = L$ and $|G : L| = p$, i.e., $G \cong \mathrm{M}_{p^n}$, $n > 3$. Then G has a normal cyclic subgroup M of order p^2 such that G/M is cyclic. As above, we must have $|G : M| = p$, a contradiction since $n > 3$. Let $G = \langle a, b \mid a^{p^m} = b^{p^n} = c^p = 1, c = [a,b], [a,c] = [b,c] = 1\rangle$ be nonmetacyclic. Write $A = \langle a, c\rangle$, $B = \langle b, c\rangle$; then $A, B \lhd G$ are abelian. Since $A, B \nleq \mathrm{Z}(G)$, we get $\mathrm{C}_G(A) = A$ and $\mathrm{C}_G(B) = B$ so $m = n = 1$, a contradiction. Now let $H \in \Gamma_1$ be nonabelian and $A < H$ abelian. If $\mathrm{C}_G(A) = A$, then $\mathrm{C}_H(A) = A$. Now let $\mathrm{C}_G(A) = \mathrm{N}_G(A)$. Then $\mathrm{C}_H(A) = H \cap \mathrm{C}_G(A) = H \cap \mathrm{N}_G(A) = \mathrm{N}_H(A)$. Thus, the hypothesis is inherited by subgroups so, by induction, $|H| = p^3$; then $|G| = p^4$. Let $A \lhd G$ be of order p^2. Since $\mathrm{C}_G(A) > A$, we get $A = \mathrm{Z}(G)$ so $\mathrm{Z}(G)$ is the unique G-invariant subgroup of order p^2. Then G is minimal nonabelian, contrary to what has been proved already.

Exercise 110. Let A be abelian of type $(4, 2, 2)$ and let G be a 2-group such that $\mathrm{c}_n(G) = \mathrm{c}_n(A)$ for all n. Prove that either $G \cong A$ or $G = \mathrm{D}_8 * \mathrm{C}_4$.

Exercise 111. Suppose that the intersection of all subgroups (normal subgroups) of index $p^n > p$ in a p-group G equals $\{1\}$. Prove that $\mathrm{dl}(G) \leq n$ $(\mathrm{cl}(G) < n)$.

Exercise 112. Let G be a noncyclic p-group, $Z \lhd G$ and G/Z are cyclic. Show that Z is a maximal cyclic subgroup of G.

Solution. Assume that $Z < Z_1 < G$, where Z_1 is cyclic. Then $Z \leq \Phi(Z_1) \leq \Phi(G)$ so $G/\Phi(G)$ is cyclic. Then G is cyclic, contrary to the hypothesis.

Exercise 113 (Janko). If G is a noncyclic p-group all of whose proper noncyclic subgroups are generated by elements of order p, then either $\exp(G) = p$ or $G \in \{\mathrm{C}_{p^2} \times \mathrm{C}_p, \mathrm{M}_{p^3}, \mathrm{D}_{2^m}, \mathrm{Q}_8\}$.

Solution (Berkovich). We use induction on $|G|$. Suppose that $\exp(G) > p$, $|G| > p^3$ and G is not a 2-group of maximal class. All noncyclic abelian subgroups of G have exponent p. There is $R \lhd G$ of type (p, p) (Lemma 1.4); then $\exp(C_G(R)) = p$ so $|G : C_G(R)| = p$ and $\exp(G) = p^2$. Let $U < G$ be cyclic of order p^2. Then $\Omega_1(RU) = RU$ (otherwise, RU has an abelian subgroup of type (p^2, p)) so $p = 2$ and $RU \cong D_8$. If $RU \leq M \in \Gamma_1$, then M is dihedral, by induction, so $M = RU$ since $\exp(G) = 4$. Thus, $|G| = 2^4$. Since G is not of maximal class, $G = MC_G(M) = MZ(G)$ (Proposition 10.17). Then G has an abelian subgroup of type $(4, 2)$, a contradiction.

Exercise 114. Let G be a nonabelian p-group and let $N \leq G' \cap Z(G)$ be of order p. Suppose that every proper nonabelian subgroup of G is generated by two elements. If all proper subgroups of G/N are abelian, then all subgroups of index p^2 in G are abelian.

Exercise 115 (Mann). Let a p-group $G = CB$, where $C > \{1\}$ is either cyclic or generalized quaternion. If $|C| \geq |B|$, then $C_G > \{1\}$.

Solution. Assume that this is false. If $x \in G$, then $CC^x \neq G$ (Ore) so $C \cap C^x > \{1\}$ and $\{1\} < \Omega_1(C) \leq \bigcap_{x \in G} C^x = C_G$.

Exercise 116 (Mann). Suppose that a p-group $G = CB_1 \dots B_n$, where all factors are pairwise permutable. If C is cyclic or generalized quaternion and $|C| \geq |B_j|$, $j = 1, \dots, n$, then $C_G > \{1\}$.

Exercise 117. Let A be an extraspecial p-group of order p^{2m+1}. Find all p-groups G such that $s_n(G) = s_n(A)$ for all n.

Exercise 118. If G is a 2-group with metacyclic $\Phi(G)$, then $\Phi(G)/\mho_2(\Phi(G))$ is abelian. (*Hint.* If $\Phi(G)/\mho_2(\Phi(G))$ is nonabelian, it is $\cong \langle a, b \mid a^4 = b^4 = 1, a^b = a^3 \rangle$.)

Exercise 119. Study the 2-groups all of whose two-generator subgroups are metacyclic.

Exercise 120. Let a p-group $G = C_1 \dots C_n$, where $C_1, \dots, C_n$ are pairwise permutable cyclic subgroups, $p^e = |C_1| \geq \dots \geq |C_n|$. Estimate $\exp(G)$.

Exercise 121. Let A be a minimal nonabelian p-group. Study the p-groups G satisfying $c_n(G) = c_n(A)$ for all n.

Exercise 122. Classify the p-groups G containing a subgroup H of order p^2 such that there is exactly one maximal chain connecting H with G.

Exercise 123. Let $n > 2$. Describe (possibly infinite) groups G provided (a) $G = \langle a, b \mid a^{2^n} = b^2, a^b = a^{-1+2^{n-1}} \rangle$. (b) $G = \langle a, b \mid a^{2^n} = b^2, a^b = a^{-1} \rangle$. (c) $G = \langle a, b \mid a^{2^n} = b^2, a^b = a^{1+2^{n-1}} \rangle$. (d) $G = \langle a, b \mid a^{2^n} = b^2, a^b = a^{-1+2^{n-1}} \rangle$, (e) $G = \langle a, b \mid a^{p^n} = b^p, a^b = a^{1+p^{n-1}} \rangle$ $(p > 2)$. (f) $G = \langle a, b \mid a^4 = b^4, a^b = a^{-1} \rangle$.

Exercise 124. Let a group X act on an abelian group A via automorphisms. Then $[a_1a_2, x] = [a_1, x][a_2, x]$ for $a_1, a_2 \in A$ and $x \in X$.

Exercise 125. Let a group X act on an abelian p-group A via automorphisms. If $[A, X, X] \leq \Omega_n(A)$, then $[\mho_n(A), X, X] = \{1\}$.

Solution. By Exercise 81 from Introduction, for $a \in A$ and $x_1, x_2 \in X$, we have $[a^{p^n}, x_1, x_2] = [[a, x_1]^{p^n}, x_2] = [a, x_1, x_2]^{p^n} = 1$, and we are done since $[\mho_n(A), X, X] = \langle [a^{p^n}, x_1, x_2] \mid a \in A, x_1, x_2 \in X \rangle$.

Exercise 126. Prove that the following conditions for a nonabelian p-group G are equivalent: (a) All minimal nonabelian subgroups are normal in G. (b) All nonabelian subgroups are normal in G.

Exercise 127 (compare with [PR, Lemma 2.12]). Let G be a p-group, $\Omega_1(G) = G$, and let $\{1\} < H \leq \mathrm{Z}(G)$ be characteristic in G. If $\mathrm{Aut}(G)$ acts transitively on the set of subgroups of order p in G/H, then $\exp(G) = \exp(H)$.

Solution. Assume that $H < G$. Let $t \in G - H$ be of order p. Set $F = \langle t, H \rangle$. Since $\bigcup_{\alpha \in \mathrm{Aut}(G)} F^\alpha = G$ and $\exp(F) = \exp(H)$, we are done.

Exercise 128. Let a p-group $G = AC > A$, where C is cyclic. Prove that C is a maximal cyclic subgroup of G.

Exercise 129. Prove that a nonmetacyclic minimal nonabelian p-group G is a product of two cyclic subgroups if and only if $p = 2$ and $G = \langle a, b \mid a^{2^m} = b^2 = c^2 = 1, m > 1, c = [a, b], [a, c] = [b, c] = 1 \rangle$.

Exercise 130. Let $G = B(4, 2) = \langle x, y \rangle$ be the free two-generator group of exponent 4. According to Burnside and Tobin, $|B(4, 2)| = 2^{12}$. Compute (a) G/G', (b) $\{\mathrm{d}(M) \mid M \in \Gamma_1\}$ and (c) $G/\mathrm{K}_3(G)$. (See §60.)

Solution. Each two-generator group of exponent 4 is an epimorphic image of G. Since the greatest two-generator abelian group of exponent 4 is of type $(4, 4)$, it follows that G/G' is abelian of type $(4, 4)$. By Schreier's theorem on generators of subgroups (see Appendix 25), all maximal subgroups of G are 3-generator. Since all maximal subgroups of the group $T = \langle x, y \mid a^4 = b^4 = c^2 = 1, c = [a, b], [a, c] = [b, c] = 1 \rangle$ are of rank 3, we get $\{\mathrm{d}(M) \mid M \in \Gamma_1\} = \{3\}$. It follows from $\mathrm{d}(G) = 2$ that $G'/\mathrm{K}_3(G)$ is cyclic and equals $\langle [x, y]\mathrm{K}_3(G) \rangle$. We have $1 = (xy)^4 \equiv x^4y^4[y, x]^6 \pmod{\mathrm{K}_3(G)}$ so that $[y, x]^2 \in \mathrm{K}_3(G)$. It follows that $G/\mathrm{K}_3(G) \cong T$.

Exercise 131. Let $\mathfrak{T}$ be the set of all maximal cyclic subgroups of a p-group G and let $H \in \Gamma_1$. Prove that $|\theta(G) - \theta(H)| \geq |\Gamma_1| - 1$, where $\theta(X)$ is the set of members of the set $\mathfrak{T}$ contained in $X \leq G$. Classify the pairs $H < G$ such that $|\theta(G) - \theta(H)| < p^2$. Consider the case where $\theta(G) - \theta(H) = p^2$.

Exercise 132. Find all p-groups G with $\mathrm{s}_n(G) = \mathrm{s}_n(\mathrm{M}_{p^m} \times \mathrm{C}_p)$ for all n.

Exercise 133. If A, B, C are pairwise distinct maximal subgroups of a noncyclic 2-group G, then $\Phi(G) = \Phi(A)\Phi(B)\Phi(C)$. Consider the similar situation for $p > 2$.

Exercise 134 (Mann). Let G be a p-group and let $x \in G$ be such that $[x, G] \le Z(G)$. If $o(x[x, G]) = p^n$ in the quotient group $G/[x, G]$, then x^{1+p^n} and x are conjugate in G.

Solution. Set $N = [x, G]$; then the equality $N = \{[x, a] \mid a \in G\}$ follows from $[x, a][x, b] = [x, ab]$ for $a, b \in G$ so there exists in G an element c such that $x^{p^n} = [x, c]$, and we get $x^{p^n+1} = x^c$.

Exercise 135 (Hall 1926; unpublished). If all characteristic abelian subgroups of a nonabelian p-group G have orders $\le p$, then G is the central product of nonabelian groups of order p^3.

Exercise 136 ([Bla12, Theorem 1]). Let G be a p-group of class c, $s \in G$ and $|C_G(s)| = p^r$. Then $|G| \le p^{rc}$.

Solution. For $1 \le i \le c$, write $H = \langle s, K_i(G)\rangle$. Every conjugate of s in H is of the form $s^h = s[s, h]$ $(h \in H)$, so the number of conjugates of s in H is at most $|H'|$. Clearly, $H' \le K_{i+1}(G)$. Thus, $|K_{i+1}(G)| \ge |H'| \ge |H : C_H(s)| \ge (|H|/p^r) \ge (|K_i(G)|/p^r)$ so $|K_i(G) : K_{i+1}(G)| \le p^r$. It follows that

$$|G| = |G : K_2(G)| \dots |K_c(G) : K_{c+1}(G)| \le p^{rc}.$$

Exercise 137. Let G be a noncyclic p-group. If $C < G$ is cyclic and $C \not\le \Phi(G)$, then C is a maximal cyclic subgroup of G.

Exercise 138. Let G be a nonabelian p-group. Prove that if $|Z(Z_2(G))| = p$, then G is extraspecial.

Solution. We have $|Z(G)| = p$. If $N \trianglelefteq G$ is such that $|N \cap Z_2(G)| \le p$, then $|N| \le p$. We have $[G', Z_2(G)] = \{1\}$. Then $G' \cap Z_2(G) \le Z(Z_2(G))$ so $|G'| = p$. It follows that $G = Z_2(G)$ so $|Z(G)| = p = |G'|$, and G is extraspecial.

Exercise 139. If G is a nonabelian p-group such that $|Z(Z_n(G))| = p$ for some $n > 1$, then $|K_n(G)| \le p$.

Exercise 140. Let G be a p-group with nonabelian Frattini subgroup $\Phi(G)$. Prove that $\Phi(G)' < G'$.

Solution. It suffices to prove that if $|G'| = p$, then $\Phi(G)$ is abelian. For $x, y \in G$, we have $1 = [x, y]^p = [x, y^p]$ so $\mho_1(G) \le Z(G)$. Then $\Phi(G) = G'\mho_1(G) \le Z(G)$.

Exercise 141. Let G be a group of order p^{n+2} and class n. If $d(G) = 3$, then G contains a subgroup of maximal class and index p.

Solution. Obviously, $G' = \Phi(G)$. We also have $|G/\mathrm{K}_3(G)| = p^4$ and $|\eta(G)/\mathrm{K}_3(G)| = p^2$ (see Theorem 1.40). Let $H/\mathrm{K}_3(G)$ be minimal nonabelian subgroup of $G/\mathrm{K}_3(G)$; then $|H/\mathrm{K}_3(G)| = p^3$ so $H\eta(G) = G$. By Theorem 1.40, $\mathrm{cl}(H) = \mathrm{cl}(G) = n$, and we are done since $|H| = p^{n+1}$.

Exercise 142 (Mann). Suppose that a p-group G contains a subgroup H such that, whenever $H \cong H_1 < G$, then $H_1 = H$. Prove that if $\mathrm{Z}(H)$ is cyclic then $\mathrm{C}_G(H)$ is either cyclic or generalized quaternion.

Exercise 143. Study the 2-groups containing only one subgroup $\cong \mathrm{Q}_{2^n}$ (D_{2^n}, SD_{2^n}, $\mathcal{H}_2 = \langle a, b \mid a^4 = b^4 = 1,\ a^b = a^3\rangle$).

Theorem 1.44. *Let A be a maximal cyclic subgroup of order $> p$ of a noncyclic p-group G. Then there exists in G a maximal cyclic subgroup B of order $> p$ such that $|A \cap B| = p$, unless $p = 2$ and G is dihedral.*

Proof. Let $A < U \le G$ be such that $|U : A| = p$. If $c_2(U) > 1$, there exists a cyclic subgroup $B_1 < U$ of order p^2 such that $B_1 \not\le A$. Then $|A \cap B_1| = p$, by the product formula. If $B_1 \le B$, where $B < G$ is maximal cyclic, then $A \cap B = A \cap B_1$, and we are done in this case.

Now let $c_2(U) = 1$ for any choice of U. Then U is dihedral (Theorems 1.10(b) and 1.17(b)) so $p = 2$. If $U = G$, there is nothing to prove. Now let $U < V \le G$ be such that $|V : U| = 2$. If $|A| = 4$, then it follows from $\mathrm{C}_G(A) = A$ that G is of maximal class (Proposition 1.8), and the theorem is true (Theorem 1.2). Now let $|A| > 4$. Write $A_1 = \Omega_2(A)$; then $A_1 \lhd V$ since A_1 is characteristic in $U \lhd V$. We have $|V : \mathrm{C}_V(A_1)| = 2$ and $A < \mathrm{C}_V(A_1)$. If $A < M \le \mathrm{C}_V(A_1)$ and $|M : A| = 2$, then M is not dihedral since $A_1 \le \mathrm{Z}(M)$, a final contradiction. □

Theorem 1.45 (Janko). *Let G be a nonabelian p-group all of whose cyclic subgroups, not contained in $\Phi(G)$, are normal in G. Then $p = 2$ and G is Dedekindian.*

Proof. We use induction on $|G|$. Let $Z \le G' \cap \mathrm{Z}(G)$ be of order p. Set $\bar{G} = G/Z$ and using the "bar convention", we get $\Phi(\bar{G}) = \Phi(G)/Z$. Let a cyclic $\bar{X} < \bar{G}$ be such that $\bar{X} \not\le \Phi(\bar{G})$. Then $X \not\le \Phi(G)$ and let $x \in X$ be such that $\langle x\rangle$ covers X/Z. Since $x \notin \Phi(G)$, $\langle x\rangle \lhd G$ so that $\langle x, Z\rangle \lhd G$ and hence $\bar{X} \lhd \bar{G}$.

Suppose that $\bar{G}$ is abelian. Then $G' = Z \cong \mathrm{C}_p$ and so $G' \le \mathrm{Z}(G)$ and $\mathrm{cl}(G) = 2$. For each $x, y \in G$, $[x^p, y] = [x, y]^p = 1$ and so $\mho_1(G) \le \mathrm{Z}(G)$. It follows that $\Phi(G) = G'\mho_1(G) \le \mathrm{Z}(G)$ and so each cyclic subgroup which is contained in $\Phi(G)$ is also normal in G. Hence each cyclic subgroup of G is G-invariant so G is Dedekindian.

Now let $\bar{G}$ be nonabelian. By induction, $p = 2$ and G/Z is Dedekindian. Hence $G = Q_1E_1$, where Q_1 and E_1 are normal in G, $Q_1 \cap E_1 = Z \le \Phi(G)$, $Q_1/Z \cong \mathrm{Q}_8$ and E_1/Z is elementary abelian. We have $\Phi(G) \le Q_1$ and $\Phi(G)/Z \cong \mathrm{C}_2$, where $\Phi(G)/Z = \mathrm{Z}(Q_1/Z) = (Q_1/Z)'$. Thus, Q_1' covers $\Phi(G)/Z$. Suppose that $Q_1' =$

$\Phi(G)$. Then $|Q_1 : Q_1'| = 4$ and so by Taussky's theorem, Q_1 is of maximal class (and order 2^4). But in that case $Q_1/Z \cong \mathrm{D}_8$, a contradiction. Hence $Q_1' < \Phi(G)$ which implies that $\Phi(G) = Q_1' \times Z \cong \mathrm{E}_4$. It follows that $\Phi(G) \leq \mathrm{Z}(G)$ and so each cyclic subgroup in $\Phi(G)$ is also normal in G. It follows that G is Dedekindian, a final contradiction since $|\Phi(G)| = 4$. □

§2

The class number, character degrees

1°. If $G > \{1\}$ is a p-group and $\chi \in \mathrm{Irr}(G)$, where $\mathrm{Irr}(G)$ is the set of irreducible complex characters of G, then $\chi(1)^2 \leq |G : \mathrm{Z}(G)|$ so $\chi(1)^2$ is a divisor (not necessarily proper) of $|G : \mathrm{Z}(G)|$. Let $\mathrm{k}(G)$ be the class number of a group G; then $\mathrm{k}(G) = |\mathrm{Irr}(G)|$. For $N \trianglelefteq G$, write $\mathrm{Irr}(G \mid N) = \mathrm{Irr}(G) - \mathrm{Irr}(G/N)$; here we consider $\mathrm{Irr}(G/N)$ as the subset $\{\chi \in \mathrm{Irr}(G) \mid N \leq \ker(\chi)\}$ of $\mathrm{Irr}(G)$. Let $\mathrm{Lin}(G) = \{\chi \in \mathrm{Irr}(G) \mid \chi(1) = 1\}$ be the set of linear characters of G; then $\mathrm{Irr}_1(G) = \mathrm{Irr}(G) - \mathrm{Lin}(G)$ is the set of nonlinear irreducible characters of G. Obviously, $\mathrm{Irr}_1(G) = \mathrm{Irr}(G \mid G')$.

Exercise 1. Let G be a group of order p^m and let $R \leq \mathrm{Z}(G)$ be of order p. Then $\mathrm{k}(G) \geq p - 1 + \mathrm{k}(G/R)$, i.e., the set $\mathrm{Irr}(G \mid R)$ contains at least $p - 1$ members.

(a) If $\mathrm{k}(G) = p - 1 + \mathrm{k}(G/R)$, then $\mathrm{Z}(G) = R$ and $\chi(1)^2 = p^{m-1}$ for all $\chi \in \mathrm{Irr}(G \mid R)$, i.e., m is odd.

(b) If m is even, then $\mathrm{k}(G) \geq p^2 - p + \mathrm{k}(G/R)$.

Solution. (a) Let $\mathrm{Irr}(G \mid R) = \{\chi_1, \dots, \chi_s\}$; then $\chi_i(1)^2 \leq |G : \mathrm{Z}(G)| \leq p^{m-1}$ ($i = 1, \dots, s$). Therefore, $(p-1)p^{m-1} = |G| - |G : R| = \sum_{i=1}^s \chi_i(1)^2 \leq sp^{m-1}$ so $s \geq p - 1$. If $s = p - 1$, then $(p-1)p^{m-1} = \sum_{\chi \in \mathrm{Irr}(G|R)} \chi(1)^2 \leq (p-1)p^{m-1}$, and we get $\chi(1)^2 = p^{m-1}$ for all $\chi \in \mathrm{Irr}(G \mid R)$ and so $|\mathrm{Z}(G)| = p$.

(b) Now suppose that m is even. Then $\chi(1)^2 \leq |G : \mathrm{Z}(G)| \leq p^{m-1}$ so $\chi(1)^2 \leq p^{m-2}$ since $m - 1$ is odd. Now (b) follows from $(p-1)p^{m-1} = (p^2 - p)p^{m-2} = |G| - |G : R| = \sum_{i=1}^s \chi_i(1)^2 \leq sp^{m-2}$.

Let $\mathrm{CL}(G) = \{K_1, \dots, K_r\}$ be the set of all G-classes, i.e., $r = \mathrm{k}(G)$, $x_i \in K_i$. Then the number of commuting ordered pairs $(a, b) \in G \times G$ equals $\sum_{i=1}^r |\mathrm{C}_G(x_i)| \cdot |K_i| = |G| \cdot \mathrm{k}(G)$. Therefore, the number of noncommuting ordered pairs $(a, b) \in G \times G$ equals $|G|^2 - |G| \cdot \mathrm{k}(G)$. Let $\varphi_2(G)$ denote the number of ordered noncommuting pairs $(a, b) \in G \times G$ such that $G = \langle a, b \rangle$ (so that, $\varphi_2(G) = 0$ if G is either abelian or not generated by two elements). Then the following equality holds (Mann):

$$(1) \qquad \sum_{H \leq G} \varphi_2(H) = |G|^2 - |G| \cdot \mathrm{k}(G).$$

Exercise 2. (a) If G is a two-generator p-group, then $\varphi_2(G) = (p^2 - 1)(p^2 - p) \cdot |\Phi(G)|^2$.

(b) If $G = \mathrm{ES}(m, p)$ is an extraspecial group of order p^{2m+1}, then the number $t = t_{m,p}$ of nonabelian subgroups of order p^3 in G equals $\frac{p^{2m}-1}{p^2-1} \cdot p^{2m-2}$.

Solution. (a) Obviously, $\varphi_2(G)$ equals the number of (ordered) minimal bases of G. By assumption, $G/\Phi(G)$ is abelian of type (p, p). If $x \in G - \Phi(G)$, $y \in G - \langle x, \Phi(G)\rangle$, then $G = \langle x, y\rangle$. The element x can be chosen in $|G| - |\Phi(G)| = |\Phi(G)|(p^2 - 1)$ ways, and y, after the choice of x, in $|G| - p|\Phi(G)| = |\Phi(G)|(p^2 - p)$ ways. Now the result follows by the combinatorial product rule.

(b) Every nonabelian two-generator subgroup of G has order p^3. Using (1) and (a) and, taking into account that $\mathrm{k}(G) = p^{2m} + p - 1$, we get

$$p^3(p-1)(p^2-1)t = p^{2m+1}(p^{2m+1} - p^{2m} - p + 1) = p^{2m+1}(p^{2m} - 1)(p - 1),$$

and the result follows.

Theorem 2.1 (P. Hall). *Let* $|G| = p^{2n+e}$, $e \in \{0, 1\}$. *Then there exists a nonnegative integer* $t = \mathrm{t}(G)$ *such that*

$$\mathrm{k}(G) = p^e + (p^2 - 1)[n + (p-1)t] = p^e + (p^2 - 1)n + (p^2 - 1)(p - 1)t.$$

Next, if $N \trianglelefteq G$, *then* $\mathrm{t}(G/N) \le \mathrm{t}(G)$.

Proof (Mann [Man7]). By Exercise 2(a) and (1), we get $|G|^2 \equiv |G|{\cdot}\mathrm{k}(G) \mod (p^2-1)(p^2-p)$ so $\mathrm{k}(G) \equiv |G| \pmod{(p^2-1)(p-1)}$. Next,

$$\begin{aligned} p^{2n+e} &= p^e + (p^{2n} - 1)p^e \\ &= p^e + (p^2-1)(p^{2n-2} + p^{2n-4} + \cdots + p^2 + 1)(p^e - 1 + 1) \\ &= p^e + (p^2-1)(p^{2n-2} + p^{2n-4} + \cdots + p^2 + 1)(p^e - 1) \\ &\quad + (p^2-1)(p^{2n-2} + p^{2n-4} + \cdots + p^2 + 1) \\ &\equiv p^e + (p^2-1)(p^{2n-2} + \cdots + p^2 + 1) \\ &\equiv p^e + (p^2-1)n \pmod{(p^2-1)(p-1)} \end{aligned}$$

since $p^{2n-2} + \cdots + p^2 + 1 \equiv n \pmod{p-1}$. Therefore,

$$\mathrm{k}(G) = p^e + (p^2-1)n + (p^2-1)(p-1)t = p^e + (p^2-1)[n + (p-1)t]$$

for some $t = \mathrm{t}(G) \in \mathbb{Z}$. It remains to prove that $\mathrm{t}(G) \ge 0$. To this end, we will use induction on $|G|$. One may assume that $n > 1$ (if $n \le 1$, then the theorem is checked easily for such small G). Let $N \le \mathrm{Z}(G)$ be of order p. Then $\mathrm{k}(G) > \mathrm{k}(G/N)$ since $\bigcap_{\chi\in\mathrm{Irr})G)} \ker \chi(1) = \ker(\rho_G) = \{1\}$ (ρ_G is the regular character of G).

Let $e = 0$; then $|G| = p^{2n}$ and $|G/N| = p^{2(n-1)+1}$ so, by what has just been proved, $\mathrm{k}(G) = 1 + (p^2-1)[n + (p-1)\mathrm{t}(G)]$ and $\mathrm{k}(G/N) = p + (p^2-1)[n - 1 +$

$(p-1)\mathrm{t}(G/N)]$ so $0 < \mathrm{k}(G)-\mathrm{k}(G/N) = 1-p+(p^2-1)[1+(\mathrm{t}(G)-\mathrm{t}(G/N))(p-1)]$. Therefore, since $\mathrm{t}(G) - \mathrm{t}(G/N) \in \mathbb{Z}$ and $1 + [\mathrm{t}(G) - \mathrm{t}(G/N)](p-1) > \frac{p-1}{p^2-1} > 0$, one obtains $\mathrm{t}(G) \geq \mathrm{t}(G/N)$ and so $\mathrm{t}(G) \geq \mathrm{t}(G/N) \geq 0$, by induction.

Let $e = 1$. In that case, $|G| = p^{2n+1}$, $|G/N| = p^{2n}$. Then, by what has been proved already, $\mathrm{k}(G) = p+(p^2-1)[n+(p-1)\mathrm{t}(G)]$ and $\mathrm{k}(G/N) = 1+(p^2-1)[n+(p-1)\mathrm{t}(G/N)]$, so $0 < \mathrm{k}(G)-\mathrm{k}(G/N) = p-1+(p^2-1)(p-1)[\mathrm{t}(G)-\mathrm{t}(G/N)]$. Therefore, we must have $\mathrm{t}(G) - \mathrm{t}(G/N) \geq 0$ so $\mathrm{t}(G) \geq \mathrm{t}(G/N) \geq 0$. □

We see that, if $N \lhd G$, then $\mathrm{t}(G) \geq \mathrm{t}(G/N)$. This allows us to control the structure of epimorphic images of G. For example, if G is not cyclic and $\mathrm{t}(G) = 0$, then $\mathrm{t}(G/G') = 0$, and so $|G : G'| = p^2$. Indeed, if G is a noncyclic abelian p-group of order p^{2n+e} such that $\mathrm{t}(G) = 0$, then by Theorem 2.1, $p^{2n+e} = |G| = \mathrm{k}(G) = p^e + (p^2-1)n$, or $p^e(p^{2n-2} + \dots + p^2 + 1) = n$ so $p^e = 1 = n$; then $e = 0$ and $|G| = p^2$, as claimed. Therefore, if G is nonabelian with $\mathrm{t}(G) = 0$ and $p = 2$, then G is of maximal class, by Taussky's theorem (moreover, then $|G| = 2^3$ or 2^4).

Remark 1. Let $\chi \in \mathrm{Irr}(G)$ be such that $\chi(1)^2 = |G : \mathrm{Z}(G)|$, where G is a finite group. Then $\chi_{\mathrm{Z}(G)} = \chi(1)\lambda$, where $\lambda \in \mathrm{Lin}(\mathrm{Z}(G))$, so $\chi(1)\chi$ is a constituent of λ^G, by reciprocity. Since $\lambda^G(1) = |G : \mathrm{Z}(G)| = \chi(1)^2 = (\chi(1)\chi)(1)$, we obtain $\lambda^G = \chi(1)\chi$. Clearly, χ vanishes outside $\mathrm{Z}(G)$.

Exercise 3. If G is a nonabelian group of order 2^{2n+e}, $e \in \{0, 1\}$ and $\mathrm{k}(G) = 2^e + 3n$, then, as we have noticed, G is of maximal class and order at most 2^4. Show that if $\mathrm{t}(G) = 0$, $|G| > p^2$ and $p > 2$, i.e., $\mathrm{k}(G) = p^e + (p^2-1)n$, then G is of maximal class.

Solution. One may assume that $n > 1$. We use induction on $|G|$. By Theorem 2.1, $\mathrm{t}(G/N) = 0$ for every $N \lhd G$ so, for $N = G'$, we have $|G : G'| = p^2$. Let $N \leq G' \cap \mathrm{Z}(G)$ be of order p. By induction, G/N is of maximal class since $n > 1$. It suffices to show that $N = \mathrm{Z}(G)$. Let $e = 1$. Then

$$|\mathrm{Irr}(G \mid N)| = \mathrm{k}(G) - \mathrm{k}(G/N) = p + (p^2-1)n - [1 + (p^2-1)n] = p - 1,$$

by Theorem 2.1, and the result follows since, by Exercise 1(a), $\mathrm{Z}(G) = N$ and $|N| = p$. Now let $e = 0$. Then $|\mathrm{Irr}(G \mid N)| = \mathrm{k}(G)-\mathrm{k}(G/N) = 1+(p^2-1)n-[p+(p^2-1)(n-1)] = p^2 - p$, by Theorem 2.1. Assume that $N < \mathrm{Z}(G)$. Then $|\mathrm{Z}(G)| = p^2$, and we get $|\mathrm{Irr}(G \mid \mathrm{Z}(G))| = p^2 - p + p - 1 = p^2 - 1$. If $|G| = p^4$, then G is minimal nonabelian and $\mathrm{k}(G) = p^3 + p^2 - p$ so $|\mathrm{Irr}(G \mid \mathrm{Z}(G))| = p^3 - p > p^2 - 1$, a contradiction. Thus, $G/\mathrm{Z}(G)$ is nonabelian. As in the solution of Exercise 1, all characters in $\mathrm{Irr}(G \mid \mathrm{Z}(G))$ have degree p^{n-1}, i.e., for every such character χ, we have $\chi(1)^2 = |G : \mathrm{Z}(G)|$. It follows that all these characters vanish on $G - \mathrm{Z}(G)$ (see Remark 1). By the Second Orthogonality Relation, if $x \in G - \mathrm{Z}(G)$, then $|\mathrm{C}_G(x)| = \sum_{\chi \in \mathrm{Irr}(G/\mathrm{Z}(G))} |\chi(x)|^2 = |\mathrm{C}_{G/\mathrm{Z}(G)}(x\mathrm{Z}(G))|$. Since $G/\mathrm{Z}(G)$ is of maximal class. then, by §9, for some $x \in G - \mathrm{Z}(G)$, the order of the centralizer of $x\mathrm{Z}(G)$ in $G/\mathrm{Z}(G)$

equals p^2. Then $|C_G(x)| = p^2$, a contradiction since $|C_G(x)| \geq |\langle x, Z(G)\rangle| > p^2$. Thus, G is of maximal class (Proposition 1.8). (In fact, $|G| \leq p^{p+2}$ [Pol].)

Proposition 2.2. *Let $H \lhd G$, $|G : H| = p$ and let s be the number of G-invariant characters in* $\mathrm{Irr}(H)$. *Then (a)* $p\mathrm{k}(G) = \mathrm{k}(H) + (p^2 - 1)s$. *(b) If, in addition, G is a p-group, then $p - 1$ divide $s - 1$.*

Proof. (a) If $\phi \in \mathrm{Irr}(H)$ is G-invariant, then $\phi^G = \chi^1 + \cdots + \chi^p$, where $\mathrm{Irr}(\phi^G) = \{\chi^1, \ldots, \chi^p\}$, $\chi^i(1) = \phi(1)$ (Clifford). Let $\mu \in \mathrm{Irr}(H)$ be not G-invariant. Then $\mu^G = \chi \in \mathrm{Irr}(G)$ and $\chi_H = \mu_1 + \cdots + \mu_p$ is the Clifford decomposition, where $\mu = \mu_1$. Since $\bigcup_{\phi \in \mathrm{Irr}(H)} \mathrm{Irr}(\phi^G) = \mathrm{Irr}(G)$, $\mathrm{Irr}((\sum_{\chi \in \mathrm{Irr}(G)} \chi)_H) = \mathrm{Irr}(H)$, we get $\mathrm{k}(G) = |\mathrm{Irr}(G)| = sp + \frac{1}{p}(|\mathrm{Irr}(H)| - s) = sp + \frac{1}{p}(\mathrm{k}(H) - s)$, and (a) follows.

(b) follows easily from Theorem 2.1. □

Exercise 4 (Burnside). Let G and H be as in Proposition 2.2. If $\mathrm{CL}(H)$ has exactly s classes that are G-classes, then the formula in Proposition 2.2(a) holds. (*Hint.* See Lemma 2.13.)

Exercise 5. Let G be a group and let $H \lhd G$ be such that G/H is cyclic of order p^n. Prove analogs of Proposition 2.2 and Exercise 4.

Proposition 2.3 ([Man7]). *The number $s = s(G)$ of nonabelian subgroups of order p^3 in a p-group G, $|G| = p^m$, $m \geq 5$, is divisible by p^2.*

Proof. By Exercise 2(a) and (1), we have

$$sp^3(p-1)(p^2-1) + \sum_{H \leq G, |H| > p^3} \varphi_2(H) = p^m \cdot [p^m - \mathrm{k}(G)]. \tag{2}$$

If $H \leq G$ is a 2-generator subgroup of order at least p^4, then $|\Phi(H)| \geq p^2$, and so p^5 divides $\varphi_2(H)$, by Exercise 2(a). Now the result follows from (2). □

Exercise 6 ([Man7]). Let G be a p-group of order p^m, $k \in \mathbb{N}$, $k > 2$, $m \geq 2k-1$. Let n_k be the number of nonabelian 2-generator subgroups of G of order p^k. Then p^{2k-4} divides $n_3 + p^2 n_4 + \cdots + p^{2k-6} n_k$.

Exercise 7 ([Kno]). If all noncentral classes of a nonabelian p-group G have size p, then $|G'| = p$.

Solution. Set $|G| = p^m$, $|G : G'| = p^a$, $|Z(G)| = p^z$. Then $\mathrm{k}(G) = p^z + \frac{p^m - p^z}{p} = p^{m-1} + p^{z-1}(p-1)$. On the other hand, $\mathrm{k}(G) = |\mathrm{Irr}(G)| \leq p^a + \frac{p^m - p^a}{p^2} = p^{m-2} + p^{a-2}(p^2-1)$. Hence $p^{m-2}(p-1) + p^{z-1}(p-1) \leq p^{a-2}(p^2-1)$ so $p^m < p^a(p+1)$. Then $p^{m-a} < p+1$ so $a = m-1$ and $|G'| = p$.

For a group G, set $\mathrm{cd}(G) = \{\chi(1) \mid \chi \in \mathrm{Irr}(G)\} = \{d_0 = 1, d_1, \ldots, d_s\}$, where $1 = d_0 < d_1 < \cdots < d_s$. If $\mathrm{Irr}(G)$ has exactly a_i characters of degree d_i, $i = 0, 1, \ldots, s$, we write $\delta(G) = \{a_0 \cdot d_0, \ldots, a_s \cdot d_s\}$ and call $\delta(G)$ the *degree vector* of G. We have $a_0 + \cdots + a_s = \mathrm{k}(G)$, $a_0 = |G : G'|$ and $|G| = a_0 + a_1 d_1^2 + \cdots + a_s d_s^2$.

Proposition 2.4 ([BZ, Lemma 3.35]). *Let G be a nonabelian p-group, $\delta(G) = \{a_0 \cdot 1, a_1 \cdot d_1, \dots, a_s \cdot d_s\}$, $1 < d_1 < \dots < d_s$. If $a_1 < p^2 - p$, then G is extraspecial.*

Proof. Let $|G| = p^m$, $a_0 = |G : G'| = p^k$, $d_i = p^{c_i}$, $i = 1, \dots, s$. Then $p^m = p^k + a_1 p^{2c_1} + \dots + a_s p^{2c_s}$ so $2c_1 \le k$ and $p^{m-2c_1} = p^{k-2c_1} + a_1 + a_2 p^{2c_2-2c_1} + \dots + a_s p^{2c_s-2c_1}$. Since $p^2 \nmid a_1$, we get $k - 2c_1 < 2$. Since $p^{k-2c_1} + a_1 \le p + a_1 < p^2$, we get $m - 2c_1 = 1$, $s = 1$ so $k = 2c_1$ and $a_1 = p - 1$. If $\chi \in \mathrm{Irr}_1(G)$, then $\chi(1)^2 = p^{2c_1} = p^{m-1}$ so $p^{m-1} = \chi(1)^2 \le |G : \mathrm{Z}(G)| \le p^{m-1}$ and $|\mathrm{Z}(G)| = p = |G'|$, and hence $\mathrm{Z}(G) = G'$. It follows that G is extraspecial. □

Proposition 2.5. *If a p-group G has a faithful irreducible character χ of degree p, then G has an abelian maximal subgroup A such that* $\mathrm{d}(A) \le p$.

Proof. There exist $A < G$ and $\lambda \in \mathrm{Lin}(A)$ such that $\chi = \lambda^G$ since G is an M-group (Appendix 2) so $|G : A| = p$ and $A \in \Gamma_1$. Since $A' \le \ker(\chi) = \{1\}$, A is abelian. Since λ is linear and $\chi(1)$ is faithful of degree p, it follows from $\{1\} = \ker(\chi) = \ker(\lambda)_G$ that $\mathrm{d}(A) \le p$ since $A/\ker(\lambda)$ is cyclic. □

Since the number of commuting pairs of elements of G is $|G|\mathrm{k}(G)$, the probability that two elements of G commute is $\mathrm{mc}(G) = \frac{|G|\mathrm{k}(G)}{|G|^2} = \frac{\mathrm{k}(G)}{|G|}$, the *measure of commutativity* of G. It follows that G is abelian if and only if $\mathrm{mc}(G) = 1$.

Lemma 2.6. *Let $H < G$. Then (a)* $\mathrm{mc}(H) \ge \mathrm{mc}(G)$ *and (b) if* $\mathrm{mc}(H) = \mathrm{mc}(G)$, *then $H' = G'$.*

Proof. (a) If $\phi \in \mathrm{Irr}(H)$, then $|\mathrm{Irr}(\phi^G)| \le |G : H|$ since $\phi^G(1) = |G : H|\phi(1)$ and, by reciprocity every irreducible constituent of ϕ^G is of degree at least $\phi(1)$. On the other hand, if $\chi \in \mathrm{Irr}(G)$ and $\phi \in \mathrm{Irr}(\chi_H)$, then $\chi \in \mathrm{Irr}(\phi^G)$. Therefore, $\mathrm{k}(G) \le \sum_{\phi \in \mathrm{Irr}(H)} |\mathrm{Irr}(\phi^G)| \le \mathrm{k}(H)|G : H|$, so $\mathrm{mc}(G) = \frac{\mathrm{k}(G)}{|G|} \le \frac{\mathrm{k}(H)}{|H|} = \mathrm{mc}(H)$.

(b) Now let $\mathrm{mc}(G) = \mathrm{mc}(H)$; then $\mathrm{k}(G) = |G : H|\mathrm{k}(H)$ so $|\mathrm{Irr}(\phi^G)| = |G : H|$ for all $\phi \in \mathrm{Irr}(H)$, hence all irreducible constituents of ϕ^G have the same degree $\phi(1)$. Taking $\phi = 1_H$, we get $G' \le H$ so G/H is abelian. Next, if $\mu, \nu \in \mathrm{Irr}(H)$ are distinct, then $\langle \mu^G, \nu^G \rangle = 0$. Therefore, if $\mathrm{Lin}(H) = \{\phi_1, \dots, \phi_n\}$, then $\mathrm{Irr}((\phi_1 + \dots + \phi_n)^G) = \mathrm{Lin}(G)$, and so $|G : G'| = |G : H|n = |G : H||H : H'| = |G : H'|$ hence $G' = H'$. □

Lemma 2.7. *If $|G : \mathrm{Z}(G)| = p^2$, then $mc(G) = \frac{p^2+p-1}{p^3}$. In particular, if G is a minimal nonabelian, then* $\mathrm{mc}(G) = \frac{p^2+p-1}{p^3}$.

Theorem 2.8. *Let G be a nonabelian p-group. Then* $\mathrm{mc}(G) \le \frac{p^2+p-1}{p^3}$ *with equality if and only if $|G : \mathrm{Z}(G)| = p^2$.*

Proof. Let $H \le G$ be minimal nonabelian; then by Lemmas 2.6(a) and 2.7, $\mathrm{mc}(G) \le \mathrm{mc}(H) = \frac{p^2+p-1}{p^3}$. Now let $\mathrm{mc}(G) = \frac{p^2+p-1}{p^3}$. By Lemma 2.7, $\mathrm{mc}(G) = \mathrm{mc}(H)$, and so $|G'| = |H'| = p$, by Exercise 1.8a and Lemma 2.6, so sizes of G-classes are 1 or p. Therefore, denoting $|G| = p^m$ and $|\mathrm{Z}(G)| = p^z$, we get

$$p^{m-1} + p^{m-2} - p^{m-3} = |G|\mathrm{mc}(G) = \mathrm{k}(G) = p^z + \frac{p^m - p^z}{p}$$
$$= p^{m-1} + p^z - p^{z-1},$$

and so $z = m - 2$. □

Let $\mathrm{T}(G) = \sum_{\chi\in\mathrm{Irr}(G)} \chi(1)$, $\mathrm{f}(G) = \mathrm{T}(G)/|G|$ (the *normalized degree* of G).

Exercise 8. Let $H < G$. Then (a) $\mathrm{f}(H) \geq \mathrm{f}(G)$, (b) if $\mathrm{f}(H) = \mathrm{f}(G)$, then $G' = H'$, (c) $\mathrm{f}(G)^2 \leq \mathrm{mc}(G)$ with equality if and only if G is abelian, (d) if G is a minimal nonabelian p-group, then $\mathrm{f}(G) = \frac{2p-1}{p^2}$.

Solution. (a) By reciprocity,

$$|G|\mathrm{f}(G) = \mathrm{T}(G) = \sum_{\chi\in\mathrm{Irr}(G)} \chi(1) \leq \sum_{\phi\in\mathrm{Irr}(H)} \phi^G(1)$$
$$= |G : H| \sum_{\phi\in\mathrm{Irr}(H)} \phi(1) = |G : H|\mathrm{T}(H) = |G|\mathrm{f}(H).$$

(b) Repeat, word for word, the proof of Lemma 2.6(b). (c) (Mann) Let $\mathrm{Irr}(G) = \{\chi_1, \dots, \chi_r\}$, $r = \mathrm{k}(G)$, $a = (\chi_1(1), \dots, \chi_r(1)) \in \mathbb{R}^r$ and $b = (1, \dots, 1) \in \mathbb{R}^r$. We have $|a \cdot b| \leq ||a|| \cdot ||b||$ with equality if and only if a and b are proportional. (d) We get $\mathrm{cd}(G) = \{1, p\}$ and $|G'| = p$ (Exercise 1.8a). If $|G| = p^m$, then $p^m \cdot \mathrm{f}(G) = \mathrm{T}(G) = |G : G'| + \frac{|G|-|G:G'|}{p^2} \cdot p = 2p^{m-1} - p^{m-2}$.

Exercise 9. Let G be a nonabelian p-group.

(a) Prove that $\mathrm{f}(G) \leq \frac{2p-1}{p^2}$.

(b) If $\mathrm{f}(G) = \frac{2p-1}{p^2}$, then $|G : \mathrm{Z}(G)| = p^2$.

Hint. (a) Let H be a minimal nonabelian subgroup of G. Use Exercise 8(a,d).

Solution. (b) Let $H < G$ be minimal nonabelian. By Exercise 8(d), we have $\mathrm{f}(H) = \frac{2p-1}{p^2} = \mathrm{f}(G)$, and so $G' = H'$ is of order p, by Exercises 8(b) and 8(a). Let $|G : \mathrm{Z}(G)| = p^{2s}$, $|G| = p^m$; then $\mathrm{cd}(G) = \{1, p^s\}$, and so $|G|\mathrm{f}(G) = 2p^{m-1} - p^{m-2} = |G : G'| + p^s \cdot \frac{|G|-|G:G'|}{p^{2s}} = p^{m-1} + p^{m-1-s}(p-1)$ and we conclude that $s = 1$.

Exercise 10. Study the pairs $N \lhd G$ of p-groups such that $N \not\leq G'$ and (a) $|N| = p$ and $\mathrm{k}(G) - \mathrm{k}(G/N) = p - 1$, (b) $|N| = p^2$ and $\mathrm{k}(G) - \mathrm{k}(G/N) = p^2 - 1$.

Exercise 11. If $\mathrm{mc}(G) > \frac{5}{8}$ ($\mathrm{f}(G) > \frac{3}{4}$), then G is abelian. Classify the groups G such that (a) $\mathrm{f}(G) = \frac{3}{4}$, (b) $\mathrm{mc}(G) = \frac{5}{8}$.

Exercise 12. If G is a p-group with cyclic derived subgroup G', then all elements of G' are commutators.

Solution. Let $G' = \langle a \rangle$ and $a = [x, y]$. If $|G'| = p$, then $a^n = [x, y]^n = [x^n, y]$ for all $n > 0$, and we are done. One may assume that $G = \langle x, y \rangle$. Let $|G'| > p$; then $G/\langle a^p \rangle$ is minimal nonabelian (Lemma 65.2(a)). By Exercise 1.69, there is $H \in \Gamma_1$ such that $|G' : H'| = p$. All elements of H' are commutators, by induction. By [BZ, Theorem 3.27], the set $G' - H'$ of all generators of G' are also commutators. and we are done.

Exercise 13. If G is a p-group, then $p - 1$ divides $\mathrm{n}(G)$, where $\mathrm{n}(G) = |\mathrm{Irr}_1(G)|$.

Exercise 14. Give a solution of the previous exercise using Theorem 2.1.

Exercise 15. Let $\exp(P) = p$. Prove that there exists a group G containing P and such that $P^\#$ is contained in a G-class.

Solution. By Cayley's theorem, one can consider P as a regular subgroup of the symmetric group $\mathrm{S}_{|P|}$ of degree $|P|$. Since all elements of $P^\#$ have the same cycle type, they are conjugate in $\mathrm{S}_{|P|}$.

Exercise 16. Given a group H, prove that there exists a group G containing H and such that any two elements of H of the same order are conjugate in G.

Let G be a nonabelian p-group and let $a_n(G)$ be the number of characters of degree p^n in $\mathrm{Irr}_1(G)$. Then, according to Mann, $p - 1$ divides $a_n(G)$ [BZ, Lemma 31.15].

The second proof of Theorem 2.1. As in Theorem 2.1, let $|G| = g = p^{2n+e}$ with $e \in \{0, 1\}$, $\mathrm{k}(G) = h$. Let a_k be the number of irreducible characters of degree p^k of G and b_k the number of conjugacy classes of size p^k, $k = 0, 1, 2, \dots$. Then $g = \sum_k a_k p^{2k} = \sum_k b_k p^k$, $\quad h = \sum_k a_k = \sum_k b_k$.
We have $h - p^e = p^e(p^{2n} - 1) + h - g$ so

$$\begin{aligned}\frac{h - p^e}{p^2 - 1} &= p^e \cdot \frac{p^{2n} - 1}{p^2 - 1} + \sum_k a_k \cdot \frac{1 - p^{2k}}{p^2 - 1} \\ &= p^e(1 + p^2 + \dots + p^{2n-2}) - \sum_k a_k(1 + p^2 + \dots + p^{2k-2}) \\ &\equiv np^e - \sum_k k a_k = n(p^e - 1) + n - \sum_k k a_k \equiv n - \sum_k k a_k \mod (p - 1).\end{aligned}$$

By Mann's result (see paragraph preceding the proof), $a_k \equiv 0 \pmod{p - 1}$ for $k > 0$. It follows that $\sum_k k a_k \equiv 0 \pmod{p - 1}$ so $\frac{h - p^e}{p^2 - 1} \equiv n \pmod{p - 1}$, or, what is the same, $h = p^e + (p^2 - 1)[n + \mathrm{t}(G)(p - 1)]$ for some integer $\mathrm{t}(G)$. As in the first proof of Theorem 2.1, one shows that $\mathrm{t}(G) \geq 0$. □

Exercise 17 (Isaacs). Let G be a nonabelian p-group. Prove that if $a_1(G) = (p-1)m$ then either p divides m or $m \equiv 1 \pmod p$. Next, $m \equiv 1 \pmod p$ if and only if $|G : G'| = p^2$ (then, in case $p = 2$, G is of maximal class).

Solution. Set $|G| = p^t$, $|G : G'| = p^s$. Then $p^t \equiv p^s + (p-1)mp^2 \pmod{p^3}$ so $p^{s-2} + (p-1)m \equiv 0 \pmod p$. If $s = 2$, we get $(p-1)m \equiv -1 \pmod p$, i.e., $m \equiv 1 \pmod p$. Conversely, if $m \equiv 1 \pmod p$, then $s = 2$.

Exercise 18. A nonabelian group G of order p^m is extraspecial if and only if $\mathrm{k}(G) = p^{m-1} + p - 1$.

We know that if G is a nonabelian p-group such that G' is a minimal normal subgroup of G, then $G/\mathrm{Z}(G)$ is elementary abelian of even rank.

Lemma 2.9 ([IsM]). *Let P be a p-group with $p \notin \mathrm{cd}(P)$ and let $L \lhd P$ be such that P/L is cyclic. Then $L' = P'$.*

Proof. It suffices to prove that $P' \leq L'$. One can assume that $L' = \{1\}$, and we have to prove that $P' = \{1\}$. Otherwise, choose $K \lhd P$ maximal such that P/K is nonabelian. Then $|(P/K)'| = p$ so P/Z is elementary abelian, where $Z/K = \mathrm{Z}(G/K)$. Since LK/K is abelian we see that $LZ/K = (LK/K)(Z/K)$ is abelian too. Also $|P/LZ| \leq p$ since P/LZ is cyclic and elementary abelian. Since P/K is nonabelian and has an abelian subgroup of index p (namely, LZ/K), it follows that $p \in \mathrm{cd}(G/K) \subseteq \mathrm{cd}(G)$, by Introduction, Theorem 17, a contradiction. Thus, P is abelian. □

Lemma 2.10 ([IsM]). *Let P be an arbitrary group and suppose that $L \leq P$ with $L' = P'$. Then $\mathrm{K}_n(P) = \mathrm{K}_n(L)$ for all $n \geq 2$.*

Proof. Working by induction on n, one may assume that $n \geq 3$. It suffices to prove that $\mathrm{K}_n(P) \leq \mathrm{K}_n(L)$. We have, by induction,

$$\mathrm{K}_n(P) = [\mathrm{K}_{n-1}(P), P] = [\mathrm{K}_{n-1}(L), P)] = [\mathrm{K}_{n-2}(L), L, P].$$

Since $\mathrm{K}_n(L) \lhd P$, it suffices to show that $[P, \mathrm{K}_{n-2}(L), L][L, P, \mathrm{K}_{n-2}(L)] \leq \mathrm{K}_n(L)$. Indeed, then also $\mathrm{K}_n(P) = [\mathrm{K}_{n-2}(L), L, P] \leq \mathrm{K}_n(L)$, the desired result. We have

$$\begin{aligned}[P, \mathrm{K}_{n-2}(L), L] &\leq [P, \mathrm{K}_{n-2}(P), L] = [\mathrm{K}_{n-1}(P), L] = [\mathrm{K}_{n-1}(L), L)] \\ &= \mathrm{K}_n(L), \\ [L, P, \mathrm{K}_{n-2}(L)] &\leq [P', \mathrm{K}_{n-2}(L)] = [L', \mathrm{K}_{n-2}(L)] = [\mathrm{K}_2(L), \mathrm{K}_{n-2}(L)] \\ &\leq \mathrm{K}_n(L).\end{aligned}$$

□

Lemma 2.11 ([IsM, Lemma 3.1]). *Let $A < G$, where A is abelian and $\mathrm{b}(G) = \max\{\chi(1) \mid \chi \in \mathrm{Irr}(G)\} = b$. Then the number of orbits in the conjugation action of A on G is at least $|G|/b$.*

Proof. Assume that G is nonabelian. By the Cauchy–Frobenius–Burnside orbit counting formula and the orthogonality relations, the number of A-orbits on G is

$$\begin{aligned}|A|^{-1}\cdot\sum_{a\in A}|\mathrm{C}_G(a)| &= |A|^{-1}\cdot\sum_{a\in A}\sum_{\chi\in\mathrm{Irr}(G)}|\chi(a)|\\ &= \sum_{\chi\in\mathrm{Irr}(G)}\langle\chi_A,\chi_A\rangle\geq\sum_{\chi\in\mathrm{Irr}(G)}\chi(1)=\mathrm{T}(G),\end{aligned}$$

where the inequality holds because A is abelian, and so χ_A is the sum of $\chi(1)$ linear characters. Now $|G|=\sum_{\chi\in\mathrm{Irr}(G)}\chi(1)^2\leq b\sum_{\chi\in\mathrm{Irr}(G)}\chi(1)=b\mathrm{T}(G)$, and thus $\mathrm{T}(G)\geq|G|/b$, proving the lemma. □

2°. If G is a p-group and $x\in G$, then the numbers $\mathrm{b}_G(x)=\log_p(|G:\mathrm{C}_G(x)|)$ and $\mathrm{b}(G)=\max\{\mathrm{b}_G(x)\mid x\in G\}$ are said the *breadth* of x and G, respectively. Note that $p^{\mathrm{b}(G)}\leq|G'|$. In the sequel, $|G|=p^n$.

The main results of this section are taken from [GMMPS].

Lemma 2.12. *If G is a p-group of order p^n, then (a) $\mathrm{b}_G(G)\leq 1\Leftrightarrow|G'|\leq p\Leftrightarrow \mathrm{k}(G)\geq p^{n-1}+p-1$, (b) $\mathrm{b}_G(G)\geq 2\Leftrightarrow|G'|\geq p^2\Leftrightarrow\mathrm{mc}(G)\leq\frac{2}{p^2}-\frac{1}{p^4}$.*

Proof. One may assume that G is nonabelian. Let $|G'|=p^k$. Then

$$(1)\qquad \mathrm{k}(G)=|\mathrm{Irr}(G)|\leq|G/G'|+\frac{|G|-|G:G'|}{p^2}\leq p^{n-2}+p^{n-k}-p^{n-k-2}.$$

Let $\mathrm{b}_G(G)=1$. Setting $|\mathrm{Z}(G)|=p^z$, we get

$$(2)\qquad \mathrm{k}(G)=p^z+\frac{p^n-p^z}{p}=p^{n-1}+p^{z-1}(p-1)\geq p^{n-1}+p-1>p^{n-1}$$

so, taking into account (1), we have $p^{n-1}<\mathrm{k}(G)<p^{n-2}+p^{n-k}$. This forces $k=1$.

Let $\mathrm{k}(G)\geq p^{n-1}+p-1$. Then by (1), $p^{n-1}+p-1\leq p^{n-2}+p^{n-k}-p^{n-k-2}$. This forces $k=1$ so $\mathrm{b}(G)=1$, completing the proof of (a).

Let $\mathrm{b}(G)\geq 2$. Then $k\geq 2$. It follows from (1) that $\mathrm{k}(G)\leq 2p^{n-2}-p^{n-4}$. Since for all primes p the inequality $2p^{n-2}-p^{n-4}<p^{n-1}+p-1$ holds and since for all p-groups G we have either $\mathrm{b}(G)\leq 1$ or $\mathrm{b}(G)\geq 2$, the proof is complete. □

Lemma 2.13 (Burnside's formula). *Suppose that H is a normal subgroup of index p in a group G. Let $\Omega=\{C_1,\dots,C_d\}$ be the set of H-classes and let G act via conjugation on the set Ω. Assume that the C_i's are labeled so that $C_i^x=C_i$, $i\leq k$, for all $x\in G$, and the remaining C_j's lie in G-orbits each of size p. Clearly, $C_1,\dots,C_k$ are each also G-conjugacy classes. We have $\mathrm{k}(G)=pk+\frac{\mathrm{k}(H)-k}{p}$.*

Proof [Grov, Proposition 3.1.4]. Let $\mathrm{k}_G(M)$ be the number of G-classes contained in a G-invariant subset M of G. Then $\mathrm{k}_G(H) = k + \frac{\mathrm{k}(H)-k}{p}$. Next we consider G-classes contained in $G - H$.

If $x \in G - H$ and $h, z \in H$, then $(hx)^z = h^z \cdot [z, x^{-1}] \cdot x \in Hx$ since $G' \le H$. Thus, H acts by conjugation on the coset Hx. In fact, each H-orbit on Hx is a G-class, i.e., $\mathrm{k}_G(Hx)$ is the number of H-orbits on Hx. To see this, take a particular element hx in some H-orbit. Since G/H is cyclic of order p, it will suffice to show that the conjugate of hx by any element of $G - H$ lies in the same H-orbit as hx. So let us conjugate hx by x^{-1}. We have $(hx)^{x^{-1}} = xh = (hx)^h$, and that element lies in the same H-orbit as hx.

Thus to complete the proof we need to count the H-orbits in each of the $p-1$ cosets Hx, $x \in G - H$. To that end we need to calculate the numbers of fixed points of elements of H (in view of the Cauchy–Frobenius–Burnside lemma). Consider first an element $a \in C_1 \cup \cdots \cup C_k$. Thus, $a^{x^{-1}} = a^b$ for some $b \in H$, and $(hx)^a = hx$, a fixed point, if and only if

$$h = h^a x^a x^{-1} = a^{-1}ha \cdot a^{-1} \cdot xax^{-1} = a^{-1}h \cdot a^{x^{-1}} = a^{-1}ha^b,$$

and this holds if and only if $a^h = a^b$. That holds if and only if $hb^{-1} \in \mathrm{C}_H(a)$, or $h \in \mathrm{C}_H(a)b$. Hence, $a \in C_1 \cup \cdots \cup C_k$ has exactly $\theta(a) = |\mathrm{C}_H(a)|$ fixed points in Hx.

Next, take $a \in C_j$, $k+1 \le j \le \mathrm{k}(H)$. If $(hx)^a = hx$, then $a^{x^{-1}} = a^h$, a contradiction since $a^h \in C_j$ but $a^{x^{-1}} \in C_j^{x^{-1}} \ne C_j$. Thus, $\theta(a)$, the number of fixed points of a in Hx, equals 0.

One may now apply the Cauchy–Frobenius Burnside lemma. Choose $a_i \in C_i$, $1 \le i \le k$. The number of G-conjugacy classes in Hx is

$$\mathrm{k}_G(Hx) = |H|^{-1} \sum_{a \in H} \theta(a) = |H|^{-1} \sum_{i=1}^{k} |C_i||\mathrm{C}_H(a_i)| = |H|^{-1} \sum_{i=1}^{k} |H| = k.$$

Thus, $\mathrm{k}(G) = k + \frac{\mathrm{k}(H)-k}{p} + (p-1)k = pk + \frac{\mathrm{k}(H)-k}{p}$. □

Lemma 2.14. *Let $|G'| \ge p^3$ and $|G : \mathrm{Z}(G)| \ge p^4$. Assume further that either* $\mathrm{mc}(G) \ge \frac{3}{p^3}$, *or $p = 2$ and $\mathrm{b}_G(x) > 1$ for all $x \in G - \mathrm{Z}(G)$ and* $\mathrm{mc}(G) \ge \frac{1}{4}$. *If all maximal subgroups of G are nonabelian, then $|M'| \ge p^2$ for all $M \in \Gamma_1$.*

Lemma 2.15. *Let $M < G$ be a maximal subgroup of a p-group G and let $A, B \le G$ be such that $[A, B] \le M'$. Then either (a) $[G, B] \le M'$ or (b) $A' \le M'$.*

Corollary 2.16. *Let $x \in G$ and $C = \mathrm{C}_G(x)$, where G is a p-group. If M is a maximal subgroup of G, then either (a) $[G, x] \le M'$ or (b) $C' \le M'$.*

Lemma 2.17. *Let G be a p-group with $|G'| \ge p^3$ and $M' = G'$ for all maximal subgroups $M < G$. Then $p \notin \mathrm{cd}(G)$ and* $\mathrm{mc}(G) < \frac{2}{p^3}$.

Theorem 2.18. *Let G be a group of order p^n, $p > 2$. If $\mathrm{mc}(G) \geq \frac{3}{p^3}$, then at least one of the following holds:*

(a) $|G'| \leq p^2$,

(b) $|G/\mathrm{Z}(G)| \leq p^3$,

(c) G has an abelian subgroup of index p.

The first proof of the following corollary was given by Mann and Scoppola (unpublished).

Corollary 2.19. *Let G be a p-group, $p > 3$, with $\mathrm{mc}(G) \geq \frac{1}{p^2}$. Then one of the following holds:*

(a) $|G'| \leq p^2$,

(b) $G/\mathrm{Z}(G)$ is of order p^3 and exponent p (then $|G'| \leq p^3$, by Theorem 21.9),

(c) G has an abelian subgroup of index p.

Suppose that G is a p-group with $\mathrm{b}(G) = 2$. Let s_0, s_1 and s_2 be the numbers of G-classes of size $1, p$ or p^2, respectively. Then $\mathrm{k}(G) = s_0 + s_1 + s_2$ and $|G| = s_0 + s_1 p + s_2 p^2 \leq (s_0 + s_1 + s_2)p^2 = \mathrm{k}(G)p^2$. It follows that $\mathrm{mc}(G) = \frac{\mathrm{k}(G)}{|G|} \geq \frac{1}{p^2}$.

Corollary 2.20. *Let G be a p-group, $p > 2$ with $\mathrm{b}(G) = 2$. Then either $|G'| = p^2$ or $G/\mathrm{Z}(G)$ is of order p^3 and exponent p.*

Corollary 2.21 ([Kno]). *If G is a p-group, $p > 2$, with $b(G) = 2$, then $|G'| \leq p^3$. (This is also true for $p = 2$.)*

§3

Minimal classes

The results of this section are taken from [LMM].

Let G be a nonabelian group of order p^n, $1 < p^{s_1} < p^{s_2} < \cdots$ the class sizes of G and u_i the number of conjugacy classes of size p^{s_i}, all i. Write $s = s_1, u = u_1$. Let $|Z(G)| = p^z$. A G-class of size p^s is said to be *minimal class*. Elements contained in minimal classes are said to be *minimal elements*. Centralizers of minimal elements are said to be *maximal centralizers*. The number s is also called the *minimal breadth*; if x is contained in a minimal class we say that it is of minimal breadth. We write $x \sim y$ to denote that elements x and y are conjugate in G.

Proposition 3.1. *Let G be a nonabelian p-group.*

(a) (Mann) For each $i \geq 1$, $p-1$ divides u_i and p^z divides $u_i p^{s_i}$.

(b) Either $z = s$ and $u \equiv p-1 \pmod{p(p-1)}$, or $s < z$ and $p(p-1)$ divides u.

Proof. (a) Let $\exp(G) = p^e$. If p does not divide $j \in \mathbb{N}$, then the map $x \mapsto x^j$ induces the permutation of the G-classes since $x^j = y^j \Leftrightarrow x = y$ $(x, y \in G)$. If K is a G-class, then $K^j = \{x^j \mid x \in K\}$ is also G-class. It follows that $K, K^2, \dots, K^{p-1}$ are distinct classes of the same size. This argument shows that $p-1$ divides u_i for all $i \geq 1$.

If $a \in G$, $x \in Z(G)$, then $C_G(a) = C_G(ax)$. It follows that the union of classes of given size is a union of some cosets of $Z(G)$. Therefore, p^z divides $u_i p^{s_i}$ for all i.

(b) It follows from the class equation $p^n = p^z + up^s + u_2 p^{s_2} + \cdots$ that $s \leq z$, and so p divides $p^{z-s} + u$. Let $z = s$. Then p divides $u + 1$, i.e., $u \equiv p-1 \pmod{p}$. In that case, we also have

$$p^{n-s} - p = [u - (p-1)] + u_2 p^{s_2 - s} + \cdots \equiv u - (p-1) \pmod{p(p-1)}$$

so that $u \equiv p - 1 \pmod{p(p-1)}$. If $z > s$, then p divides u so $p(p-1)$ divides u, by (a). □

Exercise 1. Let G be a p-group. (a) A maximal centralizer C is normal in G if and only if it is the centralizer of an element from $Z_2(G) - Z(G)$. (b) If $x \in Z_2(G)$ is of minimal breadth, then $x^p \in Z(G)$.

Solution. (a) By Grün's lemma, $[G', Z_2(G)] = \{1\}$. For $x \in Z_2(G)$, we have, therefore, $G' \leq C_G(x)$, so $C_G(x) \trianglelefteq G$. Conversely, if $C \triangleleft G$, then $Z(C) \triangleleft G$ and

$Z(C) > Z(G)$ so there exists $x \in Z(C) \cap (Z_2(G) - Z(G))$. Since $C \leq C_G(x) < G$, we must have $C_G(x) = C$ since C is a maximal centralizer.

(b) Set $C = C_G(x)$ and let $y \notin C$, but $y^p \in C$. Then, by the choice of x, $[x, y] \in Z(G)$, therefore $[x^p, y] = [x, y^p] \in [x, C] = \{1\}$. Thus $y \in C_G(x^p) - C$ hence $C_G(x^p) > C$ so $x^p \in Z(G)$.

In Proposition 3.2(b) we show that $Z_2(G)$ contains a minimal class, or, equivalently, some maximal centralizer is normal in G.

Exercise 2. Let G be a nonabelian p-group and s is the minimal breadth.

(a) Any $N \lhd G$ of order $\leq p^s$ is central so any normal subgroup of G of order p^{s+1} is abelian and belongs to $Z_2(G)$.

(b) Let $N \trianglelefteq G$ be of order p^{s+1}. Then either $N \leq Z(G)$ or $N = (N \cap Z(G)) \cup K_1 \cup \cdots \cup K_{p-1}$, where $K_1, \dots, K_{p-1}$ are minimal classes. Moreover, if $M \lhd G$ is of order $\leq p^{s2}$ and $M \nleq Z(G)$, then $M - Z(G)$ is a union of minimal classes.

Solution. (a) is obvious since each element of N has $< p^s$ conjugates in G.

(b) Let $|N| = p^{s+1}$. If $N \nleq Z(G)$, then $|N \cap Z(G)| = p^s$, by (a). If $x \in N - Z(G)$, then x has $< p^{e_2}$ conjugates in G so it belongs to a minimal class of G, and we get $N - (N \cap Z(G))$ is a union of $p - 1$ minimal classes. This argument also establishes the last assertion. (It follows from (b) and Lemma 1.1 that if $M \trianglelefteq G$ is of order p^{s+2}, then $|M'| \leq p$.)

There is an analog of Exercise 2(a) for characters. Let $\text{cd}(G) = \{1, n_1, n_2, \dots\}$, where $1 < n_1 < n_2 < \cdots$. If $H \leq G$ and $|G : H| \leq n_1$, then $G' \leq H$. Indeed, $(1_H)^G$ is the sum of linear characters of G. Exercise 2(b) also has an analog for characters. Let $H < G$ with $|G : H| \leq n_2$. Then either $G' \leq H$ or all nonlinear irreducible constituents of $(1_H)^G$ are characters of degree n_1.

Proposition 3.2. *Let G be a nonabelian p-group.*

(a) p^s is the minimal order of $N \lhd G$ such that $Z(G/N) > Z(G)/N$.

(b) Some minimal class of G is contained in $Z_2(G)$. Moreover, the number of minimal classes contained in $Z_2(G)$, is divisible by $p - 1$.

Proof. (a) Let $N \lhd G$ be of order $\leq p^s$; then $N \leq Z(G)$ (Exercise 2(a)). Let $xN \in Z(G/N)$; then all conjugates of x are contained in xN, so their number is at most $|N|$. It follows that if $|N| < p^s$, then $x \in Z(G)$ and so $Z(G/N) = Z(G)/N$. Clearly, $Z(G/N) \geq Z(G)/N$ for each $N \lhd G$. On the other hand, suppose that $z \in Z_2(G)$ is a minimal element. Set $N = [z, G] = \{[z, g] \mid g \in G\}$, If $g, h \in G$, then $[z, gh] = [z, h][z, g]^h = [z, h][z, g]$ so $N \leq G$. If $x \in G$, then, for some $z_1 \in Z(G)$, we get $N^x = [z^x, G^x] = [zz_1, G] = [z, G] = N$ so $N \trianglelefteq G$. Since $|N| = |G : C_G(z)| = p^s$, it follows that $N \leq Z(G)$ (Exercise 2(a)) and $zN \in Z(G/N)$, so $Z(G/N) > Z(G)/N$.

(b) By (a), there is $N \lhd G$ of order p^s such that $Z(G/N) > Z(G)/N$. Then there is $xN \in Z(G/N)$ such that $x \notin Z(G)$. In that case, the G-class K_x of x is contained in xN. It follows from $p^s \le |K_x| \le |xN| = p^s$ that $K_x = xN$ is a minimal class. Since $xN \le Z(G/N)$ and $N \le Z(G)$, we get $x \in K_x \subseteq Z_2(G)$, as desired. The last assertion is proved in the same way as the assertion $u_i \equiv 0 \pmod{(p-1)}$ in the proof of Proposition 3.1(a). □

Exercise 3. Let a p-group G be of maximal class and order p^n, $n > p+1$, and let G_1 be the fundamental subgroup of G. Then G has exactly $p-1$ minimal classes if and only if $Z(G_1) = Z_2(G)$.

Solution. If $x \in G$ is of minimal breadth (in our case, this breadth is 1), then $C_G(x) = G_1$ since all maximal subgroups of G other than G_1, are of maximal class (see Theorem 9.6). It follows that the set of elements of minimal breadth in G coincides with $Z(G_1) - Z(G)$. Therefore, G has exactly $p-1$ minimal classes if and only if $|Z(G_1) - Z(G)|/p = p-1$, i.e., if and only if $|Z(G_1)| = p^2$.

Exercise 4. Let G be a p-group with $|Z_2(G)| = p^2$. The group G has exactly $p-1$ minimal classes if and only if $Z(M) \le Z_2(G)$ for all $M \in \Gamma_1$.

Solution. If $a \in Z_2(G) - Z(G)$, then $|G : C_G(a)| = p$ so $s = 1$ and exactly $p-1$ minimal classes are contained in $Z_2(G)$. Now let G have exactly $p-1$ minimal classes; then all these classes are contained in $Z_2(G)$ (Proposition 3.2(b)). Let $M \in \Gamma_1$. By hypothesis, $Z_2(G)$ is the unique G-invariant subgroup of order p^2 so $Z_2(G) < M$. Assume that there is an element $x \in Z(M) - Z_2(G)$. Then $C_G(x) = M$ so x is a minimal element, contrary to what has just been said. Thus, $Z(M) \le Z_2(G)$. Now suppose that $Z(M) \le Z_2(G)$ for all $M \in \Gamma_1$. Let $x \in G - Z_2(G)$ be a minimal element. Set $K = C_G(x)$; then $K \in \Gamma_1$ since $s = 1$. In that case, the number of minimal classes in G is $> p-1$ and $Z(K) \not\le Z_2(G)$, a contradiction.

Exercise 5. If G has exactly $p-1$ minimal classes, then all these classes are contained in $Z_2(G)$ and exactly one maximal centralizer is normal in G.

Proposition 3.3. *Suppose that G has exactly $p-1$ minimal classes. Then*

(a) $Z_2(G)$ is elementary abelian, $|Z_2(G) : Z(G)| = p$, and the minimal classes are cosets of $Z(G)$ in $Z_2(G)$. Next, G has exactly one maximal centralizer.

(b) If $x \in Z_3(G) - Z_2(G)$, then the G-class of x is $xZ_2(G)$.

Proof. Since $u = p-1$, it follows from Proposition 3.1(b) that $s = z$. Now, if $x \in Z_2(G) - Z(G)$ is of minimal breadth (see Proposition 3.2(b)), then the conjugacy class of x is contained in $xZ(G)$, so its size is at most $p^z = p^s$. Thus, this class must have size p^z, and it coincides with $xZ(G)$. Since different cosets of $Z(G)$ in $Z_2(G)$ yield different classes in view of $z = s$ and there are at most $p-1$ such cosets, we must have $|Z_2(G) : Z(G)| = p$. Moreover, all those cosets are the minimal classes of G.

Therefore, if $x \in Z_3(G)-Z_2(G)$, then the class of x has size at least $p^{s+1} = |Z_2(G)|$ and it is contained in the coset $xZ_2(G)$ of size p^{s+1}. So the class of x is $xZ_2(G)$, proving (b).

Let us prove that $Z_2(G)$ is elementary abelian. By what has been proved in the previous paragraph, $Z_2(G)$ is abelian. Letting again $x \in Z_2(G)-Z(G)$, and $z \in Z(G)$, we have $z = [x, v]$ for some $v \in G$ (indeed, $xZ(G) = \{xz \mid x \in Z(G)\}$ is a G-class so for some $v \in G$ we have $x^v = xz$ or, what is the same, $z = x^{-1}x^v = [x, v]$) hence $z^p = [x^p, v] = 1$ since $x^p \in Z(G)$, by the proven part of (a). Thus, $Z(G)$ is elementary abelian, while $Z_2(G) = \langle x, Z(G)\rangle$ is abelian. Suppose that $Z_2(G)$ is not elementary abelian. Then there exists a maximal subgroup W of $Z(G)$ such that $Z_2(G)/W$ is cyclic of order p^2. By Proposition 3.2(a), $Z(G/W) = Z(G)/W$ since $|W| < p^s$. Thus, $Z_2(G/W) = Z_2(G)/W$ is cyclic, which implies that G/W is cyclic or a 2-group of maximal class, by Lemma 1.4. Since G is nonabelian, G/W is not cyclic so it is a 2-group of maximal class. In that case, G has an abelian subgroup A of index 2 and exactly one minimal class, say K. We have $|K| = |G : A| = 2$. Therefore, must be $A = Z(G) \cup K$ so the order of $Z(G)$ is 2, $W = \{1\}$. Then $|A| = 4$, $|G| = 8$ and G has exactly 3 minimal classes, a contradiction.

In our case, all minimal classes are contained in $Z_2(G)$. Let x be of minimal breadth. Then the elements $x, x^2, \dots, x^{p-1}$ belong to distinct minimal classes and have the same centralizer, say C. So $C \lhd G$ (Exercise 1(a)). Next, $C_G(x^y) = C_G(x)^y = C_G(x) = C$ so C is the unique maximal centralizer in G. □

Let G be a p-group with $z = s$. Then all G-classes, contained in $Z_2(G)-Z(G)$, are minimal. Indeed, if $x \in Z_2(G) - Z(G)$, then the coset $xZ(G)$ is a union of G-classes and contains p^s elements so it is a G-class.

Exercise 6. Let N_1 and N_2 be distinct normal subgroups of order p^s such that $Z_i/N_i = Z(G/N_i) \neq Z(G)/N_i$, $i = 1, 2$. Then $Z_1 \cap Z_2 = Z(G)$.

Solution. Let $x \in Z_1 \cap Z_2$. Then $[x, G] \le N_1 \cap N_2 < N_1$, so x has less than p^s conjugates, and hence it is central.

Exercise 7. Suppose that a nonabelian p-group G has exactly one normal subgroup of order p^s. Then either (a) $s = 1$ and $Z(G)$ is cyclic or (b) $s = z$.

Solution. If (a) holds, then G has exactly one normal subgroup of order $p = p^s$; then $Z(G)$ is cyclic. Suppose that (b) holds. Assume that N and M are normal subgroups of order p^s in G. By Exercise 2(a), $MN \le Z(G)$ so $M = N$ since $s = z$. Now suppose that G has only one normal subgroup N of order p^s; then $N \le Z(G)$. Assume that $s < z$. Then $Z(G)$ is cyclic since it has exactly one subgroup of order p^s. By Lemma 1.4, G has a normal abelian subgroup M of type (p, p) since G is not a 2-group of maximal class in view of $z > 1$. Since $M \not\le Z(G)$, we get $s = 1$, and we have case (a). It follows that if $s < z$, then $s = 1$. Since $s \le z$, it remains only the possibility $s = z$.

§4

p-groups with cyclic Frattini subgroup

In this section we study the p-groups with cyclic Frattini subgroup. As a by-product of the obtained result we give an alternate proof of Hall's classification of p-groups all of whose characteristic abelian subgroups are cyclic.

If G is a p-group, then $G/\Phi(G)$ is elementary abelian (see §1), i.e., $\Phi(G)$ is large. Therefore, it is natural to classify p-groups whose Φ-subgroups are not very complicated. In this section we consider the simplest case of the p-groups G with cyclic Φ-subgroups. It appears that, if $p > 2$, then $\Phi(G) \leq \mathrm{Z}(G)$ so $|G'| \leq p$.

A p-group G is extraspecial if and only if $G' = \mathrm{Z}(G)$ is of order p. Indeed, if $x, y \in G$, then $1 = [x, y]^p = [x, y^p]$ so $\mho_1(G) \leq \mathrm{Z}(G)$; then $G' \leq \Phi(G) = \mho_1(G)G' \leq \mathrm{Z}(G) = G'$, and we are done.

Lemma 4.1 (compare with Lemma 1.4). *Let G be a p-group, $N \trianglelefteq G$ and let $N_0 \leq N \cap \mathrm{Z}(G)$ be of order p. If N has no G-invariant abelian subgroups of type (p, p) containing N_0, it is cyclic or a 2-group of maximal class.*

Lemma 4.2. *Let G be a p-group with $|G'| = p$. Then $G = (A_1 * A_2 * \cdots * A_s)\mathrm{Z}(G)$, the central product, where $A_1, \dots, A_s$ are minimal nonabelian so $G/\mathrm{Z}(G)$ is elementary abelian of even rank. In particular, if G/G' is elementary abelian, then $|A_1| = \cdots = |A_s| = p^3$, $E = A_1 * \cdots * A_s$ is extraspecial and $G = E\mathrm{Z}(G)$.*

Proof. We use induction on $|G|$. Let $A_1 \leq G$ be minimal nonabelian. Then $|A_1'| = p = |G'|$ so $A_1' = G'$ and $A_1 \trianglelefteq G$, $A_1 = \langle x_1, y_1 \rangle$ for some $x_1, y_1 \in A_1$ (Exercise 1.8a). Set $C_1 = C_G(x_1)$, $Z_1 = C_G(y_1)$; then $C_1 \neq Z_1$. Since the cosets x_1G' and y_1G' are G-invariant of size p and are not contained in $\mathrm{Z}(G)$, they are G-classes. Therefore, $|G : C_1| = p = |G : Z_1|$. Set $N_1 = C_1 \cap Z_1$; then $N_1 = C_G(A_1)$, and so $N_1 \cap A_1 = \mathrm{Z}(A_1)$. As $|A_1 : \mathrm{Z}(A_1)| = p^2$ and $|G : N_1| = p^2$ in view of $C_1Z_1 = G$, we get $G = A_1 * N_1$ so $\mathrm{Z}(N_1) = \mathrm{Z}(G)$. If N_1 is abelian, we are done with $s = 1$. If N_1 is nonabelian, then $|N_1'| = |G'| = p$ and, by induction, $N_1 = (A_2 * \cdots * A_s)\mathrm{Z}(G)$, where $A_2, \dots, A_s$ are minimal nonabelian, and the first assertion is proven since then $G = A_1 * N_1 = (A_1 * A_2 * \cdots * A_s)\mathrm{Z}(G)$.

Now let $\exp(G/G') = p$; then $G' = \Phi(G) = A_i'$ so $A_i/G' = A_i/A_i' \cong \mathrm{E}_{p^2}$ and $|A_i| = p^3$ for all i, and $E = A_1 * \cdots * A_s$ is extraspecial since $\mathrm{Z}(E) = G' = E' = \Phi(E)$. □

Lemma 4.3. *Let E be a subgroup of a p-group G such that $|E'| = p$ and $Z(E) = \Phi(E)$. If $[G, E] = E'$, then $G = E * C_G(E)$.*

Proof. Let $d(E) = n(> 1)$; then $E = \langle x_1, x_2, \dots, x_n \rangle$. Let A be the subgroup of $\mathrm{Aut}(G)$ that induces identity on E/E'. For $\mu \in A$, we get $x_i^\mu = x_i z_i$ for some $z_i \in E'$. It follows that the action of μ on E is uniquely determined by the ordered set $\{z_1, \dots, z_n\}$. But there are exactly p^n distinct ordered sets $\{z_1, \dots, z_n\}$ with $z_i \in E'$ so $|A| \le p^n$. On the other hand, we have $\mathrm{Inn}(E) \le A$ and $|\mathrm{Inn}(E)| = |E/Z(E)| = |E/\Phi(E)| = p^n \ge |A|$ so the restriction of A on E coincides with $\mathrm{Inn}(E)$. If $g \in G$, then $\mu : e \mapsto e^g$ is an automorphism of E. By hypothesis, $E/E' \le Z(G/E')$ so μ leaves every element of E/E' invariant whence $\mu \in A = \mathrm{Inn}(E)$. It follows that, for every $g \in G$, there is $v \in E$ such that $e^g = e^v$ for all $e \in E$ so that $gv^{-1} \in C_G(E)$ and $g \in EC_G(E)$ for all $g \in G$. Thus, $G = E * C_G(E)$. □

As Janko has noticed, if, in Lemma 4.3, $Z(E) \neq \Phi(E)$, then it is possible that $E * C_G(E) < G$. Indeed, if G is extraspecial of order $\ge p^5$ and $E < G$ is maximal, then $C_G(E) < E$ and so $EC_G(E) = E < G$.

Remark 1. Let $G = EM$ be a p-group, $E, M \lhd G$. If $E \cap M = E'$ is of order p and $Z(E) = \Phi(E)$, then $G = EC_G(E)$. Indeed,

$$[G, E] = [EM, E] = [E, E][M, E] \le E'(M \cap E) = E',$$

and the result follows from Lemma 4.3.

Theorem 4.4 ([Ber3]). *Let G be a nonabelian p-group with cyclic $\Phi(G)$ of order $> p$ and write $\Phi_0 = \Omega_1(\Phi(G))$. Let $Z < G$ be cyclic of maximal order such that $\Phi(G) \le Z$; then $|Z| = p|\Phi(G)|$. Set $\Lambda_1 = \{H \le G \mid Z < H, |H : Z| = p\}$. Suppose that every $H \in \Lambda_1$ has a G-invariant abelian subgroup $T(H)$ of type (p, p) (this is a case if $p > 2$). Set $A = \langle T(H) \mid H \in \Lambda_1 \rangle$ and assume that A is nonabelian. Then $\Phi(G) \le Z(G)$, $\Phi_0 = G'$ and $G = AZ = (A_1 * \cdots * A_s)Z$, where A_i is minimal nonabelian for all i. We also have $A = Z(A)E$, where E is extraspecial, and $G = Z(A)EZ$.*

Proof. We retain the notation from the statement. We have $|G| \ge p^4$. One may assume that G has no cyclic subgroups of index p (otherwise, $G \cong M_{p^n} = \Omega_1(G)Z$). For $H \in \Lambda_1$, we have $H = ZT(H)$, where $T(H) = \Omega_1(H) \cong E_{p^2}$, $|H : Z| = p$ and $Z \cap T(H) = \Phi_0$. We get $G = \langle H \mid H \in \Lambda_1 \rangle$. Setting $A = \langle T(H) \mid H \in \Lambda_1 \rangle$, we obtain $G = \langle T(H) \mid H \in \Lambda_1 \rangle Z = AZ$. It follows that $\Phi(G) < Z$ (otherwise, $A = G$ and then $\Phi(G) = \Phi(A) = \Phi_0$ is of order p) so $|Z : \Phi(G)| = p$. We also have $\Phi(G) \le Z(H)$ $(H \in \Lambda_1)$ so $C_G(\Phi(G)) \ge \langle H \in \Lambda_1 \rangle = G$ and $\mathrm{cl}(G) = 2$. Then $\exp(G') \le \exp(G/Z(G)) = p$ so $|G'| = p$ and $G' = \Phi_0$. Since $T(H)/\Phi_0 \le Z(G/\Phi_0)$, it follows that $A/\Phi_0 \le \Omega_1(Z(G/\Phi_0))$ and, if $K/\Phi_0 < A/\Phi_0$, then $K \lhd G$. Since Φ_0 is characteristic in G, then A is also characteristic. By construction, $\exp(A/A') = p$. Suppose that A is nonabelian. Then, by Lemma 4.2,

$G = (A_1 * \cdots * A_s)Z(G)$, where $A_1, \ldots, A_s$ are minimal nonabelian of the same order p^3, $s \geq 1$. It follows from $Z(A_i) = \Phi(A_i) \leq \Phi(G)$ that $Z(A_i)$ is cyclic for $i = 1, \ldots, s$. Lemma 4.2, applied to A, implies the last assertion. □

Suppose that all characteristic abelian subgroups of a nonabelian p-group G are cyclic and $p > 2$. Then $\Phi(G)$ is cyclic so, by Theorem 4.4 and Lemma 4.2, $G = EZ(G)$, where E is extraspecial and $Z(G)$ is cyclic.

Remark 2. Let G be a 2-group of order $> 2^4$ and let $M \in \Gamma_1$ be of maximal class. Then G has at most one normal four-subgroup. Assume that this is false, and let $R, R_1 < G$ be two distinct normal four-subgroups. Then $MR = G = MR_1$ since $R, R_1 \not\leq M$. We have $M \cap R = M \cap R_1$ so, setting $S = RR_1$, we see that $M \cap S \cong C_4$ so $S \cong D_8$. By Theorem 5.4, $\Phi(G) = \Phi(M)$. Next, S centralizes $\Phi(G)$ since R and R_1 centralize $\Phi(G)$. It follows from $C_4 \cong S \cap M \leq \Phi(M) = \Phi(G) \leq C_G(S)$ that S is abelian, a contradiction.

Theorem 4.5 ([Ber3]). *Let G be a nonabelian 2-group with cyclic Frattini subgroup of order > 2 and let $\Phi_0 = \Omega_1(\Phi(G))$. Then G has a cyclic subgroup Z of order $2\cdot|\Phi(G)|$. Let the set Λ_1 be defined as in Theorem 4.4. Suppose that some $H \in \Lambda_1$ has no G-invariant abelian subgroups of type $(2,2)$. Then H is of maximal class (Lemma 1.4). Suppose that $H < G$. Let $\Lambda_2 = \{R \leq G \mid H < R, |R : H| = 2\}$. In that case, $R = HT(R)$ $(R \in \Lambda_2)$, where $T(R) \cong E_4$ is G-invariant and uniquely determined (Remark 2). We have $\Phi_0 < T(R)$ for $R \in \Lambda_2$. Let $G/H = (R_1/H) \times \cdots \times (R_s/H)$ and $A_0 = \langle T(R_i) \mid i = 1, \ldots, s\rangle$, $A = \langle T(R) \mid R \in \Lambda_2\rangle$. Then $G = A_0H = AH$ and the following holds:*

(a) $A_0 \lhd G$, A is characteristic in G, A_0 and A centralize $\Phi(G)$, $A_0 \cap H = \Phi_0$, $A \cap H \leq Z(A)$ is cyclic of order ≤ 4.

(b) A is either elementary abelian or else $A = EZ(A)$, where E is extraspecial (similar decomposition holds for A_0), $A_0/\Phi_0, A/\Phi_0 \leq Z(G/\Phi_0)$.

(c) If $A = E$, then $G = AH = EH = E * H_1$, where $H_1 < EH$ is a 2-group of maximal class, $|H_1| = |H|$ so $A = A_0$.

Proof. By Lemma 4.1, H is of maximal class. Let $H < G$. If $R \in \Lambda_2$, then $R/\Phi(G) \cong E_8$ so R is not of maximal class and R has only one G-invariant subgroup $T(R)$ of type $(2,2)$ containing Φ_0 (Lemma 4.1 and Remark 2). Then $R = HT(R)$. We have $HA = G = HA_0$, where A_0, A are defined in the statement of the theorem. Clearly, A is characteristic in G, $A_0 \lhd G$, A, A_0 centralize $\Phi(G)$, $A = EZ(A)$, where $E = \{1\}$ (so $\exp(A) = 2$) or extraspecial, by Lemma 4.3, $A/\Phi_0, A_0/\Phi_0 \leq \Omega_1(Z(G/\Phi_0))$. By the product formula, $A_0 \cap H = \Phi_0$. Set $B = A \cap H$; then B centralizes $\Phi(G)$ so it is cyclic of order ≤ 4 hence $B \leq \Phi(H) = \Phi(G)$. Since $\Phi(G)$ centralizes A, it follows that $B \leq Z(A)$, and this completes the proof of (a). Lemma 4.2 implies (b).

It remains to prove (c). By assumption, $A = E$ so $G = AH = EH$, where E is extraspecial. Since $\exp(A) = 4$ and $|H| > 8$, then $E \cap H$ is cyclic of order ≤ 4 so $E \cap H \leq \Phi(G)$. It follows that $E \cap H \leq \mathrm{Z}(E)$ so $E \cap H = \Phi_0$, and we conclude that $E = A = A_0$. We have $[G, E] = [EH, E] = [E, E][E, H] = E'$ so $G = E * \mathrm{C}_G(E)$ (Lemma 4.3). Write $H_1 = \mathrm{C}_G(E)$. By the product formula, $|H_1| = |H|$. We also have $\mathrm{cl}(H_1) = \mathrm{cl}(G) \geq \mathrm{cl}(H)$ so H_1 is of maximal class, completing the proof of (c) and thereby the theorem. □

Theorem 4.6 (P. Hall). *Let all characteristic abelian subgroups of a nonabelian p-group G be cyclic. Then $\Phi(G)$ is cyclic, $G = E * H$, the central product, where E is extraspecial and H is either cyclic or a 2-group of maximal class.*

Proof. By Proposition 1.13, $\Phi(G)$ is cyclic. By the above, it remains to consider the case $p = 2$ only. In view of Lemma 4.2, one may assume that $|\Phi(G)| > 2$ and G is neither extraspecial nor of maximal class.

(i) Let $\Phi(G) \leq \mathrm{Z}(G)$. Repeating the argument of the paragraph, following Theorem 4.4, we get $G = A\mathrm{Z}(G)$, where extraspecial A is generated by all G-invariant subgroups of type $(2, 2)$ containing $\Phi_0 = \Omega_1(\Phi(G))$ and $\mathrm{Z}(G)$ is cyclic.

(ii) Now let $\Phi(G) \not\leq \mathrm{Z}(G)$. By Theorem 4.5, $G = AH$, where A is characteristic in G, $\Omega_1(\Phi(G)) = \Phi_0 < A$, $H < G$ is a 2-group of maximal class, $|H : \Phi(G)| = 4$, $[A, \Phi(G)] = \{1\}$. Next, $A = E\mathrm{Z}(A)$, where $E = \{1\}$ or extraspecial (Lemma 4.2). Since $|A| > 2$, A is not cyclic so nonabelian; then $E > \{1\}$, i.e., E is extraspecial. If $|\mathrm{Z}(A)| = 2$, then $A = E$, $G = EH$. Suppose that $|\mathrm{Z}(A)| = 4$. Then $S = \mathrm{Z}(A)\Phi(G)$ is a characteristic abelian subgroup of G, and so S is cyclic. In that case, since $|\Phi(G)| \geq 4$, we get $\mathrm{Z}(A) \leq \Phi(G) < H$ so $|E \cap H| = 2$ and $G = AH = E\mathrm{Z}(A)H = EH$. Then

$$[E, G] = [E, EH] = [E, E][E, H] = E'E' = E'$$

so $G = E * \mathrm{C}_G(E)$ and $E \cap \mathrm{C}_G(E) = \mathrm{Z}(E)$. It follows from $\exp(G) = \exp(\mathrm{C}_G(E))$ that $\mathrm{C}_G(E)$ has a cyclic subgroup of index 2. Since $\mathrm{cl}(\mathrm{C}_G(E)) = \mathrm{cl}(G) \geq \mathrm{cl}(H) > 2$, $\mathrm{C}_G(E)$ is of maximal class. □

Theorem 4.7. *Let G be an extraspecial p-group.*

(a) G is a central product of n nonabelian groups of order p^3 so $|G| = 2^{2n+1}$.

(b) $G = AB$ is a product of maximal normal abelian subgroups A and B of order p^{n+1}, $A \cap B = \mathrm{Z}(G)$. Next, A and B are maximal abelian subgroups of G.

(c) $\mathrm{k}(G) = p^{2n} + p - 1$.

(d) All maximal abelian subgroups of G have the same order p^{n+1}. We have $\mathrm{cd}(G) = \{\chi(1) \mid \chi \in \mathrm{Irr}(G)\} = \{1, p^n\}$.

Proof. (a) Let $M \le G$ be minimal nonabelian; then $M' = G'$ so M/M' is abelian of exponent p and hence $|M| = p^3$ (Exercise 1.8a). By Lemma 4.2, $G = M_1 * \cdots * M_n$, where $M_1, \dots, M_n$ are nonabelian of order p^3 so $|G| = 2^{n+1}$.

(b) Using notations of (a), set $M_i = \langle x_i, y_i \rangle$, all i, $A = \langle x_1, \dots, x_n, \mathrm{Z}(G) \rangle$ and $B = \langle y_1, \dots, y_n, \mathrm{Z}(G) \rangle$; then $|A| = p^{n+1} = |B|$ and $G = AB$ with $A \cap B = \mathrm{Z}(G)$. Obviously, A and B are abelian and G-invariant. If $B < C < G$, where C is abelian, then $|A \cap C| > p$ and $A \cap C \le \mathrm{Z}(G)$, a contradiction. Thus, A and B are maximal abelian subgroups of G.

(c) Since all noncentral G-classes have size p, we get $\mathrm{k}(G) = \frac{|G - \mathrm{Z}(G)|}{p} + |\mathrm{Z}(G)| = p^{2n} - 1 + p$, proving (c).

(d) We have $p^{2n}(p-1) = |G| - |G : G'| = \sum_{\chi \in \mathrm{Irr}_1(G)} \chi(1)^2$ (see (b)). It follows that $\mathrm{cd}(G) = \{1, p^n\}$ so, by Ito's theorem on degrees (Introduction, Theorem 17), G has no abelian subgroups of index $< p^n$. Let $D < G$ be maximal abelian of order $p^k \le p^n$. We prove, using induction on n that this is impossible. Since $\mathrm{C}_G(D) = D$, we get $n > 3$. Let $G' < R < D$, where $R \cong \mathrm{E}_{p^2}$. Then $\mathrm{C}_G(R) = M \in \Gamma_1$ and $M = L \times U$, where $L < R$ is of order p and U is extraspecial of order p^{2n-1}. Then $D \cap U$ of order $\le p^{n-1}$ is a maximal abelian subgroup of order $\le p^{n-1}$ in U, contrary to the induction hypothesis. □

Proposition 4.8. *Let a group G of even order have no subgroups of index 2. If $P \in \mathrm{Syl}_2(G)$ is of maximal class, then all involutions are conjugate in G.*

Proof. We will prove a stronger result. Suppose that P has a subgroup Z of index 2 such that Z has only one involution. Let x be an involution in $P - Z$ (if x does not exists, there is nothing to prove). Set $|G : P| = m$, where $m > 1$ is odd. Assume that $x^g \notin Z$ for all $g \in G$. Considering the representation of G by permutations of left cosets of Z, we see that x is either represented as a product of m independent transpositions or it fixes a coset aZ. In the second case, $xaZ = aZ$, or, what is the same, $x^a \in Z$, contrary to the assumption. In the first case, however, since m is odd, G has a normal subgroup of index 2, contrary to the hypothesis. □

Proposition 4.9 (K. G. Nekrasov [BZ, Lemma 31.8]). *Let G be a 2-group. If $\Phi(G)$ is of type $(2, 2)$, then $\Phi(G) \le \mathrm{Z}(G)$.*

Proof. Assume that $\Phi(G) \not\le \mathrm{Z}(G)$. Then $\Phi(G) \cap \mathrm{Z}(G) = \langle z \rangle$, where $o(z) = 2$, $\Phi(G) = \langle a \rangle \times \langle z \rangle$ and $|G : \mathrm{C}_G(a)| = 2$. Since $\Phi(G) = \mho_1(G)$, one may assume that a is chosen so that $a = x^2$ for some $x \in G$. Now, $G = \langle b, \mathrm{C}_G(a) \rangle$ for $b \in G - \mathrm{C}_G(a)$. Put $xb = bxt$, where $1 \ne t = [x, b] \in G' \le \Phi(G)$. But x of order 4 commutes with a and z and $x^2 = a$ so $\langle x, \Phi(G) \rangle$ is abelian of type $(4, 2)$ hence $xt = tx$. We have

$$ab = x^2 b = x(xb) = x(bxt) = (xb)xt = (bxt)xt = bx^2t^2 = ba,$$

contrary to the choice of b. Thus, $\Phi(G) \le \mathrm{Z}(G)$. □

Proposition 4.10 (= Proposition 1.6). *Let G be a nonabelian 2-group. If $|G : G'| = 4$, then G is of maximal class.*

Proof. Clearly, $G' = \Phi(G)$. In view of Theorem 1.2, it suffices to prove that G has a cyclic subgroup of index 2 or, what is the same, that $\Phi(G)(= \mho_1(G))$ is cyclic. Assume that $\Phi(G)$ is not cyclic; then it contains a G-invariant subgroup H such that $\Phi(G)/H$ is of type $(2,2)$. Set $\bar{G} = G/H$. Then $|\bar{G}| = 2^4$ and $\Phi(\bar{G})$ is of type $(2,2)$. By Proposition 4.9, $\Phi(\bar{G}) \leq Z(\bar{G})$ so $\bar{G}$ is minimal nonabelian; then $|\bar{G}'| = 2$, by Exercise 1.8a, hence $|G : G'| = 8$, a contradiction. □

The following proof of Theorem 1.23 illustrates one of numerous applications of structure theorem for minimal nonabelian p-groups (Exercise 1.8a). This proof is more involved than the original one.

Theorem 4.11 (= Theorem 1.23). *If a p-group G is non-Dedekindian, then there exists a G-invariant subgroup K of index p in G' such that G/K is not Dedekindian.*

Proof. One may assume that $|G'| > p$ (otherwise, there is nothing to prove). Assuming that G is a minimal counterexample, we get $p = 2$, $|K| = 2$ so $|G'| = 4$. Then there is in G/K a subgroup $Q/K \cong Q_8$. Since Q is not of maximal class, we get $|Q'| = 2$ (Taussky) so $G' = Q' \times K \cong E_4$ and $G' \leq Z(G)$ since $Q \lhd G$. Let $A < G$ be minimal nonabelian. Assume that $|A| > 8$; then $K < A$ since all minimal nonabelian subgroups of G/K are isomorphic to Q_8. Assume that A/K is abelian; then it is of type $(4,2)$ since $d(A) = 2$. In that case, $G'/K = \mho_1(G/K) < A/K$ so $G' < A$ and $A \lhd G$. Then A/Q' is nonabelian in view of $Q' \cap K = \{1\}$, so $A/Q' \cong Q_8$. Thus, one can assume from the start that $A/K \cong Q_8$. In that case, $A = \langle x, y \mid x^4 = y^4 = 1, x^y = x^3 \rangle$ since $|\Omega_1(A)| = 4$ and $\exp(A) = 4$ (see Exercise 1.8a). All subgroups of order 2 are characteristic in A so normal in G (indeed, three epimorphic images of A of order 8, namely, D_8, Q_8 and abelian of type $(4,2)$, are pairwise non-isomorphic). Then $A/\langle y^2 \rangle \cong D_8$ so $G/\langle y^2 \rangle$ is not Dedekindian, contrary to the assumption. Thus, $|A| = 8$. Assume that $A \cong D_8$. Let $L < G'$ of order 2 be such that $L \not\leq A$. Then $AL/L \cong D_8$ so G/L is not Dedekindian, a contradiction. Thus, $A \cong Q_8$. Since all abelian subgroups of type $(2,2)$ of Dedekindian group G/A' are normal, we get $A \lhd G$. Write $C = C_G(A)$. Then G/C is a subgroup of a Sylow 2-subgroup of $\mathrm{Aut}(A) \cong S_4$, which is isomorphic to D_8. It follows that $G/C \cong E_4$ so $G = A * C$. Since $|G'| = 4$, C is nonabelian. Let $B < C$ be minimal nonabelian; then, by the above, $B \cong Q_8$. If $A \cap B > \{1\}$, then $A * B$ is extraspecial so it contains a subgroup isomorphic to D_8, contrary to what has been proved above. Thus, $A \cap B = \{1\}$. If $U < A' \times B' = G'$ is of order 2 and $A' \neq U \neq B'$, then AB/U is an extraspecial subgroup of order 2^5 of the Dedekindian group G/U, a final contradiction. □

Remarks. 3. For a nonabelian p-group G the following assertions are equivalent: (a) G' is the unique minimal normal subgroup of G, (b) all nonlinear irreducible characters of G are faithful. Obviously, (a) $\Rightarrow$ (b). Let us prove the reverse implication. Let $L < G$ be minimal normal; then G/L is abelian so $L = G'$ is a unique minimal normal subgroup of G.

4. Let G be a p-group with unique minimal normal subgroup G' and $\chi \in \mathrm{Irr}_1(G)$. Set $\chi(1) = p^n$. Then (a) $\chi_{Z(G)} = p^n\mu$, where $\mu \in \mathrm{Lin}(Z(G))$ is faithful, and $\mu^G = \chi(1)\chi$, (b) $|G : Z(G)| = p^{2n}$. (c) χ vanishes on $G - Z(G)$. (d) $|\mathrm{Irr}_1(G)| = \varphi(|Z(G)|)$, the number of faithful linear characters of $Z(G)$ (here $\varphi(*)$ is Euler's totient function). Indeed, the first assertion in (a) follows from Clifford's theorem. We have $Z(\chi) = Z(G)$ so (b) follows from [Isa1, Theorem 2.31]. Now the second assertion of (a) follows from (b), and (c) follows from [Isa1, Corollary 2.30]. If σ and τ are distinct faithful irreducible characters of G then $\sigma_{Z(G)} \neq \tau_{Z(G)}$, by (c), and (d) follows since all members of the set $\mathrm{Irr}_1(G)$ are faithful (note that $Z(G)$ is cyclic).

5. Let G be a p-group with unique minimal normal subgroup G', $|Z(G)| = p^z$, $\mathrm{cd}(G) = \{1, p^n\}$; then $|G| = p^{z+2n}$ (Remark 4). In that case, G has an abelian subgroup A of index p^n but has no abelian subgroups of index $< p^n$. We have $|A| = p^{z+n}$. Let $D < G$ be maximal abelian. Let $|D| = p^{z+s}$; then $D/Z(G)$ is elementary abelian of order p^s (recall that $\Phi(G) \leq Z(G)$). Since $A < G$ is abelian of maximal order, we get $s \leq n$. Let $x_1Z(G), \dots, x_sZ(G)$ be a minimal basis of the elementary abelian group $D/Z(G)$. Since $D = \cap_{i=1}^{s} C_G(x_i)$ has index $\leq p^s$, we get $|G : D| \leq p^s$ so $p^{z+2n} = |G| = |D||G : D| \leq p^{z+2s}$ so that $n \leq s$. Therefore, $n = s$. Thus, all maximal abelian subgroups of G have the same index p^n and $\mathrm{cd}(G) = \{1, p^n\}$. (Compare with the proof of Lemma 4.7(a).)

Exercise 1 (C. Hering). Suppose that a p-group G is not Dedekindian. Then there is $H < G$ such that $|G : N_G(H)| = p$ and $|H| \geq |G'|$.

Solution. By Theorem 4.11, there is a G-invariant $K < G'$ of index p such that G/K is not Dedekindian. Let $\langle xK \rangle$ be a nonnormal cyclic subgroup of G/K. As the order of the derived subgroup of G/K is p, we get $|G/K : C_{G/K}(xK)| = p$. Denoting $H = \langle x, K \rangle$, we obtain $|G : N_G(H)| = p$ since H is not normal in G. Since $|(G/K)'| = p$, we get $|H| \geq |G'|$.

Exercise 2. If G is a p-group and $G/\mho_2(G)$ is extraspecial, then either G is a 2-group of maximal class or $\mho_2(G) = \{1\}$.

Solution. It follows from $\exp(G/\mho_1(\mho_1(G))) = p^2$ that $\mho_2(G) \leq \mho_1(\mho_1(G))$. By hypothesis, $\mho_1(\mho_1(G)) = \mho_2(G)$ so $\mho_1(G) = \Phi(G)$ is cyclic. If $\mho_1(G)$ is of order p, we are done so we assume that $|\mho_1(G)| > p$. If G has no normal abelian subgroups of type (p, p), it is a 2-group of maximal class (Lemma 1.4). Now let $E_{p^2} \cong R \triangleleft G$ be such that $R \cap \mho_1(G) > \{1\}$; then $R \not\leq \Phi(G)$. Since $\mho_2(G) < \Phi(G)$, we see that $H = R\mho_2(G)$ is abelian of type $(|\mho_2(G)|, p)$. Next, H and $\Phi(G)$ are distinct of the same order. Set $\bar{G} = G/\mho_2(G)$. Then $\bar{H}$, $\Phi(\bar{G})$ are distinct normal subgroups of order p in the extraspecial group $\bar{G}$, a contradiction.

Exercise 3. Let G be a p-group. Prove that if $\Phi(G)$ is minimal nonabelian, then $\Phi(G)$ is metacyclic.

Exercise 4. Let G be a nonabelian p-group and $M \in \Gamma_1$. Suppose that, for every $x \in M - \mathrm{Z}(G)$, we have $|G : \mathrm{C}_G(x)| = p$. Study the structure of G.

Exercise 5. Let G be a p-group. If $F, H \in \Gamma_1$ are distinct, then $|\Phi(G) : \langle \Phi(H), \Phi(F)\rangle| \leq p$.

Exercise 6. Let $C \cong \mathrm{C}_8$ and G is the holomorph of C. Then $\Phi(G) \cong \mathrm{C}_4$ and G is not a nontrivial central product.

Exercise 7. If a p-group G is generated by normal cyclic subgroups, then $\mathrm{cl}(G) \leq 2$.

Exercise 8. If G is a p-group, $p > 2$, with abelian Frattini subgroup $\Phi(G) = Z_1 \times \dots \times Z_k$, where $Z_1, \dots, Z_k \lhd G$ are cyclic, then $\Phi(G) \leq \mathrm{Z}(G)$.

Exercise 9. Let G be an extraspecial group of order 2^{2m+1}. Then

(a) $\Omega_1(G) = G$, unless $G \cong \mathrm{Q}_8$.

(b) $\Omega_2^*(G) = \langle x \in G \mid o(x) = 4\rangle = G$, unless $G \cong \mathrm{D}_8$.

(c) $\mathrm{c}_1(G) = 2^{2m} + 2^m - 1$ if G is a central product of m groups isomorphic to D_8 and $\mathrm{c}_1(G) = 2^{2m} - 2^m - 1$ otherwise.

Exercise 10. Suppose that every subgroup of index p^2 in a p-group G is abelian. Then $|G'| \leq p^3$. (*Hint.* Use Exercises 1.8a and 1.69.)

Exercise 11. Let a p-group $G = E_1 \times \dots \times E_k$, where E_i is extraspecial of order p^{1+2n_i}, $i = 1, \dots, k$. If $\chi \in \mathrm{Irr}(G)$ is of degree p^n, where $n = n_1 + \dots + n_k$, then $G/\ker(\chi)$ is extraspecial of order p^{1+2n}.

Solution. Set $\bar{G} = G/\ker(\chi)$. We have, by [Isa1, Theorem 2.31], $|\bar{G} : \mathrm{Z}(\bar{G})| = \chi(1)^2 = p^{2n} = |G : \mathrm{Z}(G)|$. Since $\mathrm{Z}(G)$ is elementary abelian, it follows that $|\mathrm{Z}(G) : (\mathrm{Z}(G) \cap \ker(\chi))| = p$ and $|\mathrm{Z}(\bar{G})| = p$. Thus. $|\bar{G}| = p^{1+2n}$, $\ker(\chi) < \mathrm{Z}(G)$ and so $\bar{G}$ is extraspecial since $\bar{G}' = \mathrm{Z}(\bar{G})$ is of order p.

Exercise 12. Let G be a 2-group. If $\Phi(G)/[G, \Phi(G)]$ is cyclic, then $\Phi(G)$ is also cyclic. (*Hint.* Use Proposition 4.9.)

Exercise 13. Suppose that a group G is of order p^{2m+1} and $|G'| = p$. Then the following assertions are equivalent: (a) G is extraspecial. (b) G has no abelian subgroups of index p^{m-1}.

Exercise 14. Let G be an extraspecial group of order p^{2m+1}, $m > 1$, and let $M \in \Gamma_1$. Then $M = E\mathrm{Z}(M)$, where E is extraspecial maximal subgroup of M and $|\mathrm{Z}(M)| = p^2$.

§5

Hall's enumeration principle

As we saw, counting theorems are more fundamental than the corresponding existence theorems since they allow us to make strong conclusions on Ω-invariant subgroups of p-groups G, where Ω is a p-group of operators of G. Such theorems also help to study the structure of $\mathrm{Aut}(G)$ for some p-groups G. In this section we consider one of the main themes of this book — counting subgroups of given structure.

Let G be a group of order p^m and $n \le m$. Let $\mathrm{s}_n(G)$ and $\mathrm{c}_n(G)$ be the number of subgroups and cyclic subgroups of order p^n in G, respectively. Let $E = \mathrm{E}_{p^m}$ be the elementary abelian group of order p^m, $0 \le n \le m$. Set $\varphi_{m,n} = \mathrm{s}_n(E)$.

Exercise 1. Let $m \ge n$. (a) $\varphi_{m,n} = \frac{(p^m-1)\dots(p^m-p^{n-1})}{(p^n-1)\dots(p^n-p^{n-1})}$. (b) $\varphi_{m,n} = \varphi_{m,m-n}$. (c) $\varphi_{m+1,n} = \varphi_{m,n} + p^{m-n+1} \cdot \varphi_{m,n-1}$ for $1 \le n \le m$. (d) $\varphi_{m,m-1} = 1 + p + \dots + p^{m-1}$. (e) If $m > 2$, then $\varphi_{m,m-2} \equiv 1 + p \pmod{p^2}$.

Let $n, k \in \mathbb{N} \cup \{0\}$, $n \ge k$. By agreement, $0! = 1! = 1$ and $\binom{0}{k} = 0$. Set $\binom{n}{k} = \frac{n(n-1)\dots(n-k+1)}{k!} = \frac{n!}{(n-k)!k!}$. We have $\binom{n}{k} = \frac{n!}{(n-k)!k!} = \binom{n}{n-k}$.

Lemma 5.1 ([Hall, §1]). *Let $d \in \mathbb{N}$. Then*

$$(1)\ \prod_{i=1}^{d}(x - p^{i-1}) = (x-1)(x-p)\dots(x-p^{d-1}) = \sum_{i=0}^{d}(-1)^i \cdot p^{\binom{i}{2}}\varphi_{d,i} \cdot x^{d-i}.$$

Proof. We use induction on d. For $d = 1$, (1) is written as $x - 1 = \varphi_{1,0}x - \varphi_{1,1}$, and this is true since $\varphi_{1,0} = \varphi_{1,1} = 1$. By induction,

$$\prod_{i=0}^{d+1}(x - p^{i-1}) = \left(\sum_{i=0}^{d}(-1)^i p^{\binom{i}{2}}\varphi_{d,i}x^{d-i}\right) \cdot (x - p^d).$$

The coefficient of x^{d+1} in the above formula equals $\varphi_{d,0} = 1$. The coefficient of x^{d+1-i} for $1 \le i \le d$ is identical with $(-1)^i p^{\binom{i}{2}}\varphi_{d,i} - (-1)^{i-1}p^{\binom{i-1}{2}}\varphi_{d,i-1}p^d = (-1)^i p^{\binom{i}{2}}(\varphi_{d,i} + p^{d-i+1}\varphi_{d,i-1}) = (-1)^i p^{\binom{i}{2}}\varphi_{d+1,i}$, by Exercise 1(c). The coefficient of x^0 equals $-(-1)^d p^{\binom{d}{2}}p^d = (-1)^{d+1}p^{\binom{d+1}{2}}$. Formula (1) then holds for $d + 1$ if it holds for d. □

Setting $x = 1$ in (1), we obtain, for $d = 1, 2, \dots$, the following identity:

$$(2)\quad 1 - \varphi_{d,1} + p\varphi_{d,2} - p^3\varphi_{d,3} + \cdots + (-1)^i p^{\binom{i}{2}}\varphi_{d,i} + \cdots + (-1)^d p^{\binom{d}{2}} = 0.$$

Let G be a group of order p^m, $|G : \Phi(G)| = p^d$ and $\Gamma_i = \{H < G \mid \Phi(G) \leq H,\ |G : H| = p^i\}$; clearly, $|\Gamma_i| = \varphi_{d,d-i} = \varphi_{d,i}$, $i = 1, \dots, d$. Let $\mathfrak{M}$ be a set of proper subgroups of G. For $K \leq G$, denote by $\alpha(K)$ the number of members of the set $\mathfrak{M}$ that are subgroups of K. Obviously, $\alpha(G) = |\mathfrak{M}|$.

Theorem 5.2 (Hall's enumeration principle [Hall, §1]). *In the above notation,*

$$(3)\qquad \alpha(G) = \sum_{i=1}^{d} \sum_{H \in \Gamma_i} (-1)^{i-1} p^{\binom{i}{2}} \alpha(H)$$

so $\alpha(G) \equiv \sum_{H \in \Gamma_1} \alpha(H) \pmod{p}$.

Proof. Let $A \in \mathfrak{M}$ and let $D \in \Gamma_t$ be the intersection of subgroups of G containing $A\Phi(G)$ ($t \geq 1$). Then the contribution of A in the right-hand side of (3) is $\sum_{i=1}^{t}(-1)^{i-1} p^{\binom{i}{2}} \varphi_{t,i} = 1$, by (2), and this equals the contribution of A in $\alpha(G)$. □

Let us apply (3) to $G = \mathrm{E}_{p^m}$. We get, for $n \leq m$ and $i = 1, \dots, m - n$:

$$\varphi_{m,n} = \varphi_{m,1}\varphi_{m-1,n} - p\varphi_{m,2}\varphi_{m-2,n} + \cdots + (-1)^{i-1} p^{\binom{i}{2}} \varphi_{m,i}\varphi_{m-i,n} + \cdots.$$

Theorem 5.3 (Kulakoff [Kul]). *Let G be a noncyclic group of order p^m, $p > 2$ and $1 \leq n < m$. Then* $\mathrm{s}_n(G) \equiv 1 + p \pmod{p^2}$.

Proof. We use induction on m. In view of Theorem 1.10(a), one may assume that $n > 1$. Since $|\Gamma_1| = \varphi_{d,1} \equiv 1 + p \pmod{p^2}$, we may assume that $n < m - 1$. Then by (3),

$$\mathrm{s}_n(G) \equiv \sum_{H \in \Gamma_1} \mathrm{s}_n(H) - p \sum_{H \in \Gamma_2} \mathrm{s}_n(H) \pmod{p^2}.$$

Assume that $H \in \Gamma_1$ is cyclic. Then $\Gamma_2 = \{\Phi(G)\}$, $\Gamma_1 = \{H = H_1, \dots, H_p, H_{p+1}\}$, where $H_1, \dots, H_p$ are cyclic and H_{p+1} is noncyclic (Theorem 1.2). We have $\mathrm{s}_n(\Phi(G)) = \mathrm{s}_n(H_i) = 1$, $i = 1, \dots, p$, and $\mathrm{s}_n(H_{p+1}) = 1 + p$. Therefore, by (3), $\mathrm{s}_n(G) = p \cdot 1 + 1 \cdot (p + 1) - p \cdot 1 = 1 + p$. Now suppose that all $H \in \Gamma_1$ are noncyclic. Then $\mathrm{s}_n(H) \equiv 1 + p \pmod{p^2}$ ($H \in \Gamma_1$), by induction, and $p\mathrm{s}_n(H) \equiv p \pmod{p^2}$ ($H \in \Gamma_2$), by Exercise 1.9. Therefore, by (3),

$$\mathrm{s}_n(G) \equiv (p + 1)|\Gamma_1| - p|\Gamma_2| \equiv (1 + p)^2 - p \equiv 1 + p \pmod{p^2}.$$ □

Exercise 2. (a) Suppose that a 2-group G has a subgroup of maximal class and index 2. If G is not of maximal class, then $\mathrm{d}(G) = 3$.

(b) Let G be a group of order 2^m, $2 < n < m$, and let $\mathfrak{M}$ be the set of all subgroups of G of maximal class and order 2^n. Then $\alpha(G)(= |\mathfrak{M}|)$ is even.

Solution. (a) By Theorem 1.2, G has no cyclic subgroups of index 2. Let $\mathrm{d}(G) = 2$. Then $\Phi(G)$ is cyclic ($\Phi(G)$ is not of maximal class and $\Phi(G) \not\cong \mathrm{E}_4$, by Proposition 4.9). Since $\Phi(G) = \mho_1(G)$, G has a cyclic subgroup of index 2, a contradiction.

(b) We use induction on m. Let $n = m - 1$. One may assume that G has no cyclic subgroups of index 2 (otherwise, $\alpha(G) \in \{0, 2\}$, by Theorem 1.2). Let $H \in \mathfrak{M}$; then $\mathrm{d}(G) = 3$, by (a). By Exercises 1.11–1.13, $\mathrm{c}_1(H) \equiv 1 \pmod 4$. If $H \in \Gamma_1 - \mathfrak{M}$, then by Theorem 1.17(a), $\mathrm{c}_1(H) \equiv 3 \pmod 4$. By Exercise 1.9, Exercise 1(e) and (a), we have $2\sum_{F\in\Gamma_2} \mathrm{c}_1(F) \equiv 2\varphi_{3,1} \equiv 2 \pmod 4$ so, denoting $\alpha(G) = k$, we obtain

$$\mathrm{c}_1(G) \equiv k + (|\Gamma_1| - k)\cdot 3 - 2 \equiv k + (\varphi_{3,1} - k)\cdot 3 - 2 \equiv 3 - 2k \pmod 4,$$

by Exercise 1(d) and (3). Since $\mathrm{c}_1(G) \equiv 3 \pmod 4$, it follows that 4 divides $2k$, i.e., k is even. If $n < m - 1$, the result follows by induction and (3).

If G is a p-group of maximal class and $K \lhd G$ is of index $> p$, then $K \le \Phi(G)$ since G has only one normal subgroup of index p^2.

Theorem 5.4 ([Ber1]). *Let G be a 2-group of order 2^m, $2 < n < m$ and let $\alpha(M)$ be the number of subgroups of maximal class and order 2^n in $M \le G$. Then 4 divides $\alpha(G)$, unless G is of maximal class with $n = m - 1$ (in the last case, $\alpha(G) = 2$).*

Proof. We use induction on m. If G is of maximal class, then, using Theorems 1.2 and 5.2, we get $\alpha(G) = 2^{m-n}$. Assume that G is not of maximal class. Assume that $H < G$ is of maximal class and order 2^n.

(i) Let $|G| = 2^4$; then $n = 3$ and $\mathrm{cl}(G) = 2$. In that case, G has exactly three abelian subgroups of index 2 so $\alpha(G) = |\Gamma_1| - 3 = 4$. Now we assume that $m > 4$.

(ii) Let $n = m - 1$. Set $D = \mathrm{K}_3(H)$; then D is cyclic. We have $|G : D| = 2^4$ so $\{1\} < D \lhd G$. By Lemma 1.4, there is $R \lhd G$ of type $(2, 2)$. Next, $R \not\le H$ ($|H| > 2^3$ is of maximal class) so $G/(H \cap R) = (H/(H \cap R)) \times (R/(H \cap R))$; in particular, $\mathrm{d}(G) = 3$. Therefore, G/D has exactly three abelian maximal subgroups $A/D, B/D, C/D$, by the previous paragraph so A, B, C are not of maximal class. Let $F/D < G/D$ be nonabelian of order 2^3. It suffices to show that F is of maximal class (then 4 divides $\alpha(G)$). Assume that this is not the case; then, by Lemma 1.4, there is a G-invariant $L < F$ of type $(2, 2)$. As D is cyclic, $L \not\le D$. Since $H \cap L > \{1\}$ we get $D \cap L > \{1\}$ so $|DL| = 2|D|$. Then we have $LD/D = \Phi(F/D) < H/D$, and so $L < H$, a contradiction. Thus, F is of maximal class, by Lemma 1.4, so $\alpha(G) = |\Gamma_1| - 3 = 7 - 3 = 4$.

(iii) Now suppose that $n < m - 1$ and prove that $\sum_{H\in\Gamma_1} \alpha(H) \equiv 0 \pmod 4$. If $H \in \Gamma_1$ is not of maximal class, then by induction, 4 divides $\alpha(H)$. Therefore, it suffices to show that $\sum_{H\in\Gamma_1'} \alpha(H) \equiv 0 \pmod 4$, where Γ_1' is the set of all members of maximal class in the set Γ_1. By (ii), 4 divides $|\Gamma_1'|$. Since $\alpha(H) = 2^{m-1-n}$ is even and independent of $H \in \Gamma_1'$, we are done.

In view of (3), it remains to prove that $S = \sum_{H\in\Gamma_2} \alpha(H)$ is even. Suppose that $n < m - 2$. Then 2 divides S, by Exercise 2(b) and induction. Therefore, assume in what follows, that $n = m - 2$. Let Γ_2' be the set of all elements of maximal class in Γ_2. It suffices to show that 2 divides $|\Gamma_2'|$. Since 2 divides $\alpha(G)$ (Exercise 2(b)), we have to show that every normal subgroup of index 4 in G, say H, that is of maximal class, is a member of the set Γ_2'. Assume that $H \notin \Gamma_2'$. Then G/H is cyclic and $\mathrm{d}(G) \le 3$ since $\mathrm{d}(H) = 2$. Assume that $\mathrm{d}(G) = 2$. Then $\Gamma_2 = \{\Phi(G)\}$ and $\Phi(G)$ is not of maximal class (Proposition 1.3). In that case, $\alpha(G) = \sum_{K\in\Gamma_1} \alpha(K) \equiv 0 \pmod 4$ by (ii). Now let $\mathrm{d}(G) = 3$. In that case, $G' < \Phi(G)$ since $\exp(G/G') > 2$. Since $|G : H'| = 2^4$, we get $H' = G'$ so that G/G' is abelian of type $(4, 2, 2)$. Take $F \in \Gamma_2'$ (if F does not exist, we are done). We also have $F' = G'$ (compare indices!). Let $F/G' < M/G' < G/G'$, where M/G' is abelian of type $(4, 2)$. Since $G' = F' \le M'$, we get $G' = M'$. Let L be a G-invariant subgroup of index 2 in $G' = M'$. Then, if Z_1/G' and Z_2/G' are two distinct cyclic subgroups of M/G' of order 4, then Z_1/L and Z_2/L are two distinct abelian subgroups of M/L. Then $(Z_1/L) \cap (Z_2/L) = \mathrm{Z}(M/L)$ since M/L is nonabelian, hence M/L is minimal nonabelian since $\mathrm{d}(M) = 2$. This is a contradiction since M/L contains a nonabelian subgroup F/L of index 2. □

Theorem 5.5 ([Ber1]). *Let G be a 2-group of order 2^m and $1 \le n < m$. Then $\mathrm{s}_n(G) \equiv 3 \pmod 4$, unless G is either cyclic or of maximal class.*

Proof. We use induction on m. One may assume that $n < m - 1$. In view of Theorem 1.2 and Theorem 1.17(a), one may also assume that G has no cyclic subgroups of index 2 and $n > 1$. By (3), we have to prove that

$$\mathrm{s}_n(G) \equiv \sum_{H\in\Gamma_1} \mathrm{s}_n(H) - 2\sum_{F\in\Gamma_2} \mathrm{s}_n(F) \equiv 3 \pmod 4. \tag{4}$$

By Sylow we have $2\sum_{F\in\Gamma_2} \mathrm{s}_n(F) \equiv 2|\Gamma_2| \equiv 2 \pmod 4$. It remains to prove that $\sum_{H\in\Gamma_1} \mathrm{s}_n(H) \equiv 1 \pmod 4$. If $n = m - 2$, the last sum is $\equiv 3|\Gamma_1| \equiv 3 \cdot 3 \equiv 1 \pmod 4$ (Γ_1 has no cyclic members). Therefore, one may assume that $n < m - 2$. Let Γ_1' be as in Theorem 5.4. If $H \in \Gamma_1'$, then $\mathrm{s}_n(H) \equiv 1 \pmod 4$, by Exercises 1.11–1.13. Since 4 divides $|\Gamma_1'|$ (Theorem 5.4), we have $\sum_{H\in\Gamma_1'} \alpha(H) \equiv 0 \pmod 4$. Thus, it suffices to prove that $\sum_{H\in\Gamma_1-\Gamma_1'} \mathrm{s}_n(H) \equiv 1 \pmod 4$. We have $|\Gamma_1 - \Gamma_1'| \equiv 3 \pmod 4$ (Exercise 1(d) and Theorem 5.4) and, for $H \in \Gamma_1 - \Gamma_1'$, we have $\mathrm{s}_n(H) \equiv 3 \pmod 4$, by induction. Thus, $\sum_{H\in\Gamma_1-\Gamma_1'} \mathrm{s}_n(H) \equiv 3 \cdot 3 \equiv 1 \pmod 4$. □

Theorem 5.6. *(a) (Mann; see also [Ber20].) Let $S \in \mathrm{Syl}_p(G)$ be of order p^m, $0 < n < m$. Then $\mathrm{s}_n(G) \equiv \mathrm{s}_n(S) \pmod{p^2}$.*

(b) If $k \in \mathbb{N}$, $S \in \mathrm{Syl}_p(G)$, $|S| > p^k$, then the following assertions are equivalent:

(b1) $s_k(G) \not\equiv 1 + p \pmod{p^2}$;

(b2) $s_k(G) \equiv 1 \pmod{p^2}$;

(b3) *S is either cyclic or a 2-group of maximal class with* $|S| > 2^{k+1}$.

Proof. (a) For a proof, see [Ber20]. (b) follows from (a). □

For $p > 2$, Theorem 5.6(b) was proved in [Hal2, Theorem 4.6]. P. Deligne [Del] has proved that if $S \in \mathrm{Syl}_p(G)$ is noncyclic of order p^s, then the number of subgroups of order p^{s-1} in G is $\equiv 1 + p \pmod{p^2}$, a partial case of Theorem 5.6(a).

Lemma 5.7. *Let G be a nonabelian 2-generator p-group of exponent p. Then* $\mathrm{K}_3(G)$ *is a unique normal subgroup of G of index p^3.*

Proof. We have $G' = \Phi(G)$ so $|G : G'| = p^2$. Set $K = \mathrm{K}_3(G)$; then $G'/K = \mathrm{Z}(G/K)$ and $\mathrm{d}(G) = 2$ so G/K is minimal nonabelian hence $p^3 = |G : K| = |G : G'||G' : K|$ so $|G' : K| = p$. Let $M \lhd G$ be of index p^3. Since $\mathrm{cl}(G/M) = 2$, we get $M = K$. □

Theorem 5.8 ([Ber13]). *Let G be a group of order $p^m > p^3$ and exponent p, $2 < n < m$. Let $\mathfrak{M}$ denote the set of all 2-generator subgroups of order p^n in G and $\alpha(K)$ the number of elements of the set $\mathfrak{M}$ contained in $K \leq G$.*

(a) If $n = m - 1$, then $\alpha(G) \in \{0, p, p^2\}$.

(b) p divides $\alpha(G)$.

(c) Suppose that $\mathrm{d}(G) > 2$ *and G contains a subgroup H of index p such that* $\mathrm{d}(H) = 2$. *Then G has exactly $p + 1$ maximal subgroups not having two generators and their intersection has index p^2 in G.*

Proof. Assume that $\mathfrak{M} \neq \varnothing$. All elements of the set $\mathfrak{M}$ are nonabelian.

(a) Take $H \in \mathfrak{M}$ and set $D = \mathrm{K}_3(H)$. Then $D \lhd G$, H/D is nonabelian of order p^3 (Lemma 5.7). Since $\mathrm{d}(H) = 2$, we get $\mathrm{d}(G) \leq 3$. Let $H_1 \in \mathfrak{M} - \{H\}$ and $D_1 = \mathrm{K}_3(H_1)$. Then $D_1 \leq \Phi(H_1) \leq \Phi(G) < H$ and $|H : D_1| = p^3 = |H : D|$ so, by Lemma 5.7, $D_1 = D$. Let $H_2 \in \Gamma_1 - \mathfrak{M}$, i.e., $|H_2 : \Phi(H_2)| \geq p^3$. Since $\Phi(H_2) \leq \Phi(G) < H$, we get $\Phi(H_2) \leq D$, by Lemma 5.7. Then $\mathrm{d}(H_2/D) = 3$. Thus, $|\mathfrak{M}|$ equals the number of nonabelian subgroups of order p^3 in G/D, and we are done, by Exercise 1.6(a).

(b) Let $n < m - 1$. Then $\alpha(G) \equiv \sum_{H \in \Gamma_1} \alpha(H) \equiv 0 \pmod{p}$, by Theorem 5.2 and induction. (c) To prove, repeat, with small modifications, the proof of (a). □

Theorem 5.9 ([Ber13]). *Let G be a p-group of exponent p and order p^m, $1 < n < m - 1$. Then* $s_n(G) \equiv 1 + p + 2p^2 \pmod{p^3}$.

Proof. We proceed by induction on m. By (3), we have to prove that $s_n(G) \equiv 1 + p + 2p^2$. By Hall's enumeration principle,

$$s_n(G) \equiv \sum_{H \in \Gamma_1} s_n(H) - p \sum_{F \in \Gamma_2} s_n(F). \tag{5}$$

(i) Let $n = m - 2$. In that case, if $F \in \Gamma_2$ then $s_n(F) = 1$.

(i1) Suppose that $d(G) = 2$. Then $\Gamma_2 = \{\Phi(G)\}$ and $s_n(\Phi(G)) = 1$. Suppose that the set Γ_1 has a member generated by two elements. By Theorem 5.8(a), $\Gamma_1 = \{H_1, \dots, H_p, H_{p+1}\}$, $d(H_i) = 2$, $i = 1, \dots, p$, $d(H_{p+1}) > 2$. It follows that $s_n(H_i) = 1 + p$ $(i = 1, \dots, p)$, $s_n(H_{p+1}) \equiv 1 + p + p^2 \pmod{p^3}$ and $ps_n(\Phi(G)) = p$. Therefore, by (5),

$$s_n(G) \equiv p \cdot (p+1) + 1 \cdot (1 + p + p^2) - p \cdot 1 \equiv 1 + p + 2p^2 \pmod{p^3}.$$

Now suppose that the set Γ_1 has no members generated by two elements. Then $s_n(G) \equiv (1+p)(1+p+p^2) - p \equiv 1 + p + 2p^2 \pmod{p^3}$.

(i2) Let $d(G) > 2$. Suppose that the set Γ_1 has a member generated by two elements. Then the number of such members is p^2, by Theorem 5.8(c). In that case, $d(G) = 3$, so that $|\Gamma_1| = |\Gamma_2| = 1 + p + p^2$. By (5),

$$s_n(G) \equiv (1+p)p^2 + (1+p+p^2)(1+p) - p(1+p) \equiv 1 + p + 2p^2 \pmod{p^3}.$$

Now suppose that the set Γ_1 has no members generated by two elements. Since $p|\Gamma_2| \equiv p + p^2 \pmod{p^3}$, we have, by (5),

$$s_n(G) \equiv (1+p+p^2)(1+p+p^2) - (p+p^2) \equiv 1 + p + 2p^2 \pmod{p^3}.$$

(ii) Let $n < m - 2$. Then, if $H \in \Gamma_1$, then, by induction, $s_n(H) \equiv 1 + p + 2p^2 \pmod{p^3}$. If $F \in \Gamma_2$, then, by Kulakoff's theorem and Theorem 5.5, $ps_n(F) \equiv p + p^2 \pmod{p^3}$. If $d(G) = 2$, then

$$s_n(G) \equiv (1+p)(1+p+2p^2) - (p+p^2) \equiv 1 + p + 2p^2 \pmod{p^3},$$

by (5). Let $d(G) > 2$. Then $|\Gamma_2| \equiv 1 + p + kp^2 \pmod{p^3}$ for some $k \in \mathbb{N} \cup \{0\}$. If $H \in \Gamma_2$, then $ps_n(H) = p + p^2 \pmod{p^3}$. Thus, $s_n(G) \equiv (1+p+p^2)(1+p+2p^2) - (p+p^2)(1+p+kp^2) \equiv 1 + p + 2p^2 \pmod{p^3}$. □

Exercise 3. Let G be a group of order p^m and exponent p, $1 < n < m - 1$.

(a) If $s_n(G) = 1 + p + 2p^2$, then $n = m - 2$, $d(G) = 2$. and $|G/K_4(G)| = p^4$.

(b) If $n < m - 2$, then $s_n(G) \geq 1 + p + 2p^2 + p^3$.

(c) Study the structure of G satisfying $s_n(G) = 1 + p + 2p^2 + p^3$.

Exercise 4. Let G be a noncyclic group of order p^m, $m > 2$. Then $s_{m-2}(G) = 1 + p$ if and only if G is either abelian of type (p^{m-1}, p) or $G \cong M_{p^m}$.

Exercise 5. Let G be a noncyclic group of order p^m, $1 < n < m-1$. If $s_n(G) = 1+p$, then $|\Omega_1(G)| = p^2$ and $G/\Omega_1(G)$ is either cyclic or $n = 2$ and $G/\Omega_1(G) \cong Q_{2^{m-2}}$.

Exercise 6. Let G, m, n be as in Exercise 5. Describe the structure of G satisfying $s_n(G) \le 1 + p + p^2$.

Let $\mathcal{C}(G)$ be the number of maximal chains of subgroups of a p-group G. Then $\mathcal{C}(G) = \sum_{H \in \Gamma_1} \mathcal{C}(H)$. We recommend to the reader to write out the formula for the number of chief series of G (of course, this must be another formula since normality is not transitive in general).

Exercise 7. Let G be a noncyclic group of order p^m. Then $\mathcal{C}(G) \equiv 1 + (m-1)p \pmod{p^2}$, unless G is a 2-group of maximal class (then $\mathcal{C}(G) \equiv 3 \pmod 4$). If $\mathcal{C}(G) = 1 + (m-1)p$, then G is either abelian of type (p, p^{m-1}) or $G \cong M_{p^m}$. Next, $\mathcal{C}(D_{2^m}) = 2^m - 1$, $\mathcal{C}(Q_{2^m}) = 2^{m-1} - 1$, $\mathcal{C}(SD_{2^m}) = 2^{m-1} + 2^{m-2} - 1$.

Exercise 8. Find the number of maximal chains in: (a) $ES(m, p)$; (b) in $C_{p^n} \operatorname{wr} C_p$ and $C_p \operatorname{wr} C_{p^n}$; (c) in minimal nonabelian p-groups; (d) $A \times B$, where A and B are p-groups with known numbers of maximal chains.

Exercise 9. Find $\mathcal{C}(G)$, where G is an abelian p-group of given type.

Exercise 10. Let G be a metacyclic p-group of order p^m, $p > 2, m > 4$. Suppose that G has no cyclic subgroups of index p. Check that $s_{m-2}(G) = 1 + p + p^2 = s_2(G)$.

Theorem 5.10. *Let $S \in \mathrm{Syl}_p(G)$. Suppose that every maximal subgroup of S is neither cyclic nor a 2-group of maximal class. If $n > 1$, then p divides $c_n(G)$.*

Using characters, one proves (see §37)

Theorem 5.11. *If G is a p-group such that $|G : \Phi(G)| \ge p^{2k+1}$, then $c_1(G) \equiv 1 + p + \cdots + p^k \equiv \varphi_{k+1,1} \pmod{p^{k+1}}$.*

Theorem 5.12 ([Hal1, Theorem 1.61]). *Let G be a group of order p^m, $d = d(G)$, $0 \le n \le d$. Let $\mathfrak{M}$ be the set of all subgroups of index p^n in G and $\alpha(H)$ the number of elements of the set $\mathfrak{M}$ contained in $H \le G$. Then $\alpha(G) \equiv \varphi_{d,n} \pmod{p^{d-n+1}}$.*

Theorem 5.13 (see §45). *Let G be a p-group, $p > 2$. Suppose that G has no cyclic subgroups of index p. Let $\mathfrak{M}$ be the set of subgroups H of G of order p^n, $m > n \ge 4$, such that H has no cyclic subgroups of index p. Then $|\mathfrak{M}| \equiv 1 \pmod p$.*

Theorem 5.14. *Let G be a group of order p^m. If A is a subgroup of G of order p^k and $k < n < m$, then the number of subgroups of G of order p^n containing A is $\equiv 1 \pmod p$.*

To prove, use the enumeration principle.

Let G be a group of order p^m, $R \trianglelefteq G$ of order p and $k < m$. Let $\mathfrak{M}$ be a set of subgroups of order p^k in G. Set $\mathfrak{M}^+ = \{H \in \mathfrak{M} \mid R \leq H\}$. If $F \in \mathfrak{M}$, then either $R \leq F$ (in that case, obviously, $F \in \mathfrak{M}^+$) or $RF = R \times F$. Next, let $\mathfrak{X}$ be the set of all those subgroups of order p^{k+1} in G that contain R (obviously, $|\mathfrak{X}| = \mathrm{s}_k(G/R)$). For $H \leq G$, let $\alpha(H)$ be the number of those members of the set $\mathfrak{M} - \mathfrak{M}^+$ that are contained in H. If $H \in \mathfrak{X}$, then either $\alpha(H) = 0$ or $H = R \times H_1$ for each $H_1 \in \mathfrak{M} - \mathfrak{M}^+$ with $H_1 < H$. We have

$$(6) \qquad |\mathfrak{M}| = |\mathfrak{M}^+| + \sum_{H \in \mathfrak{X}} \alpha(H).$$

Indeed, let $F \in \mathfrak{M}$. If $F \in \mathfrak{M}^+$, the contribution of F in both sides of (6) equals 1. Now let $F \notin \mathfrak{M}^+$. Then $RF = R \times F$ is the unique member of $\mathfrak{X}$ containing F, and again the contribution of F in both sides of (6) equals 1, so (6) is true.

To illustrate the approach based on identity (6), we prove the following lemma which is a partial case of Theorem 1.30.

Lemma 5.15. *Suppose that a p-group G is neither cyclic nor a 2-group of maximal class. Let $R < G$ be of order p. If $k > 1$, then the number of cyclic subgroups of order p^k in G containing R, is a multiple of p.*

Proof. Let $\mathfrak{M}$ be the set of all cyclic subgroups of order p^k in G and $\mathfrak{M}^+ = \{H \in \mathfrak{M} \mid R < H\}$. Then $\mathrm{C}_G(R)$, that contains all members of the set $\mathfrak{M}^+$, is neither cyclic nor a 2-group of maximal class. Therefore, we may assume that $\mathrm{C}_G(R) = G$. Let $\mathfrak{X}$ be as in (6). Since p divides $|\mathfrak{M}|$ (Theorems 1.10(b) and 1.17(b)), it suffices to show, in view of (6), that $\sum_{H \in \mathfrak{X}} \alpha(H) \equiv 0 \pmod{p}$. Let $H \in \mathfrak{X}$. If H/R is not cyclic, then $\alpha(H) = 0$. Now let H/R be cyclic. Then H is either cyclic or abelian of type (p^k, p) so $\alpha(H) \in \{0, p\}$. □

Theorem 5.16 (= Theorem 5.3 + Theorem 5.5). *Let a p-group G be neither cyclic nor a 2-group of maximal class, $k \in \mathbb{N}$ and $p^k < p^m = |G|$. Then $\mathrm{s}_k(G) \equiv 1 + p \pmod{p^2}$.*

Proof. We proceed by induction on $|G|$. The theorem is true for $k = 1$ (Theorems 1.10(a) and 1.17(a)) and $k = m - 1$. So we assume that $1 < k < m - 1$. In view of Theorem 1.2, one may assume that G has no cyclic subgroups of index p.

Let, in (6), $\mathfrak{M}$ be the set of all subgroups of order p^k in G. Let $R \trianglelefteq G$ be of order p and $\mathfrak{M}^+$, $\mathfrak{X}$ and $\alpha(H)$ be as in (6).

(i) Let G/R be a 2-group of maximal class. We have $|\mathfrak{M}^+| = \mathrm{s}_{k-1}(G/R) \equiv 1 \pmod{4}$. Let $H \in \mathfrak{X}$; then H is noncyclic since G has no cyclic subgroups of index 2. If H/R is cyclic (the number of such H in $\mathfrak{X}$ is odd), then $\alpha(H) = 2$. Then the contribution in $\mathrm{s}_k(G)$ of all $H \in \mathfrak{X}$ such that H/R is cyclic, is $2\mathrm{c}_k(G/R) \equiv 2 \pmod{4}$. Now assume that H/R is noncyclic (the number of such H in $\mathfrak{X}$ is even). If

$R \le \Phi(H)$, then $\alpha(H) = 0$. If $R \not\le \Phi(H)$, then $H = R \times H_1$, where $\mathrm{d}(H_1) = 2$, and so $\alpha(H) = (1+2+2^2)-(1+2) = 4$ (here $1+2+2^2$ is the number of maximal subgroups of H and $1+2$ is the number of maximal subgroups of H/R). It follows that $\mathrm{s}_k(G) \equiv |\mathfrak{M}^+| + 2 \equiv 1+2 \equiv 3 \pmod 4$.

(ii) Now suppose that G/R is neither cyclic nor a 2-group of maximal class. Then, by induction, $|\mathfrak{M}^+| = \mathrm{s}_{k-1}(G/R) \equiv 1+p \pmod{p^2}$. The number of $H \in \mathfrak{X}$ such that H/R is cyclic, is a multiple of p by Theorems 1.10(b) and 1.17(b). If $H \in \mathfrak{X}$ is cyclic, then $\alpha(H) = 0$. Let $H \in \mathfrak{X}$ be such that H/R is cyclic and H is noncyclic; then H is abelian of type (p^k, p). In that case, $\alpha(H) = \mathrm{c}_k(H) = p$, and, since the number of such H is a multiple of p, their contribution in $\mathrm{s}_k(G)$ is a multiple of p^2. Now let $H \in \mathfrak{X}$ be such that H/R is noncyclic. If $R \le \Phi(H)$, then $\alpha(H) = 0$. Now suppose that $R \not\le \Phi(H)$; then $H = R \times H_1$. If $\mathrm{d}(H/R) = d (\ge 2)$ so $\mathrm{d}(H) = d+1$, then

$$\alpha(H) = (1+p+\dots+p^d) - (1+p+\dots+p^{d-1}) = p^d \equiv 0 \pmod{p^2}$$

so $\sum_{H\in\mathfrak{X}} \alpha(H) \equiv 0 \pmod{p^2}$ and $\mathrm{s}_k(G) \equiv |\mathfrak{M}^+| \equiv 1+p \pmod{p^2}$. □

Theorem 5.17 ([Fan]). *Let G be a group of order p^m and $1 \le k < m$, k is fixed. If $\mathrm{s}_k(G) \ge \mathrm{s}_k(E_{p^m})$, then one of the following holds: (a) $k = 1$ and $\exp(G) = p$, (b) $k > 1$ and $G \cong E_{p^m}$.*

Proof. We use induction on m. Let $\mathfrak{M}$ be the set of all subgroups of order p^k in G. Given $H \le G$, write $\alpha(H) = |\{L \in \mathfrak{M} \mid L \le H\}|$.

(a) Let $k = 1$. Then, $\mathrm{s}_1(G) \le \frac{p^m-1}{p-1} = \varphi_{m,1} = \mathrm{s}_1(\mathrm{E}_{p^m})$ so, by hypothesis, $\mathrm{s}_1(G) = \varphi_{m,1} = \frac{p^m-1}{p-1}$. In that case, subgroups of G of order p contain together exactly $\frac{p^m-1}{p-1} \cdot (p-1) + 1 = p^m = |G|$ distinct elements so $\exp(G) = p$, proving (a).

(b) Now we assume that $k > 1$. If $k = m-1$, then

$$\mathrm{s}_{m-1}(G) = |\Gamma_1| = 1+p+\dots+p^{\mathrm{d}(G)-1} \ge |\Gamma_1(\mathrm{E}_{p^m})| = 1+p+\dots+p^{m-1}$$

and we get $\mathrm{d}(G) = m$ so $G \cong \mathrm{E}_{p^m}$.

Next we assume that $k < m-1$. Let $R \lhd G$ be of order p. We retain the notation used in the paragraph containing (6). Then $|\mathfrak{M}^+| = \mathrm{s}_{k-1}(G/R) \le \varphi_{m-1,k-1}$ with equality if $k = 2$ and $\exp(G/R) = p$, by (a), or $k > 2$ and $G/R \cong \mathrm{E}_{p^{m-1}}$, by induction. Next, we have $|\mathfrak{X}| = \mathrm{s}_k(G/R) \le \varphi_{m-1,k}$ with equality if and only if $G/R \cong \mathrm{E}_{p^{m-1}}$, by induction. Take $H \in \mathfrak{X}$. If $R \le \Phi(H)$, then $\alpha(H) = 0$. Now let $R \not\le \Phi(H)$. Set $\mathrm{d}(H/R) = d$. Then $H = R \times H_1$ and $\alpha(H)$ equals the number of maximal subgroups of H minus the number of maximal subgroups of $H/R \cong H_1$, i.e., $\alpha(H) = (1+p+\dots+p^d) - (1+p+\dots+p^{d-1}) = p^d \le p^k$ with equality if and only if $H \cong \mathrm{E}_{p^{k+1}}$. Substituting the obtained results, we get $\mathrm{s}_k(G) \le$

$|\mathfrak{M}^+| + |\mathfrak{X}| \cdot p^k \leq \varphi_{m-1,k-1} + \varphi_{m-1,k} \cdot p^k$ with strict inequality if $R \leq \Phi(H)$ for some $H \in \mathfrak{X}$. By (6), applied to $G = \mathrm{E}_{p^m}$, we get

$$\begin{aligned}\varphi_{m,k} &= \mathrm{s}_k(\mathrm{E}_{p^m}) = \mathrm{s}_{k-1}(\mathrm{E}_{p^{m-1}}) + \mathrm{s}_k(\mathrm{E}_{p^{m-1}})[\mathrm{s}_k(\mathrm{E}_{p^{k+1}}) - \mathrm{s}_{k-1}(\mathrm{E}_{p^k})] \\ &= \varphi_{m-1,k-1} + p^k \cdot \varphi_{m-1,k}\end{aligned}$$

so, by hypothesis, $\varphi_{m,k} = \mathrm{s}_k(G)$. Then we must have, by the above argument, $\alpha(H) = p^k$ so $H \cong \mathrm{E}_{p^{k+1}}$ for all $H \in \mathfrak{X}$, as claimed. Indeed, since $\alpha(H) = p^k$, we have $\mathrm{d}(H/R) = k$ so $H/R \cong \mathrm{E}_{p^k}$. Since $\alpha(H) > 0$, R is a direct factor of H so $H \cong \mathrm{E}_{p^{k+1}}$, as was to be shown. It follows that $G/R \cong \mathrm{E}_{p^{m-1}}$ and every member of the set $\mathfrak{X}$ is isomorphic to $\mathrm{E}_{p^{k+1}}$. This is true for every choice of R. Assume that $G \not\cong \mathrm{E}_{p^m}$. Then R is a unique normal subgroup of order p in G and $\exp(G) = p$. It follows that G is extraspecial without nonabelian subgroups of order $p^3 (< p^m)$. Thus, G has no minimal nonabelian subgroups (Exercise 1.8a), a contradiction. We get $G \cong \mathrm{E}_{p^m}$. □

§6

q'-automorphisms of q-groups

1°. We assume that the reader is familiar with classical theorem of Maschke and its extension to abelian groups [BZ, Theorems 1.9 and 1.9′]. We prove Theorem 6.1, a consequence of Maschke's theorem. Theorem 6.1 is a fairly deep result on finite abelian groups with a coprime operator group. We use semidirect products instead of actions.

A p-group is said to be *homocyclic* if it is of type $(p^n, p^n, \dots, p^n)$, $n > 0$. An abelian p-group of exponent $p^e > p$ is homocyclic if and only if one of the following holds: (i) $\Phi(G)$ is homocyclic with $\mathrm{d}(\Phi(G)) = \mathrm{d}(G)$, (ii) $\mho_{e-1}(G) = \Omega_1(G)$. An abelian p-group is a direct product of homocyclic subgroups of pairwise distinct exponents.

In what follows, p and q are distinct primes and b is the least positive integer such that $q^b \equiv 1 \pmod{p}$. The rank of a p-group is its minimal number of generators.

Exercise 1. Let $Q = C_1 \times \dots \times C_d$, where $C_1, \dots, C_d$ are cyclic groups of orders $q^{m_1}, \dots, q^{m_d}$, respectively, and let $m = m_1 = \dots = m_s < m_{s+1} \le \dots \le m_d = \mu$. Set

$$F = \Omega_m(Q), \quad M = \mho_m(Q), \quad Q_1 = C_1 \times \dots \times C_s, \quad Q_2 = C_{s+1} \times \dots \times C_d.$$

Then F is homocyclic of order q^{md} and exponent q^m and M is abelian of type $(q^{m_{s+1}-m}, \dots, q^{m_d-m})$ and

(a) Every cyclic subgroup of order q^μ is a direct factor of Q.

(b) $M = \mho_m(Q_2)$.

(c) $F \cong Q/M$ is homocyclic of rank d and exponent q^m.

(d) If $Q_1 \cong L \le Q$ and $L \cap M = \{1\}$, then $Q = L \times Q_2$.

(e) If $m = \mu$ (i.e., Q is homocyclic), then any homocyclic subgroup of exponent q^m is a direct factor of Q.

(f) $\mho_{\mu-1}(Q) = \Omega_1(Q)$ if and only if Q is homocyclic.

Theorem 6.1 ([Ber22, Theorem 1]). *Suppose that Q is a normal abelian Sylow q-subgroup of a group G, R a minimal normal q-subgroup of G. Then there exists a G-invariant homocyclic q-subgroup S such that (a) $\Omega_1(S) = R$. (b) $Q = S \times S_1$, where $S_1 \lhd G$. (c) $\exp(S) = q^{e(R)}$ depends only on R. (d) If $R \le U \le Q$ and $\Omega_1(U) = R$, then $\exp(U) \le q^{e(R)} = \exp(S)$.*

Proof. We retain the notation of Exercise 1. Clearly, $R \leq Q$. In view of Maschke's theorem, we may assume that $\exp(Q) > q$.

We use induction on $|G|$. Let A be a q'-Hall subgroup of G (Schur–Zassenhaus). Since Q is abelian and $G = A \cdot Q$, a semidirect product, we have

(i) Any A-invariant subgroup of Q is G-invariant.

(ii) If $T \trianglelefteq G$ is a q-subgroup and $\Omega_1(T) = R$, then T is homocyclic. One may assume that $\exp(T) = q^r > q$. It follows from $\{1\} < \mho_{r-1}(T) \leq \Omega_1(T) = R$ and normality of $\mho_{r-1}(T)$ in G that $\mho_{r-1}(T) = R$ since R is a minimal normal subgroup of G. Hence T is homocyclic (Exercise 1(f)).

(iii) Let $T \trianglelefteq G$ be a q-subgroup of maximal order such that $\Omega_1(T) = R$. Then $\mathrm{d}(Q/T) = d - \mathrm{d}(R)$, where $d = \mathrm{d}(Q)$. Clearly, $\mathrm{d}(R) = \mathrm{d}(T)$ and, by (ii), T is homocyclic so we may assume that $T < Q$. Next, R, $\Omega_1(Q)$ and $\Omega_1(Q/T)$ are normal elementary abelian q-subgroups of G and G/T, respectively; therefore, by Maschke's theorem, there exists $L \triangleleft G$ such that $\Omega_1(Q) = R \times L$ and $\Omega_1(Q/T) = (LT/T) \times (U/T)$, where $U \triangleleft G$. In view of $\exp(L) = p$, we have

$$U \cap L = U \cap (LT \cap L) = (U \cap LT) \cap L = T \cap L = \Omega_1(T) \cap L = R \cap L = \{1\}.$$

It follows from $R \leq T \leq U$ and $U \cap L = \{1\}$ that $|R| \leq |\Omega_1(U)| \leq |\Omega_1(Q) : L| = |R|$, so that $\Omega_1(U) = R$. By the maximal choice of T, we get $U = T$. Thus, $\Omega_1(Q/T) = LT/T = LU/U \cong L$, i.e., $\mathrm{d}(Q/T) = \mathrm{d}(L) = d - \mathrm{d}(R) < d$. In particular, $T \not\leq \Phi(Q)$.

(iv) If Q is homocyclic, i.e., $Q = F$, the theorem is true. Indeed, let T be as in (iii) and $\exp(Q) = q^m$ (see Exercise 1). Since $\Phi(Q) = \Omega_{m-1}(Q)$, we have $\exp(T) = q^m$, by (iii). By (ii), T is homocyclic, so that $Q = T \times T_1$, by Exercise 1(e). By [BZ, Theorem 1.9′], $Q = T \times T_2$, where $T_2 \triangleleft G$. All other assertions of the theorem are trivial in this case.

Next we assume that Q is not homocyclic. Then $m < \mu = \log_p(\exp(Q))$, $\Omega_m(Q) = F < Q$, $\mho_m(Q) = M > \{1\}$ and $F, M \triangleleft G$.

(v) If $R \not\leq M$, then (a) and (b) are true. Indeed, $R \cap M = \{1\}$ and $R \leq F$. By (iv), the homocyclic subgroup $F = F_1 \times F_2$, where $F_1, F_2 \triangleleft AF$ and $\Omega_1(F_1) = R$. By (i), $F_1, F_2 \triangleleft G$. Since $\Omega_1(F_1) \cap M = R \cap M = \{1\}$, we get $F_1 \cap M = \{1\}$. Set $F_3 = F \cap Q_2 = \Omega_m(Q_2)$. Then $F_1F_3 = F_1 \times F_3$ is homocyclic of exponent q^m. By (iv), $F = (F_1F_3) \times F_4 = (F_1 \times F_4) \times F_3$ for some $F_4 < F$, and we have $Q = FQ_2 = [(F_1 \times F_4) \times F_3]Q_2 = (F_1 \times F_4)Q_2$. Since $(F_1 \times F_4) \cap Q_2 = (F_1 \times F_4) \cap \Omega_m(Q_2) = (F_1 \times F_4) \cap F_3 = \{1\}$, we get $Q = (F_1 \times F_4) \times Q_2 = F_1 \times (F_4 \times Q_2)$, i.e., F_1 (a normal subgroup of G) is complemented in Q. By [BZ, Theorem 1.9′], $Q = F_1 \times S_1$, where $S_1 \triangleleft G$, and (a), (b) hold with $S = F_1$. Next let $R \leq M$.

By Maschke's theorem, $\Omega_1(Q) = L_1 \times \Omega_1(M)$, where $L_1 \triangleleft G$. We have $L_1 \cap M = L_1 \cap \Omega_1(M) = \{1\}$. Let $R_1 \leq L_1$ be a minimal normal subgroup of G. Then $R_1 \cap M = \{1\}$ so that, by (v), $Q = T \times T_1$, where $T, T_1 \triangleleft G$ and $\Omega_1(T) = R_1$. It follows from $\Omega_1(T) \cap M = R_1 \cap M = \{1\}$ that $T \cap M = \{1\}$. Since $\mho_m(T) \leq$

$\mho_m(Q) \cap T = M \cap T = \{1\}$, we get $\exp(T) = q^m$. By Exercise 1(b), $R \leq M < T_1$. It follows from $AT_1 < G$ and induction that $T_1 = S \times T_2$, where $S, T_2 \lhd AT_1$ and $\Omega_1(S) = R$. By (i), $S, T_2 \lhd G$. Next, $Q = T \times T_1 = T \times (S \times T_2) = S \times (T \times T_2)$. By what has been proved already, $S, T \times T_2 \lhd G$, $\Omega_1(S) = R$, and (a), (b) hold with $S_1 = T \times T_2$ (it follows from (ii) that S is homocyclic).

Let us prove that if $Q = S \times S_1 = T \times T_1$ be decompositions of Q such that $S, S_1, T, T_1 \lhd G$ and $\Omega_1(S) = \Omega_1(T) = R$, then T and S have the same exponent (and so $S \cong T$). Assume, however, that $\exp(T) > \exp(S)$. Since S, T are homocyclic of the same rank, $|T| > |S|$. Since $|Q| = |S| \cdot |S_1|$, we have $T \cap S_1 > \{1\}$, by the product formula, and hence $\{1\} = R \cap S_1 = \Omega_1(T) \cap S_1 > \{1\}$, a contradiction. Thus, $\exp(T) = \exp(S)$, completing the proof of (c).

Let $R \leq U \leq Q$ and $\Omega_1(U) = R$. Then $U \cap S_1 = \{1\}$, and since $US_1 = U \times S_1 \leq Q = S \times S_1$, it follows that U is isomorphic to a subgroup of S. □

Corollary 6.2. *Let Q be a normal abelian Sylow q-subgroup of a group G and R a normal elementary abelian q-subgroup of G. Then $Q = S \times S_1$, where $S, S_1 \lhd G$ and $\Omega_1(S) = R$. Furthermore (M. Harris, D. Taunt), $Q = Q_1 \times \cdots \times Q_s$, where $Q_1, \dots, Q_s \lhd G$ and $\Omega_1(Q_i)$ is a minimal normal subgroup of G, $i = 1, \dots, s$.*

Proof. We proceed by induction on $|G|$. Let A be a q'-Hall subgroup of G.

Obviously, $R \leq Q$. Let R_1 be a minimal normal subgroup of G such that $R_1 \leq R$. If $R_1 = R$, the result follows from Theorem 6.1. Now assume that $R_1 < R$. By Theorem 6.1, $Q = T \times T_1$, where $T, T_1 \lhd G$ and $\Omega_1(T) = R_1$. Set $R_2 = R \cap T_1$, $G_1 = AT_1$. By induction, $T_1 = T_2 \times S_1$, where $T_2, S_1 \lhd G_1$ and $\Omega_1(T_2) = R_2$. Set $S = TT_2 = T \times T_2$. Then $\Omega_1(S) = R_1R_2 = R$, $Q = S \times S_1$ and $S, S_1 \lhd G$, completing the proof of the first assertion. Now the second assertion is obvious. □

Theorem 6.3 (= [BZ, Theorem 16″]). *Suppose that a group F acts on a group $H = A \times B$, where A is F-invariant. If $\exp(Z(A)) \cdot \exp(B/B')$ is coprime with $|F|$, then $H = A \times B_1$, where B_1 is F-invariant.*

Theorem 6.4. *Let H be a normal abelian π-Hall subgroup of G, where π is a set of primes, R a normal π-subgroup of G and let $\exp(R)$ be square free. Then there exists a normal π-subgroup S of G such that*

(a) $\Omega_1(S) = R$, where $\Omega_1(S)$ is generated by all elements of S of prime orders.

(b) $H = S \times S_1$, where $S_1 \lhd G$.

Exercise 2. Let Q be a normal abelian Sylow q-subgroup of G. Let L be a normal q-subgroup of G such that $L \cap \Phi((Q) = \{1\}$. Then $Q = L \times S$, where $S \lhd G$.

If $Q \in \mathrm{Syl}_q(G)$ is abelian and $\Omega_1(Q) \leq Z(G)$, then G is q-nilpotent, by Frobenius' normal p-complement theorem. Indeed, by Lemma 10.8, G has no minimal nonnilpotent subgroups with nonidentity normal Sylow q-subgroup.

Corollary 6.5 (Fitting). *Let Q be a normal abelian Sylow q-subgroup of a group G. Then $Q \cap \mathrm{Z}(G)$ is a direct factor of G.*

Proof. One may assume that $Q < G$. By Theorem 6.4, $Q = Q_1 \times Q_2$, where $Q_1, Q_2 \lhd G$ and $\Omega_1(Q_1) = \Omega_1(Q) \cap \mathrm{Z}(G)$. Therefore, a q'-Hall subgroup F of G centralizes Q_1 by the paragraph preceding the corollary. Since Q is abelian, this implies $\mathrm{Z}(G) \cap Q = Q_1$, and $G = Q_1 \times FQ_2 = (Q \cap \mathrm{Z}(G)) \times FQ_2$, as desired. □

Lemma 6.6 (= [BZ, Lemma 1.19]). *Suppose that $G = PQ$, $E_{p^n} \cong P \in \mathrm{Syl}_p(G)$, $E_{q^m} \cong Q \in \mathrm{Syl}_q(G)$, $Q \lhd G$. If $\mathrm{Z}(G) = \{1\}$, then $n \le \frac{m}{b}$, where b is the order of q* (mod p).

A group H is said to be an $H(p,q)$-group if $H = \mathrm{C}_p Q$, where Q is a normal homocyclic q-subgroup of H, $|\Omega_1(Q)| = q^b$ and C_p of order p acts on Q in a fixed-point-free manner (recall that b is the order of q modulo p). Then $\Omega_1(Q)$ is a minimal normal subgroup of H, and $\mathrm{C}_p \cdot \Omega_1(Q)$ is minimal nonabelian.

Theorem 6.7 (= [BZ, Theorem 1.20]). *Let $P \cong E(p^m)$ act on an abelian q-group Q, $|\Omega_1(Q)| = q^n$, and let $G = P \cdot Q$ be the natural semidirect product. If $\mathrm{Z}(G) = \{1\}$ and $n = mb$, where b is the order of q modulo p, then $G = H_1 \times \cdots \times H_m$, where the H_i's are of type $H(p,q)$ for $i = 1, \dots, m$.*

Corollary 6.8. *Let Q be a q-group, $d = \mathrm{d}(Q)$. If $P \le \mathrm{Aut}(Q)$ is abelian of exponent $p \ne q$, then $\mathrm{d}(P) \le d/b$, where b is the order of q* (mod p).

Proof. By Hall's theorem 1.15, we may consider P as a subgroup of $\mathrm{Aut}(Q/\Phi(Q)) = \mathrm{Aut}(\mathrm{E}_{q^d}) \cong \mathrm{GL}(d,q)$. Hence we may assume that $\Phi(Q) = \{1\}$. Since $\mathrm{C}_P(Q) = \{1\}$, it follows that $p \nmid |\mathrm{Z}(PQ)|$, and the result follows from Corollary 6.5 and Lemma 6.6. □

Exercise 3. Let a noncyclic group P of order p^2 act on a q-group Q, $p \ne q$. Then $Q = \langle \mathrm{C}_Q(x) \mid x \in P^\# \rangle$.

Exercise 4. Let a p'-group F act on a abelian p-group Q. If F acts irreducibly on $Q/\Phi(Q)$, then Q is homocyclic.

Exercise 5. Let d be the rank of an abelian p-group G. Is it true that if $\mathrm{GL}(d,p)$ is isomorphic to a subgroup of $\mathrm{Aut}(G)$, then G is homocyclic? Is it true that the automorphism group of a d-generator homocyclic p-group has a subgroup isomorphic to $\mathrm{GL}(d,p)$?

Exercise 6. Classify the abelian p-groups of exponent p^e with exactly e nontrivial characteristic subgroups.

Exercise 7. Let G be a homocyclic p-group of rank d. Prove that the quotient group $\mathrm{Aut}(G)/\mathrm{O}_p(\mathrm{Aut}(G)) \cong \mathrm{GL}(d,p)$.

Solution. Set $\bar{G} = G/\Phi(G)$. Consider a mapping $f : \mathrm{Aut}(G) \to \mathrm{Aut}(\bar{G})$, where, for $\alpha \in \mathrm{Aut}(G)$, $f(\alpha) = \bar{\alpha}$, the automorphism induced by α on $\bar{G}$; then f is a surjective homomorphism. If $\alpha \in \ker(f)$, then α induces the identity automorphism of $\bar{G} = G/\Phi(G)$, and so $o(\alpha)$ is a power of p (Theorem 1.15). Therefore, $\ker(f) \le \mathrm{O}_p(\mathrm{Aut}(G))$. Since $\mathrm{Aut}(G)/\ker(f) \cong \mathrm{Aut}(\bar{G}) \cong \mathrm{GL}(n, p)$ and $\mathrm{O}_p(\mathrm{GL}(n, p)) = \{1\}$, we get $\ker(f) = \mathrm{O}_p(\mathrm{Aut}(G))$.

Exercise 8. Let G be an abelian p-group. Study the structure of the factor group $\mathrm{Aut}(G)/\mathrm{O}_p(\mathrm{Aut}(G))$. If, in addition, G is homocyclic, study the structure of $\mathrm{O}_p(\mathrm{Aut}(G))$.

Exercise 9. Let G be a homocyclic p-group of rank n. Consider a mapping $f : \mathrm{Aut}(G) \to \mathrm{Aut}(\Omega_1(G))$, where, for $\alpha \in \mathrm{Aut}(G)$, $f(\alpha) = \bar{\alpha}$, the automorphism induced by α on $\Omega_1(G)$. Find $\ker(f)$ and $\mathrm{Aut}(G)/\ker(f)$.

Exercise 10. Let $P \in \mathrm{Syl}_p(G)$ be abelian of type $(a_1 \cdot p^{e_1}, \dots, a_k \cdot p^{e_k})$, $e_1 > \dots > e_k$ and $a_i \le 2$, all i. Prove that if $p > 2$ is the least prime divisor of $|G|$, then G is p-nilpotent. (*Hint.* Apply Lemma 10.8 and Corollary 6.5 to $\mathrm{N}_G(P)$.)

Exercise 11. Let G be an abelian p-group of exponent p^e, $R = \mho_{e-1}(G)$ and ϕ a p'-automorphism of G. Suppose that ϕ induces the identity automorphism on R. Show that the center of $\langle \phi, G \rangle$ contains a homocyclic group of rank α and exponent p^e, where α is the number of invariants of G that are equal to p^e.

Exercise 12. Classify the abelian p-groups G such that $\mathrm{Aut}(G)$ is (a) p-group, (b) p-nilpotent, (c) solvable.

Exercise 13. Let G be an abelian p-group all of whose invariants are pairwise distinct and let $A \le \mathrm{Aut}(G)$ be a p'-subgroup. Prove that $G = Z_1 \times \dots \times Z_s$, where $Z_1, \dots, Z_s$ are A-invariant cyclic.

Exercise 14. Let $Q \lhd G$ be a Hall subgroup. Then $\Phi(Q) = Q \cap \Phi(G)$.

Exercise 15. Let G be an abelian p-group and $T < G$ a characteristic subgroup of index p. Is it true that T is a unique characteristic subgroup of index p in G?

Exercise 16. If G is an abelian p-group and $R < G$ is of order p, then $G = C \times H$, where $\Omega_1(C) = R$. Give a proof independent of Theorem 6.1.

Exercise 17. Let p be the least prime divisor of $|G|$. If $P \in \mathrm{Syl}_p(G)$ is abelian of type $(p^{e_1}, \dots, p^{e_k})$ with $e_1 > \dots > e_k$, then G is p-nilpotent.

Exercise 18. Let G be an abelian p-group. Is it possible to classify all characteristic subgroups of G in terms of subgroups $\Omega_i(G)$ and $\mho_j(G)$ only?

Exercise 19. Let Q be a normal abelian Sylow q-subgroup of a group G, $\Omega_1(Q) = R_1 \times R_2$, where R_1 and R_2 are normal in G. Then the following assertions are equivalent: (a) $Q = S_1 \times S_2$ for some $S_1, S_2 \leq Q$ with $\Omega_1(S_i) = R_i$, $i = 1, 2$. (b) $Q = T_1 \times T_2$ for some G-invariant $T_1, T_2 \leq Q$ with $\Omega_1(T_i) = R_i$, $i = 1, 2$. (*Hint.* Suppose that (a) holds. By Corollary 6.2, Q has a G-invariant direct factor T_i of Q such that $\Omega_1(T_i) = R_i$, $i = 1, 2$. By Exercise 24, $T_i \cong S_i$, $i = 1, 2$. Therefore, $T_1 \times T_2 \cong S_1 \times S_2 = Q$ so $Q = T_1 \times T_2$.)

2°. Let G be an abelian group of order p^m and of type $(\alpha_1 \cdot p^{e_1}, \dots, \alpha_r \cdot p^{e_r})$ (i.e., G has $\alpha_i > 0$ invariants p^{e_i}, $i = 1, \dots, r$), where $e_1 > \dots > e_r$. Set $d = \alpha_1 + \dots + \alpha_r$; then $d = \mathrm{d}(G)$, $m = \alpha_1 e_1 + \dots + \alpha_r e_r$. Our aim is to find $|\mathrm{Aut}(G)|$.

Exercise 20. Let $x_1, \dots, x_d$ be generators of an abelian p-group G. If $\prod_{i=1}^{d} o(x_i) = |G|$, then $G = \langle x_1 \rangle \times \dots \times \langle x_d \rangle$. (*Hint.* We have $G = \langle x_1 \rangle \dots \langle x_d \rangle$. Using product formula, check that $\langle x_1 \rangle \cap \langle x_2, \dots x_d \rangle = \{1\}$. Use induction to show that $\langle x_2, \dots, x_d \rangle = \langle x_2 \rangle \times \dots \times \langle x_d \rangle$.)

Suppose that $\mathcal{B} = \{x_{1,1}, \dots, x_{1,\alpha_1}, x_{2,1}, \dots, x_{2,\alpha_2}, \dots, x_{r,1}, \dots, x_{r,\alpha_r}\}$ and $\mathcal{B}_1 = \{y_{1,1}, \dots, y_{1,\alpha_1}, x_{2,1}, \dots, y_{2,\alpha_2}, \dots, y_{r,1}, \dots, y_{r,\alpha_r}\}$ be two minimal (ordered!) bases of the abelian p-group G such that $o(x_{i,j}) = o(y_{i,j}) = p^{e_i}$, $i = 1, \dots, r$, $j = 1, \dots, \alpha_i$. Such bases of G we call *automorphic* since there is an $\phi \in \mathrm{Aut}(G)$ such that $x_{i,j}^{\phi} = y_{i,j}$ for all i, j. Conversely, each automorphism of G sends one basis in an automorphic one. The group $\mathrm{Aut}(G)$ acts on the set of bases automorphic to $\mathcal{B}$, regularly. Therefore, $|\mathrm{Aut}(G)|$ equals the number of bases automorphic to $\mathcal{B}$. Our solution is divided in r steps. On the i-th step we choose α_i elements, each of order p^{e_i}, $i = 1, \dots, r$. All chosen $\alpha_1 + \dots + \alpha_r = \mathrm{d}(G)$ elements, as we shall see, generate G. Since the product of orders of chosen elements equals $|G|$, it follows that they form a basis of G. It remains to compute the number of choices of all bases of G which are automorphic to $\mathcal{B}$; that number equals $|\mathrm{Aut}(G)|$. Recall that we have assumed that $e_1 > e_2 > \dots > e_r$.

Step 1. First we choose α_1 elements of maximal order p^{e_1}. All of them lie in the set $G - \Omega_{e_1-1}(G)$ (all elements in that set have the same order p^{e_1}; note also that $\Phi(G) \leq \Omega_{e_1-1}(G)$). Note that $|G : \Omega_{e_1-1}(G)| = p^{\alpha_1}$. The first element $x_{1,1}$ can be chosen in $|G - \Omega_{e_1-1}(G)|$ ways. After that choice, as the second element $x_{1,2}$ we take any element in the set $G - \langle x_{1,1}, \Omega_{e_1-1}(G) \rangle$, which contains exactly $|G| - p|\Omega_{e_1-1}(G)|$ elements. Continuing so, we can choose an α_1-th element x_{1,α_1} in $|G| - p^{\alpha_1-1}|\Omega_{e_1-1}(G)|$ ways since it lies in the set $G - \langle x_{1,1}, \dots, x_{1,\alpha_1-1}, \Omega_{e_1-1}(G) \rangle$. Modulo $\Phi(G)$, so chosen α_1 elements generate the subgroup of order p^{α_1}. Set $f_1 = m$ and $|\Omega_{e_1-1}(G)| = p^{t_1}$. We have $f_1 - t_1 = \alpha_1$. Thus, elements $x_{1,1}, \dots, x_{1,\alpha_1}$ of order p^{e_1} one can choose in

$$N_1 = (p^{f_1} - p^{t_1})(p^{f_1} - p^{1+t_1}) \dots (p^{f_1} - p^{\alpha_1-1+t_1})$$
$$= (p^{\alpha_1} - 1) \dots (p^{\alpha_1} - p^{\alpha_1-1}) p^{\alpha_1 t_1}$$

ways.

Step 2. Then we choose α_2 elements $x_{2,1}, \dots, x_{2,\alpha_2}$ of order p^{e_2}. All of them lie in the set $\Omega_{e_2}(G) - T_2$, where $T_2 = \Omega_{e_2}(\Phi(G))\Omega_{e_2-1}(G)$. We have $|\Omega_{e_2}(G)| = p^{f_2}$, where $f_2 = (\alpha_1 + \alpha_2)e_2 + \alpha_3 e_3 + \dots + \alpha_r e_r$, $|T_2| = p^{t_2}$, where $t_2 = \alpha_1 e_2 + (e_2 - 1)\alpha_2 + \alpha_3 e_3 + \dots + \alpha_r e_r$, so that $|\Omega_{e_2}(G) : T_2| = p^{f_2 - t_2} = p^{\alpha_2}$. The first element, say $x_{2,1}$, of order p^{e_2} is an arbitrary element of the set $\Omega_{e_2}(G) - T_2$ so it can be chosen in $p^{f_2} - p^{t_2} = (p^{\alpha_2} - 1)p^{t_2}$ ways. The second element, say $x_{2,2}$, is an arbitrary element of the set $\Omega_{e_2}(G) - \langle x_{2,1}, T_2 \rangle$ of cardinality $(p^{\alpha_2} - p)p^{t_2}$. Continuing so, we conclude that α_2 elements $x_{2,1}, \dots, x_{2,\alpha_2}$ of order p^{e_2} can be chosen in

$$\begin{aligned} N_2 &= (p^{f_2} - p^{t_2})(p^{f_2} - p^{1+t_2}) \dots (p^{f_2} - p^{\alpha_2 - 1 + t_2}) \\ &= (p^{\alpha_2} - 1) \dots (p^{\alpha_2} - p^{\alpha_2 - 1}) p^{\alpha_2 t_2} \end{aligned}$$

ways. Note that $\alpha_1 + \alpha_2$ elements, chosen in Steps 1 and 2, generate, modulo $\Phi(G)$, a subgroup of order $p^{\alpha_1 + \alpha_2}$.

Step 3. Elements $x_{3,1}, \dots, x_{3,\alpha_3}$ of order p^{e_3} are contained in the set $\Omega_{e_3}(G) - T_3$, where $T_3 = \Omega_{e_3}(\Phi(G))\Omega_{e_3-1}(G)$. We have $|\Omega_{e_3}(G)| = p^{f_3}$, where $f_3 = (\alpha_1 + \alpha_2 + \alpha_3)e_3 + \alpha_4 e_4 + \dots + \alpha_r e_r$, $|T_3| = p^{t_3}$, where $t_3 = (\alpha_1 + \alpha_2)e_3 + (e_3 - 1)\alpha_3 + \alpha_4 e_4 + \dots + \alpha_r e_r$ so that $|\Omega_{e_3}(G) : T_3| = p^{f_3 - t_3} = p^{\alpha_3}$. It follows that the choice of α_3 our elements can be done in

$$\begin{aligned} N_3 &= (p^{f_3} - p^{t_3})(p^{f_3} - p^{1+t_3}) \dots (p^{f_3} - p^{\alpha_3 - 1 + t_3}) \\ &= (p^{\alpha_3} - 1) \dots (p^{\alpha_3} - p^{\alpha_3 - 1}) p^{\alpha_3 t_3} \end{aligned}$$

ways. Note that $\alpha_1 + \alpha_2 + \alpha_3$ elements chosen in Steps 1, 2, and 3, generate, modulo $\Phi(G)$, the subgroup of order $p^{\alpha_1 + \alpha_2 + \alpha_3}$.

Step r. At last, we choose elements $x_{r,1}, \dots, x_{r,\alpha_r}$ of order p^{e_r}. All of them lie in the set $\Omega_{e_r}(G) - T_r$, where $T_r = \Omega_{e_r}(\Phi(G))\Omega_{e_r-1}(G)|$. We have $|\Omega_{e_r}(G)| = p^{f_r}$, where $f_r = (a_1 + \dots + \alpha_r)e_r$, and $|T_r| = p^{t_r}$, where $t_r = f_r - \alpha_r$. As above, these elements may be chosen in $N_r = (p^{\alpha_r} - 1)(p^{\alpha_r} - p) \dots (p^{\alpha_r} - p^{\alpha_r - 1})p^{\alpha_r t_t}$ ways.

The elements $x_{1,1}, \dots, x_{1,\alpha_1}, x_{2,1}, \dots, x_{2,\alpha_2}, \dots, x_{x,1}, \dots, x_{r,\alpha_r}$, in view of their choice, generate G since the subgroup generated by them, covers $G/\Phi(G)$. Since the product of their orders equals $|G|$, it follows, by Exercise 20, that

$$G = \langle x_{1,1} \rangle \times \dots \times \langle x_{1,\alpha_1} \rangle \times \langle x_{2,1} \rangle \times \dots \times \langle x_{2,\alpha_2} \rangle \times \dots \times \langle x_{r,1} \rangle \times \dots \times \langle x_{r,\alpha_r} \rangle,$$

i.e., the chosen elements form a basis of G. By the above, the number of bases of G automorphic with $\mathcal{B}$, equals $|\mathrm{Aut}(G)| = \prod_{i=1}^r N_i$, or, in implicit form, we have

Theorem 6.9. *Let G be an abelian group of type $(\alpha_1 \cdot p^{e_1}, \dots, \alpha_r \cdot p^{e_r})$. Then $|\mathrm{Aut}(G)| = p^{\sum_{i=1}^r \alpha_i t_i} \prod_{i=1}^r (p^{\alpha_i} - 1)(p^{\alpha_i} - p) \dots (p^{\alpha_i} - p^{\alpha_i - 1})$, where the numbers $t_1, t_2, \dots, t_r$ are defined in Steps 1, 2,…,r.*

§7

Regular p-groups

Theory of regular p-groups was originated in Philip Hall's fundamental paper [Hal1] and presented there in a remarkably mature form. Regular p-groups are natural generalization of abelian p-groups. All properties of regular p-groups, which we prove in what follows, are known and proved easily for abelian p-groups. There are some useful criteria for a p-group to be regular. Most popular of them are the following: (i) if $|G/\mho_1(G)| < p^p$, then G is regular (Theorem 9.8(a), (ii) if $|G'/\mho_1(G')| < p^{p-1}$, then G is regular (Theorem 9.8(c)), (iii) if $\mathrm{cl}(G) < p$, then G is regular (Theorem 7.1(b), (iv) If $\exp(G) = p$, then G is regular (Theorem 7.1(b)). In this section we prove basic properties of regular p-groups. Main results of this section, namely, Theorems 7.1 and 7.2, are due to P. Hall [Hal1]. Below we use the following commutator identities (Introduction, Exercise 13):

(1) $\qquad [x, y]^{-1} = [y, x], \quad [xy, u] = [x, u]^y [y, u], \quad [u, xy] = [u, y][u, x]^y$

and also the following important formula of Hall-Petrescu (Appendix 1):

(2) $$(xy)^n = x^n y^n c_2^{\binom{n}{2}} c_3^{\binom{n}{3}} \dots c_n^{\binom{n}{n}}, \quad c_i \in \mathrm{K}_i(\langle x, y\rangle).$$

Definition 1. A p-group G is said to be *regular* if for every $x, y \in G$, there is $c \in \mho_1(\langle x, y\rangle')$ such that $x^p y^p = (xy)^p c$.

As we see, the element c in Definition 1 is equal to a product $a_1^p \dots a_s^p$, where $a_1, \dots, a_s \in \langle x, y\rangle'$. Note that if $a, b, c \in G$ and $(ab)^p = a^p b^p c$, then $c \in H'$, where $H = \langle a, b\rangle$ (read $b^{-p} a^{-p} a^p b^p$ modulo H'). Obviously, G is regular if and only if all its two-generator subgroups are regular.

Abelian p-groups are regular. By Exercise 1.21, regular 2-groups are abelian.

Remark 1. A p-group that is not regular is called *irregular* so an irregular p-group all of whose proper subgroups are regular, is two-generator.

Let G be a regular p-group with $\exp(G') = p$; then $\mho_1(G') = \{1\}$. In that case, $x^p y^p = (xy)^p$ for any two elements $x, y \in G$. The groups G, satisfying the last condition, are called *p-abelian*. Clearly, p-abelian p-groups are regular.

Theorem 7.1. *Let G be a p-group.*

(a) Regularity is inherited by sections.

(b) If $\mathrm{cl}(G) < p$ or $|G| < p^p$ or $\exp(G) = p$, then G regular.

(c) If $\mathrm{K}_{p-1}(G)$ is cyclic, then G is regular.

Proof. (a) is obvious.

(b) By (2), we have $x^p y^p = (xy)^p c$, where $c = c_2^{\binom{p}{2}} c_3^{\binom{p}{3}} \dots c_p^{\binom{p}{p}}$, $c_i \in \mathrm{K}_i(H)$, $H = \langle x, y \rangle$, $i = 2, \dots, p$. If $\mathrm{cl}(G) < p$, then $c_p \in \mathrm{K}_p(H) = \{1\}$, and p divides $\binom{p}{i}$, so $c \in \mho_1(\langle x, y\rangle')$. If $|G| < p^p$, then $\mathrm{cl}(G) < p$. If $\exp(G) = p$, then $c = 1$.

(c) One may assume that $p > 2$; then $\mathrm{K}_{p-1}(G) \le \mathrm{K}_2(G) = G'$. Also one may assume that $G = \langle x, y\rangle$ (Remark 1) and $\mathrm{K}_{p-1}(G) > \{1\}$, by (b). Then, since every proper subgroup of $\mathrm{K}_{p-1}(G)$ is contained in $\mho_1(K_{p-1}(G))$, we get $\mathrm{K}_p(G) \le \mho_1(\mathrm{K}_{p-1}(G)) \le \mho_1(G')$. Next, p divides $\binom{p}{i}$, $i = 2, \dots, p-1$ so $c_p \in \mho_1(G')$ (see (2)). Thus, in (2), $c \in \mho_1(G')$ so G is regular. □

If $p > 2$, then metacyclic p-groups are regular (Theorem 7.1(c)). If G is the Sylow p-subgroup of the holomorph of the cyclic group of order p^n, $p > 2$, then $\mathrm{cl}(G) = n$ and G is metacyclic, by Theorem 1.2 so the class of regular p-groups is not bounded.

Exercise 1. Let G be a group. If $x^n = y^n \Rightarrow (x^{-1}y)^n = 1$ for all $x, y \in G$, then the subset $G_n = \{x \in G \mid x^n = 1\}$ is a (normal) subgroup in G.

Solution. If $x^n = y^n = 1$, then $(x^{-1})^n = y^n = 1$ so $(xy)^n = 1$. Therefore, it suffices to show that $u_1 \dots u_m \in G_n$ for all $u_1, \dots, u_m \in G_n$, i.e., $G_n \le G$. By induction on m, one may assume that $u_2 \dots u_m \in G_n$. Then $(u_1 u_2 \dots u_m)^n = 1$, and so $u_1 \dots u_m \in G_n$.

Definition 2. A p-group G is called an *R_n-group* if $x^{p^n} = y^{p^n} \Rightarrow (x^{-1}y)^{p^n} = 1$.

Exercise 2. If $G = \langle x, y\rangle$, then $G' = \langle [x, y]^g \mid g \in G\rangle$.

Solution. $G' = \langle [u_1 \dots u_m, v_1 \dots v_n] \mid u_i, v_j \in \{x, y\}\rangle$. If $m = 1$, then

$$[u_1, v_1 \dots v_n] = [u_1, v_2 \dots v_n][u_1, v_1]^{v_2 \dots v_n}$$

is a product of G-conjugates of $[x, y]^{\pm 1}$, by induction on n. If $m \ge 2$, then

$$[u_1 \dots u_m, v_1 \dots v_n] = [u_1, v_1 \dots v_n]^{u_2 \dots u_m}[u_2 \dots u_m, v_1 \dots v_n]$$

is a product of G-conjugates of $[x, y]^{\pm 1}$, by the above and induction on m.

Definition 3. A p-group G is called a *T_n-group* if $(x^{-1}y)^{p^n} = 1 \Rightarrow x^{p^n} = y^{p^n}$.

Properties R_n and T_n are inherited by sections.

Theorem 7.2. *Suppose that G is a regular p-group.*

(a) G is an R_n- and T_n-group.

(b) $\exp(\Omega_n(G)) \le p^n$.

(c) $\mho_n(G) = \{x^{p^n} \mid x \in G\}$.

(d) $|\Omega_n(G)| = |G : \mho_n(G)|$.

(e) $[x^{p^k}, y^{p^n}] = 1 \Leftrightarrow [x, y]^{p^{k+n}} = 1$.

(f) *If* $M, N \trianglelefteq G$, *then* $[\mho_r(M), \mho_s(N)] = \mho_{r+s}([M, N])$ *for* $r, s \in \mathbb{N} \cup \{0\}$.

(g) $\mathrm{K}_j(\mho_r(G)) = \mho_{rj}(\mathrm{K}_j(G))$. *In particular,* $\mho_r(G)' = \mho_{2r}(G')$.

(h) *If* $A, B \trianglelefteq G$, *we set* $[A, B; 1] = [A, B]$ *and* $[A, B; n+1] = [[A, B; n], B]$. *Then* $[\mho_r(M), N; k] = \mho_r([M, N; k])$, *where* $M, N \trianglelefteq G$.

(i) $\mathrm{K}_{j+1}(G) \leq \Omega_r(G) \Leftrightarrow \mho_r(G) \leq \mathrm{Z}_j(G)$.

(j) $[M, G] \leq \Omega_r(G) \Leftrightarrow [M, \mho_r(G)] = \{1\}$, *where* $M \trianglelefteq G$.

(k) $[\Omega_r(G), \mho_r(G)] = \{1\}$.

Proof. (a) We prove by induction on n, that the regular p-group G is an R_n- and T_n-group. By induction, all proper sections of G are R_n-groups and T_n-groups. First we prove that

(a.i) G is an R_1-group. Indeed, let $x^p = y^p$. We must prove that $(x^{-1}y)^p = 1$. One may assume that $G = \langle x, y \rangle$ is noncyclic. We have $(x^y)^p = (x^p)^y = (y^p)^y = y^p = x^p$. Let $x \in M \in \Gamma_1$; then $x, x^y \in M \triangleleft G$ and M is an R_1-group, by induction. Therefore, $(x^{-1}x^y)^p = 1$, i.e., $o([x.y]) \leq p$. By Exercise 2, $G' = \langle [x, y]^g \mid g \in G \rangle$. Also, since M is an R_1-group and $G' < M$, it follows from $\Omega_1(G') = G'$ that $\exp(G') = p$ (by induction and Exercise 1). Therefore, G is p-abelian so $x^{-p}y^p = (x^{-1}y)^p$. But $x^p = y^p$ so $(x^{-1}y)^p = 1$ hence G is an R_1-group.

(a.ii) Now let $n > 1$. We prove that, if G is an R_1- and R_{n-1}-group, then G is also an R_n-group. Suppose that $x^{p^n} = y^{p^n}$, or, what is the same, $(x^p)^{p^{n-1}} = (y^p)^{p^{n-1}}$. Then, since G is an R_{n-1}-group, we get $(x^{-p}y^p)^{p^{n-1}} = 1$ so $x^{-p}y^p \in \Omega_{n-1}(G)$, or, what is the same. $x^p \equiv y^p \pmod{\Omega_{n-1}(G)}$. Being regular, $G/\Omega_{n-1}(G)$ is an R_1-group, by (a.i). Therefore, we get $(x^{-1}y)^p \in \Omega_{n-1}(G)$. Since R_{n-1} holds, $\exp(\Omega_{n-1}(G)) \leq p^{n-1}$ (Exercise 1). Thus, $1 = ((x^{-1}y)^p)^{p^{n-1}} = (x^{-1}y)^{p^n}$ so G is an R_n-group. By induction, G is an R_{n-1}-group if $n > 1$. So, by what has just been proved, G is an R_n-group. This is true for all positive integers n.

(a.iii) We claim that G is a T_1-group. Indeed, supposing that $(x^{-1}y)^p = 1$, we have to prove that $x^p = y^p$. It follows from $x^{-1}y = (y^{-1}x)^{-1}$ that also $(y^{-1}x)^p = 1$. Since G is an R_1-group, we get $1 = (xy^{-1} \cdot x^{-1}y)^p = [x^{-1}, y]^p$. Therefore, as in (i), $\exp(G') = p$, since we may assume that $G = \langle x, y \rangle$. So since G is p-abelian, we get $x^{-p}y^p = (x^{-1}y)^p = 1$ so G is a T_1-group.

(a.iv) Let $n > 1$. We claim that if G is a T_1- and T_{n-1}-group, then G is also a T_n-group. Suppose that $(x^{-1}y)^{p^n} = 1$; then $(x^{-1}y)^p \in \Omega_{n-1}(G)$. Since the regular quotient group $G/\Omega_{n-1}(G)$ satisfies T_1, by (a.iii), we get $x^{-p}y^p \in \Omega_{n-1}(G)$. Since G is an R_{n-1}-group, we get $\exp(\Omega_{n-1}(G)) \leq p^{n-1}$, by Exercises 1, 2. Hence, we have $(x^{-p}y^p)^{p^{n-1}} = 1$. Then T_{n-1} implies $(x^p)^{p^{n-1}} = (y^p)^{p^{n-1}}$, i.e., $x^{p^n} = y^{p^n}$, and so G is a T_n-group. By induction on n, G is a T_{n-1}-group provided $n > 1$. So, by what has just been proved, G is a T_n-group. Hence, G is a T_n-group.

(b) follows from (a) and Exercise 1.

(c) We use induction on $|G|$ and n. Let $n = 1$. If G is abelian and $a \in \mho_1(G)$, then there exist $x_1, \dots, x_k \in G$ such that $a = x_1^p \dots x_k^p = (x_1 \dots x_k)^p$ is a p-th power. Now we assume that G be nonabelian. For any $x, y \in G$, we must prove that there is $z \in \langle x, y\rangle$ satisfying $x^p y^p = z^p$, i.e., the set of p-th powers in G is closed under multiplication so it is a subgroup. One may assume that $G = \langle x, y\rangle$. We have $x^p y^p = (xy)^p c$, where $c \in \mho_1(G')$. By induction in G', $c = d^p$ for $d \in G'$ since $G' < G$, and so $(xy)^p c = (xy)^p d^p$. Since $K = \langle xy, d\rangle < G$ in view of $d \in G' \leq \Phi(G)$, it follows, by induction in K, that $x^p y^p = (xy)^p c = (xy)^p d^p = z^p$ for some $z \in K$, completing case $n = 1$.

Let $n > 1$. For all $x, y \in G$, $x^{p^n} y^{p^n} = (x^p)^{p^{n-1}} (y^p)^{p^{n-1}}$, and so one can find an element $t \in \langle x^p, y^p\rangle$ such that $x^{p^n} y^{p^n} = t^{p^{n-1}}$. As above, one may assume that $G = \langle x, y\rangle$. By the case $n = 1$, we have $t = z^p$ for some $z \in G$, and so $x^{p^n} y^{p^n} = z^{p^n}$, proving (c).

(d) If $a \in b\Omega_n(G)$, then $a^{-1}b \in \Omega_n(G)$ and so, by (b), $(a^{-1}b)^{p^n} = 1$ which gives $a^{p^n} = b^{p^n}$. Therefore, all elements of a fixed coset of $\Omega_n(G)$ have the same p^n-th power so $|G : \Omega_n(G)| \leq |\mho_n(G)|$. On the other hand, if $x^{p^n} = y^{p^n}$ then $y = xz$ for some $z \in \Omega_n(G)$, by (a), so $y \in x\Omega_n(G)$. So, an element of $\mho_n(G)$ corresponds to a unique coset of $\Omega_n(G)$. This means that $|G : \Omega_n(G)| \geq |\mho_n(G)|$.

(e) Let $[x^{p^k}, y^{p^n}] = 1$. Then

$$x^{p^k} = y^{-p^n} x^{p^k} y^{p^n} = (y^{-p^n} x y^{p^n})^{p^k} \Leftrightarrow (x^{-1} y^{-p^n} x y^{p^n})^{p^k} = 1,$$

by (a). This means that $x^{-1} y^{-p^n} x \cdot y^{p^n} = (x^{-1}yx)^{-p^n} \cdot y^{p^n} \in \Omega_k(G)$ which is equivalent to $(x^{-1}yx)^{p^n} \equiv y^{p^n} \pmod{\Omega_k(G)}$, and again by (a), this is equivalent to $(y^{-1}x^{-1}yx)^{p^n} \in \Omega_k(G)$, or, in view of (b), $[x, y]^{p^{k+n}} = 1$. Similarly, $[x, y]^{p^{k+n}} = 1 \Rightarrow [x^{p^k}, y^{p^n}] = 1$.

(f) Let $M, N \trianglelefteq G$; then $[\mho_r(M), \mho_s(N)] = \langle [x^{p^r}, y^{p^s}] \mid x \in M, y \in N\rangle$. We have $[x, y]^{p^{r+s}} \equiv 1 \pmod{\mho_{r+s}([M, N])}$ for $x \in M$, $y \in N$. Therefore, by (e), $[x^{p^r}, y^{p^s}] \equiv 1 \pmod{\mho_{r+s}([M, N])}$, and so $[\mho_r(M), \mho_s(N)] \leq \mho_{r+s}([M, N])$.

On the other hand, $[x^{p^r}, y^{p^s}] \equiv 1 \pmod{[\mho_r(M), \mho_s(N)]}$, where $x \in M, y \in N$, and so, by (e), we have

$$[x, y]^{p^{r+s}} \equiv 1 \pmod{[\mho_r(M), \mho_s(N)]}.$$

Since $[M, N]/[\mho_r(M), \mho_s(N)]$ is generated by commutators of orders at most p^{r+s}, the exponent of that factor group is at most p^{r+s}, by (b). It follows that $\mho_{r+s}([M, N]) \leq [\mho_r(M), \mho_s(N)]$.

(g) For $j = 1$ the assertion is trivial. Next we proceed by induction on j. By (f) and induction, we have

$$\begin{aligned} \mathrm{K}_{j+1}(\mho_r(G)) &= [\mathrm{K}_j(\mho_r(G)), \mho_r(G)] = [\mho_{rj}(\mathrm{K}_j(G)), \mho_r(G)] \\ &= \mho_{rj+r}([\mathrm{K}_j(G), G]) = \mho_{r(j+1)}(\mathrm{K}_{j+1}(G)), \end{aligned}$$

and (g) holds for $j + 1$ if it holds for j.

(h) If $k = 1$, we have to prove that $[\mho_r(M), N] = \mho_r([M, N])$, which follows from (f) with $s = 0$. For $k > 1$, we prove this by induction on k:

$$[\mho_r(M), N; k] = [[\mho_r(M), N; k-1], N] = [[\mho_r([M, N; k-1], N])$$
$$= \mho_r([M, N; k-1], N]) = \mho_r([M, N; k]).$$

(i) The following assertions are equivalent: $K_{j+1}(G) \le \Omega_r(G)$; $\mho_r(K_{j+1}(G)) = \{1\}$ (by (b)); $[\mho_r(G), G; j] = \{1\}$ (by (h))); $\mho_r(G) \le Z_j(G)$.

(j) By (f), $[G, M] \le \Omega_r(G) \Leftrightarrow \{1\} = \mho_r([G, M]) = [\mho_r(G), M]$.

(k) We have $[G, \Omega_r(G)] \le \Omega_r(G)$. So, by (j), setting $M = \Omega_r(G)$, we get $[\Omega_r(G), \mho_r(G)] = \{1\}$. □

The standard wreath product $G = C_p \operatorname{wr} C_p$ is irregular since $\Omega_1(G) = G$ and $\exp(G) = p^2$ (Theorem 7.2(b)).

Hall has showed that if a p-group G has a subgroup A of order p^k $(k < p)$ and exponent p, then it has a normal subgroup of order p^k. Using Mann's idea, we prove this by induction on $|G|$. We have $A^G = M \triangleleft G$. There is $B \triangleleft M$ of order p^k and exponent p, by induction. Next, $B \le Z_k(M) \triangleleft G$ and $Z_k(M)$ is regular (Theorem 7.1(b)). By Theorem 7.2(b), $\exp(\Omega_1(Z_k(M)) = p$ and $B \le \Omega_1(Z_k(M))$.

In view of Theorem 7.2(c), one can define the regular p-groups as follows:

Definition 1a. A p-group G is said to be regular if for any $x, y \in G$ there exists $z \in \langle x, y \rangle'$ such that $x^p y^p = (xy)^p z^p$.

A p-group $G = G_1 \times G_2$ is regular if G_1 is regular and G_2 is p-abelian, i.e., $(ab)^p = a^p b^p$ for all $a, b \in G_2$ (O. Grün). There exist two isomorphic regular (even metacyclic) 3-groups such that their direct product is irregular (Wielandt).

Proposition 7.3. *Let G be a regular p-group, $H \le G$, $n > 0$. Then $|H : \mho_n(H)| \le |G : \mho_n(G)|$.*

Proof. By Theorem 7.2(d), $|H : \mho_n(H)| = |\Omega_n(H)| \le |\Omega_n(G)| = |G : \mho_n(G)|$. □

Exercise 3. If G is a regular p-group of exponent at least p^n, then we have $c_n(G) = \frac{|\Omega_n(G)| - |\Omega_{n-1}(G)|}{p^{n-1}(p-1)}$.

Theorems 7.4–7.8 are taken from [Man2, Man3].

Definition 4. A p-group G is termed *minimal irregular* if G is irregular but all proper sections of G are regular. (We adopt this definition only for this section.)

Theorem 7.4 (Mann). *Let G be a minimal irregular p-group of class c and exponent p^e. Then the following assertions hold:*

(a) $d(G) = 2$.

(b) $\exp(G') = p$ *(so all proper subgroups of* G *are* p*-abelian) and* G/G' *is abelian of type* (p^{e-1}, p).

(c) $Z(G) = \mho_1(G)$ *is cyclic of order* p^{e-1} *and* $\Phi(G) = Z_{c-1}(G)$. *We have* $M(G) = \langle v \in G \mid (av)^p = a^p \text{ for each } a \in G\rangle \le G'$ *(in fact,* $M(G) = G'$*, but we omit the proof of this fact).*

(d) $K_c(G) = Z(G) \cap G'$ *is a unique minimal normal subgroup of* G.

(e) *All proper sections of* G *have class less than* c.

(f) $(ab)^{p^2} = a^{p^2}b^{p^2}$ *for all* $a, b \in G$*, i.e.,* G *is* p^2*-abelian.*

(g) *If* $e > 2$*, then for all* $s = 1, \dots, e$, $\exp(\Omega_s(G)) = p^s$; $\mho_s(G)$ *consists of all* p^s*-th powers of elements of* G*; and* $|G/\Omega_s(G)| = |\mho_s(G)|$.

(h) $G \times G$ *has an irregular section of exponent* p^2.

(i) $G = \langle x, y\rangle$ *such that* $(ab)^p = a^p b^p$.

Proof. We prove assertions (a)–(f) only. If $p = 2$, then G is minimal nonabelian, and then either $|G| = 8$ or $G \cong M_{2^n}, n > 3$. Next we assume that $p > 2$.

(a) Since G is irregular, it contains elements a, b such that

$$(ab)^{-p}a^pb^p \notin \mho_1(\langle a, b\rangle')$$

so $\langle a, b\rangle$ is irregular and $G = \langle a, b\rangle$ by minimality, proving (a). In what follows we assume that $G = \langle a, b\rangle$ where a and b are chosen as above.

(b) Assume that $\exp(G') > p$. The subgroup $\mho_1(G') > \{1\}$ is normal in G. Let $T \le \mho_1(G') \cap Z(G)$ be of order p. Then G/T is regular so we have $a^pb^p = (ab)^pc^pd$, where $c \in G'$ and $d \in T \le \mho_1(G')$. In that case, we get $((ab)^{-p}a^pb^p =)c^pd \in \mho_1(G') \le \mho_1(\langle a, b\rangle')$, contrary to the choice of a, b. It follows that all proper subgroups of G are p-abelian. The last assertion we prove after proof of (c).

(c) Let $x \in G$, $u \in G' \le \Phi(G)$; then $\langle x, u\rangle < G$ so $\langle x, u\rangle$ is regular. In that case, $(xu)^p = x^pu^p = x^p$. Write $M(G) = \langle v \in G \mid (av)^p = a^p \text{ for each } a \in G\rangle$; then $M(G) \le G$. By what has just been proved, $G' \le M(G)$, and so $G/M(G)$ is abelian. We claim that $\mho_1(G) \le Z(G)$. Indeed, if $a_1, b_1 \in G$, then, since $[a_1, b_1] \in G' \le M(G)$, we get

$$b_1^{-1}a_1^pb_1 = (b_1^{-1}a_1b_1)^p = (a_1 \cdot [a_1, b_1])^p = a_1^p$$

so $\mho_1(G) \le Z(G)$.

Let $T \le Z(G)$ be of order p. Then G/T is regular and $\exp((G/T)') = p$, by (b), so G/T is p-abelian. If S is another minimal normal subgroup of G, then G/S is also p-abelian, but then G itself is p-abelian as a subgroup of $(G/T) \times (G/S)$ and so regular. Therefore, T is a unique minimal normal subgroup of G so $Z(G)$ is cyclic. Since $d(G) = 2$ we have $Z(G) \le \Phi(G)$. Thus, $\mho_1(G) \le Z(G) \le \Phi(G) = G'\mho_1(G)$ so $Z(G) = \mho_1(G)(G' \cap Z(G))$, by the modular law, hence $\mho_1(G)$ and $G' \cap Z(G)$ are incident. Here $G' \cap Z(G)$ is cyclic and of exponent p so $G' \cap Z(G) = T$ is the unique minimal normal subgroup of G. It follows that $T = G' \cap Z(G) \leq \mho_1(G)$ so $Z(G) = \mho_1(G)T = \mho_1(G)$ and $|Z(G)| = p^{e-1}$ since $Z(G)$ is cyclic. Next, $G/Z_{c-1}(G)$ is abelian (here $c = \mathrm{cl}(G)$) so $G' \leq Z_{c-1}(G)$, and since $\mho_1(G) = Z(G) \leq Z_{c-1}(G)$, we get $\Phi(G) = G'\mho_1(G) \leq Z_{c-1}(G)$. Since $G/Z_{c-1}(G)$ is not cyclic (otherwise, $G/Z_{c-2}(G)$ is abelian), it follows, in view of $d(G) = 2$, that $Z_{c-1}(G) = \Phi(G)$.

Write $\bar{G} = G/G'$. Since $\mho_1(\bar{G}) \cong \mho_1(G)/(G' \cap \mho_1(G))$ is cyclic of order p^{e-2}, the two-generator abelian group $\bar{G} = G/G'$ must be of type (p^{e-1}, p), completing the proof of (b).

(d) As $\{1\} < K_c(G) \leq Z(G) \cap G'$ is of order p, by (b) and (c), we get $K_c(G) = Z(G) \cap G'$, which is a unique minimal normal subgroup of G.

(e) If $H \in \Gamma_1$, then $Z_{c-1}(G) = \Phi(G) < H$, by (c). Then $\Phi(G) = Z_{c-1}(G) \leq Z_{c-1}(H)$ so $|H : Z_{c-1}(H)| \leq |H : \Phi(G)| = p$ implies that $H = Z_{c-1}(H)$ is of class less than c. Now (e) follows from (d).

(f) Since $G/K_c(G)$ is p-abelian we have $(xy)^p = x^p y^p z$, where $x, y \in G$, $z \in K_c(G)$. Since $x^p, y^p, z \in \mho_1(G) = Z(G)$ by (c) and in view of $|K_c(G)| = p$, we get $(xy)^{p^2} = x^{p^2} y^{p^2} z^p = x^{p^2} y^{p^2}$. □

Corollary 7.5 (Mann). *Let G be a p-group, $p > 2$. If every 2-generator subgroup H of G is such that* $\mathrm{cl}(H/\mho_1(H)) \leq p - 2$, *then G is regular.*

Proof. Suppose that G is a minimal counterexample. Then G is a minimal irregular p-group of Theorem 7.4. By Theorem 7.4(a), $d(G) = 2$, and, by Theorem 7.4(c), $\mho_1(G) = Z(G)$. By hypothesis, $\mathrm{cl}(G/Z(G)) \leq p - 2$ so $\mathrm{cl}(G) \leq p - 1$ and G is regular (Theorem 7.1(b)), contrary to the assumption. □

Remark 2 (reported by Mann). We claim that if a p-group G is such that $\bar{G} = G/Z(G)$ is absolutely regular, i.e., $|\bar{G}/\mho_1(\bar{G})| < p^p$, then G is regular. Indeed, suppose that G is a minimal counterexample, Since the hypothesis is inherited by sections of G (Proposition 7.3), G is minimal irregular. By Theorem 7.4(c), $\mho_1(G) = Z(G)$ so $G/\mho_1(G)$ is absolutely regular. Then $|G/\mho_1(G)| \leq p^{p-1}$ so $\mathrm{cl}(G) \leq p - 1$ and G is regular (Theorem 7.1(b)).

Corollary 7.6. *If all two-generator subgroups H of a p-group G satisfy $|H/\mho_1(H)| < p^{p-1}$, then G is regular.*

In the light that a direct product of two regular p-groups is not necessarily regular, the following result in unexpected.

Corollary 7.7 (Mann). *Let Z_1, Z_2 be two distinct minimal normal subgroups of a p-group G. If G/Z_i is regular for $i = 1, 2$, then G is also regular.*

Proof. We use induction on $|G|$. Let $H = K/L$ be a section of G. If $Z_1 \not\leq K$, then $K \cong KZ_1/Z_1$, as a subgroup of G/Z_1, is regular. Let $Z_1, Z_2 \leq K$. Then

$K/Z_1, K/Z_2$ are regular as subgroups of G/Z_1, G/Z_2, respectively, and so K is regular, unless $K = G$ (here we use induction). Thus, we may assume that G is minimal irregular. In that case, however, this assumption is impossible since, by Theorem 7.4(d), G has only one minimal normal subgroup. □

Theorem 7.8 (Mann). *Let G be a p-group, $p > 2$. If all subgroups of G can be generated by $\frac{1}{2}(p-1)$ elements, then G is regular.*

Proof. One may assume that $p > 3$. The hypothesis is inherited by sections. Therefore, assuming that G is a minimal counterexample, we see that G is as in Theorem 7.4. Then $2 = \mathrm{d}(G) \geq \frac{1}{2}(p-1)$ so $p = 5$ and all subgroups of G are two-generator. Now the result follows from Theorem 13.7 (however, there exists a more elementary proof). □

Theorem 7.9. *If G is a regular two-generator 3-group, then G' is cyclic.*

Proof. Suppose that G be a counterexample of minimal order. Then G' contains a G-invariant subgroup T such that $G/T \cong \mathrm{E}_{32}$ so, by induction, $T = \{1\}$. Then $G' \cong \mathrm{E}_9$ so G is 3-abelian. Since $\mathrm{C}_G(G')$ is of index 3 in G (otherwise, G is minimal nonabelian so $|G'| = 3$), one may choose two generators a, b of G so, that $b \in \mathrm{C}_G(G')$. Let $G' = \langle c \rangle \times \langle d \rangle$, where $c = [b, a]$. By Exercise 2, $c \notin \mathrm{Z}(G)$. Therefore, one may assume, that $d \in \mathrm{Z}(G)$. Then $c^a = cd^k$ for an appropriate integer $k \in \{1, 2\}$, and $c^b = c$ by the choice of b. Since G is 3-abelian, we get $a^3b^3 = (ab)^3$. On the other hand, since $ba = abc$, we get

$$\begin{aligned} a^3b^3 &= (ab)^3 = a \cdot ba \cdot ba \cdot b = a \cdot abc \cdot abc \cdot b = a^2b \cdot ac^a \cdot b^2c \\ &= a^3b^3 \cdot cc^ac = a^3b^3 \cdot c \cdot cd^k \cdot c = a^3b^3c^3d^k = a^3b^3d^k, \end{aligned}$$

so $d^k = 1$ ($k \in \{1, 2\}$). Thus, $d = 1$, a contradiction. □

Alperin [Alp1] has showed that if every two-generator subgroup of a p-group G, $p > 2$, has the cyclic derived subgroup, then G' is abelian, and so, by Theorem 7.10, regular 3-groups are metabelian.

Exercise 4. If a normal subgroup N of a p-group G, $p > 2$, contains an elementary abelian subgroup, say $H \cong \mathrm{E}_{p^{n-1}}$, of index p, then N contains a G-invariant subgroup isomorphic to H. (*Hint.* Assume that there is in N another subgroup $H_1 \cong H$. Then $N = HH_1$ is of class two so $\exp(N) = p$. Use Exercise 1.6(a).)

Exercise 5. Let G be a 2-group such that $|\Phi(G)| = 2$. Then $|G : \Omega_1(G)| \leq 4$.

Solution. Let G be nonabelian of order 2^{m+1} and $|G/\mathrm{Z}(G)| = 2^{2r}$. Then $\mathrm{cd}(G) = \{1, 2^r\}$ so the number of nonlinear irreducible characters of G equals $\mathrm{n}(G) =$

$\frac{2^{m+1}-2^m}{2^{2r}} = 2^{m-2r}$. Then, by Frobenius–Schur formula for the number of involutions [BZ, Chapter 4], we have

$$1 + c_1(G) \geq |G : G'| - 2^r |\mathrm{Irr}_1(G)| = 2^m - 2^{m-2r} \cdot 2^r = 2^m - 2^{m-r} \geq 2^{m-1}.$$

Since $|\Omega_1(G)| \geq 1 + c_1(G) \geq 2^{m-1}$, our claim follows.

Exercise 6. Let G be a regular p-group of exponent p^e and $i \leq e$. Then $\exp(\mho_i(G)) = p^{e-i}$.

Given a p-group G, let us define subgroups $\mho^k(G)$. We set $\mho^1(G) = \mho_1(G)$. If $\mho^k(G)$ has defined, we set $\mho^{k+1}(G) = \mho_1(\mho^k(G))$. Since $\exp(G/\mho^k(G)) \leq p^k$, we have $\mho_k(G) \leq \mho^k(G)$.

Exercise 7. Prove that $\mho^k(G) = \mho_k(G)$ for a regular p-group G.

Exercise 8. Suppose that a regular p-group G with $|\Omega_1(G)| = p^w$ contains an elementary abelian subgroup of order p^{w-2}. Does G contain a normal elementary abelian subgroup of order p^{w-2}?

Exercise 9. Let G be a regular p-group such that the intersection of all cyclic subgroups of G of composite orders is $> \{1\}$. Show that then $G/\Omega_1(G)$ is cyclic.

Solution. Set $H = \Omega_2(G)$; then $\exp(H) = p^2$ and $\Omega_1(H) = \Omega_1(G)$. By hypothesis, $\mho_1(H)$ is of order p. By Theorem 7.2(d), $|H : \Omega_1(H)| = |\mho_1(H)| = p$, so $G/\Omega_1(G)$ has only one subgroup of order p, namely, $H/\Omega_1(G)$. It follows that $G/\Omega_1(G)$ is cyclic since it is regular.

Exercise 10. Let G be a regular p-group with cyclic $\mho_1(G)$. Prove that $G/\Omega_1(G)$ is cyclic. Give an example of regular p-group such that $G/\mho_1(G) \not\cong \Omega_1(G)$.

Exercise 11. Prove that a group of order 3^4 and class 3 is irregular. (*Hint.* Use Theorems 7.9 and 7.1(c).)

Exercise 12. Let $G = \langle x, y \rangle$ be a p-group, $p > 2$, with cyclic G'. If $\exp(G) = p^e$ and $o(x) \geq o(y)$, then $o(x) = p^e$.

Exercise 13. Let G be a noncyclic metacyclic p-group, $p > 2$, of exponent p^e and order p^{e+s}. Then (a) $1 \leq s \leq e$ and (b) $G = AB$, where A and B are cyclic of orders p^e and p^s respectively.

Solution. (b) Let $e = s$. Take a cyclic $U < G$ of order p^e and let $N < \Omega_1(G)(= \mho_{e-1}(G))$ be of order p, $N \not\leq U$. By Theorem 7.2(c), $N \leq V < G$, where $|V| = p^e$. Since $U \cap V = \{1\}$, we get $G = UV$. Now let $e > s$. Let $A < G$ be cyclic of order p^e. By what has just been proved, $\Omega_s(G) = (A \cap \Omega_s(G))B$, where B is cyclic of order p^s and $A \cap B = \{1\}$. It follows that $G = AB$, by the product formula.

Exercise 14. Prove that a p-group, $p > 2$, with cyclic derived subgroup is regular.

Exercise 15. Suppose that any two elements of a p-group G of equal order commute. Is it true that G is abelian? (*Hint.* A minimal counterexample is minimal nonabelian.)

Exercise 16. Let G and H be regular p-groups such that $c_k(G) = c_k(H)$ for all $k \in \mathbb{N}$. Then $|\Omega_i(G)| = |\Omega_i(H)|$ for all $i \in \mathbb{N}$. If, in addition, G and H are abelian, then $G \cong H$.

Exercise 17. A minimal nonabelian p-group, $p > 2$, is generated by elements of maximal order. Consider case $p = 2$.

Exercise 18 ([Man25, Lemma 9]). Let G be a p-group of class $\leq p$, and let $x, y \in G$. Then $[x^p, y] = 1$ if and only if $[x, y^p] = 1$.

Exercise 19 (Hall). Let G be a p-group. Then the subset $M = \{a \in G \mid (ax)^p = x^p \text{ for all } x \in G\}$ is a characteristic subgroup of G.

Exercise 20 (Mann). Let $a, x \in G$, where G is a p-group of class p. Then the subgroup $\langle a, a^x \rangle$ is regular. (*Hint.* We have $\mathrm{cl}(\langle a, a^x \rangle) = \mathrm{cl}(\langle a, [a, x] \rangle) \leq \mathrm{cl}(\langle a, G' \rangle) < \mathrm{cl}(G) = p$.)

Exercise 21. Given a p-group G, set $\mathcal{M}(G) = \langle x \in G \mid C_G(x) = C_G(x^p) \rangle$. Prove [Man25] that $M = \mathcal{M}(G) \leq G$ is abelian.

Solution. Assume that there is $z \in Z_2(M) - Z(M)$ with $z^p \in Z(M)$ and x a generating element of M. Since $\mathrm{cl}(\langle x, z \rangle) \leq 2$ and $[x, z^p] = 1$, we get $[x^p, z] = [x, z]^p = [x, z^p] = 1$ so $z \in C_G(x^p) = C_G(x)$. Since x is an arbitrary generator of M, we get $z \in Z(M)$, a contradiction. Next, $\mathcal{M}(G)$ is characteristic in G.

We call $\mathcal{M}(G)$ the Mann subgroup of G. Clearly, $Z(G) \leq \mathcal{M}(G)$.

Exercise 22. The result of Exercise 21 has the following interesting consequence for p-groups, which is also due to Mann [Man25]. A p-group G that has exactly $k + 1$ distinct class sizes, is an extension of an abelian subgroup $\mathcal{M}(G)$ by a group of exponent $\leq p^k$.

Solution. Indeed, if $x \in G$ has order p^e modulo $\mathcal{M}(G)$, then $x, x^p, \dots, x^{p^e}$ have strictly increasing centralizers since $x, \dots, x^{p^{e-1}} \not\in \mathcal{M}(G)$, and so these elements belong to G-classes of pairwise distinct sizes. It follows that $e + 1 \leq k + 1$ so $e \leq k$. If $\exp(G/\mathcal{M}(G)) = p^e$, then $e \leq k$. Thus, if all noncentral classes of a p-group G have the same size, then G is an extension of an abelian group $\mathcal{M}(G)$ by a group of exponent p. If, in addition, $p = 2$, then $\Phi(G) \leq \mathcal{M}(G)$ so $\Phi(G)$ is abelian.

Exercise 23 ([Man25, Theorem 7(b)]). If $\mathcal{M}(G) \leq H \leq G$ and $\mathrm{cl}(H) \leq p$, then $\mathcal{M}(G) \leq Z(H)$.

Solution. Let a be one of the generating elements of $\mathcal{M}(G)$, and $x \in H$. We prove that $[a, x] = 1$. By induction on $o(x)$, $[a, x^p] = 1$, so, by Exercise 18, $[a^p, x] = 1$, i.e., $x \in C_H(a^p) = C_H(a)$.

Exercise 24 ([Man25, Theorem 7(c)]). $C_G(\mathcal{M}(G))$ contains all normal subgroups of G of class $< p$.

Solution. Let $N \trianglelefteq G$ be of class $< p$. Since $M = \mathcal{M}(G)$ is abelian (Exercise 21), we get $\mathrm{cl}(MN) \leq p$, by Fitting's lemma. By Exercise 23, $M \leq Z(MN)$ so $M \leq C_G(N)$.

Exercise 25. If $H < G$ is regular, then $HZ(G)$ is also regular.

Exercise 26. Let H be an regular subgroup of a p-group G and let $R \triangleleft G$ be of exponent p. If R centralizes H, then HR is regular.

§8

Pyramidal p-groups

In this section we define so called pyramidal p-groups. All regular p-groups are pyramidal and there exist irregular pyramidal p-groups. We prove for some pyramidal groups an analog of Theorem 6.1. All results of this section are taken from [Ber22, §3].

We define the upper Ω-series $\{1\} = \Omega_{(0)}(P) < \Omega_{(1)}(P) < \cdots < \Omega_{(s)}(P) < \cdots$ of a p-group P as follows:

$$\Omega_{(0)}(P) = \{1\}, \quad \Omega_{(i+1)}(P)/\Omega_{(i)}(P) = \Omega_1(P/\Omega_{(i)}(P)),\ i = 0, 1, \ldots.$$

Obviously, $\Omega_i(P) \leq \Omega_{(i)}(P)$. Next, $\Omega_{(i)}(P) = \Omega_{(i+1)}(P)$ implies $\Omega_{(i)}(P) = P$. ($\Omega_i(P) = \Omega_{i+1}(P)$ does not imply $\Omega_i(P) = P$. Indeed, let $G = \mathrm{D}_8 * \mathrm{C}_{2^n}$, $n > 2$, where $\mathrm{D}_8 \cap \mathrm{C}_{2^n}$ has order 2; then $\Omega_1(G) = \mathrm{D}_8 * \mathrm{C}_4 = \Omega_2(G) < G$.)

Set $|\Omega_{(i+1)}(P) : \Omega_{(i)}(P)| = p^{u_{i+1}}$, $|\mho_i(P) : \mho_{i+1}(P)| = p^{w_{i+1}}$, $i = 0, 1, \ldots$.

Definition 1. Let P be a p-group.

1. P is said to be *upper pyramidal* if $u_1 \geq u_2 \geq \cdots$.
2. P is said to be *lower pyramidal* if $w_1 \geq w_2 \geq \cdots$.
3. P is said to be *pyramidal* if it is upper and lower pyramidal simultaneously.
4. P is said to be *strongly pyramidal*, if all its sections are pyramidal.

The class of pyramidal p-groups is very large. Below we consider only some subclasses of pyramidal p-groups.

Regular p-groups are pyramidal (Proposition 7.3) so strongly pyramidal. Dihedral 2-groups are strongly pyramidal and semidihedral 2-groups are pyramidal. Generalized quaternion groups are lower but not upper pyramidal. Metacyclic 2-groups are lower pyramidal. Direct products of upper (lower) pyramidal p-groups have the same property. All p-groups G of maximal class with $|G| \geq p^{2p-1}$ and $|\Omega_1(G)| = p^{p-1}$ are not upper pyramidal; p-groups of maximal class are lower pyramidal. The properties 1–3 of Definition 1 are not inherited by normal subgroups and epimorphic images. Indeed, let $G = A \times B$, where A is the ordinary quaternion group and $|B| = 2$. Then G is pyramidal but G/B and A are not upper pyramidal.

Definition 2. A p-group P is said to be *generalized homocyclic* if it satisfies the following conditions (we retain the notation of Definition 1):

(i) $u_1 = u_2 = \cdots = w_1 = w_2 = \cdots$.

(ii) $\Omega_{(i+1)}(P)/\Omega_{(i)}(P)$ are abelian for all nonnegative integers i.

(iii) If H is a term of the upper or lower central series of P, then $H = \Omega_i(P)$ for some nonnegative integer i.

Any abelian generalized homocyclic p-group is homocyclic. If P is a generalized homocyclic group of exponent p^e, then $\Omega_{(i)}(P) = \Omega_i(P)$ and $\exp(\Omega_i(P)) = p^i$ for all $i \le e$. Next, $\Omega_1(P) \le Z(P)$; moreover, if $\exp(Z(P)) = p^k$, then $Z(P) = \Omega_k(P)$ (this follows from part (iii) of Definition 2). Similarly, if $\exp(P/P') = p^k$, then $P' = \mho_k(P)$ and P/P' is homocyclic.

Exercise. If G is nonabelian of order p^4 and exponent p^2, it is not generalized homocyclic. (*Hint.* We have $u_1 = u_2 = w_1 = w_2 = 2$. If $p > 2$, then G is metacyclic so $|G/G'| = p^3$ so $G' \ne \Omega_1(G)$. If $p = 2$, then $|G'| = 2$ so again $G' \ne \Omega_1(G)$.)

Sylow 2-subgroups of the Suzuki simple groups $\mathrm{Sz}(2^m)$ are generalized homocyclic (these groups have order 2^{2m} with odd $m > 1$; see §46).

Let P be a generalized homocyclic p-group of exponent p^e and $i \le e$. Then $\Omega_i(P) = \mho_{e-i}(P)$, $\exp(\Omega_i(P)) = p^i$ and $|P| = p^{u_1 e}$.

Lemma 8.1. *Let a p'-group X act on a p-group P, $p > 2$, $\exp(P) = p^e$. Let $G = X \cdot P$ be the natural semidirect product with kernel P.*

(a) Suppose that P is upper pyramidal and $\Omega_1(P)$ is a minimal normal subgroup of G. Then P, $\Omega_i(P)$ and $P/\Omega_i(P)$ are generalized homocyclic for $i > 0$. If $H < P$ is G-invariant, then $H = \Omega_i(G)$ for some $i < e$. Moreover, $\Omega_{i+1}(P)/\Omega_i(P)$ is a minimal normal subgroup of $G/\Omega_i(P)$, $i = 1, \dots, e-1$.

(b) Suppose that P is lower pyramidal and $P/\mho_1(P)$ is a minimal normal subgroup of $G/\mho_1(P)$. Then P, $P/\mho_i(P)$ and $\mho_i(P)$ are generalized homocyclic. Moreover, $\mho_i(P)/\mho_{i+1}(P)$ is a minimal normal subgroup of $G/\mho_{i+1}(P)$, $i = 1, \dots, e-1$.

Proof. We use induction on $|P|$. One may assume that $X > \{1\}$ acts nontrivially on P (otherwise, P is cyclic so homocyclic) and $e > 1$ (otherwise, P is elementary abelian so homocyclic).

(a) Let P be upper pyramidal. Since $\Omega_1(P)$ is a minimal normal subgroup of G and $[P, \Omega_1(P)] < \Omega_1(P)$, we get $\Omega_1(P) \le Z(P)$ since $[P, \Omega_1(P)] \lhd G$. It follows that $\Omega_1(P)$ is a unique minimal normal p-subgroup of G. We have $|\Omega_1(P)| = p^u$, where $u = u_1$. By hypothesis, $u_2 \le u$. Let $N/\Omega_1(P)$ be a minimal normal p-subgroup of $G/\Omega_1(P)$. Since $N/\Omega_1(P)$ is elementary abelian, it follows that $\mathrm{cl}(N) \le 2$ so N is regular, in view of $p > 2$, and $\exp(N) = p^2$ implies that $N \le \Omega_2(P)$. We have $|N/\mho_1(N)| = |\Omega_1(N)| = |\Omega_1(P)| = p^u$, by Definition 2(i) and Theorem 7.2(d). It follows that $\mho_1(N) = \Omega_1(P)$ hence $|N| = p^{2u}$. It follows from $u_2 = u$ that $N = \Omega_2(P)$. Thus,

$$\Omega_1(P/\Omega_1(P)) = \Omega_2(P)/\Omega_1(P) = N/\Omega_1(G)$$

is a minimal normal subgroup of $G/\Omega_1(P)$. By induction, $P/\Omega_1(P)$ is generalized homocyclic. Therefore, $(u =)u_1 = u_2 = \cdots = u_e$.

Let $\{1\} < H \triangleleft G$ be a p-subgroup. We claim that then $H = \Omega_i(P)$ for some i. We have $\Omega_1(P) \le H$ since $\Omega_1(P)$ is a unique minimal normal p-subgroup of G. By induction, $H/\Omega_1(P) = \Omega_j(P/\Omega_1(P))$ since $P/\Omega_1(P)$ is generalized homocyclic. Therefore, $H = \Omega_{j+1}(P)$, and the result follows with $i = j + 1$. Thus, condition (iii) of Definition 2 is fulfilled, since for H one can take any member of the upper (or lower) central series of P. Let $H < P$ be normal in G and assume that $\Omega_i(P) < H < \Omega_{i+1}(P)$. Then, by what has just has been proved, $H = \Omega_k(P)$. Since $i < k < i + 1$, we get a contradiction. Thus, $\Omega_{i+1}(P)/\Omega_i(P)$ is a minimal normal subgroup of $G/\Omega_i(P)$ so $P/\Omega_i(P)$ is generalized homocyclic.

It remains to prove that $w_1 = w_2 = \cdots = u_1$. By the previous paragraph, this is true if $e = 2$ so one may assume that $e > 2$. Since $\mho_1(P) > \{1\}$ is G-invariant and $\Omega_1(P)$ is a unique minimal normal p-subgroup of G, it follows that $\Omega_1(P) < \mho_1(P)$. Since $P/\Omega_1(P)$ is generalized homocyclic, we have, by induction, $p^{w_1} = \cdots = p^{w_{e-1}} = p^{u_2}$; since $u_2 = u_1$, our claim follows. Thus, P is generalized homocyclic. Since $\Omega_1(\Omega_i(P)) = \Omega_1(P)$ is a minimal normal subgroup of $G = X \cdot \Omega_i(P)$, in view of $\Omega_1(P) \le \mathrm{Z}(P)$, it follows, by what has just been proved, that $\Omega_i(P)$ is generalized homocyclic. The last assertion follows from the above, and (a) is proven.

(b) Let P be lower pyramidal. Since $P/\mho_1(P)$ is minimal normal p-subgroup of $G/\mho_1(P)$, it is elementary abelian, and so $\mho_1(P) = \Phi(P)$. Set $|P/\mho_1(P)| = p^w$ (obviously, $w = w_1$). Let $\mho_2(P) \le U < \mho_1(P)$, where U is a G-invariant subgroup such that $\mho_1(P)/U$ is a minimal normal subgroup of G/U. Then $\mho_1(P)/U \le \mathrm{Z}(P/U)$ so $\mathrm{cl}(P/U) \le 2$ and hence regular (Theorem 7.1(b)). We claim that $|\mho_1(P)/U| = p^w$. To prove this, one may assume without loss of generality that $U = \{1\}$. We have $|\Omega_1(P)| = |P/\mho_1(P)| = p^w$ (Theorem 7.2(d)). Since $\mho_1(P) \le \Omega_1(P)$ and $P/\mho_1(P)$ is a minimal normal subgroup of G, we get $\mho_1(P) = \Omega_1(P)$, and our claim follows. Then, by induction, $\Phi(P)$ is generalized homocyclic in general case so $\Omega_1(\Phi(G))$ is a minimal normal subgroup of $X \cdot \Phi(G)$. Assume that $e > 2$. Then $\Omega_1(P)\mho_1(P) = \Omega_1(P)\Phi(P) < P$ so $\Omega_1(P) < \Phi(P)$ since $P/\Phi(P)$ is a minimal normal subgroup of $G/\Phi(P)$, and we conclude that $\Omega_1(P) = \Omega_1(\Phi(P))$. It follows that $\Omega_1(P)$ is a minimal normal subgroup of G so P is generalized homocyclic, by (a). All other assertions in (b) now follow. □

If P is a generalized homocyclic p-group then, generally speaking, $\Omega_i(P)$ and $P/\Omega_i(P)$ are not generalized homocyclic (it is essential, in Lemma 8.1, that X acts on P in a special way).

Definition 3. A p-group P is said to be a $(*)$-group if $\Omega_1(H)$ is abelian for every section H of P.

Subgroups and epimorphic images of $(*)$-groups are $(*)$-groups. If a p-group G is a $(*)$-group, then $\exp(\Omega_k(G)) \le p^k$.

Lemma 8.2. *Let a p-group P be a $(*)$-group, $p > 2$. Suppose that a p'-group X acts on P via automorphisms in such a way that all X-invariant subgroups of P are pyramidal (this is a case if P is abelian). Then P has an X-invariant generalized homocyclic subgroup H such that $\Omega_1(H)$ is a minimal X-invariant subgroup and* $\exp(H) = \exp(P)$. *If, in addition, $\Omega_1(H) = \Omega_1(P)$, then $H = P$.*

Proof. Let $G = X \cdot P$ be the natural semidirect product. Suppose that the lemma is true for all p-groups of order $< |P|$. Since P is a $(*)$-group, $P/\mho_1(P)$ is elementary abelian so $\mho_1(P) = \Phi(P)$.

(i) Suppose that $P/\mho_1(P)$ is a minimal normal subgroup of $G/\mho_1(P)$. Then P is generalized homocyclic, by Lemma 8.1(b), and we are done with $H = P$.

(ii) Now let $P/\mho_1(P)$ be a reducible X-group. Then, since $P/\mho_1(P)$ is elementary abelian, $P = AB$, where A, B are proper X-invariant (so normal in G) subgroups of P with $A \cap B = \mho_1(P)$ (Maschke). By the remark, preceding the lemma, one of the subgroups A, B, say A, is of the same exponent as P. By induction, the X-invariant subgroup A, which is a pyramidal $(*)$-group, has an X-invariant generalized homocyclic subgroup H such that $\exp(H) = \exp(A) = \exp(P)$ and $\Omega_1(H)$ is a minimal X-invariant subgroup.

(iii) Now suppose that $\Omega_1(P) = \Omega_1(H)$. Let $\exp(P) = p^e$ and $|\Omega_1(H)| = p^u = |\Omega_1(P)|$. Then $|H| = p^{eu}$ and $|P| \leq p^{eu}$ since P is upper pyramidal. Thus, $|P| \leq |H|$ so $P = H$. □

Theorem 8.3 (Compare with Theorem 6.1). *Let a p'-group X act on a p-group P, $p > 2$, where P is a $(*)$-group, and suppose that all X-invariant subgroups of P are pyramidal and normal in P. Let $G = X \cdot P$ be the natural semidirect product, and let R be a minimal X-invariant p-subgroup of G. Then $P = S \times T$, where S and T are normal in G, S is generalized homocyclic with $\Omega_1(S) = R$.*

Proof. One may assume that $X > \{1\}$ (otherwise, P is abelian, by Theorem 1.20, and the result follows from Theorem 6.1). By Lemma 8.2, P has an X-invariant generalized homocyclic subgroup H such that $\exp(H) = \exp(P)$ and $\Omega_1(H)$ is a minimal X-invariant p-subgroup so a minimal normal subgroup of $X \cdot H$. Since $[\Omega_1(H), H] < \Omega_1(H)$ is X-invariant, we get $[\Omega_1(H), H] = \{1\}$, by minimality of $\Omega_1(H)$; then $\Omega_1(H) \leq Z(H)$. By hypothesis, $H \trianglelefteq G = X \cdot P$. Next, $\Omega_1(H)$ is a minimal normal subgroup of $G = X \cdot P$; then $\Omega_1(H) \leq Z(P)$. If $\Omega_1(H) = \Omega_1(P)$, then $H = P$, by Lemma 8.2, and the result follows with $S = H$, $T = \{1\}$.

Let $H < P$. We claim that $P = H \times H_1$, where $H_1 \lhd G$. By what has been said in the previous paragraph, $\Omega_1(H) < \Omega_1(P)$. Since, by hypothesis, $\Omega_1(P)$ is elementary abelian, it follows, by Maschke's theorem, that $\Omega_1(P) = \Omega_1(H) \times L$, where $L > \{1\}$ is X-invariant, and so $L \lhd G$. Let F be a normal p-subgroup of G maximal such that $L \leq F$ and $H \cap F = \{1\}$. Then $F > \{1\}$. We will prove that $P = H \times F$ (so that our claim is true with $S = H$ and $T = F$). It suffices to show that $P = HF$. Suppose that $HF < P$. Then $HF/F < P/F$, $HF/F \cong H$ and $\Omega_1(HF/F)(\cong \Omega_1(H))$ is an elementary abelian minimal normal subgroup of G/F. Next, $\Omega_1(P/F)$ is elementary abelian, by hypothesis, and $\Omega_1(HF/F) < \Omega_1(P/F)$, by the last part of Lemma 8.2. Therefore, by Maschke's theorem, $\Omega_1(P/F) = \Omega_1(HF/F) \times (K/F)$, where $K/F > \{1\}$ is X-invariant so normal in G/F. Since $K > F$ and

$$K \cap H \leq K \cap (HF \cap H) = (K \cap HF) \cap H \leq F \cap H = \{1\},$$

we obtain a contradiction with the choice of F. Thus, $P = HF = H \times F$.

Now we are ready to complete the proof. If $R \leq H$, the result follows with $S = H$, $T = F$. Suppose that $R \nleq H$; then $R \cap H = \{1\}$ since R is a minimal normal subgroup of G and H is G-invariant. Let D be a maximal normal p-subgroup of G, containing R and such that $D \cap H = \{1\}$. By what has been proved in the previous paragraph, $P = H \times D$. Using induction on $|P|$, we obtain $D = S \times U$, where S and U are X-invariant, S is generalized homocyclic and $\Omega_1(S) = R$. By hypothesis, S and U are normal in G. We have

$$P = H \times D = H \times (S \times U) = S \times (H \times U),$$

and the theorem holds with $T = H \times U$. □

Clearly, Theorem 8.3 generalizes Theorem 6.1.

Let X, p, P, R be as in Theorem 8.3. Let $S \leq P$ be maximal X-invariant and $\Omega_1(S) = R$. Then S is generalized homocyclic (Lemma 8.1(a)) and $S \nleq \Phi(P)$.

Corollary 8.4 (compare with Corollary 6.4). *Let X, P, G and p be as in Theorem 8.3, let $R \lhd G$ be of exponent p. Then $P = S \times T$, where S and T are normal in G and $\Omega_1(S) = R$. If S_1 is an X-invariant direct factor of P such that $\Omega_1(S_1) = R$, then $S_1 \cong S$.*

Proof. Since P is a $(*)$-group, R is elementary abelian. In view of Theorem 8.3, one may assume that R is not a minimal normal subgroup of G. Let N be a minimal normal subgroup of G contained in R. By Theorem 8.3, $P = L \times M$, where $L, M \trianglelefteq G$, $\Omega_1(L) = N$ and L is generalized homocyclic. Set $R_1 = R \cap M$. By assumption, $R_1 > \{1\}$. By induction, $M = M_1 \times T$, where $M_1, T \lhd XM$ and $\Omega_1(M_1) = R_1$. By assumption, $M_1, T \lhd G$. Set $S = L \times M_1$. Then $S \lhd G$, $\Omega_1(S) = R$ and $P = S \times T$.

Let $P = S_1 \times T_1$ be another decomposition such that S_1 and T_1 are normal in G and $\Omega_1(S_1) = R$. To fix ideas, we assume that $|T| \geq |T_1|$. Then $|S_1| \geq |S|$. Since $S_1 \cap T = \{1\}$, we have $P = S_1 \times T$ so $|S_1| = |S|$ and $|T| = |T_1|$. Therefore $S \times T = P = S_1 T = S_1 \times T$, and so $S \cong P/T \cong S_1$, completing the proof. □

Corollary 8.5. *Let X, P, G and p be as in Theorem 8.3. Then $P = P_1 \times \cdots \times P_t$, where $P_1, \dots, P_t$ are generalized homocyclic normal subgroups of G such that $\Omega_1(P_1), \dots, \Omega_1(P_t)$ are minimal normal subgroups of G.*

Problem 1. Given $k, w \in \mathbb{N}$, $k > 1$, $w > 2$, does there exist a generalized homocyclic p-group P of class k with $|P/\mho_1(P)| = p^w$?

Problem 2. Study the p-groups all of whose sections are pyramidal.

§9

On p-groups of maximal class

We prove here and in §§12,13 all those properties of p-groups of maximal class which we use in what follows. Since 2-groups of maximal class are classified, we assume in what follows that $p > 2$. We use the important paper [Man6], revising the proofs of some basic theorems on p-groups of maximal class. Main results of this section are due to Blackburn [Bla3].

Recall that a group of order p^m is of maximal class, if $\mathrm{cl}(G) = m - 1 > 1$.

Exercise 1. Let G be a p-group of maximal class and order p^m. Then

(a) $|G : G'| = p^2$, $|Z(G)| = p$, nonabelian epimorphic images of G are of maximal class.

(b) If $1 \le i < m - 1$, then G has only one normal subgroup of order p^i.

(c) If $p > 2$ and $m > 3$, then G has no cyclic normal subgroups of order p^2. (*Hint.* (a) is clear. (b) follows from (a) in the case $i = 1$ so let $m > 3$. Since $G/Z(G)$ is of maximal class, (b) follows for $i > 1$, by induction. (c) follows from Lemma 1.4 and (b).)

Lemma 9.1. *Let G be a noncyclic group of order p^m, $m > 2$. If for any $i = 1, \dots, m - 2$, G has only one normal subgroup of order p^i, then* $\mathrm{cl}(G) = m - 1$.

Proof. We use induction on m. One may assume that $m > 3$. Let $R \le Z(G)$ be of order p. In view of $G/G' \cong E_{p^2}$, G/R is nonabelian of maximal class so $|Z(G)| \le p^2$. Since R is a unique normal subgroup of order p in G, $Z(G)$ is cyclic so it suffices to show that $|Z(G)| = p$. Assume that $|Z(G)| = p^2$. By Lemma 1.4, G has a normal subgroup T of type (p, p). Then $R = Z(G) \cap T$ so $Z(G/R) = Z(G)T/R \cong E_{p^2}$ and G/R is not of maximal class, a contradiction. □

Lemma 9.2. *Let G be a p-group and let $N \lhd G$ be of order $> p$. Suppose that G/N of order $> p$ has cyclic center. If $R/N \lhd G/N$ is of order p in G/N, then R is not of maximal class.*

Proof. Let T be a G-invariant subgroup of index p^2 in N. Then $R \le C_G(N/T)$ so R/T is abelian of order p^3, and we conclude that R is not of maximal class. □

Lemma 9.3. *Let G be a p-group of maximal class and order p^m, $p > 2$, $m > 3$, and let $N \lhd G$ be of index p^3. Then* $\exp(G/N) = p$.

Proof. Assume that this is false. Let T be a G-invariant subgroup of index p in N. Set $\bar{G} = G/T$. By hypothesis, G/N has two distinct cyclic subgroups C/N and Z/N of order p^2. Then $\bar{C}$ and $\bar{Z}$ are abelian so $\bar{C} \cap \bar{Z} = \mathrm{Z}(\bar{G})$ and $\bar{G}$ is not of maximal class, a contradiction. □

Lemma 9.4. *Let $N \lhd G$ and G/N cyclic, $|N| = p^w$ and* $\exp(N) = p$. *Then (a) If G is irregular, then $w \geq p$. (b) If $w = p$ and $\Omega_1(G) = N$, then all proper subgroups of G are regular.*

Proof. (a) Since $G' < N$ and G is irregular, $\mathrm{K}_{p-1}(G)$ $(< N)$ is not cyclic (Theorem 7.1(c)). Then $|N| = |N : \mathrm{K}_{p-1}(G)||\mathrm{K}_{p-1}(G)| \geq p^{p-2} \cdot p^2 = p^p$.

(b) Let $|G : N| > p$ (otherwise, we are done, by Theorem 7.1(b)). Take $M \in \Gamma_1$ and $N \not\leq M$. Then $\Omega_1(M) = N \cap M < N$, and so M is regular by (a). Now let $N < M \in \Gamma_1$. Let D be a G-invariant subgroup of index p^2 in N. Set $\bar{G} = G/D$. Then $\bar{M}$ is abelian as a subgroup of the abelian group $\mathrm{C}_{\bar{G}}(\bar{N})$, and so $M' \leq D$. Since $|D| = p^{p-2}$, it follows that $\mathrm{cl}(M) \leq p-1$ so it is regular (Theorem 7.1(b)). □

Let G be a p-group of maximal class and order p^m, $m > 3$. For each $i = 2, \dots, m-2$, $\mathrm{K}_i(G)/\mathrm{K}_{i+2}(G)$ is a noncentral normal subgroup of order p^2 in $G/\mathrm{K}_{i+2}(G)$ so $M_i = \mathrm{C}_G(\mathrm{K}_i(G)/\mathrm{K}_{i+2}(G)) \in \Gamma_1$. The subgroup $M_2 = \mathrm{C}_G(\mathrm{K}_2(G)/\mathrm{K}_4(G))$ plays distinguished role in what follows; we denote it by G_1 and call the *fundamental subgroup* of G. In this book, if G is of maximal class, then G_1 denotes always the fundamental subgroup of G.

In the proof of Theorem 9.5 we use the following Zassenhaus identity

$$[x, y, \dots, y] = 1, \tag{1}$$

where G is a regular p-group of class $\leq p$ such that $\exp(G') \leq p$, $x, y \in G$ and y appears $p-1$ times (for the proof, see [Hup, Satz 3.9.7]). This identity is only one result, important in what follows, whose proof is omitted. We use this identity in this book only once to prove that p-groups of maximal class and order p^{p+1} are irregular.

Theorem 9.5 ([Bla3]). *Let G be a group of maximal class and order p^m, $m \leq p+1$. Then $\Phi(G)$ and $G/\mathrm{Z}(G)$ have exponent p. If $m = p+1$, then G is irregular and $|\mho_1(G)| = p$.*

Proof [Man6]. One may assume that $p > 2$. The number of distinct members in the set $\mathcal{C} = \{\mathrm{C}_G(\mathrm{K}_i(G)/\mathrm{K}_{i+2}(G)) \mid i = 2, \dots, m-2\}$ is at most $m-3 \leq (p+1)-3 = p-2$ and $|\Gamma_1 - \mathcal{C}| \geq (p+1)-(p-2) = 3$. Therefore there are two different $H, F \in \Gamma_1 - \mathcal{C}$. Let $H = \langle a, \Phi(G)\rangle$, $F = \langle b, \Phi(G)\rangle$ for $a \in H - \Phi(G)$ and $b \in F - \Phi(G)$; here a, b do not centralize $\mathrm{K}_i(G)/\mathrm{K}_{i+2}(G)$ for $i = 2, \dots, m-2$, by the choice H and F. We get $G = HF = \langle a, \Phi(G), b, \Phi(G)\rangle = \langle a, b\rangle$. Suppose that $a^p \in \mathrm{K}_i(G)$, but $a^p \notin \mathrm{K}_{i+1}(G)$ for some $i \in \{2, \dots, m-2\}$; then $o(a\mathrm{K}_{i+1}(G)) = p^2$. We have $\mathrm{K}_i(G) = \langle a^p, \mathrm{K}_{i+1}(G)\rangle$ because $|\mathrm{K}_i(G) : \mathrm{K}_{i+1}(G)| = p$. It follows that a

centralizes $K_i(G)/K_{i+2}(G)$ since $\langle a, K_{i+1}(G)\rangle/K_{i+2}(G)$ is abelian of type (p^2, p) and contains $K_i(G)/K_{i+2}(G)(= \Omega_1(\langle a, K_{i+1}(G)\rangle/K_{i+2}(G))$, contrary to the choice of a. Therefore, $a^p \in K_{m-1}(G)$, and similarly $b^p \in K_{m-1}(G)$ since $a^p, b^p \in G'$. By Theorem 7.1(b), the group $G/K_{m-1}(G)$ of order $\leq p^p$ is regular. Therefore, $\exp(G/K_{m-1}(G)) = p$ (Theorem 7.2(b)). Thus, $\mho_1(G) \leq K_{m-1}(G) = Z(G)$ so $\exp(G/Z(G)) = p$.

Let $M \in \Gamma_1$; then M is regular (Theorem 7.1(b)), and $|\mho_1(M)| \leq |\mho_1(G)| \leq p$, so $|M : \Omega_1(M)| = |\mho_1(M)| \leq p$ (Theorem 7.2(d)), and $|G : \Omega_1(M)| \leq p^2$. It follows that $G' \leq \Omega_1(M)$ so $\exp(G') = p$. Note that $\Phi(G) = G'$.

Let $c_i = [a, b, \dots, b]$ ($i-1$ times), $i = 1, \dots, m-1$ (so $c_1 = a$); then $c_i \in K_i(G)$ for all i. Since $d(G) = 2$, we get $G' = K_2(G) = \langle c_2, K_3(G)\rangle$ (Exercise 7.2). Assume that we have already proved that $K_i(G) = \langle c_i, K_{i+1}(G)\rangle$. Then $c_{i+1} = [c_i, b] \notin K_{i+2}(G)$ (otherwise, b would centralize $K_i(G)/K_{i+2}(G)$), and this implies that $K_{i+1}(G) = \langle c_{i+1}, K_{i+2}(G)\rangle$ since $|K_{i+1}(G) : K_{i+2}(G)| = p$.

It remains to prove that G is irregular if its order equals p^{p+1}. We have $K_p(G) = \langle c_p\rangle \neq \{1\}$, by the previous paragraph. Since $cl(G) = p > 2$ and $\exp(G') = p$, it follows from identity (1) that if G were regular, then $c_p = 1$, which is not the case. Then $\mho_1(G) \neq \{1\}$, by Theorem 7.1(d), so that $|\mho_1(G)| = p$ since $\mho_1(G)$ is contained in the subgroup $Z(G)$ of order p (by the above, $\exp(G/Z(G)) = p$). □

Remark. If G is a 2-generator group of exponent p, then $d(H) < p$ for each $H \in \Gamma_1$. Indeed, since $\Phi(H) \leq \Phi(G)$ and $\Phi(H) \triangleleft G$, one may assume that $\Phi(H) = \{1\}$; then H is elementary abelian. We have $G' = \Phi(G)$ so $|Z(G)| = p$ (Lemma 1.1). Then G is of maximal class so $|G| \leq p^p$ (Theorem 9.5) and hence $|H| \leq p^{p-1}$.

Definition 1. A p-group G is said to be *absolutely regular* if $|G : \mho_1(G)| < p^p$.

As Theorem 9.8(a) shows, absolutely regular p-groups are regular (the proof of this result independent of Theorem 9.6; see also Remark 7.2).

Theorem 9.6 ([Bla3]). *Let G be a group of maximal class and order p^m, $p > 2$, $m > p + 1$. Then G is irregular and*

(a) *$|G : \mho_1(G)| = p^p$. In particular, $\mho_1(G) = K_p(G)$.*

(b) *There is $G_1 \in \Gamma_1$ such that $|G_1 : \mho_1(G_1)| = p^{p-1}$.*

(c) *G has no normal subgroups of order p^p and exponent p (here the condition $m > p + 1$ is essential). Moreover, if $N \triangleleft G$ and $|G : N| > p$, then N is absolutely regular since $N < G_1$. Next, if $N \triangleleft G$ is of order p^{p-1}, then $\exp(N) = p$. In particular, G has no normal cyclic subgroups of order p^2.*

(d) *Let $Z_2 = Z_2(G)$ be a normal subgroup of order p^2 in G, $G_0 = C_G(Z_2)$. Then G_0 is regular such that $|\Omega_1(G_0)| = p^{p-1}$.*

(e) *Let $\Gamma_1 = \{M_1 = G_0, M_2, \dots, M_{p+1}\}$, where G_0 is defined in (d). Then $M_2, \dots, M_{p+1}$ are of maximal class (and so irregular; see Theorem 9.5). Thus,* *the subgroups G_1 from (b) and G_0 from (d) coincide. In what follows we call G_1 the fundamental subgroup of G.*

(f) *In this part, $m \geq 3$ (i.e., we do not assume as in other parts, that $m > p+1$). The group G has an element a such that $|C_G(a)| = p^2$, i.e., $C_G(a) = \langle a, K_{m-1}(G)\rangle$.*

(g) $[K_i(G), K_j(G)] \leq K_{i+j+1}(G)$.

Proof. (a) If $L \triangleleft G$ of index p^{p+1}, then G/L is irregular (Theorem 9.5); then G is also irregular. Since $\exp(G/L) = p^2$ and $(G/L)/Z(G/L)$ is of order p^p and exponent p (Theorem 9.5), we get $|G/\mho_1(G)| = p^p$ (here we use Exercise 1(b)).

(b) Let $\Gamma_1 = \{M_1, \dots, M_{p+1}\}$. Assume that $|M_i : \mho_1(M_i)| \geq p^p$ for all i. Then M_i has a G-invariant subgroup T_i such that $\mho_1(M_i) \leq T_i$ and $|M_i : T_i| = p^p$. By Exercise 1(b), G has only one normal subgroup of index p^{p+1} so $T_1 = T_2 = \dots = T_{p+1}$. It follows that all maximal subgroups of G/T have exponent p so that $\exp(G/T) = p$, contrary to (a). Let $M_1 = G_1$ be such that $|G_1 : \mho_1(G_1)| < p^p$. Since $|G : \mho_1(G)| = p^p$, we get $|G_1 : \mho_1(G_1)| = p^{p-1}$.

(c) Let $N \triangleleft G$ be of index $> p$; then G/N is noncyclic (since $G/G' \cong E_{p^2}$) so $N \leq \Phi(G) < G_1$, where $G_1 \in \Gamma_1$ is absolutely regular (see (b)). By Theorem 9.8(a), below, G_1 is regular so $|\Omega_1(G_1)| = |G_1 : \mho_1(G_1)| = p^{p-1}$ (Theorem 7.2(d)) hence G has no normal subgroups of order p^p and exponent p. If N is a normal subgroup of order $\leq p^{p-1}$ in G, then $N < \Phi(G) < G_1$ so $N \leq \Omega_1(G_1)$ (Exercise 1(c)) is of exponent p (Theorem 7.2). Since $p - 1 \geq 2$, G has no normal cyclic subgroups of order p^2.

(d–e) We have $\Omega_1(G_1) = K_{m-p+1}(G)$, by (b) and Theorem 7.2(d). Since all $K_i(G)$, $i > 1$, are regular, we get $K_{m-p+1}(G) = \Omega_1(K_i(G))$, $i = 2, \dots, m-p+1$.

Assume that $m = p+2$. Let $M \in \Gamma_1$ be regular. Since G has no normal subgroups of order p^p and exponent p (see (c)), it follows that $|M : \mho_1(M)| = |\Omega_1(M)| \leq p^{p-1}$ and so $|\Omega_1(M)| = p^{p-1}$ since $\Phi(G) < M$ and $|\Omega_1(\Phi(G))| = p^{p-1}$, i.e., $\mho_1(M) = K_p(G)$. Next, $M' \leq K_3(G)$ since $|G : M'| \geq p^3 = |G : K_3(G)|$ and $M' \triangleleft G$ (Exercise 1(b)). By the above, $\Omega_1(M) = K_3(G)$ (since $m = p + 2$) so $\exp(K_3(G)) = p$. By Theorem 7.2(f) with $r = 0$ and $s = 1$, we get

$$[M, K_p(G)] = [M, \mho_1(M)] = \mho_1([M, M]) \leq \mho_1(K_3(G)) = \{1\},$$

and so $M = C_G(K_p(G))$ (recall that $|Z(G)| = p$). Hence, M is a unique regular maximal subgroup of G, i.e., $M = G_1$, where G_1 is what we have called the fundamental subgroup of G. In our case, $K_p(G) = Z_2(G)$ (compare orders!). Then $M = C_G(Z_2(G)) = G_0$, where G_0 is defined in part (d). Thus, $G_1 = C_G(Z_2(G)) = G_0$.

We continue to consider the case $m = p+2$. Let $M \in \Gamma_1 - \{G_1\}$. By the previous paragraph, M is irregular so (Theorem 7.1), $\mathrm{cl}(M) = p$, i.e., M is of maximal class. Consider the central series $M > K_3(G) > K_4(G) > \dots > K_{p+2}(G) = \{1\}$ of M. This series is lower central for M since M is of maximal class. It follows that M does not centralize any factor $\mathrm{K}_i(G)/\mathrm{K}_{i+2}(G)$ for $i \geq 2$. Then $\mathrm{C}_G(\mathrm{K}_i(G)/\mathrm{K}_{i+2}(G)) = G_1$ since there is in Γ_1 only one member which is not of maximal class, and so

$$(2) \qquad [G_1, \mathrm{K}_i(G)] \leq \mathrm{K}_{i+2}(G), \ i = 2, \dots, m-2.$$

Returning to the general case $m \geq p+2$, we will prove (2) by induction on i. For $i \leq p$, (2) has proved above. Let $i > p$. Then, employing Theorem 7.2(f) and induction, we get (interpreting below G_1 as $\mathrm{K}_1(G)$),

$$\begin{aligned}[G_1, \mathrm{K}_i(G)] &= [\mathrm{K}_1(G), \mho_1(\mathrm{K}_{i-(p-1)}(G))] = \mho_1([\mathrm{K}_1(G), \mathrm{K}_{i-(p-1)}(G)]) \\ &\leq \mho_1(\mathrm{K}_{i-(p-1)+2}(G)) = \mathrm{K}_{i+2}(G).\end{aligned}$$

Since $G_1 = \mathrm{C}_G(\mathrm{K}_i(G)/\mathrm{K}_{i+2}(G))$ for all $i \geq 2$, we see that $[M, \mathrm{K}_i(G)] \not\leq \mathrm{K}_{i+2}(G)$ for every $M \in \Gamma_1 - \{G_1\}$ hence $[M, \mathrm{K}_i(G)] = \mathrm{K}_{i+1}(G)$, $i \geq 2$, and so $M > \mathrm{K}_3(G) > \cdots > \mathrm{K}_m(G) = \{1\}$ is the lower central series of M so M is of maximal class.

(f) This is trivial for $m = 3$. Let us check this for $m = 4$. Since $|G'| = p^2$, G has only one abelian subgroup A of order p^3. Take $a \in G - A$; then $\mathrm{C}_G(a) = \langle a, \mathrm{Z}(G)\rangle$ has order p^2 since $a^p \in \mathrm{Z}(G)$.

Now let $m > 4$. Let $a \in G - \mathrm{C}_G(\mathrm{K}_i(G)/\mathrm{K}_{i+2}(G))$, $i = 2, \dots, m-2$ (such an a exists since all the above centralizers coincide with G_1). By induction, we have $|\mathrm{C}_{G/\mathrm{K}_{m-1}(G)}(a\mathrm{K}_{m-1}(G))| = p^2$, and so

$$\mathrm{C}_{G/\mathrm{K}_{m-1}(G)}(a\mathrm{K}_{m-1}(G)) = \langle a\mathrm{K}_{m-1}(G), \mathrm{K}_{m-2}(G)/\mathrm{K}_{m-1}(G)\rangle.$$

Since $a \notin \mathrm{C}_G(\mathrm{K}_{m-2}(G))$, $|\mathrm{K}_{m-2}(G)| = p^2$ and $|\langle a, \mathrm{K}_{m-2}(G)\rangle| = p^3$, it follows that $\mathrm{C}_G(a) = \langle a, \mathrm{K}_{m-1}(G)\rangle$, and so $|\mathrm{C}_G(a)| = p^2$.

(g) Next, we establish, by induction on i, that

$$(3) \qquad [\mathrm{K}_i(G), \mathrm{K}_j(G)] \leq \mathrm{K}_{i+j+1}(G).$$

This has already been proved for $i = 1$ (if we interpret G_1 as $\mathrm{K}_1(G)$; see the proof of (2)). Assuming (3) for a given i, and *all* j, we get (using the Three Subgroups Lemma and the equality $[G_1, G] = \mathrm{K}_2(G)$ which is true since $G/[G_1, G]$ is abelian and $G/\mathrm{K}_3(G)$ is nonabelian)

$$\begin{aligned}[\mathrm{K}_{i+1}(G), \mathrm{K}_j(G)] &= [\mathrm{K}_i(G), G, \mathrm{K}_j(G)] \leq [\mathrm{K}_j(G), \mathrm{K}_i(G), G][G, \mathrm{K}_j(G), \mathrm{K}_i(G)] \\ &\leq [\mathrm{K}_{i+j+1}(G), G][\mathrm{K}_{j+1}(G), \mathrm{K}_i(G)] \leq \mathrm{K}_{i+j+2}(G). \qquad \square\end{aligned}$$

If $m = p+1$, G has at most two subgroups of index p that are not of maximal class: G_1 and $\mathrm{C}_G(\mathrm{Z}_2(G))$, unless both these subgroups are equal.

Exercise 2. Let G_1 be the fundamental subgroup of a p-group G of maximal class, $|G| = p^m, m > p+1$. Then $\mho_i(G) = \mho_i(G_1)$ for $i > 0$, and $\exp(G) = \exp(G_1)$.

Theorem 9.7 (Blackburn; for a more general result, see Theorem 12.9). *If $G/\mathrm{K}_{p+1}(G)$ is of maximal class then G is also of maximal class.*

If G is of order $> p^{p+2}$ and class $> p$ and every proper nonabelian epimorphic image of G is of maximal class, then G is of maximal class. Next, if, for each $i \in \{2, \dots, p+1\}$, G contains only one normal subgroup of index p^i, then G is of maximal class. These facts follow immediately from Theorem 9.7.

Theorem 9.8. *Let G be a p-group, $p > 2$.*

(a) (1st Hall regularity criterion) If G is absolutely regular, it is regular.

(b) (Hall) If G has no normal subgroups of order p^{p-1} and exponent p, it is regular.

(c) (2nd Hall regularity criterion) If $|G' : \mho_1(G')| < p^{p-1}$, then G is regular.

(d) If G is irregular, then G' has a characteristic subgroup of exponent p and order $\geq p^{p-1}$.

Proof. (a) Assume that G has a maximal subgroup M such that $|M/\mho_1(M)| \geq p^p$. Let $T/\mho_1(M)$ be a G-invariant subgroup of index p^p in $M/\mho_1(M)$. Set $\bar{G} = G/T$; then $|\bar{G}| = p^{p+1}$. If $\bar{G}$ is regular, then $|\bar{G}/\mho_1(\bar{G})| = |\Omega_1(\bar{G})| \geq |\bar{M}| = p^p$ so $G/\mho_1(G)| \geq p^p$, a contradiction. If $\bar{G}$ is irregular, it is of maximal class (Theorem 7.1(b)) so $|\bar{G}/\mho_1(\bar{G})| = p^p$ (Theorem 9.5), a contradiction again. Thus, all maximal subgroups of G are absolutely regular so regular, by induction on $|G|$. Obviously, all epimorphic images of G are absolutely regular, i.e., G is minimal irregular in sense of Theorem 7.4. By Theorem 7.4, $\mho_1(G) = \mathrm{Z}(G)$. Since $|G/\mho_1(G)| \leq p^{p-1}$, we get $\mathrm{cl}(G) < p$ so G is regular (Theorem 7.1(b)).

(b) Assume that G is irregular. By (a), $|G : \mho_1(G)| \geq p^p$. Let H be a least normal subgroup of G such that $|H : \mho_1(H)| \geq p^p$ and let F be a G-invariant subgroup of index p in H. By the choice of H, $|F : \mho_1(F)| = p^{p-1}$, and so F is regular by (a). By Theorem 7.2(d), $|\Omega_1(F)| = |F : \mho_1(F)| = p^{p-1}$ so $\Omega_1(F)$ is G-invariant of order p^{p-1} and exponent p (Theorem 7.2(b)).

(c) We use induction on $|G|$. By (a), G' is regular, and $|H' : \mho_1(H')| < p^{p-1}$ for all $H < G$ (Corollary 7.3) so H is regular, by induction. Since every proper epimorphic image of G is regular, by induction, G is a group of Theorem 7.4. Then, by Theorem 7.4(b), $\exp(G') = p$ so $|G'| < p^{p-1}$. It follows that $\mathrm{cl}(G) < p$ so G is regular (Theorem 7.1(b)).

(d) In view of (c), we may assume that G' irregular. The subgroup $H = \mathrm{Z}_{p-1}(G')$ is characteristic in G and, by Theorem 7.1(b), regular. By (b), G' has a normal subgroup N of order p^{p-1} and exponent p. Clearly, $N \leq H$ so $|\Omega_1(H)| \geq |N| = p^{p-1}$, and so $\Omega_1(H)$ is the desired subgroup (Theorem 7.2(a)). □

It follows from Theorem 9.8(a) that if a p-group $G = A_1 A_2$, where $|A_i : \mho_1(A_i)| \le p^{\frac{1}{2}(p-1)}$, $i = 1, 2$, then G is regular.

Proposition 9.9. *Suppose that a noncyclic group G of order $p^m > p^3$ has only one normal subgroup N of index p^3 and let K be a G-invariant subgroup of index p in N. Then one of the following holds:*

(*a*) $\mathrm{d}(G) = 2$, $G' < \Phi(G)$. *Then* $G \cong M_{p^4}$.

(*b*) $p > 2$, $\mathrm{d}(G) = 2$, $G' = \Phi(G)$, *G/N is nonabelian of order p^3 and exponent p. In that case, G/K is of maximal class.*

(*c*) $p = 2$, $\mathrm{d}(G) = 2$, *G is of maximal class.*

(*d*) $\mathrm{d}(G) = 3$, $N = \Phi(G) = G'$. *Then $G/K = (E/K)\mathrm{Z}(G/K)$, where E/K is nonabelian of order p^3 (and exponent p if $p > 2$) and $\mathrm{Z}(G/K) \cong \mathrm{C}_{p^2}$.*

Proof. Obviously, $|G/G'| \le p^3$. Below we use the following fact. If a minimal nonabelian p-group of order $> p^3$ has only one normal subgroup of index p^3, then $G \cong \mathrm{M}_{p^4}$.

(i) Suppose that $\mathrm{d}(G) = 2$ and $|G : G'| > p^2$. Then $N = G'$ and G/G' is abelian of type (p^2, p). In that case, G/K is minimal nonabelian so $G/K \cong \mathrm{M}_{p^4}$. Assume that $K > \{1\}$. Let $L < K$ be a G-invariant subgroup of index p. Then G/L is minimal nonabelian of order $p^5 > p^4$, a contradiction. We get a group from (a).

(ii) Suppose that $\mathrm{d}(G) = 2$ and $G' = \Phi(G)$. In that case, $G/\mathrm{K}_3(G)$ is minimal nonabelian so $N = \mathrm{K}_3(G)$. Hence, if $p = 2$, then G is of maximal class. If $p > 2$, then, as in (i), $\exp(G/N) = p$; in that case, G/K is of maximal class.

(iii) Suppose that $\mathrm{d}(G) > 2$; then $G/G' \cong \mathrm{E}_{p^3}$ so $N = G' = \Phi(G)$. Let E/K be a minimal nonabelian subgroup in G/K; then $E < G$ since $\mathrm{d}(G/K) > 2$. By Proposition 10.17, $G/K = (E/K)\mathrm{Z}(G/K)$. Since G/K has only one normal subgroup of order p, $\mathrm{Z}(G/K)$ is cyclic. Let $p > 2$ and set $E_1/K = \Omega_1(G/K)$. Then E_1/K is of order p^3 and exponent p since G/K is regular. It follows that E_1/K is nonabelian since $G/K = (E_1/K)\mathrm{Z}(G/K)$. (Sylow 2-subgroups of the Suzuki simple group $\mathrm{Sz}(2^3)$ satisfy the hypothesis of this part.) □

Theorem 9.10. *If a group G of order $p^m > p^3$ has a subgroup M of order p^{m-1} of maximal class, then G is either of maximal class or $G/G' \cong E_{p^3}$.*

Proof. We have $|G/G'| \le |G/M'| = p^3$. Suppose that G/G' is abelian of type (p, p). Then $G/\mathrm{K}_3(G)$ is nonabelian of order p^3 so $\mathrm{K}_3(G) = \mathrm{K}_2(M)$, and G is of maximal class since $\mathrm{K}_{j+1}(G) = \mathrm{K}_j(M)$ for all $j > 1$. Let $|G/G'| = p^3$; then $G' = M'$. Assume that G/G' is abelian of type (p^2, p). Let L be a G-invariant subgroup of index p in G'. Then G/L of order p^4 is minimal nonabelian, a contradiction since $M/L < G/L$ is nonabelian of order p^3. □

Theorem 9.11 ([Bla2, Theorem 2.6(i)]; Huppert; see also Corollary 36.7). *If $p > 2$, then G is metacyclic if and only if $|G : \mho_1(G)| \le p^2$.s*

Exercise 3. If G is a p-group of maximal class, $p > 3$, $|G| \geq p^{2p-3}$, then G has a normal subgroup isomorphic to $E_{p^{p-1}}$. (*Hint.* $[K_{p-1}(G), Z_{p-1}(G)] = \{1\}$.)

Exercise 4. Let G be a p-group of maximal class and order $> p^3$. Let $\alpha \in \mathrm{Aut}(G)$ with $o(\alpha) = q \neq p$, where q is a prime. Then q divides $p - 1$. (*Hint.* Fundamental subgroup G_1 is characteristic in G. Apply Maschke's theorem to G/G'.)

Exercise 5. If all subgroups of a p-group G, which have order p^3, are abelian, then $\Omega_1(G)$ is abelian.

Exercise 6. Let a p-group G of order $> p^{p+1}$ be not of maximal class and $|G/K_p(G)| = p^p$. Prove that $K_p(G)/K_{p+1}(G)$ is noncyclic.

Exercise 7. If every abelian subgroup of a nonabelian p-group G is either cyclic or of exponent p, then one of the following holds: (a) $\exp(G) = p$, (b) G is a 2-group of maximal class, (c) $p > 2$ and G is a p-group of maximal class and order $\leq p^{p+1}$.

Solution. Suppose that $\exp(G) > p$ and G is not a 2-group of maximal class. Let $A < G$ be cyclic maximal abelian; then $Z(G)$ is cyclic. Let $E_{p^2} \cong R \lhd G$ and $C = C_G(R)$. Then $\exp(C) = p$, $|G : C| = p$ so $A \cong C_{p^2}$. It follows from $C_G(A) = A$ that G is of maximal class (Proposition 1.8). Since G has no subgroups of order p^{p+1} and exponent p (Theorems 9.5 and 9.6), we get $|G| \leq p^{p+1}$.

Exercise 8. Let G be an irregular p-group and N a least normal subgroup of G such that G/N is regular. Suppose that $|G/N| = p^p$. Then G/N is of exponent p, by Theorem 9.8(a). Is G of maximal class?

Exercise 9. Let G be a group of order $p^m > p^p$, $p > 2$. If $G/K_p(G)$ is of maximal class, then G is irregular. Is G of maximal class?

Exercise 10. Let G be a 3-group of maximal class. Prove that the fundamental subgroup G_1 of G is either abelian or minimal nonabelian. (*Hint.* If G_1 is nonabelian, it is metacyclic so $|G_1'| = 3$.)

Exercise 11. Let $\phi : G \to H$ be a lattice isomorphism, where G is a p-group of maximal class and order $> p^{p+1}$. Prove that H is also of maximal class. (*Hint.* $c_1(H) = c_1(G)$. Use Theorem 13.2(a).)

If G is a group, $x \in G$ and $K_x = \{x^y \mid y \in G\}$, then $x^{-1}K_x \subseteq G'$.

Proposition 9.12 ([GLS, Lemma 10.35]). *Assume that G is a group of order p^m, $m > 2$, $x \in G$ and $G = \langle x \rangle Q$, where $Q \in \Gamma_1$ and $|C_G(x)| = p^2$. Then there is $a \in Q$ such that $[a, x, \dots, x] = z$ for some $z \in Z(G)^\#$, where x appears $m - 2$ times.*

Proof. By Proposition 1.8, G is of maximal class so $Z(G) = \langle z \rangle$ is of order p, $G' = \Phi(G)$ is of order p^{m-2}. We have $|Q : C_Q(x)| = |G : C_G(x)| = p^{m-2}$. This is the number of commutators $[a, x]$ for $a \in Q^\#$, so every element of G' has this form. Let $P = \langle x \rangle G'$; then $P \in \Gamma_1$ is of maximal class (Proposition 1.8). By induction, for some $b \in P'$, we have $[b, x, \dots, x] = z$, x appears $(m-1)-2 = m-3$ times. Since $b = [a, x]$ for some $a \in Q$, by what has just been proved, the result follows. □

If $M \trianglelefteq G$, where G is a p-group, and $Z_k(G)$ is the k-th member of the upper central series of G, then $|M \cap Z_k(G)| \geq p^k$.

Proposition 9.13. *Let $k > 1$ and $H \trianglelefteq G$, where G is a p-group. Suppose that $D = H \cap Z_k(G)$ is of maximal class. Then (a) $|H : D| \leq p$ and H is of maximal class, and (b) if, in addition, $H = G$, then $G = Z_k(G)$.*

Proof. We use induction on $|G|$. Set $C = C_H(D)$; then $C \triangleleft G$ and $C \cap D = C_H(D) \cap D = C_D(D) = Z(D)$ is of order p since D is of maximal class. As

(4) $$C \cap Z_k(G) = (C \cap H) \cap Z_k(G) = C \cap (H \cap Z_k(G)) = C \cap D = Z(D)$$

is of order p, we get $|C| = p$ so $C = Z(D) = Z(H)$. We have $C \leq \Phi(D) \leq \Phi(H)$ so $d(H) = d(H/C)$.

(a) Let $|D| = p^3$. It follows from (4) that $C = C_H(D) < D$ so H is of maximal class (Proposition 10.17) and hence $|H : D| \leq p$. Indeed, assuming that $|H : D| > p$, we get $D \leq \Phi(H)$ so D is abelian in view of $|D| = p^3$ (Burnside), a contradiction.

Now suppose that $|D| > p^3$. Then D/C is also of maximal class and $(H/C) \cap Z_{k-1}(G/C) = D/C$ so, by induction, H/C is of maximal class hence $d(H) = d(H/C) = 2$. Assume that $D < H$; then, as above, $|H : D| = |(H/C) : (D/C)| = p$. Then H is of maximal class. by Theorem 9.10.

(b) Let $H = G$. We have to prove that $G = Z_k(G)$. By (a), G is of maximal class. Assume that $D < G$. Then $D = Z_k(G) \in \Gamma_1$ since $D \not\leq \Phi(G)$: $Z(D)$ is cyclic (Proposition 1.13), and this is a final contradiction. Thus, $G = D = Z_k(G)$. □

Exercise 12. If all members of the set Γ_1, $p > 3$, are two-generator, then the p-group G is regular. (*Hint.* We have $|G/\mho_1(G)| \leq p^3$. Use Theorem 9.8(a).)

Exercise 13. Let G be a p-group of maximal class, $p > 2$, and $H < G$. Then $d(H) \leq p$. If $d(H) = p$, then $G \cong \Sigma_{p^2} \in \mathrm{Syl}_p(S_{p^2})$. Thus (Blackburn), if G is a p-group of maximal class, $p > 2$, $N \triangleleft G$ and $G/N \cong \Sigma_{p^2}$, then $N = \{1\}$.

Solution. We use induction on $|G|$. By Theorems 9.5 and 9.6, $d(H) \leq p$. Now let $d(H) = p$; then $|G| > p^p$. Let $|G| = p^{p+1}$. Then $E_{p^p} \cong H \in \Gamma_1$ and so $H = Z(G) \times K$ for some $K < H$ with $|H : K| = p$. In that case, $|G : K| = p^2$ and $K_G = \bigcap_{x \in G} K^x = \{1\}$ so $G \cong \Sigma_{p^2}$. Now let $|G| > p^{p+1}$. Let $H \leq M \in \Gamma_1$. By induction, $M \cong \Sigma_{p^2}$ so $H \cong E_{p^p}$. Since H is characteristic in M, it is normal in G, contrary to Theorem 9.6(c).

Exercise 14. Let H be a nonnormal subgroup of order p^{p-1} and exponent p in a p-group G of maximal class. Prove that $N_G(H)$ is of order p^p and exponent p.

Exercise 15. Let G be a p-group of maximal class and order $> p^3$. Then there is exactly one (i) $A \in \Gamma_1$ such that $|A : A'| > p^2$, (ii) $B \in \Gamma_1$ such that $|Z(B)| > p$.

Exercise 16. Let G be a nonabelian p-group. If for any proper normal subgroup H of G there exists $h \in H$ such that $H = \langle h^x \mid x \in G\rangle$, then G is of maximal class.

Exercise 17. Suppose that a p-group G is neither absolutely regular nor of maximal class. Show that one of the following holds:

(a) G has a characteristic subgroup of order $\geq p^p$ and exponent p.

(b) G has a characteristic subgroup H of class p such that $\Phi(H)$ is of order p^{p-1} and H is generated by all G-invariant subgroups of order p^p and exponent p containing a fixed characteristic subgroup of G of order p^{p-1} and exponent p.

Solution. If G is regular, then (a) holds. Next we assume that G is irregular. By Theorem 9.8(d), G has a characteristic subgroup R of order $\geq p^{p-1}$ and exponent p. Assume that $|R| = p^{p-1}$ and R has maximal order among characteristic subgroups of exponent p in G. Let $H = \langle M < G \mid R < M \trianglelefteq G, |M| = p^p, \exp(M) = p\rangle$; then $|H| > p^p$ (Theorem 12.1(a) and assumption), H is characteristic in G and $\mathrm{cl}(H) \leq p$ since H/R is (elementary) abelian. By assumption, H is irregular. Then $\mathrm{cl}(H) = p$ (Theorem 7.1(b)), $R = \Phi(H) = H'$ (Theorem 9.8(c)).

Exercise 18. Prove that the standard wreath product G of two groups of order p is of maximal class. (*Hint.* $\Omega_1(G) = G, \exp(G) = p^2$.)

Exercise 19. If G is a p-group of maximal class, then either $|\Omega_1(G)| = p^{p-1}$ or $|G : \Omega_1(G)| \leq p$.

Exercise 20. An irregular p-group of maximal class is generated by subgroups of maximal class and order p^{p+1}.

Exercise 21. An irregular p-group G of maximal class is generated by subgroups of order p^p not contained in G_1.

Exercise 22. Let L be a unique normal subgroup of index p^{p+1} in a p-group G. If H/L is an absolutely regular maximal subgroup of G/L and G/L is of maximal class, then H is also absolutely regular. (*Hint.* Use Theorem 12.1(a).)

Exercise 23. Let $R \triangleleft G$ be of order p, $|G| = p^m$, $m > 3$ and $3 \leq k < m$. Show that the number of subgroups of maximal class and order p^k in G that do not contain R, is divisible by p^2.

Exercise 24. Let $R \triangleleft G$ be of order p, $|G| = p^m$, $m > 3$ and $3 \leq k < m$. Show that the number of noncyclic metacyclic subgroups of order p^k in G, which do not contain R, is divisible by p^2.

Exercise 25. Let G be a nonabelian p-group.

(a) Let $Z(G)$ be cyclic and $M \in \Gamma_1$. If $x \in G - M$ is such that $x^p \in \Omega_1(Z(M))$, then $d(Z(M)) \le p$.

(b) If G is a group of exponent p with $|Z(G)| = p$. If $M \in \Gamma_1$, then $|Z(M)| \le p^{p-1}$.

Solution. (a) It suffices to show that $|\Omega_1(Z(M))| \le p^p$. Assume that this is false. Set $H = \langle x, \Omega_1(Z(M))\rangle$. If H is abelian, then $C_G(\Omega_1(Z(M))) \ge \langle x, M\rangle = G$ so $Z(G)$ is noncyclic, a contradiction. Now suppose that H is nonabelian. Then $Z(H) < \Omega_1(Z(M))$ and so $C_G(Z(H)) \ge \langle x, M\rangle = G$ hence $|Z(H)| = p$. It follows that H is of maximal class. Then $|\Omega_1(Z(M))| \le p^p$ (Theorem 9.6).

Theorem 9.14 ([Bla1, Theorem 6]). *A 3-group of maximal class is metabelian.*

Proof. In view of Exercise 10, one may assume that G_1 is minimal nonabelian. Then $\Phi(G) = G'(< G_1)$ is abelian. □

Proposition 9.15 (from Mann's book "Finite p-groups", in preparation). *For each $p^n > p^2$, there exists a p-group of maximal class and order p^n with an abelian maximal subgroup A, and $G = \langle x\rangle \cdot A$, where $o(x) = p$.*

Proof. One may assume that $n > p+1$ since Σ_{p^2} and its appropriate subgroups satisfy the asserted property. Assume also that $p > 2$,

Let $n = 1+k(p-1)$ for some $k > 1$. Let X be an abelian group, written additively, which is a direct sum of $p-1$ cyclic groups $\langle x_0\rangle, \langle x_1\rangle, \dots, \langle x_{p-2}\rangle$ of order p^k so that X is homocyclic of rank $p-1$ and exponent p^k. To attain our aim, one may assume that k is sufficiently large (for example, $k > 2$). Let the endomorphism T of X be defined by its effect on generators as follows: $x_iT = x_i + x_{i+1}$, $i = 0, 1, \dots, p-2$, where we take $x_{p-1} = -px_0 + y$ with $y \in Y = \langle p^2x_0, px_1, \dots, px_{p-2}\rangle$. Elements $x_0+x_1, x_1+x_2, \dots, x_{p-2}+x_{p-1}$ are linearly independent (write a linear dependence relation and read it modulo $\Phi(X)$) so $X = \langle XT\rangle$, and since X is finite, we get $T \in \mathrm{Aut}(X)$.

We have $x_i(T - \mathrm{I}) = x_{i+1}$, where $\mathrm{I} = \mathrm{id}_X$ $(0 \le i \le p-2)$, implying

$$X(T - \mathrm{I}) = \langle x_1, \dots, x_{p-2}, -px_0 + y\rangle = \langle px_0, x_1, \dots, x_{p-2}\rangle.$$

Thus, $|X : X(T - \mathrm{I})| = p$ so $|\ker(T - \mathrm{I})| = p$. It follows that $|C_X(T)| = p$. Since $(p^{k-1}x_{p-2})(T - \mathrm{I}) = p^{k-1}(-px_0 + y) = 0$, we get $C_X(T) = \langle p^{k-1}x_{p-2}\rangle$, and this centralizer is of order p. Let $G = \langle T\rangle \cdot X$ be a natural semidirect product. It follows that G is of maximal class. It remains to show that $T^p = \mathrm{I}$; then $G = \langle X, T\rangle$ has the stated properties. We have $x_0(T - \mathrm{I})^i = x_i$ for $i \le p-1$, and

$$x_0(T - \mathrm{I})^p = x_{p-1}(T - \mathrm{I}) = (-px_0 + y)(T - \mathrm{I}) = -px_1 + y(T - \mathrm{I}).$$

Thus,

$$\begin{aligned} x_0T^p &= x_0(\mathrm{I} + (T - \mathrm{I}))^p = x_0 + \sum_{i=1}^{p-1} \binom{p}{i} x_i - px_1 + y(T - \mathrm{I}) \\ &= x_0 + \sum_{i=2}^{p-2} \binom{p}{i} x_i + px_1 + px_{p-1} - px_1 + y(T - \mathrm{I}) \\ &= x_0 + \sum_{i=2}^{p-2} \binom{p}{i} x_i - p^2x_0 + py + y(T - \mathrm{I}) = x_0 + z + y(T + (p-1)\mathrm{I}), \end{aligned}$$

where $z \in \langle p^2x_0, px_2, \dots, px_{p-2}\rangle = Y$. Choosing y so that $y(T + (p-1)\mathrm{I}) = -z$, we obtain $x_0T^p = x_0$. Since X is a cyclic module over the ring of its endomorphisms generated by T, we get $T^p = \mathrm{I}$. Let $G = \langle t, X\rangle$, where $t^p = 1$, and let t act on X as T. Then G is the desired group for $n = 1 + kp$.

If $m > n$ and $m \equiv 1 \pmod p$, there exists a group of maximal class and order p^m with abelian subgroup of index p, by the previous paragraph. Then $G/\mathrm{K}_n(G)$ of maximal class and order p^n has abelian subgroup of index p. □

Theorem 9.16. *Let G be of maximal class and order $p^m \ge p^p$, $p > 2$. (a) The number of subgroups of order p^{p-1} and exponent p in G is $\equiv 1 \pmod{p^{m-p}}$. (b) The number of subgroups in G that are isomorphic to $\mathrm{E}_{p^{p-1}}$, is either 0 or $\equiv 1 \pmod{p^{m-p}}$. (c) If G contains a subgroup $E \cong \mathrm{E}_{p^{p-2}}$, then G has a normal subgroup isomorphic to E.*

Proof. If $m = p$, (a–c) are obvious. Next we assume that $m > p$.

(a, b) Let E be a nonnormal subgroup of order p^{p-1} and exponent p in G. Then $|G : \mathrm{N}_G(E)| = p^{m-p}$ (Exercise 14). Since G has exactly one normal subgroup of order p^{p-1} and exponent p, (a) follows.

It remains to show that G contains a normal subgroup isomorphic to $\mathrm{E}_{p^{p-1}}$ provided it has a subgroup $E \cong \mathrm{E}_{p^{p-1}}$. In view of Exercise 1.7, one may assume that $|G : E| > p^2$. If $E < G_1$, then $E = \Omega_1(G_1) \lhd G$. Now let $E < M \in \Gamma_1$. Then, by induction, there is in M a normal subgroup $E_1 \cong \mathrm{E}_{p^{p-1}}$. Then E_1 is characteristic in M so $E_1 \lhd G$. The proof of (b) is complete.

(c) We use induction on m. We assume that $p > 5$ (see Theorem 10.4) and $m > p$ (see Exercise 1.7 and Theorems 9.5 and 9.6). If $E < G_1$, then $E < \Omega_1(G_1)$, and the result follows from Exercise 1.6(a). Thus, $E < M \in \Gamma_1 - \{G_1\}$. By induction, there is $E_1 \cong \mathrm{E}_{p^{p-2}}$ which is normal in M. Then $E_1 < \Phi(M) < \Phi(G) < G_1$, so $E_1 < \Omega_1(G_1)$, and the result follows from Exercise 1.6(a). □

Exercise 26. Let G be a p-group, $p > 2$, $|G : G'| = p^2$. (a) Then G has no maximal subgroup H such that $\mathrm{d}(H) > p$. (*Hint.* Consider $G/\Phi(H)$.) (b) If, in addition, $H \in \Gamma_1$ with $\mathrm{d}(H) = p$, then $G/\Phi(H)$ is isomorphic to a Sylow p-subgroup of the symmetric group of degree p^2.

Exercise 27. Prove that all nonabelian two-generator groups of exponent 3 have order 3^3. (*Hint.* Use Theorem 9.5.)

Exercise 28. Let G be a p-group of maximal class and order $> p^{p+1}$. Show that (i) $\exp(G_1) = \exp(G)$. (ii) If $x \in G$ is of order $\geq p^3$, then $x \in G_1$.

Exercise 29. Suppose that a p-group G has a cyclic subgroup Z of index p^t, $2 < t \leq p-1$. Then G is either absolutely regular or of maximal class or $t = p-1$ and G is an L_p-group. (A p-group G is said to be an L_s-group if $\Omega_1(G)$ is of order p^s and exponent p and $G/\Omega_1(G)$ is cyclic of order $> p$; see §§17, 18.)

Proposition 9.17. *Suppose that a p-group G of maximal class, $p > 3$, has two distinct elementary abelian subgroups of order p^{p-1}. Then $|G| = p^{p+1}$.*

Proof. By Theorem 9.16(b), there is $\mathrm{E}_{p^{p-1}} \cong E \lhd G$. By Fitting's lemma, $|G| > p^p$ so G is irregular (Theorem 9.5). Assume that $|G| > p^{p+1}$. Then $E = \Omega_1(\Phi(G)) = \Omega_1(G_1)$. Let $E_1 < G$ and $E \cong E_1 \neq E$. Write $H = EE_1$. Since $H \cap G_1 = E$, we get $|H| = p^p$ so $\exp(H) = p$. By Exercise 13, H is nonabelian so $\mathrm{Z}(H) = E \cap E_1$ has index p^2 in H so $|H'| = p$. Since all subgroups of G, that contain H, are of maximal class, we get $H' = \mathrm{Z}(G)$. Let $H < F < M \leq G$, where $|F : H| = p = |M : F|$; then F and M are of maximal class. By Theorem 9.6(c), $H \ntrianglelefteq M$. Therefore, $H_1 = H^x \neq H$ for any $x \in M - F$ and $H_1 < F$. As above, $H_1' = \mathrm{Z}(G)$. Then $H/\mathrm{Z}(G)$ and $H_1/\mathrm{Z}(G)$ are two distinct abelian maximal subgroups of $F/\mathrm{Z}(G)$ so $\mathrm{cl}(F/\mathrm{Z}(G)) \leq 2$ (Fitting's lemma). In that case, $\mathrm{cl}(F) \leq 3$, a contradiction since F is of maximal class and order p^{p+1}, $p \geq 5$. Thus, $|G| = p^{p+1}$. □

It is easy to show that, in fact, the group of Proposition 9.17 is isomorphic to $\Sigma_{p^2} \in \mathrm{Syl}_p(\mathrm{S}_{p^2})$. Proposition 9.17 is not true for $p \leq 3$.

Proposition 9.18. *Let $p > 3$ and suppose that a p-group G of maximal class contains an abelian subgroup A such that $\mathrm{d}(A) = p-1$ and $\exp(A) > p$. Then $A \leq G_1$ and G_1 contains a G-invariant abelian subgroup B of order p^p with $\Omega_1(A) < B$.*

Proof. We have $|A| \geq p^p$. If $|G| = p^{p+1}$, then $A = G_1$, and we are done. Now let $|G| > p^{p+1}$. Then, by Proposition 9.17, $\Omega_1(A)$ is a unique normal elementary abelian subgroup of order p^{p-1} in G so $\Omega_1(A) = \Omega_1(G_1)$. We have $A \leq \mathrm{C}_G(\Omega_1(A)) \leq \mathrm{C}_G(\mathrm{Z}_2(G)) = G_1$. Now existence of B follows from Theorem 10.1 applied to $\Omega_1(A) < A < G_1 < G$. □

Proposition 9.19. *Let $p > 3$ and suppose that a p-group G of maximal class contains an abelian subgroup A such that $\mathrm{d}(A) = p-1$, $\exp(A) = p^k > p$ and $|A| = p^{(p-1)k-\epsilon}$, $\epsilon \in \{0, 1\}$. If $\epsilon = 0$, then $A \lhd G$. If $\epsilon = 1$, then there is an abelian $B \lhd G$ of order $|A|$.*

Proof. By Proposition 9.18, $A \leq \Omega_k(G_1)$, and we are done if $\epsilon = 0$. If $\epsilon = 1$, then $\Omega_k(G_1)$ contains $\equiv 1 \pmod p$ abelian subgroups of index p (Exercise 1.6(a)). □

In conclusion we made some additional remarks on p-groups of maximal class.

Definition 2. Let G be a group of maximal class and order p^m, $m > p + 1$. Then G is said to be (i) a $\mathcal{Q}_p$-group, if $|\Omega_1(G)| = p^{p-1}$, (ii) a $\mathcal{D}_p$-group if $\Omega_1(G) = G$, (iii) an $\mathcal{S}\mathcal{D}_p$-group if $|\Omega_1(G)| = p^{m-1}$.

It follows from Theorem 9.6 that, if G is of maximal class and order $> p^{p+1}$, it is one of the above three types. Next, Q_{2^m} is a $\mathcal{Q}_2$-group, D_{2^m} is a $\mathcal{D}_2$-group and SD_{2^m} is an $\mathcal{S}\mathcal{D}_2$-group.

Definition 3. Let G be a $\mathcal{D}_p$-group of maximal class. Then G is said to be a $\mathcal{D}_p^0$-group if $G_1 = H_p(G)$, and a $\mathcal{D}_p^1$-group if $G = H_p(G)$, where $H_p(G)$ is the Hughes subgroup of G. (D_{2^m} is a $\mathcal{D}_2^0$-group.)

Theorem 9.20. *Let G be a p-group of maximal class and order p^m, $m > p + 2$, and let $\Gamma_1 = \{G_1, G_2, \dots, G_{p+1}\}$, where G_1 is the fundamental subgroup of G. Then*

(a) If G is a $\mathcal{Q}_p$-group, then $G_2, \dots, G_{p+1}$ are $\mathcal{Q}_p$-groups.

(b) If G is a $\mathcal{D}_p^0$-group, then $G_2, \dots, G_{p+1}$ are $\mathcal{D}_p^0$-groups.

(c) G has no maximal subgroups which are $\mathcal{S}\mathcal{D}_p$-groups.

(d) If G is an $\mathcal{S}\mathcal{D}_p$-group and $\Omega_1(G) = G_2$, then $G_3, \dots, G_{p+1}$ are $\mathcal{Q}_p$-groups.

(e) If G is a $\mathcal{D}_p^1$-group, then at least two of subgroups $G_2, \dots, G_{p+1}$ are $\mathcal{D}_p$-groups.

Proof. One may assume that $p > 2$. As we know, if $i > 1$, then G_i is of maximal class and so (a) holds.

(b) We have $H_p(G) = G_1$. If $i > 1$, then $H_p(G_i) \leq G_i \cap G_1 = \Phi(G) < G_i$ so G_i is a $\mathcal{D}_p^0$-group.

(c) Let $M \in \Gamma_1$ be an $\mathcal{S}\mathcal{D}_p$-group. Then $|G : \Omega_1(M)| = p^2$ so $\Omega_1(M) = \Phi(G) < G_1$ and $|\Omega_1(M)| = |\Omega_1(G_1)| = p^{p-1}$ so $|G| = p^2|\Omega_1(M)| = p^{p+1} < |G|$, a contradiction.

(d) Clearly, G_2 is a $\mathcal{D}_p$-group. Let $i > 2$; then G_i is not an $\mathcal{S}\mathcal{D}_p$-group, by (c). Since $\Omega_1(G_i) \leq G_i \cap G_2 = \Phi(G)$ is absolutely regular, (d) follows.

(e) Let $R < G$ be of order p^p and exponent p and let $R < M \in \Gamma_1$. By (c), M is a $\mathcal{D}_p$-group. Let $x \in G - M$ be of order p; then $R_1 = \langle x, \Omega_1(G_1)\rangle$ is of order p^p and exponent p so a maximal subgroup of G containing R_1, is a $\mathcal{D}_p$-group different of M. □

§10

On abelian subgroups of p-groups

Let G be a p-group, Θ a group-theoretic property and $\mathfrak{T}$ the set of all Θ-subgroups in G. Suppose that $p \nmid |\mathfrak{T}|$. Let G act on the set $\mathfrak{T}$ via conjugation. Since the size of every G-orbit on the set $\mathfrak{T}$ is a power of p, there is a one-element G-orbit, say $\{H\} \subseteq \mathfrak{T}$; then $H \trianglelefteq G$. Moreover, if G is normal in some larger p-group W, then the set $\mathfrak{T}$ admits the action of W via conjugation and, as above, $\mathfrak{T}$ has a one-element W-orbit. We see that, in many respects, counting theorems are more fundamental than the corresponding theorems on the existence of normal subgroups.

We will show how this argument works in concrete situation. Let a p-group G contain a subgroup H of maximal class and index p. We claim that then $\mathrm{Aut}(G)$ is a π-group, where $\pi = \pi(p(p^2-1))$. Note that $\pi(\mathrm{Aut}(B)) \subseteq \pi$ for every two generator p-group B (Theorem 1.16). Therefore, the result is true if G is of maximal class. Suppose that G is not of maximal class. Let $\alpha \in \mathrm{Aut}(G)$ be of prime order $q \notin \pi$. By Theorem 12.12, $\mathrm{d}(G) = 3$ and the number of subgroups of maximal class and index p in G equals p^2. Since $q \neq p$, one of these subgroups, say H_1, is α-invariant. Since $\mathrm{Aut}(H_1)$ is a π-group, α centralizes H_1. Then $\langle\alpha\rangle$ stabilizes the chain $G > H_1 > \{1\}$. By Lemma 10.12 below, $o(\alpha)$ is a power of p, a contradiction. Similarly, if G is a 2-group of order at least 2^5, H a subgroup of maximal class and index 2 in G, then $\mathrm{Aut}(G)$ is a 2-group (the condition $|G| \geq 2^5$ is essential).

If $A < G$ is maximal normal abelian, G is a p-group, then $\mathrm{C}_G(A) = A$. This property allows us to control the structure of G provided all its maximal abelian normal subgroups are small. Some results in this direction were proved and used in [FT] and [Tho3]. For example, if $p > 2$ and $A < G$ is maximal normal elementary abelian, then $\Omega_1(\mathrm{C}_G(A)) = A$. Alperin [Alp3] has generalized that result as follows: Let A be a maximal normal abelian subgroup of exponent $\leq p^n$ of a p-group G, where $p^n > 2$; then $\Omega_n(\mathrm{C}_G(A)) = A$. We will prove the following stronger result:

Theorem 10.1. *Let $A < B \leq G$, where A, B are abelian subgroups of a p-group G, $|B : A| = p$, $\exp(B) = p^n$ and $p^n > 2$. Let $\mathfrak{A}$ be the set of all abelian subgroups T of G such that $A < T$, $|T : A| = p$ and* $\exp(T) \leq p^n$. *Then* $|\mathfrak{A}| \equiv 1 \pmod p$.

Proof. We use induction on $|G|$. If $T \in \mathfrak{A}$, then $T \leq \mathrm{C}_G(A)$ so we assume that $G = \mathrm{C}_G(A)$. We also assume that $A > \{1\}$ (Sylow) and $|G : A| > p$ (Exercise 1.6(a)).

Let $\mathfrak{A}'$ be the set of G-invariant elements in the set $\mathfrak{A}$. Since $|\mathfrak{A}| \equiv |\mathfrak{A}'| \pmod p$, it suffices to prove that $|\mathfrak{A}'| \equiv 1 \pmod p$. Let $A < B \leq M \in \Gamma_1$ and $B \in \mathfrak{A}$. Then, by induction, the number of elements of the set $\mathfrak{A}$ contained in M is $\equiv 1 \pmod p$, and so M contains $T \in \mathfrak{A}'$. Set $D = \langle T \mid T \in \mathfrak{A}' \rangle$. Then $D/A \le \Omega_1(Z(G/A))$, and so $cl(D) \le 2$. By construction, $\Omega_n(D) = D$. Therefore, if $x, y \in D$, with $o(x), o(y) \le p^n$, we get

$$(xy)^{p^n} = x^{p^n} y^{p^n} [y, x]^{\binom{p^n}{2}} = [y, x]^{\binom{p^n}{2}} = 1$$

since $p^n > 2$ insures that p divides $\binom{p^n}{2}$ and $[y, x]^p = [y^p, x] = 1$ in view of $y^p \in A \le Z(G)$. Hence $\exp(D) \le p^n$. If H/A is a subgroup of order p in D/A, then H is abelian, $H \lhd G$ and $\exp(H) \le \exp(D) \le p^n$, i.e., $H \in \mathfrak{A}'$. Therefore, one may assume that $G = D$; then G/A is elementary abelian. We have $|\mathfrak{A}'| = c_1(G/A) \equiv 1 \pmod p$. □

Corollary 10.2. *Let G be a p-group, $N \trianglelefteq G$ and let A be a maximal G-invariant abelian subgroup of N with* $\exp(A) = p^n$, $p^n > 2$. *Then* $\Omega_n(C_N(A)) = A$.

Proof. Assume that $A < \Omega_n(C_N(A))$. Then there is $x \in C_N(A) - A$ of order $\le p^n$ such that $x^p \in A$. In that case, $\langle x, A \rangle$ is abelian of exponent p^n and $|\langle x, A \rangle : A| = p$. By Theorem 10.1, the number of abelian subgroups of exponent p^n and order $p|A|$ between A and N is $\equiv 1 \pmod p$. It follows that among these subgroups there is one, say B, that is normal in G, contrary to the choice of A. □

Remark 1. Isaacs [Isa7] has proved that if $p^n > 2$, G is a p-group and $N \lhd G$ is such that $Z(N)$ contains all G-invariant abelian subgroups of N of exponent dividing p^n, then $\Omega_n(N) \le Z(N)$. Let us prove this. Set $A = \Omega_n(Z(N))$. Then A is a maximal G-invariant abelian subgroup of exponent dividing p^n in N, and so $\Omega_n(N) = \Omega_n(C_N(A)) = A$, by Corollary 10.2. It follows that $\Omega_n(N) = A \le Z(N)$.

Lemma 10.3. *Let G be a p-group, $p > 2$, $N \trianglelefteq G$ and let $E_{p^2} \cong A \le N$ be G-invariant. If N has a subgroup $B \cong E_{p^3}$, then N has a G-invariant subgroup $B_1 \cong E_{p^3}$ such that $A < B_1$.*

Proof. Since $p^2 \nmid |\mathrm{Aut}(A)|$, it follows that $A < AC_B(A)$ and $AC_B(A)$ is elementary abelian of order at least p^3 in N. Let D/A be a subgroup of order p in $AC_B(A)/A$; then $D \cong E_{p^3}$, and the result now follows from Corollary 10.2. □

Let $\mathfrak{E}_k(X)$ be the set of subgroups isomorphic to E_{p^k} in a p-group X. Set $e_k(X) = |\mathfrak{E}_k(X)|$. Let $\mathfrak{E}'_k(X)$ be the set of X-invariant members in the set $\mathfrak{E}_k(X)$, and write $e'_k(X) = |\mathfrak{E}'_k(X)|$.

Theorem 10.4. *If $e_3(G) > 0$ for a p-group G, $p > 2$, then $e_3(G) \equiv 1 \pmod p$.*

Proof. We have to prove that $e_3(G) \equiv 1 \pmod p$. By Theorem 5.2,

(1) $$e_3(G) \equiv \sum_{H \in \Gamma_1} e_3(H) \pmod p.$$

Assume that the theorem holds for all proper subgroups of G.

Take $H \in \Gamma_1$. By induction, either $e_3(H) = 0$ or else $e_3(H) \equiv 1 \pmod{p}$. If $e_3(H) \equiv 1 \pmod{p}$ for all $H \in \Gamma_1$, then by (1), $e_3(G) \equiv |\Gamma_1| \equiv 1 \pmod{p}$. Now we assume that $e_3(H) = 0$ for some $H \in \Gamma_1$. By Exercise 1.8, H has no subgroup L of order p^4 and exponent p. By assumption, G has a subgroup $E_0 \cong \mathrm{E}_{p^3}$. Let $E_0 \le F \in \Gamma_1$. By induction, $e_3(F) \equiv 1 \pmod{p}$, and so there is $E \cong \mathfrak{E}'_3(G)$.

Since $e_3(G) \equiv e'_3(G) \pmod{p}$, it suffices to prove that $e'_3(G) \equiv 1 \pmod{p}$. One may assume that there is $E_1 \in \mathfrak{E}'_3(G) - \{E\}$ (otherwise, there is nothing to prove). Set $D = EE_1$. By Fitting's lemma (Introduction, Theorem 21), $\mathrm{cl}(D) \le 2$ so $\exp(D) = p$ since $p > 2$. Considering $D \cap H$ and taking into account that H has no subgroups of order p^4 and exponent p, we get $|H \cap D| = p^3$ so $|D| = p^4$. By Exercise 1.6(a), $e_3(D) \equiv 1 \pmod{p}$ so the number of members of the set $\mathfrak{E}'_3(G)$ in D is $\equiv 1 \pmod{p}$; therefore, one may assume that G has a normal subgroup $E_2 \cong \mathrm{E}_{p^3}$ such that $E_2 \not\le D$. Suppose that $(\mathrm{E}_{p^2} \cong) E \cap E_1 < E_2$. Then $E \cap E_1 \le \mathrm{Z}(EE_1E_2)$ and $EE_1E_2/(E \cap E_1) \cong \mathrm{E}_{p^3}$. In that case, EE_1E_2 is of order p^5 and class 2, and so $\exp(EE_1E_2) = p$. Then $H \cap EE_1E_2$ is of order p^4 and exponent p, contrary to what has been said above. Therefore, $E \cap E_1 \not\le E_2$. Since $|EE_2| = |E_1E_2| = p^4$, we get $|E \cap E_2| = p^2 = |E_1 \cap E_2|$. Since $E \cap E_2$, $E_1 \cap E_2$ are different maximal subgroups of E_2 in view of $E \cap E_1 \not\le E_2$, we get $E_2 = (E \cap E_2)(E_1 \cap E_2) < EE_1 = D$, contrary to the choice of E_2. □

Theorem 10.5. *If $e_4(G) > 0$ for a p-group G, $p > 2$, then $e_4(G) \equiv 1 \pmod{p}$.*

Proof. Suppose that G is a counterexample of minimal order. Then $|G| > p^5$ (Exercise 1.6(a)). By Theorem 10.4, there is $A \in \mathfrak{E}'_3(G)$.

A. Let G have no elementary abelian subgroups of order p^4 containing A. Then

(i) A is a maximal elementary abelian subgroup of G. Therefore,

(ii) If $K < G$ is of exponent p, then $\mathrm{C}_K(A) \le A$.

Note that a Sylow p-subgroup of $\mathrm{Aut}(A) (\cong \mathrm{GL}(3, p))$ is nonabelian of order p^3 and exponent p. Therefore, in view of (ii), the following two assertions are true:

(iii) If $K < G$ and $K \cong \mathrm{E}_{p^4}$, then $|A \cap K| = p^2$. Next, $\mathfrak{E}'_5(G) = \varnothing$.

(iv) G has no subgroups of order p^7 and exponent p. If $K \le G$ is of order p^6 and exponent p, then $A < K$ and K/A is nonabelian.

Let $E < M \in \Gamma_1$, where $E \in \mathfrak{E}_4(G)$. By induction, $|\mathfrak{E}_4(M)| \equiv 1 \pmod{p}$ so that $\mathfrak{E}'_4(G) \ne \varnothing$. Next we assume that $E \lhd G$ and there is $E_1 \in \mathfrak{E}'_4(G) - \{E\}$. Set $K = EE_1$; then $\exp(K) = p$. By (iv), $|K| \le p^6$. Assume that $|K| = p^6$. Then $A < K$, by (iv). Since $\mathrm{C}_K(A) = A$, by (ii), we get $E \cap E_1 = \mathrm{Z}(K) < A$. Since $K/(E \cap E_1)$ is elementary abelian (Lemma 1.11), it follows that K/A, as an epimorphic image of $K/(E \cap E_1)$, is elementary abelian of order p^3, contrary to (iv). Thus, $|K| = p^5$. Hence,

(v) If $B, B_1 \in \mathfrak{E}'_4(G)$ are different, then $|BB_1| = p^5$, i.e., $|B \cap B_1| = p^3$. Moreover, $B \cap B_1 = \mathrm{Z}(BB_1)$ since BB_1 is nonabelian, by (iii).

Thus, $K = EE_1$ is nonabelian of order p^5 and exponent p, $E \cap E_1 = Z(K)$ is of order p^3. Therefore, by (ii),

(vi) $A \not\leq K$.

There is $E_2 \in \mathfrak{E}'_4(G)$ such that $E_2 \not\leq K$ (otherwise, $e'_4(G) \equiv e'_4(K) \equiv 1 \pmod p$, by Theorem 10.1, applied to the pair $Z(K) < K$). Then, by (v), $|E \cap E_2| = p^3 = |E_1 \cap E_2| = |E \cap E_1|$. Assume that $E_2 \cap E \neq E_2 \cap E_1$. Then $E_2 = (E_2 \cap E)(E_2 \cap E_1) \leq EE_1 = K$, a contradiction. Hence $E_2 \cap E = E_2 \cap E_1$. Then $C_G(E_2 \cap E) \geq EE_1 = K$, and so $E_2 \cap K = E \cap E_1 = Z(K)$. Set $L = KE_2 = EE_1E_2$. Then $|L| = p^6$. Since $L/Z(K) \cong E_{p^3}$, we get $cl(L) = 2$, and since $\Omega_1(L) = L$, we get $\exp(L) = p$ since $p > 2$. By (iv), $A < L$. Then $C_L(A) > A$ since $|Z(L)| = p^3 = |A|$, contrary to (ii).

B. Every member of the set $\mathfrak{E}'_3(G)$ is contained in some member of the set $\mathfrak{E}'_4(G)$ (see Theorem 10.1). Set $\mathfrak{E}'_3(G) = \{A_1, \dots, A_r\}$ and $\mathfrak{E}'_4(G) = \{B_1, \dots, B_s\}$. We have to prove that $s \equiv 1 \pmod p$. Suppose that A_i is contained in α_i elements of the set $\mathfrak{E}'_4(G)$, and B_j contains β_j elements of the set $\mathfrak{E}'_3(G)$. By the double counting,

$$\alpha_1 + \dots + \alpha_r = \beta_1 + \dots + \beta_s. \tag{2}$$

We have $\beta_j \equiv \frac{p^4-1}{p-1} = 1 + p + p^2 + p^3 \equiv 1 \pmod p$, $(j = 1, \dots, s)$. By Theorem 10.1, $\alpha_i \equiv 1 \pmod p$ for $i = 1, \dots, r$. Therefore, reading (2) modulo p, one obtains $r \equiv s \pmod p$. Since $r \equiv 1 \pmod p$, by Theorem 10.4, we get $s \equiv 1 \pmod p$. □

Moreover [KonJ], if $p > 2$, then $e_5(G) \equiv 1 \pmod p$ provided $e_5(G) > 0$ (but there exists G such that $e_6(G) = 2$ [JonK]; similar example was given by G. A. Miller).

Corollary 10.6. *Let N be a normal subgroup of a p-group G, $p > 2$. Suppose that N has no G-invariant subgroups isomorphic to E_{p^3}. Then N has no subgroups isomorphic to E_{p^3}.*

Corollary 10.7. *Let N be a normal subgroup of a p-group G, $p > 2$. Suppose that N has no G-invariant subgroups isomorphic to E_{p^4}. Then N has no subgroups isomorphic to E_{p^4}.*

Recall that a nonnilpotent group G is said to be *minimal nonnilpotent* if all its proper subgroups are nilpotent. Given $a > 1$ and $n > 1$, a prime p is said to be a *Zsigmondy prime* for a pair $\langle a, n\rangle$, if p divides $a^n - 1$ but $p \nmid a^i - 1$ for all $0 < i < n$.

Lemma 10.8 (O. Yu. Schmidt, Yu. A. Golfand; see also Theorem A.22.1). *Let S be a minimal nonnilpotent group. Then $S = P \cdot Q$, where $p \neq q$ are primes, $P \in Syl_p(S)$ is cyclic, $Q = S' \in Syl_q(S)$ is either elementary abelian or special, $|P : (P \cap Z(S))| = p$. If $q > 2$, then $\exp(Q) = q$. Next, p is a Zsigmondy prime for the pair $\langle q, b\rangle$, where $b = \log_q(|Q : (Z(G) \cap Q)|)$. If Q is nonabelian, then $\log_q(|Q/Z(Q)|)$ is even and $Z(Q) = Q \cap Z(G)$.*

Remarks. 2. Let us prove [FT, Lemma 8.4(ii)]. Let a p-group G, $p > 2$, have no normal subgroups isomorphic to E_{p^3} and $\alpha \in \mathrm{Aut}(G)^{\#}$. If $o(\alpha) = q$ is a prime, $q \neq p$, then as we will prove, q divides $p^2 - 1$. Let $W = \langle\alpha\rangle \cdot G$ be the natural semidirect product. Take in W a minimal nonnilpotent subgroup $S = \langle\alpha\rangle \cdot P$, where $P = S \cap G$. By Theorem 10.4, P has no subgroups isomorphic to E_{p^3}. Since P is elementary abelian or special of exponent p, we get $d(P) = 2$, and we are done (Theorem 1.16).

3. Let us prove [FT, Lemma 8.5]. Suppose that $P \in \mathrm{Syl}_p(G)$, where $p > 2$ is the smallest prime divisor of $|G|$. If P has no elementary abelian subgroups of order p^3, then G is p-nilpotent. Assume that G is a counterexample. Then G has a p-closed minimal nonnilpotent subgroup S, p divides $|S|$, by Frobenius' normal p-complement theorem. If $\pi(S) = \{p, q\}$, then $q > p + 1$, and so $q \nmid p^2 - 1$, contrary to Remark 2.

Using Lemma 10.8 and number theoretic [FT, Lemma 5.1], it is easy to prove the following

Proposition 10.9 ([FT, Lemma 8.8]). *Suppose that Q is a q-group, $q > 2$, α an automorphism of Q of prime order p. If $p \equiv 1 \pmod q$ and Q has a maximal subgroup Q_0 that has no normal subgroups isomorphic to E_{q^3}, then $p = 1 + q + q^2$ and $Q \cong E_{q^3}$.*

Theorem 10.10. *Let G be a nonmetacyclic group of order p^m, $p > 2$, $3 < n < m$. Then the number of metacyclic subgroups of order p^n in G is a multiple of p, unless G is a 3-group of maximal class, and then the above number is equal* 1.

Proof. We use induction on $|G|$. Let $\mathfrak{M}(H)$ denote the set of all metacyclic subgroups of order p^n in $H \leq G$. One may assume that $\mathfrak{M}(G) \neq \varnothing$. By Hall's enumeration principle (see Theorem 5.2),

$$|\mathfrak{M}(G)| \equiv \sum_{G \in \Gamma_1} |\mathfrak{M}(H)| \pmod p \tag{3}$$

By (3), one may assume that $|\mathfrak{M}(H)| \not\equiv 0 \pmod p$ for some $H \in \Gamma_1$. Then H is either metacyclic or a 3-group of maximal class, by induction. If G is metacyclic, then $|\mathfrak{M}(G)| = s_n(G) \equiv 1 \pmod p$ (Sylow).

(i) Let G be a 3-group of maximal class and let $H \in \mathfrak{M}(G)$; then $\exp(H) = 3^e$, where $e \geq 2$ since $n > 3$. All subgroups of order 3^n, that are not contained in G_1, are of maximal class (see Theorem 9.6). By Theorems 9.6 and 9.11, G_1 is metacyclic, and now $|\mathfrak{M}(G)| = |\mathfrak{M}(G_1)| \equiv 1 \pmod 3$, by Sylow's theorem. Next we assume that G is neither metacyclic nor a 3-group of maximal class. If $n = m - 1$, then the set Γ_1 has no members which are 2-groups of maximal class (otherwise, by Theorem 12.12(b), $G/K_3(G)$ is of order 3^4 and exponent 3; then $|\mathfrak{M}(G)| = 0$, contrary to the assumption). Thus, our H is metacyclic so absolutely regular.

(ii) Let $n = m - 1$. In that case, by Theorems 7.2 (if G is regular) and 12.1(b) (if G is irregular; then $p = 3$), $G = HR$, where $R = \Omega_1(G)$ is of order p^3 and exponent p. Then $G/R \cong H/(H \cap R)$ is metacyclic of order $\geq p^3$ so of exponent $> p$. If $F \in \Gamma_1$ is a 3-group of maximal class, then, by Theorem 12.12(b), $G/\mathrm{K}_3(G)$ is of order 3^4 and exponent 3, a contradiction: the metacyclic H cannot be a member of the set Γ_1. If $L \in \Gamma_1$ and $R \not\leq L$, then $|\Omega_1(L)| = |L \cap R| = p^2$ so, by Theorems 12.1(a) and 9.5 and what has just been proved, L is metacyclic. The number of such L is $|\Gamma_1| - |\Gamma_1(G/R)| \equiv 0 \pmod{p}$.

(iii) $n < m - 1$. By Theorem 13.6 and (i), the contribution of all members of the set Γ_1 which are 3-groups of maximal class in the right-hand side of (3), is a multiple of 3^2. By (ii), the contribution of all members of the set Γ_1, which are metacyclic, in the right-hand side of (3) is a multiple of p. If $H \in \Gamma_1$ is neither metacyclic nor a 3-group of maximal class, then p divides $|\mathfrak{M}(H)|$, by induction. If $H \in \Gamma_1$ is metacyclic, then $|\mathfrak{M}(H)| \equiv 1 \pmod{p}$. Since the number of metacyclic members of the set Γ_1 is a multiple of p, by (ii), we get, by (3), $|\mathfrak{M}(G)| \equiv 0 \pmod{p}$. □

If $G = G_0 > G_1 > \cdots > G_n = \{1\}$ is a chain of G-invariant subgroups of a group of G then $A = \{\alpha \in \mathrm{Aut}(G) \mid (xG_i)^\alpha = xG_i \text{ for all } x \in G_{i-1}\}$ is a subgroup of $\mathrm{Aut}(G)$, the *stability group* of the chain.

Lemma 10.11. *Let $G = G_0 > G_1 > \cdots > G_n = \{1\}$ be a chain of normal subgroups of an arbitrary finite group G and A the stability group of that chain. Let p be a prime such that $p \nmid |\mathrm{Z}(G_{i-1}/G_i)|$ for $i = 2, \ldots, n$. Then $p \nmid |A|$.*

Proof. We use induction on n. Assume that A has an element α of order p. Let $W = \langle\alpha\rangle \cdot G$ be the natural semidirect product. By induction, α induces the identity on G/G_{n-1} so $H = \langle\alpha\rangle \cdot G_{n-1} \lhd W$. By hypothesis, α centralizes G_{n-1} so that $H = \langle\alpha\rangle \times G_{n-1}$. Assume that $\langle\alpha\rangle$ is not characteristic in H. Then there exists $\phi \in \mathrm{Aut}(H)$ such that $\phi(\langle\alpha\rangle) = \langle\beta\rangle \neq \langle\alpha\rangle$. Obviously, $\langle\beta\rangle \lhd H$ and $\langle\alpha, \beta\rangle$ is a subgroup of $\mathrm{Z}(H)$ of order p^2. Then $\langle\alpha, \beta\rangle \cap G_{n-1}$ is a subgroup of $\mathrm{Z}(G_{n-1})$ of order p, contrary to the assumption. Thus, $\langle\alpha\rangle$ is characteristic in H. Then $\langle\alpha\rangle \lhd W$ so α centralizes G and hence $\alpha = \mathrm{id}_G$, a contradiction. □

Lemma 10.12 ([FT, Lemma 8.1]). *If G is a π-group, then the stability group A of a chain $G = G_0 > G_1 > \cdots > G_n = \{1\}$ of G-invariant subgroups is a π-group.*

Proof. We use induction on n. Suppose that A has an element α of order p for some $p \in \pi'$. By induction, α centralizes G_1. Let $x \in G$; then $x^\alpha = xy$ with $y \in G_1$. In that case, $x = x^{\alpha^p} = xy^p$ so $y^p = 1$. It follows from $p \in \pi'$ that $y = 1$ so $x^\alpha = x$ for all $x \in G$ and $\alpha = \mathrm{id}_G$, a contradiction. □

Lemma 10.13. *Let G be a π-group and suppose that D is a subgroup generated by all elements of G of orders 4 and p for all $p \in \pi$. Let α be a π'-automorphism of G. If α centralizes D, then $\alpha = \mathrm{id}_G$.*

This follows easily from Lemma 10.8.

Lemma 10.14 ([FT, Lemma 8.12]). *Let G be a p-group, let $U \leq G$ be abelian and let α be a p'-automorphism of G. If α centralizes $C_G(U)$, then $\alpha = \mathrm{id}_G$.*

Corollary 10.15. *Let α be a p'-automorphism of a p-group G and U an abelian α-admissible subgroup of G. Let $\epsilon = 1$ if $p > 2$ and $\epsilon = 2$ if $p = 2$. If α centralizes $\Omega_\epsilon(C_G(U))$, then $\alpha = \mathrm{id}_G$.*

Proof. Obviously $C_G(U)$ is α-invariant, and α centralizes $C_G(U)$ by Lemma 10.13. Now the result follows from Lemma 10.14. □

Corollary 10.16 ([Bla6]). *Let p, G and ϵ be as in Corollary 10.15. Let U be a maximal among abelian subgroups of G of exponent p^n, where $n \geq \epsilon$. If a p'-automorphism α of G centralizes U, then $\alpha = \mathrm{id}_G$.*

Proof. Let x be an element of $C_G(U)$ of order at most p^n. Since $\langle U, x\rangle$ is abelian of exponent at most p^n, it follows that $x \in U$ by the maximal choice of U. Thus, $\Omega_n(C_G(U)) \leq U$, and the result follows from Corollary 10.15. □

The following proposition which characterizes p-groups of maximal class, is cited many times in the book.

Proposition 10.17. *Let G be a p-group, $B \leq G$ nonabelian of order p^3 and $C_G(B) < B$. Then G is of maximal class.*

Proof. Assume that $|G| \geq p^4$ and the proposition has been proved for groups of order $< |G|$. It is known that a Sylow p-subgroup of $\mathrm{Aut}(B)$ is nonabelian of order p^3. Now, $C_G(B) = Z(B) = Z(G)$. Therefore, by N/C-Theorem, $N_G(B)/Z(G)$ is nonabelian of order p^3. If $x \in G - C_G(B)$ centralizes $N_G(B)/Z(G)$, then x normalizes B so $x \in N_G(B)$, a contradiction. Thus, $C_G(N_G(B)/Z(G)) < N_G(B)/Z(G)$ so, by induction, $G/Z(G)$ is of maximal class. Since $|Z(G)| = p$, we are done. □

Remarks. 4. (See also §83.) Let G be a p-group and $N \leq \Phi(G)$ be a G-invariant nonabelian subgroup of order p^4. Then one of the following holds: (a) $N = M \times C$, where M is nonabelian of order p^3 and $|C| = p$, (b) $p > 2$ and $N = \langle x, y \mid x^{p^2} = y^{p^2} = 1, x^y = x^{p+1}\rangle$ is metacyclic. Indeed, $Z(N)$ is noncyclic (Proposition 1.13). Suppose that $d(N) = 2$. Then N is metacyclic (Theorem 44.13) and $\exp(N) = p^2$ so N is as in (b) if $p = 2$. Assume that $p > 2$; then $L = \langle x^2 y^2\rangle$ is characteristic in N so normal in G (indeed, $L < N$ is the unique subgroup of order 2 that is maximal cyclic in N). Then N/L is nonabelian and $Z(N/L)$ is cyclic, which is impossible. If $d(N) = 3$, then N, not being minimal nonabelian, is as in (a) since $Z(N)$ is noncyclic.

5. We claim that a p-group G is of maximal class if it has a subgroup H such that $N = N_G(H)$ is of maximal class. To prove this, one may assume that $N < G$

and $|N| > p^3$ (otherwise, $C_G(H) = H$ and G is of maximal class, by Proposition 1.8). We use induction on $|G|$. Since $Z(G) < N$, we get $Z(G) = Z(N)$ so $|Z(G)| = p$. Since $N < G$, H is not characteristic in N so $|N : H| = p$. Then $N_{G/Z(G)}(H/Z(G)) = N/Z(G)$ is of maximal class, and so, by induction, $G/Z(G)$ is also of maximal class. Since $|Z(G)| = p$, the claim follows.

Proposition 10.18. *Let A be a subgroup of a p-group G such that $C_G(A)$ is metacyclic. If $|A| \le p^2$ and $\exp(A) = p$, then G has no normal subgroups of order p^{p+1} and exponent p.*

Proof. Suppose that $D \trianglelefteq G$ is of order p^{p+1} and exponent p. We have $C_D(A) > \{1\}$. Then $H = AC_D(A) \le C_G(A)$ is of exponent p and metacyclic, by hypothesis, so its order is p^2. We have $C_{AD}(H) = H$. Therefore, AD is of maximal class, by Proposition 1.8, contrary to Theorems 9.5 and 9.6. □

Proposition 10.19. *Let G be a metacyclic p-group containing a nonabelian subgroup B of order p^3. Then (a) if $p = 2$, then G is of maximal class, (b) if $p > 2$, then $|G| = p^3$, i.e., $G = B$.*

Proof. If G is of maximal class, then either $p = 2$ or $G = B$, Assume that G is not of maximal class. Then $C_G(B) \not\le B$ (Proposition 10.17) so $d(BF) > 2$, a contradiction. □

Exercise 1. Let a p-group G of order p^m, $m > 4$, be neither abelian nor minimal nonabelian. Let any nonabelian subgroup of G of order $> p^3$ have exactly one abelian maximal subgroup. Prove that G is of maximal class. (*Hint.* By induction, all proper nonabelian subgroups of G are of maximal class. Use Theorem 13.5.)

Exercise 2. Let G be a metacyclic 2-group. If $|\Omega_1(G)| \ne 2^2$, then G is either cyclic or a 2-group of maximal class.

Proposition 10.20. *Let G be a nilpotent group and G/G' abelian of rank d. Suppose that $|G'|^d \le |G/Z(G)|$. If $\alpha \in \mathrm{Aut}(G)$ induces the identity automorphism on G/G', then $\alpha \in \mathrm{Inn}(G)$.*

Proof. Let $G/G' = \langle x_1G'\rangle \times \cdots \times \langle x_dG'\rangle$. Since $G' \le \Phi(G)$, it follows that $G = \langle x_1, \dots, x_d\rangle$. Let A be the set of all automorphisms of G which induce identity on G/G'; then $\mathrm{Inn}(G) \le A \le \mathrm{Aut}(G)$. If $\alpha \in A$, then, for $i = 1, \dots, d$, we obtain $x_i^\alpha = x_i y_i$, where $y_i \in G'$. Obviously, α is uniquely determined by elements $y_1, \dots, y_d$. There are $|G'|^d$ distinct d-sequences $\{y_1, \dots, y_d\}$ of elements in G'. Hence $|A| \le |G'|^d$. On the other hand, by hypothesis, $|A| \le |G'|^d \le |G/Z(G)| = |\mathrm{Inn}(G)| \le |A|$. This proves that $A = \mathrm{Inn}(G)$. (Thus, central automorphisms of extraspecial and minimal nonabelian p-groups are inner.) □

Corollary 10.21. *Let $G < W$, where G is nilpotent and such that G/G' is abelian of rank d and $|G'|^d \le |G/Z(G)|$. If $[G, W] = G'$, then $W = G * C_W(G)$.*

Proof. It follows from $[G, W] \leq G'$ that $G \lhd W$. Let $w \in W$. Conjugation by w induces an automorphism α in G. If $g \in G$, then by hypothesis, $[w, g] = y \in G'$, so that $(g^{-1})^\alpha = w^{-1}g^{-1}w = yg^{-1}$. Hence α induces identity on G/G'. Then, by Proposition 10.20, $\alpha \in \mathrm{Inn}(G)$. It follows that for each $w \in W$, there exists $u \in G$ such that $w^{-1}gw = u^{-1}gu$ for all $g \in G$. Thus, $wu^{-1} \in \mathrm{C}_W(G)$, and so $w \in \mathrm{C}_W(G)G$ for each $w \in W$ or, what is the same, $W = \mathrm{C}_W(G) * G$. □

Propositions 10.20 and Corollary 10.21 hold provided G is extraspecial or minimal nonabelian.

A p-group G is said to be *generalized regular* if, whenever $x^p = y^p$ $(x, y \in G)$, then $(x^{-1}y)^p = 1$; then $\exp(\Omega_1(G)) = p$ (Exercise 7.1).

Proposition 10.22. *Let a p'-group Q act on a generalized regular p-group G. If Q acts trivially on $\mho_1(G)$, then $[G, Q] \leq \Omega_1(G)$.*

Proof. For all $x \in G$, $y \in Q$, we have $x^p = (x^p)^y = (x^y)^p$. Therefore, $[x, y]^p = (x^{-1}x^y)^p = 1$ since G is generalized regular, so that $[G, Q] \leq \Omega_1(G)$. □

Proposition 10.23. *Let G be a group of maximal class and order $p^m > p^{p+1}$. Set $\bar{G} = G/\mathrm{Z}(G)$. Let $\bar{D} < \bar{G}$ of order p^2 be such that $\mathrm{C}_{\bar{G}}(\bar{D}) = \bar{D}$. Then (a) D is nonabelian of order p^3 and $\mathrm{C}_G(D) < D$. (b) D has exactly p subgroups R of order p^2 such that $\mathrm{C}_G(R) = R$. (c) If R is from (b) and $x \in R - \mathrm{Z}(G)$, then $|\mathrm{C}_G(x)| = p^2$.*

Proof. Since $\bar{G}_1$ is not of maximal class, $\bar{D} \not\leq \bar{G}_1$. One may assume that $p > 2$. If $u \in G - D$ centralizes D, then $\bar{u}$ centralizes $\bar{D}$ and $\bar{u} \notin \bar{D}$, a contradiction. Thus, $\mathrm{C}_G(D) \leq D$. Since $\mathrm{Z}(G) < D$ and $\mathrm{Z}(\bar{G}) < \bar{D}$, we get $\mathrm{Z}_2(G) < D$. Since $m > p+1$, we get $\mathrm{C}_G(\mathrm{Z}_2(G)) = G_1$ (Theorem 9.6). Assume that D is abelian. Then $D < \mathrm{C}_G(\mathrm{Z}_2(G)) = G_1$, a contradiction. Thus, D is nonabelian, completing the proof of (a). Now (b) follows from Exercise 26 and (c) follows from (b). □

Proposition 10.24. *Let R be a subgroup of order p of a nonabelian p-group G. If there is only one maximal chain connecting R with G, then either $\mathrm{C}_G(R) \cong E_{p^2}$ (then G is of maximal class, by Proposition 1.8) or $G \cong M_{p^{n+2}}$.*

Proof. If $R \lhd G$, then G/R is cyclic so G is abelian, a contradiction. It follows that $\mathrm{C}_G(R)$ is noncyclic so $\mathrm{C}_G(R) = R \times Z$, where $Z \cong \mathrm{C}_{p^n}$. Assume that $n > 1$. Set $\Omega_1(\mathrm{C}_G(R)) = U$; then $U \cong \mathrm{E}_{p^2}$. By hypothesis, $\mathrm{N}_G(U)/U$ is cyclic so $\mathrm{N}_G(U)$ is either abelian or isomorphic to $\mathrm{M}_{p^{n+2}}$. Since $U = \Omega_1(\mathrm{N}_G(U))$ is characteristic in $\mathrm{N}_G(U)$, we get $\mathrm{N}_G(U) = G$ so $U \lhd G$. In that case, the nonabelian group $G \cong \mathrm{M}_{p^{n+2}}$. Now let $n = 1$. In that case, any subgroup of G, properly containing U, is of maximal class (Proposition 1.8). Let $U < B < G$. Then $\mathrm{N}_G(B)$ is of maximal class so $|\mathrm{N}_G(B) : B| = p$ (Theorem 9.6(c)). Thus, there is only one maximal chain connecting R and G. □

Proposition 10.25. *Suppose that $\{1\} \leq A < B \leq G$, where G is a p-group, $p > 2$, B elementary abelian of order p^k, $2 \leq k \leq 4$. Then the number of elementary abelian subgroups of G of order p^k that contain A is congruent to* $1 \pmod p$.

Denote by $\mathfrak{E}^*(G)$ the set of all maximal elementary abelian subgroups of a p-group G. If $E \in \mathfrak{E}^*(G)$, then $\Omega_1(C_G(E)) = E$ so, if E is of order p^2 and $p > 2$, then $C_G(E)$ is metacyclic (see Theorem 13.7).

Theorem 10.26. *Suppose that G is a p-group, $E \in \mathfrak{E}^*(G)$ is of order p^2 and $E \ntrianglelefteq G$. Then, for each subgroup S of order p in E such that $S \nleq Z(G)$, we have $C_G(S) = S \times Q$, where Q is cyclic or generalized quaternion. Next, G has no normal subgroups of order p^{p+1} and exponent p.*

Proof. By hypothesis, $Z(G)$ is cyclic so $E = Z \times S$, where $Z = \Omega_1(Z(G))$ and $|S| = p$. If G has no normal abelian subgroups of type (p, p), it is a 2-group of maximal class (Lemma 1.4), and the theorem is true. Now let $E_{p^2} \cong F \triangleleft G$; then $F \nleq Z(G)$ so $|G : C_G(F)| = p$. Since $E \in \mathfrak{E}^*(G)$ and $Z < F$, we have $S \nleq C_G(F)$ and $G = S \cdot C_G(F)$, a semidirect product. Let Q be the centralizer of S in $C_G(F)$. Then, by the modular law, $C_G(S) = S \times Q$. Since $E \in \mathfrak{E}^*(G)$, we see that Q has no abelian subgroups of type (p, p) so it is either cyclic or generalized quaternion. Assume that $U \trianglelefteq G$ is of order p^{p+1} and exponent p. Set $H = SU$; then $E < H$ since $Z < U$. Then $C_H(S) = E \cong E_{p^2}$ since $\exp(C_H(S)) = p$ and S is not contained in a subgroup isomorphic to E_{p^3}, so H is of maximal class (Proposition 1.8), contrary to Theorems 9.5 and 9.6. □

Corollary 10.27 ([BG, Lemma 5.2, Theorem 5.3]). *Suppose that G is a p-group, $p > 2$, $E \nleq Z(G)$ and $E \in \mathfrak{E}^*(G)$ is of order p^2. Set $Z = \Omega_1(Z(G))$, $W = \Omega_1(Z_2(G))$ and $T = C_G(W)$. Then (a) $E \nleq T$, (b) $|Z| = p$ and $W \cong E_{p^2}$, (c) T is characteristic of index p in G, (d) $E = Z \times S$, where $|S| = p$. Furthermore, $G = S \cdot T$ and $C_G(S) = S \times C_T(S)$, where $C_T(S)$ is cyclic.*

Exercise 3. If G is a p-group, then one of the following holds: (a) $|\Omega_1(G)| \leq p^2$, (b) The set $\mathfrak{E}^*(G)$ has no members of order p^2, (c) G has a subgroup S of order p such that $C_G(S) = S \times Q$, where Q is either cyclic or generalized quaternion.

Exercise 4. Let $F \in \mathfrak{E}^*(G)$ be of order p^2 and let $T = C_G(F)$ be of index p in a p-group G. Then $\Omega_1(T) = F$ and either (a) $\Omega_1(G) = F$ or (b) G has a subgroup S of order p such that $S \nleq T$ and $C_G(S) = S \times Q$, where $Q = C_T(S)$ has no abelian subgroups of type (p, p).

Exercise 5. Let G be a nonabelian p-group. Suppose that an element $x \in G$ of order p^2 is not contained in the set $S = \bigcup_{A<G} A$, where A runs over all noncyclic abelian subgroups of G. Then G is of maximal class.

Proposition 10.28. *A nonabelian p-group G is generated by minimal nonabelian subgroups.*

Proof. We use induction on $|G|$. Let K be generated by all minimal nonabelian subgroups of G. Assume that $K < G$. Then $K \in \Gamma_1$, by induction, and all other maximal subgroups of G are abelian. Since the number of abelian maximal subgroups of G is 0, 1 or $p + 1$ (Exercise 1.6(a)), we get a contradiction. □

Exercise 6 ([Ber31]). Let G be a nonabelian group and let $K = K(G)$ be the subgroup generated by all minimal nonabelian subgroups of G. Then G/K is abelian.

Solution (Isaacs). One may assume that $|G|$ is not prime power. Suppose that G/K is nonabelian. Assume that $K = K(G) \not\leq \Phi(G)$. Then there is a maximal subgroup M of G such that $MK(G) = G$. By induction, $M/K(M)$ is abelian, and $K(M) \leq M \cap K(G) < K(G)$. Thus $G/K(G) \cong M/(M \cap K(G))$, and this is abelian as an epimorphic image of $M/K(M)$. We can thus assume that $K(G) \leq \Phi(G)$. Let $\{1\} < P \in \mathrm{Syl}(K(G))$; then $P \lhd G$ but is not complemented in G and thus P is not Sylow in G (Schur–Zassenhaus). Let $P < S \in \mathrm{Syl}(G)$. Then $K(S) \leq S \cap K(G) < S$, and thus S is abelian (Proposition 10.28). Since $K(G)$ is nilpotent with all abelian Sylow subgroups, it is abelian, a contradiction.

Mann [Man25] proved the following unexpected and nice result: If G is a p-group and the subgroup $\mathcal{M}(G)$ is generated by all elements $x \in G$ such that $\mathrm{C}_G(x^p) = \mathrm{C}_G(x)$, then $\mathcal{M}(G)$ is abelian (see the paragraph following Exercise 7.20). Repeating, word for word, his argument, we prove the following

Proposition 10.29. *Let p divides $|G|$, where G is a finite group. Let $A = \mathcal{M}(G)$ be the subgroup of G, generated by all elements $x \in G$ such that p divides $o(x)$ and $\mathrm{C}_G(x^p) = \mathrm{C}_G(x)$. Then $p \nmid |\mathrm{Z}_2(A) : \mathrm{Z}(A)|$. (In particular, if G is a p-group, then $\mathrm{Z}_2(A) = \mathrm{Z}(A)$ so $A = \mathrm{Z}(A)$ is abelian.)*

Proof. Suppose that p divides $|\mathrm{Z}_2(A) : \mathrm{Z}(A)|$. Then we can find an element $z \in \mathrm{Z}_2(A) - \mathrm{Z}(A)$ such that $z^p \in \mathrm{Z}(A)$. Let x be any of the generating elements of A. In that case, $\mathrm{cl}(\langle z, x\rangle) \leq 2$ so $[z, x^p] = [z^p, x] = 1$, and hence $z \in \mathrm{C}_G(x^p) = \mathrm{C}_G(x)$. Therefore, $z \in \mathrm{Z}(A)$, a contradiction. □

Proposition 10.30 (Bozikov–Janko). *Let G be a noncyclic p-group and $Z < G$ cyclic of order p^n, $n > 1$. Then G contains a cyclic subgroup Z_1 of order p^n such that $|Z \cap Z_1| = p^{n-1}$, unless $p = 2$, $n > 2$ and G is of maximal class (in that case G is isomorphic to Q_{2^m} or SD_{2^m} with $m > 2$) or $p = 2$ and G is dihedral.*

Proof (Berkovich). The theorem holds for $\mathrm{M}_{p^{n+1}}$ and the abelian group of type (p^n, p). The theorem is not true for 2-groups mentioned in the statement. So one may assume that Z is not contained in a subgroup of G that is either abelian of type (p^n, p) or isomorphic to $\mathrm{M}_{p^{n+1}}$. We also assume that G is not a 2-group of maximal class. Then there is in G a normal subgroup $R \cong \mathrm{E}_{p^2}$. Set $H = RZ$. If $R \cap Z > \{1\}$, then, by assumption, $p = 2$ and $H \cong \mathrm{D}_8$ so $n = 2$. Since, by assumption, Z is not contained in abelian subgroup of type $(4, 2)$, we get $\mathrm{C}_G(H) < H$ so G is of maximal class (Proposition 10.17), contrary to the assumption. Now let $R \cap Z = \{1\}$. Then $C_R(Z) > \{1\}$ so Z is contained in abelian subgroup of type (p^n, p), contrary to the assumption. □

Exercise 7. Let p divides $|G|$. Suppose that $L \leq G$ is either elementary abelian p-subgroup or cyclic of order 4 if $p = 2$, and each such L satisfies $L \leq Z(N_G(L))$. Prove that G is p-nilpotent. (*Hint.* Use Lemma 10.8.)

Exercise 8 (Janko). If G is a nonabelian p-group, then $G = A \cup (\bigcup M)$, where $A < G$ is any maximal normal abelian and M runs over all minimal nonabelian subgroups of G.

Exercise 9. Suppose that $A < B \leq G$, where A, B are abelian subgroups of a p-group G and $|B : A| = p$. Then the number $a(G)$ of abelian subgroups of G of order $p|A|$, containing A, is $\equiv 1 \pmod{p}$.

Solution. We may assume that $C_G(A) = G$ so $A \leq Z(G)$. Then $a(G) = c_1(G/A) \equiv 1 \pmod{p}$ (Sylow), and we are done.

Exercise 10. Let $H < G$, where G is a p-group. If all subgroups of order $p|H|$ of G containing H, are of maximal class, then G is also of maximal class.

Solution. Set $N = N_G(H)$. In view of Remark 5, one may assume that $|N : H| > p$. Let $D < H$ be N-invariant of index p^2 (D exists since $|H| > p$). Set $C = C_N(H/D)$; then $C > H$. Let $F/H \leq C/H$ be of order p; then F is not of maximal class, a contradiction.

Exercise 11. If $H < G$ is nonabelian of order p^3, G a p-group and $N_G(H)$ metacyclic, then $p = 2$ and G is of maximal class. (Use Proposition 10.19 and Remark 10.5.)

Exercise 12. Let a p-group G be neither cyclic nor a 2-group of maximal class and let Z be a cyclic subgroup of order p^n is G. Then the number of cyclic subgroups of G of order p^{n+k}, containing Z, is a multiple of p.

Exercises 13–16 are taken from Mann's letter.

Exercise 13. Let $A < G$ be abelian of index p in a nonabelian p-group G.

(a) If G is regular, then $|G/Z(G)| \leq p^{r+1}$, where $r = d(A)$, so $cl(G) \leq r + 1$.

(b) If $r = d(A) \leq p - 2$, then $|G : Z(G)| \leq p^{r+1}$.

Solution. (a) We have $Z(G) < A$. If $\exp(A) = p$, then $|G| = p^{r+1}$, and we are done. Now let $\exp(A) > p$. Set $H = \langle x^p \mid x \in G - A \rangle$; then $H \leq Z(G) \cap \mho_1(G)$. Set $\bar{G} = G/H$. Then all elements in $\bar{G} - \bar{A}$ have order p. Since $\bar{G} = \langle \bar{G} - \bar{A} \rangle$ and $\bar{G}$ is regular, we get $\exp(\bar{G}) = p$ so $H = \mho_1(G)$. We have $\mho_1(A) \leq \mho_1(G) \leq H$ so

$$|G : Z(G)| \leq |G : H| \leq |G : \mho_1(A)| = |G : A||A : \mho_1(A)| = p^{r+1},$$

proving the first inequality. Next, $cl(G/Z(G)) \leq r$ so $cl(G) \leq r + 1$.

(b) Since $|G : \mho_1(G)| \le |G : \mho_1(A)| = p^{r+1} \le p^{p-1}$ we get (a) $\Rightarrow$ (b) (Theorem 9.8(a)).

Exercise 14. Let $A \in \Gamma_1$, where G is a regular p-group. Set $H = \langle x^p \mid x \in G - A\rangle$. Then $H = \mho_1(G)$.

Let $A = G_1 < G$, where G_1 is the fundamental subgroup of an irregular p-group G of maximal class. Set $H = \langle x^p \mid x \in G - A\rangle$. Then $H \le \mathrm{Z}(G)$ (see Theorem 13.19) so $H < \mho_1(G)$. Hence, $\mathrm{cl}(G)$ is unbounded.

Exercise 15 (see Exercise 13(a)). Let $A \triangleleft G$, where G is a regular p-group and $\exp(G/A) = p^f$. Set $H = \langle x^{p^f} \mid x \in G - A\rangle$. Then $H = \mho_f(G)$.

Solution. We have $H \triangleleft G$. Set $\bar{G} = G/H$. We have $o(\bar{x}) \le p^f$ for all $\bar{x} \in \bar{G} - \bar{H}$ so $\Omega_f(\bar{G}) = \bar{G}$ and $\exp(\bar{G}) = p^f$ (Theorem 7.2), and we have $\mho_f(G) \le H$. The reverse inclusion is obvious.

Exercise 16. Let G be a regular p-group, $A \triangleleft G$ abelian, $G/A \cong \mathrm{C}_{p^f}$ and $r = \mathrm{d}(A)$. Then $|G : \mathrm{Z}(G)| \le p^{(r+1)f}$.

Solution. We have $H = \mho_f(G) \le \mathrm{Z}(G)$. Let $L/A \le G/A$ be of order p; then $\Omega_1(G) = \Omega_1(L)$. Clearly, $|G/\mho_1(G)| = |\Omega_1(G)| = |\Omega_1(L)| \le p|\Omega_1(A)| = p^{r+1}$. Since G is pyramidal (see §8), we get $|G/\mho_f(G)| \le p^{(r+1)f}$.

Exercise 17. Let G be a group of order 2^4 and exponent > 2. Then $\mathrm{c}_1(G) \le 11$ with equality if and only if $G \cong \mathrm{D}_8 \times \mathrm{C}_2$.

Exercise 18. Let G be a non-elementary abelian group of order 2^5. Then $\mathrm{c}_1(G) \le 23$. (*Hint.* Use Taussky's theorem and Frobenius–Schur formula for the number of involutions [BZ, Chapter 4].)

Exercise 19. If G is a group of order 2^5 and exponent > 2. If $\mathrm{c}_1(G) = 23$, then $G \cong \mathrm{D}_8 \times \mathrm{E}_4$.

Exercise 20. Let G be a group of order 2^m, $m > 5$ and $G \ncong \mathrm{E}_{2^m}$. Show that $\mathrm{c}_1(G) \le 3 \cdot 2^{m-2} - 1$ with equality if and only if $G \cong \mathrm{D}_8 \times \mathrm{E}_{2^{m-3}}$. (By Exercises 17 and 19, the result also holds for $m = 4, 5$; it is also holds for $m = 3$.)

Exercise 21. Find all groups of order 2^5 containing exactly 19 involutions.

Exercise 22. Does there exist a group of order 2^m containing exactly $3 \cdot 2^{m-2} - 5$ involutions for each $m > 5$?

Exercise 23. Let $m > 5$ and let $\mathcal{T}$ be the set of all groups G of order 2^m with $\mathrm{c}_1(G) < 3 \cdot 2^{m-2} - 5$. Find $\max\{\mathrm{c}_1(G) \mid G \in \mathcal{T}\}$.

Exercise 24. Let G be a group of order 3^4 and class 3, let a 3-group G_0 be lattice isomorphic with G via ϕ. Is it true that $\mathrm{cl}(G_0) = 3$? (Answer is: 'yes'.)

Remark 6. We will clear up the subgroup and normal structure of groups G of order p^4. We will show that one of the following holds: (i) G is abelian of one of the following types: (p^4), (p^3, p), (p^2, p^2), (p^2, p, p), (p, p, p, p) (five groups). (ii) G is minimal nonabelian (three groups, by Exercise 1.8a). (iii) G is of maximal class (if $p = 2$, there is three groups). (iv) $G = M \times C$, where M is nonabelian of order p^3 and $|C| = p$ (two groups). (v) $G = M * C_{p^2}$, where M is nonabelian of order p^3, $\Omega_1(M) = M$ (one type). Indeed, suppose that G is not of types (i)–(iii). Then it has a minimal nonabelian subgroup M of order p^3. By Proposition 10.17, $G = M\mathrm{Z}(G)$. If $\mathrm{Z}(G)$ is noncyclic, we get two groups from (iv). Now let $\mathrm{Z}(G)$ be cyclic and M metacyclic. Then $|G'| = p$ and $\Phi(G) = G' = \mho_1(G)$. Let $p > 2$. Since G is regular, we get $|\Omega_1(G)| = |G/\mho_1(G)| = p^3$ and $\exp(\Omega_1(G)) = p$. We have $G = \Omega_1(G)\mathrm{Z}(G)$, so $\Omega_1(G)$ is nonabelian, and case $p > 2$ is complete. Now let $p = 2$. Then, according to Appendix 16, there is exactly one group of type (v). We have 11 groups of order p^4, which are not of maximal class so there are 14 groups of order 2^4.

Proposition 10.31. *Let G be a p-group of order $p^m > p^4$ with exactly one noncyclic abelian subgroup A of order p^3. Then one of the following holds: (a) G is abelian of type (p^{m-1}, p), (b) $G \cong M_{p^m}$, (c) $p = 2$ and $G = \langle a, b \mid a^{2^{m-2}} = 1, b^4 = a^{2^{m-3}}, a^b = a^{-1}\rangle$.*

Proof. Obviously $A \triangleleft G$ and G is not a 2-group of maximal class. Assume that G is of maximal class, $p > 2$. Let $\mathrm{E}_{p^2} \cong U \triangleleft G$. Then $\mathrm{C}_G(U)/U$ has only one subgroup of order p so $\mathrm{C}_G(U)/U$ is cyclic. In that case, $\mathrm{C}_G(U)$ is abelian of type (p^n, p). Then $\Phi(\mathrm{C}_G(U)) \triangleleft G$ is cyclic of order $> p$, a contradiction. Thus, G is not of maximal class.

(i) If $B < G$ is nonabelian of order p^3, then $\mathrm{C}_G(B) < B$ (otherwise, $B * \mathrm{C}_G(B)$ has two distinct noncyclic abelian subgroups of order p^3). Then G is of maximal class (Lemma J(c)), contrary to the previous paragraph.

(ii) Let $U \le G$ be minimal nonabelian. Then $U \cong \mathrm{M}_{p^n}$, $n > 3$ (see (i)) so $A < U$ and $\Omega_1(U) = \Omega_1(A) \cong E_{p^2}$; then A is abelian of type (p^2, p) and $\Omega_1(A) \triangleleft G$.

(iii) Assume that there is $x \in G - A$ of order p. Then $B = \langle x, \Omega_1(A)\rangle$ is of order p^3. By (i), B is abelian, a contradiction since $B \neq A$. Thus, $\Omega_1(G) = \Omega_1(A)$.

(iv) Assume that there is $y \in G - A$ of order p^2. Write $Y = \langle y \rangle$. Set $H = \Omega_1(A)Y$; then $|H| = p^3$, by (iii), so H is abelian and $H \neq A$, a contradiction.

Thus, $\Omega_2(G) = A$ so G is one of groups (a), (b), (c) (Lemma 42.1). □

A nonabelian 2-group G is called *generalized dihedral* if it has a nontrivial subgroup A of exponent > 2 such that all elements of the set $G - A$ are involutions. Then A has index 2 in G, $G = \langle b \rangle \cdot A$ and $a^b = a^{-1}$ for all $a \in A$ so A is abelian and A is characteristic in G; A is called the *kernel* of G.

Theorem 10.32 (Janko). *If all nonabelian subgroups of a nonabelian 2-group G are generated by involutions, then G is generalized dihedral.*

Proof. If G is minimal nonabelian, then $\mathrm{d}(G) = 2$ so it is generated by *two* involutions. It follows that G is dihedral. Since $|G'| = 2$, we get $G \cong \mathrm{D}_8$.

We use induction on $|G|$ and suppose that G is not minimal nonabelian. Then there is a nonabelian $M_1 \in \Gamma_1$. By induction, M_1 is generalized dihedral with kernel, say A. By the above, A is characteristic in M_1 so normal in G, $|G : A| = 4$. It follows from $\Omega_1(G) = G$ that $G/A \cong \mathrm{E}_4$. Let $M_1/A, M_2/A, M_3/A$ be all subgroups of order 2 in G/A. Since $A \not\leq \mathrm{Z}(M_1)$, one may assume, that M_2 is also nonabelian so M_2 is generalized dihedral with kernel A, by induction. In this case, $G = (M_1 - A) \cup (M_2 - A) \cup M_3$ is a partition. It follows that all elements in the set $G - M_3$ are involutions so G is generalized dihedral with kernel M_3. □

Theorem 10.33 (Berkovich). *(a) All minimal nonabelian subgroups of a nonabelian 2-group G are dihedral if and only if G is generalized dihedral.*

(b) Suppose that a 2-group is neither abelian nor minimal nonabelian. If all proper nonabelian subgroups of G are generated by involutions, then $\Omega_1(G) = G$.

Proof. (a) Let $H \leq G$ be nonabelian. By Proposition 10.28, H is generated by subgroups $\cong \mathrm{D}_8$ so involutions, and we are done, by Theorem 10.32.

(b) In that case, all minimal nonabelian subgroups of G are dihedral, and now the result follows from (a). □

It appears that the analog of Theorem 10.32 also holds for $p > 2$ (see Mann's comments to #115 in 'Research problems and themes I').

Denote by $\mathcal{MA}_k(G)$ the set of all $\mathcal{A}_1$-subgroups (= minimal nonabelian subgroups) U of G such that $U = \Omega_k(U) = U$. Let $\langle \mathcal{MA}_k(G)\rangle = \langle H \mid H \in \mathcal{MA}_k(G)\rangle$ and let Σ_{p^2} denote a Sylow p-subgroup of the symmetric group of degree p^2.

Theorem 10.34. *Let a nonabelian p-group $G = \Omega_k(G)$. Then*

(a) $\mathcal{MA}_k(G) \neq \varnothing$,

(b) $\Omega_k(F) = F$, $\Omega_k(H) = H$ for some distinct $F, H \in \Gamma_1$,

(c) if $k = 1$, then $G = \langle \mathcal{MA}_1(G)\rangle$, unless G has a subgroup $\cong \Sigma_{p^2}$, $p > 2$.

Lemma 10.35. *Let $E < G$, where $G = \Omega_k(G)$ is a nonabelian p-group and $\Omega_k(E) = E$. Then (a) $E \leq M \in \Gamma_1$, where $\Omega_k(M) = M$, (b) $\Omega_k(F) = F$, $\Omega_k(H) = H$ for two distinct $U, V \in \Gamma_1$.*

Proof. (a) Let $E \leq M < G$, where $\Omega_k(M) = M$ and $|M|$ is as large as possible. Since $M^G < G$ and $\Omega_k(M^G) = M^G$, we get $M^G = M$, by the choice of M, so $M \lhd G$. Let $x \in G - M$ be such that $o(x)$ is as small as possible; then $o(x) \leq p^k$ and one may assume that $x^p \in M$. Since $\Omega_k(\langle x, M\rangle) = \langle x, M\rangle$, we get $\langle x, M\rangle = G$ since $|\langle x, M\rangle| > |M|$ so $M \in \Gamma_1$.

(b) Let $x \in G$ be of order $\leq p^k$. Then, by (a), $x \in U \in \Gamma_1$ with $\Omega_k(U) = U$. If $y \in G - U$ with $o(y) \leq p^k$, then $y \in V \in \Gamma_1$ with $\Omega_k(V) = V$, and $U \neq V$. □

Proof of Theorem 10.34. We use induction on $|G|$.

To facilitate the general case, we first consider case $p^k = 2$. If $x, y \in G$ are noncommuting involutions (x, y exist since $G = \Omega_1(G)$ is nonabelian), then $\langle x, y\rangle$ is dihedral so contains a subgroup $A \cong D_8$. Then A is an $\mathcal{A}_1$-subgroup and $\Omega_1(A) = A$, proving (a). By Lemma 10.35(a), $A \leq U \in \Gamma_1$, where $\Omega_1(U) = U$. By induction, $U = \langle \mathcal{M}\mathcal{A}_1(U)\rangle$. Let $y \in G - U$ be an involution. Then $y \in V \in \Gamma_1$, where $\Omega_1(V) = V$, proving (b). Assume that (c) is false. Then, by induction, V must be elementary abelian (otherwise, by Theorem 10.28, V contains an $\mathcal{A}_1$-subgroup L such that $L \not\leq U$, and then $\langle \mathcal{M}\mathcal{A}_1(G)\rangle \geq UL = G$). We have $V = \langle V - Z(G)\rangle$ so there is (an involution) $z \in V - Z(G)$ such that $z \notin U$. Since z does not centralize U, there is an involution $w \in U$ such that $zw \neq wz$. Then $L = \langle w, z\rangle$ is dihedral so all its minimal nonabelian subgroups are dihedral of order 8 and generate L. Since $L \not\leq U$, there exists in L an $\mathcal{A}_1$-subgroup B such that $B \not\leq U$. Then $\langle \mathcal{M}\mathcal{A}_1(G)\rangle \geq L\langle \mathcal{M}\mathcal{A}_1(U)\rangle = UL = G$. Next we assume that $p^k > 2$.

(a) We have to prove that G has an $\mathcal{A}_1$-subgroup generated by elements of orders $\leq p^k$. One may assume that all proper subgroups of G, generated by elements of orders $\leq p^k$, are abelian (otherwise, the result follows by induction). Let $E < G$ be an abelian subgroup of exponent $\leq p^k$ of maximal possible order. Then $\Omega_k(E^G) = E^G < G$ so, by assumption, E^G must be abelian hence $E^G = E$, by the choice of E, and so $E \lhd G$. Let $x \in G - E$ be of minimal possible order; then $o(x) \leq p^k$, by hypothesis, and one may assume that $x^p \in E$. Set $F = \langle x, E\rangle$. Then F is nonabelian and $\Omega_k(F) = F$ so we must have $F = G$, by assumption. Thus, $E \in \Gamma_1$ and $Z(G) < E$. Let $\langle x, Z(G)\rangle \leq T \in \Gamma_1$, where $\Omega_k(T) \leq p^k$ (T exists, by Lemma 10.35(a)); then $T \neq E$. By assumption, T is abelian; then $G = ET$ and $E \cap T = Z(G)$ is of index p^2 in G. Since $p^k > 2$, then $\exp(G) \leq p^k$ (indeed, if $a, b \in G$ and $o(a), o(b) \leq p^k$, then $(ab)^{p^k} = a^{p^k} b^{p^k} [b, a^{p^k(p^k-1)/2}] = 1$ since $G/Z(G)$ is abelian of type (p, p) so $\mathrm{cl}(G) = 2$ and $\exp(G') = p$); in that case, every $\mathcal{A}_1$-subgroup, contained in G, also has exponent $\leq p^k$, and the proof of (a) is complete.

(b) follows from Lemma 10.35(b).

(c) In this part, $k = 1$, and we may assume that $p > 2$ (see the paragraph preceding (a)). One may assume that G is not an $\mathcal{A}_1$-group; then $|G| \geq p^4$. By (b), there are distinct $U, V \in \Gamma_1$ such that $\Omega_1(U) = U$ and $\Omega_1(V) = V$. If U and V are both nonabelian, then, by induction, $U = \langle \mathcal{M}\mathcal{A}_1(U)\rangle$ and $V = \langle \mathcal{M}\mathcal{A}_1(V)\rangle$ so, since $G = UV$, we get

$$\langle \mathcal{M}\mathcal{A}_1(G)\rangle = \langle \mathcal{M}\mathcal{A}_1(UV)\rangle \geq \langle \mathcal{M}\mathcal{A}_1(U)\rangle\langle \mathcal{M}\mathcal{A}_1(V)\rangle = UV = G.$$

In view of (a) and Lemma 10.35(a), one may assume that U is nonabelian. Next let V be abelian (otherwise, there is nothing to prove); then $\exp(V) = p$ and $Z(G) < V$. Assume that G has no subgroup isomorphic to Σ_{p^2}.

Since $\langle V - \mathrm{Z}(G)\rangle = V$, the set $V - \mathrm{Z}(G)$ has an element $y \notin U$ (of order p); then y does not centralize U so there is $x \in U$ of order p such that $xy \neq yx$. Set $K = \langle x, y\rangle$; then $\Omega_1(K) = K$ is nonabelian. Assume that $K < G$. Then $\langle \mathcal{MA}_1(K)\rangle = K$, by induction, so, by Theorem 10.28, there is $L \in \mathcal{MA}_1(K)$ such that $L \not\leq U$ since $K \not\leq U$. In that case, $G = UL$ is generated by members of the set $\mathcal{MA}_1(G)$ since $\langle \mathcal{MA}_1(U)\rangle = U$, by induction. Thus, one may assume that $K = G$, i.e., $G = \langle x, y\rangle$ so $\mathrm{d}(G) = 2$. Since $\Omega_1(G) = G$, we get $G' = \Phi(G)$ so $|G : G'| = p^2$. In that case, as we know, G is of maximal class. Let $V = \mathrm{Z}(G) \times E$, where $E < V$ and $|G : E| = p^2$. Since $E_G = \{1\}$, G is isomorphic to a subgroup of Σ_{p^2}. If $\exp(G) = p$, then $G = \langle \mathcal{MA}_1(G)\rangle$ (Theorem 10.28). Now we assume $\exp(G) > p$. Then $G(= \Omega_1(G))$ is irregular so $p^{p+1} = |\Sigma_{p^2}| \geq |G| \geq p^{p+1}$ whence $G \cong \Sigma_{p^2}$. □

Definition 1. Given a positive integer k, a subgroup H of a p-group G is said to be k-quasi-maximal in G, if $H \triangleleft G$ and $G = \langle x\rangle H$ for some $x \in G - H$ with $o(x) = p^k$.

Recall that $\Omega_k^*(G) = \langle x \in G \mid o(x) = p^k\rangle$.

Definition 2. Given a positive integer k, a p-group G is said to be $\mathcal{V}_k$-group if it is nonabelian, $\Omega_k^*(G) = G$, $G/\mathrm{Z}(G)$ is abelian of rank 2 and exponent $\leq p^k$.

Theorem 10.36. *Suppose that $k > 1$ and a nonabelian p-group $G = \Omega_k^*(G)$.*

(a) G has a $\mathcal{V}_k$-subgroup,

(b) either G is a $\mathcal{V}_k$-group or G has two distinct k-quasi-maximal subgroups, say A and B, such that $\Omega_k^*(A) = A$ and $\Omega_k^*(B) = B$.

Proof. We proceed by induction on $|G|$.

Let $X < G$ be such that $\Omega_k^*(X) = X$ (X exists since $\exp(G) \geq p^k$) and $|X|$ is as large as possible. Then $X^G < G$ and $\Omega_k^*(X^G) = X^G$ so $X^G = X$, i.e., $X \triangleleft G$. Let $a \in G - X$ be of order p^k and $Y = \langle a, X\rangle$. Since $\Omega_k^*(Y) = Y$, we get $Y = G$, by the choice of X, so X is a k-quasi-maximal subgroup of G.

(a) Assume that G has no proper $\mathcal{V}_k$-subgroups. Then, by induction, all k-quasi-maximal subgroups of G are abelian. By the previous paragraph, every element of G of order p^k is contained in a k-quasi-maximal subgroup generated by elements of order p^k; let X be one of such subgroups. Therefore, we have $G = \langle a, X\rangle$, where $o(a) = p^k$. Let $a \in Y$, where Y is a k-quasi-maximal subgroup of G such that $\Omega_k^*(Y) = Y$. We have $G = XY$ so, since X and Y are abelian, we get $\mathrm{cl}(G) = 2$ (Fitting's lemma) and $X \cap Y \leq \mathrm{Z}(G)$. Since G/X and G/Y are cyclic of order $\leq p^k$, it follows that $G/(X \cap Y)$ is a subgroup of abelian group $(G/X) \times (G/Y)$. Since G is nonabelian, $G/\mathrm{Z}(G)$ is abelian of rank 2 and exponent $\leq p^k$. In that case, G itself is a $\mathcal{V}_k$-group, completing the proof of (a).

(b) was proved in the previous paragraph. □

Exercise 25. Let $B < G$ be nonabelian of order p^3, G a p-group and $\mathrm{C}_G(B) < B$. Then $\mathrm{Z}_2(G) \leq B$.

Exercise 26. Let $B < G$ be nonabelian of order p^3, G a p-group and $\mathrm{C}_G(B) < B$. Then each maximal subgroup $K \neq \mathrm{Z}_2(G)$ of B satisfies $\mathrm{C}_G(K) = K$.

Exercise 27. Let G be a group of order $p^m > p^{p+1}$, let $H < G$ be not normal and let $\mathrm{N}_G(H)$ be of maximal class. Then G is of maximal class and $H \not\leq G_1$, the fundamental subgroup of G. If $K \neq H \cap G_1$ is a maximal subgroup of H, then $\mathrm{N}_G(K) = H$.

Solution. By Remark 10.5, G is of maximal class. Write $\mathrm{N}_G(H) = N$. Since H is not characteristic in N, we get $|N : H| = p$ (Theorem 9.6). Then $H \not\leq G_1$ since $|\mathrm{Z}(G_1)| > p$, and so $H \cap G_1$ is maximal in H. Let $K \neq H \cap G_1$ be maximal in H and assume that $\mathrm{N}_G(K) > H$. Let $H < F \leq \mathrm{N}_G(K)$, where $|F : H| = p$; then $F = N$ (compare orders!) so $F = N$. It follows that $K \lhd F = N$. Since $|N : K| = |F : K| = |F : H||H : K| = p^2 > p$ and N is of maximal class, we get $K = \Phi(F) < \Phi(G) < G_1$ so $K = H \cap G_1$, a contradiction.

§11

On the power structure of a p-group

All results of this section are due to Mann [Man5]; his paper yields axiomatic treatment of some important properties of p-groups.

Let X be a p-group and let n run over all positive integers. We are interested in the following statements:

(1) $\mho_n(X) = \{x^{p^n} \mid x \in X\}$.

(2) $\exp(\Omega_n(X)) \leq p^n$.

(3) $|\mho_n(X)| = |X : \Omega_n(X)|$.

Definition 1. A p-group G is said to be a $\mathcal{P}_i$-group ($i = 1, 2, 3$), if all sections of G satisfy (i). G is said to be a $\mathcal{P}$-group, if it is a $\mathcal{P}_i$-group for $i = 1, 2, 3$.

Regular p-groups are $\mathcal{P}$-groups. The nonabelian (so irregular) 2-group M_{2^n} is a $\mathcal{P}$-group. Here we study some interrelations between properties $\mathcal{P}_1$, $\mathcal{P}_2$, $\mathcal{P}_3$ and $\mathcal{P}$. Results of this section, as we shall see, have powerful potential.

Here we construct, following [Man5], an irregular group G of order p^{p+1}, $p > 2$, satisfying $|\Omega_1(G)| = p^p$. Start with the direct product $H = \langle a_0\rangle \times \langle a_1\rangle \times \cdots \times \langle a_{p-2}\rangle$, where $o(a_0) = p^2, o(a_1) = \cdots = o(a_{p-2}) = p$. Denote $a_{p-1} = a_0^{kp}$, where $k \not\equiv -1 \pmod p$. Define an automorphism σ of H by $a_i^\sigma = a_i a_{i+1}, i = 0, 1, \dots, p-2$. Then $\sigma^p = \mathrm{id}_H$ and $a_{p-1}^\sigma = a_{p-1}$. Let $G = \langle b\rangle \cdot H$ be the semidirect product with kernel H, where b induces σ on H and $b^p = 1$. Then $|G| = p^{p+1}, Z(G) = \langle a_{p-1}\rangle$ is of order p so G is of maximal class. It is easy to check that $\Omega_1(G) = \langle \Phi(G), b\rangle$ is of order p^p and exponent p. Obviously, G is a $\mathcal{P}$-group.

In what follows we use freely the following obvious fact. If $\{1\} < N \trianglelefteq G$, where G is a p-group, then $N \cap \Omega_1(G) > \{1\}$.

Theorem 11.1. *Let G be a p-group. If each section of G of exponent p^2 is a $\mathcal{P}_i$-group ($i = 1, 2$) or a $\mathcal{P}$-group, then G satisfies the same properties.*

Proof. Let $\exp(G) = p^e$. One may assume that $e > 2$ (otherwise, there is nothing to prove). Then the theorem holds for all proper sections of G.

(a) Let $i = 1$ and let each section S of G of exponent p^2 be a $\mathcal{P}_1$-group, i.e., $\mho_1(S) = \{x^p \mid x \in S\}$. We have $\exp(\mho_1(G)) > p$. Let $z \in \mho_1(\mho_1(G)) \cap Z(G)$ be of order p. By induction in $\mho_1(G)$, $z = u^p$ for some $u \in \mho_1(G)$. Also by induction applied to $G/\langle z\rangle$, given $a, b \in G$, we have $a^p b^p = c^p z^j = c^p(u^p)^j = c^p(u^j)^p$ for some $c \in G$, $j \in \mathbb{Z}$. As $u \in \mho_1(G) \leq \Phi(G)$, we have $G > K = \langle c, u\rangle$, so again by induction in K, $c^p(u^j)^p = d^p$ for some $d \in K$. Thus, $a^p b^p = d^p$ so $\mho_1(G) = \{x^p \mid x \in G\}$.

Now let $n > 1$ and $e > n$. We have $\mho_n(G) \leq \mho_{n-1}(\mho_1(G))$ because of the estimate $\exp(G/\mho_{n-1}(\mho_1(G))) \leq p^n$, so induction in $\mho_1(G)$ yields: if $a \in \mho_n(G)$, then $a = b^{p^{n-1}}$ for some $b \in \mho_1(G)$. By the previous paragraph, $b = c^p$ for some $c \in G$; then $a = (c^p)^{p^{n-1}} = c^{p^n}$ so $\mho_n(G) = \{x^{p^n} \mid x \in G\}$. Thus, G is a $\mathcal{P}_1$-group.

(b) Let $i = 2$ and let each section S of G of exponent p^2 satisfy $\mathcal{P}_2$, i.e., $\exp(\Omega_1(S)) = p$. Let $a, b \in G$ be of order p, and let $Z \leq Z(G)$ be of order p. Setting $H = \langle a, b, Z\rangle$, we get $\exp(H/Z) \leq p$, by induction, hence $\exp(H) \leq p^2$, i.e., H is a $\mathcal{P}_2$-group, and so $\exp(H) = p$ since $\Omega_1(H) = H$, whence $o(ab) \leq p$. Thus, $\exp(\Omega_1(G)) = p$.

Now let $n > 1$. By the previous paragraph, $\Omega_{n-1}(G/\Omega_1(G)) = \Omega_n(G)/\Omega_1(G)$ so $\exp(\Omega_n(G)/\Omega_1(G)) \leq p^{n-1}$, by induction, hence, by the previous paragraph again, we have $\exp(\Omega_n(G)) \leq \exp(\Omega_n(G)/\Omega_1(G)) \cdot \exp(\Omega_1(G) \leq p^n$, i.e., G is a $\mathcal{P}_2$-group.

(c) Suppose that every section of G of exponent p^2 is a $\mathcal{P}$-group. Since G is a $\mathcal{P}_i$-group, $i = 1, 2$, by (a) and (b), it remains to prove that G is also a $\mathcal{P}_3$-group. One may assume that $e > n$ (otherwise, there is nothing to prove); then $\mho_1(G) > \{1\}$. Let $\{1\} < N = \Omega_1(G) \cap \mho_n(G) (\triangleleft G)$; then $\exp(N) = p$ and each element of N is a p^n-th power since G is a $\mathcal{P}_i$-group for $i = 1, 2$. Set $H/N = \Omega_n(G/N)$; then $\Omega_n(G) \leq H$; moreover, $\Omega_n(G) < H$ since H has an element of order $p^{n+1} > \exp(\Omega_n(G))$. Let us prove that $H = \Omega_{n+1}(G)$ and $N = \mho_n(H)$. Indeed, let $x \in G$ be of order p^{n+1}. Then $x^{p^n} \in \Omega_1(G) \cap \mho_n(G) = N$ so $o(xN) = p^n$. This means that $xN \in H/N$ so $x \in H$. It follows that $H \geq \Omega_{n+1}(G)$. The reverse inclusion is also true, by definition of H. Since G/N is a $\mathcal{P}_2$-group, $\exp(H/N) = p^n$ so $N \geq \mho_n(H)$. The reverse inclusion is also true since every element of N is a p^n-th power of some element of G (really, that element is contained in H since $H = \Omega_{n+1}(G)$). So, by induction,

$$\begin{aligned} |G : \mho_n(G)| &= |G/N : \mho_n(G/N)| = |\Omega_n(G/N)| = |H/N| = |\Omega_{n+1}(G) : N| \\ &= |\Omega_{n+1}(G) : \mho_n(\Omega_{n+1}(G))| = |\Omega_n(\Omega_{n+1}(G))| = |\Omega_n(G)|. \qquad \square \end{aligned}$$

Proposition 11.2. *If a $\mathcal{P}_1$-group G has a subgroup E of exponent $\leq p^n$ and index p^k, then $|\mho_n(G)| \leq p^k$.*

Proof. We use induction on $|G|$. One may assume that G is noncyclic and $\exp(G) > p^n$.

(a) Let $k = 1$. Take $Z \leq Z(G)$ of order p. If $Z \not\leq E$, then $G = E \times Z$ and $|\mho_n(G)| = |\mho_n(E)| = 1 < p$. Now let $Z \leq E$. In that case, $\exp(E/Z) = p^{n-\mu}$, where $\mu = 0$ or 1. Then, by induction, $|\mho_n(G/Z)| \le |\mho_{n-\mu}(G/Z)| \le p$, hence $|\mho_n(G)| \le p^2$. Assume that $|\mho_n(G)| = p^2$. We have $\mho_n(G) \le \Phi(G)(< E)$, and there exists a subgroup A in E such that $\mho_n(G) \le \Phi(G) \le A < E$ and $|G : A| = p^2$. If $a \in G - A$, then $M = \langle a, A\rangle \in \Gamma_1$, $a^{p^n} \in \mho_n(M)$ and $|\mho_n(M)| \le p$, by induction, since $|M : A| = p$ and $\exp(A) \le \exp(E) \le p^n$. If $M/A = M_1/A, \dots, M_{p+1}/A$ are all subgroups of order p in G/A, then all p^n-th powers in G are contained in the set $U = \bigcup_{i=1}^{p+1} \mho_n(M_i)$ in view of $\bigcup_{i=1}^{p+1} M_i = G$, and each of subgroups $\mho_n(M_i)$ has order 1 or p, and at least one of which, $\mho_n(E)$, is trivial since $E \in \{M_1, \dots, M_{p+1}\}$). Then $|U| < p^2 = |\mho_n(G)|$, which is not the case since all elements of $\mho_n(G)$ are p^n-th powers, by hypothesis. Thus, $|\mho_n(G)| \le p$, completing case $k = 1$.

(b) Now let $k > 1$ and $E < M \in \Gamma_1$. Then $|M : E| = p^{k-1}$ so $|\mho_n(M)| \le p^{k-1}$, by induction. Set $\bar{G} = G/\mho_n(M)$. Then $\exp(\bar{M}) \le p^n$ and $|\bar{G} : \bar{M}| = p$. Therefore, by (a), we have $|\mho_n(\bar{G})| \le p$, so $|\mho_n(G)| \le |\mho_n(\bar{G})||\mho_n(M)| \le p^k$. □

Exercise 1. If G is a p-group such that $|\mho_1(G)| = p$, then G is a $\mathcal{P}_1$-group.

Solution. We have $\exp(G) = p^2$. If $Z = \langle z\rangle < G$ is of order p^2, then $\langle z^p\rangle = \mho_1(G)$ so $\mho_1(G) = \{x^p \mid x \in G\}$. If H is a section of G, then $|\mho_1(H)| \le p$.

For $\Theta \in \{\mathcal{P}_1, \mathcal{P}_2, \mathcal{P}_3, \mathcal{P}\}$, we term a p-group G a *minimal non-Θ-group* if G does not enjoy the property Θ, but all of its proper sections do (this definition we apply only in this section; compare with definition of minimal irregular p-groups in §7).

Theorem 11.3. *Let G be a minimal non-$\mathcal{P}_1$-group. Then* (a) $\mathrm{d}(G) = 2$ and $\exp(G) = p^2$, (b) $\exp(\Phi(G)) = p$, (c) $\mathrm{Z}(G) = \mho_1(G) \cong \mathrm{E}_{p^2}$, (d) if $M \in \Gamma_1$, then $|\mho_1(M)| \le p$.

Proof. (a) Let G be a counterexample of minimal order. Since G is not a $\mathcal{P}_1$-group, there exist $a, b \in G$ such that $a^p b^p$ is not a p-th power. By minimality, $G = \langle a, b\rangle$. By Theorem 11.1, $\exp(G) = p^2$, completing the proof of (a).

(b) Let $\exp(\Phi(G)) \ge p^2$ and let $Z < \mho_1(\Phi(G)) \cap \mathrm{Z}(G)$ be of order p. Since G/Z is a $\mathcal{P}_1$-group, we get $a^p b^p = c^p z$ for some $c \in G$ and $z \in Z^\#$. Since $\Phi(G)$ is a $\mathcal{P}_1$-group, we get $z = u^p$ for some $u \in \Phi(G)$. Since $G \ne \langle c, u\rangle$, it follows by induction in $\langle c, u\rangle$ that $c^p z = c^p u^p = v^p$ for some $v \in \langle c, u\rangle$; then $a^p b^p = c^p u^p = v^p$, contrary to the choice of a and b.

(c,d) It follows from $\mathrm{d}(G) = 2$ that $\mathrm{Z}(G) \le \Phi(G)$, and so $\exp(\mathrm{Z}(G)) = p$, by (b). If $Z \le \mathrm{Z}(G)$ is of order p, then, as above, we have $a^p b^p = c^p z$ for some $c \in G$ and $z \in Z^\#$. Thus, $z = c^{-p} a^p b^p \in \mho_1(G)$, so $Z \le \mho_1(G)$ and $\mathrm{Z}(G) \le \mho_1(G)$ since $\exp(\mathrm{Z}(G)) = p$.

Every $x \in G$ is contained in some $M \in \Gamma_1$. By Proposition 11.2, $|\mho_1(M)| \le p$ since $\exp(\Phi(G)) = p$ and $|M : \Phi(G)| = p$, in view of $\mathrm{d}(G) = 2$, proving (d). Thus, $\mho_1(M) \le \mathrm{Z}(G)$. As $x^p \in \mho_1(M)$, we have $\mho_1(G) \le \mathrm{Z}(G)$ so $\mho_1(G) = \mathrm{Z}(G)$ since the reverse inclusion holds. Moreover, choosing M such that $|\mho_1(M)| = p$ and applying Proposition 11.2 to $G/\mho_1(M)$, we get $|\mho_1(G/\mho_1(M))| \leq p$ so $|\mho_1(G)| \leq p^2$. Equality holds, in view of Exercise 1. This completes the proof of (c). □

Corollary 11.4. *A $\mathcal{P}_2$-group G is a $\mathcal{P}_1$-group.*

Proof. Suppose that G is a counterexample of minimal order. Then G is a minimal non-$\mathcal{P}_1$-group since all proper sections of G are $\mathcal{P}_2$-groups so $\mathcal{P}_1$-groups, by induction. By Theorem 11.3(c) and its proof, $Z(G)$ has exactly $p+1$ subgroups of order p and at least one of them does not consist of p-th powers. By Theorem 11.3(d), $|\mho_1(M)| \leq p$ for every $M \in \Gamma_1$. Let $\Gamma_1 = \{M_1, \dots, M_{p+1}\}$ (see Theorem 11.3(a)). By what has just been said, $Z = \mho_1(M_i) \in \{\mho_1(M_j), \{1\}\}$ for some $i \neq j$. In the both cases, $Z \leq \Phi(M_i) \leq \Phi(G) < M_j$. Since M_i/Z and M_j/Z are distinct maximal subgroups of exponent p in a $\mathcal{P}_2$-group G/Z, it follows that $\exp(G/Z) = p$, and so G is a $\mathcal{P}_1$-group, by Exercise 1, a contradiction. □

Theorem 11.5. *If a p-group G is a minimal non-$\mathcal{P}_2$-group, then*

(a) $d(G) = 2$, $Z(G) \leq \Phi(G)$, $\exp(G) = p^2$, $\Omega_1(G) = G$ *so* $G' = \Phi(G)$.

(b) *G has a maximal subgroup, which is of exponent p so that* $\exp(\Phi(G)) = p$.

(c) *$Z(G) = \mho_1(G)$ is of order p so G is a $\mathcal{P}_1$-group.*

(d) *G is a minimal non-$\mathcal{P}$-group.*

Proof. (a,b) Assume that $\exp(G) > p^2$. Then every section of G of exponent p^2 is a $\mathcal{P}_2$-group; in that case, G is a $\mathcal{P}_2$-group (Theorem 11.1), a contradiction. Therefore, $\exp(G) = p^2$. As G is not a $\mathcal{P}_2$-group, there exist elements $a, b \subset G$ of order p such that $o(ab) = p^2$. By minimality, $G = \langle a, b\rangle$ so $d(G) = 2$, $\Omega_1(G) = G$ hence $\Phi(G) = G'$ (Exercise 2) and $Z(G) \leq \Phi(G)$, proving (a). Let $\langle a\rangle^G = M$ be the normal closure of $\langle a\rangle$; then $M < G$ and $\Omega_1(M) = M$, so $\exp(M) = p$ since M is a $\mathcal{P}_2$-group. Since $G = \langle M, b\rangle = M\langle b\rangle$ and $b^p = 1$, we have $|G : M| = p$, and so $M \in \Gamma_1$; in that case, $\exp(\Phi(G)) \leq \exp(M) = p$, completing the proof of (b).

(c) Let $Z \leq Z(G)$ be of order p; then $G/Z = \langle aZ, bZ\rangle$ is of exponent p since $G/Z(G)$ is a $\mathcal{P}_2$-group and $o(a) = o(b) = p$. In that case, $Z = \mho_1(G)$ so G is a $\mathcal{P}_1$-group (Exercise 1). Thus, Z is a unique minimal normal subgroup of G so $Z(G)$ is cyclic. Since $Z(G) \leq \Phi(G)$, we get $\exp(Z(G)) = p$, by (b); then $|Z(G)| = p$.

(d) Let H be a proper section of G. We have only to show that $|H : \Omega_1(H)| = |\mho_1(H)|$. This is true if $\exp(H) = p$. If $\exp(H) = p^2$, the equality follows from (b) and (c). □

Theorem 11.6. *A p-group G is a $\mathcal{P}_2$-group if and only if all its sections H satisfy $|\mho_n(H)| \leq |H : \Omega_n(H)|$ for all $n \in \mathbb{N}$. Thus, a $\mathcal{P}_3$-group is also a $\mathcal{P}_2$-group.*

Proof. Suppose that G is a $\mathcal{P}_2$-group; then it is a $\mathcal{P}_1$-group (Corollary 11.4). If $|H : \Omega_n(H)| = p^k$, then, since $\exp(\Omega_n(H)) \leq p^n$, it follows from Proposition 11.2 that

$|\mho_n(H)| \le p^k$ $(= |H : \Omega_n(H)|)$ (take into account that H is a $\mathcal{P}_2$-group so a $\mathcal{P}_1$-group, by Corollary 11.4). Conversely, if G is not a $\mathcal{P}_2$-group, some section H of G is a minimal non-$\mathcal{P}_2$-group, which, by Theorem 11.5(a,c), satisfies $|\mho_1(H)| = p$ and $H = \Omega_1(H)$ so $|H : \Omega_1(H)| = 1 < |\mho_1(H)|$. □

Corollary 11.7. *A $\mathcal{P}_3$-group is a $\mathcal{P}$-group.*

Proof. Indeed, a $\mathcal{P}_3$-group G is a $\mathcal{P}_2$-group (Theorem 11.6), and so a $\mathcal{P}_1$-group (Corollary 11.4). □

Exercise 2. Let a p-group G be a $\mathcal{P}$-group. (a) If $H < G$, then $|H/\mho_n(H)| \le |G/\mho_n(G)|$ (see Proposition 7.3). (b) G is pyramidal (see §8).

Theorem 11.8. *If G is a minimal non-$\mathcal{P}$-group, then (a) G is two-generator of exponent p^2. (b) either $\Omega_1(G) = G$ or $\Omega_1(G) = \Phi(G)$. In any case, $\Phi(G) = G'$ and this subgroup has exponent p. (c) $Z(G) = \mho_1(G)$ is of order p. (d) If H is a proper section of G, then $\mathrm{cl}(H) < \mathrm{cl}(G)$.*

Theorem 11.9. *If E is a normal subgroup of exponent p^n in a $\mathcal{P}_1$-group G, then $[\mho_n(G), E] = \{1\}$. In particular, $[\mho_m(G), \mho_n(G)] \le \mho_{m+n}(G)$ and, in a $\mathcal{P}_2$-group G, $[\mho_n(G), \Omega_n(G)] = \{1\}$.*

Corollary 11.10. *If G is a $\mathcal{P}_2$-group, $a, b \in G$ and $b^{p^n} = 1$, then $\langle a^{p^n} \rangle = \langle (ab)^{p^n} \rangle$.*

Proposition 11.11. *If G is a $\mathcal{P}$-group, then $\mho_n(G') \le [\mho_n(G), G]$.*

Theorem 11.12. *Let G be a $\mathcal{P}_i$-group, $i = 1, 2, 3$, and let H be a group of exponent p. Then $G \times H$ is a $\mathcal{P}_i$-group.*

Theorem 11.13. *Let G be a group of exponent p^e and $C \cong C_{p^e}$. If $G \times C$ is a $\mathcal{P}_1$-group, then G is regular.*

§12

Counting theorems for p-groups of maximal class

In this section we prove a number of counting theorems for irregular p-groups of maximal class. Since the 2-groups of maximal class are classified, we will confine to the case $p > 2$. We retain the notation of §§5, 9. Our exposition is based on some results of §9 and Theorem 12.1 which is the most important consequence of Blackburn's theory of p-groups of maximal class.

Recall that a p-group G is absolutely regular if $|G/\mho_1(G)| < p^p$.

Theorem 12.1 (Blackburn [Bla5]). *(a) If a p-group G is neither absolutely regular nor of maximal class, it has a normal subgroup of order p^p and exponent p.*

(b) If a non-absolutely regular p-group G has an absolutely regular subgroup H of index p, then G is either irregular of maximal class or else $\Omega_1(G)$ is of order p^p and exponent p and $G = H\Omega_1(G)$.

We claim that Theorem 12.1(b) $\Rightarrow$ Theorem 12.1(a). Indeed, suppose that G is neither absolutely regular nor of maximal class and that Theorem 12.1(b) is true. We must to prove that then G has a normal subgroup of order p^p and exponent p. In view of Lemma 1.4, we may assume that $p > 2$. Next, in view of Theorem 7.2, we may assume that G is irregular. Let H be a normal absolutely regular subgroup of G of maximal order. If $|G : H| = p$, then, since $|G : \mho_1(G)| \geq p^p$ (Theorem 9.8(a)), it follows that $|H : \mho_1(H)| = p^{p-1}$. In that case, by Theorem 12.1(b), $\exp(\Omega_1(G)) = p$, $G = H\Omega_1(G)$, $|\Omega_1(G)| = p^p$, and Theorem 12.1(a) follows. Now suppose that $|G : H| > p$. Let $D < H$ be a G-invariant subgroup of index p^2 in H and let B/H be a G-invariant subgroup of order p in $C_G(H/D)$. Since B/D is abelian of order p^3, B is not of maximal class and, by the choice of H, B is not absolutely regular. By Theorem 12.1(b), $B = H\Omega_1(B)$, where $\Omega_1(B)$ is of order p^p and exponent p and $\Omega_1(B)$ is characteristic in B so normal in G.

Lemma 12.2 (= Exercise 9.28). *Let G be a p-group of maximal class, $k > 2$ and* $\exp(G) \geq p^k$. *Then $\Omega_k^*(G) \leq G_1$. In particular,* $\exp(G) = \exp(G_1)$.

Lemma 12.3. *Let a group G be of maximal class and order $p^m > p^{p+1}$, $p > 2$. Then (a) $c_1(G) \equiv 1 + p + \cdots + p^{p-2} \pmod{p^p}$. (b) $c_2(G) \equiv p^{p-2} \pmod{p^{p-1}}$. (c) If $n > 2$, then $p^{p-1} \mid c_n(G)$. (d) The number of subgroups of order p^{p-1} and exponent p in G is $\equiv 1 \pmod{p^{m-p}}$.*

Proof. Let $\Gamma_1 = \{M_1 = G_1, M_2, \dots, M_{p+1}\}$, where G_1 is the fundamental subgroup of G. Then for all $n \in \mathbb{N}$ we have

$$(1) \qquad \mathrm{c}_n(G) = \mathrm{c}_n(M_1) + \dots + \mathrm{c}_n(M_{p+1}) - p \cdot \mathrm{c}_n(\Phi(G)).$$

Since $|\Omega_1(M_1)| = p^{p-1}$ (Theorem 9.6), we have $\mathrm{c}_1(M_1) = 1 + p + \dots + p^{p-2}$. Next, since $|\Omega_2(M_1)| \geq p^{p+1}$ (see Theorem 9.6) and $M_1 = G_1$ is regular, we have $\mathrm{c}_2(M_1) = \frac{|\Omega_2(M_1)| - |\Omega_1(M_1)|}{p(p-1)} \equiv p^{p-2} \pmod{p^{p-1}}$. Now let $n > 2$ and $p^n \leq \exp(M_1)(= \exp(G))$. We have $|\Omega_{n-1}(M_1)| = p^{(p-1)(n-1)}$ so

$$(2) \qquad \mathrm{c}_n(M_1) = \frac{|\Omega_n(M_1)| - |\Omega_{n-1}(M_1)|}{(p-1)p^{n-1}} \equiv 0 \pmod{p^{(p-2)(n-1)}}$$

and we conclude that $\mathrm{c}_n(M_1) \equiv 0 \pmod{p^{p-1}}$ since $p > 2$ and $n > 2$.

(a) First assume that $m = p + 2$; then $\exp(G) = p^2$ since $p > 2$. Assume that for some $i > 1$, a subgroup M_i has a subgroup, say R, of order p^p and exponent p. Then, by Theorem 9.6(e), M_i has exactly p such subgroups since R is not normal in G and $\exp(M_i) = p^2$. This M_i has exactly one absolutely regular subgroup of index p so $\mathrm{c}_2(M_i) = p^{p-2}$. For such i, we have

$$\mathrm{c}_1(M_i) = \mathrm{c}_1(\Phi(M_i)) + p \cdot p^{p-1} \equiv \mathrm{c}_1(\Phi(M_i)) = 1 + p + \dots + p^{p-2} \pmod{p^p}.$$

In any case, $|\Omega_1(\Phi(G))| = p^{p-1} = |\Omega_1(G_1)|$ so $\mathrm{c}_2(\Phi(G)) = \frac{p^p - p^{p-1}}{p(p-1)} = p^{p-2}$, $\mathrm{c}_2(G_1) = \frac{p^{p+1} - p^{p-1}}{p(p-1)} = (p+1)p^{p-2}$. If, for some $j > 1$, M_j has no subgroups of order p^p and exponent p, then $|\Omega_1(M_j)| = p^{p-1}$, and so $\mathrm{c}_1(M_j) = 1 + p + \dots + p^{p-2}$, $\mathrm{c}_2(M_j) = \frac{p^{p+1} - p^{p-1}}{p(p-1)} = (p+1)p^{p-2} \equiv 0 \pmod{p^{p-2}}$. Thus,

$$\mathrm{c}_1(G) \equiv (p+1)\varphi_{p-1,p} - p\varphi_{p-1,1} \equiv 1 + p + \dots + p^{p-2} \pmod{p^p},$$
$$\mathrm{c}_2(G) \equiv (p+1)p^{p-2} - p \cdot p^{p-2} \equiv p^{p-2} \pmod{p^{p-1}}.$$

Now let $m > p + 2$. Then for $i > 1$ we have, by induction, $\mathrm{c}_1(M_i) \equiv \varphi_{p-1,1} \pmod{p^p}$ (this also holds for $i = 1$) so substitution in (1) yields the result.

(b) Let us find $\mathrm{c}_2(G) \pmod{p^{p-1}}$. The case $m = p + 2$ was considered in (a) so let $m > p + 2$. Then, by induction and what has been proved already, $\mathrm{c}_2(G) = \sum_{i=1}^{p+1} \mathrm{c}_2(M_i) - p\mathrm{c}_2(\Phi(G)) \equiv (p+1)p^{p-2} - p \cdot p^{p-2} \equiv p^{p-2} \pmod{p^{p-1}}$.

(c) Let $n \geq 3$. Then, by Lemma 12.2, $\mathrm{c}_n(G) = \mathrm{c}_n(G_1)$ and, by (2), p^{p-1} divides this number. (This is not true for $p = 2$.)

(d) This was proved in Theorem 9.16(a). □

Exercise 1. If a p-group G satisfies $\mathrm{c}_1(G) = \varphi_{p,1}$, then $|\Omega_1(G)| = p^p$.

Lemma 12.4. *If G is a p-group of maximal class, then $\Omega_2(G) = G$.*

Proof. One may assume that $p > 2$ and $\exp(G) > p^2$ so $|G| \geq p^{2p-1}$. We use induction on $|G|$. If $M \in \Gamma_1$ is irregular, then, by induction, $\Omega_2(M) = M$. Since the set Γ_1 has exactly p irregular members, the result follows. □

Given a p-group G, $S = \Omega_1(\mathrm{Z}(G))$ is its socle. Set $|S| = p^s$. Let Δ_i be the set of all subgroups of order p^i in S, $i = 0, 1, \dots, s$. Let $\mathfrak{N}$ be a set of nonidentity normal subgroups of G. Given $K \in \Delta_i$, let $\nu(K)$ be the number of members of $\mathfrak{N}$ containing K.

Theorem 12.5 (enumeration principle for normal subgroups). *Let $\mathfrak{N}$ be a set of nonidentity normal subgroups of a p-group G. Then*

$$\nu(\{1\}) = |\mathfrak{N}| = \sum_{i=1}^{s} \sum_{K \in \Delta_i} (-1)^{i-1} p^{\frac{1}{2}i(i-1)} \nu(K). \tag{3}$$

Proof. If $H \in \mathfrak{N}$ and $|H \cap S| = p^j$ ($j > 0$ since $H > \{1\}$), then the contribution of H in the right-hand side of (3) is

$$\varphi_{j,1} - p\varphi_{j,2} + \cdots + (-1)^{i-1} p^{i(i-1)/2} \varphi_{j,i} + \cdots + (-1)^{j-1} p^{j(j-1)/2} \varphi_{j,j} = 1,$$

by (5.2), and (3) is proved. □

Theorem 12.6. *Suppose that a group G of order p^m is not of maximal class and $m > n > p+1$. Let $\mathfrak{N}$ be the set of normal subgroups D of G such that G/D is of maximal class and order p^n. Then p divides $\nu(\{1\}) = |\mathfrak{N}|$.*

Proof. We retain the notation of Theorem 12.5. One may assume that $\mathfrak{N} \neq \varnothing$. We proceed by induction on m. By (3),

$$|\mathfrak{N}| = \nu(\{1\}) \equiv \sum_{H \in \Delta_1} \nu(H) \pmod{p}. \tag{4}$$

If $H \in \Delta_1$ and G/H is not of maximal class, then p divides $\nu(H)$, by induction. If G/H is of maximal class, it contains only one normal subgroup of index p^n (Exercise 9.1) so $\nu(H) = 1$. Thus, in view of (4), we have to prove that p divides the number of minimal normal subgroups of G whose quotient groups are of maximal class. One can assume that there exists $D \in \Delta_1$ such that G/D is of maximal class. Since G is not of maximal class, $|\mathrm{Z}(G)| = p^2$. By Lemma 4.1, there is $R \lhd G$ such that $D < R$ and $R \cong \mathrm{E}_{p^2}$. Then $R = \mathrm{Z}(G)$ so $\mathrm{Z}(G)$ equals the socle of G. Since $\mathrm{cl}(G) = m-2$, there is at most one $U \in \Delta_1$ such that G/U is not of maximal class (see Lemma 1.11). Thus, it suffices to show that there exists at least one such subgroup because $|\Delta_1| = p+1$. Assume that this is false, i.e., G/K is of maximal class for all $K \in \Delta_1$. Then all epimorphic images of G of order p^{m-1} are of maximal class. By Theorem 9.7, $G/\mathrm{K}_{p+1}(G)$ is not of maximal class. Therefore, by what has just been said, $\mathrm{K}_{p+1}(G) = \{1\}$ so $\mathrm{cl}(G) = p$. However, $\mathrm{cl}(G) = m-2 > p$, a contradiction. □

Exercise 2. Let G be a nonabelian p-group, $|G| > p^4$. Prove that G is either minimal nonabelian or of maximal class provided all members of the set Γ_1 are either abelian or of maximal class.

Exercise 3. Study the irregular p-groups containing a maximal regular subgroup which is absolutely regular.

Exercise 4. If all maximal subgroups of an irregular p-group G are either absolutely regular or of maximal class, then G is of maximal class.

Exercise 5 ([Bla5, Theorem 2.1]). Let G be a p-group and suppose that G has a proper subgroup H of index p^r, $r < p$, such that $|H : \mho_1(H)| \leq p^{p-r}$. Then one of the following holds: (a) G is absolutely regular, (b) G is of maximal class, (c) $\Omega_1(G)$ is of order p^p and $G = H\Omega_1(G)$.

Exercise 6 ([Bla5, Corollary 2.2]). Suppose that a group G of order p^m has a cyclic subgroup H of index p^{p-1}. Then one of the following holds: (a) G is absolutely regular, (b) G is a 2-group of maximal class, (c) G is a p-group of maximal class and order p^{p+1}, $p > 2$, (d) $E = \Omega_1(G)$ is of order p^p and $G = HE$.

Exercise 7. Let a p-group G be neither absolutely regular nor of maximal class. Suppose that all subgroups of order p^p and exponent p of G are of maximal class. Prove that $G = A\Omega_1(G)$, where $A \in \Gamma_1$ and $|\Omega_1(G)| = p^p$.

Solution. Let $R \lhd G$ be abelian of type (p, p). Set $C = C_G(R)$ and let $R < A \leq C$, where $A \in \Gamma_1$. If A has no G-invariant subgroup of order p^p and exponent p, it is absolutely regular and $G = A\Omega_1(G)$ with $|\Omega_1(G)| = p^p$ (Theorem 12.1(b)). If A has a G-invariant subgroup M of order p^p and exponent p, then $\exp(RM) = p$ so one may assume that $R < M$, and so M is not of maximal class, a contradiction.

Let G be a group of order p^m and exponent p^e. Then the number $\mathrm{ce}(G) = m - e$ is said to be the *coexponent* of G [SanW]. It is known (see Theorem 1.2) that if $p > 2$ and $\mathrm{ce}(G) = 1$, then $\mathrm{cl}(G) \leq 2$. We will prove the following

Theorem 12.7 (compare with [SanW, Theorems A and B]). *Let $p > 2$ and G a p-group. Then* $\mathrm{cl}(G) \leq 2 \cdot \mathrm{ce}(G)$. *If* $\mathrm{cl}(G) = 2 \cdot \mathrm{ce}(G)$*, then G is metacyclic.*

Proof. We use induction on $|G|$. Let $|G| = p^m$, $\exp(G) = p^e$ and G nonabelian. In view of Theorem 1.2, one may assume that $\mathrm{ce}(G) = m - e > 1$.

Let $R \lhd G$ be of order p^w and exponent p, where w is as large as possible. Then $w \geq 2$ (Lemma 1.4). Let $\mathrm{ce}(G/R) = t$ and let Z be a cyclic subgroup of G of order p^e such that $|G/R : (ZR/R)| = p^t$ (then $|G : Z| = p^{m-e} = p^{\mathrm{ce}(G)}$); in that case, $|R \cap Z| \leq p$. Since $|RZ : Z| \in \{p^w, p^{w-1}\}$, we get

$$p^{m-e} = p^{\mathrm{ce}(G)} = |G : Z| = |G : RZ||RZ : Z| \in \{p^{t+w}, p^{t+w-1}\},$$

i.e., $\mathrm{ce}(G) \in \{t + w - 1, t + w\}$. By induction, $\mathrm{cl}(G/R) \leq 2t$. We have $\mathrm{cl}(G) \leq \mathrm{cl}(G/R) + w \leq 2t + w$ since $|R| = p^w$.

(i) Assume that $\mathrm{cl}(G) > 2\mathrm{ce}(G)$. Then $2t + w > 2(t + w - 1) = 2t + 2w - 2$ so $w < 2$, a contradiction. Thus, $\mathrm{cl}(G) \leq 2\mathrm{ce}(G)$.

(ii) Let $\mathrm{cl}(G) = 2\mathrm{ce}(G)$. Then $2t + w \geq 2t + 2w - 2$ so $w = 2$ (Lemma 1.4), and, by Theorems 12.1(a) and 9.11, G is either metacyclic or a 3-group of maximal class. Let G be a 3-group of maximal class and $m > 3$. We have $m - 1 = \mathrm{cl}(G) = 2\mathrm{ce}(G)$ so m is odd. Next, G has a metacyclic subgroup G_1 of index 3 and exponent 3^e such that $|\Omega_i(G_1)| = 3^{2i}$, $i = 1, \dots, e$; then $|G_1| = 3^{2e}$ and $m = 2e + 1$. It follows that $\mathrm{ce}(G) = e + 1$. Then $m - 1 = \mathrm{cl}(G) = 2(e + 1) = m + 1$, a contradiction. □

Theorem 12.8 (P. J. Sanders). *If G is regular, then* $\mathrm{cl}(G) \leq \mathrm{ce}(G) + 1$.

Theorem 12.9 (compare with Theorem 9.7). *Suppose that a p-group G has only one normal subgroup L of index p^{p+1}. If G/L is of maximal class then G is also of maximal class.*

Proof. Assume that $L \not\leq G'$; then $|G/G'| \geq p^3$. Since $p + 1 \geq 3$, there is $M \triangleleft G$ of index p^{p+1} such that $M \leq G'$ and $|(G/M) : (G'/M)| \geq p^3$; then $M \neq L$, a contradiction. Thus, $L < G' \leq \Phi(G)$ so that $|G : G'| = p^2$ and $\mathrm{d}(G) = 2$. One may assume that $|G| > p^{p+1}$ and $p > 2$ (Taussky). Since G/L is irregular, one has $\exp(G/L) = p^2$ (Theorem 9.5). Then G/L has an absolutely regular subgroup H/L of index p so $\exp(H/L) = p^2$. Assume that H is not absolutely regular. Then $|H/\mho_1(H)| \geq p^p$ so $H/\mho_1(H)$ has a G-invariant subgroup $S/\mho_1(H)$ of index p^p. We have $|G/S| = p^{p+1}$ so $S = L$, a contradiction since $\exp(H/L) = p^2 > p = \exp(H/S)$. Thus, H is absolutely regular. Assume that G is not of maximal class.

(i) Let $|G| = p^{p+2}$. Then $|L| = p$ and $|\mathrm{Z}(G)| = p^2$. By Lemma 4.1, there is $R \triangleleft G$ of type (p, p) such that $L < R$. Since $|\mathrm{Z}(G/L)| = p$, we get $\mathrm{Z}(G) = R \cong \mathrm{E}_{p^2}$. In that case, $\mathrm{Z}(G)$ has exactly $p + 1 > 1$ subgroups of order p that are normal and have index p^{p+1} in G, a contradiction.

(ii) Let $|G| > p^{p+2}$. Then, by Theorem 12.1(b) which is independent of this theorem, $G = H\Omega_1(G)$ and $\Omega_1(H) = H \cap \Omega_1(G)$, where $\Omega_1(G)$ is of order p^p and exponent p so $|\Omega_1(H)| = p^{p-1}$ and $G/\Omega_1(H) = (H/\Omega_1(H)) \times (\Omega_1(G)/\Omega_1(H))$, $|\Omega_1(G)/\Omega_1(H)| = p$. Then $|G/G'| \geq p^3 > p^2$, a contradiction. □

Theorem 12.10. *Let G be a group of order $p^m > p^p$ and let $r > 2$. Suppose that $G/\mathrm{K}_r(G)$ is of maximal class. If N is a G-invariant subgroup of index p in $\mathrm{K}_r(G)$, then G/N is also of maximal class. If, in addition, $r \leq p$ (so $p > 2$), then $G/\mathrm{K}_r(G)$ is of exponent p.*

Proof. Assume that G/N is not of maximal class. Then $\mathrm{Z}(G/N)$ is of order p^2. In that case, $\mathrm{K}_{r-1}(G/N) = \mathrm{Z}(G/N)$ so $\mathrm{K}_r(G) \leq N$, which is not the case. The second assertion follows from Theorem 9.5. □

Corollary 12.11. *If, for $i = 2, \dots, p + 1$, a nonabelian p-group G has exactly one normal subgroup of index p^i, then G is of maximal class.*

Recall that if G is a nonabelian p-group, then the subgroup $\eta(G)$ is defined as follows: $\eta(G)/\mathrm{K}_3(G) = \mathrm{Z}(G/\mathrm{K}_3(G))$. Clearly, $G' \leq \eta(G)$ and $\eta(G)$ is characteristic in G. By Theorem 1.40, if $G = H\eta(G)$ for some $H < G$, then $\mathrm{cl}(H) = \mathrm{cl}(G)$.

Theorem 12.12. *Suppose that a group G of order p^m, $m > 3$, contains a subgroup H of maximal class and index p. If G is not of maximal class, then*

(a) (Theorem 9.10) $\mathrm{d}(G) = 3$.

(b) Set $\nu = m - 2$ if $m \leq p + 1$ and $\nu = p$ if $m > p + 1$. Then $G/\mathrm{K}_\nu(G)$ is of order $p^{\nu+1}$ and, if $m > 4$, it is of exponent p.

(c) Exactly p^2 maximal subgroups of G are of maximal class. If, in addition, $p > 2$ and $m > 4$, then the remaining $p + 1$ maximal subgroups of G have no two generators and their intersection $\eta(G)$ has index p^2 in G.

Proof. (b, c) By (a), $\Phi(G) = G' = H'$ so that $|G/\mathrm{K}_\nu(G)| = p^{\nu+1}$ since H is of maximal class so $\mathrm{K}_i(G) = \mathrm{K}_i(H)$, $i > 1$. Since $G/G' \cong \mathrm{E}_{p^3}$ and $|G/\mathrm{K}_3(G)| = p^4$, it follows that $\eta(G)$ is of index p^2 in G. If $\eta(G) \not\leq F$ and $F \in \Gamma_1$, then $F\eta(G) = G$ and again we get $\mathrm{K}_i(G) = \mathrm{K}_i(F)$, all $i > 1$ (Theorem 1.40) so F is of maximal class. It follows from the structure of (nonabelian) $G/\mathrm{K}_3(G)$ that it contains exactly p^2 nonabelian subgroups $M_1/\mathrm{K}_3(G), \dots, M_{p^2}/\mathrm{K}_3(G)$ (these subgroups are exactly those that do not contain $\eta(G)/\mathrm{K}_3(G)$) so, by the above, $M_1, \dots, M_{p^2}$ are of maximal class; the intersection of the remaining $p + 1$ maximal subgroups of $G/\mathrm{K}_3(G)$ equals $\eta(G)/\mathrm{K}_3(G)$ since $G/\eta(G) \cong \mathrm{E}_{p^2}$. This proves (c) apart of the assertion on exponent and generators since the $p + 1$ maximal subgroups of G containing $\eta(G)$, are not of maximal class: indeed, $\eta(G)/\mathrm{K}_3(G)$ is a central factor of G of order p^2.

(i) Let $m > p + 1$. Then $\nu = p$ and $G/\mathrm{K}_p(G)$ is regular of order p^{p+1} and class $p - 1$. Since M_i is of maximal class and irregular, $M_i/\mathrm{K}_p(G) = M_i/\mathrm{K}_p(M_i)$ is of order p^p and exponent p, $i = 1, \dots, p^2$ (Theorems 9.5 and 9.6). Since the regular group $G/\mathrm{K}_p(G)$ is generated by subgroups $M_i/\mathrm{K}_p(G)$ of exponent p, $i = 1, \dots, p^2$, we get $\exp(G/\mathrm{K}_p(G)) = p$ (Theorem 7.2(b)). Other $p + 1$ maximal subgroups of G, say $T_1, \dots, T_{p+1}$, satisfy, for $p > 2$, $|T_j/\mathrm{K}_3(G)| = p^3$ and $\exp(T_i/\mathrm{K}_3(G)) = p$ so $\mathrm{d}(T_i) \geq 3$. This completes the proof of (c) in the case $m > p + 1$.

(ii) Now let $m \leq p + 1$; then $p > 3$ since $m > 4$. Here $\nu = m - 2 \leq p - 1$. Then $G/\mathrm{K}_\nu(G)$ is regular of order $p^{\nu+1} = p^{m-1} \leq p^p$. Since M_i is of maximal class and order p^{m-1} and $\mathrm{K}_{m-2}(M_i) = \mathrm{K}_{m-2}(G) = \mathrm{K}_\nu(G)$, we see that $M_i/\mathrm{K}_\nu(G)$, $i = 1, \dots, p^2$, is of exponent p (Theorem 9.5); then $\Omega_1(G/\mathrm{K}_\nu(G)) = G/\mathrm{K}_\nu(G)$ so $\exp(G/\mathrm{K}_\nu(G)) = p$. Again, as in (i), the remaining $p + 1$ maximal subgroups $T_1, \dots, T_{p+1}$ of G satisfy $\mathrm{d}(T_j) > 2$, provided $p > 2$ and $m > 4$, and $\eta(G) = \bigcap_{i=1}^{p+1} T_i$ has index p^2 in G. □

Proof of Theorem 12.1(a) independent of Theorem 12.1(b). In view of Lemma 1.4, one may assume from the start that $p > 2$. Let $|G| = p^m$. We have to prove that if G has no normal subgroups of order p^p and exponent p, it is either absolutely regular or of maximal class. If G is regular, it is absolutely regular since $p^p > |\Omega_1(G)| = |G/\mho_1(G)|$ (Theorem 7.2(d)). Next we assume that G is irregular; then $m > p$ and $\exp(G) > p$ (Theorem 7.1(b)). One may assume that $p > 2$ (Lemma 1.4) and $m > p+1$ (Theorem 7.1(b)). Let G be a counterexample of minimal order. We have $\mho_1(G) > \{1\}$. Let $N \leq \mho_1(G) \cap \mathrm{Z}(G)$ be of order p. There are two cases to consider. If $N \not\leq G'$, then $|G/G'| = p^3$ so G/N is neither absolutely regular (Theorem 9.8(a)) nor of maximal class. There are two cases to consider.

(i) Suppose that G/N has no normal subgroups of order p^p and exponent p. By induction, G/N is either absolutely regular or irregular of maximal class and $|G/N| \geq p^{p+1}$. By Theorem 9.8(a), G/N is not absolutely regular so G/N is irregular of maximal class, and we have $\mathrm{d}(G) = \mathrm{d}(G/N) = 2$ since $N < \Phi(G)$. By the previous paragraph, $N < G'$.

Then $\mathrm{cl}(G) = m - 2$ and $\mathrm{Z}(G) \cong \mathrm{E}_{p^2}$ since G has a normal abelian subgroup of type (p, p) containing N so coinciding with $\mathrm{Z}(G)$ (Lemma 4.1). Let $H/N < G/N$ be absolutely regular of index p (Theorems 9.5, 9.6). Then H is regular (Remark 7.2) so $\Omega_1(H) \triangleleft G$ has exponent p. It follows that $|\Omega_1(H)| < p^p$ so H is absolutely regular. We have $|(G/N)/\mho_1(G/N)| = |G/\mho_1(G)| = p^p$ (Theorem 9.8(a)). We also have $|G : G'| = p^2$ and so $G' < H$ is absolutely regular. Then $K/N = \Omega_1(H/N)$ is of order p^{p-1} and exponent p. On the other hand, K is of order p^p and $\exp(K) = p^2$. Since $\mathrm{d}(G) = 2$, we get $\mathrm{Z}(G) < \Phi(G) < H$. Let $\mathrm{Z}(G) = N \times S$. Since G/S is not absolutely regular (Remark 7.2), we get $S < \mho_1(G)$ so $\mathrm{Z}(G) \leq \mho_1(G)$. We have $\mathrm{c}_1(\mathrm{Z}(G)) = p + 1$ so, since $\mathrm{cl}(G) = m - 2$, one may assume that $\mathrm{cl}(G/S) = m - 2$. Assume that G/S has a normal subgroup M/S of order p^p and exponent p. Then $M \in \Gamma_1$ (Theorem 9.6(c)) so $\mathrm{Z}(G) < M$ and M is regular. In that case, $|\Omega_1(M)| = |M/\mho_1(M)| \geq p^p$ so $\Omega_1(M)$ contains a G-invariant subgroup of order p^p and exponent p, a contradiction. Thus, G/S has no normal subgroups of order p^p and exponent p so G/S is also of maximal class. Clearly, $\mathrm{Z}(G) \leq \Omega_1(H)$. It follows that $\mathrm{Z}(G) < K$. Since G/S has no normal subgroups of order p^{p-1} and exponent p^2, we get $\exp(K/S) = p$. Then $\exp(K) = p$ since K of order p^p is isomorphic to a subgroup of $(K/N) \times (K/S)$, and this is a contradiction.

(ii) Next suppose that G/N has a normal subgroup K/N of order p^p and exponent p. If K is regular, then $|\Omega_1(K)| = |K : \mho_1(K)| \geq |K : N| \geq p^p$ (Theorem 7.2(d)), a contradiction since, in our case, $\Omega_1(K)$ is of exponent p (Theorem 7.2(b)) and normal in G. Hence, K is irregular of order p^{p+1}, and therefore it is of maximal class (Theorem 7.1(b)). Now let $A \triangleleft G$ contain K and A is of maximal class and $|A|$ is as large as possible. We shall prove that $A = G$, thus completing the proof. Since a p-group of maximal class A/N contains a normal subgroup K/N of order p^p and exponent p, we get $|A| \leq p|K| = p^{p+2}$.

Assume that we have $A < G$. Let B/A be a normal subgroup of G/A of order p. Then $|B| = p|A| \leq p^{p+3}$ and B is not of maximal class, by the choice of A. By Theorem 12.12(a,b), $\mathrm{d}(B) = 3$ and $B/\mathrm{K}_p(B)$ is of order p^{p+1} and exponent p. By Theorem 12.12(c), B has exactly $p + 1$ maximal subgroups $M_1, \dots, M_{p+1}$ of ranks at least 3, and these subgroups are not absolutely regular since $B/\mathrm{K}_p(B)$ is of order p^{p+1} and exponent p and all $M_i > \mathrm{K}_p(B)$ in view of $\mathrm{K}_p(B) \le \Phi(B)$. Since $B \triangleleft G$, one of these $p+1$ subgroups, say $M = M_1$, is also normal in G. If M is regular, then, by Theorem 7.2, $\Omega_1(M)$ is of order $|M/\mho_1(M)| \ge p^p$ and exponent p, which is not the case. Thus, M is irregular. In that case, since $\mathrm{d}(M) > 2$, we get $|M| = p^{p+2}$ so $|B| = p^{p+3}$ and $\mathrm{cl}(M) = p$. It follows that $|M : M'| = p^3$, $M' = \Phi(M)$ (Theorem 12.12(a)) and $|M : \mathrm{K}_3(M)| = p^4$. In that case, $|M : \eta(M)| = p^2$. Let T be a maximal subgroup of M such that $\eta(M) \not\le T$ (recall that $\mathrm{d}(M) = 3$); then $T\eta(M) = M$. In that case, by Theorem 1.40, $\mathrm{cl}(T) = \mathrm{cl}(M)(= p)$ so T is of maximal class since $|T| = p^{p+1}$. Therefore, by Theorem 12.12(c), $M/\mathrm{K}_p(M)$ is of order p^{p+1} and exponent p. By Theorem 5.8(b), we can choose a $G/\mathrm{K}_p(M)$-invariant subgroup $T_1/\mathrm{K}_p(M)$ of index p in $M/\mathrm{K}_p(M)$ such that $|(T_1/\mathrm{K}_p(M)) : (T_1/\mathrm{K}_p(M))'| > p^2$. In that case, $|T_1 : T_1'| \ge p^3$ so T_1 is not of maximal class. Since $|T_1| = p^{p+1}$, it follows that T_1 is regular (Theorem 7.1). Since $M/\mathrm{K}_p(M)$ is of order p^{p+1} and exponent p, T_1 is not absolutely regular. Then $\Omega_1(T_1)$ is a G-invariant subgroup of order $\ge p^p$ and exponent p (Theorem 7.2), completing the proof. □

The above proof is easier than original one due to Blackburn.

Proposition 12.13. *Let G be a p-group. If $A \in \Gamma_1$ is absolutely regular and $M < G$ is irregular of maximal class, then G is of maximal class.*

Proof. Assume that G is not of maximal class. One may assume that every subgroup of G that contains M as a subgroup of index p, is not of maximal class. Let $M < L \le G$, where $|L : M| = p$. By Theorem 12.12(c), $L/\mathrm{K}_p(L)$ is of order p^{p+1} and exponent p so L has no absolutely regular maximal subgroups. This is a contradiction since $A \cap L$ is a maximal absolutely regular subgroup of L. □

Proposition 12.14 ([Bla5, Theorem 2.1]). *Theorem 12.1(b) is true.*

Proof. Let H be an absolutely regular subgroup of index p in an irregular p-group G and suppose that G is not of maximal class. Then G has a normal subgroup E of order p^p and exponent p, by Theorem 12.1(a), and $G = HE$, by the product formula, so $G/E \cong H/(H \cap E)$ is absolutely regular. It remains to prove that $E = \Omega_1(G)$. If this is not so, there is $x \in G - E$ of order p. Then $B = \langle x, E\rangle$ is of order p^{p+1}, and, since $B \not\le H$, we get $|H \cap B| = p^p$. It follows that $\exp(H \cap B) = p^2$ since $H \cap B(< H)$ is absolutely regular, so B is irregular since $\Omega_1(B) = B$ (Theorem 7.2(b)). We conclude that $\mathrm{cl}(B) = p$ so B is of maximal class. Then G is also of maximal class, by Proposition 12.13. □

The proof of Theorem 12.1 is complete.

If a p-group G is regular and $|\Omega_1(G)| = p^w$, then w is said to be the *width* of G.

Exercise 8. A p-group G is absolutely regular of width w if and only if $\Omega_2(G)$ is regular of width w.

Corollary 12.15. *Suppose that a p-group G contains an absolutely regular subgroup of index p, $|G| > p^{p+1}$. If G is not of maximal class, then all members of the set Γ_1 are not of maximal class.*

Proposition 12.16. *Let an irregular p-group G be not of maximal class, $|G| = p^m > p^{p+1}$. Suppose that all nonabelian members of the set Γ_1 are either absolutely regular or of maximal class. Then one of the following holds: (a) $p = 2$, $G = DZ(G)$ is of order 16, $|D| = 8$. (b) $E = \Omega_1(G)$ is elementary abelian of order p^p, G/E is cyclic and $C_G(E)$ is maximal in G. (c) G is minimal nonabelian 2-group.*

Proof. Assume that the set Γ_1 has no abelian members. Let $R \triangleleft G$ be abelian of type (p, p); then any maximal subgroup H of G such that $R < H \leq C_G(R)$, is absolutely regular, and so we get $|\Omega_1(G)| = p^p$ (Theorem 12.1(b)). It follows that all members of the set Γ_1 containing $\Omega_1(G)$, are of maximal class so G is also of maximal class (Exercise 13.10), contrary to the hypothesis.

Assume that there is an abelian $A \in \Gamma_1$. Since G is irregular, A is a unique abelian member of the set Γ_1, provided $p > 2$. Now let $p = 2$ and let the set Γ_1 have another abelian member B. Then $|G'| = 2$ so $\mathrm{cl}(G) = 2$. If $M \in \Gamma_1$ is of maximal class, then $|M| = 8$, and we get case (a). In what follows we assume that A is a unique abelian member of the set Γ_1. If M does not exist, G is a minimal nonabelian 2-group. If $m = 4$, then $p = 2$ (by hypothesis, $m > p + 1$) and G is a group from (a). Next assume that $m > 4$.

Assume, in addition, that there is an absolutely regular $H \in \Gamma_1$. By Theorem 12.1(b), $G = EH$, where $E = \Omega_1(G)$ is of order p^p and exponent p. By Corollary 12.15, the set Γ_1 has no members of maximal class. Then all members of the set Γ_1, containing $\Omega_1(G)$, are abelian. If $G/\Omega_1(G)$ is cyclic, then $C_G(\Omega_1(G)) \in \Gamma_1$, and we get case (b). Now assume that $G/\Omega_1(G)$ is noncyclic. Then $\Omega_1(G) \leq Z(G)$ so $\mathrm{cl}(G) = 2$; then $p = 2$ since G is irregular. Then H is cyclic so $G/\Omega_1(G)$ is cyclic, contrary to assumption.

Now assume that all nonabelian members of the set Γ_1 are of maximal class. In that case, in view of Theorem 12.12(c), we get case (a). □

Exercise 9. (i) Let G be a group of order p^m, $m > n \geq 3$, $p > 2$. Suppose that G is not metacyclic. Then p divides the number of $N \triangleleft G$ such that G/N is metacyclic of order p^n. (ii) If $3 < n < m$ and G nonmetacyclic, then the number of $N \triangleleft G$ such that G/N is metacyclic of order 2^n, is even. (*Hint.* (ii) See §47.)

Problems

Problem 1. Suppose that a p-group G of maximal class has an abelian subgroup of index p^4. Is it true that G has a normal abelian subgroup of index p^4? (For $p \in \{2, 3\}$, the answer is 'yes'.)

Problem 2. Is it true that there exists a p-group containing a regular subgroup of index p^2, $p > 2$, but not containing a normal regular subgroup of index p^2?

Problem 3. Let H be a subgroup of maximal class and order p^4 in a p-group G. Suppose that $C_G(H) = Z(H)$. Study the normal structure of G.

Problem 4. Study the irregular p-groups with maximal regular subgroup of order p^{p+1}.

Problem 5. Study the irregular p-groups with abelian maximal regular subgroup.

§13

Further counting theorems

Here we prove the main counting theorems of this book. Our exposition is based on results of §§7, 9, 12. Some particular cases were proved in §§1, 5 by means of elementary arguments. Many proofs are new.

Lemma 13.1 (= Theorem 12.12(c)). *Let a p-group G be not of maximal class, $p > 2$ and there is $H \in \Gamma_1$ of maximal class. Then* $\mathrm{d}(G) = 3$ *and the set Γ_1 has exactly $p + 1$ members, that are not two-generator, and their intersection $\eta(G)$ (see Theorem 1.40) has index p^2 in G.*

Definition. A p-group G is said to be *thin* if it is either absolutely regular (i.e., $|G/\mho_1(G)| < p^p$) or irregular of maximal class.

Recall that $\varphi_{n,1} = 1 + p + \cdots + p^{n-1}$. The following Theorem 13.2 is a deep generalization of Theorems 1.10 and 1.17.

Theorem 13.2. *If G is a group of order p^m which is not thin, then (a)* $\mathrm{c}_1(G) \equiv \varphi_{p,1}$ (mod p^p). *(b) If $k > 1$, then p^{p-1} divides* $\mathrm{c}_k(G)$.

Proof. We proceed by induction on $|G|$. By hypothesis and Theorem 12.1(a), G has a normal subgroup R of order p^p and exponent p.

In view of Theorem 1.17, one may assume that $p > 2$. If $\exp(G) = p$, then $\mathrm{c}_1(G) = \frac{p^m-1}{p-1} = 1 + p + \cdots + p^{m-1} \equiv \varphi_{p,1} \pmod{p^p}$ since $m \geq p$, and $\mathrm{c}_k(G) = 0$ for $k > 1$. Now we assume that $\exp(G) \geq p^k$. By Theorem 7.2, one may assume that G is irregular so, since G is not thin, we get $m > p + 1$ (Theorems 7.1(b) and 9.5).

Suppose that G/R is cyclic. Then it follows from Lemma 9.4 that $|\Omega_1(G)| \in \{p^p, p^{p+1}\}$ and $\exp(\Omega_1(G)) = p$ so $\Omega_s(G)$ is of exponent p^s and order p^{p+s-1} or p^{p+s} for every s with $p^s \leq \exp(G)$; then $\mathrm{c}_s(G) = \frac{|\Omega_s(G)|-|\Omega_{s-1}(G)|}{\varphi(p^s)}$. It follows that $\mathrm{c}_1(G) \equiv \varphi_{p,1} \pmod{p^p}$, and, if $s > 1$, then $\mathrm{c}_s(G) \equiv 0 \pmod{p^{p-1}}$.

Now suppose that G/R is not cyclic. Then there is $D/R \triangleleft G/R$ such that $G/D \cong \mathrm{E}_{p^2}$. Let $H_1/D, \ldots, H_{p+1}/D$ be all distinct subgroups of order p in G/D. Then

$$\text{(1)} \qquad \mathrm{c}_s(G) = \sum_{i=1}^{p+1} \mathrm{c}_s(H_i) - p \cdot \mathrm{c}_s(D),$$

for all $s \in \mathbb{N}$ (see the proof of Theorem 1.10). By construction, all H_i are not absolutely regular. If all H_i are not of maximal class, then, by induction, we have, for all i, $c_1(H_i) \equiv \varphi_{p,1} \pmod{p^p}$ and $p \cdot c_1(D) \equiv p + \cdots + p^{p-1} \pmod{p^p}$ and, for $k > 1$, $c_k(H_i) \equiv 0 \pmod{p^{p-1}}$, $p \cdot c_k(D) \equiv 0 \pmod{p^{p-1}}$ (the congruences for $k > 1$ are true also in the case where D is of maximal class, by Lemma 12.3(a–c)). Substituting these congruences in (1), we obtain the desired results.

Suppose that H_1 is of maximal class; then $|H_1| = p^{p+1}$ since $R < H_1$, so $|G| = p^{p+2}$. Then, by Theorem 12.12(c) (recall that $p > 2$), one can choose $p + 1$ distinct maximal subgroups $F_1, \dots, F_{p+1}$ in G such that $d(F_i) > 2$ and $L = \bigcap_{i=1}^{p+1} F_i$ has index p^2 in G. (in fact, $L/\mho_1(G) = \eta(G/\mho_1(G))$; see also Theorem 1.40 and Lemma 13.1). By Theorem 7.1, F_i (of order p^{p+1}) is regular. In that case, we have equality (1) with F_i instead of H_i and L instead of D. Then, with $k > 1$, we have, since regular subgroup F_i is not absolutely regular for all i:

$$c_1(F_i) \equiv \varphi_{p,1} \pmod{p^p}, \qquad p \cdot c_1(L) \equiv p + \cdots + p^{p-1} \pmod{p^p},$$
$$c_k(F_i) \equiv 0 \pmod{p^{p-1}}, \qquad p \cdot c_k(L) \equiv 0 \pmod{p^{p-1}}.$$

Substituting these congruences in (1), we complete the proof. □

As noticed in §1, Theorem 13.2(a) follows from Theorem 13.2(b), in view of identity $|G| = 1 + \sum_{i=1}^{e} \varphi(p^i) c_i(G)$, there $p^e = \exp(G)$.

Remark 1. Using Theorem 9.8, one can give the proof, independent of Theorem 12.1(a), of the following weaker assertions on irregular p-groups G (Hall): $c_1(G) \equiv \varphi_{p-1,1} \pmod{p^{p-1}}$, $c_n(G) \equiv 0 \pmod{p^{p-2}}$, $(n > 1)$ (for an alternate approach, see the proof of Theorem 13.4).

Corollary 13.3. *Suppose that an irregular p-group G is not thin and $k < p$. Then it has a normal subgroup M of order p^k and exponent p, and M is contained in $n_M \equiv 1 \pmod{p}$ subgroups of order p^{k+1} and exponent p.*

Proof. The existence of M follows from Theorem 9.8(d). If $x \in G - M$ is of order p, then $\langle x, M\rangle$ has order $p^{k+1} (\le p^p)$ so its exponent is p, by Theorems 7.1(b) and 7.2(b); x exists by Theorem 12.1(a). In $\langle x, M\rangle$, but outside M, there are exactly p^k subgroups of order p. Then $n_M \cdot p^k + \varphi_{k,1} = c_1(G) \equiv \varphi_{p,1} \pmod{p^p}$, by Theorem 13.2(a) so $n_M \cdot p^k \equiv p^k + \cdots + p^{p-1} \pmod{p^p}$ and $n_M \equiv 1 \pmod{p}$. □

For a regular p-group G, set $p^{\mu(G)} = |\Omega_1(G)|$. Now let G be irregular. Among all subgroups of G of exponent p we take one which has maximal order $p^{t(G)}$. Next, among all irregular subgroups of G we take one which has minimal order $p^{r(G)}$ and set $\mu(G) = \min\{t(G), r(G) - 2\}$. Let $e_k(G)$ be the number of subgroups of order p^k and exponent p in G.

Theorem 13.4. *Let G be a p-group. Then:*

(a) $e_k(G) \equiv 1 \pmod{p}$, $k = 1, \dots, \mu(G)$.

(b) *The number of two-generator subgroups of order* p^k, $k = 3, \dots, \mu(G) - 1$, *and exponent* p *in* G *is a multiple of* p.

(c) $c_1(G) \equiv \varphi_{\mu(G),1} \pmod{p^{\mu(G)}}$.

(d) *If* $n > 1$, *then* $p^{\mu(G)-1} \mid c_n(G)$.

Proof. We consider only the case $k = \mu(G)$ in (a) (the case where $k < \mu(G)$ is treated in the same way with help of Theorem 5.2). Thus, we assume, for $k < \mu(G) = s$, that $e_k(G) \equiv 1 \pmod p$. The assertion is true for regular G, by Theorem 7.2(b); next we assume that G is irregular. Then G has a subgroup A of order p^s (where $s = \mu(G)$) and exponent p. If $A < H \in \Gamma_1$, then $\mu(H) \geq \log_p(|A|) = \mu(G)$ so, by induction, $e_k(H) \equiv 1 \pmod p$. Therefore, one may assume that $A \lhd G$. Suppose that $x \in G - A$ is of order p (if x does not exist, we are done). Since, by hypothesis $K = \langle x, A\rangle$ is regular, in view of $|K| = p^{\mu(G)+1} < p^{r(G)}$, we have $\exp(K) = p$, by Theorem 7.2(b). If $M \in \Gamma_1$, then $|M \cap K| \geq p^{\mu(G)} = p^s$ so, by induction, $e_s(M) \equiv 1 \pmod p$ since $\mu(M) \geq s = \mu(G)$). Now the result follows, by Theorem 5.2. □

If G is irregular, then $\mu(G) \geq p - 1$ so $e_{p-1}(G) \equiv 1 \pmod p$ (Hall).

Theorem 13.5. *If a* p*-group* G *is not thin, then* $e_p(G) \equiv 1 \pmod p$.

Proof. By hypothesis, $|G| > p^p$. We use induction on $|G|$. Let $R \lhd G$ be of order p^{p-1} and exponent p (Theorem 12.1(a)). Then the number of subgroups of order p^p and exponent p containing R, is $\equiv 1 \pmod p$ (Corollary 13.3). Let K be a nonnormal subgroup of order p^{p-1} and exponent p in G and set $N = N_G(K)$; then N is not of maximal class (Remark 10.5) so K is not characteristic in N hence N is not thin. Obviously, every subgroup L of order p^p and exponent p containing K is also contained in N. By Corollary 13.3, the number of such L is $\equiv 1 \pmod p$. Let $\mathfrak{M}_{p-1} = \{R_1, \dots, R_u\}$ ($\mathfrak{M}_p = \{S_1, \dots, S_v\}$) be the set of all subgroups of order p^{p-1} (of order p^p) and exponent p in G so that $u = e_{p-1}(G)$, $v = e_p(G)$. We have to prove that $v \equiv 1 \pmod p$. Let α_i be the number of all elements of the set $\mathfrak{M}_p$ containing $R_i \in \mathfrak{M}_{p-1}$ and let β_j be the number of all maximal subgroups in $S_j \in \mathfrak{M}_p$ (all of them are contained in $\mathfrak{M}_{p-1}$), $i = 1, \dots, u$, $j = 1, \dots, v$. By double counting,

$$\alpha_1 + \dots + \alpha_u = \beta_1 + \dots + \beta_v. \tag{2}$$

As we have noticed, $\alpha_i \equiv 1 \pmod p$, all i. By Sylow, $\beta_j \equiv 1 \pmod p$, all j. By Theorem 13.4(a), $u \equiv 1 \pmod p$ so, by (2), $v \equiv u \equiv 1 \pmod p$. □

Exercise 1. If a p-group G of order p^m is not thin and $p > 3$, then: (a) The number of 2-generator subgroups of order p^{p-1} and exponent p in G is divisible by p. (b) $e_{p-1}(G) \equiv 1 + p \pmod{p^2}$. (c) If $1 < k < p - 1$, then $e_k(G) \equiv 1 + p + 2p^2 \pmod{p^3}$.

Exercise 2. Let G be a p-group. If $M \lhd G$ has no G-invariant subgroups of order p^p and exponent p, it is thin.

Remark 2. If an irregular p-group G has noncyclic center, it has a characteristic subgroup of order $> p^{p-1}$ and exponent p (compare with Theorem 9.8(d)). Indeed, G has a normal subgroup R of order p^p and exponent p (Theorem 12.1(a)). One can choose R so that $|R \cap Z(G)| > p$. Then $R \leq Z_{p-1}(G)$. Since $Z_{p-1}(G)$ is regular (Theorem 7.1), we get $\Omega_1(Z_{p-1}(G)) = p$, and we are done.

Theorem 13.6 ([Ber3, Theorem 3]; [Bla8]). *Let a group G of order p^m be not thin, $n \in \mathbb{N}$ and $m > n \geq p + 1$. Denote by $\alpha(G)$ the number of subgroups of maximal class and order p^n in G. Then p^2 divides $\alpha(G)$.*

Proof. We use induction on m. For $n = m - 1$ the assertion of the theorem is a part of Theorem 12.12(c). In the sequel we assume that $n < m - 1$; then $m \geq p + 3$. In view of Theorem 5.4, one may assume that $p > 2$ and $\alpha(G) > 0$.

By Hall's enumeration principle,

$$\alpha(G) \equiv \sum_{H \in \Gamma_1} \alpha(H) - p \sum_{H \in \Gamma_2} \alpha(H) \pmod{p^2}. \tag{3}$$

The first sum in (3) is divisible by p^2. Indeed, if $H \in \Gamma_1$ is of maximal class, then $\alpha(H) = p^{m-1-n}$ does not depend on H since $m - 1 > p + 1$. Since the set Γ_1 has exactly p^2 elements of maximal class (Theorem 12.12(c)), the contribution of elements of maximal class in the first sum is divisible by p^2. The contribution of the remaining elements of the set Γ_1 is divisible by p^2, by induction.

It remains to prove that p divides $\sum_{H \in \Gamma_2} \alpha(H)$. By what has just been said and (3), p divides $\alpha(G)$ always, and so by induction, the theorem is true if $n < m-2$ (then every member of the second sum in (3) is a multiple of p, by induction or by Theorem 9.6). In what follows, we assume that $n = m - 2$. We may also assume that there exists $H \in \Gamma_2$ which is of maximal class. It suffices to show that the number of such H is divisible by p. In that case, $d(G) \leq 4$ since $d(H) = 2$. Since $Z(H)$ is cyclic, $H \neq \Phi(G)$, by Proposition 1.13, and so we get $|G : \Phi(G)| \geq p^3$.

(i) Suppose that $|G : \Phi(G)| = p^4$. Then, if $K < G$ is of maximal class and index p^2 in G (in that case, $|K| = p^n$), we have $|\Phi(K)| = p^{n-2} = |\Phi(G)|$ so $\Phi(K) = \Phi(G)$ since $\Phi(K) \leq \Phi(G)$, and hence $K \in \Gamma_2$. Thus, by the previous paragraph, p divides $\sum_{H \in \Gamma_2} \alpha(H)$, and we are done.

(ii) It remains to consider the case where $|G : \Phi(G)| = p^3$. Let $\alpha'(G)$ be the number of normal subgroups of maximal class and index p^2 in G. Since p divides $\alpha(G) - \alpha'(G)$, we see that p divides $\alpha'(G)$. It remains to show that every normal subgroup T of maximal class and index p^2 in G belongs to the set Γ_2; indeed, then $\sum_{H \in \Gamma_2} \alpha(H) (= \alpha'(G))$ is a multiple of p. Therefore, assume that G has a normal subgroup T of maximal class and index p^2 such that $T \notin \Gamma_2$; then $G/T \cong C_{p^2}$

so $G' < \Phi(G)$; in that case, G/G' is abelian of type (p^2, p, p). One may assume that there exists $H \in \Gamma_2$ of maximal class (otherwise, we are done); then $G' = H'$ (compare indices!). Let M/H be a subgroup of order p in G/H such that $M/G'(= M/H')$ is abelian of type (p^2, p) (such an M exists since G/G' is abelian of type (p^2, p, p) so it contains only one subgroup of type (p, p, p); on the other hand, G/H has exactly $p + 1 > 1$ subgroups of order p); then $M \in \Gamma_1$. Since $G' = H' \leq M' \leq G'$, we have equalities elsewhere. Let L be a G-invariant subgroup of index p in G'; then H/L is nonabelian of order p^3 since H is of maximal class. Let us consider the group M/L. We see that $(M/L)' = G'/L$ is of order p so $\mathrm{d}(M/L) = \mathrm{d}(M/G') = 2$ and hence M/L is minimal nonabelian. This is a contradiction since M/L contains a nonabelian subgroup H/L of index p. Thus, T does not exist so all normal subgroups of maximal class and index p^2 in G are members of the set Γ_2, as was to be shown. □

Theorem 13.6 is true for $n = m - 1 \geq 3$, by Theorem 12.12(c).

The following theorem plays important role in the Odd Order Paper [FT] (for more elementary proof, see Theorem 69.4).

Theorem 13.7 ([Bla5, Theorem 4.1(iii)]). *Let G be a p-group, $p > 2$. Suppose that G has no normal elementary abelian subgroups of order p^3. Then one of the following holds:*

(a) G is metacyclic.

(b) G is a 3-group of maximal class not $\cong \Sigma_{32}$.

(c) $G = EH$, where $E = \Omega_1(G)$ is nonabelian of order p^3 and exponent p, H is cyclic of index p^2 in G, $\mathrm{Z}(G) \leq H$ is cyclic, $|G : \mathrm{Z}(G)| \leq p^3$, $|H : \mathrm{C}_G(E)| \leq p$.

Proof. (See also proof of Theorem 69.4.) By Theorem 10.4, G has no subgroups isomorphic to E_{p^3} so G has no subgroups of order p^4 and exponent p.

(i) Suppose that G has no normal subgroups of order p^3 and exponent p. If G is irregular, it is a 3-group of maximal class (Theorem 12.1(a)). If G is regular, it is metacyclic (Theorem 9.11) since $p^2 \geq |\Omega_1(G)| = |G/\mho_1(G)|$ (Theorem 7.2).

(ii) Next we assume that G has a normal subgroup E of order p^3 and exponent p; then E is nonabelian. Since $\exp(\Omega_1(\mathrm{Z}(G))E) = p$, we get $\Omega_1(\mathrm{Z}(G)) < E$ so $\mathrm{Z}(G)$ is cyclic. Set $C = \mathrm{C}_G(E)$; then C is normal in G since $E \lhd G$, and $E \cap C = \mathrm{Z}(E)$ is of order p. Assume that C has a G-invariant abelian subgroup A of type (p, p). Then $\mathrm{cl}(EA) = 2$ so $\exp(EA) = p$ and $|EA| \geq p^4$, a contradiction. Then C is cyclic (Lemma 1.4). Next, G/C is isomorphic to a subgroup of a Sylow p-subgroup of $\mathrm{Aut}(E)$ that is nonabelian of order p^3 and exponent p (see §33). Since $|E \cap C| = p$, it follows that $|G : C| \geq p^2$. If $|G : C| = p^2$, we get $G = E * C$.

Let $R < E$ be G-invariant of order p^2; then $R \cong \mathrm{E}_{p^2}$. Set $L = \mathrm{C}_G(R)$. Then $E \not\leq L$ since E is nonabelian, and so $L \in \Gamma_1$. Since $R \leq \mathrm{Z}(L)$, we have $\Omega_1(L) = R$ (otherwise, L has an elementary abelian subgroup of order p^3). Therefore, by

Theorems 12.1(a), 7.2(d) and 9.11, L is metacyclic. If G is irregular, then, by Theorem 12.1(b), $E = \Omega_1(G)$ since L is absolutely regular in view of $p > 2$. The last equality is also true if G is regular by what has been said in the first paragraph.

It remains to consider the case where $|G : C| = p^3(= |\mathrm{Aut}(E)|_p)$; then G/C is nonabelian of order p^3 and exponent p so, if $x \in G - CE$; then $x^p \in C$. Set $H = \langle x, C\rangle$ and $T = \Omega_1(H)(\leq \Omega_1(G) = E)$. Assume that $|T| > p$; then $T \cong \mathrm{E}_{p^2}$ and $H = CT$, by the product formula. We have $T = H \cap E$ since $E = \Omega_1(G)$. By the choice of x, $G = HE$. On the other hand, $G = EH = ETC = EC < G$, and this is a contradiction. Thus, T is of order p so H is cyclic of index p^2 in G. We have $\mathrm{C}_G(C) \geq EH = G$ so $C = \mathrm{Z}(G) < H$. □

Corollary 13.8. *Let G be a p-group, $p > 2$, and let $N \trianglelefteq G$ have no G-invariant subgroups $\cong \mathrm{E}_{p^3}$. Then N is one of the groups (a)–(c) of Theorem 13.7.*

We offer a new proof of the following Huppert's Theorem 9.11: If $p > 2$, then G is metacyclic if and only if $|G : \mho_1(G)| \leq p^2$. By Theorem 9.8(a), G is regular so $|\Omega_1(G)| = |G/\mho_1(G)| = p^2$, and the result follows from Theorem 13.7.

Here we offer a variant of the proof of Theorem 13.7. Assume that G is not a group from conclusion of Theorem 13.7. By Lemma 1.4, G has a normal abelian subgroup R of type (p, p). Set $H = \mathrm{C}_G(R)$; then $\Omega_1(H) = R$, by Theorem 10.4, so H is absolutely regular, by Theorem 12.1(a). Then H is metacyclic, by Theorems 7.2(d) and 9.11. Since G is not metacyclic, we conclude that $H \in \Gamma_1$. If G has no normal subgroup of order p^3 and exponent p, it is a 3-group of maximal class, by Theorem 12.1(a). Let G have a normal subgroup E of order p^3 and exponent p; then E is nonabelian. In that case, $E = \Omega_1(G)$, by Theorem 12.1(b). Continuing, as in the proof of Theorem 13.7, we complete the proof.

Exercise 3. Let $p > 2$, $n > 3$, $H \cong M_{p^n} \in \Gamma_1$, where G is a p-group. If G does not split over H, then G is metacyclic.

Exercise 4. Let H be an abelian maximal subgroup of type (p, p^n) of a p-group G, $p > 2$, $n > 2$. If G does not split over H, then G is metacyclic. (*Hint.* We have $\Omega_1(G) = \Omega_1(H)$. Apply Theorem 13.7. If G is a 3-group of maximal class, then $\Omega_1(\Phi(H)) \cong \mathrm{C}_{p^2}$ is normal in G, a contradiction.)

Exercise 5. Let H be an absolutely regular maximal subgroup of a p-group G, $p > 2$. Suppose that G does not split over H. Prove that G is either absolutely regular or of maximal class. (*Hint.* Use Theorem 12.1(b).)

Theorem 13.9. *Suppose that a group G of order p^m is not thin. If $m > n \geq p$, then the number of absolutely regular subgroups of order p^n in G is a multiple of p.*

Proof. If $p = 2$, the result follows from Theorem 1.17(b). So we assume that $p > 2$. If $n = p$, the result follows from Theorem 13.5 and Sylow's theorem. Thus, in what follows we assume that $m > n > p > 2$ and that G has an absolutely regular subgroup

H of order p^n. If $X \le G$, we write $\alpha(X)$ for the number of absolutely regular subgroups of order p^n in G.

(i) Suppose that $n = m - 1$. Then $G = H\Omega_1(G)$, where $\Omega_1(G)$ is of order p^p and exponent p (Theorem 12.1(b)). Let $\Omega_1(G) \not\le M \in \Gamma_1$. Then $\Omega_1(M) = M \cap \Omega_1(G)$ is of order p^{p-1} so M is either absolutely regular or irregular of maximal class (Theorem 12.1(a)). By Theorem 12.13, M is not irregular of maximal class. It follows that the number of absolutely regular subgroups of index p in G equals $\varphi_{\mathrm{d}(G),1} - \varphi_{\mathrm{d}(G/\Omega_1(G)),1} \equiv 0 \pmod{p}$.

(ii) Next we assume that $n < m - 1$. We use induction on m. We have

$$\alpha(G) \equiv \sum_{H \in \Gamma_1} \alpha(H) \pmod{p}. \tag{4}$$

If $H \in \Gamma_1$ is of maximal class, then $|H| > p^n \ge p^{p+1}$, and so $\alpha(H) \equiv 1 \pmod{p}$ (this follows from Theorem 9.6), so the contribution of all such H in the right-hand side of (4) is divisible by p, by Theorem 13.6. If $H \in \Gamma_1$ is absolutely regular, then $\alpha(H) \equiv 1 \pmod{p}$, by Sylow's theorem, and the contribution of all such H in the right-hand side of (4) is a multiple of p, by (i). If $H \in \Gamma_1$ is not thin, then p divides $\alpha(H)$, by induction. It follows then from (4) that p divides $\alpha(G)$. □

The following theorem supplements Theorem 13.7.

Theorem 13.10. *Suppose that a group G of order p^m is a normal subgroup of a p-group W, where p is odd and $m \ge 6$. Next, suppose that all W-invariant subgroups of G of order p^r have two generators, where r is fixed and $4 \le r \le m - 2$. Then G is either metacyclic or a 3-group of maximal class.*

Proof. In view of Theorem 5.8(a), $\exp(G) > p$.

(a) First we prove this for $r = 4$. Suppose that G is neither metacyclic nor a 3-group of maximal class. Then G has a subgroup of order p^3 and exponent p (Theorems 12.1(a), 7.2 and 9.11). Then G has a W-invariant subgroup E of order p^3 and exponent p since the number of such subgroups in G is $\equiv 1 \pmod{p}$, by Theorems 13.4(a) and 13.5. Let N be a W-invariant subgroup of order p in E. By hypothesis, G/N has no W-invariant elementary abelian subgroups of order p^3. It follows from Corollary 13.8 and Theorem 13.7 that one of the following holds: (i) G/N is metacyclic, (ii) G/N is a 3-group of maximal class, or (iii) $G/N = \Omega_1(G/N)(C/N)$, where $\Omega_1(G/N)$ is nonabelian of order p^3 and exponent p and C/N is cyclic of order $\ge p^3$ since $m - 1 \ge 5$.

(i) Suppose that G/N is metacyclic. Then G/E is also metacyclic, so its exponent $> p$ since $m \ge 6$. Since $\exp(\mathrm{Aut}(E))_p = p$ (see §33), we get $\mathrm{C}_G(E) \not\le E$, and $\mathrm{C}_G(E)$ is W-invariant since G and E are. If A/E is a W/E-invariant subgroup of order p in $E\mathrm{C}_G(E)/E$, then A is a W-invariant subgroup of order $p^4 = p^r$ in G with $\mathrm{d}(A) \ge 3$, contrary to the hypothesis. (Let us prove that $\mathrm{d}(A) \ge 3$. This is the case if

$E \cong \mathrm{E}_{p^3}$ since then A is abelian. Now suppose that E is nonabelian. Obviously, $\mathrm{Z}(A)$ is of order p^2 so $A = E\mathrm{Z}(A)$. It follows that $A/(A \cap \mathrm{Z}(E)) \cong \mathrm{E}_{p^3}$.)

(ii) Suppose that G/N is a 3-group of maximal class. Since G is not of maximal class, we get $|\mathrm{Z}(G)| = 3^2$. It is easy to show using Lemma 1.4 that $\mathrm{Z}(G)$ is of type $(3, 3)$ so one may assume that $\mathrm{Z}(G) < E$; then E is elementary abelian. Since the center of a Sylow 3-subgroup of the holomorph of E is of order 3 and $|\mathrm{Aut}(E)|_3 = 3^3 \le |G : E|$, it follows that $\mathrm{C}_G(E) > E$. If A/E is a W-invariant subgroup of order p in $\mathrm{C}_G(E)/E$, then A is abelian W-invariant of order $3^4 = 3^r$ and $\mathrm{d}(A) > 2$, a contradiction.

(iii) Suppose that $G/N = \Omega_1(G/N)(C/N)$, where $\Omega_1(G/N)$ is nonabelian of order p^3 and exponent p and C/N is cyclic of order $\ge p^3$ (recall that $|G/N| \ge p^5$). Then G/E is metacyclic so its exponent is greater than p. As above, $E\mathrm{C}_G(E) > E$, and we obtain a contradiction as in (i). Thus the theorem is true for $r = 4$.

(b) If $4 < r \le m - 2$, we use induction on r. Since $\exp(G) > p$, we get $\mho_1(G) > \{1\}$. Let N be a W-invariant subgroup of order p in $\mho_1(G)$. By induction, G/N is either metacyclic or a 3-group of maximal class.

(i) Suppose that G/N is metacyclic. Then $|G : \mho_1(G)| = |(G/N) : \mho_1(G/N)| \le p^2$ so G, by Theorem 9.11, is metacyclic.

(ii) Suppose that G/N is a 3-group of maximal class. Since $m \ge r + 2 \ge 7$, we get $\mathrm{cl}(G) \ge 5$. Then $|G : \mho_1(G)| = 3^3$ so $\mho_1(G) > \mathrm{K}_4(G)$. One can take $N < \mathrm{K}_4(G)$. Then $G/\mathrm{K}_4(G)$ is of maximal class as a nonabelian epimorphic image of G/N. In that case, G is also of maximal class, by Theorem 9.7. □

Lemma 13.11. *Suppose that A is a normal abelian subgroup of order p^k in a group G of order p^n. If $\mathrm{C}_G(A) = A$, then $n \le \binom{k+1}{2}$.*

Proof. Set $\mathrm{d}(A) = d$. Then $|\mathrm{Aut}(G)|$ divides $|\Phi(A)|^d \cdot |\mathrm{GL}(d, p)|$, and so $n \le k + (k - d)d + \frac{1}{2}d(d-1) \le \binom{k+1}{2}$. Indeed, $2\left(\binom{k+1}{2} - k - (k-d)d - \frac{1}{2}d(d-1)\right) = (k-d)(k-d-1) \ge 0$. □

I am indebted to Mann for the proof of the following

Theorem 13.12 ([Mil2]). *Suppose that G is a group of order p^n, $n, k \in \mathbb{N}$ and $n > \binom{k}{2}$. If $a_k(G)$ is the number of abelian subgroups of order p^k in G, then $a_k(G) \equiv 1 \pmod{p}$.*

Proof. We use induction on k. One may assume that $k > 1$. Let $a'_k(G)$ be the number of normal abelian subgroups of order p^k in G. Since p divides $a_k(G) - a'_k(G)$, it suffices to show that $a'_k(G) \equiv 1 \pmod{p}$. By induction, $a'_{k-1}(G) \equiv 1 \pmod{p}$; then G has a normal abelian subgroup A of order p^{k-1}. By Lemma 13.11, $\mathrm{C}_G(A) > A$ since $k - 1 + \binom{k-1}{2} = \binom{k}{2} < n$. The number of normal abelian subgroups of G of order p^k containing A is equal to the number of G/A-invariant subgroups of order p in

$C_G(A)/A$, and the last number is $\equiv 1 \pmod p$ (Sylow). The combinatorial argument in the end of the proof of Theorem 13.5 yields $a'_k(G) \equiv 1 \pmod p$. □

Theorem 13.13 ([Mil1]). *If p is odd, $k > 1$ and Sylow p-subgroups of G are not cyclic, then the number of cyclic subgroups of order p^k in G is divisible by p.*

Exercise 6. Let G be a nonabelian p-group, $p > 2$. Prove that if the centralizer of each $x \in G - Z(G)$ is metacyclic, then G is one of the groups of Theorem 13.7.

Exercise 7. Let G be a p-group. Then (a) if $G/K_3(G) \cong M_{p^n}$, then $K_3(G) = \{1\}$, (b) if $n > 2$ and $G/K_n(G) \in \{Q_{2^n}, SD_{2^n}\}$, then $K_n(G) = \{1\}$, (c) If $n > 2$ and $G/K_n(G) \cong D_{2^n}$, then G is a 2-group of maximal class.

Exercise 8. Study the p-groups G, $p > 2$, with $|G/K_3(G)| = p^3$.

Exercise 9. Study the p-groups G such that $G/K_3(G)$ is extraspecial.

Exercise 10. (a) Let $H < G$, where G is a p-group. If every subgroup of G of order $p|H|$ containing H is of maximal class, then G is also of maximal class.

(b) Let A be a proper absolutely regular subgroup of a p-group G, $p > 2$, $\exp(A) > p$, such that whenever $A < B \leq G$ with $|B : A| = p$, then $\Omega_1(B) = B$. Then G is of maximal class. If, in addition, $|A| > p^p$, then $|G : A| = p$.

Solution. (a) We have $|H| > p$. Set $N = N_G(H)$. In view of Remark 10.5 and hypothesis, one may assume that $|N : H| > p$ (otherwise, there is nothing to prove). Let $D < H$ be N-invariant of index p^2. Set $C = C_N(H/D)$; then $C > H$. Let $F/H \leq C/H$ be of order p; then F is not of maximal class, a contradiction.

(b) Let G be a counterexample of minimal order; then $N_G(A)$ is not of maximal class (Remark 10.5) so one may assume that $N_G(A) = G$. If $B/A \leq G/A$ is of order p, then B is irregular since $\exp(B) \geq \exp(A) > p$ and $\Omega_1(B) = B$ (Theorem 7.2). Assume that B is not of maximal class. Since B is also not absolutely regular, we get $B = A\Omega_1(B)$, where $\exp(\Omega_1(B)) = p$ (Theorem 12.1(b)) so $\Omega_1(B) < B$, contrary to the hypothesis. Thus, every subgroup of G of order $p|A|$ containing A is of maximal class so G is of maximal class, by (a). In our case, A is a maximal regular subgroup of G; then either $|G : A| = p$ or $|A| = p^p$.

Exercise 11. Let G be a p-group. If $\Omega_1(G)$ is isomorphic to a Sylow p-subgroup of S_{p^2}, then either $G = \Omega_1(G)$ or SD_{2^4}. (*Hint.* Use Theorem 13.5.)

Exercise 12. Let G be a nonabelian group of order p^m, $m > 4$, and suppose that G has a proper nonabelian subgroup. Suppose that all nonabelian maximal subgroups of G are of maximal class. Prove that then G is also of maximal class.

Exercise 13. Let $H < G$ be p-groups, $|G : H| > p^k$. Suppose that all subgroups of G of order $p^k|H|$, containing H, are of maximal class. Is G of maximal class?

Exercise 14. Let G be a p-group and let T be generated by all elementary abelian subgroups of G of order p^3 and $x \in G - T$ is of order p. Prove that $C_T(x)$ is cyclic or generalized quaternion.

Exercise 15. If $\Omega_1(G)$ is irregular of maximal class, then G is also of maximal class. (*Hint.* Use Theorem 13.5.)

Proposition 13.14. *(a) If $\Omega_1(G)$ is thin, then G is thin.*

(b) A p-group G is of maximal class if and only if $\Omega_2(G)$ is of maximal class.

Proposition 13.15. *Let G be a noncyclic group of order $p^m > p^4$. Prove that if G has only one abelian subgroup A of order p^3, then G is a 2-group of maximal class. (Hint. Use Proposition 2.3 and Theorems 5.3 and 5.5.)*

Proposition 13.16. *Suppose that a nonabelian p-group G has a cyclic subgroup U of order p^2 such that $C_G(U)$ is cyclic. Then G is of maximal class.*

Theorem 13.17 ([Man13]). *Let $G > \{1\}$ be a metacyclic p-group, $p > 2$. Then G is such, that all extensions of G by a group of order p split, precisely when G is the Sylow p-subgroup of the holomorph of C_{p^k}, excepting $p^k = 9$.*

Proposition 13.18. *Let $M < G$ be of maximal class, where G is a p-group.*

(a) Set $D = \Phi(M)$, $N = N_G(M)$ and $C = C_N(M/D)$. Let t be the number of subgroups $K \le G$ of maximal class such that $M < K$ and $|K : M| = p$. Then $t = c_1(N/M) - c_1(C/M)$. If G is not of maximal class, then $t \equiv 0 \pmod p$.

(b) Suppose, in addition, that M is irregular and G is not of maximal class and $k \in \mathbb{N}$ is fixed. Then the number t of subgroups $L < G$ of maximal class and order $p^k|M|$ such that $M < L$, is a multiple of p.

Proof. (a) Note that $|N : C| \le p$. First assume that $M < C$. If K/M is a subgroup of order p in C/M, then K/D is abelian of order p^3 so K is not of maximal class. Let K/M be a subgroup of order p (in N/M) not contained in C/M. Then K/D is nonabelian of order p^3. Since $D = \Phi(M) \le \Phi(K)$ and K/D is nonabelian, it follows that $d(K) = d(K/D) = 2$ so K is of maximal class, by Theorem 12.12(a). Now assume that G is not of maximal class. Then N is also not of maximal class (Remark 10.5). If $C = N$, then $t = 0$. If $C = M$, then N/D is nonabelian of order p^3 so N is of maximal class (Theorem 12.12(a) again), contrary to the assumption. Now let $M < C < N$; then the numbers of subgroups of order p in C/M and N/M are $\equiv 1 \pmod p$ so the number of subgroups $L/M < N/M$ of order p not contained in C/M, is a multiple of p; since L, by the above, is of maximal class, we get: p divides t.

(b) If $k = 1$, the assertion follows from (a). Now let $k > 1$. We proceed by induction on k. Let $\mathfrak{N} = \{P_1, \dots, P_u\}$ be the set of subgroups of maximal class and order $p^{k-1}|M|$ in G containing M. By induction, $u \equiv 0 \pmod p$. Let $\{V_1, \dots, V_a\}$

and $\{W_1, \dots, W_b\}$ be the sets of subgroups of maximal class and order $p|P_1|$ in G containing P_i and P_j, respectively, $i \neq j$. By (a), a and b are multiples of p. Assume that $X \in \{V_1, \dots, V_a\} \cap \{W_1, \dots, W_b\}$. Then P_i and P_j are subgroups of index p in X so $X = P_i P_j$. Since X is of maximal class, we get $d(X) = 2$. It follows that $M \leq P_i \cap P_j = \Phi(X)$, a contradiction since M is irregular and $\Phi(X)$ is absolutely regular (Theorems 9.5 and 9.6). Thus, $\{V_1, \dots, V_a\} \cap \{W_1, \dots, W_b\} = \varnothing$. Let us show that, in this way, we have counted all wanted subgroups. Let $\mathfrak{M}_i = \{V_{i,1}, \dots, V_{i,a_i}\}$ be the set of all subgroups of maximal class and order $p|P_i|$ in G with $P_i < V_{i,r}$ for all $r \in \{1, \dots, a_i\}$, $i \in \{1, \dots, u\}$. Suppose that $V < G$ is of maximal class with $M < V$ and $|V : M| = p^k$. By Theorem 9.6, any maximal subgroup P of V with $M < P$ is of maximal class so $P \in \mathfrak{N}$. It follows that $V \in \bigcup_{i=1}^{u} \mathfrak{M}_i$, and our claim follows. Thus. $t = \sum_{i-1}^{u} |\mathfrak{M}_i| \equiv 0 \pmod{p}$. □

Remark 3. Let G be a p-group of maximal class and let $H < G$ be such that $|H/\mho_1(H)| > p^{p-1}$. Then one of the following holds: (i) H is irregular of maximal class, (ii) $|H| = p^p$. Indeed, by Theorems 9.5 and 9.6, irregular subgroups of G are of maximal class and regular subgroups of G are either absolutely regular or of order p^p. If, in addition, $p > 2$ and $d(H) > p - 1$, then, as we know, $G \cong \Sigma_{p^2}$.

Let $M \in \Gamma_1$, where G is a p-group of maximal class, $|G| > p^4$. Let M_1 be the fundamental subgroup of M. Then $|G : M_1| = p^2$ and $M_1 \lhd G$ so $M_1 = \Phi(G) < G_1$, and we get $M_1 = G_1 \cap M$, where G_1 is the fundamental subgroup of G.

Theorem 13.19. *Let G be a group of maximal class of order p^m and exponent p^e, $m > p + 1$. Then*

(a) If $3 \leq k \leq e$, then $\Omega_k^*(G) \leq G_1$, where $\Omega_k^*(G) = \langle x \in G \mid o(x) = p^k \rangle$, i.e., every element in $G - G_1$ has order $\leq p^2$. If $m > p + 1$, then $\exp(G_1) = p^e$.

(b) If $x \in G - G_1$, then $x^p \in Z(G)$ so $H_p(G/Z(G)) = G_1/Z(G)$.

(c) If $H < G$, $H \nleq G_1$ and $|H| > p^p$, then H is of maximal class.

(d) If $H < G$ is irregular, then $\Omega_1(G_1) < H$.

(e) If $H < G$ is of order p^p and is not contained in G_1, then $\Omega_1(G_1) < H$.

(f) If $p > 2$ and $L < G$ is of order p^{p-1}, then either $L < G_1$ or $Z(G) < L$.

(g) If $L < G$ is of order p^{p-1} and $L \nleq G_1$, then $\Omega_1(G_1)$ normalizes L.

Proof. To prove these assertions, we use induction on m. One may assume that $p > 2$ since all assertions are easily checked in the case $p = 2$.

(a) Let $a \in G - G_1$ with $o(a) > p^2$. If $a \in M \in \Gamma_1$, then $M \neq G_1$ so M is of maximal class (Theorem 9.6). By the paragraph preceding the theorem, $M \cap G_1 = M_1$ is the fundamental subgroup of M so, by induction, $a \in M_1 < G_1$, contrary to the choice of a. Hence, if $e > 2$, then $\exp(G_1) = p^e$ and $\Omega_k^*(G) \leq G_1$ for $k > 2$. (It follows that $\Omega_2(G) \geq \langle G - G_1 \rangle = G$; this is also true for $m \leq p + 1$.)

(b) By (a), $o(x) \le p^2$. Assume that $o(x) = p^2$. Let $x \in M \in \Gamma_1$; then M is of maximal class. If $m \le p+2$, then $x^p \in \mho_1(M) = Z(M) = Z(G)$. If $m > p+2$, then, by induction, $x^p \in Z(M) = Z(G)$.

(c) If $|G| = p^{p+2}$, the result follows, by Theorem 9.6(e). Now let $|G| > p^{p+2}$. Let $H < M \in \Gamma_1$; then $M(\neq G_1)$ is of maximal class and $M \cap G_1 = \Phi(G) = M_1$ is the fundamental subgroup of M. Then, by induction on m, H is of maximal class.

(d,e) Assume that $\Omega_1(G_1) \not\le H$. Let R be the least G-invariant subgroup of $\Omega_1(G_1)$ not contained in H. Then $|RH| = p|H| > p^p$ so RH is of maximal class, by (c). Then $R \le \Phi(RH) < H$, a contradiction.

(f) Assume that $Z(G) \not\le L$ and $L \not\le G_1$; then $F = L \times Z(G)$ is of order p^p. Let $F < H \le G$, where $|H : F| = p$. Then $|H| = p^{p+1}$ and $H \not\le G_1$ so H is of maximal class, by (c). In that case, $Z(H) = Z(G)$ so that $L_H = \{1\}$, and we conclude that $H \cong \Sigma_{p^2} \in \mathrm{Syl}_p(\mathrm{S}_{p^2})$. By Remark 3, $H = G$. Then $L < F = G_1$, a contradiction (by Fitting's lemma, all members of the set $\Gamma_1 - \{G_1\}$, where $G \cong \Sigma_{p^2}$, are of maximal class).

(g) Let $L < H < G$, where $|H : L| = p$; then $H \not\le G_1$. In that case, $\Omega_1(G_1) < H$, by (e), and $L \lhd H$. □

Supplement 1 to Theorem 13.19. *Suppose that G is a group of maximal class and order p^m, $m > p+1$. If $H < G$, then one of the following holds:*

(a) $H \le G_1$,

(b) $\Omega_1(G_1) \le H$,

(c) $|H\Omega_1(G_1)| = p^p$.

Proof. Suppose that $H \not\le G_1$, $\Omega_1(G_1) \not\le H$ and $|H\Omega_1(G_1)| > p^p$. Then $|H| \le p^{p-1}$. By Theorem 13.19(c), $H\Omega_1(G_1)$ is irregular of maximal class so, by (d), $\Omega_1(G_1) \le \Phi(H\Omega_1(G_1))$ so $H\Omega_1(G_1) = H$, a contradiction. □

Supplement 2 to Theorem 13.19 = Theorem 9.16. *Suppose that G is a group of maximal class and order p^m, $m > p-1$. If G contains a subgroup $E \cong \mathrm{E}_{p^k}$, $k \in \{p-2, p-1\}$, then G contains a normal subgroup isomorphic to E.*

Let G be a p-group such that $\bar{G} = G/Z(G)$ is generated by two distinct cyclic subgroups $\bar{U}$ and $\bar{V}$. Then $\bar{U} \cap \bar{V} = \{1\}$. Indeed, U and V are abelian so $C_G(U \cap V) \ge UV = G$, and we have $Z(G) \le U \cap V \le Z(G)$.

Proposition 13.20. *Let G be a 2-group such that $\bar{G} = G/Z(G)$ is of maximal class. Then $\bar{G} \cong \mathrm{D}_{2^n}$ for some $n > 2$ and $|G'| = 2^{n-1}$. If, in addition, $n > 3$, then $G' \cong \mathrm{C}_{2^{n-1}}$.*

Proof. If a non-dihedral 2-group G is of maximal class, then it is generated by two cyclic subgroups U and V with $U \cap V > \{1\}$, implying the first assertion.

To prove the second assertion, we use induction on n. Let $\bar{U}$ and $\bar{V}$ be different maximal subgroups of $\bar{G}$ and $\bar{V}$ is cyclic; then V is abelian. If $n = 3$, then, by Lemma 1.1, $|U'| \leq 2$ so $|G'| \leq 2|U'V'| \leq 4$ (Exercise 1.69(a)).

Now let $n > 3$. Then $U/Z(U)$ is dihedral. By induction, $|U'| \leq 2^{n-2}$. Then $|G'| \leq 2|U'V'| \leq 2^{n-1}$, by Exercise 1.69(a) again. Since $|(G/Z(G))'| = 2^{n-2}$, we get $|G'| \geq 2^{n-2}$. Assume that $|G'| = 2^{n-2}$. Then $G' \cap Z(G) = \{1\}$, which is not the case. Thus, $|G'| = 2^{n-1}$. Assume that G' is not cyclic. Then $G' = U_1 \times V_1$, where $U_1 \cong C_{2^{n-2}}$, $|V_1| = 2$ and $U_1 \cap Z(G) = \{1\}$. In that case, $\Phi(G') = \Phi(U_1) \triangleleft G$ so $\Omega_1(U_1) \leq Z(G)$, contrary to the equality $U_1 \cap Z(G) = \{1\}$. □

Remarks. 4. Let G be of order $p^m > p^k \geq p^{p+1}$, and $M \in \Gamma_1$. Suppose that $H < G$ is of maximal class and order p^k such that $H \not\leq M$, and all such H are of maximal class. We claim that then G is of maximal class. Indeed, if $m = k + 1$, then all members of the set $\Gamma_1 - \{M\}$ are of maximal class. In that case, the number of members of the set Γ_1 that are of maximal class, is not a multiple of p^2; then G is of maximal class, by Theorem 13.6. Now let $m > k + 1$. Let $R \neq M \cap H$ be a maximal subgroup of H. Then all subgroups of G of order $p|R| = p^k$, containing R, are of maximal class, and the claim follows from Exercise 10.

5. Let G be a p-group of maximal class and order $p^m > p^{p+1}$ and G_1 its fundamental subgroup. Then, if $x \in G - G_1$, then $|C_G(x)| = p^2$ [Bla3]. It follows that all abelian subgroups of G of order $> p^2$ are contained in G_1. Moreover, if $U < G$ is of order $> p^2$ and $U \not\leq G_1$, then U is of maximal class. Indeed, if $x \in U - G_1$, then, by what has just been said, $|C_U(x)| = p^2$, so U is of maximal class (Proposition 1.8). In particular, if $R < G$ is of order p^p and exponent p, then R is of maximal class. If there are in the set Γ_1 exactly k members of maximal class containing subgroups of order p^p and exponent p, then G contains exactly kp^{m-p-1} such subgroups. In that case, $c_1(G) = 1 + p + \cdots + p^{p-2} + kp^{m-2}$. We do not use these results in what follows.

Exercise 16 (see Proposition 13.18). Let $p < k < m$ and let G be a group of order p^m. Suppose that $A < G$ is absolutely regular of order p^k such that $|A : \mho_1(A)| = p^{p-1}$. If the number of thin subgroups of G of order p^{k+1} that contain A is not divisible by p, then G is also thin.

Exercise 17. Let G be a p-group that is not of maximal class. If $|\Omega_1(G)| = p^{p+1}$, then $\exp(\Omega_1(G)) = p$. (*Hint.* Use Theorem 13.5.)

Exercise 18. Let G be a p-group, $p > 2$, with $|G| > p^{p+2}$ and let $|\Omega_2(G)| = p^{p+2}$. Then the following holds:

(a) G has a normal elementary abelian subgroup E of order p^3.

(b) G/E is either absolutely regular or irregular of maximal class.

(c) Let G/E be irregular of maximal class. Then $p = 3$, $\Omega_1(G/E) \cong E_{3^2}$ and E is a unique normal subgroup of G isomorphic to E_{3^3}. Next, E is a maximal normal subgroup of exponent 3 in G.

(d) If $M \triangleleft G$ is of order p^{p+1} and exponent p, then G/M is cyclic.

(e) If $p > 3$, then G/E is absolutely regular.

(f) If $p = 3$ and G/E is irregular of order $\geq 3^5$, then $E \leq Z(\Omega_2(G))$ so $\mathrm{cl}(\Omega_2(G)) \leq 2$, $E = \Omega_1(G)$.

Exercise 19. Let $\mathfrak{M}$ be the set of all irregular p-groups of maximal class and order $\geq p^{p+1}$, $p > 2$. Then $p^a = \max\{|A| \mid A < G \in \mathfrak{M},\ A' = \{1\},\ A \not\leq G_1\} < p^p$. Give a proof independent of Remark 5.

Exercise 20. Suppose that a p-group G of order $> p^4$ is neither metacyclic nor of exponent p, $p > 2$. Prove that if all nonmetacyclic subgroups of G are generated by elements of order p, then $p = 3$, G is of maximal class and every irregular member of the set Γ_1 has two distinct nonabelian subgroups of order 3^3 and exponent 3. (If $|G| = p^4$, then again G is a 3-group of maximal class. For $p = 2$, see §84.)

Solution. I am indebted to Mann for discussion of this exercise. There is in G a maximal metacyclic subgroup H of exponent $> p$. Let $H < K \leq G$, where $|K : H| = p$. Then $\Omega_1(K) = K$ so K is irregular. If, for every choice, K is of maximal class, then G is also of maximal class (Exercise 10). If K is not of maximal class, then $K = H\Omega_1(K)$, where $\Omega_1(K)(< K)$ is of order p^3 and exponent p (Theorem 12.1(b)), a contradiction.

Exercise 21. Suppose that a subgroup of maximal class H of order $> p^{p+1}$ is normal in a p-group G and G/H is cyclic of order $> p$. Then G has only one normal subgroup of order p^p and exponent p. (This is not true for $|H| = p^{p+1}$.)

Solution. Set $\Phi = \Phi(H)$, $C = C_G(H/\Phi)$; then $|G : C| \leq p$ and C/Φ is abelian of rank 3 since $|G : H| > p$ (Theorem 12.12(a)). By Theorem 13.5, C contains a G-invariant subgroup R of order p^p and exponent p (Theorem 13.5). Since $|H| > p^{p+1}$, we get $R \not\leq H$. Let H_1 be the fundamental subgroup of H. Since G/H is cyclic, we get $\Omega_1(G) \leq HR$ and $|HR| = p|H|$ so $H \cap R = H_1 \cap R$ has order p^{p-1} whence $|H_1R| = p|H_1| = |H|(> p^{p+1})$. Next, $\Omega_1(H_1R) = R$ (Theorem 12.1(b)) since H_1R is not thin. Assume that $R_1 < G$ is another normal subgroup of order p^p and exponent p. Then $R_1H = RH$ since $R_1 < \Omega_1(G) \leq RH$, and $R \cap R_1 = H \cap R = H_1 \cap R$ is of order p^{p-1}. Put $S_1 = RR_1$. Since H has no normal subgroup of order p^p and exponent p, the subgroup $T = H \cap S_1$ is of order p^p and exponent p^2. By Theorem 13.5, G has at least $p+1$ normal subgroups of order p^p and exponent p. Let $R_2 \triangleleft G$ be of order p^p and exponent p such that $R \neq R_2 \neq R_1$. We have $R \cap R_2 = R \cap R_1 = R \cap H$. Assume that $R_2 \not\leq RR_1$. Set $S_2 = RR_2$. Then $H \cap S_2 = T$ since H has only one normal subgroup of order p^p. We have $S_1 = TR_1 = TR$ and $S_2 = TR_2 = TR$ so $S_1 = S_2$. Then $R_2 < S_1$, contrary to the choice of R_2. Thus, all normal subgroups of G of order p^p and exponent p are contained in S_1. Since $\exp(T) = p^2$ and $\Omega_1(S_1) = S_1$, we conclude that the subgroup S_1 of order p^{p+1} is of maximal class. It follows that $\mathrm{d}(S_1) = 2$. Since

$\geq p+1$ normal in G subgroup of order p^p and exponent p are contained in S_1, we get $\exp(S_1) = p$, a final contradiction.

Exercise 22. Let $H \lhd G$, where G is a p-group and let G/H of order $> p$ have cyclic center. If K/H is a normal subgroup of order p in G/H, then K is not of maximal class.

Exercise 23 (= Theorem 69.3). If a p-group G, $p > 2$, has no normal subgroups of order p^3 and exponent p, it is either metacyclic or a 3-group of maximal class.

Exercise 24 (compare with [Mil6]). Let G be a p-group, $p > 2$ and $c_n(G) = p$. Then (a) G is either abelian of type (p^m, p) or (b) $G \cong M_{p^{m+1}}$, $m \geq n$, or (c) $p = 3$, $n = 2$, G is a 3-group of maximal class and order 3^4 with $c_1(G) = 1 + 3 + 3^3$.

Exercise 25. Given $n > 1$, classify the non-absolutely regular p-groups G containing exactly p^{p-2} cyclic subgroups of order p^n, $n > 1$. Show that one of the following holds: (a) $p = 2$, $n = 2$, G is dihedral, (b) $p = 2$, $n > 2$, G is any 2-group of maximal class, (c) $p > 2$, $n = 2$ and G is of maximal class and order p^{p+1} such that it has exactly one absolutely regular subgroup of index p.

Exercise 26. Let G be a p-group of order $> p^{p+1}$. If G is not thin, then one of the following holds: (a) G has a subgroup E of order p^{p+1} and exponent p, (b) there is $H \in \Gamma_1$ such that $|\Omega_1(H)| = p^p$.

Solution. By Theorem 12.1(a), G has a normal subgroup R of order p^p and exponent p. Let D be a G-invariant subgroup of index p^2 in R. Set $C = C_G(R/D)$. If an element $x \subset C - R$ has order p, then $E = \langle x, R \rangle$ is of class at most $p-1$ so regular of order p^{p+1}; it follows that then $\exp(E) = p$. Now let (a) be not true. Then, if $R/D < H/D \leq C/D$ and H/D is maximal in G/D (the equality $C = G$ is possible), then $\Omega_1(H) = R$ and (b) holds.

Exercise 27. Suppose that a p-group G of order $> p^4$, $p > 2$, has exactly one nonmetacyclic maximal subgroup. Then $\Omega_1(G)$ is of order p^3 and exponent p and $G/\Omega_1(G)$ is cyclic, i.e., G is an L_3-group (see §§17,18).

Exercise 28. Classify the irregular p-groups of order $> p^{p+1}$ containing only one maximal subgroup which is not absolutely regular.

Exercise 29 (inspired by Janko's letter at Nov. 9, 2004). Classify the pairs of p-groups $H < G$ such that H is of order $> p^2$ and has a cyclic subgroup of index p and $H \cap Z = \{1\}$ for each cyclic $Z < G$ not contained in H.

Solution. Let $H < M \leq G$, where $|M : H| = p$. Then all elements in $M - H$ have order p so M is irregular since $\Omega_1(M) = M$ and $\exp(M) = \exp(H) > p$. Let $p = 2$. Then H is abelian. If H is cyclic, it is a unique cyclic subgroup of its order in M so M is dihedral. In that case, G is is of maximal class (Exercise 10) so dihedral.

Let $p > 2$, $n > 2$ and let H be noncyclic. Then H is either abelian of type (p^n, p) or $\mathrm{M}_{p^{n+1}}$. By Theorem 12.1(b), either $M = HE$, where $E = \Omega_1(M)$ is of order p^3 and exponent p or M is a 3-group of maximal class. Since all elements of the set $M - H$ have order p, in the first case we get $E = \Omega_1(M) = M$, a contradiction. If M is a 3-group of maximal class, then $\Phi(H)$ is cyclic and normal in M so $n = 2$ since M has no normal cyclic subgroups of order 3^2. In that case, G is (a 3-group) of maximal class, by Exercise 10(a). Then H is characteristic in M. Assume that $M < G$. Then $\mathrm{N}_G(H) > M$. Let $M < N$, where $|N : M| = 3$; then $H = \Phi(N)$ since N is of maximal class. If N_1 is the fundamental subgroup of N, it is metacyclic of order 3^4 and exponent 3^2. If $Z < N_1$ is cyclic of order 3^2 such that $Z \not\leq H$, then $H \cap Z > \{1\}$, a contradiction. Thus, $M = G$ is of order 3^4. In that case, H is the unique metacyclic member of the set Γ_1.

Exercise 30. Let $H < G$ be nonnormal of order p^p and exponent p and suppose that a p-group G is not of maximal class. Suppose that H^G, the normal closure of H in G, is of maximal class. Prove that then G has a normal subgroup F of order p^p and exponent p such that $|HF| = p^{p+1}$ and $H \cap F \triangleleft G$. (*Hint.* Use Corollary 13.3 and Theorem 13.19(e).)

Exercise 31. Let p^p be the maximal order of subgroups of exponent p in a p-group G. Then either $|\Omega_1(G)| = p^p$ or the intersection of all subgroups of order p^p and exponent p in G has order p^{p-1}.

Exercise 32. Let G be a p-group, $p > 2$. Suppose that $|\Omega_1(\mathrm{Z}(G))| = p^n$. Let $\mathcal{E}_k$ be the set of elementary abelian subgroups of order p^k in G. Then (a) if $k \leq n$, then $|\mathcal{E}_k| \equiv 1 \pmod{p}$ and (b) if $\Omega_1(\mathrm{Z}(G)) < \Omega_1(G)$, then $|\mathcal{E}_{n+1}| \equiv 1 \pmod{p}$.

Exercise 33. Let H be a noncyclic subgroup of a p-group G, $p > 2$. Prove that if $\mathrm{N}_G(H)$ is metacyclic then G is either metacyclic or a 3-group of maximal class. (*Hint.* Prove that G has no normal subgroups of order p^3 and exponent p.)

Exercise 34 (Jonah–Konvisser). Let G be a nonmetacyclic p-group of order $> p^4$, $p > 2$. Prove that the number of abelian subgroups of type (p^2, p) in G is divisible by p.

Exercise 35. Let $|G| = p^m$, $m > p+2$, and $M \in \Gamma_1$. Suppose that G has a subgroup H of maximal class and order $p^k > p^{p+1}$ such that $H \not\leq M$, and all such subgroups are of maximal class. Is it true that G is of maximal class?

Let $\mathfrak{M}_n(G)$ be the set of subgroups of maximal class and order p^n in a p-group G, and write $\mu_n(G) = |\mathfrak{M}_n(G)|$. By Theorem 12.12(a), if $m > 3$, then either G is of maximal class or $\mu_{m-1}(G) \equiv 0 \pmod{p^2}$. Therefore, it is natural to classify the p-groups satisfying $\mu_n(G) = p^2$ for $n \geq 3$.

Exercise 36. Let G be a p-group of order p^m, $3 \leq n < m$ and $\mu_n(G) = p^2$. Take $S \in \mathfrak{M}_n(G)$ and set $N = \mathrm{N}_G(S)$, $D = \langle A \mid A \in \mathfrak{M}_n(G)\rangle$. Then one of the following holds: (a) $G = D$ is of maximal class, $m = n + 2$ and, if $H \in \Gamma_1$ is irregular, then $\mu_n(H) = p$. (b) $D = N$, $|D| = p^{n+1}$, $\mathrm{d}(D) = 3$, $p^{n+2} \le |G| \le p^{n+3}$.

Proposition 13.21. *Let G be a group of maximal class and order p^m, $m > p + 1$, and let $n \ge p - 1$. Given $K \le G$, let $\alpha_n(K)$ be the number of subgroups $H \le K$ of order p^n and such that $|H : \mho_1(H)| = p^{p-1}$. Then $\alpha_n(G) \equiv 1 \pmod p$.*

Remark 6. Let a p-group G satisfy the following conditions: (i) G contains a proper abelian subgroup A of order $\ge p^3$. (ii) Whenever $A < H \le G$ and $|H : A| = p$, then $|\mathrm{Z}(H)| = p$. Then: (a) G is of maximal class, (b) $|G : A| = p$. Indeed, let H be as in (ii). Then H is of maximal class (Lemma 1.4 and induction), and (a) follows from Exercise 10 since H is arbitrary. Using induction, one may assume that $|G : A| = p^2$. Let H be as above. Then A is characteristic in H (Fitting's lemma) so $A \lhd G$, and we get $A = \Phi(G)$. Then all maximal subgroups of G are of maximal class which is not a case since $\mathrm{C}_G(\mathrm{Z}_2(G)) \in \Gamma_1$ is not of maximal class.

Exercise 37. Let G be an irregular p-group.

(a) If any subgroup of G is either absolutely regular or generated by elements of order p, then G is of maximal class.

(b) Let A be a maximal absolutely regular subgroup of G. Suppose that $\exp(A) > p$ and every subgroup B of G of order $p|A|$, containing A, is generated by elements of order p. Then G is of maximal class.

Exercise 38. Let G be an irregular p-group. Suppose that every maximal subgroup of G is either thin or of exponent p. Prove that $|G| \le p^{p+2}$, unless G is of maximal class. (*Hint.* If G is not of maximal class and $|G| > p^{p+2}$, there is in G a maximal subgroup of exponent p.)

Exercise 39. If a non-absolutely regular p-group G of order $> p^{p+1}$ has > 0 however $\le p$ non-absolutely regular maximal subgroups, then it is ether an L_p-group (see §§17,18) or of maximal class.

Exercise 40. If a 2-group G has only one proper subgroup isomorphic to D_8, then $G \cong \mathrm{SD}_{16}$.

Exercise 41. The following conditions for a p-group G are equivalent: (a) $\mathrm{c}_1(G) = \mathrm{s}_2(G) = p + 1$, (b) $\mathrm{c}_1(G) = p + 1 = \mathrm{c}_2(G) + 1$, (c) G is either an L_2-group or $p = 2$ and $G = \langle a, b \mid a^{2^{m-2}} = b^8 = 1, a^b = a^{-1}, a^{2^{m-3}} = b^4, m > 4\rangle$.

Exercise 42. Let G be of maximal class and order $> p^{p+1}$. Prove that if $H < G$ is of order $> p^2$, then either $H \le G_1$ or H is of maximal class.

Solution. Let $|H| = p^k$. The result is known if $k = 3$ [Bla3]. Assuming that $k > 3$, we use induction on k. Then all maximal subgroups of H which $\ne H \cap G_1$, are of maximal class, by induction. Then the set $\Gamma_1(H)$ contains exactly $|\Gamma_1(H)| - 1 \not\equiv 0 \pmod{p^2}$ members of maximal class so H is of maximal class, by Theorem 12.12(c).

Proposition 13.22. *Let a p-group G of order $> p^{p+2}$ be not thin. If $R \lhd G$ is of order p, then one of the following holds: (a) The number of abelian subgroups of type (p, p) in G, containing R, is $\equiv 1 + p + \cdots + p^{p-2} \pmod{p^{p-1}}$, (b) The number of cyclic subgroups of order p^2 in G, containing R, is a multiple of p^{p-1}.*

Recall that $e_k(G)$ is the number of subgroups of order p^k and exponent p in G.

Theorem 13.23. *Let G be a p-group of order $> p^{p+3}$ with $e_p(G) = p + 1$, and let $R_1, \dots, R_{p+1}$ be all its subgroups of order p^p and exponent p. Set $H = \Omega_1(G)$. Then one of the following holds:*

(a) H is of order p^{p+1} and exponent p and $\mathrm{d}(H) = 2$.

(b) $|H| = p^{p+2}$, $\exp(H) = p^2$, $\mathrm{d}(H) = 3$, $\bigcap_{i=1}^{p+1} R_i = \Phi(H)$. Assume that $R = R_1 \lhd G$. Then

(b1) $\Gamma_1(H) = \{M_1, \dots, M_{p^2}, T_1, \dots, T_{p+1}\}$, where $\Gamma_1(H)$ is the set of maximal subgroups of H, $M_1, \dots, M_{p^2}$ are of maximal class, $T_1, \dots, T_{p+1}$ are regular with $|\Omega_1(T_i)| = p^p$. Exactly one of subgroups T_i, say T_1, is normal in G.

(b2) $RR_i \not\leq \Phi(G)$ so $H \not\leq \Phi(G)$.

(b3) If $H \not\leq M \in \Gamma_1$, then $e_p(M) = 1$. In particular, M is not of maximal class.

Next assume that $R = R_1$ is a unique normal subgroup of order p^p and exponent p in G. Put $N = \mathrm{N}_G(R_2)$; then $|G : N| = p$.

(b4) $R < T_1 \leq \Phi(G)$ so, if $M \in \Gamma_1$ does not contain H, then $\Omega_1(M) = R$.

(b5) $RR_2, \dots, RR_{p+1}$ are distinct G-conjugate subgroups of maximal class and order p^{p+1} with $e_p(RR_i) = 2$ for $i = 2, \dots, p+1$. Next, the above p subgroups and T_1 are the only maximal subgroups of H containing R.

(b6) $T_2, \dots, T_{p+1}$ are conjugate in G. One can choose numbering so that $\Omega_1(T_i) = R_i$ for all i.

(b7) Let $K \in \Gamma_1(H)$ be of maximal class. Assume that $K < L < G$ but $H \not\leq L \not\leq N$. Then L is of maximal class and order p^{p+2} and $e_p(L) = e_p(K) \in \{0, p\}$.

(b8) If $K \in \Gamma_1(H)$ is of maximal class and $0 < e_p(K) < p$, then K is not normal in G.

(b9) Suppose that there is $K \in \Gamma_1(H)$ with $e_p(K) = p$; then $K \lhd G$ is of maximal class. In that case, H contains exactly $p-1$ maximal subgroups L such that $e_p(L) = 0$, and all these L are G-invariant. Exactly $p^2 - p$ maximal subgroups of H of class p are not normal in G and their normalizers are all equal to N.

(b10) If $M \in \Gamma_1(H)$ is of maximal class, then $\mathrm{C}_G(M)$ is cyclic.

Proof. Since the set $\{R_i\}_1^{p+1}$ of cardinality $p+1$ is G-invariant, one may assume that $R = R_1 \lhd G$. Since $|G| > p^{p+1}$, G is not of maximal class. Let D be a G-invariant subgroup of order p^{p-1} and exponent p. If $x \in G - D$ is of order p, then $\langle x, D\rangle = R_i$ for some i (here we use Theorem 7.1(b)) so all elements of G of order p lie in the set $\bigcup_{i=1}^{p+1} R_i$; then $\Omega_1(G) = \langle R_1, \dots, R_{p+1}\rangle$. We also conclude that $D = \bigcap_{i=1}^{p+1} R_i$ so D is the unique G-invariant subgroup of order p^{p-1} and exponent p in G. If G has a subgroup of order p^{p+1} and exponent p, then that subgroup contains all R_i so coincides with $\Omega_1(G)$, and G is as stated in part (a). Next we assume that $\exp(\Omega_1(G)) > p$.

Set $N = \mathrm{N}_G(R_2)$. Then, since R_2 has at most p conjugates, we get $|G : N| \le p$ so N is normal in G. In any case, all $R_i < N$. Indeed, $R < N$ since $|RR_2| = p^{p+1}$ so R normalizes R_2. Our claim is obvious if $R_2 \lhd G$. If R_2 is not normal in G, then $R_2, \dots, R_{p+1}$ are conjugate in G, and again $R_i < N$ for $i > 1$ since $N \lhd G$. Since $R_i \lhd N$ for all $i > 1$, $R_s R_t < G$ for all s and t.

By assumption, $|\Omega_1(G)| > |\bigcup_{i=1}^{p+1} R_i| = p^{p+1}$. Then RR_i, $i > 1$, being of order p^{p+1} and exponent $> p$, is irregular hence of maximal class. One may assume that $R_3 \not\le RR_2$. Set $H = RR_2R_3$; then $|H| = p^{p+2}$, and so $H/D \cong \mathrm{E}_{p^3}$ and $\mathrm{d}(H) = 3$ since $\mathrm{d}(RR_2) = 2$ and $|H : RR_2| = p$; then H is not thin. By Theorem 13.5, $e_p(H) \equiv 1 \pmod p$ so $e_p(H) = p+1$ since $e_p(H) > 1$; then $H = \Omega_1(G)$.

Since H is not of maximal class, the set $\Gamma_1(H)$ of maximal subgroups of H is such as given in (b1) (Theorem 12.12(c)). Next we use the notation as in (b1).

Assume that $RR_i \le \Phi(G)$ $(i > 1)$. Then $|R \cap \mathrm{Z}(\Phi(G))| > p$ (indeed, every G-invariant subgroup of order p^2 of $\Phi(G)$ is contained in $\mathrm{Z}(\Phi(G))$) so $\mathrm{Z}(R) \le \mathrm{Z}(RR_i)$, a contradiction since RR_i is of maximal class. Then also $H \not\le \Phi(G)$, proving (b2).

Suppose that $(\Omega_1(G) =)H \not\le M \in \Gamma_1$. By Theorem 12.1(b), M is not absolutely regular. Since $e_p(M) < e_p(H) = p+1$, it follows that $e_p(M) \le p$ so either $|\Omega_1(M)| = p^p$ or M is of maximal class. Assume that M is of maximal class. Since $|M| > p^{p+2}$, M has no normal subgroups of order p^p and exponent p. Write $F = M \cap H (\lhd G)$; then $|F| = p^{p+1}$. Assume that $F = T_i$ for some i (see (b1)). Since T_i is not absolutely regular, $\Omega_1(T_i) \lhd G$ is of order p^p and exponent p, a contradiction. If $F = M_j$ for some $j \le p^2$ (see (b1)), then $|M : F| = p$ so $|M| = p^{p+2}$ (Theorem 9.6(c)), contrary to the hypothesis. Thus, $|\Omega_1(M)| = p^p$ and, if R is the unique normal subgroup of G of order p^p and exponent p, then $R \le \Phi(G)$.

Next we assume that $R_2 \ntrianglelefteq G$; then R is a unique normal subgroup of G of order p^p and exponent p and $R_2, \dots, R_{p+1}$ are conjugate in G; then $R \le \Phi(G)$.

Since $\mathrm{d}(RR_2) = 2$ and $\exp(RR_2) > p$, not all p conjugates of R_2 are contained in RR_2 so RR_2 is not normal in G. Then $\mathrm{N}_G(RR_2) = N = \mathrm{N}_G(R_2)$ and $RR_2, \dots, RR_{p+1}$ is a class of p conjugate subgroups of G. Thus, if $i > 1$, we have $e_p(RR_i) = 2$ for all $i > 1$, and (b5) is proved.

We have $\exp(T_i) = p^2$ for all i, by (b1). Since $H/\mathrm{K}_p(H)$ is of order p^{p+1} and exponent p (Theorem 12.12(b)), we get $|T_i/\mho_1(T_i)| = p^p$ for all i so, taking into account that T_i is regular, we get $|\Omega_1(T_i)| = |T/\mho_1(T_i)| = p^p$. Since $T_1 \lhd G$ (see (b1)), we get $\Omega_1(T_1) = R$. Since $H/R \cong \mathrm{E}_{p^2}$, R is contained in exactly $p+1$ maximal subgroups of H, namely, in $T_1, RR_2, \dots, RR_{p+1}$. If $i > 1$, then T_i is not normal in G since $R \neq \Omega_1(T_i)$, so one may assume that $\Omega_1(T_i) = R_i$ for all i since $T_2, \dots, T_{p+1}$ are conjugate in G.

Let $(\Omega_1(G) =)H \not\leq M \in \Gamma_1$ (see (b2)). Then $\Omega_1(M) = R$ so $\Omega_1(\Phi(G)) = R$. As we have noticed $H \cap M = T_1$. Thus, $T_1 \leq \Phi(G)$, and the proof of (b4) is complete.

Let $K \in \Gamma_1(H)$ be of maximal class. Assume that $K < L < G$ but $H \not\leq L$. Then $L \cap H = K \triangleleft L$ so $e_p(L) = e_p(K) \leq p$ and L is of maximal class (Theorem 13.5). Then $R \not\leq K$ since $|L| > p^{p+1}$. Since K is irregular of maximal class, we get $|L| = p^{p+2}$ (Theorem 9.6(c)). In that case, $e_p(K) \in \{0, p\}$ since L has no normal subgroups of order p^p and exponent p, and this completes the proof of (b7).

Let $K \in \Gamma_1(H)$ be of maximal class and $1 \leq e_p(K) < p$. Then K is not normal in G. This is clear if $R < K$, by (b5). If $R \not\leq K$ and $K \triangleleft G$, then all subgroups of order p^p and exponent p in K are normal in G, a contradiction since R is a unique G-invariant subgroup of order p^p and exponent p. This proves (b8).

Assume that $K \in \Gamma_1(H)$ and $e_p(K) = p$. Then $R \not\leq K$ (see (b5)) and $R_i R_j = K$ for distinct $i, j > 1$. Since the set $\{R_2, \dots, R_{p+1}\}$ is G-invariant, we have $K \triangleleft G$. If $i > 1$, then R_i is contained in exactly $p-1$ maximal subgroups of H of maximal class distinct of K and T_i. Therefore, the set $\Gamma_1(H)$ contains exactly $p(p-1)$ pairwise distinct members M of maximal class different of K and such that $e_p(M) > 0$. All remaining $p-1$ members L of class p ($L \neq K$) of the set $\Gamma_1(H)$ satisfy $e_p(L) = 0$, and all these L are G-invariant since they constitute a G-invariant set. The proof of (b9) is complete.

Let us prove (b10). Assume that $\mathrm{C}_G(M)$ has a subgroup $L \neq \mathrm{Z}(M)$ of order p. Then $M \times L = H$ (compare orders). Since G has no subgroups of order p^{p+1} and exponent p, we get $|\Omega_1(M)| = p^{p-1}$, and then $\Omega_1(H) = \Omega_1(M) \times L < H$, a contradiction. Thus, $\mathrm{C}_G(M)$ has only one subgroup of order p so it is cyclic or generalized quaternion. Assume that $\mathrm{C}_G(M)$ is generalized quaternion; then $p = 2$. Let $\mathrm{Q}_8 \cong M_1 \leq \mathrm{C}_G(M)$; then $\mathrm{c}_1(M * M_1) \in \{11, 19\}$, a contradiction since $\mathrm{c}_1(G) = 7$. Thus, $\mathrm{C}_G(M)$ is cyclic. □

Theorem 13.24. *Let G be a p-group. Then the number of irregular subgroups of maximal class in the set Γ_2 is a multiple of p.*

Proof. Let Γ_2' be the set of all irregular members of maximal class in the set Γ_2. We may assume that $\Gamma_2' \neq \varnothing$; then G is not of maximal class (indeed, the Φ-subgroup of a p-group of maximal class is absolutely regular), $\mathrm{d}(G) \leq 4$ and $|G| \geq p^{p+3}$ (Theorems 9.5 and 9.6). Since $\Phi(G) \notin \Gamma_2'$ in view of the center of each member of the set Γ_2' is of order p (see Proposition 1.13), we get $\mathrm{d}(G) > 2$. Let $\mathfrak{M}$ be the set of all (irregular) subgroups of maximal class and index p^2 in G; then $\Gamma_2' \subseteq \mathfrak{M}$. By Theorem 13.6, $|\mathfrak{M}|$ is a multiple of p^2 so the number of normal subgroups of maximal class and index p^2 in G is a multiple of p. Therefore, we may assume that G has a normal subgroup H of maximal class such that $G/H \cong C_{p^2}$.

Assume that $d(G) = 4$. Let $L \in \mathfrak{M}$. Since $|G : \Phi(L)| = |G : L||L : \Phi(L)| = p^4 = |G : \Phi(G)|$ and $\Phi(L) \le \Phi(G)$, we get $\Phi(L) = \Phi(G)$ so $L \in \Gamma'_2$. In that case, $\mathfrak{M} = \Gamma'_2$ so $H \in \Gamma'_2$, a contradiction. Thus, $d(G) = 3$. It follows that G/G' is abelian of type (p^2, p, p) and $G' = H'$.

Let $F \in \Gamma'_2$. Then $G' = F'$ (compare indices in G). Set $T/G' = \Omega_1(G/G')$; then $T/G' \cong E_{p^3}$. Since $G/F \cong E_{p^2}$, G/F contains a subgroup M/F of order p such that $M \ne T$. We have $M' = F' = G'$ since $F' \le M'$, and so M/G' is abelian of type (p^2, p). Let L be a G-invariant subgroup of index p in G'. Then $F/L < M/L$ is nonabelian of order p^3. On the other hand, M/L is minimal nonabelian since $d(M) = d(M/L) = 2$ and $(M/L)'$ is of order p (see, for example, Lemma 65.2(a)), a contradiction. Thus, H does not exist, and we conclude that p divides $|\Gamma'_2|$. □

Proposition 13.25. *Let G be a nonabelian p-group of order $> p^3$ and exponent $> p > 2$, all of whose nonnormal abelian subgroups are cyclic of the same order p^ξ. Then $|G'| = p$ and one and only of the following holds:*

(a) $\xi = 1$. *In that case, one of the following assertions is true:*

  (a1) $G = \Omega_1(G) * Z$, *where $Z = Z(G)$ is cyclic and $|\Omega_1(G)| = p^3$.*

  (a2) $G \cong M_{p^n}$.

(a) $1 < m \le \xi$, $G = \langle a, b \mid a^{p^m} = b^{p^\xi} = 1, a^b = a^{p^{m-1}}\rangle$ *is a metacyclic minimal nonabelian group.*

Blackburn has posed the following problem: (i) Classify the p-groups G having a subgroup Z of order p contained in a unique abelian subgroup of G of order p^2. We consider more general problem. (ii) Classify the p-groups G having a subgroup Z of order p contained in only one abelian subgroup of type (p, p).

Proposition 13.26. *Suppose that a p-group G has a subgroup Z of order p such that G has only one abelian subgroup of type (p, p) containing Z. Then one of the following holds:*

(a) $|G : C_G(Z)| \le p$ *and* $\Omega_1(C_G(Z)) \cong E_{p^2}$. *If $p > 2$, then $C_G(Z)$ is metacyclic.*

(b) *There is in G a normal subgroup $V \cong E_{p^2}$ such that $T = C_G(V)$ has index p in G, $G = Z \cdot T$ (a semidirect product) and $C_G(Z) = Z \times Q$, where $Q = C_T(Z)$ is either cyclic or generalized quaternion.*

(c) $p = 2$ *and G is either dihedral or semidihedral.*

Proof. If G has no normal abelian subgroups of type (p, p), it is as in (c). Now suppose that G has a normal abelian subgroup V of type (p, p).

Suppose that $Z < V$. Then $T = C_G(V)$ satisfies $\Omega_1(T) = V$ and $|G : T| \le p$. In particular, if $p > 2$, then T is metacyclic, by Theorems 10.4 and 13.7, and G is as in (a).

Next suppose that $Z \not\leq V$. Set $T = \mathrm{C}_G(V)$. Since $Z \not\leq T$, we get $|G : T| = p$ and so $G = Z \cdot T$, a semidirect product. Therefore, by the modular law, $\mathrm{C}_G(Z) = Z \times Q$, where $Q = \mathrm{C}_G(Z) \cap T = \mathrm{C}_T(Z)$. By hypothesis, Q has no abelian subgroups of type (p, p) so Q is either cyclic or generalized quaternion, and G is as in (b). □

Thus, problem (ii) is reduced to the following two problems:

(ii1) Classify the 2-groups with exactly three involutions (see §82).

(ii2) (Blackburn) Classify, for $p > 2$, the p-groups G, containing a subgroup Z of order p such that $\mathrm{C}_G(Z) = Z \times Q$, where Q is cyclic (see [Bla13]).

Proposition 13.27. *Let G be a group of exponent $p^e > p$. If $\mathrm{H}_p(G) \leq H < G$ and H is either absolutely regular or of maximal class, then G is of maximal class.*

Proof. Since each regular subgroup of exponent p^e coincides with its H_p-subgroup, G is irregular. Let H be absolutely regular. Then each subgroup B of G of order $p|H|$, containing H, is generated by elements of order p; then G is of maximal class (Exercise 10(b)).

Now suppose that H is of maximal class but not absolutely regular. Since $\exp(H) = \exp(G) > p$, we get $|H| \geq p^{p+1}$ so H is irregular. Assume that $|H| > p^{p+1}$. Then H_1, the fundamental subgroup of H, is characteristic in H so normal in G. Let $H_1 < F < G$, where $|F : H_1| = p$ and $F \neq H$. Then $H_1 = \mathrm{H}_p(F)$ so F is of maximal class, by the previous paragraph. Thus, all subgroups of G of order $p|H_1|$ that contain H_1, are of maximal class. Then G is also of maximal class (Exercise 10(a)). Now let $|H| = p^{p+1}$. Assume that G is not of maximal class. In view of Theorem 13.18(b), one may assume that $|G : H| = p$. By Theorem 12.12(c), the set Γ_1 contains exactly $p + 1$ regular members $T_1, \dots, T_{p+1}$. Since $\Omega_1(T_i) = \langle T_i - H\rangle = T_i$, we get $\exp(T_i) = p$ for all i. By Theorem 12.12(c), $G = \bigcup_{i=1}^{p+1} T_i$ so $\exp(G) = p$, a contradiction. □

Theorem 13.28. *Let G be a group of order p^m and exponent $> p$ such that $H = \Omega_2^*(G) < G$ is of maximal class. Then G is also of maximal class and $|G : H| = p$.*

Proof. We have $\exp(H) > p$. By Proposition 13.19(a), if $n > 2$, then $\Omega_n^*(G)$ is not of maximal class. If $p = 2$, then $\mathrm{c}_2(G) = \mathrm{c}_2(H)$ is odd; then G is of maximal class so semidihedral and $H \in \Gamma_1$ is generalized quaternion (Theorem 1.17(b)). Assume that $p > 2$.

(a) Let H be regular. Then, by Theorem 9.5, $|H| \leq p^p$, H is absolutely regular and $|H : \Omega_1(H)| = p$. Set $|\Omega_1(H)| = p^k$, $k \leq p - 1$. Then $\mathrm{c}_2(G) = \mathrm{c}_2(H) = p^{k-1}$ so, by Theorems 13.2(b), 7.1 and 7.2, either G is absolutely regular or $k = p - 1$ and G is irregular of maximal class.

(a1) Let G be regular; then $\mathrm{c}_2(G) = p^{k-1}$, $k \leq p - 1$ and $|\Omega_1(G)| = p^k$ so G is absolutely regular and $H = \Omega_2(G)$. It follows that $G/\Omega_1(G)$ is cyclic. The argument in the proof of Lemma 9.4(b) shows that H is not of maximal class, a contradiction. Thus, G is irregular.

(a2) Let G be irregular of maximal class. By Theorem 9.6(c), $H \in \Gamma_1$ so $|G| = p|H| \le p^{p+1}$ and hence $|G| = p^{p+1}$. In that case, H is a unique maximal subgroup of G of exponent p^2.

Next we assume that H is irregular (of maximal class).

(b) Assume that $|H| > p^{p+1}$. Then $c_2(G) = c_2(H) \equiv p^{p-2} \pmod{p^{p-1}}$ (Lemma 12.3(c)) so G is also of maximal class (Theorem 13.2(b)). Then $H \in \Gamma_1$.

Next we assume that $|H| = p^{p+1}$.

(c) Let G be of maximal class. Then $|G : H| = p$ so $|G| = p^{p+2}$. Let G_1 be the fundamental subgroup of G; then $H \neq G_1$ (Theorem 9.5), $\exp(G_1) = p^2$ so $\Omega_2^*(G_1) = G_1 \not\le H$, a contradiction.

(d) Thus, G is not of maximal class. Since $c_2(H) = c_2(G) \equiv 0 \pmod{p^{p-1}}$ (Theorem 13.2(b)), it follows that $\Omega_1(H)$ is of order p^p and exponent p, all other maximal subgroups of H are absolutely regular. We want to obtain a contradiction. Let $H < M \le G$, where $|M : H| = p$. Since $\Omega_1(H) \lhd M$, we conclude that M is not of maximal class. To obtain a contradiction, we may assume that $M = G$; then $H \in \Gamma_1$ so $|G| = p^{p+2}$. In that case, $G/K_p(G)$ is of order p^{p+1} and exponent p, $d(G) = 3$, $\Gamma_1 = \{M_1 = H, M_2, \dots, M_{p^2}, T_1, \dots, T_{p+1}\}$, where $M_1, \dots, M_{p^2}$ are of maximal class, $T_1, \dots, T_{p+1}$ are regular with $d(T_i) > 2$ for all i (Theorem 12.12(c)). Since $\exp(G) = p^2$, it follows from $\Omega_2^*(T_i) \le H$ that $\exp(T_i) = p$ for all i so $H \cap T_i = \Omega_1(H)$ for all i. Thus, $\Omega_1(H)$ is contained in $p+2$ pairwise distinct members $H, T_1, \dots, T_{p+1}$ of the set Γ_1, a contradiction, since $G/\Omega_1(H) \cong E_{p^2}$ has exactly $p+1$ maximal subgroups. □

Proposition 13.29. *If G is a p-group of exponent $> p$ such that $H = \Omega_2^*(G)$ is a proper absolutely regular subgroup of G, then G is either absolutely regular or irregular of maximal class. Let G be not absolutely regular. If $p = 2$, then $G \cong D_8$. Now let $p > 2$. If $H \le G_1$, then all elements of the set $G - G_1$ have the same order p; if $H \not\le G_1$, then* $\exp(G_1) = p$ *and* $|G| = p^{p+1}$.

Proof. We have $\exp(H) = p^2$ (Theorem 7.2) and $c_2(G) = c_2(H)$. Let $S = \Omega_1(H)$ and $|S| = p^k$; then $k \le p-1$. In that case, setting $|H| = p^h$, we get $c_2(H) = \frac{|H - \Omega_1(H)|}{\varphi(p^2)} = \frac{p^h - p^k}{p(p-1)} = p^{k-1}(1 + p + \dots + p^{h-k-1}) \not\equiv 0 \pmod{p^{p-1}}$ so G is either absolutely regular or irregular of maximal class (Theorem 13.2(b)).

Suppose that G is irregular of maximal class. If $H \le G_1$, then all elements in $G - G_1$ are of order p (Theorem 13.19(b)). In particular, if $p = 2$, then $G \cong D_8$. Now suppose that $p > 2$. If $H \not\le G_1$, then $H \in \Gamma_1, \exp(G_1) = p$ so $|G| = p^{p+1}$. □

It is interesting to classify the p-groups G satisfying $|\Omega_2^*(G)| = p^{p+1}$.

Theorem 13.30. *Let G be an irregular p-group of order $> p^{p+1}$. If $K = \Omega_1(G) < G$ is of maximal class, then one of the following holds:*

(a) If K is irregular, then G is of maximal class and $|G : K| = p$.

(b) *If K is regular, then $p > 2$, K is of order p^p and all maximal subgroups of G not containing K, are absolutely regular.*

Proof. Assume that K is irregular. We have $e_p(G) = e_p(K) > 1$ and $e_p(K) \not\equiv 1 \pmod p$, so G is of maximal class (Theorem 13.5). In that case, $|G : K| = p$ (Theorem 9.6(c)).

Now let K be regular. Then $\exp(K) = p$ so $p > 2$ since K is nonabelian. Since G is irregular, we get $|K| \geq p^{p-1}$. If $|K| = p^{p-1}$, then G is of maximal class (Theorem 12.1(a)). In that case, $K \leq \Phi(G)$ so $|Z(K)| > p$, a contradiction. Since the order of regular p-group of maximal class is at most p^p, it remains to consider the case $|K| = p^p$. If G is of maximal class, then $|G| = p^{p+1}$ (Theorem 9.6(e)). Next assume that G is not of maximal class. By Lemma 1.4, K has a G-invariant $R \cong E_{p^2}$. Setting $C_G(R) = M$, we get $K \not\leq M$ so $|G : M| = p$. Then $\Omega_1(M) = K \cap M$ is of order p^{p-1} so M is absolutely regular since it is not of maximal class in view of $R \leq Z(M)$. Now let $F \in \Gamma_1$ and suppose that F is of maximal class. Since $M \in \Gamma_1$ is absolutely regular, then G is of maximal class (Theorem 12.13), a contradiction. Taking, from the start, $F \not\geq K$, we see that F is absolutely regular. □

Theorem 13.31. *If an irregular p-group $G = \Omega_1(G)$ possesses only one nonabelian subgroup, say R, of order p^p and exponent p, then $G \cong \Sigma_{p^2}$.*

§14

Thompson's critical subgroup

The following theorem is due essentially to Thompson (originally, it was proven for p-groups) but we formulate it in a more general form. The theorem shows that every solvable group has a subgroup enjoying a number of remarkable properties which allows us to control the structure of the whole group.

Theorem 14.1 (compare with [FT, Lemma 8.2]). *A solvable group G has a characteristic subgroup B such that:*

(a) *B is nilpotent of class ≤ 2.*

(b) *$Z(B)$ is a maximal characteristic abelian subgroup of G.*

(c) *$B/Z(B)$ is generated by all minimal normal subgroups of $G/Z(B)$ contained in $C_G(Z(B))/Z(B)$.*

(d) *$C_G(B) = Z(B)$.*

(e) *Let $\sigma \in \mathrm{Aut}(G)$ and $(o(\sigma), |B|) = 1$. If $\sigma(x) = x$ for all $x \in B$, then $\sigma = \mathrm{id}_G$.*

Proof. Assume that G is nonabelian. Let A be a maximal characteristic abelian subgroup of G. Since the socle of G is a nonidentity characteristic abelian subgroup of G, $A > \{1\}$. Then $C_G(A)$ is characteristic in G. For $C_G(A) = A$, we set then $B = A$, and this B satisfies conditions (a)–(d).

Now suppose that $C_G(A) > A$. Let $\mathcal{M}$ be the set of all minimal normal subgroups of G/A contained in $C_G(A)/A$. Write $B/A = \langle R/A \mid R/A \in \mathcal{M}\rangle$. We claim that B is the desired subgroup. Clearly, $B(> A)$ is characteristic in G since A and $C_G(A)$ are characteristic. Since $Z(B) \geq A$ and $Z(B)$ is a characteristic abelian subgroup of B and so of G then, by the maximal choice of A, we get $Z(B) = A$, proving (b). Since B/A is abelian, B is nilpotent of class 2 in view of $B > A$. This proves (a).

Next, $C_G(B) \cap B = Z(B)(= A)$. Assume that $C_G(B) > A$; then $C_G(B) \not\leq B$. Let L/A be a minimal normal subgroup of G/A contained in $C_G(B)/A$; then $L \not\leq B$. Since $L \leq C_G(B) \leq C_G(A)$, we conclude that $L \leq B$, contrary to the choice of L. This completes the proof of assertions (a)–(d).

It remains to prove (e). Now we do not assume that $B > A$. Let $\sigma \in \mathrm{Aut}(G)$ with $(o(\sigma), |B|) = 1$ and such that $\sigma_B = \mathrm{id}_B$. Let $W = \langle\sigma\rangle \cdot G$ be the natural semidirect product of G and $\langle\sigma\rangle$; W is solvable since G and $\langle\sigma\rangle$ are solvable. Clearly, $B \triangleleft W$. We have $C_W(B) = A \times \langle\sigma\rangle$ and this subgroup is normal in W. It follows from $(|A|, o(\sigma)) = 1$ that $\langle\sigma\rangle$ is characteristic in $\mathrm{C}_W(B)$ so $\langle\sigma\rangle \lhd W$. We conclude that $W = G \times \langle\sigma\rangle$ so $\sigma = \mathrm{id}_G$. □

Definition 1. Let G be a solvable group. The subgroup B of Theorem 14.1 is called a *Thompson critical subgroup* of G.

Exercise 1 (P. Hall, 1926, unpublished dissertation). Suppose that all nonidentity characteristic abelian subgroups of a nonabelian p-group G have order p. Then G is extraspecial.

Solution. $\mathrm{Z}(\Phi(G))$ is of order p so $\Phi(G)$ is cyclic, and we get $|\Phi(G)| = p$. Since $|\mathrm{Z}(G)| = p$, we get $G' = \Phi(G) = \mathrm{Z}(G)$ so G is extraspecial.

Exercise 2. Classify the solvable groups all of whose nonidentity characteristic abelian subgroups have prime orders.

Recall that a p-group G is said to be *special* if it is nonabelian and $\mathrm{Z}(G) = G' = \Phi(G)$; then, by Exercise 1.19, also $\exp(G') = p$.

Exercise 3. Suppose that all nonidentity characteristic abelian subgroups of a nonabelian p-group G have the same order. Prove that then Thompson critical subgroups of G are either elementary abelian or special.

Solution. Let A be a nonidentity characteristic abelian subgroup of G. By hypothesis, A is elementary abelian and maximal abelian characteristic subgroup of G so $A = \mathrm{Z}(G)$. Let $A < T \le G$, where T is a Thompson critical subgroup of G. We claim that T is special. Indeed, $\mathrm{Z}(T) = A$. Since $T' \le \mathrm{Z}(T) = A$, we get $T' = A$. Since T/A is elementary abelian, $\Phi(T) \le A$, and so $\Phi(T) = A$.

In general, a nonabelian Thompson critical subgroup of a p-group G is not necessarily special (example: G is a minimal nonabelian group of order p^5 and exponent p^2). If T is a nonabelian Thompson critical subgroup of a p-group G, then $\exp(T') = \exp(T/\mathrm{Z}(T)) = p$ (use Exercise 1.19).

Theorem 14.2 (Thompson). *Suppose that a group A acts on a nonabelian p-group P in such a way that $[P, A] = P$ and A centralizes all abelian characteristic subgroups of P. Then* $\mathrm{cl}(P) = 2$. *If, in addition, $p \nmid |A|$, then P is special.*

Proof. Let L be an abelian characteristic subgroup of P. Then $[L, A] = \{1\}$ so $[L, A, P] = \{1\}$ and $[P, L, A] \le [L, A] = \{1\}$. By the Three Subgroups Lemma, $[A, P, L] = \{1\}$. Since, by hypothesis, $[A, P, L] = [P, L]$ we obtain $L \le \mathrm{Z}(P)$. Thus,

(i) All characteristic abelian subgroups of P are contained in $\mathrm{Z}(P)$.

Assume that $\mathrm{cl}(P) = c > 2$. Set $T = \mathrm{K}_{c-1}(P)$; then $T \not\le \mathrm{Z}(P)$. It follows from $2(c-1) = c - 1 + (c-1) \ge c - 1 + 2 = c + 1$ that

$$[T, T] = [\mathrm{K}_{c-1}(P), \mathrm{K}_{c-1}(P)] \le \mathrm{K}_{2c-2}(P) \le \mathrm{K}_{c+1}(P) = \{1\},$$

and so T is a characteristic abelian subgroup of P. Therefore, by (i), $T \leq Z(P)$, contrary to what has just been said. Since P is nonabelian, we have

(ii) $\mathrm{cl}(P) = 2$ so $P' \leq Z(P)$.

Next, assuming, in addition, that $p \nmid |A|$, we prove that P is special.

Set $\bar{P} = P/P'$. By hypothesis, $[\bar{P}, A] = \bar{P}$ so that A acts on $\bar{P}$ without fixed points, by Fitting's lemma (see Corollary 6.5), so, by (i), $Z(P) \leq P'$. Since the reverse inclusion holds, by (ii), we get

(iii) $P' = Z(P)$.

It follows from (ii) and Exercise 1.19 that $\exp(P') \leq \exp(P/P')$. Assume that $\exp(P/P') = p^m > p$. Then $2m - 2 = m + (m-2) \geq m$. Therefore, for $x, y \in P$, we have (see Exercise 1.18)

$$[x^{p^{m-1}}, y^{p^{m-1}}] = [x, y]^{p^{2m-2}} = [x, y^{p^{2m-2}}] = 1,$$

since $y^{p^{2m-2}} \in \langle y^{p^m} \rangle \leq P' = Z(P)$. It follows that $\mho_{m-1}(P)$ is an abelian characteristic subgroup of P. By (i), $\mho_{m-1}(P) \leq Z(P) = P'$ so $\exp(P/P') \leq p^{m-1}$, contrary to the assumption. Thus, $\exp(P/P') = p$ so $Z(P) = P' = \Phi(P)$, and P is special. □

Theorem 14.3 (Hall–Higman [HH]). *Suppose that a π'-group Q acts on a π-group G. Let R be a subgroup of Q acting on G nontrivially. Next, suppose that R acts trivially on all proper Q-invariant subgroups of G. Then G is either an elementary abelian or a special p-group. Next, R acts on $\Phi(G)$ trivially, and Q acts on $G/\Phi(G)$ irreducibly.*

Proof. Let $W = Q \cdot G$ be the natural semidirect product of Q and G; then W is π-separable. By [BZ, Appendix C, Lemma 1(b)], W has a $\pi' \cup \{p\}$-Hall subgroup H_p for each $p \in \pi$ such that $Q < H_p$. The intersection $G \cap H_p$ is a Q-invariant Sylow p-subgroup of G. One may assume that R does not centralizes $G \cap H_p$ for some p; then G is a p-group, by hypothesis. In that case, by Theorem 1.15, R acts on $G/\Phi(G)$ nontrivially, i.e., $[R, G] \not\leq \Phi(G)$. Hence Q acts on $G/\Phi(G)$ irreducibly (otherwise, if $G/\Phi(G) = (U/\Phi(G)) \times (V/\Phi(G))$, where both factors are nontrivial and R-invariant, by Maschke's theorem, then R centralizes $UV = G$). It follows that $[Q, G] = G$. By hypothesis, R acts on $\Phi(G)$ trivially. If G is abelian, it is of exponent p, by Fitting's lemma (see Corollary 6.5). If G is nonabelian, the result follows from Theorem 14.2 applied to the pair $\{R, G\}$: indeed, the p'-group R centralizes all characteristic abelian subgroups of the p-group G. □

Corollary 14.4. *For each odd prime p, a p-group $G > \{1\}$ has a characteristic subgroup D of class at most 2 and exponent p such that every nontrivial p'-automorphism of G induces a nontrivial automorphism of D.*

Proof. Let $\alpha \in \mathrm{Aut}(G)$, $o(\alpha) = q$, where $q \neq p$ is a prime, let T be a Thompson critical subgroup of G and $D = \Omega_1(T)$. Then D is characteristic in G since T is, and $\exp(D) = p$ by Theorem 14.1 and Exercise 1.19. Let $\beta = \alpha_T$ be the restriction of α to T. By Theorem 14.1(e), $o(\beta) = q$. Assume that β centralizes D. Then $H = \langle \beta, T \rangle$ has no minimal nonnilpotent subgroups, by Lemma 10.8, so β centralizes T; then H is nilpotent, which is not the case. Thus, α does not centralize D. □

Corollary 14.5. *A nonidentity 2-group G has a characteristic subgroup D of class at most 2 and exponent 2 or 4 such that every nontrivial p'-automorphism α of G induces a nontrivial automorphism of D.*

§15

Generators of p-groups

In this section we estimate the number of generators of p-groups in terms of ranks of some their normal abelian subgroups.

Exercise. Let $p^n > 2$ and let G be a p-group such that $\Omega_n(G) \leq Z(G)$. Then $\Omega_n(G/\Omega_1(G)) \leq Z(G/\Omega_1(G))$,

Solution. Let $A/\Omega_1(G)$ be a maximal normal abelian subgroup of $G/\Omega_1(G)$ of exponent $\leq p^n$; then $\mathrm{cl}(A) \leq 2$ and $\exp(A') \leq p$. For $a \in A$ and $g \in G$ we write $a^g = ab$ for some $b \in A$. We compute

$$a^{p^n} = (a^{p^n})^g = (a^g)^{p^n} = (ab)^{p^n} = a^{p^n} b^{p^n} [b,a]^{\binom{p^n}{2}} = a^{p^n} b^{p^n},$$

since p divides $\binom{p^n}{2}$ in view of $p^n > 2$, and so $b^{p^n} = 1$, and we conclude that $b \in \Omega_n(G) \leq Z(G)$. We have $\exp(A) \leq \exp(\Omega_1(G)) \exp(A/\Omega_1(G)) \leq p^{n+1}$ so $a^p \in \Omega_n(G) \leq Z(G)$, and we get $a^p = (a^p)^g = (a^g)^p = (ab)^p = a^p b^p$. It follows that $b^p = 1$ hence $b \in \Omega_1(G)$. Thus, $A/\Omega_1(G) \leq Z(G/\Omega_1(G))$.

Theorem 15.1 (Thompson). *Let G be a p-group, $p > 2$, such that $\Omega_1(G) \leq Z(G)$. Then $d(G) \leq d(Z(G))$. In particular, $d(U) \leq d(Z(U)) \leq d(G)$ for $U < G$.*

Proof. (Blackburn) We use induction on $|G|$. Set $Z = \Omega_1(G)$ and $Z \cong E_{p^n}$. First we prove that $d(G) \leq n$. Let A/Z be a maximal normal abelian subgroup of G/Z of exponent p. Then $\mathrm{cl}(A) \leq 2$ and so, since $A' \leq Z \leq Z(A)$, we get $\exp(A') \leq \exp(A/Z(A)) \leq \exp(A/Z) = p$. By Exercise 1, $A/Z \leq Z(G/Z)$. By Corollary 10.2, $\Omega_1(G/Z) = A/Z$. Since A is regular (Theorem 7.1), it follows by Theorem 7.2(d) that $|A/Z| \leq |A/\mho_1(A)| = |\Omega_1(A)| = |Z| = p^n$, and so $d(A/Z) \leq n$. By induction, $d(G/Z) \leq d(Z(G/Z)) = d(A/Z) \leq n$. If $Z \leq \Phi(G)$, we have $d(G) = d(G/Z)$, and in that case all is done for $U = G$. Let $Z \not\leq \Phi(G)$. In that case, there exists a maximal subgroup M in G such that $Z \not\leq M$. Then $G = Z_1 \times M$ for some subgroup Z_1 of order p in Z, and $|\Omega_1(M)| = p^{n-1}$, $\Omega_1(M) = \Omega_1(Z(M))$. By induction, $d(G) = 1 + d(M) \leq 1 + d(Z(M)) \leq 1 + (n-1) = n$. The last assertion now follows. □

Definition. A maximal rank of abelian normal subgroups of a p-group G is said to be the *normal rank* of G and denoted by $r_n(G)$.

Theorem 15.2 (Thompson). *Suppose that G is a p-group, $p > 2$. Then each subgroup of G can be generated by $\frac{1}{2}k(k+1)$ elements, where $k = \mathrm{r}_n(G)$.*

Proof. Let B be a normal elementary abelian subgroup of G of rank k. Set $C = C_G(B)$. Then $\Omega_1(C) = B$, by Theorem 10.1. By Theorem 15.1, $\mathrm{d}(V) \le k$ for all $V \le C$. Let $U \le G$. Then $UC/C \le G/C$ so $U/(U \cap C) \cong UC/C$ is isomorphic to a subgroup of $\mathrm{Aut}(B)$. Since $B \cong \mathrm{E}_{p^k}$, we get $\mathrm{Aut}(B) \cong \mathrm{GL}(k,p)$ so $|U/(U\cap C)| \le |\mathrm{GL}(k,p)|_p = p^{\frac{1}{2}k(k-1)}$. By Theorem 15.1, applied to the pair $U \cap C \le C$, we get $\mathrm{d}(U \cap C) \le k$. Therefore,

$$\mathrm{d}(U) \le \mathrm{d}(U/(U\cap C)) + \mathrm{d}(U \cap C) \le \frac{1}{2}k(k-1) + k = \frac{1}{2}k(k+1). \qquad \square$$

Theorem 15.3 ([Man1]). *If G is a 2-group such that $\Omega_2(G) \le \mathrm{Z}(G)$, then $\mathrm{d}(G) \le \mathrm{d}(\mathrm{Z}(G))$. Next, every subgroup U of G can be generated by $\mathrm{d}(\mathrm{Z}(G))$ elements.*

Proof. Let $P = \Omega_1(G) = \Omega_1(\mathrm{Z}(G))$. Then $|P| = 2^n$, where $n = \mathrm{d}(\mathrm{Z}(G))$, and so $|\Omega_2(G)| \le 2^{2n}$. Let A/P be maximal among normal abelian subgroups of exponent at most 4 in G/P; then $\exp(A) \le 8$ and $A' \le P \le \mathrm{Z}(A)$ so $\mathrm{cl}(A) \le 2$, and we get $\mho_1(A) \le \mathrm{Z}(A)$ and $\exp(A') \le \exp(P) \le 2$. By Exercise 1, $A/P \le \mathrm{Z}(G/P)$. Then by Corollary 10.2, $\Omega_2(G/P) \le A/P \le \mathrm{Z}(G/P)$. By induction, $\mathrm{d}(G/P) \le \mathrm{d}(\mathrm{Z}(G/P))$.

Let $B = \Omega_2(G)$; then $B/P = \Omega_1(\mathrm{Z}(G/P))$. If A/P is as above, then $A/P \le \mathrm{Z}(G/P)$ so $\mathrm{d}(A/P) = \mathrm{d}(B/P)$. Then, by induction, $\mathrm{d}(G/P) \le \mathrm{d}(\mathrm{Z}(G/P)) = \mathrm{d}(B/P) \le n$. If $P \le \Phi(G)$, then $\mathrm{d}(G) = \mathrm{d}(G/P)$, and we are done. Let $P \not\le \Phi(G)$ and let $M \in \Gamma_1$ be such that $P \not\le M$; then $G = M \times P_1$ for some $P_1 < P$ of order 2. It follows that $\mathrm{Z}(G) = \mathrm{Z}(M) \times P_1$. Then $\Omega_2(M) = M \cap \Omega_2(G) \le \mathrm{Z}(M)$, $\mathrm{d}(\mathrm{Z}(M)) = n-1$ so we get, by induction in M, $\mathrm{d}(G) \le 1 + \mathrm{d}(M) \le 1 + \mathrm{d}(\mathrm{Z}(M)) = 1 + (n-1) = n$. The last assertion now follows. $\square$

Let A be an abelian 2-group of exponent ≤ 4 with $\mathrm{d}(A) = k$. Then $|A| \le 2^{2k}$ and $|\Phi(A)| \le 2^k$. By Theorem 1.16, $|\mathrm{Aut}(A)|_2 \le 2^{\binom{k}{2}}|\Phi(A)|^k = 2^{k^2+\binom{k}{2}}$.

Theorem 15.4 ([Man1]). *Let G be a 2-group, $k = \mathrm{r}_n(G)$. Then each subgroup of G can be generated by $k^2 + \frac{1}{2}k(k+1) = \frac{1}{2}k(3k+1)$ elements.*

Proof. Let A be maximal among normal abelian subgroups of G of exponent at most 4 and $\mathrm{d}(A) = k$ so $|A| \le 2^{2k}$. Suppose that $C = C_G(A)$; then $C \trianglelefteq G$ and $\Omega_2(C) = A \le \mathrm{Z}(C)$ (Theorem 10.1). Let $H \le G$. Then $H/(H \cap C)$ is isomorphic to a subgroup of $\mathrm{Aut}(A)$, so by the remark, preceding the theorem, $|H/(H\cap C)| \le 2^{k^2+\frac{1}{2}k(k-1)}$. Let $B = H \cap C$; then $\Omega_2(B) = B \cap A \le \mathrm{Z}(B)$, so by Theorem 15.3, $\mathrm{d}(B) \le \mathrm{d}(\mathrm{Z}(B)) \le \mathrm{d}(\mathrm{Z}(A)) = k$. Thus, $\mathrm{d}(H) \le \mathrm{d}(H/B) + \mathrm{d}(B) \le k^2 + \frac{1}{2}k(k-1) + k = k^2 + \frac{1}{2}k(k+1)$. $\square$

Remark (Mann). Let G be a p-group, p odd. We claim that if $|\Omega_1(G)| = p^n$, then $|G : \mho_1(G)| \le p^{n^2}$. Let $C = C_G(\Omega_1(G))$. Then G/C is an automorphism group of $\Omega_1(G)$, and therefore its order is at most $p^{n(n-1)/2}$. We have $\Omega_1(C) \leq Z(C)$ so $d(C) \leq d(\Omega_1(Z(C)))(= d(Z(C)))$ (Theorem 15.1). Here $d(Z(C)) \leq n$ so $d(C) \leq n$. Write $H = C/\mho_1(C)$. Let A be a maximal normal abelian subgroup of H. Then $A \cong E_{p^k}$, $k \leq n$ (Theorem 15.1 applied to A), and $H/A = H/C_H(A)$ is a group of automorphisms of A, so again we have $|H/A| \leq p^{n(n-1)/2}$. Combining the obtained three inequalities, we get $|G/\mho_1(C)| \leq p^{n^2}$. Since $\mho_1(C) \leq \mho_1(G)$, we are done.

§16

Classification of finite p-groups all of whose noncyclic subgroups are normal

If in a p-group all cyclic subgroups are normal, then all subgroups are normal and we call such a group Dedekindian. A Dedekindian p-group G is either abelian or $G = Q \times A$, where $Q \cong Q_8$ is the quaternion group and A is elementary abelian (see §1). Here we study non-Dedekindian p-groups all of whose noncyclic subgroups are normal. Such groups have been considered by D. S. Passman [Pas], but he omitted 2-groups of order $\leq 2^7$. In fact here lies the main difficulty. Here we refine, improve and extend the arguments of Passman so that all p-groups will be included. It turns out that we get in addition five exceptional 2-groups: one group of order 2^6, three groups of order 2^5 and one group of order 2^4. First we prove the following auxiliary result. This solves Problem 733.

Lemma 16.1 ([Pas]). *Let G be a p-group all of whose noncyclic subgroups are normal. Let H be any nonnormal subgroup in G. Then H is a maximal cyclic subgroup, $|G : \mathrm{N}_G(H)| = p$ and $\mathrm{N}_G(H)/H$ is either cyclic or $p = 2$ and $\mathrm{N}_G(H)/H \cong Q_8$ is quaternion.*

Proof. By assumption, H is cyclic. Assume that H is not maximal cyclic and let $K > H$ be a maximal cyclic subgroup containing H. Let $L > K$ be a subgroup of G such that $|L : K| = p$. Since L is noncyclic, we have $\mathrm{d}(L) = 2$ and $L \trianglelefteq G$. Thus $|L : \Phi(L)| = p^2$ and $\Phi(L) = \Phi(K)$. But $\Phi(L) \triangleleft G$ and $H \leq \Phi(L)$ and so $H \triangleleft G$, a contradiction. Thus, H is a maximal cyclic subgroup in G.

Set $N = \mathrm{N}_G(H)$ so that $\{1\} \neq N/H$ is Dedekindian and so N/H is Dedekindian. If N/H has two distinct subgroups N_1/H and N_2/H of order p, then $H = N_1 \cap N_2 \triangleleft G$, a contradiction. Thus N/H has only one subgroup of order p and so N/H is either cyclic or $p = 2$ and $N/H \cong \mathrm{Q}_8$.

Set $M/H = \Omega_1(N/H)$ so that $|M : H| = p$ and $M \trianglelefteq G$. All G-conjugates of H are contained in M. Since H is not normal in G, there are exactly p-conjugates with H since $\mathrm{d}(M) = 2$ so $|G : N| = p$. □

Theorem 16.2 (Janko–Bozikov). *Let G be a non-Dedekindian p-group all of whose noncyclic subgroups are normal. Then G is one of the following groups:*

(i) *G is metacyclic minimal nonabelian and G is not isomorphic to Q_8.*

(ii) $G = G_0 * Z$, *the central product of a nonabelian group* G_0 *of order* p^3 *with a cyclic group* Z, *where* $G_0 \cap Z = Z(G_0)$ *and if* $p = 2$, *then* $|Z| > 2$.

(iii) $p = 2$ *and* $G = Q \times Z$ *where* $Q \cong Q_8$ *and* Z *is cyclic of order* > 2.

(iv) G *is a group of order* 3^4 *and maximal class with* $\Omega_1(G) = G' \cong E_9$.

(v) $G = \langle a, b \mid a^8 = b^8 = 1,\ a^b = a^{-1},\ a^4 = b^4 \rangle$, *where* $|G| = 2^5$, $G' \cong C_4$, $Z(G) \cong C_4$, $G' \cap Z(G) \cong C_2$ *and* $\Omega_2(G)$ *is abelian of type* $(4, 2)$.

(vi) $G \cong Q_{16}$, *the generalized quaternion group of order* 2^4.

(vii) $G = D_8 * Q_8$, *an extraspecial* 2*-group of order* 2^5.

(viii) $G = \langle a, b, c \mid a^4 = b^4 = [a, b] = 1,\ c^2 = a^2,\ a^c = ab^2,\ b^c = ba^2 \rangle$, *where* G *is the minimal non-metacyclic group of order* 2^5, G *is a special* 2*-group with* $G' = \Omega_1(G) \cong E_4$.

(ix) $G = \langle a, b, c, d \rangle$ *where*

$$a^4 = b^4 = [a, b] = 1, \quad c^2 = a^2 b^2, \quad a^c = a^{-1}, \quad b^c = a^2 b^{-1}, \quad d^2 = a^2,$$
$$a^d = a^{-1} b^2, \quad b^d = b^{-1}, \quad [c, d] = 1.$$

Here G *is a special group of order* 2^6 *with* $G' = \Omega_1(G) \cong E_4$ *in which every maximal subgroup is isomorphic to the minimal non-metacyclic group of order* 2^5 *(from (viii)).*

Conversely, all the above groups satisfy the assumptions of the theorem.

Proof. Let G be a non-Dedekindian p-group all of whose noncyclic subgroups are normal. In particular, G is nonabelian. If G has no normal elementary abelian subgroups of order p^2, then G is of maximal class. It follows that $G \cong Q_{16}$ is as in (vi).

In what follows we assume that G has a normal abelian subgroup W of type (p, p). Since each subgroup of G/W is normal, G/W is Dedekindian and so G/W.

Suppose that $p = 2$ and there is a normal four-subgroup W such that G/W is nonabelian Dedekindian. In that case, $\bar{G} = G/W = \bar{Q} \times \bar{A}$, where $\bar{Q}$ is quaternion and $\bar{A}$ is elementary abelian. By Lemma 16.1, $d(G) \le 4$ and so $|\bar{A}| \le 4$ which implies $2^5 \le |G| \le 2^7$.

First we consider the case $|G| = 2^5$ so that $G = Q$ and $G/W \cong Q_8$. Set $S/W = \Phi(G/W)$ so that S is abelian since $|G : C_G(W)| \le 2$. Suppose that S is elementary abelian. If $s \in S$ is such that $s \notin Z(G)$, then $C_G(s)/\langle s \rangle$ contains the four-subgroup $S/\langle s \rangle$, contrary to Lemma 16.1. Hence $S \le Z(G)$. Since G is not Dedekindian, there is $g \in G - S$ such that $g^2 \in S - W$ and $\langle g \rangle \ntrianglelefteq G$. But then $N_G(\langle g \rangle) = \langle g \rangle \times W$ and $N_G(\langle g \rangle)/\langle g \rangle \cong E_4$, contrary to Lemma 16.1. We have proved that S is abelian of type $(4, 2)$. If $g \in G - S$, then $g^2 \in S - W$ and so $o(g) = 8$. We have proved that $\Omega_2(G) = S$ is of order 8. By Lemma 42.1,

$$G = \langle a, b \mid a^8 = b^8 = 1,\ a^4 = b^4,\ a^b = a^{-1} \rangle$$

since $M_{2^{n+1}}$ $(n > 2)$ does not possess a factor-group isomorphic to Q_8. Thus, G is as in (v).

Now we consider the case $|G| > 2^5$. We have $G = QA$, $Q \cap A = W$, Q and $A \lhd G$, $Q/W \cong Q_8$, and A/W is elementary abelian of order 2 or 4. In particular, $\exp(G) \leq 8$. Suppose that u is an involution in G with $u \notin Z(G)$. By Lemma 16.1, $|G : C_G(u)| = 2$ and $C_G(u)/\langle u\rangle$ is either cyclic of order ≤ 8 or $C_G(u)/\langle u\rangle \cong Q_8$. In any case, $|G| \leq 2^5$ which is a contradiction. We have shown that $\Omega_1(G) \leq Z(G)$ and so $W \leq Z(G)$. Assume that $E = \Omega_1(G)$ is of order ≥ 8. Since G is not Dedekindian, G possesses a cyclic subgroup Z of composite order which is not normal in G. Set $\langle z\rangle = Z \cap E$ so that z is a central involution. Let $e, f \in E - \langle z\rangle$ such that $\langle e, f, z\rangle \cong E_8$. In this case $S_1 = \langle Z, e\rangle$ and $S_2 = \langle Z, f\rangle$ are normal in G. But $S_1 \cap S_2 = Z$ and so $Z \lhd G$, a contradiction. It follows that $\Omega_1(G) = W \leq Z(G)$. Set $T/W = \Phi(Q/W)$ so that T is abelian of type $(4, 2)$. For each $x \in Q - T$, $x^2 \in T - W$ and so $o(x) = 8$ and $\Omega_2(Q) = T$ is of order 8. By Lemma 42.1, Q is a metacyclic group isomorphic to the group of part (v) of our theorem and so $Z(Q) \cong C_4$, contrary to the fact that $W \leq Z(G)$.

In what follows we may assume that G has a normal abelian subgroup W of type (p, p) and for each such W, G/W is abelian. This gives that $G' \leq W$ and so G' is elementary abelian of order p or p^2. Also, G has no abelian subgroups of type (p, p, p). Indeed, if $E \cong E_{p^3}$, then, considering maximal subgroups of E, we know that they are normal in G and each of them contains G' which would imply $G' = \{1\}$.

(i) Assume that $Z(G)$ is cyclic. Set $Z = \Omega_1(Z(G))$ so that $|Z| = p$ and let J be another subgroup of order p. Then J is not normal in G and $W = Z \times J$ is a normal abelian subgroup of type (p, p). If $N = N_G(J)$, then $|G : N| = p$ and $N/J \neq \{1\}$ is either cyclic or $p = 2$ and $N/J \cong Q_8$ (Lemma 16.1).

First suppose that $N/J \cong Q_8$ so that $|G| = 2^5$. We have $W = Z(N)$ and since G/N acts faithfully on W, we get $Z = Z(G)$. If X/J is a maximal subgroup of N/J, then $X/J \cong C_4$ and so X is abelian. Hence N has at least three abelian maximal subgroups which implies $|N'| = 2$ and $N' < W$ so that $N' = Z = Z(G)$. Let $n \in N - W$ so that $n^2 \in W - J$. Suppose that $n^2 \notin Z$ so that $\langle n^2\rangle$ is not normal in G. By Lemma 16.1, $\langle n^2\rangle$ must be a maximal cyclic subgroup in G, which is not the case. Thus, $n^2 \in Z$ and so $\Phi(N) = N' = Z$. Let Q^* be a maximal subgroup of N which does not contain J so that $Q^* \cong Q_8$, $N = J \times Q^*$ and $Q^* \lhd G$. Suppose that there is $x \in G - N$ such that $o(x) \leq 4$. Then $x^2 \in W$ and $D = \langle W, x\rangle \cong D_8$. There is $i \in D - N$ such that i is a noncentral involution. By Lemma 16.1 , $|G : C_G(i)| = 2$ and the fact that $[i, J] \neq \{1\}$ gives that $C_N(i)$ covers N/J so that $C_N(i) = Q_1 \cong Q_8$. We get $G = Q_1 * D$ is extraspecial of order 2^5 which is as in (vii). Suppose that all elements in $G - N$ are of order > 4 so that $\Omega_2(G) = N = J \times Q^*$. Since $C_G(Q^*) \cap N = W$, we have $\exp(C_G(Q^*)) \leq 4$ and so $C_G(Q^*) = W$ which gives $G/W \cong D_8$, contrary to the fact that G/W is abelian.

We may assume that $N/J \neq \{1\}$ is cyclic so that N is abelian. Since J is maximal cyclic (Lemma 16.1), N is noncyclic and so $N = J \times A$, where A is cyclic of order

$\geq p^2$ (since we may assume that $|G| > p^3$) and so $Z < A$ because $\Omega_1(\mho_1(N))(< A)$ is central in G. By the proof of Lemma 16.1, $N/Z(G) \cong G'$.

Now assume that there exists an element x of order p in $G - N$. Then $\langle Z, x\rangle$ is abelian of type (p, p) so that $\langle Z, x\rangle$ is normal in G and $G' \leq W \cap \langle Z, x\rangle = Z$ which implies $G' = Z$ and $|G'| = p$. Since $N/Z(G) \cong G'$, we have $|N : Z(G)| = p$. But $J \not\leq Z(G)$ and so we may set $N = Z(G) \times J$, where $Z(G)$ is cyclic. Since $\langle W, x\rangle = G_0$ is nonabelian of order p^3, we get $G = Z(G) * G_0$ so G is as in (ii).

Now suppose that there exist no elements of order p in $G - N$ and consider the abelian group G/W. If G/W is cyclic, then the fact that $\Omega_1(N) = J \times Z = W \cong E_{p^2}$ implies that G has a cyclic subgroup of index p and so G is metacyclic. Now, $G' \leq W$ and the commutator group of a metacyclic group is cyclic and so $|G'| = p$. By Lemma 65.2(a), G is minimal nonabelian which gives case (i). We may assume that G/W is noncyclic and so G/W is abelian of type (p^a, p) with $a \geq 1$. Let $G \geq R > W$ be such that R/W is abelian of type (p, p). Suppose $p = 2$ and let $x \in R - N$. Then $x^2 \in W$ and $[W, x] \neq 1$ so that $\langle W, x\rangle \cong D_8$. But then there are involutions in $\langle W, x\rangle - N$, a contradiction. Hence $p > 2$. If R (of order p^4) is regular, then $p^2 = |\Omega_1(R)| = |G/\mho_1(R)|$. Thus, $\mho_1(R) = W$ and so there is $x \in R - N$ such that $x^p \notin Z$. By Lemma 16.1, $x^p \in Z(G)$ which implies $W \leq Z(G)$, a contradiction. It follows that R is irregular and so $p = 3$, R is of class 3 and so $R' = G' = W$. Finally, we know that $N/Z(G) \cong E_{p^2}$ (Lemma 16.1) and $R \cap N$ is abelian of type (p^2, p). Thus, if $a \geq 2$, then $R \cap N$ contains a central element of order p^2, contrary to the fact that R is of maximal class. Hence $a = 1$, $R = G$ and we have obtained the case (iv) of our theorem.

(ii) Assume that $Z(G)$ is noncyclic. Set $Z = Z(G)$ so that $W = \Omega_1(Z) = \Omega_1(G) \cong E_{p^2}$ because G has no elementary abelian subgroups of order p^3. Moreover, $G' \leq W$ so G is of class 2. For any $x, y \in G$, $[x^p, y] = [x, y]^p = 1$ and so $\Phi(G) \leq Z$ and G/Z is elementary abelian.

Suppose that either $p > 2$ or $p = 2$ and $\{x \in G \mid x^2 \in G'\} \leq Z$. We show that under this assumption the map $x \Rightarrow x^p$ is a one-to-one map from G/W into Z. Indeed, this is clear for $p > 2$ since in that case $x \Rightarrow x^p$ $(x \in G)$ is a homomorphism with kernel $\Omega_1(G) = W$. Let $p = 2$. If $x^2 = y^2$, then

$$(xy^{-1})^2 = xy^{-1}xy^{-1} = x^2(x^{-1}y^{-1}xy)y^{-2} = x^2[x, y]y^{-2} = [x, y].$$

Hence, by our assumption, $xy^{-1} \in Z$ so x and y commute and $(xy^{-1})^2 = x^2y^{-2} = 1$, $xy^{-1} \in W = \Omega_1(G)$. Thus this fact follows. This gives $|G/W| \leq Z$ and so $|G| \leq |Z||W|$. Since G is nonabelian, we have $G/Z \cong E_{p^2}$ and the map $x \Rightarrow x^p$ is onto which implies $Z = \Phi(G)$. Each maximal subgroup of G is abelian and so G is minimal nonabelian. Since $\Omega_1(G) = W \cong E_{p^2}$, G is metacyclic as in part (i) of the theorem.

In the rest of the proof we may assume that $p = 2$ and $\{x \in G \mid x^2 \in G'\} \not\leq Z$. Hence there is $x \in G - Z$ with $x^2 \in G' \leq W$ and $o(x) = 4$. Suppose there is $y \in Z$

with $y^2 = x^2$. Then $(xy^{-1})^2 = 1$ so $xy^{-1} \in W \leq Z$ and $x \in Z$, a contradiction. This implies that $Z = Z_1 \times \langle x^2 \rangle$, where Z_1 is cyclic. Since $\Omega_2(G) \nleq Z(G)$, Theorem 1.21 implies that G has a non-normal cyclic subgroup H of order 4. Lemma 16.1 implies that H is maximal cyclic in G, $|G : N| = 2$ with $N = N_G(H)$ and N/H is either cyclic or quaternion.

First suppose $N/H \cong Q_8$ so that $|G| = 2^6$. Set $S/H = \Phi(N/H)$ so that S is abelian of type $(4, 2)$ because $|N : C_N(H)| \leq 2$ and H is maximal cyclic. We have $\Omega_1(S) \cong E_4$ which implies that $\Omega_1(S) = W = \Omega_1(G) \leq Z(G)$. On the other hand, $Z(G) \leq N$ and since $H \nleq Z(G)$, we get $W = Z(G)$. Since $G/Z(G)$ is elementary abelian and $d(G) \leq 4$, we have $\Phi(G) = W$ and $\exp(G) = 4$.

Suppose in addition that G possesses a subgroup $Q = \langle k, l \rangle \cong Q_8$ so that $Q \lhd G$. Since $G' \leq W$, no element y in G induces an outer automorphism on Q (otherwise, we may assume that $k^y = l$ and then $[k, y] = k^{-1}l$, where $o(k^{-1}l) = 4$). It follows that $L = C_G(Q)$ covers G/Q (noting that $\mathrm{Aut}(Q_8) \cong S_4$) and so $G = Q * L$ with $Q \cap L = Z(Q) = \langle z \rangle$ and $k^2 = l^2 = z$. We have $W = Z(G) \leq L$ and since $Z(L) \leq Z(G)$, we get $Z(L) = \Omega_1(L) = W$. Also, $\Phi(G) = \Phi(Q)\Phi(L)$ which gives $\Phi(L) = W$. The subgroup L is obviously minimal nonabelian (of exponent 4) and L is metacyclic since $\Omega_1(L) \cong E_4$. We get $L = \langle a, b \mid a^4 = b^4 = 1, a^b = a^{-1} \rangle$, where $W = Z(L) = \langle a^2, b^2 \rangle$ and a^2 and b^2 are the only involutions in W which are squares in L. If $c \in L$ is such that $c^2 = z$, then $o(kc) = 2$ and $kc \notin L$, a contradiction. It follows that $a^2 \neq z$ and $b^2 \neq z$ which implies $a^2b^2 = z$. We have $\langle bk, a \rangle = R \cong Q_8$ since $(bk)^2 = b^2k^2 = b^2z = a^2$ and $(bk)^a = ba^2k = (bk)a^2 = (bk)^{-1}$. Also, $R \cap Q = \{1\}$ and R is normal in G so that $G = Q \times R$. On the other hand, $(bk)^l = (bk)z$ and so $z \in R$, a contradiction. We have proved that a quaternion group is not a subgroup of G.

Let X be any minimal nonabelian subgroup of G. Since X is not isomorphic to Q_8, $|\Omega_1(X)| \leq 4$, and $\exp(X) = 4$, X is isomorphic to the group $\langle a, b \mid a^4 = b^4 = 1, a^b = a^{-1} \rangle$. By Theorem 57.3, G is the special group of part (ix).

In what follows we assume that N/H is cyclic which implies $d(G) \leq 3$.

(ii1) Suppose that $|G'| = 2$. Set $H = \langle h \rangle \cong C_4$ and let $N \geq K > H$ be such that $|K/H| = 2$. Then K is normal in G (since K is noncyclic) and if $g \in G - N$, then $h^g = hi$ with an involution $i \in K - H$ and $i \in G'$ so that $G' = \langle i \rangle$. In that case $H \leq Z(N)$ which gives that N is abelian. By Lemma 1.1, $N/Z \cong G'$ and so $G/Z \cong E_4$. We know that there is an element $x \in G - Z$ such that $x^2 = i$ and so $\langle x \rangle$ is normal in G. Now, $Z_1 \cap \langle x \rangle = \{1\}$ and so $(Z_1 \times \langle x \rangle)/\langle x \rangle$ has index 2 in $G/\langle x \rangle$. If $G/\langle x \rangle$ is cyclic, then G is metacyclic which together with $|G'| = 2$ implies that G is minimal nonabelian and we have obtained groups of part (i) of our theorem. If $G/\langle x \rangle$ is noncyclic, then $G/\langle x \rangle$ is abelian of type $(2^n, 2)$, $n \geq 1$ (since we may assume $|G| \geq 2^4$), and so there is $y \in G - (\langle x \rangle \times Z_1)$ such that $y^2 \in \langle x \rangle$. But $\langle x \rangle \nleq Z(G)$ and so $y^2 \in \langle i \rangle$ which together with $x^y = x^{-1}$ implies that $\langle x, y \rangle \cong Q_8$ and so $G = \langle x, y \rangle \times Z_1$ with $|Z_1| > 2$. We have obtained the groups stated in part (iii) of our theorem.

(ii2) Now suppose $|G'| = 4$ so that $G' = W = \Omega_1(Z) = \Omega_1(G)$. By Lemma 1.1, $|G/Z| = 8$ and so $G/Z \cong \mathrm{E}_8$ which implies $\mathrm{d}(G) = 3$ and so $Z = \Phi(G)$.

Suppose that G possesses a subgroup $\langle x, y\rangle = Q \cong Q_8$, where we set $\mathrm{Z}(Q) = \langle z\rangle$. Then $Q \triangleleft G$ and again there is no element in G inducing an outer automorphism on Q (since $G' \cong \mathrm{E}_4$). It follows that $G = Q\mathrm{C}_G(Q)$ with $Q \cap \mathrm{C}_G(Q) = \langle z\rangle$. We get $\Phi(G) \geq \langle z\rangle$ and $\Phi(G) \leq C = \mathrm{C}_G(Q)$ and so $|C : \Phi(G)| = 2$. Since $Z = \Phi(G)$, C is abelian and so $G' = \langle z\rangle$, a contradiction. Hence Q_8 is not a subgroup of G.

Assume that $Z > W = G'$ and so G/W is abelian of type $(2^m, 2, 2,)$, $m > 1$. Set $M/W = \Omega_1(G/W)$ so that $M/W \cong \mathrm{E}_8$. Set $Z_0 = M \cap Z$, where $|Z_0 : W| = 2$ and Z_0 is abelian of type $(4, 2)$. Let $v \in Z_0 - W$ so that $o(v) = 4$ and $v^2 = u \in W$. Let R/W be a complement of Z_0/W in M/W. We have $|R| = 2^4$ and $R/W \cong \mathrm{E}_4$. Suppose there is $y \in R$ such that $y^2 = u$. Then $o(yv) = 2$ and $yv \notin R$, a contradiction. It follows that R is nonabelian (otherwise, $R \cong \mathrm{C}_4 \times C_4$ and in that case u is a square in R) and since Q_8 is not a subgroup of R, we get that R is minimal nonabelian with $\Omega_1(R) = W$ and $\exp(R) = 4$ so that $R = \langle a, b \mid a^4 = b^4 = 1, a^b = a^{-1}\rangle = \mathcal{H}_2$ and we must have $a^2 \neq u$ and $b^2 \neq u$ so that $a^2b^2 = u$. Then we get $(bv)^2 = b^2v^2 = b^2u = a^2$ and $a^{bv} = a^{-1}$ so that $\langle a, bv\rangle \cong \mathrm{Q}_8$, a contradiction.

We have proved that $Z = W$ and so $G' = W = \mathrm{Z}(G) = \Phi(G) = \Omega_1(G) \cong \mathrm{E}_4$ and so G is a special group of order 2^5. Since Q_8 is not a subgroup of G and $\Omega_1(G) \cong \mathrm{E}_4$, each minimal nonabelian subgroup of G is isomorphic to the above group H_2. By Theorem 57.3, G is the minimal non-metacyclic group of order 2^5 and so G is in (viii).

It is an easy exercise to show that all the groups of our theorem satisfy the assumptions of this theorem and we are done. □

Classification of p-groups all of whose nonnormal subgroups have the same order follows from Theorem 16.2 [Ber26, Zap].

§17

Counting theorems for regular p-groups

In this section we prove a few counting theorems for regular p-groups.

Given a p-group G, set $p^{w(G)} = |G/\mho_1(G)|$. If $w(G) < p$, then G is regular (Theorem 9.8(a)).

Definition 1. A p-group G is said to be an L_s-*group* if (L1) $\Omega_1(G)$ is of order p^s and exponent p; (L2) $G/\Omega_1(G)$ is cyclic of order $> p$ (so $\exp(G) > p^2$). Write $\mathrm{L}_{s,n}(G)$ for the set of all L_s-subgroups of order p^n in G.

An L_2-group has a cyclic subgroup of index p so it is either abelian of type (p^n, p) or $\cong \mathrm{M}_{p^{n+1}}$ for some $n > 2$. If G is an L_s-group, then $\mathrm{cl}(G) \le s$.

Definition 2. A 2-group G is said to be a U_2-*group* if (U1) G has a normal four-subgroup R (a *kernel* of G); (U2) G/R is of maximal class; (U3) If T/R is a cyclic subgroup of index 2 in G/R, then $\Omega_1(T) = R$. Write $\mathrm{U}_n(G)$ for the set of all U_2-subgroups of order p^n in G (obviously, $n \ge 5$).

The subgroup T of Definition 2 is an L_2-group. By Lemma 18.1, a U_2-group G has only one kernel. U_2-groups are classified in §67.

Throughout this and the following section, $\mathcal{M}$ usually denotes the set of proper subgroups of G which must be counted. For $F \le G$, let $\alpha(F) = |\{H \in \mathcal{M} \mid H \le F\}|$ be the number of members of the set $\mathcal{M}$ contained in F; then $\alpha(G) = |\mathcal{M}|$.

Lemma 17.1. *Suppose that a p-group G is neither cyclic nor a 2-group of maximal class, $|G| = p^m$, $n < m$ is fixed. Let $H < G$ be cyclic and $\mathcal{M} = \{C < G \mid H < C, |C| = p^n > p, C$ is cyclic $\}$. Then p divides $|\mathcal{M}|$.*

For the proof, see Lemma 5.15.

Lemma 17.2 (see Theorem 13.19). *Suppose that G is a p-group of maximal class, $|G| = p^m > p^{p+1}$ and $H < G$ is some largest (by inclusion) regular subgroup. Then either $|G : H| = p$ or H is of order p^p and exponent $\le p^2$.*

Lemma 17.3. *Let G be a p-group and $H < G$. (a) (Remark 10.5) If $\mathrm{N}_G(H)$ is of maximal class, then G is also of maximal class. (b) Suppose that G is irregular and H is a largest regular subgroup of G. If $|H| = p^p$, then G is of maximal class.*

Proof. (b) Let $H < F \leq G$, where $|F : H| = p$. Then F is of maximal class so $\mathrm{N}_G(H)$ is also of maximal class since $F \leq \mathrm{N}_G(H)$ (Exercise 13.10), and we are done, in view of (a). □

Lemma 17.4. *Let G be a group of order p^m.*

(a) Suppose that $m > p + 1$ and $D \lhd G$ is of order p^p and exponent p. If G/D is cyclic, then $\exp(\Omega_1(G)) = p$. If $H/D < G/D$ is of order p, then H is regular.

(b) Let G be an L_w-group. If G is irregular, we suppose that $w = p$. Then $\mho_1(G)$ is cyclic.

(c) If G is an L_p-group, then all proper subgroups of G are regular.

(d) Every L_p-group of exponent p^n is generated by elements of order p^n.

(e) If G is an irregular L_p-group, then $\mathrm{Z}(G)$ is cyclic.

(f) Let $R \lhd G$ be of order p^k and exponent p, $k \leq p$. Suppose that G/R is cyclic of order $> p$ and $R < \Omega_1(G)$. If $Z < G$ is cyclic of order p^{m-k} and $G = RZ$, then $\exp(G) = p^{m-k}$

Proof. (a) follows from Exercise 13.10(a).

(b) We have $\exp(G) = p^{m-w+1}$. If G is regular, the result follows from Theorem 7.2(d). If G is irregular, then $|G : \mho_1(G)| \geq p^p$ (Theorem 9.8(a)) so $|\mho_1(G)| \leq p^{m-p}$. Since $\exp(G) = p^{m-p+1} \leq p \cdot \exp(\mho_1(G))$, $\mho_1(G)$ must be cyclic so $|G/\mho_1(G)| = p^p$.

(c) coincides with Theorem 9.4(b).

(d,e,f) Let $\exp(G) = p^n$, $n > 3$. Let $\Omega_1(G) < H \in \Gamma_1$; then $\Omega_{n-1}(G) = H$ is of exponent p^{n-1} so all elements of the set $G - H$ have the same order p^n. In that case $\langle G - H\rangle = G$, and (d) is proved. Let us prove (e). Assume that $\mathrm{Z}(G)$ is noncyclic and $L = \Omega_1(\mathrm{Z}(G))$. Then G/L is absolutely regular, and the result follows from Remark 7.2. (f) is obvious. □

Until the end of this section, G is a regular p-group.

Conjecture A(s). *Let G be a regular group of order p^m, $w = w(G)$, $s \in \{1, \dots, w\}$, $m > n > s + 1$. If p^{w-s+1} does not divide $|\mathrm{L}_{s,n}(G)|$, then G is an L_w- group and $|\mathrm{L}_{s,n}(G)| \equiv p^{w-s} \pmod{p^{w-s+1}}$.*

In particular, if A(s) is true, then p^{w-s} must divide $|\mathrm{L}_{s,n}(G)|$ always. By Theorem 7.2, A(1) and A(w) are true. Let us prove that A(1) holds. Note that L_1-group is cyclic of order $\geq p^3$. Let G be regular with $|\Omega_1(G)| = p^w$ and $G/\Omega_1(G)$ is noncyclic of exponent $\geq p^2$. Then

$$|\mathrm{L}_{1,n}| = \mathrm{c}_n(G) = \frac{|\Omega_n(G) - \Omega_{n-1}(G)|}{(p-1)p^{n-1}} \equiv 0 \pmod{p^w}$$

since $|\Omega_{n-1}(G)| \geq p^{w+n-1}$ in view of $|\Omega_1(G/\Omega_{n-2}(G)| \geq p^2$.

Reduction Theorem. *Let $s \in \mathbb{N}$ be fixed. Suppose that* A(*s*) *is true for all regular* L_w*-groups, where $w \geq s$. Then* A(*s*) *is true for all regular p-groups.*

Proof. One may assume that $1 < s < w = w(G)$. Let G be a regular p-group with $w = w(G)$. We have to prove that $|L_{s,n}| \equiv 0 \pmod{p^{w-s+1}}$.

We use induction on $|G|$. Set $D = \Omega_1(G)$, $\mathcal{M} = L_{s,n}(G)$, $\alpha(G) = |\mathcal{M}|$; then $|D| = p^w$, where $w = w(G)$, and $\exp(D) = p$ (Theorem 7.2(b)). Supposing that G is not an L_w-group, we have to prove that $\alpha(G) \equiv 0 \pmod{p^{w-s+1}}$. By hypothesis, G/D is not cyclic; therefore, G/D has a normal subgroup T/D such that $G/T \cong E_{p^2}$. Let $F_1/T, \dots, F_{p+1}/T$ be all subgroups of order p in G/T. We may assume that $\alpha(G) > 0$. Let $H \in \mathcal{M}\ (= L_{s,n}(G))$. Then $\Omega_1(H) = H \cap D < D = \Omega_1(T)$ since $s < w$, and so HT/T is cyclic as an epimorphic image of a cyclic group $HD/D \cong H/\Omega_1(H)$. Hence, $H < F_j$ for some $j \in \{1, \dots, p+1\}$ ($H < F_j$ since $\Omega_1(H) < \Omega_1(F_j) = D$).

It follows from $\alpha(G) > 0$ that $\exp(G) > p^2$ since the exponent of every member of the set $\mathcal{M}$ is $> p^2$. Then $D < T$ (since $\exp(T) \geq \frac{1}{p}\exp(G) > p = \exp(D)$), and so, by induction, p^{w-s} divides $\alpha(T)$ (it is possible that $\alpha(T) = 0$, for example, if $\exp(T) = p^2$); then p^{w-s+1} divides $p\alpha(T)$. Assume that $p^{w-s+1} \nmid \alpha(F_j)$ for some $j \in \{1, \dots, p+1\}$. Then, by induction, F_j is an L_w-subgroup, so that F_j/D is a cyclic subgroup of index p in the noncyclic group G/D. By Theorem 1.2, G/D has exactly p cyclic subgroups $F_1/D, \dots, F_p/D$ and one noncyclic subgroup F_{p+1}/D of index p (if $p = 2$, then G/D is abelian). Then $F_1, \dots, F_p$ are L_w-groups, and so, by assumption, $\alpha(F_i) \equiv p^{w-s} \pmod{p^{w-s+1}}$ for $i \in \{1, \dots, p\}$. Therefore, p^{w-s+1} divides $\sum_{i=1}^{p} \alpha(F_i)$. Since F_{p+1} is not an L_w-group, the number p^{w-s+1} divides $\alpha(F_{p+1})$, by induction. Hence, p^{w-s+1} divides $\alpha(G)$ since $\alpha(G) = \sum_{i=1}^{p+1} \alpha(F_i) - p\alpha(T)$. □

In view of the Reduction Theorem, in the proof of Conjecture A(s), one can confine to L_w-groups. If X is a regular L_w-group of order p^m and exponent p^e, then $w(X) = m - e + 1$.

Lemma 17.5. *Let G be a regular L_w-group of order p^m, $w = w(G) > 1$. For $k \in \{1, \dots, d-1 = d(G) - 1\}$, let $\Gamma_k^1 = \{H \in \Gamma_k \mid \exp(H) = \exp(G)\}$ and $t_k = |\Gamma_k^1|$. Then $t_k = p^k \cdot \varphi_{d-1,k}$, where $\varphi_{m,n} = s_n(E_{p^m})$.*

Proof. Set $D = \Omega_1(G)$. Since G/D is cyclic, we have $G' < D$. Since cyclic $\mho_1(G) \leq \Phi(G) < \Omega_{m-w}(G) \in \Gamma_1$ and $\exp(\Omega_{m-w}(G)) = p^{m-w}$, it follows that $p^{m-w} = \exp(\Phi(G)) < \exp(G) = p^{m-w+1}$. Let $\mathcal{N} = \{C_1, \dots, C_r\}$ be the set of all cyclic subgroups of order $p^{m-w+1} = \exp(G)$ in G. Then

$$r = c_{m-w+1}(G) = \frac{|G - \Omega_{m-w}(G)|}{p^{m-w}(p-1)} = \frac{p^m - p^{m-1}}{p^{m-w}(p-1)} = p^{w-1}.$$

Next, $DC_i = G$ and $|G : C_i\Phi(G)| = p^{d-1}$, $i = 1, \dots, r$. Therefore, the number of members of the set Γ_k containing $C \in \mathcal{N}$, equals the number of subgroups of index

p^k in $G/C\Phi(G)$, i.e., $\varphi_{d-1,k}$. Let $\Gamma_k^1 = \{A_1, \dots, A_{t_k}\}$. Then, for $j \in \{1, \dots, t_k\}$, we have $w(A_j) = (m-k) - (m-w+1) + 1 = w-k$, so $c_{m-w+1}(A_j) = p^{w(A_j)-1} = p^{w-k-1}$. By what has just been proved, the number of pairs $C < A$ with $C \in \mathcal{N}$, $A \in \Gamma_k^1$ equals $p^{w-1}\phi_{d-1,k}$. On the other hand, that number is also equal to $t_k \cdot p^{w-k-1}$. Thus, $p^{w-1}\varphi_{d-1,k} = t_k \cdot p^{w-k-1}$, and so $t_k = p^k \cdot \varphi_{d-1,k}$. □

It follows from the Reduction Theorem and Lemma 17.5 that Theorem A(w-1) is true. Indeed, let $\alpha(G) = |\mathrm{L}_{w-1,n}(G)|$, where G is an L_w-group. One may assume that $G = \langle H \mid H \in \mathrm{L}_{w-1,n}(G)\rangle$. In that case, $p^{m-w+1} = \exp(G) = p^{n-(w-1)+1}$ so $n = m-1$. It follows that then $\mathrm{L}_{w-1,m-1}(G) = \Gamma_1^1$ and $\alpha(G) = t_1 = p \cdot \phi_{d-1,1} \equiv p \pmod{p^2}$.

Theorem A($w-2$). *Conjecture* A($w-2$) *is true so p^3 divides $|\mathrm{L}_{w-2,n}(G)|$, unless G is an L_w-group (in the last case, $|\mathrm{L}_{w-2,n}(G)| \equiv p^2 \pmod{p^3}$). Of course, we suppose that $w > 2$.*

§18

Counting theorems for irregular p-groups

In this section we retain the notation and agreements of §17.

Definition 1. Let $k \in \mathbb{N}$. A p-group G is said to be k*-tame* if

(a) G has a subgroup of order p^k and exponent p and

(b) whenever $H < G$ is of order p^{k+1} and H has a subgroup of index p and exponent p, then H is regular.

We write $\chi(G) = \max\{k \in \mathbb{N} \mid G \text{ is } k\text{-tame}\}$.

A p-group G satisfies $\chi(G) = 1$ if and only if one of the following holds: (i) G is cyclic, (ii) G is generalized quaternion, (iii) G has a subgroup isomorphic to D_8.

If G is irregular, then $\chi(G) \geq p - 1$ (Theorems 7.1 and 9.8). If G is a regular p-group and $|\Omega_1(G)| = p^w$, then $\chi(G) = w = w(G)$. If G is an irregular p-group of maximal class, then $\chi(G) = p - 1$ (Theorems 9.5, 9.6, 9.8(d) and 7.1).

Definition 2. A 2-group G is said to be a U_s-group if (U_s1) G has a normal elementary abelian subgroup R of order 2^s; (U_s2) G/R is a 2-group of maximal class; (U_s3) If T/R is a cyclic subgroup of index 2 in G/R, then $\Omega_1(T) = R$, i.e., T is an L_s-group (see §17). The subgroup R is called the *kernel* of the U_s-group G. (For related definitions, see §§19, 36.)

If G is a U_s-group of order 2^{n+s}, then $\exp(G) = 2^n$.

Let us consider the following general counting

Conjecture B(s). *Suppose that G is an irregular p-group of order p^m, $s \in \{1, \dots, p\}$ and $m > n > s + 1$. If $p^{p-s+1} \nmid |\mathrm{L}_{s,n}(G)|$, then one of the following holds:*

(a) G is an L_p-group and $|\mathrm{L}_{s,n}(G)| \equiv p^{p-s} \pmod{p^{p-s+1}}$.

(b) $p = 2$ and G is a U_2-group.

(c) G is a p-group of maximal class.

Given a p-group G, let $e_n(G)$ denote the number of subgroups of order p^n and exponent p in G and $\mathrm{sol}_n(G) = \{x \in G \mid x^{p^n} = 1\}$. In the sequel $|G| = p^m$. Until the end of this section we retain the notation introduced above.

Lemma 18.1. *Suppose that G is a U_s-group with kernel R, $|G| = 2^m$. Then all G-invariant elementary abelian subgroups of G are contained in R.*

For a proof, see paragraph preceding Proposition 36.15.

Theorem 18.2 ([Ber16]). *Let G be a p-group, $\chi = \chi(G) > 1$.*

(a) $\mathrm{e}_k(G) \equiv 1 \pmod{p}$ *for every natural* $k \leq \chi$.

(b) *Let $R \lhd G$ be of order p^k ($k \leq \chi$) and exponent p. If $S \leq G$ is of exponent p, then* $\exp(RS) = p$.

(c) *If $H < G$ is of order $p^{1+s} < p^{1+\chi}$ and H has a subgroup of index p and exponent p, it is regular so G is also k-tame for all $k < \chi$.*

(d) $\mathrm{c}_1(G) \equiv 1 + p + \cdots + p^{\chi-1} \pmod{p^\chi}$.

(e) $\mathrm{e}_{\chi-1}(G) \equiv 1 + p \pmod{p^2}$.

(f) *If $\chi > 3$ and $n \in \{2, \ldots, \chi - 2\}$, then* $\mathrm{e}_n(G) \equiv 1 + p + 2p^2 \pmod{p^3}$.

(g) *If $\chi > 3$ and $n \in \{3, \ldots, \chi - 1\}$, then p divides the number of two-generator subgroups $H < G$ such that $|H| = p^n$ and* $\exp(H) = p$.

(h) *If $n > 1$, then $p^{\chi-1}$ divides $\mathrm{c}_n(G)$ and p^χ divides* $|\mathrm{sol}_1(G)|$.

(i) *If* $\exp(G) \geq p^n$, *then $p^{\chi+n-1}$ divides* $|\mathrm{sol}_n(G)|$.

(j) *If $0 < n < \chi$ and $R_0 < G$ is of order p^n and exponent p, then* $|\{R \leq G \mid R_0 < R,\ |R| = p^{1+n},\ \exp(R) = p\}| \equiv 1 \pmod{p}$.

(k) *If $n > 2$ for $p > 2$ and $n > 3$ for $p = 2$ and* $\exp(G) \geq p^n$, *then p^χ divides $\mathrm{c}_n(G)$ and $p^{\chi+n}$ divides $|\mathrm{sol}_n(G)|$, unless G is either an L_χ- or U_χ-group.*

(l) *If $\chi + 2 \leq n < m$, then p divides $|\mathrm{L}_{\chi,n}(G)|$, unless G is an L_χ- or U_χ-group.*

Proof. Suppose that the theorem is proved for all proper subgroups of G.

(a) There is $R < G$ of order p^χ and exponent p. If $R \leq H \in \Gamma_1$, then $\chi(H) \geq \chi$, so by induction, $\mathrm{e}_\chi(H) \equiv 1 \pmod{p}$. This means that H has a G-invariant subgroup of order p^χ and exponent p so one may assume that $R \lhd G$. If $\Omega_1(G) = R$, then (a), (d–g) are correct (see §5). Now let $x \in G - R$ be of order p. Set $L = \langle x, R\rangle$; then L is regular (see Definition 1) so $\exp(L) = p$ (Theorem 7.2(b)). If $H \in \Gamma_1$, then $|H \cap L| \geq p^\chi$ so $\chi(H) \geq \chi$, and, by induction, $e_\chi(H) \equiv 1 \pmod{p}$. By Theorem 5.2, we get $\mathrm{e}_k(G) \equiv \sum_{H\in\Gamma_1} \mathrm{e}_k(H) \equiv |\Gamma_1| \equiv 1 \pmod{p}$, proving (a).

Our argument also yields the following result:

($*$) If $H \in \Gamma_1$ and $\chi(H) < \chi$, then $|\Omega_1(G)| = p^\chi$, $\exp(\Omega_1(G)) = p$, $\Omega_1(H) = H \cap \Omega_1(G)$, $\chi(H) = \chi - 1$.

It remains to prove the last equality in ($*$). Indeed, set $R = \Omega_1(G)$ and let $\Omega_1(H) < L < H$, where $|L : \Omega_1(H)| = p$; then $\exp(L) > p$. In that case, $|LR : R| = p$, by the product formula, so LR is regular, and we are done.

(b) Let $R \lhd G$ be of order p^χ and exponent p and S a subgroup of exponent p in G. We claim that $\exp(RS) = p$. Let $x = uv \in RS - R$, where $u \in R$ and $v \in S$. Then $\langle x, R\rangle = \langle v, R\rangle$ is regular so of exponent p since $o(v) = p$. It follows that $o(x) \leq p$ for all $x \in RS - R$ so $\exp(RS) = p$.

Now let $k < \chi$ and let $R \triangleleft G$ be of order p^k and exponent p. By (a), there is $M \triangleleft G$ of order p^χ and exponent p. By what we have just been proved, $RM \triangleleft G$ is of exponent p. Let $R < T < RM$, where T is a normal subgroup of G of order p^χ. By the previous paragraph, $\exp(TS) = p$ so $\exp(RS) = p$.

(c) Let $s < \chi$ and let $H < G$ of order p^{1+s} be such that H has a subgroup F of index p and exponent p. We have to prove that H is regular. Let $R \triangleleft G$ be of order p^χ and exponent p. By (b), $\exp(RF) = p$. Let $H < L \le RH$, where $|L| = p^{1+\chi}$. Then L has a subgroup $L \cap RF$ of index $\le p$ and exponent p so L is regular. Then $H < L$ is also regular.

(d) By (a), there is $R \triangleleft G$ of order p^χ and exponent p. One may assume that $R < \Omega_1(G)$ (otherwise, there is nothing to prove). Assume that G/R is cyclic. Let M/R be a subgroup of order p in G/R. By ($\mathrm{T}_\chi 2$), M is regular so $M = \Omega_1(G)$, and (d) follows. Now let G/R be not cyclic. Then G/R has a normal subgroup H/R such that $G/H \cong \mathrm{E}_{p^2}$. Let $H_1/R, \dots, H_{p+1}/H$ be all subgroups of order p in G/H. We have $c_1(G) = c_1(H_1) + \dots + c_1(H_{p+1}) - pc_1(H)$. It follows from $R < H_i$ and $R \le H$ that $\chi(H_i) \ge \chi$ and $\chi(H) \ge \chi$. Now the result follows by induction from the formula for $c_1(G)$.

(e) As in (d), one may assume that $|\Omega_1(G)| > p^\chi$. Then for all $H \in \Gamma_1$, we have $\chi(H) \ge \chi$ so, by induction, $e_{\chi-1}(H) \equiv 1 + p \pmod{p^2}$, and the result follows, by Theorem 5.2.

(f) and (g) are proved as previous two parts with help of Theorem 5.8.

To prove (h) and (i), we have to repeat, word for word, the proof of Theorem 13.2. To prove (j), we have to apply (a) and the second part of (h).

(k) Let $R \triangleleft G$ be of order p^χ and exponent p.

If G/R is cyclic and $\Omega_1(G) > R$ (i.e., G is not an L_χ-group), then $\Omega_1(G)$ is of order $p^{\chi+1}$ and exponent p so $|\mathrm{sol}_n(G)| = p^{n+\chi}$ and $c_n(G) = p^\chi$ for $n > 2$. Next we assume that G/R is not cyclic.

Let $p = 2$ and let G/R be of maximal class but G be not a U_χ-group. Then, if T/R is a cyclic subgroup of index 2 in G/R, then $|\Omega_1(T)| = 2^{\chi+1}$, $|\mathrm{sol}_n(T)| = 2^{\chi+n}$ and $c_n(T) = 2^\chi$. Since, in view of $n > 3$, all elements of G of order 2^n are contained in T, we are done. Next we assume that G/R is not a 2-group of maximal class.

Let D/R be a normal subgroup of G/R such that $G/D \cong \mathrm{E}_{p^2}$ and let $M_1/D, \dots, M_{p+1}/D$ be all subgroups of order p in G/D. We confine to the number $c_n(G)$ only. Then (see the proof of Theorem 13.2)

$$c_n(G) = \sum_{i=1}^{p+1} c_n(M_i) - pc_n(D). \tag{1}$$

All groups in (1) are χ-tame, by (c). One may assume that $p^\chi \nmid c_n(M_1)$. By induction, M_1 is either an L_χ- or U_χ-group.

(1k) Suppose that M_1 is an L_χ-group; then $R = \Omega_1(M_1)$. Since G/R is neither cyclic nor 2-group of maximal class, one may assume that M_i/R is cyclic for $i = 1, \dots, p$, and M_{p+1}/R is abelian noncyclic (Theorem 1.2; this is true also if G/R is abelian). Let S/R be a G-invariant subgroup of order p in M_1/R. Then $S/R \le \Phi(M_1/R) \le \Phi(M_i/R) < G_i/R$ for $i = 1, \dots, p$, and it follows that $M_1, \dots, M_p$ are L_χ-groups. Then $c_n(M_i) = p^{\chi-1}$ and p^χ divides $c_n(M_{p+1})$, by induction. Clearly, p^χ divides $pc_n(D)$. It follows from (1) that p^χ divides $c_n(G)$. (All this true for $n > 2$.)

(2k) Suppose that M_1 is a U_χ-group; then $p = 2$ and M_1/R is a group of maximal class. By the above, G/R is not of maximal class. Then G/R has exactly four subgroups $M_1/R, \dots, M_4/R$ of index 2 that are of maximal class (see §13); the number of maximal subgroups in G/R is seven; let $M_1/R, \dots, M_7/R$ be all maximal subgroups of G/R (see Theorem 5.4). By the second paragraph of (k), $c_n(M_i) = 2^{\chi-1}$, $i = 1, 2, 3, 4$, since these M_i are U_χ-subgroups. By induction, 2^χ divides $c_n(M_t)$, $t = 5, 6, 7$. Next, 2^χ divides $2c_n(D)$. Then, by (1), 2^χ divides $c_n(G)$.

Assertion for $\mathrm{sol}_n(G)$ is proved similarly.

(l) If G is an L_χ-group, then $|L_{\chi,n}(G)| = 1$.

Now let G be a U_χ-group, then $p = 2$ and $|L_{\chi,n}(G)| = 1$ if either $n - \chi > 2$ (in that case, T contains all L_χ-subgroups of G) or G/D is dihedral, where D is the kernel of G. If $n - \chi = 2$ and G/D is generalized quaternion, then $|L_{\chi,n}(G)| = c_2(G/D) = 1 + 2^{m-\chi-2}$. If $n - \chi = 2$ and G/R is semidihedral, then $|L_{\chi,n}(G)| = c_2(G/D) = 1 + 2^{m-\chi-3}$. In both these cases, $|L_{\chi,n}(G)| \equiv 1 \pmod 2$. In what follows we assume that G is neither an L_χ- nor U_χ-group. By the enumeration principle,

$$(2) \qquad |L_{\chi,n}(G)| = \alpha(G) \equiv \sum_{H \in \Gamma_1} \alpha(H) \pmod p,$$

where $\alpha(H) = |L_{\chi,n}(H)|$. One may assume that $p \nmid \alpha(H)$ for some $H \in \Gamma_1$. Then, by induction, H is either an L_χ- or U_χ-group.

(1l) Suppose that $H \in \Gamma_1$ is an L_χ-group. Then $\Omega_1(H) = R \lhd G$, $|R| = p^\chi$ and $\exp(R) = p$. If G/R is cyclic then $\Omega_1(G) = R$ and so G is an L_χ-group, a contradiction. Let G/R be noncyclic. If $p = 2$ and G/R is of maximal class, then G is a U_χ-group, which is not the case. Then, by Theorem 1.2, G/R has exactly p cyclic subgroups $H_1/R = H/R, \dots, H_p/R$ of index p. Since H_1 is an L_χ-group then, as before, $H_2, \dots, H_p$ are also L_χ-groups. The remaining maximal subgroup H_{p+1}/R is not cyclic, so H_{p+1} is neither an L_χ-group nor U_χ-group hence p divides $\alpha(H_{p+1})$, by induction. For $i = 1, \dots, p$, we have $\alpha(H_i) = 1$. Thus, if S is a normal subgroup of G of order p^χ and exponent p, then S is contained in p or 0 maximal subgroups of G which are L_χ-groups. Using this information and double counting, it is easy to prove that p divides $|L_{\chi,n}(G)|$.

(2l) Now let $H \in \Gamma_1$ be a U_χ-group. Then $G/\Omega_1(H)$ has exactly four subgroups of index 2 which are of maximal class, and their inverse images are U_χ-groups. As above, using induction, we prove that $|L_{\chi,n}(G)|$ is even, completing the proof. □

Exercise 1. For a 2-group G of order $> 2^4$ the following conditions are equivalent: (a) G is a U_2-group with $\mathrm{d}(G) = 3$, (b) G' is cyclic and $G/G' \cong \mathrm{E}_8$.

Exercise 2. If G is an L_s-group, then $|G : \mho_1(G)| \leq p^s$.

Theorem B(p). *Suppose that G is an irregular p-group of order p^m, $m > n > p+1$, $\mathcal{M} = \mathrm{L}_{p,n}(G)$, $\alpha(G) = |\mathcal{M}|$. Then one of the following holds:*

(a) G is an L_p-group, $\alpha(G) = 1$.

(b) $p = 2$, G is a U_2-group, $\alpha(G)$ is odd.

(c) $\alpha(G) \equiv 0 \pmod p$.

Proof. Suppose that the theorem has proved for all proper subgroups of G and that $\mathcal{M} \neq \varnothing$. By the enumeration principle,

$$\alpha(G) \equiv \sum_{H \in \Gamma_1} \alpha(H) \pmod p. \tag{3}$$

If G is an L_p-group, then $\alpha(G) = 1$. The p-groups of maximal class has no L_p-subgroups (Theorem 9.6). Suppose that $p = 2$ and G is a U_2-group with kernel D. Let $H \in \mathcal{M}$; then $\exp(H) > 4$. Assume that $D \nleq H$. Then $H/(H \cap D) \cong HD/D \leq G/D$. In that case, H is abelian since all abelian subgroups of exponent > 2 in G/D are cyclic, and this is a contradiction. Thus, $D < H$ so $|H \cap T| > 4$, where T/D is cyclic of index 2 in G/D. In that case, $\alpha(G) = \mathrm{c}_{n-2}(G/D)$ is odd.

Next we assume that G is neither an L_p- nor U_2-group. Assume that $p \nmid \alpha(G)$. Then, by (3), there is $H \in \Gamma_1$ such that $p \nmid \alpha(H)$ so, by induction, H is either an L_p- or U_2-group.

(i) Let H be an L_p-group. Then $D = \Omega_1(H) \lhd G$. Assume that G/D is cyclic. Since $n > p + 1$, we get $\Omega_1(G) = D$ so G is an L_p-group, which is not the case.

Assume that $p = 2$ and G/D is a 2-group of maximal class. We claim that then G is a U_2-group. Indeed, let T/D be a cyclic subgroup of index 2 in G/D. Let K/D be a unique subgroup of order 2 in H/D; then $K/D \leq \Phi(H/D) \leq \Phi(G/D) < T/D$. Since K is abelian of type $(4, 2)$, it follows that $\Omega_1(T) = D$, and our claim follows.

Now let G/G be neither cyclic nor a 2-group of maximal class. Then G/D has exactly $p + 1$ maximal subgroups $H_1/D = H/D, \dots, H_p/D$ and H_{p+1}/D and the first p of them are cyclic, H_{p+1}/D is noncyclic and abelian (see Theorem 1.2). As above, $H_1, \dots, H_p$ are L_p-groups. Since H_{p+1} is not an L_p-group, p divides $\alpha(H_{p+1})$, by induction. Since $\alpha(H_i) = 1$ for $i = 1, \dots, p$, then p divides $\alpha(G)$, by (3).

(ii) Let H be a U_2-group with kernel D; then $p = 2$ and $D \lhd G$. In view of (i), one may assume that the set Γ_1 has no elements that are L_2-groups.

Assume that G/D is of maximal class. We claim that then G is a U_2-group. Indeed, let T/D be a cyclic subgroup of index 2 in G/D. It is enough to prove that $\Omega_1(T) = D$. If not, $\Omega_1(T)$ is elementary abelian of order 8 (Lemma 17.4(a)). Since G/D has only one normal subgroup of order 2, it follows that $\Omega_1(T/D) < H/D$ so H is not a U_2-group, a contradiction.

Thus, G/D is not of maximal class. Then, by Theorem 5.4, G/D has exactly four subgroups of maximal class and index 2: $H_1/D = H/D, \dots, H_4/D$. As above, H_i are U_2-groups so $\alpha(H_i)$ is odd for $i = 1,2,3,4$. Let $\Gamma_{1,D} = \{H_1, \dots, H_7\}$ be the set of all maximal subgroups of G containing D (similarly, we can define the set $\Gamma_{1,R}$ for every G-invariant four-subgroup R). By induction, $\alpha(H)$ is even if $H \in \Gamma_1$ is not an U_2-subgroup. Let $F \in \Gamma_1$ be an U_2-subgroup and $D \not\leq F$. Let R be the kernel of F; then $R \lhd G$. As above, $DR/R \leq H_1/D$ so H_1 is not a U_2-group, which is a contradiction. It follows that F does not exist. Thus, if $H \in \Gamma_1$, then $\alpha(H)$ is odd if H is a U_2-subgroup and even otherwise. Therefore, $\alpha(G)$ is even by (3), completing the proof. □

Corollary 18.3. *Conjecture* $\mathrm{B}(p)$ *is true.*

Theorem B($p-1$). *Suppose that G is an irregular p-group of order p^m, $m > n > p$, $\mathcal{M} = \mathrm{L}_{p-1,n}(G)$, $\alpha(G) = |\mathcal{M}|$. Then one of the following holds:*

(a) $\alpha(G) \equiv 0 \pmod{p^2}$.

(b) *G is an L_p-group, $\alpha(G) \equiv p \pmod{p^2}$.*

(c) *$p = 2$, G is a U_2-group, $\alpha(G) \equiv 2 \pmod 4$.*

(d) *G is a 3-group of maximal class, $n = 4$ and $\alpha(G) = 3$ if $m = 6$ and $\alpha(G) = 12$ if $m > 6$.*

Proof. Suppose that the theorem has proved for all proper subgroups of G. Since Theorem B(1) is true (see Theorems 1.10(a) and 1.17(a)), we assume that $p > 2$.

Let G be an L_p-group. All proper subgroups of G are regular (Lemma 17.4(c)) so $\exp(\Omega_k(G)) = p^k$ for $p^k < \exp(G)$. If $H \in \mathcal{M}$, then $\exp(H) = p^{n-(p-1)+1} = p^{n-p+2}$, $H \in \Omega_{n-p+2}(G)$ so one may assume that $\Omega_{n-p+2}(G) = G$. Then, $p^{m-p+1} = \exp(G) = p^{n-p+2}$ so $n = m-1$, i.e., $\mathcal{M} \subset \Gamma_1$. If $H \in \Gamma_1$ and $\Omega_1(G) \not\leq H$, then $|\Omega_1(H)| = p^{p-1}$, H is an L_{p-1}-subgroup, and so $H \in \mathcal{M}$. Then $\alpha(G) = -1 + \varphi_{\mathrm{d}(G),1} = p + p^2 + \dots + p^{\mathrm{d}(G)-1}$ since only one maximal subgroup of G, namely $\mho_1(G)\Omega_1(G)$, has exponent $< p^{m-p+1}$. If $\Omega_1(G) < H \in \Gamma_1$, then $\alpha(H) = 0$ since $\exp(H) < \exp(G)$. Since $\mathrm{d}(G) > 1$, it follows that $\alpha(G) = -1 + \varphi_{\mathrm{d}(G),1} \equiv p \pmod{p^2}$, proving (b). Next we assume that G is not an L_p-group.

Suppose that G is of maximal class. If $H \in \mathcal{M}$ and $\exp(H) = p^k > p^2$, then H is generated by elements of order p^k since it is regular; then $H \leq G_1$, where G_1 is the fundamental subgroup of G (Theorem 13.19), and we get $\alpha(G) = \alpha(G_1) = c_{k-1}(G_1/\Omega_1(G_1))$ (this number depends only on n, by Theorem 9.6). Since G_1 is not an L_{p-1}-group (by Theorem 9.6, every epimorphic image of G of order $\geq p^p$ is not absolutely regular), we get $\alpha(G_1) \equiv 0 \pmod{p^2}$, unless $p = 3$. Suppose that $p = 3$ and that $3^2 \nmid \alpha(G_1)$. Then $k = 3$, $|\Omega_3(G_1)| \in \{3^5, 3^6\}$ and $\alpha(G) \in \{3, 12\}$. This completes the proof of (d). Next we assume that G is not of maximal class.

For $i = 1, 2$, we write

$$\Gamma_{i,1} = \{H \in \Gamma_i \mid H \text{ is an } \mathrm{L}_p\text{-group}\}, \quad \Gamma_{i,2} = \{H \in \Gamma_i \mid H \text{ is of maximal class}\},$$
$$\Gamma_{i,3} = \{H \in \Gamma_i \mid H \text{ is absolutely regular}\}, \quad \Gamma_{i,4} = \Gamma_i - (\Gamma_{i,1} \cup \Gamma_{i,2} \cup \Gamma_{i,3}).$$

Let $n = m - 1$. Then, by Theorem 12.1(b), $G = H\Omega_1(G)$, where $H \in \mathcal{M}$ and $|\Omega_1(G)| = p^p$. Since H is an L_{p-1}-group, then $G/\Omega_1(G) \cong H/\Omega_1(H)$ is cyclic so G is an L_p-group. In what follows we assume that $n < m - 1$.

(i) By induction, p^2 divides $\sum_{H \in \Gamma_{1,4}} \alpha(H)$.

(ii) We claim that p^2 divides $\sum_{H \in \Gamma_{1,2}} \alpha(H)$. Indeed, let $H \in \Gamma_{1,2}$, i.e., H is of maximal class. Then, by what has been proved already, $\alpha(H) \pmod{p^2}$ depends only on n. Since p^2 divides $|\Gamma_{1,2}|$, by Theorem 13.6, our claim follows.

(iii) We claim that p^2 divides $\sum_{H \in \Gamma_{1,1}} \alpha(H)$. Indeed, if $H \in \Gamma_{1,1}$, i.e., H is an L_p-group, then by the result of the second paragraph of the proof, $\alpha(H) \equiv p \pmod{p^2}$. Since p divides $|\Gamma_{1,1}|$ by Theorem B(p), we are done.

(iv) We claim that p^2 divides $\sum_{H \in \Gamma_{1,3}} \alpha(H)$. Indeed, suppose that $H \in \Gamma_{1,3}$. Then (Theorem 12.1(b)) $G = HR$, where $R = \Omega_1(G)$ is of order p^p and exponent p, $H \cap R = \Omega_1(H)$. Since G is not an L_p-group, $H/(H \cap R) \cong G/R$ is noncyclic. Then, by Theorem 1.10(b), $\alpha(H) = c_{n-p+1}(H/(H \cap R)) = c_{n-p+1}(G/R)$ is a multiple of p that does not depend on the choice of H. Then $H \in \Gamma_{1,3}$ if and only if $R \not\leq H$. Therefore, p divides $|\Gamma_{1,3}|$, and our claim follows.

It follows from (i)–(iv) that

(v) p^2 divides $\sum_{H \in \Gamma_1} \alpha(H)$. Since $n < m-1$ and $n \geq p+1$, we have $m \geq p+3$. By the enumeration principle, it remains to show that

$$\sum_{H \in \Gamma_2} \alpha(H) \equiv 0 \pmod{p}. \tag{4}$$

Assume that (4) is false. Then there is $H \in \Gamma_2$ such that $p \nmid \alpha(H)$. By Theorem A($p-1$), H is an L_{p-1}-group and so $\alpha(H) = 1$. Then $\Omega_1(H) = D = \Omega_1(\Phi(G))$ (see Theorem 9.8(d)) so H/D is cyclic of order $p^{n-(p-1)} > p$.

Suppose that $H = \Phi(G)$ (or, what is the same, $\mathrm{d}(G) = 2$). Then $H/D = \langle x^p D\rangle$ for some $x \in G$. Hence, $T/D = \langle x, H\rangle/D$ is cyclic of index p in G/D, and T is an L_{p-1}-group since H is. By Theorem 12.1(b), $R = \Omega_1(G)$ is of order p^p and exponent p and $G/R \cong T/D$ is cyclic so G is an L_p-group.

Let $\Phi(G) < H$. Since $\Omega_1(H) = D \leq \Phi(G)$, by Theorem 9.8(d), and $\mho_1(H) \leq \Phi(H) \leq \Phi(G)$, it follows that $\Phi(G) = \mho_1(H)\Omega_1(H)$ has index p in H (see Lemma 17.4(b)). Next, $\mho_1(H) \lhd G$. Since $\mho_1(H)$ is cyclic of composite order, $\Gamma_{2,2} = \varnothing$: p-groups of maximal class and order $> p^3$, $p > 2$, have no normal cyclic subgroups of order p^2. If $F \in \Gamma_2$ and F is an L_{p-1}-group, then $D < F$ and F/D is cyclic. On the other hand, every cyclic subgroup of G/D of order $|H/D|$ contains $\Phi(G/D)$, and so its inverse image is an L_{p-1}-group. Hence, the set Γ_2 contains $c_{m-2-(p-1)}(G/D)$ members that are L_{p-1}-groups. Since p divides $c_{m-2-(p-1)}(G/D)$ (G/D is regular nonmetacyclic of exponent $> p$ and $m-2-(p-1) \geq 2$), it follows that the number of L_{p-1}-subgroups in the set Γ_2 is divisible by p, and their contribution in $\sum_{H\in\Gamma_2} \alpha(H)$ is divisible by p, so (4) is true, completing the proof. □

We omit a fairly complicated proof of the following

Theorem B($p-2$). *Suppose that G is a group of order p^m, $m > n > p-1$, $\mathcal{M} = L_{p-2,n}(G)$, $\alpha(G) = |\mathcal{M}|$. Then one of the following holds:*

(a) G is an L_p-group, $\alpha(G) \equiv p^2 \pmod{p^3}$.

(b) $p = 3 = n$, G is a 3-group of maximal class, $\alpha(G) \equiv 3^2 \pmod{3^3}$.

(c) $\alpha(G) \equiv 0 \pmod{p^3}$.

Lemma 18.4 (= Proposition 4.9). *Let G be a 2-group. If the subgroup $\Phi(G)$ is of type $(2,2)$, then $\Phi(G) \leq Z(G)$.*

Exercise 3. Suppose that a 2-group G of order $> 2^4$ is such that G' is cyclic and G/G' is abelian of type $(4,2)$. Then Aut(G) is a 2-group and (a) $\Phi(G)$ is abelian, (b) G has a cyclic subgroup of index 4, (c) G is a U_2-group.

Proposition 18.5. *Suppose that a 2-group G is not of maximal class. Then, for all i, the number of elements of maximal class in the set Γ_i is a multiple of 4.*

Proof. Assume that the set Γ_i^0 of members of maximal class in the set Γ_i is nonempty. In view of Theorem 5.4, we may assume that $i > 1$. Since $\Phi(G)$ is not of maximal class (Proposition 1.13), we have $d = d(G) > i$.

(i) Suppose that $|\Phi(G)| = 2$. Then $i = d-2$ since groups of maximal class are two-generator. If L is a nonabelian subgroup of order 8 in G, then $\Phi(L) = \Phi(G)$ and $L \in \Gamma_i$. Hence we must only prove that the number of nonabelian subgroups of order 8 is a multiple of 4. This is true, by Proposition 2.3.

(ii) Suppose that $|\Phi(G)| > 2$ and $i = d-1$; then $d > 2$. Assume that $\Phi(G)$ is noncyclic. Let $U < \Phi(G)$ be G-invariant such that $\Phi(G)/U \cong E_4$; then $\Phi(G)/U \leq Z(G/U)$ (Lemma 18.4) so Γ_i^0 is empty, a contradiction. Thus, $\Phi(G)$ is cyclic. Let A be a subgroup of order 4 in $\Phi(G)$; then $|G : C_G(A)| = 2$ since $\Gamma_i^0 \neq \varnothing$. Let $T/\Phi(G) < G/\Phi(G)$ be of order 2 such that $T \not\leq C_G(A)$. Then $C_G(A)/\Phi(G)$ has $3+4k$ subgroups of order 2. Since $G/\Phi(G)$ has $3+4k_1$ subgroups of order 2, then the number of subgroups of order 2 in $G/\Phi(G)$ which are not contained in $C_G(A)/\Phi(G)$, equals $4(k_1-k) \equiv 0 \pmod 4$. Thus we can choose T in $4(k_1-k)$ ways. Now, by the choice, $A \not\leq Z(T)$, and T contains the cyclic subgroup $\Phi(G)$ of index 2. If $|\Phi(G)| = 4$, then T is of maximal class. If $|\Phi(G)| > 4$, then $A \leq \Phi(\Phi(G))$ so T is of maximal class by Theorem 1.2 since T/A is noncyclic. Thus, 4 divides $|\Gamma_i^0|$.

(iii) Suppose that $|\Phi(G)| > 2$ and $d - 1 > i$. Then $i = d - 2$, $d > 3$ and $\Phi(G)$ is cyclic since $\Gamma_i^0 \neq \varnothing$. Let D be a subgroup of index 2 in $\Phi(G)$. Then by (i), G/D has $4k$ nonabelian subgroups of order 8, and let L/D be one of them. Because $D = \Phi(\Phi(G)) \leq \Phi(L)$, it follows that $\Phi(L) = \Phi(G)$. Since $\Phi(L) = \mho_1(L)$, there is in L a cyclic subgroup of index 2. Since L/D is nonabelian, it follows from Theorem 1.2 that L is of maximal class, and the proof is complete. □

Remark 1. Let G be a U_s-group of order 2^m with kernel $R \cong \mathrm{E}_{2^s}$, $s > 1$. Let T/R be a cyclic subgroup of index 2 in G/R. Then, if $k > 3$, we have $\mathrm{c}_k(G) = \mathrm{c}_k(T) = 2^{s-1}$. Now let $k = 3$. Set $|G/R| = 2^{n+1}$, where $n + 1 = m - s$. If G/R is dihedral, then all elements in $G - T$ have order ≤ 4 so $\mathrm{c}_3(G) = \mathrm{c}_3(T) = 2^{s-1}$. Let $G/R \cong \mathrm{Q}_{2^{n+1}}$, $n \geq 2$. Then all elements in $G - T$ have order 8 so $\mathrm{c}_3(G) = \mathrm{c}_3(T) + \frac{|G-T|}{\varphi(8)} = 2^{s-1} + 2^{m-3} \equiv 2^{s-1} \pmod{2^s}$. Now let $G/R \cong \mathrm{SD}_{2^{n+1}}$, $n \geq 3$. Let $M/R \cong \mathrm{Q}_{2^n}$ be maximal in G/R. Then $\mathrm{c}_3(G) = \mathrm{c}_3(M) = 2^{s-1} + 2^{m-4} \equiv 2^{s-1} \pmod{2^s}$. If $N/R \cong \mathrm{D}_{2^n}$ is maximal in G/R, then $\mathrm{c}_2(G) = \mathrm{c}_2(N) \equiv 2^{s-1} \pmod{2^3}$.

A subgroup H of a p-group G is said to be *k-good* ($k \in \mathbb{N}$) if $\exp(H) = p$ and $\exp(\Omega_1(\langle x, H\rangle)) = p$ for every $x \in G$ of order p^k. If H is k-good in G and $H < F < G$, then H is k-good in F. Let $N(p^k, G) = |\mathrm{sol}_k(G)|$ be the number of solutions of $x^{p^k} = 1$ in G.

Theorem 18.6. *Let $n > 1$ and $k > 2$. Suppose that a p-group G of exponent $\geq p^k$ has a k-good normal subgroup R of order p^n. Then p^{n+k} divides $N(p^k, G)$ and p^n divides $\mathrm{c}_k(G)$, unless G is an L_n- or U_n-group.*

Proof. Suppose that G is a counterexample of minimal order.

(i) Suppose that G/R is cyclic. Since G is not an L_n-group and $\exp(\Omega_1(G)) = p$ (in fact, $\Omega_1(G) \leq RC$, where $C < G$ is cyclic of order p^k), it follows that $|\Omega_1(G)| = p^{n+1}$. Hence, $N(p^k, G) = |\Omega_k(G)| = p^{n+k}$, $\mathrm{c}_k(G) = p^n$. and G is not a counterexample. Thus, G/R is not cyclic.

(ii) Suppose that G/R is a 2-group of maximal class. Let $T/R < G/R$ be cyclic of index 2. Since G is not a U_n-group, we get $\Omega_1(T) \cong \mathrm{E}_{2^{n+1}}$. It follows from the structure of G/R that all elements outside T satisfy $x^8 = 1$. Since $k > 2$, we have $N(2^k, G) = N(2^k, T) + |G - T|$. Next, $N(2^k, T) = 2^{n+k}$ so 2^{n+k} divides $N(2^k, G)$ since $|G - T| = \frac{1}{2}|G|$ is divisible by 2^{n+k}, and G is not a counterexample. Thus G/R is not a 2-group of maximal class.

It follows from (i) that G/R has a normal subgroup H/R such that G/H is of type (p, p). Let $M_1/H, \dots, M_{p+1}/H$ be all subgroups of order p in G/H. Then

$$N(p^k, G) = N(p^k, M_1) + \cdots + N(p^k, M_{p+1}) - p \cdot N(p^k, H). \tag{5}$$

We may assume that $\exp(G) \geq p^{k+1}$; then $|G| \geq p^{n+k+1}$ since $\exp(G/R) \geq p^k$ and G/R is noncyclic, by (i). Next, $|H| \geq p^{n+k-1}$ and $\exp(H) \geq p^k$. Since R

is k-good in H, it follows that p^{n+k} divides $pN(p^k, H)$ (in fact, this is true if H is L_n- or U_n-group; if not, this follows by induction). Therefore, by assumption, $p^{n+k} \nmid N(p^k, M_i)$ for some i. By induction, M_i is an L_n- or U_n-group.

Suppose that M_i is an L_n-group. In view of Theorem 1.2 and (ii), one may assume that $M_1/R, \dots, M_p/R$ are cyclic and M_{p+1}/R is noncyclic abelian with cyclic subgroup of index p. By induction, $N(p^k, M_{p+1}) \equiv 0 \pmod{p^{n+k}}$. It is easy to check that $M_1, \dots, M_p$ are L_n-groups. By (i), $\sum_{i=1}^{p} N(p^k, M_i) = p^{n+k}$. It follows from (5) that p^{n+k} divides $N(p^k, G)$.

Let M_i be a U_n-group. We may assume that $i = 1$. Let T_1/R be a cyclic subgroup of index 2 in M_1/R. By definition, $\Omega_1(T_1) = R$. By Theorem 5.4, we may assume that $M_1/R, \dots, M_4/R$ are of maximal class and three remaining maximal subgroups $M_5/R, M_6/R, M_7/R$ of G/R are neither cyclic nor of maximal class so their inverse images are neither L_n- nor U_2-groups. As above, the first four M_i's are U_n-groups. If $j > 4$, then, by induction, 2^{n+k} divides $N(2^k, M_j)$. If $i \le 4$, then $N(2^k, M_i) \equiv 2^{n+k-1} \pmod{2^{n+k}}$, and so 2^{n+k} divides $N(2^k, G)$, by (5). To prove the last assertion on $\mathrm{c}_k(G)$, we have to repeat, word for word, the previous part of the proof. □

Remark 2. Let G be a 2-group and let $H \in \Gamma_1$ be of maximal class. Then H has a G-invariant cyclic subgroup T of index 2. We claim that T is contained in exactly two subgroups of maximal class and order $2|T|$. One may assume that G has no cyclic subgroups of index 2 (otherwise, G is of maximal class, by Theorem 1.2, and we are done since then $T = \Phi(G)$). Let $U < T$ be of index 4. Since H/U is nonabelian, G/U is not metacyclic. Indeed, otherwise $\exp(G/U) = 8$ so G/U has a cyclic subgroup F/U of index 2. Since $U < \Phi(T) \le \Phi(F)$, it follows that F is cyclic, a contradiction. In particular, $G/T \cong \mathrm{E}_4$. Let $H/T = H_1/T, H_2/T, H_3/T < G/T$ be distinct of order 2. Since G/U (of order 2^4) has an abelian subgroup of index 2, one may assume that H_3/U is abelian. It remains to show that H_2 is of maximal class. Since $T/U \not\le \mathrm{Z}(G/U)$, it follows that H_2/T is nonabelian. Since H_2 has a cyclic subgroup T of index 2, it follows that H_2 is of maximal class (Theorem 1.2).

Corollary 18.7 ([Ber14]). *Conjecture* B(1) *is true. In other words, if an irregular p-group G is not of maximal class, $k > 2$, then $\mathrm{c}_k(G) \equiv 0 \pmod{p^p}$, unless G is an L_p- or U_2-group.*

Proof. Assume that G is neither an L_p- or U_2-group. We proceed by induction on $|G|$. By Theorem 12.1(a), there is $R \lhd G$ of order p^p and exponent p. If G/R is cyclic, then $\Omega_1(G)$ is of order p^{p+1} and exponent p and $\mathrm{c}_k(G) = p^p$. In what follows we assume that G/R is not cyclic. Then G/R contains a normal subgroup T/R such that $G/T \cong \mathrm{E}_{p^2}$. Let $H_1/T, \dots, H_{p+1}/T$ be all maximal subgroups of G/T. We have $\mathrm{c}_k(G) = \sum_{i=1}^{p+1} \mathrm{c}_k(H_i) - p\mathrm{c}_k(T)$. One may assume that $\exp(G) \ge p^k$. We claim that $\mathrm{c}_k(T) \equiv 0 \pmod{p^{p-1}}$. Assume that this is false; then $\exp(T) \ge p^k$. In that case, T is irregular of maximal class (Theorem 13.2(b)) so $|T| = p^{p+1}$ and

$\exp(T) = p^2 < p^k$, a contradiction. It follows that $pc_k(T) \equiv 0 \pmod{p^p}$. It remains to prove that

$$(6) \qquad \sum_{i=1}^{p+1} c_k(H_i) \equiv 0 \pmod{p^p}.$$

Assume that (6) is not true. Then $p^p \nmid c_k(H_i)$ for some i. We may assume that $i = 1$. Taking into account that $R < H_1$, we must consider the following two possibilities: (i) H_1 is an L_p-group, (ii) H_1 is a U_2-group.

(i) Suppose that H_1 is an L_p-group; then $\Omega_1(H_1) = R$. Assume that $H_1/R, \dots, H_p/R$ are cyclic and H_{p+1}/R is abelian of type (p^n, p). Since $k > 2$, $K/R := \Omega_1(H_1/R) \le \Phi(G/R) < H_i/R$ so $R = \Omega_1(H_1) = \Omega_1(K) = \Omega_1(H_i)$, and we conclude, that H_i is an L_p-group for $i = 2, \dots, p$. It follows that for the same i we have $c_k(H_i) = p^{p-1}$. By induction, $c_k(H_{p+1}) \equiv 0 \pmod{p^p}$ so (6) is true. It remains to consider the case where G/R is a 2-group of maximal class (Theorem 1.2). Since $\Omega_1(H_1) = R$ and H_1/R is a cyclic subgroup of index 2 in G/R, we conclude that G is a U_2-group, contrary to the hypothesis.

(ii) Now suppose that H_1 is a U_2-group; then $p = 2$. Since G is not a U_2-group, we conclude that G/R is not of maximal class. Let T/R be a G-invariant cyclic subgroup of index 2 in H_1/R. By Remark 2, one may assume that H_2/R is of maximal class and H_3/R is not of maximal class. It follows from $\Omega_1(T) = R$ (indeed, $T < H_1$) that H_2 is a U_2-subgroup. Clearly, H_3 is not an L_2-subgroup since G/R has no cyclic subgroups of index 2. We have, by induction, $c_k(H_3) \equiv 0 \pmod 4$. Since $c_k(H_i) \equiv 2 \pmod 4$ $(i = 1, 2)$, by Remark 1, we get $c_k(G) \equiv c_k(H_1) + c_k(H_2) + c_k(H_3) \equiv 2 + 2 + 0 \equiv 0 \pmod 4$, so (6) is true. □

Thus, Conjecture B(s) is true for 2- and 3-groups and all s.

Exercise 4. Suppose that a p-group G has a 2-good normal subgroup R of order $p^n > p$ and exponent p. Then p^n divides $N(p, G)$.

Solution. Let $x \in G - R$ be an element of order p. Then the subgroup $\Omega_1(\langle x, R\rangle) = p$ since R is 2-good (check that $\exp(\langle x, R\rangle) = p$). Hence, such element x produces $p^n(p-1)$ solutions of equation $y^p = 1$ not contained in R. We see that the set of solutions of $y^p = 1$ in $G - R$ is a disjoint union of the sets $\langle x, R\rangle - R$ of cardinality $p^n(p-1)$ and R so p^n divides $N(p, G)$.

Corollary 18.8. *Let a 2-group G possess an elementary abelian subgroup E of order 8, $k > 3$. Then one of the following holds:*

(a) $c_k(G) \equiv 0 \pmod 8$,

(b) G is either an L_3*- or* U_3*-group.*

Exercise 5. Let G be a p-group and $H \triangleleft G$ be of order p^p and exponent p. Show that H is k-good in G for all $k > 2$.

Exercise 6. Let G be a p-group and $H \triangleleft G$ be of exponent p, G/H is cyclic of order $> p$, $|H| < p^{2p-1}$ and $|H'| < p^{p-1}$. Then $\exp(\Omega_1(G)) = p$.

Solution. Let T be a G-invariant subgroup of H containing H' and such that $|H : T| = p^p$ if $|H : H'| > p^p$ and $T = H'$ if $|H : H'| \le p^p$. Note that $\exp(\mathrm{Aut}(H/T))_p = p$ (see the paragraph, preceding Exercise 1.42). Therefore, setting $C = C_G(H/T)$, we get $|G : C| \le p$. The group C/T is abelian as an extension of the central subgroup H/T by the cyclic group C/H. It follows that $\mathrm{cl}(C) \le p-1$ so C is regular (Theorem 7.1). Then $\exp(\Omega_1(C)) = p$. Since $C > H$ in view of $|G : C| \le p < |G : H|$, we get $\Omega_1(G) \le C$.

Exercise 7. Let H be an absolutely regular normal subgroup of a p-group G such that G/H is cyclic of order $> p$. Prove that $\exp(\Omega_1(G)) = p$. (*Hint.* Use Theorem 12.1(b).)

Exercise 8. Let G be a U_2-group of order 2^{n+3} with kernel R. Then

(a) G' is cyclic of index 8 in G so $|G'| = 2^n$.

(b) If $R \not\le \Phi(G)$, then $\Phi(G) = G'$.

(c) If $R < \Phi(G)$, then $\Phi(G)$ is abelian of type $(2^n, 2)$.

(d) G has a cyclic subgroup of index 4.

(e) G is metacyclic if and only if $\mathrm{d}(G) = 2$ and G has a normal cyclic subgroup of index 4.

(f) If G is metacyclic and G/R is dihedral, then $R \le \mathrm{Z}(G)$.

(g) If G has a nonnormal cyclic subgroup of index 4, then $\mathrm{d}(G) = 2$.

(h) G has exactly two normal cyclic subgroups of index 8. Moreover, if $n + 1 \ge k > 3$, then $c_k(G) = 2$.

(i) Suppose that $R \not\le \Phi(G)$. Then exactly four maximal subgroups of G, not containing R, are of maximal class. In that case, G/R is dihedral.

(j) $\mathrm{Aut}(G)$ is a 2-group.

Solution. Let $T/R < G/R$ be cyclic of index 2.

(a) Since G is not of maximal class, $R \not\le G'$ (Taussky). Since R centralizes G', RG', as a subgroup of T, is abelian of type $(2^n, 2)$ so G' is cyclic. We have $|(G/R) : (G/R)'| = 4$ and $(G/R)' = G'R/R = \Phi(G/R)$. If $R \le \Phi(G)$, we get $\Phi(G) = RG'$ so, since $|R \cap G'| = 2$, we get $|G : G'| = 8$.

(b) In that case, $|G : G'| = 8 = |G : \Phi(G)|$ so $\Phi(G) = G'$.

(c) In that case, $\Phi(G) = RG' < T$ and R centralizes $\Phi(G)$. Since $\Omega_1(T) = R$ and $\Phi(G) < T$, $\Phi(G)$ is abelian of type $(2^n, 2)$.

(d) Since $\Phi(G) = \mho_1(G)$ and every set of generators of $\Phi(G)$ contains an element of order 2^n, we get $\exp(G) = 2^{n+1}$.

(e) If a cyclic subgroup Z of index 4 is normal in G, then $Z > G'$ and $Z \neq \Phi(G)$ (otherwise, G contains a cyclic subgroup of index 2) so G/Z is cyclic and G is metacyclic. If G has a normal cyclic subgroup C such that G/C is cyclic, then $|G : C| = 4$ since G/G' is abelian of type $(4, 2)$.

(f) Let L/R be a nonnormal subgroup of order 2 in G/R. Since G is not of maximal class, L must be abelian (Proposition 10.19) so $L \leq C_G(R)$. Since G/R is generated by nonnormal subgroups of order 2, it follows that $R \leq Z(G)$.

(g) Let $Z < G$ be nonnormal cyclic of index 4. The isomorphism $G/Z_G \cong D_8$ is trivial. The second assertion follows from $Z_G \leq \Phi(Z) \leq \Phi(G)$.

(h) If Z is a normal cyclic subgroup of order 2^n in G or a cyclic subgroup of G of order $2^k > 2^3$, then $RZ < T$, and the result follows since $c_n(T) = 2$.

(i) We have $d(G) = 3$. Then $G/\eta(G) \cong E_{p^2}$ so there is $H \in \Gamma_1$ such that $G = H\eta(G)$. Then $cl(H) = cl(G) = n + 1$ so H is of maximal class. By Theorem 13.5, there is in Γ_1 exactly four members of maximal class.

(j) Let $d(G) = 3$. Assume that $\mu \in Aut(G)$ is of prime order $p > 2$. Then there is $H \in \Gamma_1$ of maximal class such that $H^\mu = H$. Since $|H| > 8$, we get $\mu_H = id_H$. Since μ stabilizes the chain $G > H > \{1\}$, we get $\mu = id_G$, a contradiction. This also true if $R < \Phi(G)$ since then G/G' is abelian of type $(4, 2)$.

Exercise 9. For a 2-group G of order $> 2^4$ the following conditions are equivalent: (a) G is a U_2-group with $d(G) = 2$, (b) G' is cyclic and G/G' is abelian of type $(4, 2)$.

§19

Some additional counting theorems

1°. We begin with the following

Definition. A p-group G is said to be a $\mathrm{U}^{(p)}$-group, if it satisfies the following conditions: ($1\mathrm{U}^{(p)}$) G has a normal subgroup R of order p^p and exponent p. ($2\mathrm{U}^{(p)}$) G/R is a group of maximal class and order $> p^{p+1}$. ($3\mathrm{U}^{(p)}$) If $T/R < G/R$ is the fundamental subgroup, then $\Omega_1(T) = R$. Let $\mathrm{U}_n^{(p)}(G)$ denote the set of all $\mathrm{U}^{(p)}$-subgroups of order p^n in G.

If G and R are as in the definition (R is said to be the *kernel* of G), then R contains all G-invariant subgroups of exponent p. Indeed, if $R_1 \lhd G$ is minimal such that $\exp(R_1) = p$ and $R_1 \not\leq R$, then $|RR_1/R| = p$ so $RR_1/R < \Phi(G/R) < T/R$, a contradiction since $\Omega_1(T) = R$. The U_2-groups of order $> 2^5$ (see §§17,18) are $\mathrm{U}^{(2)}$-groups.

Remark. Let G be a $\mathrm{U}^{(p)}$-group with kernel R. Assume that $R \not\leq \Phi(G)$. Then $G = RM$ with some $M \in \Gamma_1$ and M has no G-invariant subgroups of order p^p and exponent p. Then M is of maximal class (Theorems 12.1(a) and 13.5) so $\mathrm{d}(G) = 3$ and the set Γ_1 contains exactly p^2 members of maximal class (Theorem 13.6).

Theorem $\mathrm{U}^{(p)}$. *Let G be a p-group of order p^m, $2p+1 < n < m$. Set $\mathcal{M} = \mathrm{U}_n^{(p)}(G)$, $\alpha(G) = |\mathcal{M}|$. Then (a) p divides $\alpha(G)$, and (b) p^2 divides $\alpha(G)$, unless G is a $\mathrm{U}^{(p)}$-group and $n = m - 1$.*

Proof. We use induction on $|G|$. One may assume that $\alpha(G) > 0$.

Let G be a $\mathrm{U}^{(p)}$-group with kernel R. It is easy to see that if $R < L < G$ and L/R is of maximal class and order $> p^{p+1}$, then H is a $\mathrm{U}^{(p)}$-group. The group G/R has exactly p^i subgroups of maximal class and index p^i, where $m - p - i \geq p + 1$; so, to compute $\alpha(G)$, it suffices to show that $R < H$ for every $H \in \mathcal{M}$. First suppose $n = m - 1$. Since the kernel R_1 of H is characteristic in H so normal in G, we get $R = R_1 < H$. Now let $n < m - 1$ and let $H < M \in \Gamma_1$. Then M is not of maximal class (Theorem 9.6) so $R < M$, by Theorem 13.5. Since M/R is not absolutely regular (indeed, $H/R < M/R$), M is a $\mathrm{U}^{(p)}$-group. By induction in M, $R < H$. Then p divides $\alpha(M)$, and the result follows, by Theorem 5.2. Next we assume that G is not a $\mathrm{U}^{(p)}$-group.

(i) Let $n = m - 1$, i.e., $\mathcal{M} \subset \Gamma_1$. Take $H \in \mathcal{M}$. Let D be the kernel of H. Since D is characteristic in H, we get $D \triangleleft G$.

Assume that G/D is of maximal class and let T/D be its fundamental subgroup. We will prove that then G is a $\mathrm{U}^{(p)}$-group. To this end, we have to show that $\Omega_1(T) = D$. The intersection $T_1/D = (T/D) \cap (H/D)$ is absolutely regular of index p in H/D hence it is the fundamental subgroup in H/D, and so $\Omega_1(T_1) = D$ since H is an $\mathrm{U}^{(p)}$-group. Since $\Omega_1(T/D) = \Omega_1(T_1/D)$ and $\Omega_1(T/D) \leq \Phi(H/D) < T_1/D$, we get $\Omega_1(T) = \Omega_1(T_1) = D$ so G is a $\mathrm{U}^{(p)}$-group.

Now suppose that G/D is not of maximal class. By Theorem 13.6, the number of subgroups of maximal class and index p in G/D equals p^2 since $\mathrm{d}(G) = 3$ (Theorem 12.12(a))). Let F/D be a subgroup of maximal class and index p in G/D. We claim that F is a $\mathrm{U}^{(p)}$-group. Let T_1/D be the fundamental subgroup of F/D. To prove our claim, one must show that $\Omega_1(T_1) = D$. We have $\Omega_1(T_1/D) \leq \Phi(F/D) \leq \Phi(G/D) < H/D$, so $\Omega_1(T_1/D) \leq \Omega_1(T/D)$, where T/D is the fundamental subgroup of H/D. It follows that $\Omega_1(T_1) \leq \Omega_1(T) = D$, as claimed. Thus, the set Γ_1^D of all members of the set Γ_1 containing D, has exactly p^2 members which are $\mathrm{U}^{(p)}$-groups; denote that set $\mathcal{M}_D$. Suppose that $D_1 = D, D_2, \ldots, D_k$ are normal subgroups of G of order p^p and exponent p. Then $\Gamma_1^{D_i} \cap \Gamma_1^{D_j} = \varnothing$ $(i \neq j)$ since each U^p-subgroup has only one normal subgroup of order p^p and exponent p (see the Remark). It follows that $|\mathcal{M}| = |\sum_{i=1}^k |\mathcal{M}_{D_i}| \equiv 0 \pmod{p^2}$.

(ii) If $n < m - 1$, the result follows by induction on $|G|$ with help of Theorem 5.2. □

2°. By Theorem B(1), if $n > 2$, then, as a rule, p^p divides $\mathrm{c}_n(G)$. For $n = 2$, the situation is more complicated.

Theorem 19.1. *Suppose that G is an irregular group of order p^m, $p > 2$, $m > p + 3$ and $\mathrm{c}_2(G) < p^p$. Let F be a normal abelian subgroup of type (p, p) in G and $\mathrm{C}_G(F) = L$. If G is neither an L_p-group nor a 3-group of maximal class, then the following assertions hold: (a) $\exp(L) = p$, $|G : L| = p$ so $\exp(G) = p^2$, (b) $|\mathrm{Z}(G)| = p$ and $\mathrm{Z}(G) = \mho_1(G)$, (c) $p > 3$ and $m \leq 2p$.*

Proof. (i) We claim that if $p > 3$, then G is not of maximal class. Indeed, let G be a p-group of maximal class and G_1 its fundamental subgroup. Taking into account that G_1 is absolutely regular and the properties of p-groups of maximal class (see Theorem 9.6), we get

$$\mathrm{c}_2(G_1) = \frac{|\Omega_2(G_1) - \Omega_1(G_1)|}{p(p-1)} \geq \frac{p^{p-1+3} - p^{p-1}}{p(p-1)} = p^{p-2}(p^2 + p + 1) > p^p,$$

a contradiction. If $p = 3$, then $\mathrm{c}_2(G_1) = \frac{3^4 - 3^2}{3(3-1)} = 12 < 3^3$, so some 3-groups of maximal class satisfy the hypothesis. In what follows we assume that G is not of maximal class. Then, by Theorem 12.1(a),

(ii) There is $R \triangleleft G$ of order p^p and exponent p.

(iii) If $\Omega_1(G) = R$, then G is an L_p-group. Indeed, assume that G/R contains a subgroup H/R of type (p, p). Then

$$c_2(H) = \frac{|H - R|}{p(p-1)} = \frac{p^{p+2} - p^p}{p(p-1)} = p^{p-1}(p+1) > p^p,$$

a contradiction. In our case, therefore, G/R is cyclic, as was to be shown.

(iv) We claim that $\Omega_1(G) > R$ and G/R is not cyclic. The first part of that assertion follows from (iii) since G is not an L_p-group. Suppose that G/R is cyclic. Then $|\Omega_1(G)| = p^{p+1}$ so $|\Omega_2(G)| = p^{p+2}$, and we get $c_2(G) = \frac{p^{p+2}-p^{p+1}}{p(p-1)} = p^p$, a contradiction.

(v) If $C < G$ is cyclic of order p^2, then $C \cap R > \{1\}$ and $\mho_1(CR) = \mho_1(C)$. Assume that $C \cap R = \{1\}$. Then $\Omega_1(CR)$ is of order p^{p+1} and exponent p (Lemma 17.4(a)). In that case, as in (iv), $c_2(CR) = p^p$, a contradiction. Suppose that $C \cap R > \{1\}$. Then $|CR| = p^{p+1}$, and the second assertion is obvious.

(vi) If $C < G$ is cyclic of order p^2, then $t = c_1(N_G(C)) < p^p$. Indeed, let C_0 be a subgroup of order p in C. Take a subgroup Z of order p in $N_G(C)$ distinct of C_0. Then $|CZ| = p^3$, $c_1(CZ) = p + 1$, $c_2(CZ) = p$. We see that p subgroups of order p in CZ, which $\neq C_0$, produce $p - 1$ cyclic subgroups of order p^2 in $N_G(C)$ which $\neq C$. Let Z_1 be a subgroup of order p in $N_G(C)$ such that $Z_1 \not\leq CZ$. Then $CZ \cap CZ_1 = C$ and the new p subgroups of order p in CZ_1 produce the new $p - 1$ cyclic subgroups of order p^2 (contained in CZ_1). We shall continue this process until all subgroups of order p in $N_G(C)$ will be exhausted. Therefore,

$$\frac{t-1}{p} \cdot (p-1) \leq c_2(N_G(C)) - 1 \leq c_2(G) - 1 < (p-1)p^{p-1},$$

since p^{p-1} divides $c_2(G)$ (Theorem 13.2(b)). Thus, $t \leq p^p$. Moreover, by Kulakoff's Theorem 5.3, $t \equiv 1 + p \pmod{p^2}$, and so $t < p^p$, as was to be shown.

(vii) We claim that $Z(G)$ is cyclic, and this is true even for $m > p + 2$. Indeed, assume that $Z(G)$ has a subgroup $T = Z_1 \times Z_2$, where subgroups Z_1, Z_2 are of order p. Since $\exp(TR) = p$ and $TR \triangleleft G$, one may assume that $T < R$. Let $Z_1 \not\leq C$, where $C < G$ is cyclic of order p^2. Since $C \cap R > \{1\}$ by (v), it follows that $|CR| = p^{p+1}$, and so CR/Z_1 is absolutely regular (since $\exp(CR/Z_1) = p^2$ and $|CR/Z_1| = p^p$). Let f be the number of absolutely regular subgroups of G/Z_1 of order p^p, containing R/Z_1, and let K/Z_1 be one of them; then K is regular since $\mathrm{cl}(K) \leq p - 1$. By Theorem 7.2, $|\Omega_1(K)| = R$, and so $c_2(K) = \frac{|K-\Omega_1(K)|}{p(p-1)} = p^{p-1}$. It follows from $p^p > c_2(G) \geq f \cdot p^{p-1}$ that $f < p$; in particular, $p \nmid f$. Then G/Z_1 is either absolutely regular or of maximal class and $f = 1$ (Theorem 13.5).

Suppose that G/Z_1 is absolutely regular. Then G is regular (Remark 7.2) so $\Omega_1(G) = R$. In that case, by (iii), G is an L_p-group, contrary to the hypothesis.

Let G/Z_1 be of maximal class. Applying the previous argument to G_1/Z_1, where G_1/Z_1 is the fundamental subgroup of G/Z_1, we obtain that G_1/R is cyclic, which is impossible, since every epimorphic image of G/Z_1 of order p^p is of prime exponent.

(viii) Let $E_{p^2} \cong F \triangleleft G$ and $L = C_G(F)$; then $|G : L| = p$, by (vii). Assume that $\exp(L) > p$. Since $|L| > p^{p+2}$ and $c_2(L) \le c_2(G) < p^p$, it follows from (vii) that L is regular. Since $|L \cap R| \ge p^{p-1}$, we get $|\Omega_1(L)| \ge p^{p-1}$. Since L is regular, $|\Omega_1(L)| \le p^p$ and $\Omega_2(L) \le p^{p+1}$, by Theorem 7.2 (otherwise $c_2(L) \ge p^p$).

Assume that $|\Omega_1(L)| = p^{p-1}$. Then, by Theorem 12.1(b), $|\Omega_1(G)| = p^p$ (since L is absolutely regular), contrary to (iv).

Hence, $|\Omega_1(L)| = p^p$. One may assume that $R = \Omega_1(L)$. It follows from $|\Omega_2(L)| = p^{p+1}$ that L/R is cyclic so L is an L_p-subgroup. By Lemma 17.4(b), $\mho_1(L)$ is cyclic, and its index in L is p^p. Since $|L| > p^{p+2}$, we get $|\mho_1(L)| > p^2$. Since $\Omega_2(\mho_1(L)) \triangleleft G$ is cyclic of order p^2, we get $c_1(G) < p^p$, by (vi). Then, by Theorem 13.2(a), $c_1(G) = 1 + p + \cdots + p^{p-1}$ so $\Omega_1(G) = R$, contrary to (iv). Thus, $\exp(L) = p$ so $\exp(G) = p^2$.

(ix) G has no normal cyclic subgroups of order p^2, by (vi).

(x) By (vii) and (ix), $Z = Z(G)$ is of order p. We claim that $Z = \mho_1(G)$. Assume that this is false. Then $\exp(G/Z) = p^2$, by (viii). We claim that G/Z is irregular, otherwise,

$$c_2(G) > c_2(G/Z) = \frac{p^{m-1} - p^{m-2}}{p(p-1)} = p^{m-3} > p^p,$$

a contradiction. Since $\exp(G) = \exp(G/Z) = p^2$, it follows that G has a cyclic subgroup C of order p^2 that does not contain Z. Set $T = CR$. By (v), $|T| = p^{p+1}$. Since $Z \ne \mho_1(C) = \mho_1(T)$ (Lemma 17.4(b)), it follows that $|Z(T)| \ge p^2$ (recall that $Z < R < T$), and so T is regular. In that case, $c_2(T) = p^{p-1}$. Since $Z \ne \mho_1(T)$, the quotient group T/Z is absolutely regular. Let f be the number of absolutely regular subgroups H/Z of order p^p in G/Z such that $R/Z < H/Z$ (by Remark 7.2, all such subgroups H are regular). By Corollary 13.3, p divides f (here we use the existence of L; see (viii)). Since $f > 0$ (in view of the existence of T/Z), it follows that $f \ge p$. Since $c_2(G) \ge f \cdot p^{p-1} \ge p^p$, we obtain a contradiction. Thus, $Z = \mho_1(G)$.

(xi) We claim that $m \le 2p$ and $p > 3$. Indeed, let $C_{p^2} \cong C < G$. Set $N = N_G(C)$. Then $|G : N| \le c_2(G) < p^p$ so $|G : N| \le p^{p-1}$. Since $N \cap L$ is of exponent p, by (viii), and normalizes C, it follows that $|N \cap L| \le p^p$, by (vi). Therefore, $|N| \le p^{p+1}$. Thus,

$$p^m = |G| = |N| \cdot |G : N| \le p^{p+1} \cdot p^{p-1} = p^{2p},$$

and so $m \le 2p$. Then $2p \ge m > p + 3$, and so $p > 3$. □

Problem. Suppose that an irregular p-group G has a subgroup of index p and exponent p. Study the structure of G if $c_2(G) \le p^p$.

§20

Groups with small abelian subgroups and partitions

In this section we consider the p-groups G such that whenever A is an abelian subgroup of G, then AZ/Z is cyclic, where $Z = \mathrm{Z}(G)$. They are called, according to [Hei1], p-groups with small abelian subgroups. There are many interesting p-groups with small abelian subgroups (many such groups were constructed in [Hei1]). Main result of this section, which is due to Mann [Man8], depends on some properties of groups admitting nontrivial partitions.

Definition 1. Let $\Sigma = \{T_1, \dots, T_k\}$ be a set of nonidentity subgroups of a group $G > \{1\}$. We say that Σ is a *partition* of G if $G^\# = \bigcup_{i=1}^k T_i^\#$ is the set-theoretic partition. A partition Σ is *nontrivial* if $|\Sigma| > 1$. Subgroups $T_1, \dots, T_k$ are called *components* of Σ.

Definition 2. Let G be a group. A subgroup $\mathrm{H}_p(G) = \langle x \in G \mid o(x) \neq p\rangle$ is called the H_p-*subgroup* of G.

Lemma 20.1. *Let Σ be a nontrivial partition of a p-group G of exponent $> p$.*

(a) (Baer) Suppose that $T \in \Sigma$ be such that $z \in (T \cap \mathrm{Z}(G))^\#$ is of order p. If $B \in \Sigma - \{T\}$, then $\exp(B) = p$. Next, $\exp(\mathrm{Z}(G)) = p$, Σ has exactly one component of exponent $\neq p$, say U, and $\mathrm{H}_p(G)\mathrm{Z}(G) \leq U$.

(b) Let G be a p-group and $T \in \Sigma$ is either cyclic of composite order or a 2-group of maximal class. Then $p = 2$, T is cyclic, $|G : T| = 2$ and G is dihedral.

Proof. (a) Assume that $x \in B \in \Sigma - \{T\}$ is of order $> p$. Then $1 \neq (xz)^p = x^p \in B^\#$ so $xz \in B$, whence $1 \neq z \in B \cap T$, a contradiction. Thus, $\exp(B) = p$ so $\mathrm{H}_p(G) \leq T$. If $\exp(\mathrm{Z}(G)) > p$, then $\mathrm{Z}(G) \leq T$, by what has just been said, and $\Omega_1(G) \not\leq T$ since $T < G$. Let $y \in \mathrm{Z}(G)^\#$ be of order $> p$. Then, if $u \in G$ is of order p, then $o(uy) > p$ so $uy \in T$, and we conclude that $u \in T$, a contradiction since $\Omega_1(G) = G$. Thus, $\exp(\mathrm{Z}(G)) = p$. If $u \in T$ is of order $> p$ and $v \in \mathrm{Z}(G)^\#$, then $o(uv) > p$ so $uv \in T$, and we conclude that $v \in T$. Thus, $\mathrm{Z}(G) < T$.

(b) By (a), $\{1\} < \mathrm{Z}(G) \leq T$ and all elements in the set $G - T$ have order p. Let $|G| = p^n$ and $\exp(T) = p^t$. Then $\mathrm{c}_t(G) = \mathrm{c}_t(T) \equiv 1 \pmod p$ so G is a 2-group of maximal class (Theorems 1.10(b) and 1.17(b)), and since all elements in $G - T$ are involutions, $|G : T| = 2$ and G is dihedral (then T is cyclic). □

Lemma 20.2. *Let G be a p-group such that $G/Z(G) = (U_1/Z(G))(U_2/Z(G))$, where $U_1/Z(G)$ and $U_2/Z(G)$ are cyclic. Then $U_1 \cap U_2 = Z(G)$.*

Proof. Since U_1 and U_2 are abelian, $(Z(G) \leq) U_1 \cap U_2 \leq Z(G)$. □

Lemma 20.3 (reported by Mann). *Let G be a p-group such that $G/Z(G)$ is extraspecial. Then $G/Z(G)$ is either of order p^3 and exponent $p > 2$ or $\cong D_8$.*

Proof. Write $Z = Z(G)$. Assume that $|G/Z| = p^{2n+1}$, $n > 1$. Then (see §4) there exist elements $x_1, y_1, \dots, x_n, y_n \in G - Z$ such that $G = \langle x_1, y_1, \dots, x_n, y_n, z, Z\rangle$, where $\langle zZ(G)\rangle = Z(G/Z)$, and

$$G/Z = (\langle x_1, y_1, Z\rangle/Z) * \cdots * (\langle x_n, y_n, Z\rangle/Z)$$

is a central product, $|\langle x_i, y_i, Z\rangle/Z| = p^3$ and $[x_i, y_i] \in \langle zZ\rangle$. We have $Z_2(G) = \langle z, Z\rangle$. We can select $x_1, y_1, \dots, x_n, y_n$ in such a way that $[x_i, y_i^{-1}] = zu_i$, where $u_i \in Z$, all i. Let $i \neq j$. By the Hall–Witt commutator identity,

$$[x_i, y_i^{-1}, x_j]^{y_i}[y_i, x_j^{-1}, x_i]^{x_j}[x_j, x_i^{-1}, y_i]^{x_i} = 1.$$

The second and third factors are equal to 1 since $[y_i, x_j^{-1}], [x_j, x_i^{-1}] \in Z$, so the first factor equals $1 = [z, x_j]^{y_i}$ since $[x_i, y_i] = zu_i'$ with $u_i' \in Z$. Thus, $z \in C_G(x_j)$ for all j. Similarly, $[z, y_j] = 1$ for all j, and so $z \in Z$, which is a contradiction. Thus, $n = 1$, i.e., $|G/Z| = p^3$, and the result follows from Lemma 20.2. □

Exercise 1. Suppose that G is a nonabelian p-group such that $G/Z(G)$ is abelian of type $(p^{n_1}, \dots, p^{n_k})$, where $n_1 \geq n_2 \geq \cdots \geq n_k$, $k > 1$. Then $n_1 = n_2$.

Theorem 20.4 ([Man8]). *Suppose that G is a nonabelian p-group and A/Z cyclic for each abelian $A < G$, where $Z = Z(G)$. Then G/Z is either elementary abelian or dihedral or nonabelian of order p^3 and exponent $p > 2$.*

Proof. Let $x \in G - Z$ be such that $x^p \in Z$. Set $C = C_G(x)$ and let $a \in C - Z\langle x\rangle$ with $a^p \in Z$. Then $T = \langle a, x, Z\rangle$ is abelian and $T/Z = \langle xZ\rangle \times \langle aZ\rangle \cong E_{p^2}$ is not cyclic, contrary to the hypothesis. This means that $\langle xZ\rangle$ is the only subgroup of order p in C/Z, i.e., C/Z is cyclic or generalized quaternion. Let $y \in G - Z$ be such that $y^p \in Z$. Set $C_1 = C_G(y)$. By the above, C_1/Z has only one subgroup of order p. If $z \in (C \cap C_1) - Z$ with $z^p \in Z$, then $\langle zZ\rangle \leq \langle xZ\rangle \cap \langle yZ\rangle$. Therefore $C_G(z) \geq \langle C, C_1\rangle$, and so $C = C_1$. Hence G/Z has a partition Σ which consists of some quotient groups $C_G(x)/Z$ $(o(xZ) = p)$ that are cyclic or generalized quaternion. By Lemma 20.2, G/Z is not generalized quaternion. so, by Lemma 20.1(b), G/Z is of exponent p or dihedral.

Let $\exp(G/Z) = p$ and $z \in Z_2(G) - Z$. Then $A = C_G(z) = \langle x, Z\rangle (\leq Z_2(G))$ since A/Z is cyclic of order p. It follows from $[G', Z_2(G)] = \{1\}$ and $C_G(A) = A \leq Z_2(G)$ that $G' \leq A$ so G/A is elementary abelian. If $Z_2(G)/Z$ is noncyclic, then G/Z is elementary abelian. Otherwise, $|Z_2(G)/Z| = p$. In the last case, $G/Z(G)$ is extraspecial so, by Lemma 20.3, G/Z is nonabelian of order p^3 and exponent p. □

Exercise 2 ([Man8]). Let $J = \mathrm{J}(G)$ be the Thompson subgroup of a p-group G generated by abelian subgroups of G of maximal order and $Z = \mathrm{Z}(G)$. If J/Z is cyclic, then either $J = G$ or else $p = 2$ and G/Z is dihedral, $|G : J| = 2$.

Solution. By hypothesis, J is abelian. It follows that $J/\mathrm{Z}(G)$ is the unique cyclic subgroup of its order in $G/\mathrm{Z}(G)$ so $G/\mathrm{Z}(G)$ is either cyclic or a 2-group of maximal class (Theorems 1.10(b) and 1.17(b)). In the first case, $J = G$. In the second case, by Lemma 20.2, G/Z is dihedral, and, clearly, $|G : J| = 2$.

§21

On the Schur multiplier and the commutator subgroup

1°. Let $G > \{1\}$ be a finite group. I. Schur (see, for example, [BZ, Chapter 6]) has showed that there exists a pair $M < \Gamma$ of finite groups such that (M1) $M \leq \mathrm{Z}(\Gamma) \cap \Gamma'$. (M2) $\Gamma/M \cong G$. (M3) If a pair $M_0 < \Gamma_0$ satisfies (M1) and (M2), then $|\Gamma| \geq |\Gamma_0|$.

A group M is determined uniquely up to isomorphism and is called the *Schur multiplier* of the group G (we write $M = \mathrm{M}(G)$ and $p^{m(G)} = |\mathrm{M}(G)|$ if G is a p-group). However, the group Γ is not defined uniquely. A group Γ, satisfying (M1–M3), is called a *representation group* of the group G. For details, see [BZ, Chapter 6] or the books [Kar1–3].

In this section we prove some simple facts on Schur multipliers of p-groups.

Lemma 21.1. *If G is cyclic, then* $\mathrm{M}(G) = \{1\}$.

Lemma 21.2. *If G is a p-group, then* $\mathrm{M}(G)$ *is also a p-group.*

Lemma 21.3. *Let G be a 2-group of maximal class. If* $\mathrm{M}(G) > \{1\}$, *then G is dihedral,* $|\mathrm{M}(G)| = 2$ *and every group of maximal class and order* $2|G|$ *is a representation group of G.*

Lemma 21.4. *If $G \cong \mathrm{M}_{p^n}$ (see Theorem 1.2), then* $\mathrm{M}(G) = \{1\}$.

Lemma 21.1 is trivial. Lemma 21.2 follows from Schur–Zassenhaus theorem. Lemmas 21.3 follows from Taussky's theorem. The representation group Γ of the group of Lemma 21.4 is minimal nonabelian so $|\Gamma'| = p$.

Lemma 21.5 (Wiegold, Berkovich). *Let G be a p-group and $|G/\mathrm{Z}(G)| = p^n$. Then there exists an integer $s \geq 0$ such that $|G'| = p^{\binom{n}{2}-s}$ and $|(G/\mathrm{Z}(G))'| \leq p^{1+s}$. Next, if $|(G/\mathrm{Z}(G))'| = p^{1+s}$, then* $\exp(\mathrm{Z}_2(G)/\mathrm{Z}(G)) = p$.

Proof. We proceed by induction on n. If G is abelian, then $n = 0$ and one may take $s = 0$. Now let G be nonabelian. Then $G/\mathrm{Z}(G)$ is noncyclic and $n > 1$.

Let $z_0 \in \mathrm{Z}_2(G) - \mathrm{Z}(G)$ be fixed. By Grün's lemma, the mapping $\phi : G \to [G, z_0]$, defined by $x \mapsto [x, z_0]$, is a homomorphism of G into $\mathrm{Z}(G)$. Put $N = \mathrm{im}(\phi)$ and $|N| = p^t$. Then $N \leq \mathrm{Z}(G)$, $\ker(\phi) = \mathrm{C}_G(z_0) \geq \langle z_0, \mathrm{Z}(G)\rangle > \mathrm{Z}(G)$, and therefore, $|G : \mathrm{C}_G(z_0)| \leq p^{n-1}$. Since $N \cong G/\ker(\phi) = G/\mathrm{C}_G(z_0)$, we get $t \leq n - 1$.

Set $p^b = |G/N : Z(G/N)|$. Since $N \neq z_0N \in Z(G/N)$ and $Z(G)/N < Z(G/N)$, we have $b \leq n-1$. Note, that $N = [z_0, G] \leq G'$, and so $|G'| = |N| \cdot |G'/N|$. It follows, by induction in G/N, that $|G'/N| \leq p^{\binom{b}{2}}$, and therefore

$$|G'| \leq p^{\binom{b}{2}+t}. \tag{1}$$

Because of $b \leq n-1$ and $t \leq n-1$, it follows from (1) that $|G'| \leq p^{\binom{n}{2}}$. Thus, there exists an integer $s \geq 0$ such that $|G'| = p^{\binom{n}{2}-s}$, proving the first assertion. Owing to the last formula and (1), we have

$$\begin{aligned} t &\geq \log_p |G'| - \binom{b}{2} = \frac{1}{2}n(n-1) - s - \frac{1}{2}b(b-1) \\ &\geq \frac{1}{2}n(n-1) - s - \frac{1}{2}(n-1)(n-2) = n-1-s, \end{aligned}$$

and therefore $p^{n-1-s} \leq p^t = |N| = |G : C_G(z_0)|$. Because $G/C_G(z_0) \cong N \leq Z(G)$, we see that $G/C_G(z_0)$ is abelian, and so $G'Z(G) \leq C_G(z_0)$. Therefore,

$$\begin{aligned} |G : G'Z(G)| &= |(G/Z(G)) : (G'Z(G)/Z(G))| \\ &= |(G/Z(G)) : (G/Z(G))'| \geq |G : C_G(z_0)| = |N| = p^t \geq p^{n-1-s}, \end{aligned}$$

and so $|(G/Z(G))'| \leq \frac{|G/Z(G)|}{p^{n-1-s}} = \frac{p^n}{p^{n-1-s}} = p^{1+s}$, proving the second assertion.

Assume that $|(G/Z(G))'| = p^{1+s}$. Since $G'Z(G) \leq C_G(z_0)$ for all $z_0 \in Z_2(G) - Z(G)$, it follows that

$$p^t = |N| = |G : C_G(z_0)| \leq |G : G'Z(G)| \leq |(G/Z(G)) : (G/Z(G))'| = p^{n-1-s},$$

and so $t \leq n-1-s$, and we conclude that $t = n-1-s$.

Assume, by the way of contradiction, that $\exp(Z_2(G)/Z(G)) > p$. Then there exists $z_0 \in Z_2(G) - Z(G)$ such that $z_0^p \notin Z(G)$. Because $z_0N \in Z(G/N)$, where $N = [G, z_0]$, we see that $|G/N : Z(G/N)| \leq p^{n-2}$. Therefore, by induction, $|(G/N)'| = |G'/N| \leq p^{\frac{1}{2}(n-2)(n-3)}$ and $|G'| \leq |N| \cdot |G'/N| \leq p^{n-1-s+\frac{1}{2}(n-2)(n-3)}$, or $\frac{1}{2}n(n-1) - s \leq n-1-s+\frac{1}{2}(n-2)(n-3)$. It follows that $(n-2)(n-1) \leq (n-2)(n-3)$, and we conclude that $n = 2$. Then $Z_2(G) = G$ so $G/Z(G)$ is cyclic, G is abelian, which is impossible. □

Corollary 21.6 ([Gre]). *If G is a group of order p^n, then $m(G) \leq \binom{n}{2}$.*

Proof. Let Γ be a representation group of G and let $M(G) \cong M \leq \Gamma' \cap Z(\Gamma)$ be such that $\Gamma/M \cong G$. Then $|\Gamma/Z(\Gamma)| \leq |G| = p^n$, and so, by Lemma 21.5, $|\Gamma'| \leq p^{\binom{n}{2}}$. Since $M \leq \Gamma'$, the result follows. □

Lemma 21.7 (Schur; see [BZ, Exercise 6.35]). *Let G be an abelian p-group of order p^n with invariants $\{p^{e_i}\}_1^d$, $e_1 \leq \cdots \leq e_d$; then $n = e_1 + \cdots + e_d$. In that case,*

$$m(G) \leq (d-1)e_1 + (d-2)e_2 + \cdots + e_{d-1}$$
$$= (n - e_d) + (n - e_d - e_{d-1}) + \cdots + (n - e_d - \cdots - e_2).$$

If $e_d = 1$, i.e., $G \cong \mathrm{E}_{p^d}$, then $m(G) = \frac{1}{2}n(n-1)$. If $d > 1$ and $e_d > 1$, then $m(G) \leq (n-2) + (n-3) + \cdots + (n-d) < (n-2) + \cdots + 1 = \frac{1}{2}(n-1)(n-2)$. In particular, if $m(G) = \frac{1}{2}n(n-1) - s$, $s \leq 2$, then (a) if $s = 0$, then $G = \mathrm{E}(p^n)$; (b) if $s = 1$, then $G = C_{p^2}$; (c) if $s = 2$, then G is abelian of type (p^2, p).

Theorem 21.8 ([Ber12]). *Let G be a group of order p^n. Then $m(G) = \frac{1}{2}n(n-1)$ if and only if G is elementary abelian.*

Proof. Let Γ be a representation group of G and let a subgroup $M \leq \Gamma' \cap \mathrm{Z}(G)$ be such that $M \cong \mathrm{M}(G)$ and $\Gamma/M \cong G$. Using Lemma 21.5, we get $p^{\frac{1}{2}n(n-1)} = |M| \leq |\Gamma'| \leq p^{\frac{1}{2}n(n-1)}$, and so $M = \Gamma'$. This means that $G = \Gamma/M$ is abelian. Now $G \cong \mathrm{E}(p^n)$, by Lemma 21.7(a). □

If G is a p-group with $|G'| = |\mathrm{Z}(G)| = p$, it is extraspecial.

Theorem 21.9 ([Ber12]). *Let G be a p-group and $|G/\mathrm{Z}(G)| = p^n$. If $|G'| = p^{\frac{1}{2}n(n-1)}$, then $G/\mathrm{Z}(G)$ is either elementary abelian or nonabelian of order p^3 and exponent $p > 2$.*

Proof. We use the same notation as in the proof of Lemma 21.5. Our assumption says that $s = 0$. Hence, by Lemma 21.5, $|(G/\mathrm{Z}(G))'| \leq p$.

Assume that $G/\mathrm{Z}(G)$ is abelian. Take $z_0 \in G - \mathrm{Z}(G)$. If $z_0^p \not\in \mathrm{Z}(G)$, then, $\frac{1}{2}n(n-1) = \log_p |G'| \leq \frac{1}{2}(n-2)(n-3) + (n-2) = \frac{1}{2}(n-1)(n-2)$ (Lemma 21.5), a contradiction. Thus, if $G/\mathrm{Z}(G)$ is abelian, it is of exponent p.

Now assume that $|(G/\mathrm{Z}(G))'| = p$. We claim that then $G/\mathrm{Z}(G)$ is extraspecial. Obviously, it suffices to prove that $|\mathrm{Z}_2(G)/\mathrm{Z}(G)| = p$ (see the paragraph preceding the theorem). We have $\exp(\mathrm{Z}_2(G)/\mathrm{Z}(G)) = p$ (Lemma 21.5). Assume that $\mathrm{Z}_2(G)/\mathrm{Z}(G)$ has two distinct subgroups $A/\mathrm{Z}(G)$ and $B/\mathrm{Z}(G)$ of order p. Then we may write $A = \langle z_0, \mathrm{Z}(G)\rangle$ and $B = \langle y_0, \mathrm{Z}(G)\rangle$. By inequality (1), we have $b = n - 1 = t$ for every $z_0 \in \mathrm{Z}_2(G) - \mathrm{Z}(G)$. Hence $A = C_G(z_0)$ and $B = C_G(y_0)$ and G/A and G/B are abelian of order p^{n-1}. But then $G' \leq A \cap B = \mathrm{Z}(G)$, which is a contradiction since $G/\mathrm{Z}(G)$ is nonabelian. Thus $|\mathrm{Z}_2(G)/\mathrm{Z}(G)| = p$, as was to be shown.

Let us prove that $G/\mathrm{Z}(G)$ is nonabelian of order p^3 and exponent p. We have $|G/\mathrm{Z}(G)| = p^3$ (Lemma 20.3). Assume that $\exp(G/\mathrm{Z}(G)) = p^2$. Then, if $G/\mathrm{Z}(G)$ has two distinct cyclic subgroups $A/\mathrm{Z}(G)$ and $B/\mathrm{Z}(G)$ of index p, then A, B are abelian, and so $A \cap B = \mathrm{Z}(G)$ has index p^2 in G, a contradiction. Thus, $G/\mathrm{Z}(G)$ is nonabelian of exponent p or dihedral, $|G/\mathrm{Z}(G)| = p^3$. If $G/\mathrm{Z}(G)$ is dihedral of order

8, then, by hypothesis, $|G'| = 2^3$; however, $|G| = 2 \cdot |G'| \cdot |Z(G)|$, since G has an abelian subgroup of index 2 (see Lemma 1.1), and so $|G'| = 2^2$, a contradiction. □

Lemma 21.10 ([Kar3, Theorem 11.8.23]). *If $G = \mathrm{ES}(m, p)$ is an extraspecial group of order p^{1+2m}. If $m > 1$, then $m(G) = 2m^2 - m - 1$. If $m = 1$ and $\mathrm{M}(G) > \{1\}$, then either $m(G) = 1$ and $G = \mathrm{D}_8$ or $\mathrm{M}(G) \cong \mathrm{E}_{p^2}$ and $\exp(G) = p$.*

Theorem 21.11 ([Ber12]). *Let G be a group of order p^n. Then the following conditions are equivalent: (a) $m(G) = \frac{1}{2}n(n-1) - 1$. (b) G is either cyclic of order p^2 or nonabelian of order p^3 and exponent $p > 2$.*

Proof. By Lemmas 21.1 and 21.10, if G is from (b), then $m(G) = \frac{1}{2}n(n-1) - 1$.

Let $m(G) = \frac{1}{2}n(n-1) - 1$ and let Γ be a representation group of G. Then there exists a subgroup $M \in \Gamma' \cap Z(G)$ such that $\Gamma/M \cong G$ and $M \cong \mathrm{M}(G)$. In that case, $G' \cong \Gamma'/M$ and $|\Gamma : Z(\Gamma)| \le |\Gamma : M| = |G| = p^n$, so that $\log_p |\Gamma'| = \frac{1}{2}n(n-1) - s \ge m(G) = \frac{1}{2}n(n-1) - 1$ (Lemma 21.5), and so $s \le 1$.

(i) Assume that $M < Z(\Gamma)$. Then $|\Gamma : Z(\Gamma)| \le p^{n-1}$ and, by Lemma 21.5, we have $\frac{1}{2}(n-1)(n-2) \ge \log_p |\Gamma'| \ge \log_p |M| = m(G) = \frac{1}{2}n(n-1) - 1$, and so $n \le 2$. In this case, by hypothesis, $\mathrm{M}(G) = \{1\}$ so G is cyclic of order p^2 (indeed, $\mathrm{M}(\mathrm{E}_{p^2}) \cong \mathrm{C}_p$, by Lemma 21.7).

(ii) Next, suppose that G is noncyclic and $M = Z(\Gamma)$. By Lemma 21.7, G is not elementary abelian. Therefore, $n > 2$. If G is abelian then, as in Lemma 21.7, $d > 1$, $e_d > 1$ imply $m(G) = (n - e_d) + \cdots + (n - e_d - \cdots - e_2) \le (n-2) + \cdots + 1 = \frac{1}{2}(n-1)(n-2) < \frac{1}{2}n(n-1) - 1$ (since, by assumption, $n \ge 3$), a contradiction.

(iii) Now assume that G is nonabelian. Then $M < \Gamma'$. Therefore, $\frac{1}{2}n(n-1) - s = \log_p |\Gamma'| > m(G) = \frac{1}{2}n(n-1) - 1$, so that $s = 0$. Then, by Theorem 21.9, $G \cong \Gamma/Z(\Gamma) = \Gamma/M$ is nonabelian of order p^3 and exponent $p > 2$. □

Definition. A group H is called *capable* if there is a group G such that $G/Z(G) \cong H$.

Exercise 1. Let $H = \langle x_1, \dots, x_r \rangle$ and suppose that H has an element $u \ne 1$ such that $u \in \langle x_i \rangle$ for $i = 1, \dots, r$. Then H is not capable. Hence, extraspecial p-groups generated by elements of order p^2 are not capable.

Solution. Assume that there exists G such that $G/Z(G) \cong H$. Let $u = vZ(G)$, $X_i = \langle x_i Z(G) \rangle$ and Y_i is the inverse image of X_i in G for all i. Then all Y_i are abelian and contain v. Therefore, $v \in \bigcap_{i=1}^r Y_i$ so $C_G(v) \ge \langle Y_1, \dots, Y_r \rangle = G$ and $v \in Z(G)$, a contradiction.

Exercise 2. If $H = F \times \mathrm{C}_{p^n}$, where $\exp(F) < p^n$, then H is not capable.

Hint. Let $x_1, \dots, x_r$ be all elements of order p^n in H, $\mathrm{C}_{p^n} = \langle c \rangle$. If $x_i \ne c$, then $\langle x_i \rangle \cap \langle c \rangle > \{1\}$ (otherwise, $\exp(H/\langle c \rangle) \ge p^n$). Then $\langle x_1 \rangle \cap \cdots \cap \langle x_r \rangle > \{1\}$. In view of Exercise 1, it remains to show that $\langle x_1, \dots, x_r \rangle = H$. Let $f \in F$. Then elements of order p^n generate $\langle f, c\rangle = \langle f\rangle \times \langle c\rangle$ so $f \in \langle x_1, \dots, x_r\rangle$, and we get $H = \langle x_1, \dots, x_r\rangle$. Use Exercise 1.

Let a p-group H be not capable, let Γ be a representation group of H and let $M \le \mathrm{Z}(\Gamma) \cap \Gamma'$ be such that $M \cong \mathrm{M}(H)$ and $\Gamma/M \cong H$; then $M < \mathrm{Z}(\Gamma)$.

Exercise 3. Let G be an abelian p-group. If two largest invariants of G are not equal, it is not capable. (*Hint.* Use Exercise 2.)

Exercise 4. Let $\mathcal{H} = \langle a, b \mid a^4 = b^4 = 1, a^b = a^{-1}\rangle$. Show that the Schur multiplier of $\mathcal{H}$ is nontrivial. (*Hint.* Check that $\Gamma = \langle x, y \mid x^8 = y^4 = 1, x^y = x^{-1}\rangle$ is an epimorphic image of a representation group of $\mathcal{H}$.)

2°. The proofs of presented below results see in [Kar2].

Theorem 21.12 (N. Blackburn). *Let $\Sigma_{p^n} \in \mathrm{Syl}_p(\mathrm{S}_{p^n})$. Then $\mathrm{M}(\Sigma_{p^n}) \cong \mathrm{E}_{p^s}$, where $s = \frac{1}{12}(p-1)(n-1)n(2n-1)$ if $p > 2$ and $s = \frac{1}{6}n(n^2-1)$ if $p = 2$. In particular, $|\mathrm{M}(\Sigma_{p^2})| = p^{\frac{1}{2}(p-1)}$ for $p > 2$ and C_2 for $p = 2$.*

Theorem 21.13 (D. L. Johnson). *A noncyclic group of exponent p has nontrivial Schur multiplier. If G is a noncyclic p-group with $\mathrm{M}(G) = \{1\}$, then $\Omega_1(G) < G$.*

Proposition 21.14 (B. H. Neumann). *Let $G = \langle a, b \mid a^{p^m} = b^{p^n} = 1, b^{-1}ab = a^{1+p^{m-n}}\rangle$, where $m > n > 0$ if p is odd, and $m - 1 > n > 0$ if $p = 2$. Then $|G| = p^{m+n}$ and $\mathrm{M}(G) = \{1\}$.*

3°. We do not assume in Theorem 21.15 that G is finite.

Theorem 21.15 (Schur). *Let $|G : \mathrm{Z}(G)| = m$. Then, for $x \in G$, the map $x \mapsto x^m$ is homomorphic and $|G'|$ is an m-number so the multiplier of finite group is finite.*

Proof. Let $V = V_{G \to \mathrm{Z}(G)}$ be the transfer homomorphism of G into $\mathrm{Z}(G)$. Then, for any $x \in G$, we have $V(x) = \prod_{i=1}^r s_i x^{\rho_i} s_i^{-1}$, $r \in \mathbb{N}$, $\sum_{i=1}^r \rho_i = m$, $s_i x^{\rho_i} s_i^{-1} \in \mathrm{Z}(G)$. It follows from the last inclusion that $s_i x^{\rho_i} s_i^{-1} = x^{\rho_i}$ and $V(x) = x^m$.

Since $V(G) = \mathrm{im}(V)$ is a subgroup of $\mathrm{Z}(G)$, it is abelian and the kernel K of V contains G'. Hence $y^m = V(y) = 1$ for all $y \in G'$. It remains to show that G' is finite. Let $L = G' \cap \mathrm{Z}(G)$. Since $[x_1, x_2] = [x_1', x_2']$ whenever $x_i \equiv x_i' \pmod{\mathrm{Z}(G)}$, $i = 1, 2$, G' is finitely generated. But $|G'/L| = |\mathrm{Z}(G)G'/\mathrm{Z}(G)|$ which is finite and an m-number. By Schreier's theorem, L is a finitely generated abelian group whose elements z all satisfy $z^m = 1$. Thus, $|L|$ is an m-number, and so $|G'|$ is an m-number as well. (Recall that $a \in \mathbb{N}$ is an m-number if $\pi(a) \subseteq \pi(m)$.) Let Γ be a representation group of a finite group G and let $\mathrm{M}(G) \cong M \le \Gamma' \cap \mathrm{Z}(G)$ be such that $\Gamma/M \cong G$. Since $\Gamma/\mathrm{Z}(\Gamma)$ is finite, Γ' is finite so $\mathrm{M}(G)$ is finite. □

Theorem 21.16 (Baer). *Let H be a normal subgroup of a finite group G, let K be a subgroup of $H \cap \mathrm{Z}(G)$, and let $|H : K| = m$. Then $\exp([H, G])$ divides m.*

Proof. Let $V = V_{H \to K}$ be the transfer of H to K. Then $V(x) = x^m$ for $x \in H$ (see the proof of Theorem 21.15). Let $y \in G$. Then

$$V([x,y]) = V(x^{-1} \cdot x^y) = V(x^{-1})V(x^y) = x^{-m}(x^y)^m = x^{-m}(x^m)^y$$
$$= [x^m, y] = 1,$$

since $x^m \in K \in Z(G)$. But $V([x,y]) = [x,y]^m$. Hence $[x,y]^m = 1$. All generators $[x,y]$ of $[H,G]$ belong to $\ker(V)$, so $\exp([H,G])$ divides m. □

4°. Items 21.17–21.21 are taken from [Isa18].

Theorem 21.17. *Let G be a finite capable group. Then there is a finite group H such that $H/Z(H) \cong G$.*

Proof. Since G is capable, there is, by definition, a possibly infinite group H such that $H/Z(H) \cong G$. Let T be a transversal for the cosets $Z(H)$ in H, and let $K = \langle T \rangle$. Then $|T| = |H : Z(H)| = |G| < \infty$ so K is finitely generated. We have $Z(H)K = H$ so $Z(K) = K \cap Z(H)$. It follows that

$$K/Z(K) \cong K/(K \cap Z(H)) \cong KZ(H)/Z(H) \cong H/Z(H) \cong G.$$

We can therefore replace H by K; then H is finitely generated.

By Schreier's theorem (see Appendix 25), the abelian group $Z(H)$ is finitely generated since $|H : Z(H)| = |G| < \infty$, and thus we can write $Z(H) = D \times F$, where D is finite and F is torsion free. Certainly, $Z(H)/F \leq Z(H/F)$, and we claim that the equality holds there. To see this, let $hF \in Z(H/F)$, so that $[H,h] \leq F \leq Z(H)$. It follows that the map $x \mapsto [x,h]$ defines a homomorphism from H into F. Since F lies in the kernel of this homomorphism and $|H : F| < \infty$, it follows that the image of this homomorphism is a finite subgroup of the torsion free group F. It follows that $[H,h] = \{1\}$ and $h \in Z(H)$, as required. Now H/F is finite and $(H/F)/Z(H/F) = (H/F)/(Z(H)/F) \cong H/Z(H) \cong G$. □

Proposition 21.18 (Isaacs–Mann). *Let G be a p-group such that $|G'| = p$ and $|G : Z(G)| > p^2$. Then G is not capable.*

Proof. Assume that $G = H/Z(H)$. By Lemma 4.2, we can decompose G in a central product of two nonabelian subgroups (take into account here that $|G : Z(G)| > p^2$). One can thus write $H = XY$, where $[X,Y] \leq Z(H)$ and $X'Z(H) = Y'Z(H) = H'Z(H)$. Since $[X,Y,Y] = \{1\} = [Y,X,Y]$, the Three Subgroups Lemma yields that $\{1\} = [Y,Y,X] = [Y',X]$ so $[H',X] = [X'Y',X] = \{1\}$. We have similarly $[H',Y] = \{1\}$. Consequently, $H' \leq Z(H)$, which is a contradiction since $\mathrm{cl}(H) > 2$, by assumption. □

Given $n = \prod_{i=1}^k p_i^{\alpha_i}$, the prime decomposition of $n > 1$, let $\lambda(n) = \sum_{i=1}^k \alpha_i$.

Theorem 21.19 ([Isa18]). *Let $A \le G$ be abelian of index n and let $|G'| = m$. Then $|G : \mathrm{Z}(G)| \le nm^{\lambda(n)}$.*

Proof. It is clear that we can choose a subset X in G such that $G = \langle X, A\rangle$ and $|X| \le \lambda(|G : A|) = \lambda(n)$. The size of a conjugacy class of G does not exceed $|G'| = m$, i.e., $|G : \mathrm{C}_G(x)| \le m$ for each $x \in G$ so $|G : \mathrm{C}_G(X)| \le m^{|X|} \le m^{\lambda(n)}$. Since A is abelian and $G = \langle A, X\rangle$, we see that $A \cap \mathrm{C}_G(X) \le \mathrm{Z}(G)$ and thus

$$|G : \mathrm{Z}(G)| \le |G : A||A : (A \cap \mathrm{C}_G(X))| \le n|G : \mathrm{C}_G(X)| \le nm^{\lambda(n)},$$

since $|A : (A \cap \mathrm{C}_G(X))| = |A\mathrm{C}_G(X) : \mathrm{C}_G(X)| \le |G : \mathrm{C}_G(X)|$. □

Proposition 21.20 ([Isa18]). *Let G be a p-group with cyclic G'. If $p = 2$ and $|G'| > 2$, assume, in addition, that $|G' \cap \mathrm{Z}(G)| > 2$. Then $|G : \mathrm{Z}(G)| \ge |G'|^2$.*

Proof. Since G' is cyclic, we have $G' = \langle [a, b]\rangle$ for some $a, b \in G$. Then $G' = X'$, where $X = \langle a, b\rangle$. Since $X\mathrm{Z}(G) \le G$, we have

$$|X : \mathrm{Z}(X)| \le |X : (X \cap \mathrm{Z}(G))| = |X\mathrm{Z}(G) : \mathrm{Z}(G)| \le |G : \mathrm{Z}(G)|$$

so, taking into account our aim, there is no loss of generality to assume that $G = X$, and thus we have $G = \langle a, b\rangle$. Write $A = \langle \mathrm{Z}(G), a\rangle$ and $B = \langle \mathrm{Z}(G), b\rangle$. Then both A and B are abelian, and since $G = \langle A, B\rangle$, we get $A \cap B = \mathrm{Z}(G)$.

Consider the inner automorphism σ of G induced by a, and note that $o(\sigma) = |\langle a\rangle : (\langle a\rangle \cap \mathrm{Z}(G))|$. Since $[G, \sigma] = G'$ is cyclic (indeed, $G/[G, \sigma]$ is abelian), σ fixes all elements of order p in G', and also, by hypothesis, σ fixes all elements of order 4 in G'. By Lemma 1.32 and the product formula, $|A : \mathrm{Z}(G)| = |\langle a\rangle \mathrm{Z}(G) : \mathrm{Z}(G)| = |\langle a\rangle : (\langle a\rangle \cap \mathrm{Z}(G))| = o(\sigma)$, and that number is a multiple of $\exp([G, \sigma]) = |[G, \sigma]| = |G'|$ so we have $|A : \mathrm{Z}(G)| \ge |G'|$. Similarly, $|B : \mathrm{Z}(G)| \ge |G'|$, and thus, since $A \cap B = \mathrm{Z}(G)$, we get

$$|G : \mathrm{Z}(G)| = \frac{|AB|}{|\mathrm{Z}(G)|} = \frac{|A||B|}{|A \cap B||\mathrm{Z}(G)|} = |A : \mathrm{Z}(G)||B : \mathrm{Z}(G)| \ge |G'|^2. \qquad □$$

Theorem 21.21 ([Isa18]). *Let G be a p-group with cyclic G'. If $p = 2$ and $|G'| > 2$, assume, in addition, that $|G' \cap \mathrm{Z}(G)| > 2$. Then there exist subgroups X and Y of G such that $G = X * Y$, $X' = G'$ and $|G : Y| = |G'|^2$.*

§22

On characters of p-groups

1°. Below we consider a generalization of CM-groups (see [BZ, Chapter 9]).

Lemma 22.1. *Let G be a p-group of order p^m, $\exp(Z(G)) = p^s$. Then G has a normal subgroup N such that $\ker(\chi) = N$ for at least $\varphi(p^s) = (p-1)p^{s-1}$ irreducible characters χ of G.*

Proof. (Similar argument is due to Zhmud) Let $C \leq Z(G)$ be cyclic of order p^s and N a maximal normal subgroup of G such that $N \cap C = \{1\}$; then $\Omega_1(C)N/N$ is the unique minimal subgroup of G/N so $\mathrm{Irr}(G/N)$ has a faithful character. Let k be the number of faithful irreducible characters of G/N. We claim that $k \geq (p-1)p^{s-1}$. One may assume that $N = \{1\}$. Then $Z(G) \geq C$ is cyclic. Let $\chi \in \mathrm{Irr}(G)$ be faithful. We have $\chi(1)^2 \leq |G : Z(G)| \leq p^{m-s}$. But the sum of squares of the degrees of faithful irreducible characters of G equals $|G| - |G/\Omega_1(C)| = p^m - p^{m-1} = (p-1)p^{m-1}$ so $kp^{m-s} \geq (p-1)p^{m-1}$ hence $k \geq (p-1)p^{s-1}$. □

Definition 1. If n is the least natural number such that every normal subgroup of G is the kernel of at most n irreducible characters of G, then G is said to be a CM_n-group, and we write $\gamma(G) = n$.

It follows from Lemma 22.1 that if $G > \{1\}$ is a p-group then $\gamma(G) \geq p-1$.

Theorem 22.2. *Let G be a group of order p^m. Then the following assertions are equivalent:*

(a) *G is a CM_{p-1}-group.*

(b) *$|G/\ker(\chi)| = p\chi(1)^2$ for every nonprincipal irreducible character χ of G.*

Proof. (i) Assume that G is a CM_{p-1}-group. Take $\chi \in \mathrm{Irr}(G)$. To prove (b), one may assume that χ is faithful. Then $Z(G)$ is cyclic, and by Lemma 22.1, $|Z(G)| = p$. Let k be the number of faithful irreducible characters of G. We have $\chi(1)^2 \leq p^{m-1}$ so $kp^{m-1} \geq (p-1)p^{m-1}$, and hence $k \geq p-1$. By assumption, however, $k \leq p-1$ so $k = p-1$ and $\chi(1)^2 = p^{m-1}$, proving (b).

(ii) Assume that $|G/\ker(\chi)| = p\chi(1)^2$ for every nonprincipal $\chi \in \mathrm{Irr}(G)$. Let $N = \ker(\chi) < G$. We have to prove that N is the kernel of exactly $p-1$ irreducible characters of G. One may assume that $N = \{1\}$. By assumption, every faithful irreducible character of G has degree $p^{\frac{1}{2}(m-1)}$ so $|Z(G)| = p$. If k is the number of faithful irreducible characters, then $kp^{m-1} = p^m - p^{m-1}$ so $k = p-1$. □

Definition 2. A normal subgroup of G that is a kernel of some $\chi \in \mathrm{Irr}(G)$, is said to be a *kernel* in G. The normal closure of a cyclic subgroup $\langle a \rangle$ in G is said to be an *antikernel* in G. Let $j(G)$ be the number of kernels in G and $j^*(G)$ the number of antikernels in G.

By Zhmud's theorem [BZ, Chapter 9], $j(G) = j^*(G)$. Therefore, if a p-group G is a CM_{p-1}-group and $|\mathrm{Irr}(G)| = \mathrm{k}(G) = k$, then

$$j(G) = \frac{k-1}{p-1} + 1 = j^*(G). \tag{1}$$

Indeed, every kernel $\neq G$ is the kernel of exactly $p-1$ irreducible characters of G. It follows that $j(G) = \frac{k-1}{p-1} + 1$ since G is the kernel of exactly one irreducible character of G; the second equality in (1) is true, by Zhmud's theorem. It follows from (1) the following

Proposition 22.3. *Let a p-group G be a CM_{p-1}-group and $A \neq \{1\}$ an antikernel of G. Let $\{K_1, \dots, K_t\}$ be the set of G-classes such that $\langle K_i \rangle = A$, $i = 1, \dots, t$. Then $t = p - 1$. Conversely, if for every antikernel $A \neq \{1\}$ of a p-group G, there exists exactly $p-1$ conjugacy classes of G such that every one of them generate A, then G is a CM_{p-1}-group.*

Proof. Let $A_1, \dots, A_{j^*(G)-1}$ be all nonidentity antikernels of G, and suppose that there are exactly t_i G-classes every one of which generate A_i. Then $t_i \geq p-1$ since $x_i, \dots, x_i^{p-1}$ belong to different G-classes for each $x \in G^{\#}$. We have, by the above and (1) that

$$(p-1)(j^*(G)-1) \leq \sum_{i=1}^{j^*(G)-1} t_i = k-1 = (p-1)(j(G)-1) = (p-1)(j^*(G)-1).$$

It follows that $t_i = p-1$ for all $i < j^*(G)$.

Conversely, let $t_i = p-1$ for $i = 1, \dots, j^*(G)-1$. Then $k - 1 = (j^*(G) - 1)(p-1) = (j(G)-1)(p-1)$ so every $\neq G$ kernel in G is the kernel of exactly $p-1$ irreducible characters of G (see Lemma 22.2), i.e., G is a CM_{p-1}-group. □

Let $A = \langle x \rangle^G$ be an antikernel in G. Then $A/[A,G]$ is an antikernel in $G/[A,G]$. Since $A/[A,G] \leq \mathrm{Z}(G/[A,G])$, the subgroup $A/[A,G]$ is cyclic, so $\varphi(|A/[A,G]|)$ generators of that subgroup are representatives of distinct $G/[A,G]$-classes. Thus, there are at least $\varphi(|A/[A,G]|)$ classes generating A.

2°. Let $\mathrm{cd}(G) = \{d_0 = 1, d_1, \dots, d_s\}$ and let $\mathrm{Irr}(G)$ have exactly a_i characters of degree d_i, $i = 0, \dots, s$. Then $\bar{\delta}(G) = \{a_0 \cdot 1, a_1 \cdot d_1, \dots, a_s \cdot d_s\}$ is called the *degree vector* of G. We have $\mathrm{k}(G) = a_0 d_0 + a_1 d_1 + \dots + a_s d_s$. Write $\mathrm{n}(G) = |\mathrm{Irr}_1(G)|$, where $\mathrm{Irr}_1(G) = \mathrm{Irr}(G) - \mathrm{Lin}(G)$; then $\mathrm{n}(G) = a_1 + \dots + a_s$.

Exercise 1. (a) Given a p-group G, $p-1$ divides $\mathrm{n}(G) = |\mathrm{Irr}_1(G)|$.

(b) Classify the p-groups having at most p^2 nonlinear irreducible characters.

Solution. (a) If $|G| = p^m$, $|G : G'| = p^k$ and the degree vector of G is $\{p^k \cdot 1, a_1 \cdot p^{c_1}, \dots, a_r \cdot p^{c_r}\}$, then $\mathrm{n}(G) = a_1 + \dots + a_r$ and $a_1 + \dots + a_r \equiv a_1 p^{2c_1} + \dots + a_r p^{2c_r} = |G| - |G : G'| = p^m - p^k \equiv 0 \pmod{p-1}$.

Exercise 2. Let G be a group, $N \trianglelefteq G$, $x \in G$. Prove that $|\mathrm{C}_{G/N}(xN)| \le |\mathrm{C}_G(x)|$. (*Hint.* Use the Second Orthogonality Relation.)

Theorem 22.4 (Mann). *A p-group G has exactly $p^2 - p$ nonlinear irreducible characters of minimal degree, say d, if and only if one of the following holds:*

(a) $|G'| = p$ and $|\mathrm{Z}(G)| = p^2$.

(b) $|G : G'| = p^3$, $|G' : \mathrm{K}_3(G)| = p$, and if H is a maximal subgroup of G then $H' = G'$ or $H' = \mathrm{K}_3(G)$. In that case, $d = p$, there always exist maximal subgroups H satisfying $H' = \mathrm{K}_3(G)$, and maximal subgroups satisfying $H' = G'$ exist if and only if $\mathrm{d}(G) = 3$.

Theorem 22.5 (Isaacs–Passman [IsP1]). *If G is a nonabelian p-group with $\mathrm{cd}(G) = \{1, p\}$, then one and only one of the following holds: (a) G has an abelian subgroup of index p, (b) $G/\mathrm{Z}(G)$ is of order p^3 and exponent p.*

Lemma 22.6. *Let G be a nonabelian group of order p^m and $\bar{\delta}(G) = \{a_0 \cdot 1, a_1 \cdot d_1, \dots, a_s \cdot d_s\}$ the degree vector of G. Then (a) (Mann) $p - 1$ divides a_i for $i = 1, \dots, s$. (b) If $\mathrm{T}(G) = (\sum_{\chi \in \mathrm{Irr}(G)} \chi(1)) = p^{m-1}$, then $p = 2$.*

Proof. (a) Let $\chi \in \mathrm{Irr}_1(G)$, $\chi(1) = d > 1$, and let $x \cdot \ker(\chi) \in \mathrm{Z}(G/\ker(\chi))$ be of order p. If R is a representation of G with character χ, then $R(x) = \epsilon \cdot \mathrm{I}$ is a scalar $\chi(1) \times \chi(1)$ matrix. Then the number of algebraic conjugates of χ is divisible by the number of algebraic conjugates of ϵ, and the latter is $p - 1$ since ϵ is a p-th root of 1. Summing over the set of all characters of degree d, we obtain the desired result.

(b) Let $\{p^k \cdot 1, a_1 \cdot p^{c_1}, \dots, a_s \cdot p^{c_s}\}$ be the degree vector of G. By (a), $a_i = (p-1)a_i'$, where $a_i' \in \mathbb{N}$, $i = 1, \dots, s$. Let $0 < c_1 < \dots < c_s$. It follows from

$$|G| = p^m = p^k + (p-1)[a_1' p^{2c_1} + \dots + a_s' p^{2c_s}],$$
$$\mathrm{T}(G) = p^{m-1} = p^k + (p-1)[a_1' p^{c_1} + \dots + a_s' p^{c_s}]$$

that $|G| - \mathrm{T}(G) = p^{m-1}(p-1) = (p-1)[a_1' p^{c_1}(p^{c_1} - 1) + \dots + a_s' p^{c_s}(p^{c_s} - 1)]$, and so $p^{m-1} = a_1' p^{c_1}(p^{c_1} - 1) + \dots + a_s' p^{c_s}(p^{c_s} - 1) \equiv 0 \pmod 2$ hence $p = 2$. □

3°. The following two nice theorems are due to Mann. Recall that $\mathrm{Irr}(\chi)$ is the set of all irreducible constituents of a character χ of a group G.

Theorem 22.7 (Mann). *A nonabelian group G has a faithful irreducible character χ such that $\mathrm{Irr}(\chi^2) \subseteq \mathrm{Lin}(G)$ if and only if $|G'| = 2$ and $\mathrm{Z}(G)$ is cyclic.*

Proof. (a) Let $\chi \in \mathrm{Irr}(G)$ be faithful and $\mathrm{Irr}(\chi^2) \subseteq \mathrm{Lin}(G)$; then χ is nonlinear. If $\lambda \in \mathrm{Irr}(\chi^2)$, then $\langle \chi^2, \lambda\rangle = \langle \chi, \lambda\bar{\chi}\rangle = 1$ so $\chi = \lambda\bar{\chi}$. Therefore,

(i) $\chi^2 = \lambda_1 + \dots + \lambda_k$, where $\lambda_1, \dots, \lambda_k$ are pairwise distinct linear characters of G such that $\lambda_i\bar{\chi} = \chi$ for each i. In particular, $k = \chi(1)^2$.

(ii) χ vanishes outside $\mathrm{Z}(G)$. Indeed, take $x \in G$. Applying (i), we obtain $\chi^2 \cdot \bar{\chi} = \sum_{i=1}^k \lambda_i \bar{\chi} = k\chi$, whence, $\chi(\chi\bar{\chi} - k \cdot 1_G) = 0$. Hence, if $\chi(x) \neq 0$ then $|\chi(x)|^2 = k = \chi(1)^2$ so $|\chi(x)| = \chi(1)$ and, since χ is faithful, we get $x \in \mathrm{Z}(\chi) = \mathrm{Z}(G)$.

(iii) We claim that $G' \le \mathrm{Z}(G)$. Indeed, if $x \in G'$ then, by (i), $\chi(x)^2 = \lambda_1(x) + \dots + \lambda_k(x) = k \neq 0$, and so $x \in \mathrm{Z}(G)$, by (ii).

(iv) $|G'| = 2$. Indeed, by Clifford's theorem, $1_{G'} \notin \mathrm{Irr}(\chi_{G'})$ so

$$0 = |G'|\langle \chi_{G'}, 1_{G'}\rangle = \sum_{x\in G'} \chi(x) = \chi(1) + \sum_{x\in (G')^\#} \chi(x). \tag{2}$$

We claim that $\chi(x) = -\chi(1)$ for $x \in (G')^\#$. Indeed, since $\lambda_i\bar{\chi} = \chi$, we have $\bar{\chi}(x) = \chi(x)$, i.e., $\chi(x) \in \mathbb{R}$. Therefore, by (ii), $\chi(x)^2 = |\chi(x)|^2 = \chi(1)^2$, so that $|\chi(x)| = \chi(1)$. Since χ is faithful, we have $\chi(x) = -\chi(1)$. Therefore, by (2), $\chi(1) - \chi(1)(|G'| - 1) = 0$, hence $|G'| = 2$.

(v) $\mathrm{Z}(G)$ is cyclic since $\chi \in \mathrm{Irr}(G)$ is faithful.

(vi) $|G : \mathrm{Z}(G)| = \chi(1)^2$ $(= k)$. Indeed, by (ii), $|G| = \sum_{x\in G} |\chi(x)|^2 = \sum_{x\in \mathrm{Z}(G)} |\chi(x)|^2 = |\mathrm{Z}(G)|\chi(1)^2$.

(b) Now let G be a group with cyclic center and derived subgroup of order 2. Then G is nilpotent and $\exp(G/\mathrm{Z}(G)) = 2$. There exists a faithful $\chi \in \mathrm{Irr}(G)$. Then $|G : \mathrm{Z}(G)| = \chi(1)^2$ [BZ, Lemma 3.37] and χ vanishes outside $\mathrm{Z}(G)$. Let $\lambda \in \mathrm{Lin}(G)$ be such that $\mathrm{Z}(G) \le \ker(\lambda)$. Then $\langle \chi^2, \lambda\rangle = |G|^{-1} \sum_{x\in \mathrm{Z}(G)} \chi(x)^2\lambda(x) = |G|^{-1} \sum_{x\in \mathrm{Z}(G)} \chi(x)^2 = |G|^{-1}|\mathrm{Z}(G)|\langle \chi_{\mathrm{Z}(G)}, \overline{\chi_{\mathrm{Z}(G)}}\rangle$.

If $\chi_{\mathrm{Z}(G)} = \chi(1)\mu$, then, $\langle \chi^2, \lambda\rangle = |G|^{-1}|\mathrm{Z}(G)|\chi(1)^2\langle \mu, \bar{\mu}\rangle = \langle \mu, \bar{\mu}\rangle$. As $G/\mathrm{Z}(G)$ is an elementary abelian 2-group, the linear characters of G, whose kernels contain $\mathrm{Z}(G)$, can take values -1 and 1 only. Therefore, $\langle \chi^2, \lambda\rangle = \langle \mu, \bar{\mu}\rangle = \langle \mu, \mu\rangle = 1$, and since $|G/\mathrm{Z}(G)| = \chi(1)^2$, it follows that $\mathrm{Irr}(\chi^2) \subseteq \mathrm{Lin}(G)$. □

Theorem 22.8 (Mann). *A nonabelian group G has a faithful irreducible character χ such that* $\mathrm{Irr}(\chi\bar{\chi}) \subseteq \mathrm{Lin}(G)$ *if and only if* $\mathrm{Z}(G)$ *is cyclic and* $G' \le \mathrm{Z}(G)$.

Exercise 3. Suppose that G is nonabelian of exponent p with $|\mathrm{Z}(G)| = p$, and let $M \in \Gamma_1$. Then $|\mathrm{Z}(M)| \le p^{p-1}$.

Solution. Take $x \in G - M$ and assume that $|\mathrm{Z}(M)| > p$. Set $H = \langle x, \mathrm{Z}(M)\rangle$. Then $\mathrm{Z}(H) = \mathrm{Z}(G)$ so $|\mathrm{Z}(H)| = p$ and H is of maximal class hence $|H| \le p^p$ (Theorem 9.5). Thus, $|\mathrm{Z}(M)| = \frac{1}{p}|H| \le p^{p-1}$.

Exercise 4. Suppose that a nonabelian group G of exponent p has a faithful irreducible character χ of degree p^n. Then $|G| \le p^k$, where $k \le p^n + \frac{p^{n-1}-1}{p-1}$.

Solution. Let $n = 1$. Then G has an abelian subgroup of index p [BZ, Chapter 18] and $|Z(G)| = p$ (since $Z(G)$ is cyclic) so $|G : G'| = p|Z(G)| = p^2$, and G is of maximal class. By Theorem 9.5, $|G| \le p^p$, proving our assertion for $n = 1$. Now let $n > 1$. There are $M \in \Gamma_1$ and $\mu \in \mathrm{Irr}(M)$ such that $\chi = \mu^G$, $\mu(1) = p^{n-1}$. Let $\chi_M = \mu_1 + \cdots + \mu_p$ be the Clifford's decomposition, $\mu_1 = \mu$. Then $M/\ker(\mu_i)$ has a faithful irreducible character μ_i of degree p^{n-1} so, by induction, $|M/\ker(\mu_i)| \le p^t$, where $t \le p^{n-1} + \frac{p^{n-2}-1}{p-1}$. Since M is a subgroup of the direct product of groups $M/\ker(\mu_i)$, $i = 1, \dots, p$, we get $|G| = p^k$, where $k \le 1 + pt = 1 + p(p^{n-1} + \frac{p^{n-2}-1}{p-1}) = p^n + \frac{p^{n-1}-1}{p-1}$. (According to Mann's letter, Herzog and Praeger have proved in J. Algebra **43**, 1 (1976), 216–220, that if G is a finite subgroup of exponent e in $\mathrm{GL}(n, F)$, $\mathrm{char}(F) = 0$, then $|G| \le e^n$. This improves the obtained estimate.)

Exercise 5. Let G be a nonabelian group and $\chi \in \mathrm{Irr}(G)$ faithful. Study the situation where $\mathrm{Irr}(\chi^m \bar{\chi}^n) \subseteq \mathrm{Lin}(G)$, in detail.

4°. An element $x \in G$ is said to be *nonvanishing* [INW], if $\chi(x) \neq 0$ for all $\chi \in \mathrm{Irr}(G)$. Obviously, all elements of $Z(G)$ are nonvanishing.

Proposition 22.9 ([INW]). *If G is a p-group, then the set of nonvanishing elements of G coincides with $Z(G)$.*

Lemma 22.10. *Let $H \lhd G$, where G is not necessarily a p-group, and suppose that all elements of the coset Hg are conjugate in G. If χ is a character of G and $1_H \notin \mathrm{Irr}(\chi_H)$, then $\chi(g) = 0$.*

Proof. Let $\mathcal{X}$ be a matrix representation which affords χ, and write $M = \sum_{h \in H} \mathcal{X}(h)$. If $h_1 \in H$, then $M\mathcal{X}(h_1) = \sum_{h \in H} \mathcal{X}(hh_1) = M$, and we see that each row of the matrix M is the fixed vector for the restricted representation $\mathcal{X}_H$. Since by hypothesis, this representation has no principal constituents, it follows that each row of M is the zero vector, and thus $M = 0$. We conclude that $M\mathcal{X}(g) = 0$. But $M\mathcal{X}(g)$ is the sum of the conjugate matrices $\mathcal{X}(hg)$ as h runs over H. Its trace, therefore, is exactly $|H|\chi(g)$, and the result follows. □

Lemma 22.11. *Let G be a p-group, $x \in G$ a nonvanishing element, and suppose that $M \le G$. If $H = [M, x]$ is normal in G and centralized by M, then $[M, x] = \{1\}$.*

Proof. Since H is centralized by M, the map $m \mapsto [m, x]$ is a homomorphism of M onto H, and thus if $h \in H$, we can write $h^{-1} = [\mu, x]$ for some $\mu \in M$. Then $h = [x, \mu] = x^{-1}x^{\mu}$, and we have $xh = x^{\mu}$. All elements of the coset xH are thus G-conjugate, and since $H \lhd G$, it follows from Lemma 22.10 that $\chi(x) = 0$ if $\chi \in \mathrm{Irr}(G)$ and $H \not\le \ker(\chi)$. Since x is nonvanishing, however, this never happens, and thus $H \le \ker(\chi)$. We conclude that $H = \{1\}$ since $\bigcap_{\chi \in \mathrm{Irr}(G)} \ker(\chi) = \{1\}$. □

Proof of Proposition 22.9. We will prove the following more general result: If G is supersolvable, then all the nonvanishing elements of G lie in $\mathrm{Z}(F)$, where $F = \mathrm{F}(G)$ is the Fitting subgroup of G. (It is proved in [INW], that the above assertion is also true for solvable G.) Let N be a minimal normal subgroup of G. Working by induction on $|G|$, we apply induction to the group $\bar{G} = G/N$. Let $x \in G$ be nonvanishing. Then $\bar{x}$ is a nonvanishing element of $\bar{G}$ so, by induction, $\bar{x}$ centralizes $\mathrm{F}(\bar{G}) \geq \bar{F}$, and thus $[F, x] \leq N$. Since $N \lhd G$ is of prime order and $[F, x] \leq N$, we see that $[F, x] \lhd G$. Also, $N \leq \mathrm{Z}(F)$, and thus F centralizes $[F, x]$. It now follows from Lemma 22.11 (with $M = F$) that $[F, x] = \{1\}$, and thus $x \in \mathrm{C}_G(F) = \mathrm{Z}(F)$. □

5°. Now we consider the p-groups G with $|\mathrm{cd}(G)| = 2$.

Theorem 22.12 ([Ban]). *Let G be a p-group with* $\mathrm{cd}(G) = \{1, p^k\}$. *Then $\Phi(G)$ is abelian.*

Proof. Let $H < G$ and $|G : H| \leq p^k$. Since all irreducible constituents of (reducible character) $(1_H)^G$ are linear, we get $G' \leq H$ so $H \lhd G$.

Let $\chi \in \mathrm{Irr}_1(G)$; then $\chi = \lambda^G$, where $\lambda \in \mathrm{Lin}(H)$ and $H < G$. As $|G : H| = p^k$, we have $G' \leq H$, by the above. By Clifford's theorem, $\mathrm{Irr}(\chi_H) \subseteq \mathrm{Lin}(H)$ so $H/\ker(\chi_H)$ is abelian.

We claim that G/H is elementary abelian, where H is as above. It suffices to assume that $|G'| = p$. Then $G/\mathrm{Z}(G)$ is elementary abelian of order $\chi(1)^2 = p^{2k}$, $|G : H| = \chi(1) = p^k$. Let us show, retaining assumption $|G'| = p$, that H is abelian. Indeed, $G' \not\leq H \cap \ker(\chi) = \ker(\chi_H)$ so $G' \cap \ker(\chi_H) = \{1\}$. Therefore, H is abelian since H/G' and $H/\ker(\chi_H)$ are abelian. Since $\chi(1) = p^k = |G : H|$, H is a maximal abelian subgroup of G, by Ito's theorem on degrees, so $\mathrm{Z}(G) < H$. Then $G/\mathrm{Z}(G) \cong \mathrm{E}_{p^{2k}}$, and our claim follows. Thus, $\Phi(G) \leq H$ so $\Phi(G)$ is abelian.

Now we omit the assumption $|G'| = p$. If $\mu \in \mathrm{Irr}(\Phi(G))$, then $\mu \in \mathrm{Irr}(\tau_{\Phi(G)})$ for some $\tau \in \mathrm{Irr}(G)$. We have $\tau = \sigma^G$ for some $\sigma \in \mathrm{Lin}(F)$, $F < G$. By the above, $\Phi(G) \leq F$. Then $\tau_{\Phi(G)}$ is a sum of linear constituents (Clifford) so μ is linear. It follows that $\Phi(G)$ is abelian since μ is arbitrary. □

Exercise 6 ([Man12]). Let G be a p-group, $\chi = \lambda^G$, where $\chi \in \mathrm{Irr}(G)$ is faithful, $\lambda \in \mathrm{Lin}(H)$, $H \trianglelefteq G$ (such χ is said to be *normally monomial*). Then all normal subgroups of G which linearly induce χ are abelian, and have maximal order among all abelian subgroups of G. Conversely, if H is an abelian subgroup of G of maximal order (not necessarily normal) then χ is linearly induced from H. All faithful irreducible normally monomial characters of G have the same degree, which is the maximal degree of all irreducible characters of G.

Exercise 7 ([Man12]). Two normally monomial irreducible characters with the same kernel have the same degree.

6°. Next we consider some properties of characters of minimal degree (see also Appendix 10). Given a nonabelian group G, the number $d = \min\{n \in \mathrm{cd}(G) - \{1\}\}$ is called the *minimal degree* of G. Irreducible characters of degree d are called *minimal characters*.

Exercise 8 ([Man12]). If a nonabelian p-group G has a faithful irreducible character which is minimal, then all nonlinear irreducible characters of G have the same degree, and either G contains a unique abelian subgroup of maximal order, which is then normal and has an abelian factor group, or $|G'| = p$.

Theorem 22.13 ([Ban]). *Let G be a nonabelian p-group with minimal degree d. Then there is a G-invariant subgroup N of index p in G' with* $\mathrm{cd}(G/N) = \{1, d\}$.

Theorem 22.14 ([Man12]). *Let G be a p-group with minimal degree $d > 1$.*

(a) d^2 equals the minimal index of subgroups K satisfying $[K, x] \neq G'$ for all $x \in G$.

(b) Given K as in (a), we have $\mathrm{cl}(G/[K, G]) = 2$, *and all the nonlinear irreducible characters of $G/[K, G]$ are minimal characters of G.*

(c) If a minimal character of G is linearly induced from $H < G$, then H contains K as in (a), and any K is contained in an appropriate H. Both types of subgroups (K and H) contain $\Phi(G)$.

(d) For a minimal character χ of G the following are equivalent:

(d1) χ is obtained from a subgroup K as above.

(d2) $\mathrm{K}_3(G) \leq \ker(\chi)$.

(d3) χ is linearly induced for more than one subgroup.

For these characters χ we have $K = \mathrm{Z}(\chi)$.

Exercise 9 ([Man12]). For a p-group G with exactly a_1 minimal characters of degree p^c, one of the following holds: (a) $a_1 = p - 1$ and G is extraspecial [BZ, Lemma 3.35]; (b) $|G : G'| = p^{2c}$ and $a_1 \equiv -1 \pmod{p^2}$; (c) $|G : G'| = p^{2c+1}$ and $a_1 \equiv -p \pmod{p^2}$, (d) $|G : G'| \geq p^{2c+2}$ and $a_1 \equiv 0 \pmod{|G : G'|/p^{2c}}$.

Theorem 22.15 ([Man12]). *The following conditions on a p-group G are equivalent: (a) G has exactly $p^2 - 1$ irreducible characters of degree p. (b) $|G : \mathrm{K}_4(G)| = p^4$ and $C' = \mathrm{K}_4(G)$, where $C = C_G(G'/\mathrm{K}_4(G))$. (c) G is either of order p^4 and class 3, or $|G : \mathrm{K}_5(G)| = p^5$ and $G/\mathrm{K}_5(G)$ has no abelian subgroups of index p.*

Exercise 10. Let G be a p-group and $S = \Omega_1(\mathrm{Z}(G))$ is of order p^k. Prove that G has a faithful character τ such that $|\mathrm{Irr}(\tau)| = k$.

Solution. The subgroup $S \cong \mathrm{E}_{p^k}$ has maximal subgroups $S_1, \dots, S_k$ such that $S_1 \cap \dots \cap S_k = \{1\}$. Let $M_i \lhd G$ be maximal such that $M_i \cap S = S_i$; then $SM_i/M_i \cong S/S_i$ is the unique minimal normal subgroup of G/M_i. Since $(M_1 \cap \dots \cap M_k) \cap S = S_1 \cap \dots \cap S_k = \{1\}$, it follows that $M_1 \cap \dots \cap M_k = \{1\}$. Let $\chi_i \in \mathrm{Irr}(G/M_i)$ be faithful, all i. Then $\tau = \chi_1 + \dots + \chi_k$ is the required character.

7°. Theorem 22.16 improves Jordan's theorem in case of p-groups [BZ, Chapter 18].

Theorem 22.16 ([Pas]). *Suppose that a p-group G has a faithful irreducible character χ of degree p^n. Set $j_p(n) = 1 + p + \cdots + p^{n-1}$ for $n > 1$ and $j_p(0) = j_p(1) = 1$. Then G contains an abelian subgroup of index $\le p^{j_p(n)}$.*

Proof. We may assume that $n > 0$. If $n = 1$, then G has an abelian subgroup of index p [BZ, Theorem 18.1] so let $n > 1$. We proceed by induction on n. Since G is an M-group, $\chi = \lambda^G$, where λ is a linear character of a subgroup F of index p^n in G. Let $F < H \in \Gamma_1$. Then $\lambda^H = \mu_1 \in \mathrm{Irr}(H)$. Let $\chi_H = \mu_1 + \cdots + \mu_p$ be the Clifford decomposition. We have $\mu_i(1) = p^{n-1}$ for all i.

By induction, $H/\ker(\mu_i)$ contains an abelian subgroup $A_i/\ker(\mu_i)$ of index $\le p^{j_p(n-1)}$. We have $\{1\} = \ker(\chi_H) = \bigcap_{i=1}^p \ker(\mu_i)$. Therefore, H is isomorphic to a subgroup of the direct product of groups $H/\ker(\mu_i)$, $i = 1, \dots, p$. By the above, $|\prod_{i=1}^p H/\ker(\mu_i) : \prod_{i=1}^p A_i/\ker(\mu_i)| \le (p^{j_p(n-1)})^p = p^{pj_p(n-1)}$. Since $H \le \prod_{i=1}^p H/\ker(\mu_i)$ and the last group contains an abelian subgroup, say T, of index $\le p^{pj_p(n-1)}$, then H contains an abelian subgroup $H \cap T$ of index $\le p^{pj_p(n-1)}$, by the product formula. Thus, $|G : (H \cap T)| = |G : H||H : (H \cap T)| \le p \cdot p^{pj_p(n-1)} = p^{j_p(n)}$. □

If A is an abelian normal subgroup of a p-group G, then $\mathrm{b}(G) = \max\{\chi(1) \mid \chi \in \mathrm{Irr}(G)\}$ divides $|G : A|$. We show that the equality $\mathrm{b}(G) = |G : A|$ is possible and the group G/A may be arbitrary. Indeed, let a regular wreath product $G = Z \,\mathrm{wr}\, H$, where H is a regular permutation group and $|Z| = p$. We must to prove that $\mathrm{b}(G) = |G : B| = |H|$, where B is the base subgroup of the wreath product. Let $B = Z_1 \times \cdots \times Z_{|H|}$, where $Z_1, \dots, Z_{|H|}$ (of order p) regularly permuted by H. It is known that $\mathrm{Z}(G)$ equals the diagonal subgroup of B so it is a unique minimal normal subgroup of G. Set $L = Z_2 \times \cdots \times Z_{|H|}$ and let μ be a nonprincipal linear character of B/L. Set $\chi = \mu^G$. Then $\ker(\chi) = L_G = \{1\}$ so χ is faithful. We have $\chi(1) = |G : B| = |H|$. It remains to prove that $\chi \in \mathrm{Irr}(G)$. To this end, it is enough to show that the inertia group $\mathrm{I}_G(\mu) = B$. Take $h \in H^\#$ and assume that $\mu^h = \mu$. Then $\mu(hxh^{-1}) = \mu(x)$ for all $x \in B$. However, if x is a generator of Z_1, then hxh^{-1} is a generator of Z_j for some $j > 1$, so $hxh^{-1} \in L$. It follows that $\mu(x) = \mu(hxh^{-1}) = 1$, contrary to the choice of μ. Thus, $\chi(1) \in \mathrm{Irr}(G)$.

8°. The following eight lemmas and theorems are taken from Kazarin's letter at Nov. 11, 2000. Given $H \lhd G$, we write $\mathrm{Irr}(G \mid H) = \mathrm{Irr}(G) - \mathrm{Irr}(G/H)$.

Lemma 22.17. *Let $H \le G'$ be G-invariant of order p^2, G is a group of order p^m. Then $|\mathrm{Irr}(G \mid H)| \equiv 0 \pmod{p^2 - 1}$. In particular, if $|G'| = p^{2n}$, then $\mathrm{n}(G) \equiv 0 \pmod{(p^2 - 1)}$ and $\mathrm{n}(G) \ge (p^2 - 1) \cdot n$. If $|G'| = p^{2n+1}$, then $\mathrm{n}(G) \equiv 0 \pmod{(p - 1)}$ and $\mathrm{n}(G) \ge (p^2 - 1) \cdot n + (p - 1)$.*

Proof. If $\mathrm{Irr}(G \mid H) = \{\chi^1, \dots, \chi^t\}$, then

$$p^{m-2}(p^2-1) = |G| - |G/H| = \chi^1(1)^2 + \dots + \chi^t(1)^2 \equiv t \pmod{(p^2-1)},$$

completing the proof of the first assertion and the first half of the second one (by induction). If $|G'| = p^h$ and $\mathrm{Irr}_1(G) = \{\chi^1, \dots, \chi^{\mathrm{n}(G)}\}$, then

$$p^{m-h}(p^h-1) = |G| - |G/G'| = \sum_{i=1}^{\mathrm{n}(G)} \chi^i(1)^2 \equiv \mathrm{n}(G) \pmod{p^2-1},$$

so $p-1$ divides $\mathrm{n}(G)$. Using induction, we complete the proof. □

Exercise 11. (a) If $G = A \times B$, then $\mathrm{n}(G) = \left(\frac{\mathrm{k}(A)}{\mathrm{n}(A)} + \frac{\mathrm{k}(B)}{\mathrm{n}(B)} - 1\right)\mathrm{n}(A)\mathrm{n}(B)$ if $\mathrm{n}(A)\mathrm{n}(B) \neq 0$. If $\mathrm{n}(A) = 0$, then $\mathrm{n}(G) = |A|\mathrm{n}(B)$.

(b) If G is a p-group with $|G'| = p$, then $\mathrm{n}(G) = \frac{p-1}{p} z_0$, where $z_0 = |\mathrm{Z}(G)|$.

Solution. (b) Let $|G| = p^m$; then $\mathrm{cd}(G) = \{1, p^r\}$. If $\chi \in \mathrm{Irr}_1(G)$, then $p^{2r} = \chi(1)^2 = |G : \mathrm{Z}(G)|$. We have $p^m = |G| = |G : G'| + \mathrm{n}(G)p^{2r} = p^{m-1} + \mathrm{n}(G)p^{2r}$ so that $\mathrm{n}(G) = \frac{p^m - p^{m-1}}{p^{2r}} = \frac{p^m - p^{m-1}}{p^m}|\mathrm{Z}(G)| = \frac{p-1}{p} z_0$.

Let $\mathrm{k}_G(M)$ be the number of G-classes having nonempty intersection with $M \subseteq G$. In particular, $\mathrm{k}_G(G) = \mathrm{k}(G)$. Let T be a transversal of G' in G, $1 \in T$. Since any subset xG' is G-invariant for $x \in G$, we get $\mathrm{n}(G) + |G : G'| = \mathrm{k}(G) = \sum_{x \in T} \mathrm{k}_G(xG')$ hence (Isaacs–Passman)

$$(*) \qquad \mathrm{n}(G) = k_G(G') - 1 + \sum_{x \in T^\#} (\mathrm{k}_G(xG') - 1) \geq \mathrm{k}_G(G') - 1$$

with equality if and only if $\mathrm{k}_G(xG') = 1$ for all $x \in G - G'$.

Lemma 22.18. *Let G be a nonabelian p-group. Then: (a) If $G' \not\leq \mathrm{Z}(G)$, then $\mathrm{n}(G) \geq |\mathrm{Z}(G)|$; (b) If $\mathrm{Z}(G) < G'$, then $\mathrm{n}(G) > |\mathrm{Z}(G)|$; (c) If $G' \leq \mathrm{Z}(G)$, then $\mathrm{n}(G) \geq |\mathrm{Z}(G)| - |\mathrm{Z}(G) : G'|$.*

Proof. (a) Set $m = |G' \cap \mathrm{Z}(G)|$. For every $z \in \mathrm{Z}(G)$, the coset zG' contains m central G-classes (their union equals $z(G' \cap \mathrm{Z}(G))$) and at least one noncentral G-class. Therefore, $\mathrm{k}_G(zG') - 1 \geq m$. The number of distinct cosets zG' ($z \in \mathrm{Z}(G)$) equals $|\mathrm{Z}(G)G' : G'| = \frac{|\mathrm{Z}(G)|}{|G' \cap \mathrm{Z}(G)|} = \frac{|\mathrm{Z}(G)|}{m}$. Then, by $(*)$, we get $\mathrm{n}(G) \geq m \cdot \frac{|\mathrm{Z}(G)|}{|G' \cap \mathrm{Z}(G)|} = m \cdot \frac{|\mathrm{Z}(G)|}{m} = |\mathrm{Z}(G)|$. (b) follows from $(*)$.

(c) Let $G' \leq \mathrm{Z}(G)$. Then, for $z \in \mathrm{Z}(G)$, we have $\mathrm{k}_G(zG') - 1 = |G'| - 1$. Therefore, by $(*)$, we get $\mathrm{n}(G) \geq \frac{|\mathrm{Z}(G)|}{|G'|}(|G'| - 1) = |\mathrm{Z}(G)| - |\mathrm{Z}(G) : G'|$. □

Lemma 22.19. *Let G be a p-group, $|G'| = p^2$, $|G| = p^m$, $|Z(G)| = p^\beta$. If $G' \not\leq Z(G)$, then $n(G) = (p-1)(2p^{\beta-1} + \lambda)$, where $\lambda \equiv 2(-1)^\beta \pmod{p+1}$ and $0 < \lambda \leq p^{m-4}(p+1) - 2p^{\beta-1}$. If $G' \leq Z(G)$, then $n(G) = p^{\beta-2}(p^2-1) + \lambda(p-1)$, where $\lambda \equiv 0 \pmod{p+1}$ and $0 \leq \lambda \leq (p^{m-4} - p^{\beta-2})(p+1)$.*

Exercise 12 (= [BZ, Lemma 31.7] (Isaacs–Passman)). Let G be a p-group with $|G'| = p^r$, $r > 0$, and suppose that every nonlinear irreducible character of G has degree $\geq p^r$, i.e., the minimal degree of G is $\geq p^r$. Then $G' \leq Z(G)$.

Solution. Let $a \in G - Z(G)$ and let $\Omega = \{a = a_1, \dots, a_k\}$ be the G-class containing $a = a_1$. Since $1 < k \leq |G'| = p^r$, the character θ of the permutation representation $g \mapsto \binom{a_1, \dots, a_k}{a_1^g, \dots, a_k^g}$ $(g \in G)$ has degree k and involves 1_G (Burnside). In that case, θ is the sum of linear characters so $G' \leq \ker(\theta) \leq C_G(\Omega)$. Since Ω is an arbitrary G-class contained in $G - Z(G)$ and $\langle G - Z(G) \rangle = G$, we conclude that $G' \leq Z(G)$.

Lemma 22.20. *Let G be a p-group, $|G'| = p^2$ and $n(G) = p^2 - 1$. Then either (i) $cl(G) = 3$ and $|G| \in \{p^4, p^5\}$, or (ii) $G' = Z(G) \cong E_{p^2}$ and $G/R \cong ES(m, p)$ for each $R < G'$ of order p (here $ES(m, p)$ is an extraspecial group of order p^{2m+1}), and G is special.*

Proof. Assume that $|G| = p^4$ and G is not of maximal class. Then $|G'| = p^2 = |Z(G)|$ so $G' = Z(G)$ and G is minimal nonabelian; in that case, $|G'| = p$, a contradiction. Thus, G is of maximal class. Then $cd(G) = \{1, p\}$, in view of $\chi(1)^2 \leq |G : Z(G)|$ for all $\chi \in Irr(G)$, so $n(G)p^2 + p^2 = |G| = p^4$ so $n(G) = p^2 - 1$.

Next we assume that $|G| \geq p^5$. Let $R \leq G' \cap Z(G)$ be of order p. Then $|G'/R| = p$ and $G'/R \leq Z(G/R)$ so $\bar{z}_0 = |Z(G/R)| \in \{p, p^2\}$ (Exercise 11(b)).

(i) Suppose that $\bar{z}_0 = p$. Then $G/R \cong ES(m, p)$, $cd(G/R) = \{1, p^m\}$, $|Irr(G \mid R)| = n(G) - |Irr_1(G/R)| = (p^2 - 1) - (p - 1) = p^2 - p$ so $|G| - |G/R| = p^{2m+1}(p-1) = p^{2m}(p^2 - p)$. It follows that every character in $Irr(G \mid R)$ has degree p^m so $cd(G) = \{1, p^m\}$. Since $|G| > p^4$, we get $m > 1$; then, by Exercise 12, $G' \leq Z(G)$. Next, by Lemma 22.18(c), $G' = Z(G)$. It is easy to see that $G' \cong E_{p^2}$.

(ii) Now let $\bar{z}_0 = |Z(G/R)| = p^2$. Then $n(G/R) = \frac{p-1}{p} p^2 = p^2 - p$ (Lemma 22.18(c)). Therefore, $|Irr(G \mid R)| = p^2 - 1 - (p^2 - p) = p - 1$ so G/R is extraspecial. Setting $|G| = p^n$, we have $|G| - |G/R| = (p-1)p^{n-1}$. Let $Irr(G \mid R) = \{\chi^1, \dots, \chi^{p-1}\}$ with $p^{k_1} = \chi^1(1) \leq \dots \leq \chi^{p-1}(1) = p^{k_{p-1}}$; then $(p-1)p^{n-1} = p^{2k_1} + \dots + p^{2k_{p-1}}$. It follows that $2k_1 = \dots = 2k_{p-1} = n - 1$ so $n = 2m + 1$. Then $cd(G/R) = \{1, p^{m-1}\}$ so $cd(G) = \{1, p^{m-1}, p^m\}$.

If $G' \not\leq Z(G)$, then $m - 1 = 1$ (Exercise 12) so $n = 2m + 1 = 5$, $|G| = p^5$.

Now let $m > 2$; then $G' \leq Z(G)$. In that case, for every subgroup R of G' of order p, $n(G/R) = p^2 - p$. Then $n(G) = (p^2 - p) + p(p-1) > p^2 - 1$, a contradiction. □

Theorem 22.21. *Let G be a p-group with $n(G) = p^2 + p - 2$. Then one of the following holds: (a) $p = 2$ and $|G'| = 2$, $|Z(G)| = 8$; (b) $p > 2$, $|G'| = p^3$ and $|G| = p^5$; (c) $|G'| = p^3$ and $|G| = p^7$.*

Proof. We have $|G'| < p^4$ (use Lemma 22.17).

(i) If $|G'| = p$, then (Exercise 11(b)) $\mathrm{n}(G) = \frac{p-1}{p} z_0 = p^2+p-2 = (p-1)(p+2)$ so $z_0 = p(p+2)$ is a power of p. It follows that $p = 2$ and $z_0 = 2(2+2) = 8$.

(ii) If $|G'| = p^2$, then $\mathrm{n}(G) \equiv 0 \pmod{p^2-1}$ (Lemma 22.17), and so $\mathrm{n}(G) \neq p^2+p-2$.

(iii) Thus $|G'| = p^3$ (Lemma 22.17). Let R be a subgroup of order p in $G' \cap \mathrm{Z}(G)$. Then $|(G/R)'| = p^2$ and $\mathrm{n}(G/R) = p^2-1$; moreover, all $p-1$ characters in $\mathrm{Irr}(G \mid R)$ have the same degree, say p^m. Setting $|G| = p^n$, we get $|G| - |G/R| = p^{n-1}(p-1) = (p-1)p^{2m}$, i.e., $n = 2m+1$. By Lemma 22.20, either $|G/R| = p^4$ or $\mathrm{cd}(G/R) = \{1, p^{m-1}\}$. If $m-1 \geq 3$, then $G' \leq \mathrm{Z}(G)$ and so, by $(*)$, $\mathrm{n}(G) \geq \mathrm{k}_G(G') \geq |G'| - 1 = p^3 - 1 > p^2+p-2$, a contradiction. Hence, $m \leq 3$, i.e., $|G| \in \{p^5, p^7\}$. □

Corollary 22.22. *If G is a p-group with $|G'| = p^3$ and $|G| \notin \{p^5, p^7\}$, then $\mathrm{n}(G) \geq p^2+2p-3$ (see Lemma 22.17).*

Theorem 22.23. *Let G be a 2-group with $|G'| = 4$. If $\mathrm{n}(G) \leq 5$, then $\mathrm{n}(G) = 3$ and one of the following holds: (a) G is special of order 2^{2m+2}, $\mathrm{cd}(G) = \{1, 2^m\}$; (b) G is of maximal class and order 2^4; (c) $|G| = 2^5$, $\mathrm{cd}(G) = \{1, 2, 2^2\}$.*

Theorem 22.24. *Let G be a 2-group with $|G'| = 8$. If $\mathrm{n}(G) \leq 5$, then $|G| = 2^6$.*

Thus, if $|G| = p^n$ and $\mathrm{n}(G) = 3$, then one of the following holds: (a) $p = 2, n = 4$ and G is of maximal class, or (b) $p = 2$, $n = 5$, $|G'| = 4$, $\{z_0, z_1\} = \{2, 6\}$, or (c) G is special with $n = 2m+2$, $|G'| = 4$ and $\{z_0, z_1\} = \{4, 0\}$. Here $z_i = |\{x \in G \mid |G : \mathrm{C}_G(x)| = p^i, i = 0, 1, \dots, \log_p(|G'|)$.

According to the letter of Mann, if G is a 2-group with $\mathrm{n}(G) = 4$, then $|G'| = 2$ and $|\mathrm{Z}(G)| = 8$. However, the proof of this is not trivial.

Remarks. 1. Let G be the holomorph of the cyclic group L of order 8. Then $G/L \cong \mathrm{E}_4$ and $\Phi(G) = G' = \mho_1(L)$ is cyclic of order 4, $\mathrm{Z}(G) = \Omega_1(L)$. By Ito's theorem on degrees (Introduction, Theorem 17), we get $\mathrm{cd}(G) \subseteq \{1, 2, 4\}$. By Theorem 22.5, $4 \in \mathrm{cd}(G)$ so $\mathrm{cd}(G) = \{1, 2, 4\}$ and $\mathrm{Irr}_1(G)$ has exactly two characters of degree 2 and one character of degree 4 so $\mathrm{n}(G) = 3$. This is a group of Lemma 22.18(b). The Sylow 2-subgroup G of the 2-closed minimal nonnilpotent group of order $5 \cdot 2^6$ satisfies $G' = \mathrm{Z}(G)$ and $\mathrm{n}(G) = 3$.

2. (Janko). Below we present a 2-group G of order 2^6 with $\mathrm{n}(G) = 5$. Let

$$G = \langle u, y, l \mid u^4 = y^4 = l^4 = [u, y] = 1, u^l = u^3 y, y^l = u^2 y \rangle.$$

Here $A = \langle u, y \rangle \lhd G$ is abelian of type $(4, 4)$. The group G is a split extension of A with $\langle l \rangle = \mathrm{C}_4$, $\mathrm{Z}(G) = \langle y^2 \rangle$ is of order 2, $G' = \langle u^2, y \rangle$ is abelian of type $(4, 2)$, $\Omega_1(A) = \langle u^2, y^2 \rangle \cong \mathrm{E}_4$. Set $B = A\langle l^2 \rangle$. All elements in $B - A$ are involutions. All elements in $G - B$ are of order 4. The conjugacy classes in G with representatives: 1

(of size 1), y^2 (of size 1), u^2 (of size 2), three conjugacy classes in $A - \Omega_1(A)$ (each of size 4), three conjugacy classes of involutions in $B - A$ with representatives: l^2 (of size 4), l^2y (of size 4), l^2u (of size 8), four conjugacy classes of elements of order 4 in $G - B$ (each of size 8). Hence $\mathrm{k}(G) = 13$. It follows from $|G : G'| = 8$ that $\mathrm{n}(G) = \mathrm{k}(G) - |G : G'| = 13 - 8 = 5$. (Mann showed that if $\mathrm{n}(G) = 5$, then $|G| = 2^6$; this result is also obtained by Kazarin's student E. Chankov.)

3. Write $\mathrm{Irr}_{(k)}(G) = \{\chi \in \mathrm{Irr}(G) \mid \chi(1) = k\}$. Let $G = A * B$, where A and B are 2-groups of maximal class and orders 2^m and 2^n, respectively, $|G| = 2^{m+n-1}$. Then $\mathrm{cd}(G) \subseteq \mathrm{cd}(A \times B) = \{1, 2, 4\}$. Let $r = |\mathrm{Irr}_{(2)}(G)|$ and $s = |\mathrm{Irr}_{(4)}(G)|$. Let us find r and s. We have $G \cong (A \times B)/Z$, where Z is the diagonal subgroup of $\mathrm{Z}(A) \times \mathrm{Z}(B)$, $|G : G'| = 2^4$, $|\mathrm{Z}(G)| = 2$. It follows from Lemma 1.1 that G has no abelian subgroups of index 2 so G has no faithful irreducible characters of degree 2 [BZ, Theorem 18.1]. Thus, r is the number of irreducible characters of degree 2 of the group $G/\mathrm{Z}(G) \cong (A/\mathrm{Z}(A)) \times (B/\mathrm{Z}(B)$, i.e., $r = 4(2^{m-3} - 1) + 4(2^{n-3} - 1) = 2^{m-1} + 2^{n-1} - 8$. We have $|G| = 2^{m+n-1} = 16 + 4r + 16s = 16 + 2^{m+1} + 2^{n+1} - 32 + 16s$ so $s = 2^{m+n-5} - 2^{m-3} - 2^{n-3} + 1$. In particular, if $m = 4$, $n = 3$, then $r = 4$, $s = 2$ and so $\mathrm{n}(G) = r + s = 6$.

4. E. Chankov proved that, if G is a p-group with $\mathrm{n}(G) = 4$, then either $G \cong \mathrm{ES}(m, 5)$, $p = 5$ or $|G| = 2^{2n+3}$, $|\mathrm{Z}(G)| = 8$ and $|G'| = 2$. He also classified all p-groups with $\mathrm{n}(G) = 6$.

Exercise 13. Let $\mathcal{H} = \langle a, b \mid a^4 = b^4 = 1, a^b = a^3\rangle$ be nonabelian metacyclic of exponent 4 and $Q = \langle c, d \mid c^4 = d^2 = 1, c^d = c^3\rangle \cong \mathrm{D}_8$. Let $G = \mathcal{H} * Q$ be the central product with amalgamated subgroup $\mathcal{H} \cap Q = \langle b^2\rangle = \langle c^2\rangle$. Find $\mathrm{cd}(G)$ and $\mathrm{n}(G)$.

Exercise 14. Does there exist a 2-group G with $\mathrm{n}(G) = 11$?

Theorem 22.25 (inspired by Ito's letter at Jan. 3, 2005). *A p-group G with a faithful irreducible character of degree p^n has the derived length at most $n + 1$. The above estimate is best possible.*

Proof. It is known (see [BZ, Theorem 18.1]) that if G has a faithful irreducible character of degree p, then it has an abelian subgroup of index p. Let $\chi \in \mathrm{Irr}(G)$ be faithful of degree p^n, $n > 1$. We want to prove that $\mathrm{dl}(G)$, the derived length of G, does not exceed $n + 1$, and $n + 1$ is the best possible estimate. We proceed by induction on n. If G is a 2-group of maximal class, then $n = 1$, by Ito's theorem on degrees, and $\mathrm{dl}(G) = 2$. In what follows we assume that G is not a 2-group of maximal class. Then G has a normal subgroup R of type (p, p). We have $|G : \mathrm{C}_G(R)| \leq p$ so there exists in G a maximal subgroup M such that $R < M \leq \mathrm{C}_G(R)$. In particular, $\mathrm{Z}(M)$ is noncyclic. Therefore, the Clifford's decomposition of χ_M, which is faithful on M, is of the form $\chi_M = \mu_1 + \cdots + \mu_p$, where $\mu_1, \ldots, \mu_p$ are pairwise distinct G-conjugate irreducible characters of M. Since χ is faithful, we have $\bigcap_{i=1}^p \ker(\mu_i) = \ker(\chi_M) \leq M \cap \ker(\chi) = \{1\}$. By induction, $\mathrm{dl}(M/\ker(\mu_i)) \leq n$

for $i = 1, \dots, p$ since $\mu_i(1) = p^{n-1}$. It follows that M is isomorphic to a subgroup of the direct product $(M/\ker(\mu_1)) \times \cdots \times (M/\ker(\mu_p))$ so $\mathrm{dl}(M) \le n$. Since G/M is abelian, we get $\mathrm{dl}(G) \le n+1$. (Ito wrote that his proof of the estimate $\mathrm{dl}(G) \le n+1$ coincides with the above argument.)

It remains to prove that the above estimate is best possible. Let $G = \Sigma_{p^{n+1}} \in \mathrm{Syl}_p(\mathrm{S}_{p^{n+1}})$, where $\mathrm{S}_{p^{n+1}}$ is the symmetric group of degree p^{n+1}. Then $G = U \,\mathrm{wr}\, C$, the standard wreath product, where the 'passive' factor $U \cong \Sigma_{p^n}$ and the 'active' factor $C \cong \mathrm{C}_p$. Let $U_1, \dots, U_p$ be coordinate subgroups of the base B of G; then $B = U_1 \times \cdots \times U_p$. By induction, U has a faithful irreducible character of degree p^{n-1} and, as we know, the derived length of U equals n. Set $B_1 = U_2 \times \cdots \times U_p$. Let $\phi \in \mathrm{Irr}(B/B_1)$ be a faithful irreducible character of degree p^{n-1} (ϕ exists since $B/B_1 \cong U$). We claim that $\chi = \phi^G$ is a faithful irreducible character of G of degree p^n. We have $\ker(\chi) = (B_1)_G = \bigcap_{x \in G} B_1^x = \{1\}$. Indeed, $\mathrm{Z}(G) = \langle z_1 z_1^c \dots z_1^{c^{p-1}} \rangle$ is a unique minimal normal subgroup of G; here z_1 is a generator of $\mathrm{Z}(U_1)$ and c a generator of $\mathrm{C}_p = C$. Since $\mathrm{Z}(G) \not\le B_1$, we get $(B_1)_G = \{1\}$, as asserted. Now assume that χ is reducible. Then, by Clifford theory, $\chi = \mu_1 + \cdots + \mu_p$, where $\mu_1, \dots, \mu_p$ are pairwise distinct irreducible characters of G of the same degree $\phi(1) = p^{n-1}$. In that case, by reciprocity, $\chi_B = p \cdot \phi$ and $\ker(\chi) \cap B = \ker(\chi_B) = \ker(\phi) = B_1$ is not normal in G, a contradiction. Thus, $\chi \in \mathrm{Irr}(G)$ is faithful of degree p^n and the derived length of our G equals $n+1$. □

§23

On subgroups of given exponent

Let G be a group of order p^m and $m \leq 1 + (p-1)k$. According to Hall [Hal2, Theorem 2.64], if $\Omega_k(G) = G$, then $\exp(G) \leq p^k$. In this section we improve this result.

Lemma 23.1. *Suppose that G is a group of maximal class and order p^m. If $m > p+1$, then there exists $H < G$ such that $H \not\leq G_1$, where G_1 is the fundamental subgroup of G, and $|H| = p^{p+1}$. Moreover, H is of maximal class.*

Proof. This follows from Theorems 9.6 and 13.19(c). □

Lemma 23.2. *Suppose that A, B are two normal subgroups of a p-group G, $|A| \leq p^n$ and $|B| \leq p^n$. Then $\mathrm{cl}(AB) \leq n$. This estimate is best possible.*

The first assertion is trivial. Let G be a group of maximal class and order p^{n+1} and let $A, B \in \Gamma_1$ be distinct. Then $G = AB$ and $\mathrm{cl}(G) = n$.

Lemma 23.3. *Let G be a p-group.*

(a) If $R \lhd G$ of order p^n is such that G/R is cyclic of order at least p^2 and $H/R \lhd G/R$ is of order p, then $\mathrm{cl}(H) \leq n-1$.

(b) If $\Omega_2(G) < G$, then $\Omega_2(G)$ is not of maximal class.

Proof. (a) See Lemma 17.4(c). (b) follows from Theorems 9.5, 9.6 and 13.2. □

Lemma 23.4 (= Corollary 13.3). *Let the p-group G be neither absolutely regular nor of maximal class and take $L \lhd G$ of order p^s and exponent p, $0 < s < p$. Let $\mathfrak{M}$ denote the set of all normal subgroups of G of order p^{s+1} and exponent p containing L. Then $|\mathfrak{M}| \equiv 1 \pmod{p}$.*

Example 1. Let $\mho^1(G) = \mho_1(G)$, $\mho^2(G) = \mho_1(\mho^1(G))$, $\mho^3(G) = \mho_1(\mho^2(G))$, and so on. We have $\exp(G/\mho^i(G)) \leq p^i$ so $\mho_i(G) \leq \mho^i(G)$. We will show that the strong inequality is possible. Let $G = B(4,2)$ be the 2-group of maximal order such that its exponent is 4 and $\mathrm{d}(G) = 2$. It is known (Burnside, Tobin) that $|G| = 2^{12}$. Obviously, $\mho_2(G) = \{1\}$. Since $\mathrm{d}(G) = 2$, we have $|G : \mho_1(G)| = 4$. By Schreier's theorem (Appendix 25), $\mathrm{d}(\mho_1(G)) \leq 1 + (\mathrm{d}(G) - 1)|G : \mho_1(G)| = 5$. Since $|\mho_1(G)| = 2^{10}$, $\mho_1(G)$ is not elementary abelian so $\mho^2(G) = \mho_1(\mho_1(G)) > \{1\} = \mho_2(G)$. (If G is regular, then $\mho^i(G) = \mho_i(G)$, all i.)

Example 2. Let G be an absolutely regular p-group. If $|G| > p^{(p-1)k}$, then $\exp(G) > p^k$. Now let G be a p-group of maximal class and order p^m, $m > p+1$. If $m-1 = (p-1)k$, then $\exp(G) = p^k$. If $m-1 > (p-1)k$, then $\exp(G) > p^k$.

The proof of our generalization of [Hal2, Theorem 2.64] is based on the following

Lemma 23.5. *Let G be a p-group. If $|G| = p^m \leq p^{kp}$, then $\mho^{k-1}(G)$ is either absolutely regular or of order p^p and exponent p; the same is true for $\mho_{k-1}(G)$. If, in addition, $m < kp$, then the above two subgroups are absolutely regular.*

Proof. Assume that $\mho^{k-1}(G)$ is not absolutely regular. One may assume that $k > 1$. Then $\mho^j(G)$ is not absolutely regular for $j \leq k-1$ so $|\mho^{i-1} : \mho^i(G)| \geq p^p$ for $i = 1, \dots, k$. In that case,

$$|G : \mho^{k-1}(G)| = \prod_{i=1}^{k-1} |\mho^{i-1}(G) : \mho^i(G)| \geq p^{p(k-1)}$$

so $|\mho^{k-1}(G)| \leq p^p$. In that case, if $\mho^{k-1}(G)$ is of exponent $> p$, it is absolutely regular. In view of $\mho_{k-1}(G) \leq \Omega_1(\mho^{k-1}(G))$, we are done (Theorem 7.1). □

Remarks. 1. Let $k > 1$ and let G be a group of order p^{1+pk} and exponent p^k. We will prove that $\mho_{k-1}(G)$ is regular of order $\leq p^{p+1}$ and exponent $\leq p$. Assume that $\mho^{k-1}(G)$ is not absolutely regular. Then, as in Lemma 23.5, $|\mho^{k-1}(G)| \leq p^{p+1}$ and $\mho_{k-1}(G)$ is generated by elements of order $\leq p$. Therefore, it suffices to prove that $\mho_{k-1}(G)$ is regular. Assume that this is false. Then it is of maximal class and order p^{p+1}. Let R be a G-invariant subgroup of $\mho_{k-1}(G)$ of order p^2. Since $k > 1$, we have $\mho_{k-1}(G) \leq \Phi(G)$. Then $\mho_{k-1}(G) \leq \Phi(G) \leq C_G(R)$, and $R \leq Z(\mho_{k-1}(G))$, a contradiction. The same argument shows that $\mho^{k-1}(G)$ is regular.

2. Let $k > 1$, $p > 2$ and let G be a group of order p^{2+pk} and exponent p^k. We claim that $\mho^{k-1}(G)$ is regular of order p^{p+2} at most. As in Lemma 23.5, $|\mho^{k-1}(G)| \leq p^{p+2}$. Therefore, it suffices to prove that $\mho^{k-1}(G)$ is regular. Assume that this is false. Then $\mho^{k-1}(G)$ has a G-invariant subgroup R of order p^3 and exponent p (Theorems 13.7 and 13.5). Since $p^2 \nmid \exp(\mathrm{Aut}(R))$, we have $\exp(G/C_G(R)) \leq p$ so $\mho_1(G) \leq C_G(R)$. Since $\mho^{k-1}(G) \leq \mho_1(G)$, it follows that $R \leq Z(\mho^{k-1}(G))$ so $\mathrm{cl}(\mho^{k-1}(G)) \leq p-1$ and $\mho^{k-1}(G)$ is regular (Theorem 7.1(b)).

Theorem 23.6. *For $k > 1$ and a p-group G, one of the following assertions holds: (a) $\exp(\Omega_k(G)) \leq p^k$; (b) G is of maximal class; (c) G has a subgroup of order $p^{2+(p-1)k}$ and exponent $\leq p^k$.*

Proof. Set $|G| = p^m$. Suppose that G is a counterexample with minimal $|G| + k$. This means that $\exp(\Omega_k(G)) > p^k$ so G is irregular, by Theorem 7.2, G is not of maximal class and G has no subgroups of order $p^{2+(p-1)k}$ and exponent $\leq p^k$. Next, $\exp(G) \geq p^{k+1} > p^2$ so every maximal subgroup of G has exponent $\geq p^k$.

Suppose that $H \in \Gamma_1$ is of maximal class. By Theorem 12.1(a), G has a normal subgroup R of order p^p and exponent p. Assume that $R < H$; then $|H| = p^{p+1}$ (Theorem 9.6) so $|G| = p^{p+2}$, $k = 2$. Since G is not of maximal class, we get $\exp(G/\mathrm{K}_{p+1}(G))| = p$ (Theorem 12.12(b)) so $\exp(G) = p^2 < p^3$, a contradiction. Thus, $R \not\leq H$ so $G = RH$ and $|H \cap R| = p^{p-1}$. By Theorem 12.12(b) again, $\exp(G) = \exp(H)$ so $\exp(H) \geq p^{k+1}$. Then there is $F \leq H$ of maximal class, of exponent p^k and order $p^{1+(p-1)k}$ (see Example 1). Then RF is of order $p^{2+(p-1)k}$ and exponent p^k since $RF/R \cong F/(F \cap R)$ is of exponent p^{k-1}, so G is not a counterexample.

Thus, the set Γ_1 has no members of maximal class. Then, by induction, we have $\exp(\Omega_k(L)) = p^k$ for every $L \in \Gamma_1$ since L has no subgroups of order $p^{2+(p-1)p}$ and exponent $\leq p^k$ and so, by assumption, $|\Omega_k(L)| < p^{2+(p-1)k}$. Next, every $L \in \Gamma_1$ is not absolutely regular (otherwise, by Theorem 12.1(b), $G = L\Omega_1(G)$ satisfies (a)). Thus,

(i) If $L \in \Gamma_1$, then L is neither absolutely regular nor of maximal class.

(ii) $\exp(\Omega_k(M)) = p^k$ for all $M \in \Gamma_1$.

Let $H \in \Gamma_1$ be such that $|\Omega_k(H)| \geq |\Omega_k(L)|$ for all $L \in \Gamma_1$. Set $D = \Omega_k(H)$. By (ii), $\exp(D) = p^k$ so $|D| < p^{2+(p-1)k}$. It follows that if $M < G$ is of exponent $\leq p^k$, then $|M| \leq |D|$. As G is a counterexample, there exists $z \in G - D$ with $o(z) \leq p^k$. Since $z^p \in H$ and $o(z^p) < p^k$, we get $z^p \in D$. Set $L = \langle z, D\rangle$; then $|L| = p|D|$. Assume that $L < G$. Then $L \leq F \in \Gamma_1$, and $\Omega_k(F) \geq L > D$, contrary to the choice of H. Thus, $L = G$ so $|G : D| = p$ hence $H = D$. In particular, $\exp(H) = p^k$. Thus

(iii) If $H \in \Gamma_1$ is chosen as in the previous paragraph (i), then $\exp(H) = p^k$, $\exp(G) = p^{k+1}$, $|H| \leq p^{1+(p-1)k}$, and hence $|G| \leq p^{2+(p-1)k}$. On the other hand, $H < \Omega_k(G)$ so $G = \Omega_k(G)$.

If $U \in \Gamma_1 - \{H\}$, then $H \cap U \leq \Omega_k(U)$, by (iii). If $x \in G - H$ is of order $\leq p^k$ (see (iii)), then $x \in L \in \Gamma_1 - \{H\}$, and, by (ii), $L = \langle x, L \cap H\rangle$ is of exponent p^k since $\Omega_k(L) = L$. Thus,

(iv) The set Γ_1 has two distinct members of exponent p^k: H and L.

By (iv), $\Omega_1(\mho_{k-1}(L)) = \mho_{k-1}(L)$ so, by Lemma 23.5, $\exp(\mho_{k-1}(L)) = p$, $|\mho_{k-1}(L)| \leq p^{p-1}$ since $1 + k(p-1) < pk$ and the same is true for H. Next, $\mho_{k-1}(H), \mho_{k-1}(L) \triangleleft G$. Set $C = \mho_{k-1}(L)\mho_{k-1}(H)$. By Lemma 23.2, $\mathrm{cl}(C) < p$ so C is regular of exponent p (Theorems 7.1(b) and 7.2(b)). We define $T < G$ as follows. If $|C| \geq p^p$, set $T = C$. If $|C| < p^p$, let T be a G-invariant subgroup of order p^p and exponent p in G such that $C < T$ (see Lemma 23.4). Assume that $T \not\leq H$. Then $G = HT$ and $G/T \cong H/(H \cap T)$ is of exponent p^{k-1} since $\mho_{k-1}(H) \leq T \cap H$. In that case $\exp(G) \leq p^k$, contrary to (i). Thus, $T \leq H$. Similarly, $T \leq L$. We have $\exp(H/T) = p^{k-1} = \exp(L/T)$ so

$$G = HL, \;\; G/T = (H/T)(L/T), \;\; |G/T| \leq p^{1+(p-1)(k-1)}, \;\; \exp(G/T) = p^k.$$

Thus, G/T does not satisfy condition (c) of the theorem for $k-1$. We get $\Omega_{k-1}(G/T) \geq (H/T)(L/T) = G/T$ (thus, G/T does not satisfy condition (a) of the theorem for $k-1$) so G/T is irregular since $\exp(G/T) = p^k > p^{k-1}$ (Theorem 7.2(b)). It follows that $k > 2$ (otherwise, $|G/T| \leq p^p$ and G/T is regular). Assume that G/T is of maximal class. Since $\exp(G/T) = p^k$, we get $|G/T| \geq p^{2+(p-1)(k-1)}$ (see Example 2), contrary to the inequality given in the displayed line. Hence, G/T does not satisfy condition (b) of the theorem. Thus, the theorem is not true for $k-1$. However, $k-1 > 1$ and $|G/T| + (k-1) < |G| + k$, contrary to the induction hypothesis. □

Corollary 23.7. *Let $k > 1$ and let G be a p-group of order $\leq p^{2+(p-1)k}$. If $G = \Omega_k(G)$, then $\exp(G) \leq p^k$, unless G is of maximal class and order $p^{2+(p-1)k}$.*

Exercise 1. Classify the 2-groups without (a) subgroups of order 2^4 and exponent 4, (b) normal subgroups of order 2^4 and exponent 4.

Exercise 2. Suppose that a p-group G has exactly one maximal subgroup M such that $\exp(M) = \exp(G)$. Prove that $d(G) = 2$.

§24

Hall's theorem on normal subgroups of given exponent

1°. Probably, the following remarkable 'conditionless' structure theorem is one of the deepest consequences of Hall's theory of regular p-groups.

Theorem 24.1. *Let $H > \{1\}$ be a normal subgroup of a p-group G. Then there exists in H a chain $\mathcal{C} : \{1\} = L_0 < L_1 < \cdots < L_n = H$ of G-invariant subgroups with the properties ($i = 1, \dots, n$):*

(a) *L_i/L_{i-1} is of order $\leq p^{p-1}$ and exponent p, and*

(b) *either the order of L_i is exactly $p^{i(p-1)}$, or else $L_i = \Omega_i(H)$.*

A chain $\mathcal{C}$, having properties (a) and (b), is said to be a $(p-1)$-*admissible Hall chain* in H. The length of $\mathcal{C}$ is $\geq \log_p(\exp(H))$ since $\exp(L_i) \leq p^i$ for all i.

Our proof of Theorem 24.1 is based on the following partial case of Theorem 23.6 for which we offer an independent proof.

Theorem 24.2 (Hall). *Let $k \in \mathbb{N}$ and let G be a p-group. If G has no subgroups of order $p^{(p-1)k+1}$ and exponent $\leq p^k$, then* $\exp(\Omega_k(G)) \leq p^k$.

Theorems 24.1 and 24.2 are trivial for $p = 2$.

Set $\mho^0(G) = G$, $\mho^1(G) = \mho_1(G)$, $\mho^{i+1}(G) = \mho_1(\mho^i(G))$, $i = 1, 2, \dots$ (see §23, Example 1). As we have noticed in §23, Example 1, $\mho_i(G) \leq \mho^i(G)$. The subgroups $\mho^i(G)$ are characteristic in G. These subgroups control the structure of subgroups $\mho_i(G)$.

To facilitate the proof of Theorem 24.1 and subsequent results, we first prove Lemma 24.3(e,f), Theorem 24.2 and assertions of Remarks 1–4.

Almost all parts of Lemma 24.3 are proved in the previous sections.

Lemma 24.3. *Let G be a p-group.*

(a) *(Theorem 9.8(d)) If G is irregular, it has a characteristic subgroup of order $\geq p^{p-1}$ and exponent p. If G is an arbitrary p-group and $H \triangleleft G$ has a subgroup of order $p^k \leq p^{p-1}$ and exponent p, then H has a G-invariant subgroup of order p^k and exponent p.*

(b) *(Theorem 18.1) Let $W \trianglelefteq G$ have a subgroup of order p^{p-1} and exponent p, let $R < W$ be G-invariant of order $p^k < p^{p-1}$ and exponent p. Then there exists a G-invariant subgroup $H < W$ of order p^{p-1} and exponent p such that $R < H$.*

(c) *(i) (Theorem 7.1) p-groups of class < p are regular. (ii) (Theorem 9.8(a)) If* $|G/\mho_1(G)| < p^p$, *then G is regular (such G is called absolutely regular). (iii) (Theorem 7.2(b,d)) If G is regular, then* $\exp(\Omega_n(G)) \le p^n$ *and* $|\Omega_n(G)| = |G/\mho_n(G)|$ *for* $n \in \mathbb{N}$.

(d) *Suppose that G is neither absolutely regular nor of maximal class. (i) (Theorem 13.5) The number of subgroups of order* p^p *and exponent p in G is* $\equiv 1 \pmod p$. *(ii) (Corollary 13.3) If* $A < W$ *be a G-invariant subgroup of order* $p^a < p^p$ *and* $W \triangleleft G$ *is neither absolutely regular nor of maximal class, then, for* $b \in \{a+1, \dots, p\}$, *there is in W a G-invariant subgroup H of order* p^b *and exponent p containing A.*

(e) *Suppose that* $H \trianglelefteq G$, *where* $|H| \le p^{(p-1)e}$ *and* $\exp(H) = p^e$. *Then there is a chain* $\{1\} = T_0 < T_1 < \cdots < T_e = H$ *of length e of G-invariant subgroups such that*

$$p^{p-1} \ge |T_1/T_0| \ge |T_2/T_1| \ge \cdots \ge |T_e/T_{e-1}|,$$
$$\exp(T_i/T_{i-1}) = p, \quad i = 1, \dots, e.$$

If $|H| = p^{(p-1)e}$, *then* $|T_i/T_{i-1}| = p^{p-1}$ *for all i.*

(f) *Suppose that* $H \trianglelefteq G$, *where* $|H| = p^{(p-1)e}$ *and* $\exp(H) \le p^e$. *Then there exists a chain* $\{1\} = T_0 < T_1 < \cdots < T_e = H$ *of length e of G-invariant subgroups such that* $|T_i/T_{i-1}| = p^{p-1}$ *and* $\exp(T_i/T_{i-1}) = p$ *for* $i = 1, \dots, e$.

Proof. (e) If H is absolutely regular, $\{1\} < \Omega_1(H) < \cdots < \Omega_e(H) = H$ is the desired chain. Now assume that H is not absolutely regular. We use induction on $|H|$ and e. The subgroups $\mho^{e-1}(H)$ and $L = \mho_{e-1}(H)$ are absolutely regular and $\exp(L) = p$, $|L| \le p^{p-1}$ (Lemma 23.5). By (b), $L \le U \le H$, where U is G-invariant of order p^{p-1} and exponent p. Since $|H/U| \le p^{(p-1)(e-1)}$ and $\exp(H/U) = p^{e-1}$, there is, by induction, a chain $U/U = T_1/U < \cdots < T_e/U = H/U$ of G-invariant subgroups such that $p^{p-1} \ge |T_i/T_{i-1}|$ and $\exp(T_i/T_{i-1}) = p$ for $i = 1, \dots, e$ (here $T_0 = \{1\}$). Then $\{1\} = T_0 < T_1 < \cdots < T_e = H$ is the desired chain.

(f) In view of (e), one may assume that $\exp(H) < p^e$ so H is not absolutely regular (Example 23.2). Then H has a G-invariant subgroup T_1 of order p^{p-1} and exponent p, by (c). Since $|H/T_1| = p^{(p-1)(e-1)}$ and $\exp(H/T_1) \le \exp(H) \le p^{e-1}$, there is, by induction, a chain $T_1/T_1 < T_2/T_1 < \cdots < T_e/T_1 = H/T_1$ of G-invariant subgroups such that T_{i+1}/T_i is of order p^{p-1} and exponent $p, i = 1, \dots, e-1$. Then $\{1\} = T_0 < T_1 < \cdots < T_e = H$ is the desired chain. □

Remarks. Let G be a p-group and $k, j \in \mathbb{N}$.

1. Let $\exp(\Omega_k(G)) \le p^k$ and let $G/\Omega_k(G)$ be regular. We claim that

$$\exp(\Omega_{k+j}(G)) \le p^{p+j} \quad \text{and} \quad \Omega_j(G/\Omega_k(G)) = \Omega_{k+j}(G)/\Omega_k(G).$$

Indeed, set $H = \Omega_k(G)$ and $F/H = \Omega_j(G/H)$. If $x \in F$, then $x^{p^j} \in H$ (Lemma 24.3(c)(iii)) so $o(x) \le p^{k+j}$ and $F \le \Omega_{i+j}(G)$. Now suppose that $y \in G$ with

$o(y) \le p^{k+j}$. Then $y^{p^j} \in H$ so $yH \in F/H$ and $y \in F$, and we conclude that $\Omega_{i+j}(G) \le F$.

2. Let $H \trianglelefteq G$ and $\exp(\Omega_k(H)) = p^k$, let $H/\Omega_k(H)$ be absolutely regular and $|\Omega_k(H)| \le p^{(p-1)k}$. Let $\{1\} = L_0 < L_1 < \cdots < L_k = \Omega_k(H)$ be a $(p-1)$-admissible Hall chain in $\Omega_k(H)$ existing by Lemma 24.3(e). For a nonnegative integer s, put $L_{k+s}/L_k = \Omega_s(H/L_k)$. We claim that $\{1\} = L_0 < L_1 < \cdots < L_k < L_{k+1} < \cdots < H$ is a $(p-1)$-admissible Hall chain in H. Indeed, $\Omega_i(\Omega_k(H)) = \Omega_i(H)$ for $i \le k$, and we are done, by Remark 1.

3. Let $M \trianglelefteq G$ and $\Omega_j(G/M) \le H/M$ for some $H \le G$. Then $\Omega_j(G) \le H$. Indeed, if $x \in G$ with $o(x) \le p^j$, then $o(xM) \le p^j$ so $xM \le \Omega_j(G/M) \le H/M$.

4. Let $H \trianglelefteq G$ and let $F_0 < H$ be a G-invariant subgroup of order p. Let H/F_0 be of order $p^{(p-1)e}$ and exponent $\le p^e$. We claim that there is in H a $(p-1)$-admissible Hall chain of length $e+1$ with last index equal p. We proceed by induction on e. Set $\bar{G} = G/F_0$. By Lemma 24.3(f), there is a $(p-1)$-admissible Hall chain $\{\bar{1}\} = \bar{F}_0 < \bar{F}_1 < \cdots < \bar{F}_e = \bar{H}$ in $\bar{H}$. By induction, there is a $(p-1)$-admissible Hall chain $\{1\} = L_1 < \cdots < L_{e-1} < F_{e-1}$ in F_{e-1} such that $|F_{e-1}/L_{e-1}| = p$. Then H/L_{e-1} is of order p^p so regular, and H/F_{e-1} is of order p^{p-1} and exponent p so $\Omega_1(H/L_{e-1})$ is of order $\ge p^{p-1}$ and exponent p (Lemma 24.3(c)(iii)). Let L_e/L_{e-1} be a G-invariant subgroup of order p^{p-1} in $\Omega_1(H/L_{e-1})$ (see Lemma 24.3(c) again). Then $\{1\} = L_0 < L_1 < \cdots < L_{e-1} < L_e < H$ is the desired chain.

Proof of Theorem 24.2. Suppose that G is a minimal counterexample. Then we have $\exp(\Omega_k(G)) > p^k$ so G is irregular (Lemma 24.3(c)) and $\exp(M) \ge p^k$ for all $M \in \Gamma_1$. By induction, $\exp(\Omega_k(H)) \le p^k$ for all $H < G$.

Let $k = 1$ and let $R \trianglelefteq G$ be of exponent p of maximal order. Since G has no subgroups of order $p^{(p-1)1+1} = p^p$ and exponent p, we get $|R| = p^{p-1}$ (Lemma 24.3(a)). If $x \in G - R$ is of order p, then $S = \langle x, R\rangle$ is of order $p^p = p^{(p-1)1+1}$ and exponent p (Lemma 24.3(c)), a contradiction. Thus, $R = \Omega_1(G)$, and the theorem is true for $k = 1$.

Now let $k > 1$. Obviously, G has a noncyclic subgroup of order $p^{k+1} = p^{k(2-1)+1}$ so $p > 2$. Let $A < G$ be a subgroup of maximal order among subgroups of exponent $\le p^k$; then $|A| \le p^{(p-1)k}$, by assumption. Let $A \le M \in \Gamma_1$; then $\Omega_k(M) = A$, whence $A \triangleleft G$ and $\exp(A) = p^k$ since $\exp(M) \ge p^k$. There is $g \in G - A$ with $o(g) \le p^k$. Then $g^p \in M$ so $g^p \in \Omega_{k-1}(M) \le A$. Set $B = \langle g, A\rangle$; then $|B| = p|A| > |A|$. If $B \le F \in \Gamma_1$, then $B \le \Omega_k(F)$ has exponent p^k, contrary to the choice of A. Thus, $B = G$ and $|G : A| = p$, $|G| \le p^{(p-1)k+1} < p^{pk}$ and $\exp(G) = p^{k+1}$. It follows from Lemma 23.5 that $\mho_{k-1}(G)$ is absolutely regular of exponent p^2 since it is generated by elements of orders $\le p^2$ and has an element of order p^2.

By Lemma 23.5, $\mho_{k-1}(A)$ is of exponent p so of order $\le p^{p-1}$ since it is absolutely regular. Let $x_1, \dots, x_n$ be all elements of orders $\le p^k$ in $G - A$; the set of these elements is G-invariant. Then $L = \langle x_1^{p^{k-1}}, \dots, x_n^{p^{k-1}}, \mho_{k-1}(A)\rangle$ is a G-invariant subgroup contained in $\Omega_1(\mho_{k-1}(G))$ so $\exp(L) = p$ and $|L| \le p^{p-1}$ (Lemmas

24.3(c) and 23.5). By Lemma 24.3(b), $L \leq H \lhd G$, where H is of order p^{p-1} and exponent p. If $H \not\leq A$, then $G = AH$ and $G/(H \cap A) = (A/(H \cap A)) \times (H/(H \cap A))$ is of exponent p^{k-1} so $\exp(G) = p^k$, a contradiction. Now let $H < A$. Set $\bar{G} = G/H$. Then

$$\bar{G} = \langle \bar{x}_1, \bar{A} \rangle, \quad \exp(\bar{A}) = p^{k-1}, \quad |\bar{A}| \leq p^{(p-1)(k-1)}, \quad o(\bar{x}_1) \leq p^{k-1}.$$

We also have $|\bar{G}| \leq p^{(p-1)(k-1)+1}$ and $\Omega_{k-1}(\bar{G}) = \bar{G}$. Then, by induction, $\exp(\bar{G}) = p^{k-1}$ so $\exp(G) \leq p^k$, a final contradiction. □

Remarks. 5. Let G be a p-group such that $\Omega_k(G) = G$. If A is a maximal among $X < G$ satisfying $\Omega_k(X) = X$, then $|G : A| = p$. Since $A^G < G$ and $\Omega_k(A^G) < G$, we get $A^G = A$ so $A \lhd G$. Let $y \in G - A$ be of minimal order; then $o(y) \leq p^k$ and, assuming that $y^p \in A$, we get $\Omega_k(\langle y, A \rangle) = \langle y, A \rangle > A$ so $G = \langle y, A \rangle > A$ and $|G : A| = p$.

6. Let G be a p-group and let $A < G$ be maximal among subgroups of G of exponent $\leq p^k$. We claim that if $|A| \leq p^{(p-1)k}$, then $A = \Omega_k(G)$. Assume that this is false. Assume, in addition, that A is not normal in G. Then A is not characteristic in $N = \mathrm{N}_G(A)$ so $A < \Omega_k(N)$. Take $x \in N - A$ of minimal order with $o(xA) = p$ in N/A. Then $o(x) \leq p^k$, $x^p \in A$ so $B = \langle x, A \rangle$ is of order $p|A| \leq p^{(p-1)k+1}$ and $\Omega_k(B) = B$. Then $\exp(B) \leq p^k$ (Theorem 24.2), a contradiction. Thus, $N = G$, and arguing, as above, we get a contradiction. Thus, $A = \Omega_k(G)$.

Proof of Theorem 24.1. Set $\exp(H) = p^e$. We may assume that $e > 1$, $p > 2$ and H is not absolutely regular. We proceed by induction on $|H|$. Indeed, if H is absolutely regular, then $\{1\} < \Omega_1(H) < \cdots < \Omega_e(H) = H$ is a *unique* $(p-1)$-admissible Hall chain in H. Next, if $p = 2$, then any part of a chief series of G, lying below H, is a Hall chain in H.

Let $F_0 \leq H \cap \mathrm{Z}(G)$ be of order p and set $\bar{G} = G/F_0$. By induction, there is in $\bar{H}$ a $(p-1)$-admissible Hall chain $\{\bar{1}\} = \bar{F}_0 < \bar{F}_1 < \cdots < \bar{F}_n = \bar{H}$. We have $\exp(\bar{F}_i) \leq p^i$ so $\exp(F_i) \leq p^{i+1}$ for all i. Let i_0 be the greatest value of i such that $|\bar{F}_i| = p^{(p-1)i}$. In view of Remark 4, one may assume that $i_0 < n$; then $|\bar{F}_{i_0+1}| < p^{(p-1)(i_0+1)}$; we also have $\Omega_{i_0+1}(\bar{H}) = \bar{F}_{i_0+1}$ since our chain in $\bar{H}$ satisfies (b). It follows that $\Omega_{i_0+1}(H) \leq F_{i_0+1}$ (Remark 3) so

$$(*) \qquad \Omega_{i_0+1}(H) = \Omega_{i_0+1}(F_{i_0+1}).$$

Since $|F_{i_0+1}| \leq p^{(p-1)(i_0+1)}$, then $\exp(\Omega_{i_0+1}(F_{i_0+1})) \leq p^{i_0+1}$ (Theorem 24.2) so in view of $(*)$, we get

$$(**) \qquad \exp(\Omega_{i_0+1}(H)) \leq p^{i_0+1}.$$

Next, by the choice of i_0, we have $|\bar{F}_{i_0+1}/\bar{F}_{i_0}| < p^{p-1}$, and we conclude that $\bar{H}/\bar{F}_{i_0}$ has no G-invariant subgroups of order p^{p-1} and exponent p (indeed, if $\bar{U}/\bar{F}_{i_0}$

is a such subgroup, then $\exp(\bar{U}) \le p^{i_0+1}$ so $\bar{U} \le \Omega_{i_0+1}(\bar{H}) = \bar{F}_{i_0+1}$, which is a contradiction). Thus, $\bar{F}_{i_0+1}/\bar{F}_{i_0} = \Omega_1(\bar{H}/\bar{F}_{i_0})$ whence $\bar{H}/\bar{F}_{i_0}$ is absolutely regular (Lemma 24.3(a)) so $H/F_{i_0}(\cong \bar{H}/\bar{F}_{i_0})$ is also absolutely regular.

Assume that $i_0 = 0$. Then $|\bar{F}_1| < p^{p-1}$ so $\Omega_1(\bar{H}) = \bar{F}_1$, by (b), and $|F_1| = |F_0||\bar{F}_1| \le p^{p-1}$. In that case, F_1 must be of order p^{p-1} and exponent p (otherwise, H is absolutely regular, by Lemma 24.3(a)). Then $\Omega_1(H) = F_1$ and $H/\Omega_1(H)$ is absolutely regular (see the previous paragraph). By Remark 2, there is a $(p-1)$-admissible Hall chain in H.

Now we assume that $i_0 > 0$; then $\bar{F}_1$ is of order p^{p-1} and exponent p so $|F_1| = p^p$ and $\exp(F_1) \le p^2$. We also have $\exp(F_{i_0+1}) \le |F_0| \exp(\bar{F}_{i_0+1}) \le p \cdot p^{i_0+1} = p^{i_0+2}$ and, according to this, we have to consider separately the following three possibilities:

(i) $\exp(F_{i_0+1}) < p^{i_0+1}$,

(ii) $\exp(F_{i_0+1}) = p^{i_0+1}$, and

(iii) $\exp(F_{i_0+1}) = p^{i_0+2}$.

(i) Suppose that $\exp(F_{i_0+1}) < p^{i_0+1}$; then, by $(*)$, $F_{i_0+1} = \Omega_{i_0+1}(F_{i_0+1}) = \Omega_{i_0+1}(H)$. It follows from the last equality that $\exp(H) < p^{i_0+1}$ so $F_{i_0+1} = \Omega_{i_0+1}(H) = H$. Then, by Remark 4, there exists in F_{i_0} a $(p-1)$-admissible Hall chain $\{1\} = L_0 < L_1 < \cdots < L_{i_0} < F_{i_0}$ satisfying

$$|F_{i_0} : L_{i_0}| = p, \quad |L_{i_0}| = p^{(p-1)i_0},$$
$$|H/L_{i_0}| = |H/F_{i_0}||F_{i_0}/L_{i_0}| \le p^{p-2} \cdot p = p^{p-1},$$

and H/L_{i_0} is regular of exponent $\le p^2$.

If $\exp(H/L_{i_0}) = p$, then $\{1\} = L_0 < L_1 < \cdots < L_{i_0} < H$ is the desired chain.

Now we assume that $\exp(H/L_{i_0}) = p^2$. By Lemma 24.3(c)(iii), $U/L_{i_0} = \Omega_1(H/L_{i_0})$ is of exponent p and index $|F_{i_0}/L_{i_0}| = p$ in H/L_{i_0} since $\exp(H/F_{i_0}) = p$. Therefore, $\{1\} = L_0 < L_1 < \cdots < L_{i_0} < U$ is a $(p-1)$-admissible Hall chain in U. Let $W/L_{i_0} = \mho_1(H/L_{i_0})$; then $|W/L_{i_0}| = |(H/L_{i_0}) : (U/L_{i_0})| = p$ (part (iii) of Lemma 24.3(c)). Since $\exp(H/W) = p$ and $|H/W| < p^{p-1}$, we get $\mho_1(H) < W$ ($<$ since $|H/\mho_1(H)| \ge p^p$: H is not absolutely regular). Therefore, there exists a G-invariant subgroup T_{i_0} satisfying $\mho_1(H) < T_{i_0} < W$ and $|T_{i_0}| = p^{(p-1)i_0}$ (recall that $p^{(p-1)i_0} = |L_{i_0}| < |H|$ and $|H : \mho_1(H)| \ge p^p > |H : W|$). We have $\exp(T_{i_0}) \le \exp(H) \le p^{i_0}$, so there exists in T_{i_0} a $(p-1)$-admissible Hall chain $\{1\} = T_0 < T_1 < \cdots < T_{i_0}$ and all indices of that chain equal p^{p-1} (Lemma 24.3(f)). Since $|H/T_{i_0}| \le p^{p-1}$ and $\exp(H/T_{i_0}) = p$, $\{1\} = T_0 < T_1 < \cdots < T_{i_0} < H$ is the desired chain.

(ii) Let $\exp(F_{i_0+1}) = p^{i_0+1}$; then $F_{i_0+1} = \Omega_{i_0+1}(H)$ in view of $(*)$. Since $H/\Omega_{i_0+1}(H)$, as an epimorphic image of H/F_{i_0}, is absolutely regular and $|\Omega_{i_0+1}(H)| \le p^{(p-1)(i_0+1)}$, there is the desired chain in H (Lemma 24.3(e) and Remark 2).

(iii) Suppose that $\exp(F_{i_0+1}) = p^{i_0+2}$. Then, by (**), $\exp(\Omega_{i_0+1}(H)) = p^{i_0+1}$. We have $F_{i_0} \leq \Omega_{i_0+1}(H)$ and H/F_{i_0} is absolutely regular so $H/\Omega_{i_0+1}(H)$ is absolutely regular and, in addition, $|\Omega_{i_0+1}(H)| \leq p^{(p-1)(i_0+1)}$. Therefore, there is a $(p-1)$-admissible Hall chain in H, by Lemma 24.3(e) and Remark 2. □

Let $H \trianglelefteq G$, where G is a p-group, $\exp(H) = p^e$, $|H| = p^m$ and let $\mathcal{C} : \{1\} = L_0 < L_1 < \cdots < L_n = H$ be a $(p-1)$-admissible Hall chain in H. Let, in addition, $m \geq (p-1)e$. Assume that for some $i \leq e$, we have $|L_i| < p^{(p-1)i}$; then $n > e$ and, by Theorem 24.1, $L_e = \Omega_e(H) = H$ so $n = e$, a contradiction. Thus, for all $i \leq e$, we must have $|L_i| = p^{(p-1)i}$.

Supplement 1 to Theorem 24.1. *Let a fixed $k < p$ and let H be a normal subgroup of a p-group G. Then there is in H a chain $\{1\} = L_0 < L_1 < \cdots < L_n = H$ of G-invariant subgroups with the properties ($i = 1, \ldots, n$):*

(a) L_i/L_{i-1} is of order $\leq p^k$ and exponent p, and

(b) either the order of L_i is exactly p^{ik}, or else $L_i = \Omega_i(H)$.

Supplement 2 to Theorem 24.1. *Let H be a regular normal subgroup of a p-group G and let $k \in \mathbb{N}$ be fixed. Then there is in H a chain $\{1\} = L_0 < L_1 < \cdots < L_n = H$ of G-invariant subgroups with the properties ($i = 1, \ldots, n$):*

(a) L_i/L_{i-1} is of order $\leq p^k$ and exponent p, and

(b) either the order of L_i is exactly p^{ik}, or else $L_i = \Omega_i(H)$.

To prove the above supplements, it suffices to repeat the proof of Theorem 1. In Supplement 2, one can replace regularity by property $\mathcal{P}_3 = \mathcal{P}$ from §11.

Remark 7. An irregular p-group G of maximal class is a $\mathcal{P}$-group if and only if $p > 2$, $|G| = p^{p+1}$ and $|\Omega_1(G)| = p^p$. If all subgroups of order p^{p+1} of an irregular p-group G of maximal class are $\mathcal{P}$-groups, then $|G| = p^{p+1}$.

As the proof of Theorem 24.1 shows, if $\{1\} = L_0 < L_1 < \cdots < L_n = H$ is a $(p-1)$-admissible Hall chain in $H \trianglelefteq G$, then $|L_1 : L_0| \geq |L_2 : L_1| \geq |L_n : L_{n-1}|$.

Theorem 24.4. *Let $k > 1$. Suppose that a p-group G has no subgroups of order $p^{(p-1)k+2}$ and exponent $\leq p^k$. Then one of the following holds: (a) $\exp(\Omega_k(G)) \leq p^k$, (b) G is of maximal class and order $\geq p^{(p-1)k+2}$.*

Lemma 24.5. *Suppose that G is a group of order $p^{(p-1)k+2}$ and $\Omega_k(G) = G$. Then either $\exp(G) \leq p^k$ or G is of maximal class.*

Proof. Suppose that G is a minimal counterexample. Then $\exp(\Omega_k(G)) > p^k$ so G is irregular, and G is not of maximal class. By Theorem 24.2, G has a subgroup A of order $p^{(p-1)k+1}$ and exponent $\leq p^k$. It follows from $\exp(G) > p^k$ that $\exp(A) = p^k$ since $|G : A| = p$, and then $\exp(G) = p^{k+1}$. By Theorem 12.1(b), A is not absolutely regular. Since G is not of maximal class, we get $k > 1$.

Assume that A is of maximal class; then A is irregular (Theorem 9.5). In that case, $\exp(G) = \exp(A) = p^k$ (Theorem 12.12(c)), a contradiction.

Since $|G| = p^{(p-1)k+2} \le p^{pk}$, the subgroup $\mho_{k-1}(G)$ is absolutely regular (Lemma 23.5) in view of $\exp(\mho_1(G)) > p$. Since $\exp(A) = p^k$ so $\mho_{k-1}(A) \le \Omega_1(\mho_{k-1}(G))$, we get $\exp(\mho_{k-1}(A)) = p$ and $|\mho_{k-1}(A)| \le p^{p-1}$, by Lemma 24.3(c)(iii). Let $x_1, \dots, x_n$ be all elements of orders $\le p^k$ in $G - A$. Set $T = \langle x_1^{p^{k-1}}, \dots, x_n^{p^{k-1}}, \mho_{k-1}(A)\rangle$. Then the G-invariant subgroup $T \le \Omega_1(\mho_{k-1}(G))$ so $\exp(T) = p$ and $|T| \le p^{p-1}$ (Lemma 24.3(c)). By Lemma 24.3(d)(ii), $T < H < A$, where H is a G-invariant subgroup of order p^p and exponent p. Set $\bar{G} = G/H$. Then $|\bar{G}| = p^{-p}|G| = p^{(p-1)(k-1)+1}$, $\Omega_{k-1}(\bar{G}) = \bar{G}$ since $\bar{G} = \langle \bar{x}_1, \bar{A}\rangle$, $\exp(\bar{A}) = p^{k-1}$ and $o(\bar{x}) \le p^{k-1}$. By Theorem 24.2, $\exp(\bar{G}) = p^{k-1}$ so $\exp(G) = p^k$, a final contradiction. □

Proof of Theorem 24.4. If G is of maximal class and exponent $> p^k$, its order is $\ge p^{(p-1)k+2}$ and it satisfies the hypothesis (Theorems 9.6 and 13.19). Suppose that G is a counterexample of minimal order. Then G is not of maximal class, $\exp(G) \ge \exp(\Omega_k(G)) \ge p^{k+1}$ so G is irregular and all its maximal subgroups have exponent $\ge p^k$. Then, by Theorem 24.2, G has a (proper) subgroup A of order $p^{(p-1)k+1}$ and exponent $\le p^k$. Since A is maximal among subgroups of G of exponent $\le p^k$, we get $\exp(A) = p^k$. In view of Lemma 24.5, $|G| > p^{(p-1)k+2}$.

Now let $H \in \Gamma_1$ be of maximal class; then $\exp(G) = \exp(H)$ (Theorem 12.12(b)). Let $R \lhd G$ be of order p^p and exponent p (Lemma 24.3(d)(i)). Assume that $R < H$. Then $|H| = p^{p+1}$ (Theorem 9.6), and so, by Theorem 12.12(b), $\exp(G) = \exp(H) = p^2 < p^{k+1}$, a contradiction. Now let $R \not\le H$; then $G = RH$ and $G/R \cong (H(R\cap H))\times(R/(R\cap H))$ has a subgroup B/R of order $p^{(p-1)(k-1)+1}$ and exponent p^{k-1}, by Theorems 9.5 and 9.6. Then $\exp(B) \le p^k$ and $|B| = p^{(p-1)(k-1)+1+p} = p^{(p-1)k+2} > |A|$, a contradiction. Thus, H does not exist.

The hypothesis is inherited by subgroups. Therefore, if $M \in \Gamma_1$, then we have $\exp(\Omega_k(M)) = p^k$ since M is not of maximal class, by the previous paragraph. If we take, from the start, M so that it contains A, we get $A = \Omega_k(M)$ so $A \lhd G$. By assumption, there is $x \in G - A$ with $o(x) \le p^k$; then $o(x) < p^k$ so $x^p \in \Omega_{k-1}(M) \le A$. Set $B = \langle x, A\rangle$. Then $|B| = p|A| = p^{(p-1)k+2}$, $\exp(B) = p^{k+1}$, by the choice of A, and $\Omega_k(B) = B$ so, by Lemma 24.5, B must be of maximal class. By the previous paragraph, $|G : B| > p$. Let $B < M < G$, where M is maximal in G. Then, by induction, $\exp(\Omega_k(M)) = p^k$, a contradiction since $B \le \Omega_k(M)$ and $\exp(B) > \exp(\Omega_k(M))$. □

Theorem 24.6. *Let G be a p-group and $k > 1$. Suppose that G has a proper subgroup A of order $p^{(p-1)k+1}$ which is maximal among subgroups of G of exponent $\le p^k$. Then one of the following holds: (a) $\Omega_k(G) = A$, (b) A and G are of maximal class.*

Proof. Suppose that $\exp(\Omega_k(G)) > p^k$; then G is irregular. It follows that $\exp(A) = p^k$.

First suppose that $A \lhd G$. Let $x \in G - A$ be of minimal order. Then $o(x) \le p^k$, by assumption, and one may assume that $x^p \in A$ so $B = \langle x, A\rangle$ has order $p^{(p-1)k+2}$ and exponent p^{k+1}, and $\Omega_k(B) = B$. In that case, by Lemma 24.5, B is of maximal class. It follows from Example 23.2 that A is also of maximal class. Now let $A < D \le G$ be such that $|D : A| = p$. Since $\exp(D) > p^k = \exp(A)$, it follows from Theorem 12.12(b) that D must be of maximal class. Thus, all subgroups of G of order $p|A|$, containing A, are of maximal class so G is also of maximal class, by Exercise 13.10.

Now suppose that $A \ntrianglelefteq G$. Set $N_G(A) = N$. Since $N < G$, A is not characteristic in N so $\Omega_k(N) > A$ and, by the previous paragraph, N is of maximal class. Then, by Remark 10.5, G is also of maximal class. By Theorem 9.6, A is also of maximal class. □

Proposition 24.7. *Let G be a group of order $p^{(p-1)k+3}$, $k > 2$. Suppose that $\Omega_k(G) = G$ and $\exp(G) > p^k$. Then one of the following holds:*

(a) G is of maximal class.

(b) G has a maximal subgroup A with $\exp(A) = p^k$, A has a G-invariant subgroup H of order p^p and exponent p such that G/H and A/H are of maximal class.

Proof. We have $|G| = p^{(p-1)k+3} \le p^{kp}$ since $k \ge 3$. It follows that the subgroup $\mho_{k-1}(G)$ is of order $\le p^p$ (Lemma 23.5) so it is regular. Suppose that G is not of maximal class. Then, by Theorem 24.4, there is $A \in \Gamma_1$ with $\exp(A) = p^k$. By Example 23.2, A is neither absolutely regular nor of maximal class. By Lemma 23.5, $\mho_{k-1}(A)$ is of order $\le p^{p-1}$ and exponent p since $|A| = p^{(p-1)k+2} < p^{pk}$ and $\mho_{k-1}(A) \le \Omega_1(\mho_{k-1}(A))$. Let $x_1, \dots, x_n$ be all elements of order $\le p^k$ in $G - A$. Set $T = \langle x_1^{p^{k-1}}, \dots, x_n^{p^{k-1}}, \mho_{k-1}(A)\rangle$; then $T \lhd G$, $T < A$ since $x^p \in A$ for all $x \in G$, $T \le \Omega_1(\mho_{k-1}(G))$ so T is of order $\le p^p$ and exponent p. By Lemma 24.3(d)(ii), $T \le H < A$, where H is a G-invariant subgroup of order p^p and exponent p. Set $\bar{G} = G/H$. We have

$$\exp(\bar{A}) = p^{k-1}, \quad |\bar{G}| = p^{(p-1)(k-1)+2}, \quad \Omega_{k-1}(\bar{G}) = \bar{G}, \quad \exp(\bar{G}) > p^{k-1}.$$

Then $\bar{G}$ is of maximal class, by Lemma 24.5. It follows from Theorem 9.6 that $\bar{A}$ is also of maximal class. □

Corollary 24.8. *Let G be a group of order p^{3p}. If $\Omega_3(G) = G$, then one of the following holds:*

(a) $\exp(G) \le p^3$.

(b) G is of maximal class; then $\exp(G) = p^4$ if $p > 2$ and $\exp(G) = p^5$ if $p = 2$.

(c) G has a maximal subgroup A with $\exp(A) = p^3$, A has a G-invariant subgroup H of order p^p and exponent p such that G/H and A/H are of maximal class.

Proposition 24.9. *Let $k > 3$, $p > 2$ and let G be a p-group containing a normal subgroup A of order $p^{(p-1)k+2}$ and exponent $\le p^k$ which is maximal among subgroups* *of G of exponent* $\le p^k$*. Then one of the following holds:*

(a) $\Omega_k(G) = A$.

(b) $|G/A| = p$*, there is a G-invariant* $R < A$ *of order* p^p *and exponent p such that* G/R *and* A/R *are of maximal class.*

Proof. Assume that $\Omega_k(G) > A$. Then, if $A < U \le G$, where $|U : A| = p$, then $\exp(U) = p^{k+1}$ so $\exp(A) = p^k$. By Example 23.2, A is neither absolutely regular nor of maximal class.

By Lemma 23.5, $\mho_{k-1}(A)$ is of order $\le p^{p-1}$ and exponent p since $|A| < p^{kp}$ in view of $k > 3$. Then, by Corollary 13.3, $\mho_{k-1}(A) < R < A$, where R is a G-invariant subgroup of order p^p and exponent p. Set $\bar{G} = G/R$. We have $|\bar{A}| = p^{(p-1)(k-1)+1}$ and $\exp(\bar{A}) = p^{k-1}$. Clearly, $\bar{A}$ is maximal among subgroups of exponent p^{k-1} in $\bar{G}$. It follows from Theorem 24.6 that either $\Omega_{k-1}(\bar{G}) = \bar{A}$ or $\bar{G}$ and $\bar{A}$ are of maximal class. In the second case, $|G : A| = p$ (indeed, each normal subgroup of $\bar{G}$ of index $> p$ is not of maximal class since its center is of order $> p$).

It remains to consider the possibility $\Omega_{k-1}(\bar{G}) = \bar{A}$. Then $\Omega_{k-1}(G) \le A$, by Remark 3. By assumption, there exists an element $x \in G - A$ such that $o(x) \le p^k$ and $x^p \in A$. Since $\Omega_{k-1}G) \le A$, we get $o(x) = p^k$. Set $B = \langle x, A \rangle$; then $|B| = p|A| = p^{(p-1)k+3}$ since $A \triangleleft G$, $\Omega_k(B) = B$ and $\exp(B) = p^{k+1}$, by the choice of A. Since a maximal subgroup A of B is neither absolutely regular nor of maximal class, B is not of maximal class as well (Theorem 9.6). Therefore, by Proposition 24.7, there is in A a B-invariant subgroup K of order p^p and exponent p such that A/K and B/K are of maximal class. We have $\Omega_2(B/K) = B/K$ so $\Omega_3(B) = B$. In that case, $B \le \Omega_3(G) \le \Omega_{k-1}(G) \le A$, since $k > 3$, and this is a contradiction. Thus, $\Omega_k(G) = A$. □

Proposition 24.10. *Suppose that G is a p-group of order* $\le p^{pk}$ *such that* $\Omega_k(G) = G$ *and all sections of G of order* p^{p+1} *are* $\mathcal{P}$*-groups. Then* $\exp(G) \le p^k$.

Proof. Suppose that G is a counterexample of minimal order; then we have $k > 1$, $\exp(\Omega_k(G)) > p^k$ so G is irregular. In that case, by Remark 5, there is $A \in \Gamma_1$ such that $\Omega_k(A) = A$; then $G = \langle y, A \rangle$, where $y \in G - A$ with $o(y) \le p^k$. Since A satisfies the hypothesis, we get $\exp(A) \le p^k$, by induction, so $\exp(G) = p^{k+1}$ and $\exp(A) = p^k$.

If G is of maximal class, then G is of order p^{p+1} (Remark 7); then $\exp(G) = p^2 < p^{k+1}$, a contradiction.

Assume that A is of maximal class. Then $|A| = p^{p+1}$ (Remark 7) so $\exp(A) = p^2$. In that case, by the previous paragraph and Theorem 12.12(b), $\exp(G) = \exp(A) = p^2 < p^{k+1}$, and G is not a counterexample.

Now assume that A is absolutely regular. Then, by Theorem 12.1(b), $G = A\Omega_1(G)$, where $\Omega_1(G)$ is of order p^p and exponent p, whence $\exp(\Omega_k(G)) = p^k$, contrary to the assumption. In what follows we assume that A is neither absolutely regular nor of maximal class.

By Lemma 23.5, $\mho_{k-1}(A)$ is of order $\leq p^{p-1}$ and exponent p since $|A| < p^{pk}$. Next, by the same lemma, $\mho_{k-1}(G)$ is absolutely regular since it contains an element of order p^2. Let $x_1, \dots, x_n$ be all elements of orders $\leq p^k$ in $G - A$ and set $T = \langle x_1^{p^{k-1}}, \dots, x_n^{p^{k-1}}, \mho_{k-1}(A)\rangle$. Then $T < A$, $T \lhd G$, and $T \leq \Omega_1(\mho_{k-1}(G))$ so T is of exponent p and order $\leq p^{p-1}$. Therefore, by Corollary 13.3, $T < H < A$, where H is a G-invariant subgroup of order p^p and exponent p. Set $\bar{G} = G/H$; then $\bar{G} = \langle \bar{x}_1, \bar{A}\rangle$, where $o(\bar{x}_1) \leq p^{k-1}$, $\exp(\bar{A}) = p^{k-1}$ so $\Omega_{k-1}(\bar{G}) = \bar{G}$, and $|\bar{G}| \leq p^{p(k-1)}$. Obviously, $\bar{G}$ satisfies the hypothesis with $k-1$ instead of k. Then, by induction, $\exp(\bar{G}) \leq p^{k-1}$ so $\exp(G) \leq p^k$ and G is not a counterexample. □

Corollary 24.11. *Let $k \in \mathbb{N}$ and let A be a subgroup of a p-group G which is maximal among subgroups of G of exponent $\leq p^k$. Suppose that all sections of the subgroup $\Omega_k(G)$ of order p^{p+1} are $\mathcal{P}$-groups. Then, if $|A| < p^{kp}$, then $\Omega_k(G) = A$.*

Proof. Assume that G is a counterexample of minimal order; then $A < \Omega_k(G)$. Assume that $A \lhd G$. If $x \in G - A$ is of minimal order and $x^p \in A$, then $o(x) \leq p^k$. Set $B = \langle x, A\rangle$; then $|B| \leq p^{kp}$ and $B = \Omega_k(B) \leq \Omega_k(G)$. By Proposition 24.10, $\exp(B) = p^k$, contrary to the choice of A. Now assume that A is not normal in G. Then $N = N_G(A) < G$. By induction, $A = \Omega_k(N)$ so A is characteristic in N. It follows that $N = G$, contrary to the assumption. □

Lemma 24.12 (Hall). *Let $p > 2$. For any sequence (1) of G-invariant subgroups of H which satisfy (a), but not necessarily (b), we have: if $g \in G$ and $x \in L_i$, then $(gx)^{p^i} = g^{p^i}$.*

Exercise 1. Let H be a subgroup of the p-group G such that $(gh)^{p^k} = g^{p^k}$ for all $g \in G$ and $h \in H$. Then the number of solutions of $x^{p^k} = a$ in G ($a \in G$) is a multiple of $|H|$.

Exercise 2. The characteristic subgroup $\bar{\Omega}_k$ of the p-group G, which is the product of all possible L_k that have occurred in some sequence satisfying the conditions of Theorem 24.1 with $G = H$, is of exponent at most p^k. Moreover, $[\bar{\Omega}_k, \mho_k(G)] = \{1\}$.

§25

On the lattice of subgroups of a group

Let $\mathcal{L}(G)$ be the lattice of subgroups of a group G. Two operations in $\mathcal{L}(G)$ are defined as follows: $(A, B) \mapsto A \cap B$ (intersection) and $(A, B) \mapsto \langle A, B \rangle$ (union).

Two groups G and H are said to be lattice isomorphic via lattice isomorphism $\varphi : \mathcal{L}(G) \to \mathcal{L}(H)$, if φ is a bijection of $\mathcal{L}(G)$ onto $\mathcal{L}(H)$ such that, for every $A, B \leq G$, we have $(A \cap B)^\varphi = A^\varphi \cap B^\varphi$ and $\langle A, B \rangle^\varphi = \langle A^\varphi, B^\varphi \rangle$.

There exists a fairly close connection between the group and the lattice of its subgroups. We prove that if G is a p-group and $\mathcal{L}(G_1) \cong \mathcal{L}(G)$, then, as a rule, G_1 is also a p-group. We also show that lattice isomorphisms respect some other group theoretic properties.

In what follows, φ denotes a lattice isomorphism of two groups G and G_1. The following lemma is an easy exercise.

Lemma 25.1. *Let $\varphi : \mathcal{L}(G) \to \mathcal{L}(G_1)$ be a lattice isomorphism.*

(a) $\Phi(G)^\varphi = \Phi(G^\varphi) = \Phi(G_1)$ and $\mathcal{L}(G/\Phi(G)) \cong \mathcal{L}(G_1/\Phi(G_1))$.

(b) If $G \cong \mathrm{C}_{p^m}$, then $G^\varphi \cong \mathrm{C}_{q^m}$ for some prime q. Next, $\mathcal{L}(\mathrm{C}_{p^m}) \cong \mathcal{L}(\mathrm{C}_{q^m})$.

(c) If p, q are distinct primes and $G = \mathrm{C}_{pq}$, then $G^\varphi \cong \mathrm{C}_{rs}$ for primes $r \neq s$.

(d) If G is elementary abelian of order p^2, then either $G^\varphi \cong G$ or G^φ is isomorphic to a nonabelian group of order qp, where a prime q divides $p - 1$.

(e) If G is a nonabelian group of order qp, p, q are distinct primes, $p > q$, then either $G_1 \cong \mathrm{E}_{p^2}$ or G is a nonabelian group of order rp, where a prime r divides $p - 1$.

Lemma 25.2 (O. Ore). *Suppose that $G \cong \mathrm{C}_m$ and G_1 are lattice isomorphic. Then G_1 is also cyclic.*

Proof. We use induction on $|G|$. Assume that G_1 is a minimal counterexample. Then G_1 is a minimal noncyclic group. Since $G/\Phi(G)$ and the noncyclic $G_1/\Phi(G_1)$ are lattice isomorphic, we have $\Phi(G) = \{1\} = \Phi(G_1)$; then G has no elements of order p^2 for all prime p. It follows that G_1 is either E_{p^2} or nonabelian of order pq, $p \neq q$. Then, however, G and G_1 have different numbers of maximal subgroups. □

Let p be a prime. A group G is said to be a $\mathcal{P}$-group [Suz2] if it is either an elementary abelian p-group or $G = Q \cdot P$ is a Frobenius group, where Q is of prime order $q \neq p$, $P \in \mathrm{Syl}_p(G)$ is elementary abelian and the action of Q on P is scalar. We omit an easy proof of the following

Lemma 25.3. *Let $G = Q \cdot P \cong \mathrm{C}_q \cdot \mathrm{E}_{p^{n-1}}$ be a $\mathcal{P}$-group, p and q are distinct primes, $q < p$. Then G and $G_1 \cong \mathrm{E}_{p^n}$ are lattice isomorphic.*

Lemma 25.4. *Let $G \cong \mathrm{E}_{p^n}$, $n > 1$. Then either $G_1 = G^\varphi$ is an elementary abelian p-group or a $\mathcal{P}$-group.*

Proof. If G_1 is a p-group, it is elementary abelian since $\Phi(G_1) = \{1\}$ (Lemma 25.1(a)). In what follows we assume that G_1 is not a p-group. We use induction on n. In view of Lemma 25.1(c), we may assume that $n > 2$. By Lemma 25.1(a), $\Phi(H_1) = \{1\}$ for all $H_1 \leq G_1$ so all elements of G_1 have prime orders and all its nilpotent subgroups are elementary abelian. Let $q \neq p$ be a prime divisor of $|G_1|$ and let $Z_1 < G_1$ be of order q. Let $Z < G$ be such that $Z^\varphi = Z_1$ and let $Z < E < G$ be of order p^2. By Lemma 25.1(c), $E_1 = E^\varphi$ is a nonabelian group of order qp and q divides $p-1$. By Lemma 25.1(b,d), $q^2 \nmid |G_1|$. By Lemma 25.1(b–d), Burnside's theorem on normal p-complement and the structure of Frobenius complements, $P_1 \in \mathrm{Syl}_p(G_1)$ is normal and $|G_1 : P_1| = q$. By the above, P_1 is elementary abelian, and $G_1 = Z_1 \cdot P_1$ is a Frobenius group with kernel P_1, by Lemma 25.1(d). If L_1 is a subgroup of order p in P_1, then $\lambda(\langle Z_1, L_1\rangle) = 2$ since, if $L < G$ is such that $L^\varphi = L_1$, then ZL is abelian of type (p,p). It follows that all subgroups of P_1 are Z_1-invariant. It is a standard fact that then a generator of Z_1 induces a scalar transformation on P_1, i.e., G_1 is a $\mathcal{P}$-group. □

Remarks. 1. Let $G = F \cdot H$ be a semidirect product with kernel H, $(|F|, |H|) = 1$. It is known that $\Phi(H) \leq \Phi(G)$. We claim, that $\Phi(H) = H \cap \Phi(G)$. Without loss of generality, we may assume that $\Phi(H) = \{1\}$; then it suffices to show that $D = H \cap \Phi(G) = \{1\}$. Assume that this is false: $D > \{1\}$ (note that $D \triangleleft G$). It follows from $\Phi(D) \leq \Phi(H) = \{1\}$ that D is abelian of square free exponent. Let K be a subgroup of H minimal such that $KD = H$. Then $K \cap D \leq \Phi(K)$ and $K \cap D \triangleleft H$ so $K \cap D \leq \Phi(H) = \{1\}$ whence $H = K \cdot D$ is a semidirect product. Let p divides $|D|$, $P_1 \in \mathrm{Syl}_p(K)$, $P_2 \in \mathrm{Syl}_p(D)$. Then $P_1P_2 \in \mathrm{Syl}_p(H) = \mathrm{Syl}_p(G)$ and $P_1 \cap P_2 = \{1\}$. By Gaschütz's theorem [Hup, Hauptsatz 1.17.4], there exists $T < G$ such that $G = T \cdot P_2$ with $T \cap P_2 = \{1\}$. As $P_2 \leq D \leq \Phi(G)$, we get $G = T$ so $P_2 = \{1\}$. Thus, $\{1\} = D = H \cap \Phi(G)$, as was to be shown.

2. Let G be nonsolvable but all its proper subgroups are solvable. Let N be a maximal normal subgroup of G; then G/N is nonabelian simple since N is solvable. If H is a maximal subgroup of G, then $HN < G$ since HN is solvable. It follows that $N < H$ so $N = \Phi(G)$ and $G/\Phi(G)$ is nonabelian simple.

We are ready to give a new proof of the following

Theorem 25.5 ([Suz2]). *If G is a noncyclic p-group and $\mathcal{L}(G) \cong \mathcal{L}(G_1)$, then G_1 is either a p-group or a $\mathcal{P}$-group.*

Proof. We use induction on $|G|$. Let φ be an isomorphism of $\mathcal{L}(G)$ onto $\mathcal{L}(G_1)$. In view of Lemmas 25.1 and 25.4, we may assume that G is neither cyclic nor elementary abelian. Then $\Phi(G) > \{1\}$ so $\Phi(G_1) > \{1\}$ (Lemma 25.1(a)). Assume that G_1 is not a p-group. Then $G_1/\Phi(G_1)$ is not a p-group (Schur–Zassenhaus) so it is a $\mathcal{P}$-group (Lemmas 25.1(a) and 25.4). Set $|G_1/\Phi(G_1)| = qp^n$, where p and q are primes and q divides $p-1$; then $\mathrm{d}(G) = n+1$. In that case, $G_1 = Q_1 \cdot P_1$, where $P_1 = G_1' \in \mathrm{Syl}_p(G_1)$ and $Q_1 \in \mathrm{Syl}_q(G_1)$ (this follows from properties of Φ-subgroups [Gas1]). By Lemma 25.1(c), G_1 has no elements of order pq so $|Q_1| = q$ and G_1 is a Frobenius group with kernel P_1 of index q. By Remark 1, $\Phi(G_1) = \Phi(P_1)$. Set $P = P_1^{\varphi^{-1}}$; then P is maximal in G. If M is a maximal subgroup of G and $M \neq P$, then M is elementary abelian, by induction, since M^φ is not a p-group (indeed, P_1 is the unique maximal subgroup of G_1 which is a p-group). It follows that the set Γ_1 has $(1+p+\cdots+p^n)-1$ elementary abelian members. Since G is not elementary abelian, we get $n = 1$ (Exercise 1.6(a)). Then P_1 is cyclic and G_1 has exactly $1+|P_1|$ subgroups of prime order so $\mathrm{c}_1(G) = 1+|P_1|$ (here $\mathrm{c}_1(G_1)$ is the number of subgroups of G_1 of prime orders) and hence $|P_1| = p$, by Kulakoff's Theorem 5.3. Then $|G| = p^2$, contrary to the assumption. □

Recall that if $n = \prod_{i=1}^k p^{\alpha_i}$ is a standard prime decomposition, then $\lambda(n) = \sum_{i=1}^k \alpha_i$. Next, $\lambda(G) = \lambda(|G|)$.

Lemma 25.6. *Let G be a minimal nonnilpotent group such that $\lambda(G) > 2$ and $G' \in \mathrm{Syl}_q(G)$. Then $G_1 = G^\varphi$ is also minimal nonnilpotent (here φ is the lattice isomorphism) and $G_1' \in \mathrm{Syl}_q(G_1)$.*

Proof. By Theorem 22.5, G_1 is not a p-group. Let $G = P \cdot Q$, where $Q = G' \in \mathrm{Syl}_q(G)$, $P \in \mathrm{Syl}_p(G)$ and let $b \in \mathbb{N}$ be the least integer such that p divides $q^b - 1$. Then (see Theorem A.22.1) $G/\Phi(G)$ is of order pq^b with maximal subgroup of order p and this quotient group has exactly $1+q^b$ maximal subgroups; then $G_1/\Phi(G_1)$ has a maximal subgroup of prime order, say r. It follows that $|G_1/\Phi(G_1)| = rs^t$, where r, s are distinct primes and $t \in \mathbb{N}$ is least such that r divides $s^t - 1$.

Suppose that $b > 1$. Then $t > 1$ and $G_1/\Phi(G_1)$ has a normal r-complement, by Burnside's theorem, and then $G_1 = P_1 \cdot Q_1$, where $P_1 \in \mathrm{Syl}_r(G_1)$, $Q_1 \in \mathrm{Syl}_s(G_1)$. Since G_1 has exactly $1+s^t$ maximal subgroups we get $s^t = q^b$, i.e., $s = q$ and $t = b$. By Remark 1, $Q_1 \cap \Phi(G_1) = \Phi(Q_1)$. Suppose that $\Phi(Q_1) > \{1\}$. Then $Q^\varphi = Q_1$, by Theorem 25.5. It follows from properties of minimal nonnilpotent groups that $|\Phi(Q_1)| = |\Phi(Q)| < q^b$, so by Sylow's theorem, $P_1\Phi(Q_1)$ is nilpotent and P_1 centralizes $\Phi(Q_1)$. Since all Sylow r-subgroups generate G_1, it follows that $\Phi(Q_1) \leq \mathrm{Z}(G_1)$ and then all maximal subgroups of G_1 are nilpotent. Now suppose that $\Phi(Q_1) = \{1\}$; then $\Phi(G_1) = \Phi(P_1)$ and $|P_1 : \Phi(P_1)| = r$, and again, as it easy to see, G_1 is minimal nonnilpotent.

Suppose that $b = 1$. Then $|Q| = q$ so $|Q_1| = q$ and $|P_1 : \Phi(G_1)| = r$. As in the last sentence of the previous paragraph, G_1 is minimal nonnilpotent. □

Now we are ready to give a new proof of the following classical

Theorem 25.7 ([Suz2]). *Let a group G be solvable and φ an isomorphism from $\mathcal{L}(G)$ onto $\mathcal{L}(G_1)$. Then the group G_1 is also solvable.*

Proof. Let G_1 be a counterexample of minimal order. Then G_1 is nonsolvable but all its proper subgroups are solvable, by induction so $G_1/\Phi(G_1)$ is nonabelian simple, by Remark 2. As $G_1/\Phi(G_1)$ and $G/\Phi(G)$ are lattice isomorphic, by Lemma 25.1(a), we get $\Phi(G_1) = \{1\}$ so G_1 is nonabelian simple. Let $K \triangleleft G$ be of prime index, say r, and set $K_1 = K^\varphi$.

Let q be the least prime divisor of $|G_1|$ and let $S_1 = P_1 \cdot Q_1$ be a minimal nonnilpotent subgroup of G_1, where $Q_1 = S_1' \in \mathrm{Syl}_q(S_1)$ and $P_1 \in \mathrm{Syl}_p(S_1)$ (by Frobenius' normal p-complement theorem, such an S_1 exists). Set $S = S_1^{\varphi^{-1}}$. Since $|Q_1| > q$, we have $\lambda(S_1) > 2$ so $\lambda(S) > 2$ and S is minimal nonnilpotent (Lemma 25.6). Assume that $S \not\leq K$. Then $G = KS$ since $K \triangleleft G$ is of prime index r. It follows that $r = p$, and so $S' = Q = Q_1^{\varphi^{-1}} \leq K_1^{\varphi^{-1}} = K$. This inclusion is also true if $S \leq K$. It follows that $S_1' = Q_1 \leq K_1$ for every choice of S. Thus, if q is the minimal prime divisor of $|G_1|$ and D_1 is generated by normal Sylow q-subgroups of all minimal nonnilpotent subgroups of G_1, then $D_1 \leq K_1 (< G_1)$. Since $D_1 > \{1\}$ is normal in G_1, it follows that G_1 is not simple, a final contradiction. □

Proposition 25.8. *Let G and G_0 be groups of order p^m and let $\mathcal{L}(G) \cong \mathcal{L}(G_0)$ via φ.*

(a) If G is metacyclic then G_0 is also metacyclic.

(b) If $\Omega_1(G) = G$ and $N \triangleleft G$, then $N_1 = N^\varphi \triangleleft G_0$.

(c) If G is of maximal class and order $p^m > p^{p+1}$, then G_0 is also of maximal class. If, in addition, $N \triangleleft G$, then $N^\varphi \triangleleft G_0$.

(d) If $G \in \mathrm{Syl}_p(S_{p^n})$, $n > 1$, then $G_0 \cong G$. Next, $\mathrm{K}_i(G)^\varphi = \mathrm{K}_i(G_0)$ for $i > 1$.

Proof. (a) If $p > 2$, then $|G_0/\mho_1(G_0)| = |G/\mho_1(G)| \leq p^2$ so G_0 is metacyclic, by Theorem 9.11. If $p = 2$, the result follows since G_0 and all its maximal subgroups are two-generator since the same is true for G (see Theorem 43.3).

(b) Let Z_0 be a subgroup of order p in G_0 and $Z = Z_0^{\varphi^{-1}}$, $Z \not\leq N$. Since $K = \langle Z, N\rangle$ has order $p|N|$, $K_0 = K^\varphi$ has order $p|N| = p|N_0|$. Since $K_0 = \langle Z_0, N_0\rangle$, Z_0 normalizes N_0. It follows that $G_0 = \Omega_1(G_0)$ normalizes N_0.

(c) We have $\mathrm{c}_1(G_0) = \mathrm{c}_1(G) \equiv 1 + p + \cdots + p^{p-2} \pmod{p^p}$ so either G_0 is absolutely regular or of maximal class (see Theorem 13.2(a)). Since $|G_0/\mho_1(G_0)| = |G/\mho_1(G)| = p^p$, G_0 is not absolutely regular. The remaining assertions in (c) follow from Proposition 25.9, below.

(d) It is known that G is transitive on the set $\{1, \dots, p^n\}$. Let H be the stabilizer of a point in G and let $H_0 = H^\varphi$. We have $|G_0 : H_0| = |G : H| = p^n$. It remains to show, that if $N_0 = \bigcap_{x_0 \in G_0} H_0^{x_0}$, then $N_0 = \{1\}$: indeed, then G_0 is a subgroup of S_{p^n} so, since $|G_0| = |G|$, we have $G \cong G_0$, by Sylow's theorem. Let $N = N_0^{\varphi^{-1}}$; then $N < H$. Since $\Omega_1(G) = G$ we have $\Omega_1(G_0) = G_0$ so $N \lhd G$, by (b). In our case, $N \leq H_G = \bigcap_{x \in G} H^x = \{1\}$, so $N = \{1\}$. It follows that $N_0 = N^\varphi = \{1\}$. □

Proposition 25.9. *Suppose that a p-group G contains a regular subgroup H of exponent p^e and index p such that $\exp(G/\mho_1(H)) = p$. Let φ be a lattice isomorphism of G onto a p-group G_1. Then there exists a chief series $\{1\} = H^0 < H^1 < \cdots < H^m = G$ such that $(H^i)^\varphi \trianglelefteq G_1$, all i. Moreover, the last assertion is also true if G is regular or of maximal class.*

Proof. (i) In view of Proposition 25.8(b), we may assume that $e > 1$; then, by hypothesis, $\mho_1(H) = \mho_1(G) > \{1\}$. We have $\exp(G) = \exp(G_1) = p^e$. It suffices to prove that G has a normal subgroup $N \leq \mho_1(H)$ of order p such that $N_1 = N^\varphi \lhd G_1$, and then apply induction to G/N and G_1/N_1. Set $H_1 = H^\varphi$, $F = \Omega_{e-1}(H)$, $F_1 = \Omega_{e-1}(H_1) = F^\varphi$. Then $F \lhd G$ and $F_1 \lhd G_1$. Next, $\exp(H/F) = p$ so $\mho_1(G) = \mho_1(H) \leq F$, and we get $\exp(G/F) = p$. Similarly, $\exp(G_1/F_1) = p$. Let D/F be a G/F-invariant subgroup of order p in H/F and let $D_1 = D^\varphi$. By Proposition 25.8(b), $D_1/F_1 \lhd G_1/F_1$ since G/F and G_1/F_1 are of exponent p. By construction, $\exp(D) = p^e = \exp(D_1)$. Set $N = \mho_{e-1}(D)$. Since $D \leq H$ is regular, it follows that $p = |D : F| = |D : \Omega_{e-1}(D)| = |\mho_{e-1}(D)|$, i.e., $N = \mho_{e-1}(D)$ is of order p. Setting $N_1 = \mho_{e-1}(D_1)$, we see that N_1 is a characteristic subgroup of D_1 of order p, so $N_1 < H_1$ is normal in G_1. Since $N_1 = N^\varphi$, we are done.

(ii) Let G be regular of exponent p^e. We may assume that G is not cyclic and $e > 1$. Set $F = \mho_1(G)$, $F_1 = F^\varphi$. Let $D/F \lhd G/F$ be of order p and $N = \mho_{e-1}(D)$; then N is a characteristic subgroup of D of order p so $N \lhd G$. Set $D_1 = D^\varphi$, $N_1 = \mho_{e-1}(D_1)$; then $N_1 = N^\varphi \lhd G_1$, by construction. Applying induction to G/N and G_1/N_1, we prove the proposition in this case.

(iii) Let G be of maximal class. In view of (ii), we may assume that G is irregular, i.e., $|G| > p^p$. Then G contains an absolutely regular subgroup H of index p such that $\mho_1(H) = \mho_1(G)$, and the result follows from (i). □

Exercise 1. Let $G = \mathrm{GL}(2,3)$ and let G_1 be lattice isomorphic to G. Prove that $G_1 \cong G$.

Hint. We have $\Phi(G) = \mathrm{Z}(G)$ and $G/\Phi(G) \cong S_4$. It follows that $G_1/\Phi(G_1) \cong S_4$. If $P \in \mathrm{Syl}_2(G)$, then P is semidihedral. It follows that a Sylow 2-subgroup of G_1 is also semidihedral, i.e., G_1 is a covering group of S_4 with semidihedral Sylow 2-subgroup. It follows that $G_1 \cong G$.

Exercise 2. Study the p-groups which are lattice isomorphic to minimal nonabelian p-groups.

Exercise 3. Let A be an abelian 2-group of exponent 4 and let a 2-group G be lattice isomorphic to A. Show that $G \cong A$.

Hint. Since $\mathrm{c}_i(G) = \mathrm{c}_i(A)$, $i = 1, 2$, it suffices to show that G is abelian. Assuming that G is a counterexample of minimal order, we see that G is minimal nonabelian so $\mathrm{d}(G) = 2$, and we get $\mathrm{d}(A) = 2$. Then A is a subgroup of the abelian group of type $(4, 4)$, and it is easy to see that $|G| = |A| = 16$. By Proposition 25.8(a), G is metacyclic. Since there is only one nonabelian metacyclic group of order 16 and exponent 4, we get $G = \langle a, b \mid a^4 = b^4 = 1, a^b = a^{-1}\rangle$. It is easy to check that the involution a^2b^2 is not a square. Since all involutions are squares in A (Theorem 6.1), we get a contradiction.

Exercise 4. Let $p > 2$. Prove that the abelian group of type (p^2, p^2), $p > 2$, is lattice isomorphic to the nonabelian metacyclic group of order p^4 and exponent p^2.

Exercise 5. Let G be a nonabelian Dedekindian 2-group and G_0 is lattice isomorphic with G. Prove that $G_0 \cong G$.

Exercise 6. Let $G = \mathrm{ES}(m, p)$ be an extraspecial group of order p^{1+2m} and exponent p^2. Classify all groups G_1 that are lattice isomorphic with G.

A. Caranti [Ca] has showed that if a group G of order $p^n > p^3$ is of maximal class and G_0 is lattice isomorphic with G and G_0 is not of maximal class, then $n \leq p$. [Ca, Theorem 12] contains the list of all G satisfying the foregoing condition.

Problems

Problem 1. Classify the p-groups that are lattice isomorphic with minimal nonabelian p-groups.

Problem 2. Classify the p-groups that are lattice isomorphic with $\mathcal{A}_2$-groups (see §§65, 71).

Problem 3. Study the p-groups that are lattice isomorphic with special p-groups.

Problem 4. Classify the groups that are lattice isomorphic with Frobenius groups.

§26

Powerful p-groups

Below we follow closely to [DdSMS] and [LubM].

1°. Here we consider the case $p > 2$ (the case $p = 2$ is considered in 2°).

Definition 1. Let $p > 2$. A p-group G is said to be *powerful* if $\mho_1(G) = \Phi(G)$. (Since $\Phi(G) = G'\mho_1(G)$, this is equivalent to saying that $G' \leq \mho_1(G)$.)

Epimorphic images of powerful p-groups are powerful. A product G of pairwise permutable cyclic p-subgroups is powerful since $G/\mho_1(G)$ is elementary abelian. If a group of exponent p is powerful, it must be abelian. We prove some important properties of powerful p-groups. It is surprising that one can said so much on groups satisfying such 'not very restrictive condition'.

Exercise 1. All subgroups of a p-group G are powerful if and only if $\langle x, y\rangle$ is metacyclic. for all $x, y \in G$. (*Hint.* Use Theorems 9.11 and 69.1.)

Exercise 2. If G is a regular p-group, then $\mho_1(G)$ is powerful.

Solution. It suffices to show that $\mho_1(G)' \leq \mho_1(\mho_1(G))$. By Theorem 7.2(f), $\mho_1(G)' = \mho_2(G')$. But, if X is a p-group, then $\exp(X/\mho_1(\mho_1(X))) \leq p^2$ so $\mho_2(X) \leq \mho_1(\mho_1(X))$. It follows that $\mho_1(G)' = \mho_2(G') \leq \mho_1(\mho_1(G')) \leq \mho_1(\mho_1(G))$.

Proposition 26.1. *Let a p-group $G = P_1 \dots P_k$, where $P_1, \dots, P_k$ are powerful (we do not suppose that the factors are pairwise permutable). If $\mathrm{d}(P_1) + \dots + \mathrm{d}(P_k) = \mathrm{d}(G)$, then G is also powerful. In particular, if a p-group G is a product of $\mathrm{d}(G)$ cyclic subgroups, it is powerful.*

Proof. Set $\bar{G} = G/\mho_1(G)$; then $\mathrm{d}(G) = \mathrm{d}(\bar{G})$. It suffices to prove that $\bar{G}$ is elementary abelian, i.e., $|\bar{G}| = p^{\mathrm{d}(\bar{G})}$. We get $\bar{G} = \bar{P}_1 \dots \bar{P}_k$ and $|\bar{G}| = |\bar{P}_1 \dots \bar{P}_k| \leq p^{\mathrm{d}(\bar{P}_1)+\dots+\mathrm{d}(\bar{P}_k)} = p^{\mathrm{d}(\bar{G})}$ in view of $|\bar{P}_i| = p^{\mathrm{d}(\bar{P}_i)}$. Since $|\bar{G}| \geq p^{\mathrm{d}(\bar{G})}$, we get $|\bar{G}| = p^{\mathrm{d}(\bar{G})}$ so $\bar{G}$ is elementary abelian. □

Definition 2. A subgroup N of G is *powerfully embedded* in G if $[N, G] \leq \mho_1(N)$ (or, what is the same, if $N \trianglelefteq G$ and $N/\mho_1(N) \leq \mathrm{Z}(G/\mho_1(N))$.

Thus, G is powerful if and only if G is powerfully embedded in G. If N is powerfully embedded in G, then N is powerful since $N/\mho_1(N)$ is abelian as a subgroup of $Z(G/\mho_1(G))$. A subgroup N of exponent p is powerfully embedded in a p-group G if and only if it is contained in $Z(G)$. A cyclic normal subgroup is powerfully embedded in a p-group G.

Lemma 26.2. *If N is powerfully embedded in G and $K \lhd G$, then NK/K is powerfully embedded in G/K.*

Proof. Let ϕ be the natural epimorphism $G \to \bar{G} = G/K$. Then $N^\phi = NK/K$, $\mho_1(N)^\phi = \mho_1(N^\phi)$. We have to prove that $[N^\phi, G^\phi] \leq \mho_1(N^\phi)$. Applying ϕ to $[N, G] \leq \mho_1(N)$, we obtain the desired inclusion. □

Lemma 26.3. *Let $K, N \trianglelefteq G$ and $K \leq \mho_1(N)$. Then N is powerfully embedded in G if and only if N/K is powerfully embedded in G/K.*

Proof. In view of Lemma 26.2, it remains to prove that if N/K is powerfully embedded in G/K, then N is powerfully embedded in G. By assumption, $[N/K, G/K] \leq \mho_1(N/K) = \mho_1(N)/K$, and so $[N, G] \leq \mho_1(N)$. □

Lemma 26.4. *Let N be powerfully embedded in G and $x \in G$. Then $H = \langle N, x\rangle$ is powerful.*

Proof. We have $N/\mho_1(N) \leq Z(H/\mho_1(N)$ so $H/\mho_1(N)$ is abelian since H/N is cyclic. Then $\mho_1(H) \geq \mho_1(N) \geq H'$ so H is powerful. □

If M, N are powerfully embedded in G, then

$$[MN, G] = [M.G][N.G] \leq \mho_1(M)\mho_1(N) \leq \mho_1(MN)$$

so MN is powerfully embedded in G.

Lemma 26.5. *Suppose that N is powerfully embedded in G and $S < G$ is such that $N = S^G$. Then $N = S$.*

Proof. Assume that $S < N$. Without loss of generality, one may assume that S is maximal in N. Then $S/\Phi(N) < N/\Phi(N) \leq Z(G/\Phi(N))$, and we conclude that $S \lhd G$. Then $S^G = S < N$, contrary to the hypothesis. □

Lemma 26.6. *Let $N < W \leq G$, where N, W are normal in G, and suppose that N is not powerfully embedded in W. Then there exists a normal subgroup J in G such that $\mho_1(N)[N, W, W] \leq J < M = \mho_1(N)[N, W]$, $|M : J| = p$.*

Proof. We have $[N, W] \nleq \mho_1(N)$. Then $\mho_1(N) < M = \mho_1(N)[N, W]$. Since $M \lhd G$, there exists a normal subgroup J of G such that $\mho_1(N) \leq J < M$ and $|M : J| = p$. Since $[N, W, W] \leq [M, G] \leq J$, it follows that $\mho_1(N)[N, W, W] \leq J$. □

As noticed in [DdSMS], the point of Lemma 26.6 is that in order to establish that N is powerfully embedded in W, where $N < W$ are normal subgroups of a p-group G, we can factor out a suitable J and thereby reduce to the case where $\exp(N) = p$ and $|[N, W]| = p$. This technique is illustrated in the proof of the following

Proposition 26.7. *If N is powerfully embedded in G, then $\mho_1(N)$ is also powerfully embedded in G.*

Proof. It is given that $[N, G] \le \mho_1(N)$, and we may assume that $\mho_1(\mho_1(N)) = \{1\}$. It is enough to prove that $\mho_1(N) \le \mathrm{Z}(G)$. To this end, one may assume, without loss of generality, that $[\mho_1(N), G, G] = \{1\}$. We have $[N, G, G, G] \le [\mho_1(N), G, G] = \{1\}$, and so $[N, G, G] \le \mathrm{Z}(G)$. Then, by Grün's lemma, for given $x \in N$ and $g \in G$, the map $w \mapsto [x, g, w]$ is a homomorphism from G into $\mathrm{Z}(G)$. Then $\prod_{j=0}^{p-1}[x, g, x^j] = \prod_{j=0}^{p-1}[x, g, x]^j = [x, g, x]^{\binom{p}{2}}$. Hence

$$\begin{aligned}
[x^p, g] &= [x, g]^{x^{p-1}}[x, g]^{x^{p-2}} \dots [x, g] \\
&= \prod_{j=p-1}^{0} [x, g][x, g, x^j] \\
&= [x, g]^p \prod_{j=0}^{p-1}[x, g, x^j] \qquad \text{since } [x, g, x^j] \in \mathrm{Z}(G) \text{ for each } j \\
&= [x, g]^p [x, g, x]^{\binom{p}{2}} = 1
\end{aligned}$$

since $\mho_1([N, G]) = \{1\}$. Thus $[\mho_1(N), G] = \{1\}$. □

In particular, if G is a powerful p-group, then $\Phi(G)$ is powerfully embedded in G; moreover, $\mho^k(G)$ is powerfully embedded in G for all $k \in \mathbb{N}$ (see §§23, 24; recall that $\mho^1(G) = \mho_1(G)$, $\mho^{i+1}(G) = \mho_1(\mho^i(G))$). However, we do not know, in the case under consideration, if $\mho_k(G)$ is powerfully embedded in G for $k > 1$.

It follows from Proposition 26.7 that if N is powerfully embedded in G, then $N/\mho^{p-1}(N)$ is regular (since its class is at most $p - 1$; see Theorem 7.1(b)), $\mathrm{d}(\mho^i(N)) \le \mathrm{d}(N)$. It remains to prove only the last assertion for $i = 1$. By Proposition 26.7, $N/\mho^2(N)$ is of class two at most; therefore (Theorem 7.3)

$$p^{\mathrm{d}(N)} = |N : \mho_1(N)| \ge |\mho_1(N) : \mho^2(N)| = p^{\mathrm{d}(\mho_1(N))}.$$

Set $P_1(G) = G$, $P_{i+1}(G) = \mho_1(P_i(G))[P_i(G), G]$ for $i \ge 1$. Then, for a p-group G, we have $P_2(G) = \mho_1(P_1(G))[P_1(G), G] = \mho_1(G)[G, G] = \mho_1(G)G' = \Phi(G)$.

Lemma 26.8. *Let G be a powerful p-group.*

(a) *For each $i \ge 1$, $P_i(G)$ is powerfully embedded in G and $P_{i+1}(G) = \mho_1(P_i(G)) = \Phi(P_i(G))$. (In particular, $P_3(G) = \Phi(\Phi(G))$.)*

(b) *For each* $i \geq 1$, *the map* $x \mapsto x^p$ *induces a homomorphism from* $P_i(G)/P_{i+1}(G)$ *onto* $P_{i+1}(G)/P_{i+2}(G)$.

Proof. (a) By hypothesis, $G = P_1(G)$ is powerfully embedded in G. Suppose that $P_i(G)$ is powerfully embedded in G for some $i \geq 1$. Then

$$P_{i+1}(G) = \mho_1(P_i(G))[P_i(G), G] \leq \mho_1(P_i(G))\mho_1(P_i(G)) = \mho_1(P_i(G)) \leq P_{i+1}(G),$$

and so $P_{i+1}(G) = \mho_1(P_i(G))$. Then, by Proposition 26.7, $P_{i+1}(G)$ is powerfully embedded in G. Therefore, by Definition 1, $\mho_1(P_i(G)) = \Phi(P_i(G))$, proving (a).

(b) By (a), $P_i(G)$ is powerful, $P_{i+1}(G) = \Phi(P_i(G)) = P_2(P_i(G))$. Next, $P_{i+2}(G) = \mho_1(\mho_1(P_i(G))) = P_3(P_i(G))$. So, changing notation, we may assume that $i = 1$; and then replacing G by $G/P_3(G)$, we may assume that $P_3(G) = \{1\}$. Then $G' \leq \Phi(G) = P_2(G) \leq Z(G)$, so for $x, y \in G$ we have (see Exercise 1.18) $(xy)^p = x^p y^p [y, x]^{p(p-1)/2}$. Since $p > 2$, we get $[y, x]^{p(p-1)/2} \in \mho_1(P_2(G)) = P_3(G) = \{1\}$, and so $(xy)^p = x^p y^p$. Since $\mho_1(P_2(G)) = P_3(G) = \{1\}$ and $\mho_1(G) = P_2(G)$, this shows that $x \mapsto x^p$ induces a homomorphism from $G/P_2(G)$ onto $P_2(G)/P_3(G)$. □

Lemma 26.9. *If a powerful p-group* $G = \langle a_1, \ldots, a_d \rangle$, *then* $\mho_1(G) = \langle a_1^p, \ldots, a_d^p \rangle$.

Proof. Let $\theta : G/P_2(G) \to P_2(G)/P_3(G)$ be the homomorphism given in the previous lemma. Then

$$P_2(G)/P_3(G) = \theta(G/P_2(G)) = \langle \theta(a_1 P_2(G)), \ldots, \theta(a_d P_2(G)) \rangle = \langle a_1^p, \ldots, a_d^p \rangle P_3(G).$$

Since $P_3(G) = \Phi(P_2(G))$, we get $\mho_1(G) = \langle a_1^p, \ldots, a_d^p \rangle$. □

Proposition 26.10. *If G is a powerful p-group, then every element of* $\mho_1(G)$ *is a p-th power.*

Proof. We proceed by induction on $|G|$. Let $g \in \mho_1(G) = P_2(G)$. By Lemma 26.8(b), there exists $x \in G$ and $y \in P_3(G)$ such that $g = x^p y$. Put $H = \langle \mho_1(G), x \rangle$. Since $\mho_1(G) = P_2(G)$ is powerfully embedded in G, by Lemma 26.8(a), it follows from Lemma 26.4 that H is powerful. Also, $g \in \mho_1(H)$ since $y \in P_3(G) = \mho_1(\mho_1(G)) \leq \mho_1(H)$. We may assume that G is not cyclic; then $H < G$. Then g is a p-th power in H, by induction. □

Let G be a powerful p-group. Then $\mho_1(G) = \{x^p \mid x \in G\}$, by Proposition 26.10. Since $\mho_1(G)$ is also powerful (Proposition 26.7), we get $\mho^2(G) = \{y^p \mid y \in \mho_1(G)\} = \{x^{p^2} \mid x \in G\}$, and so on. In general, we have $\mho^k(G) = \{x^{p^k} \mid x \in G\}$.

Let us summarize the main features of the lower P-series in a powerful p-group.

Theorem 26.11. *Let $G = \langle a_1, \dots, a_d \rangle$ be a powerful p-group.*

(a) $P_i(G)$ is powerfully embedded in G.

(b) $P_{i+k}(G) = P_{k+1}(P_i(G)) = \mho_k(P_i(G))$ for each $k \geq 0$.

(c) $P_i(G) = \mho_{i-1}(G) = \langle a_1^{p^{i-1}}, \dots, a_d^{p^{i-1}} \rangle$.

(d) The map $x \mapsto x^{p^k}$ induces a surjective homomorphism $P_i(G)/P_{i+1}(G) \to P_{i+k}(G)/P_{i+k+1}(G)$ for each i and k.

Proof. (a) = Lemma 26.8(a). We also observed that $P_{i+1}(G) = \mho_1(P_i(G)) = P_2(P_i(G))$. It follows from Proposition 26.10 that $P_{i+1}(G) = \{x^p \mid x \in P_i(G)\}$ and then by induction $P_i(G) = \{x^{p^{i-1}} \mid x \in G\}$ (see also paragraph following Proposition 26.10). Since $P_i(G) < G$, this implies that $P_i(G) = \mho_{i-1}(G)$. Similarly, repeated applications of Lemma 26.9 show that $P_i(G) = \langle a_1^{p^{i-1}}, \dots, a_d^{p^{i-1}} \rangle$, proving (c). Part (d) follows from Lemma 26.8(b). Finally, taking $P_i(G)$ in place of G and $k + 1$ in place of i, in (c), we get

$$P_{k+1}(P_i(G)) = \mho_k(P_i(G)) = \{x^{p^k} \mid x \in P_i(G)\} = \{y^{p^{i-1+k}} \mid y \in G\} = P_{i+k}(G),$$

and (b) is proved. □

Corollary 26.12. *If $G = \langle a_1, \dots, a_d \rangle$ is a powerful p-group, then $G = \langle a_1 \rangle \dots \langle a_d \rangle$, i.e., G is a product of its d cyclic subgroups $\langle a_i \rangle$ which are not necessarily pairwise permutable.*

Proof. Say $P_e(G) > P_{e+1}(G) = \{1\}$. Using induction on e, we may suppose that $G = \langle a_1 \rangle \dots \langle a_d \rangle P_e(G)$. But $P_e(G) = \langle a_1^{p^{e-1}}, \dots, a_d^{p^{e-1}} \rangle$ and $P_e(G) \leq \mathrm{Z}(G)$; the second inclusion follows since $P_e(G)$ is powerfully embedded in G and $\{1\} = P_{e+1} = \Phi(P_e(G))$. Then

$$G = (\langle a_1 \rangle \langle a_1^{p^{e-1}} \rangle) \dots (\langle a_d \rangle \langle a_d^{p^{e-1}} \rangle) = \langle a_1 \rangle \dots \langle a_d \rangle. \qquad \square$$

Theorem 26.13. *If H is a subgroup of a powerful p-group G, then $\mathrm{d}(H) \leq \mathrm{d}(G)$.*

Proof. We use induction on $|G|$. Let $d = \mathrm{d}(G)$ and put $m = \mathrm{d}(P_2(G))$ (here $P_2(G) = \Phi(G)$). By the remark after Proposition 26.7, $m \leq d$. By Lemma 26.8(a), $P_2(G)$ is powerful so, by induction, $\mathrm{d}(K) \leq m$, where $K = H \cap P_2(G)$.

Now the map $\theta : G/P_2(G) \to P_2(G)/P_3(G)$, induced by $x \mapsto x^p$, is an epimorphism (by Lemma 26.8(b)), and $\dim_{\mathrm{GF}(p)}(\ker(\theta)) = d - m$ (we consider $G/P_2(G)$ as a vector GF(p)-space). Hence $\dim_{\mathrm{GF}(p)}(\ker(\theta) \cap HP_2(G)/P_2(G)) \leq d - m$, whence, denoting $e = \dim_{\mathrm{GF}(p)}(HP_2(G)/P_2(G))$, we get

$$\dim_{\mathrm{GF}(p)}(\theta(HP_2(G)/P_2(G))) \geq \dim_{\mathrm{GF}(p)}(HP_2(G)/P_2(G)) - (d - m) = e - (d - m) = m - (d - e).$$

Let $h_1, \dots, h_e \in H$ be such that $HP_2(G) = \langle h_1, \dots, h_e \rangle P_2(G)$. Since $\Phi(K) = \Phi(H \cap \Phi(G)) \le \Phi(\Phi(G)) = P_3(G)$, the subspace $K/\Phi(K)$, spanned by the cosets of $h_1^p, \dots, h_e^p$, has dimension at least $\dim_{\mathrm{GF}(p)}(\theta(HP_2(G)/P_2(G))) \ge m - (d - e)$. Since $\mathrm{d}(K) \le m$, we can find $d - e$ elements $y_1, \dots, y_{d-e}$ of K such that $K = \langle h_1^p, \dots, h_e^p, y_1, \dots, y_{d-e} \rangle \Phi(K)$. Then $K = \langle h_1^p, \dots, h_e^p, y_1, \dots, y_{d-e} \rangle$, and so, by the modular law,

$$\begin{aligned} H &= H \cap \langle h_1, \dots, h_e \rangle P_2(G) = \langle h_1, \dots, h_e \rangle (H \cap P_2(G)) \\ &= \langle h_1, \dots, h_e \rangle K = \langle h_1, \dots, h_e, y_1, \dots, y_{d-e} \rangle. \end{aligned}$$

Thus $\mathrm{d}(H) \le d$, as required. □

Definition 3. Let $G > \{1\}$ be a p-group, $r \in \mathbb{N}$. Let $\mathrm{V}(G, r)$ denote the intersection of the kernels of all homomorphisms of G into $\mathrm{GL}(r, p)$.

Obviously, $\mathrm{V}(G, r)$ is a characteristic subgroup of G. Recall that $\mathrm{UT}(r, p)$ is the upper unitriangular group of dimension r over $\mathrm{GF}(p)$. Since the image of any homomorphism of a p-group G into $\mathrm{GL}(r, p)$ is a p-group and it is conjugate to the subgroup of $\mathrm{UT}(r, p)$, we could equally well define $V(G, r)$ as the intersection of the kernels of all homomorphisms of G into $\mathrm{UT}(r, p)$. Note that $g \in G$ belongs to $V(r, G)$ if and only if g acts trivially in every k-dimensional representation of G over $\mathrm{GF}(p)$, $k \le r$.

Definition 4. The number $\mathrm{r}(G) = \max\{\mathrm{d}(H) \mid H \le G\}$ is said to be the *sectional rank* of a p-group G (so $\mathrm{r}(G) = \mathrm{d}(G)$ provided G is powerful).

Given $r \in \mathbb{N}$, define the integer $\mu(r)$ by $2^{\mu(r)-1} < r \le 2^{\mu(r)}$.

Lemma 26.14. *(a) The group* $\mathrm{UT}(r, p)$ *has a series, of length* $\mu(r)$*, of normal subgroups, with elementary abelian factors. (In particular,* $\mho_{\mu(r)}(\mathrm{UT}(r, p)) = \{1\}$*.)*

(b) If G *is a* p*-group, then* $G/V(G, r)$ *has a series such as in (a).*

Proof. (a) We proceed by induction on r. The result is trivial if $r = 1$. If $r > 1$, put $s = [r/2]$, the integer part of $r/2$. Then the elements of $\mathrm{UT}(r, p)$ have the form $x = \left(\begin{smallmatrix} A & B \\ 0 & C \end{smallmatrix}\right)$, with $A \in \mathrm{UT}(s, p)$ and $C \in \mathrm{UT}(r - s, p)$. The mapping which sends x to matrix $\left(\begin{smallmatrix} A & 0 \\ 0 & C \end{smallmatrix}\right)$, is a homomorphism from $\mathrm{UT}(r, p)$ into $\mathrm{UT}(s, p) \times \mathrm{UT}(r - s, p)$, and its kernel consisting of matrices $\left(\begin{smallmatrix} I_s & B \\ 0 & I_{r-s} \end{smallmatrix}\right)$, is an elementary abelian p-group. Now the result follows by induction.

(b) follows from (a), since $G/\mathrm{V}(G, r)$ is isomorphic to a subgroup of the direct product of finitely many copies of $\mathrm{UT}(r, p)$. □

Lemma 26.15. *Let* G *be a* p*-group and* $r \in \mathbb{N}$*. Put* $V = \mathrm{V}(G, r)$*. If* $N \trianglelefteq G$*,* $\mathrm{d}(N) \le r$*, and* $N \le V$*, then* N *is powerfully embedded in* V*.*

Proof. We use induction on $|N|$. Suppose that $[N, V] \not\leq \mho_1(N)$. In view of Lemma 26.6, we may assume that $\mho_1(N) = \{1\}$ and $|[N, V]| = p$. There exists $M \lhd G$ with $[N, V] \leq M < N$ and $|N : M| = p$. By induction, $N/[N, V]$ is elementary abelian so $\mathrm{d}(M/[N, V]) = \mathrm{d}(N/[N, V]) - 1 \leq r - 1$, and we get $\mathrm{d}(M) \leq r$ in view of $|[N, V]| = p$. Hence, by induction, $[M, V] \leq \mho_1(M) \leq \mho_1(N) = \{1\}$. Thus, $M \leq \mathrm{Z}(V)$, and so $M \leq \mathrm{Z}(N)$; since N/M is cyclic (of order p), we conclude that N is abelian. Since $\exp(N) = p$, it follows that N is elementary abelian. By Definition 3, since $|N| \leq p^r$, we must have $[N, V] = \{1\}$, contrary to the initial assumption. □

Theorem 26.16. *Let G be a p-group with $r = \mathrm{r}(G)$. Then $V = V(G, r)$ is a powerful subgroup of G, $|G : V| \leq p^{r\mu(r)}$.*

Proof. By Lemma 26.14(b), there is a series of normal subgroups running from G to V, of length at most $\mu(r)$, with each factor elementary abelian of order $\leq p^r$ so $|G : V| \leq p^{r\mu(r)}$. By hypothesis, $\mathrm{d}(V) \leq r$. Then V is powerfully embedded in V, by Lemma 26.15, hence V is powerful. □

2°. In this subsection we consider 2-groups only.

Note that if G is a 2-group then $G' \leq \Phi(G) = \mho_1(G)$. Therefore, powerful 2-groups are defined as follows:

Definition 5. A 2-group G is said to be *powerful* if $G/\mho_2(G)$ is abelian.

The group M_{2^n}, $n > 3$, is powerful. Next, $G = M * \mathrm{C}_{2^n}$, where M is nonabelian of order 8, $n > 2$ and $|G| = 2^{n+2}$, is powerful. A metacyclic 2-group G such that G/G' is abelian of type $(2^m, 2^n)$, $m \geq n > 1$, is powerful.

All results proved above for powerful p-groups, $p > 2$, apart from Lemma 26.6, Lemma 26.15 and Theorem 26.16, are true for powerful 2-groups as well (with obvious modifications).

Definition 6. Let G be a 2-group and $N \trianglelefteq G$. Then N is *powerfully embedded* in G if $[N, G] \leq \mho_2(N)$.

Obviously, G is powerful if and only if it is powerfully embedded in itself.

If N is powerfully embedded in G and $H = \langle a, N \rangle$, then H is powerful. Indeed, $N/\mho_2(N) \leq \mathrm{Z}(\mho_2(N))$ and H/N is cyclic. Then $H/\mho_2(N)$ is abelian and so $H' \leq \mho_2(N)$. But $\mho_2(N) \leq \mho_2(H)$ and we are done. Moreover, the subgroup AN, where N is as above and A is abelian, is powerful.

Theorem 26.17. *If M and N are powerfully embedded in a 2-group G so are $[N, G]$, $\mho_1(N)$, $[M, N]$, and MN.*

Proof. To prove that $K \trianglelefteq G$ is powerfully embedded, we can always assume that $\mho_2(K) = \{1\}$, and try to show that $[K, G] = \{1\}$. Moreover, if $[K, G] \neq \{1\}$, then we can work in the quotient group G/X, where X is a G-invariant subgroup of index 2 in $[K, G]$ and we may assume $X = \{1\}$; then $|[K, G]| = 2$. In any case, one may assume that $[K, G, G] = \{1\}$. In particular, $K \leq Z_2(G)$.

(a) Suppose that $K = [N, G]$. Let $a \in N$, $b \in G$ so that $[a, b] \in K \leq Z_2(G)$ and $\mathrm{cl}(\langle a, [a, b]\rangle) \leq 2$. We compute

$$(a^4)^b = (a^b)^4 = (a[a, b])^4 = a^4[a, b]^4[a, b, a]^6 = a^4$$

since $[a, b]^4 \in \mho_2(K) = \{1\}$ and $[a, b, a]^6 \in \mho_1([K, G]) = \{1\}$ (here $6 = \binom{4}{2}$). It follows that $a^4 \in Z(G)$, and so $\mho_2(N) \leq Z(G)$. But N is powerfully embedded in G and so $K = [N, G] \leq \mho_2(N) \leq Z(G)$.

(b) Suppose that $K = \mho_1(N)(= \Phi(N))$. Let $a \in N$, $b \in G$. Since $K \leq Z_2(G)$ and $a^2 \in K$, we get $\mathrm{cl}(\langle a^2, b\rangle) \leq 2$ and $1 = [a^2, b]^2 = [a^4, b]$, where we have used our assumption that $\mho_1([K, G]) = 1$. Thus, $\mho_2(N) \leq Z(G)$. Since N is powerfully embedded in G, $[N, G] \leq \mho_2(N) \leq Z(G)$ and so $N \leq Z_2(G)$. Thus $\mathrm{cl}(\langle a, b\rangle) \leq 2$ which implies $[a^2, b] = [a, b]^2$. Since $\mho_2(N)$ is abelian,

$$\begin{aligned}\mho_1(\mho_2(N)) &= \langle (a_1^4 \dots a_k^4)^2 \mid a_i \in N\rangle = \langle a_1^8 \dots a_k^8 \mid a_i \in N\rangle \\ &= \mho_3(N) \leq \mho_2(\mho_1(N)) = \mho_2(K) = \{1\}\end{aligned}$$

and so $\mho_2(N)$ is elementary abelian. But $[N, G] \leq \mho_2(N)$ and so $[N, G]$ is elementary abelian too. Hence $[a^2, b] = [a, b]^2 = 1$ and so $K = \mho_1(N) \leq Z(G)$.

(c) Suppose that $K = [M, N]$. Let $a \in M$, $b \in N$; then $K \leq Z_2(G)$ implies that $\mathrm{cl}(\langle a, [a, b]\rangle) \leq 2$. We compute $(a^4)^b = (a^b)^4 = (a[a, b])^4 = a^4[a, b]^4[a, b, a]^6 =$ 1 since $\mho_2(K) = \{1\}$ and $\mho_1([K, G]) = \{1\}$. Hence $[\mho_2(M), N] = \{1\}$ and so also $[M, G, N] = \{1\}$ (noting that M is powerfully embedded in G). Similarly, $\{1\} = [N, G, M] = [G, N, M]$. By the Three Subgroups Lemma, $[K, G] = [M, N, G] = \{1\}$.

(d) Suppose that $K = MN$. We have assumed $\mho_2(K) = \mho_2(MN) = \{1\}$ and so also $\mho_2(M) = \mho_2(N) = \{1\}$. Since M and N are powerfully embedded in G, we get $[M, G] \leq \mho_2(M) = \{1\}$, $[N, G] \leq \mho_2(N) = \{1\}$ and so $M \leq Z(G)$ and $N \leq Z(G)$. Hence $K = MN \leq Z(G)$ and we are done. □

Corollary 26.18. *If G is a powerful 2-group, then $K_i(G)$, $G^{(i)}$ (the i-th derived subgroup of G) and $\Phi(G) = \mho_1(G)$ are powerfully embedded in G. If $K_{i+1}(G) \leq H \leq K_i(G)$ for some $i \geq 2$, then H is powerful.*

Proof. We have to prove only the last statement. We have (since $K_{i+1}(G)$ is powerfully embedded in G)

$$\mho_2(H) \geq \mho_2(K_{i+1}(G)) \geq [G, K_{i+1}(G)] = K_{i+2}(G) \geq [K_i(G), K_i(G)] \geq H',$$

where we still have to show that for $i \geq 2$, $[K_i(G), K_i(G)] \leq K_{i+2}(G)$. Indeed, we have $[K_{i-1}(G), K_i(G), G] \leq [K_{i+1}(G), G] = K_{i+2}(G)$ and $[K_i(G), G, K_{i-1}(G)] \leq$

$[K_{i+1}(G), K_{i-1}(G)] \leq K_{i+2}(G)$, and so, by the Three Subgroups Lemma, we obtain $[G, K_{i-1}(G), K_i(G)] \leq K_{i+2}(G)$ which gives $[K_i(G), K_i(G)] \leq K_{i+2}(G)$. □

Theorem 26.19. *If G is a powerful 2-group, then $\mho_j(G)$ is powerfully embedded in G and $\mho_1(\mho_j(G)) = \mho_{j+1}(G)$.*

Proof. By Theorem 26.17, $\mho_1(G)$ is powerfully embedded in G. Suppose that we have already proved that $\mho_j(G)$ $(j \geq 1)$ is powerfully embedded in G. We want to show that $\mho_1(\mho_j(G)) = \mho_{j+1}(G)$. For that purpose we may assume that $\mho_{j+1}(G) = 1$, and then show that $\mho_1(\mho_j(G)) = \{1\}$, i.e., $\mho_j(G)$ is elementary abelian. Let N be a minimal normal subgroup of G contained in $\mho_j(G)$. By induction (applied to the factor group G/N), $\mho_1(\mho_j(G)) \leq N$ and so $\mho_2(\mho_j(G)) = \{1\}$. But $\mho_j(G)$ is powerfully embedded in G, so $[\mho_j(G), G] = \{1\}$ and so $\mho_j(G) \leq Z(G)$ and $\mho_j(G)$ is abelian. Since $\mho_{j+1}(G) = \{1\}$, $\mho_j(G)$ is generated by elements of order 2 and so $\mho_j(G)$ is elementary abelian. Hence $\mho_1(\mho_j(G)) = \{1\}$ and we are done. Hence, $\mho_1(\mho_j(G)) = \mho_{j+1}(G)$ and so, by Theorem 26.17, $\mho_{j+1}(G)$ is powerfully embedded in G. □

Corollary 26.20. *If G is a powerful 2-group, then $\mho_i(\mho_j(G)) = \mho_{i+j}(G)$.*

Proof. By Theorem 26.19 and induction on i,

$$\mho_{i+1}(\mho_j(G)) = \mho_1(\mho_i(\mho_j(G))) = \mho_1(\mho_{i+j}(G)) = \mho_{i+j+1}(G). \quad \square$$

Corollary 26.21. *If G is a powerful 2-group, then $\mho_{2i}(G) \geq K_{i+1}(G)$.*

Proof. We use induction on i. By definition, $\mho_2(G) \geq \mho_1(G) = \Phi(G) = K_2(G)$. Using Corollary 26.20, we have (since $K_{i+1}(G)$ is powerfully embedded in G)

$$\mho_{2i+2}(G) = \mho_2(\mho_{2i}(G)) \geq \mho_2(K_{i+1}(G)) \geq [K_{i+1}(G), G] = K_{i+2}(G). \quad \square$$

Proposition 26.22. *If M and N are powerfully embedded in a 2-group G, then $\mho_k(MN) = \mho_k(M)\mho_k(N)$.*

Proof. Obviously, $\mho_1(M)\mho_1(N) \leq \mho_1(MN)$. On the other hand, in the group $G/\mho_1(M)\mho_1(N)$ both M and N are central and of exponent 2 (since $[G, M] \leq \mho_2(M) \leq \mho_1(M)$ and $[G, N] \leq \mho_2(N) \leq \mho_1(N)$), so we also have $\mho_1(MN) \leq \mho_1(M)\mho_1(N)$, and the result holds for $k = 1$. By induction,

$$\begin{aligned}\mho_{k+1}(MN) &= \mho_1(\mho_k(MN)) = \mho_1(\mho_k(M)\mho_k(N)) \\ &= \mho_1(\mho_k(M))\mho_1(\mho_k(N)) = \mho_{k+1}(M)\mho_{k+1}(N),\end{aligned}$$

and we are done. □

Proposition 26.23. *If G is a powerful 2-group, then each element of $\mho_i(G)$ can be written as a^{2^i} for some $a \in G$, i.e., $\mho_i(G) = \{x^{2^i} \mid x \in G\}$.*

Proof. First we deal with the case $i = 1$. Let $x \in \mho_1(G)$. Since $G/\mho_2(G)$ is abelian, x is a square mod$(\mho_2(G))$ and so $x \in b^2\mho_2(G)$ for some $b \in G$. Then $x \in \mho_1(H)$, where $H = \langle b, \mho_1(G)\rangle$ since $\mho_2(G) = \mho_1(\mho_1(G))$. Now, $H \neq G$ (unless G is cyclic) and H is powerful by the remark following Definition 6. So $x = a^2$ for $a \in H$, and we are done in the case $i = 1$.

We have $\mho_{i+1}(G) = \mho_1(\mho_i(G))$ so, if $x \in \mho_{i+1}(G)$, then there is $b \in \mho_i(G)$ such that $x = b^2$. By induction, there is $a \in G$ such that $b = a^{2^i}$ so $x = a^{2^{i+1}}$. □

Proposition 26.24. *Let $G = \langle a_1, \dots, a_d\rangle$ be a powerful 2-group. Then $\mho_i(G) = \langle a_1^{2^i}, \dots, a_d^{2^i}\rangle$.*

Proof. For $i = 1$ we may assume $\mho_1(\mho_1(G)) = \mho_2(G) = \{1\}$ and so (since G is powerful) G is abelian and then $\mho_1(G) = \langle a_1^2, \dots, a_d^2\rangle$. In general, by induction on i, $\mho_{i+1}(G) = \mho_1(\mho_i(G)) = \mho_1(\langle a_1^{2^i}, \dots, a_d^{2^i}\rangle) = \langle a_1^{2^{i+1}}, \dots, a_d^{2^{i+1}}\rangle$. □

Theorem 26.25. *Let $G = \langle a_1, \dots, a_d\rangle$ be a powerful 2-group. Then, there holds $G = \langle a_1\rangle \dots \langle a_d\rangle$.*

Proof. Let $2^e = \exp(G)$. Then $\mho_{e-1}(G) = \langle a_1^{2^{e-1}}, \dots, a_d^{2^{e-1}}\rangle$ is a central subgroup in G (this follows from Theorem 26.17). Applying induction to $G/\mho_{e-1}(G)$, we get $G = \langle a_1\rangle \dots \langle a_d\rangle\mho_{e-1}(G) = \langle a_1\rangle \dots \langle a_d\rangle\langle a_1^{2^{e-1}}\rangle \dots \langle a_d^{2^{e-1}}\rangle = \langle a_1\rangle \dots \langle a_d\rangle$. □

Theorem 26.26. *If H is a subgroup of a powerful 2-group G, then $\mathrm{d}(H) \le \mathrm{d}(G)$.*

Proof. The proof is almost the same as the proof of the corresponding Theorem 26.13 for $p > 2$ with some slight changes. Because of the importance of this theorem we shall give it explicitly.

We will use induction on $|G|$. The map $\theta : \; G/\mho_1(G) \longrightarrow \mho_1(G)/\mho_2(G)$, induced by $x \to x^2$, is an epimorphism. Indeed, $G/\mho_2(G)$ is abelian and so for $x, y \in G$, $(xy)^2 \equiv x^2y^2 \pmod{\mho_2(G)}$, and $G = \langle x_1, \dots, x_d\rangle$ implies $\mho_1(G) = \langle x_1^2, \dots x_d^2\rangle$, where $\mho_1(\mho_1(G)) = \mho_2(G)$. Let $d = \mathrm{d}(G)$ and put $m = d(\mho_1(G))$ so that $m \le d$. By induction (since $\mho_1(G)$ is powerful), $\mathrm{d}(K) \le m \le d$, where we set $K = H \cap \mho_1(G)$. Also, $\dim_{\mathrm{GF}(2)}(\ker(\theta)) = d - m$. So, $\dim_{\mathrm{GF}(2)}(\ker(\theta) \cap H\mho_1(G)/\mho_1(G)) \le d - m$, whence, denoting $e = \dim_{\mathrm{GF}(2)}(H\mho_1(G)/\mho_1(G))$, we get

$$\begin{aligned}\dim_{\mathrm{GF}(2)}(\theta(H\mho_1(G)/\mho_1(G))) &\ge \dim_{\mathrm{GF}(2)}(H\mho_1(G)/\mho_1(G)) - (d - m)\\ &= e - (d - m) = m - (d - e).\end{aligned}$$

Let $h_1, \dots, h_e \in H$ be such that $H\mho_1(G) = \langle h_1, \dots, h_e\rangle\mho_1(G)$ holds so that $\langle h_1, \dots, h_e\rangle$ covers H/K. Since $\Phi(K) = \Phi(H \cap \mho_1(G)) \le \mho_1(\mho_1(G) = \mho_2(G))$, the subspace of $K/\Phi(K)$, spanned by the cosets of $h_1^2, \dots, h_e^2$, has dimension at least $\dim_{\mathrm{GF}(2)}(\theta(H\mho_1(G)/\mho_1(G))) \ge m - (d - e)$.

Since $d(K) \le m$, we can find $d - e$ elements $y_1, \dots, y_{d-e}$ of K such that

$$K = \langle h_1^2, \dots, h_e^2, y_1, \dots, y_{d-e} \rangle \Phi(K) = \langle h_1^2, \dots, h_e^2, y_1, \dots, y_{d-e} \rangle.$$

But $\langle h_1, \dots, h_e \rangle$ covers H/K and so $H = \langle h_1, \dots, h_e, y_1, \dots, y_{d-e} \rangle$ which implies $d(H) \le d$. □

Thus, the inequality $r(G) \le d(G)$ holds for all powerful p-groups G, p is arbitrary.

Definition 7. A metacyclic 2-group H is called *ordinary metacyclic* (with respect to A) if H has a cyclic normal subgroup A such that H/A is cyclic and H centralizes $A/\mho_2(A)$ (in that case, $H/\mho_2(A)$ is abelian).

Proposition 26.27. *A two-generator 2-group is powerful if and only if it is ordinary metacyclic.*

Proof. (i) Let G be a powerful 2-group with $d(G) = 2$. Then $G/\mho_2(G)$ is a two-generator abelian group and so $G/\mho_2(G)$ is metacyclic. By a result of N. Blackburn (see Corollary 44.9), G is also metacyclic. Let A be a cyclic normal subgroup of G with G/A cyclic. If G centralizes $A/\mho_2(A)$, then G is ordinary metacyclic and we are done. Suppose that G does not centralize $A/\mho_2(A)$. Let $G = \langle A, g \rangle$ so that g inverts $A/\mho_2(A) \cong C_4$ (since $x \mapsto x^{-1}$ is a unique automorphism of the cyclic group of order 4). If $|G : A| = 2$, then $G/\mho_2(A)$ (being nonabelian of order 8) is isomorphic to D_8 or Q_8. In that case $\exp(G/\mho_2(A)) = 4$ and so $\mho_2(G) = \mho_2(A)$, contrary to our assumption that G is powerful (in which case $G/\mho_2(G)$ must be abelian). Hence $|G : A| \ge 4$ and so $|G| \ge 2^4$. Assume that $\langle g \rangle \cap A \le \mho_2(A)$. Since g^2 centralizes $A/\mho_2(A)$, $Y = \mho_2(A)\langle g^2 \rangle$ is normal in G and $G/Y \cong D_8$. But then $\mho_2(G) \le Y$ and $G/\mho_2(G)$ is nonabelian, a contradiction. It follows that $\langle g \rangle \cap A \not\le \mho_2(A)$. Since g inverts $A/\mho_2(A)$, $\langle g \rangle \not\ge A$ and so $\langle g \rangle \cap A = \mho_1(A)$. It follows that $\langle g \rangle$ is a cyclic subgroup of index 2 in G. Since G/A is cyclic of order > 2, G is not of maximal class and therefore $G \cong M_{2^n}$, $n \ge 4$ (Theorem 1.2). But this group is ordinary metacyclic with respect to $\langle g \rangle$, where $o(g) \ge 8$.

(ii) Suppose that G is ordinary metacyclic with respect to a normal cyclic subgroup A. Then $[G, A] \le \mho_2(A)$ and G/A is cyclic which implies that $G/\mho_2(A)$ is abelian and so G is powerful since $\mho_2(A) \le \mho_2(G)$. □

Theorem 26.28 ([Wil1]). *Let P be a* Q_8*-free powerful* 2*-group. Then P is also* D_8*-free so modular.*

Proof. We use induction on $|P|$. Let $2^n = \exp(P)$ and let $x \in P$ with $o(x) = 2^n$. We may presume that $n \ge 3$, P being abelian otherwise. As P is powerful, $\mho_{n-2}(P)$ is powerfully embedded in G and so $[\mho_{n-2}(P), P] = \{1\}$ and $\mho_{n-2}(P) \le Z(P)$. Set $x^{2^{n-2}} = y$ and $z = y^2$; then $y \in Z(P)$. Obviously, $P/\langle z \rangle$ is powerful and so, by induction, $P/\langle z \rangle$ is D_8-free. Suppose there are subgroups U and V of P, where U is a normal subgroup of V with $V/U \cong D_8$. Since $P/\langle z \rangle$ is D_8-free, $z \notin U$

and $z \notin P - V$. Thus $z \in V - U$ and therefore $Z(V/U) = (\langle z \rangle U)/U$. Since $z = y^2$, $y \in Z(P)$ so $y \in P - V$, we get $\langle V, y \rangle / U \cong C_4 * D_8$. On the other hand, $C_4 * D_8 \cong C_4 * Q_8$ and G is not Q_8-free, a contradiction. □

Theorem 26.29. *All subgroups of a 2-group G are powerful if, and only if, G is D_8-free and Q_8-free.*

Proof. If all subgroups of G are powerful, then G is obviously D_8-free and Q_8-free. Suppose that G is D_8-free and Q_8-free and let $H \leq G$. We have to show that $X = H/\mho_2(H)$ is abelian. We have $\exp(X) \leq 4$ and suppose that X is nonabelian. Let X_0 be a minimal nonabelian subgroup of X. Then X_0 is metacyclic (Theorem 26.27) so $X_0 = \langle a, b \mid a^4 = b^4 = 1, a^b = a^{-1} \rangle$, But $X_0/\langle b^2 \rangle \cong D_8$, a contradiction. Thus, X_0 does not exist. Hence X is abelian. □

3°. Below, p is arbitrary.

Proposition 26.30. *(a) Powerful p-groups of exponent p^2 are of class at most 2. (For $p = 2$, this follows from the definition.)*

(b) [Man24] If G is a powerful p-group of exponent p^e, then the map $x \mapsto x^{p^{e-1}}$ is an endomorphism of G.

If $p > 2$ and G is powerful of exponent p^2, then $\Phi(\Phi(G)) = \{1\}$ and $\Phi(G)$ is powerfully embedded in G so $\Phi(G) \leq Z(G)$. This proves part (a).

The following theorems (see [Man24]) supplement the results of §11.

Definition 8. Let G be a p-group.

1. G is called *order closed*, if in all sections H of G, $\exp(\Omega_k(H)) \leq p^k$ for all $k \in \mathbb{N}$. (Such groups are called $\mathcal{P}_2$-groups in §11.)
2. G is called *power closed* if all elements in $\mho_k(H)$ are p^k-th powers for all sections H of G and all $k \in \mathbb{N}$. (Such groups are called $\mathcal{P}_1$-groups in §11.)
3. G is called a P-group, if all its sections H satisfy $|\Omega_k(H)| = |H : \mho_k(H)|$ for all $k \in \mathbb{N}$. (Such groups are called $\mathcal{P}_3$-groups in §11.)

Groups satisfying all above defined properties are called $\mathcal{P}$-groups (see §11).

Theorem 26.31. *Let G be a power closed p-group, in which $\exp(\Omega_1(G)) = p$ and $|\Omega_1(G)| = |G : \mho_1(G)|$. Then G is a $\mathcal{P}$-group.*

Theorem 26.32. *Let G be a p-group in which $|\mho_1(G)| = p$, $\Omega_1(G)$ has exponent p and is a maximal subgroup of G. Then G is a $\mathcal{P}$-group.*

Theorem 26.33. *Let G be a power closed p-group and $\epsilon = 1$ if $p > 2$ and $\epsilon = 2$ if $p = 2$. Then $\mho_\epsilon(G)$ is a $\mathcal{P}$-group.*

Theorem 26.34. *Let G be a power closed p-group, possessing a chief series $\{N_i\}$ (with $N_0 = \{1\}$), such that for each i, the subgroup $\Omega_1(G/N_i)$ has exponent p. Then G is order closed.*

Proposition 26.35. *Let G be a power closed p-group with $\exp(\Omega_1(G)) = p$. Then for each $\mathcal{P}$-subgroup K of G, the subgroup $K\Omega_1(G)$ is also a $\mathcal{P}$-group.*

Theorem 26.36 ([Wil1]). *If $p > 2$ and G is a powerful p-group, then $\Omega_i(G) = \{x \in G \mid o(x) \leq p^i\}$ and the class of $\Omega_i(G)$ is $i + 1$ at most for all i.*

Theorem 26.37 ([HL]). *If G is a powerful p-group, then $|\mho_i(G)| = |G : \Omega_i(G)|$ for all i.*

Theorem 26.38 ([Arg]). *If $p > 2$ and a p-group G satisfies $\mathrm{K}_{p-1}(G) \leq \mho_1(G)$), then $\mho_1(G) = \{x^p \mid x \in G\}$.*

Exercise 3. Let N be a normal subgroup of a p-group G such that $N/N' \leq \mathrm{Z}(G/N)$. If $S < G$ is such that $S^G = N$, then $S = N$.

§27

p-groups with normal centralizers of all elements

In 'Research problems and themes' was posed the problem to classify finite p-groups with the title property and there we noticed that p-groups of class ≤ 2 have that property. We prove here that p-groups of class ≤ 2 are "almost" all p-groups with such a property.

Theorem 27.1 (Rocke [Rock]). *Let G be a p-group in which the centralizer of each element is normal. Then either* $\mathrm{cl}(G) \leq 2$ *or G is a* 3*-group of class* 3.

Proof (Bozikov–Janko [BozJ3]). Suppose that a p-group G satisfies:

($*$) For each $x \in G$, $\mathrm{C}_G(x)$ is normal in G.

Take $H \leq G$. If $h \in H$, then $\mathrm{C}_H(h) = \mathrm{C}_G(h) \cap H$ and so $\mathrm{C}_H(h) \trianglelefteq H$ so H has also the property ($*$).

Take $x, y \in G$. Since $\mathrm{C}_G(x) \trianglelefteq G$, $[x, y] \in \mathrm{C}_G(x)$. Similarly, $[x, y] \in \mathrm{C}_G(y)$. Hence $[x, y] \in \mathrm{Z}(\langle x, y\rangle)$ and so $\langle x, y\rangle/\langle [x, y]\rangle$ is abelian. This implies that $\langle x, y\rangle' = \langle [x, y]\rangle$ and so $\mathrm{cl}(\langle x, y\rangle) \leq 2$. We have proved that each two-generator subgroup of G is of class ≤ 2. Consequently, $(xy)^p = x^p y^p [y, x]^{p(p-1)/2}$ and so, if $p > 2$, the group G is regular.

Suppose that the theorem is false. Then $p \neq 3$ and $\mathrm{cl}(G) \geq 3$. Using induction on $|G|$, we may assume that each proper subgroup of G is of class ≤ 2. By the previous paragraph, $\mathrm{d}(G) \geq 3$.

Since $G' \not\leq \mathrm{Z}(G)$, there exist elements $a, b \in G$ such that $[a, b] \notin \mathrm{Z}(G)$. In that case $M = \langle a, b, \Phi(G)\rangle$ is maximal in G. Otherwise, there are two distinct maximal subgroups X, Y of G containing $\langle a, b\rangle$ and $\mathrm{C}_G([a, b]) \geq \langle X, Y\rangle = G$ (since X and Y are of class ≤ 2), a contradiction. Suppose that $c \in G - M$ is such that $G = \langle M, c\rangle = \langle a, b, c, \Phi(G)\rangle = \langle a, b, c\rangle$ and therefore $\mathrm{d}(G) = 3$. Since $\mathrm{C}_G([a, b]) = M$, we have $[a, b, c] \neq 1$.

It is easy to see that $\mathrm{cl}(G) = 3$. Suppose that $x \in G$ and $g \in G'$, so that $g \in \Phi(G)$ and therefore $|G : \langle \Phi(G), x, g\rangle| \geq p^2$. It follows that there are two distinct maximal subgroups X and Y of G which contain $\langle x, g\rangle$ and so $\mathrm{C}_G([x, g]) \geq \langle X, Y\rangle = G$. It follows that $\mathrm{K}_3(G) = [G, G'] \leq \mathrm{Z}(G)$, and so $\mathrm{cl}(G) = 3$.

We now use the Hall–Witt's identity:

$$[a, b^{-1}, c]^b [b, c^{-1}, a]^c [c, a^{-1}, b]^b = 1, \tag{1}$$

where a, b, c are the above elements in G. Since $\mathrm{cl}(G) = 3$, this simplifies into:

$$[a, b^{-1}, c][b, c^{-1}, a][c, a^{-1}, b] = 1. \tag{2}$$

Now, $\mathrm{cl}(\langle a, b\rangle) \leq 2$, and so $[a, b^{-1}] = [a, b]^{-1}$. Also, $\mathrm{cl}(\langle [a, b], c\rangle) \leq 2$ and so $[[a, b]^{-1}, c] = [a, b, c]^{-1}$. Hence $[a, b^{-1}, c] = [a, b, c]^{-1}$ and, similarly, $[b, c^{-1}, a] = [b, c, a]^{-1}$ and $[c, a^{-1}, b] = [c, a, b]^{-1}$. Thus, from (2) we obtain

$$[a, b, c]^{-1}[b, c, a]^{-1}[c, a, b]^{-1} = 1, \tag{3}$$

and the fact that $\mathrm{K}_3(G) = [G, G']$ is abelian gives also

$$[a, b, c][b, c, a][c, a, b] = 1. \tag{4}$$

Since $\mathrm{cl}(G/\mathrm{K}_3(G)) = 2$, we get $[ab, c] = [a, c][b, c]s$ with $s \in \mathrm{K}_3(G) \leq \mathrm{Z}(G)$. On the other hand, $\mathrm{cl}(\langle ab, c\rangle) \leq 2$, and so $[ab, c]$ commutes with ab. This implies

$$\begin{aligned} 1 = [[a, c][b, c], ab] &= [[a, c], ab]^{[b,c]}[[b, c], ab] \\ &= [[a, c], ab][[b, c], ab] \\ &= [[a, c], b][[a, c], a]^b[[b, c], b][[b, c], a]^b \\ &= [a, c, b][b, c, a], \end{aligned}$$

where we have used the facts that $[[a, c], ab] \in \mathrm{Z}(G)$, $[[b, c], a] \in \mathrm{Z}(G)$, $[[a, c], a] = 1$ and $[[b, c], b] = 1$ (noting that the subgroups $\langle a, c\rangle$ and $\langle b, c\rangle$ are of class ≤ 2).

We have proved that $[a, c, b][b, c, a] = 1$ which is equivalent with

$$[c, a, b] = [b, c, a], \tag{5}$$

since $[a, c, b] = [[c, a]^{-1}, b] = [c, a, b]^{-1}$.

Similarly, we get (replacing in (5) a, b, c with b, c, a in that order)

$$[a, b, c] = [c, a, b]. \tag{6}$$

Putting (5) and (6) in (4), we get finally $[a, b, c]^3 = 1$. Since G is not a 3-group, this gives $[a, b, c] = 1$ which contradicts to our choice of elements a, b, c.

Assume now that G is a 3-group with the property $(*)$. Let $x, y, z \in G$ and set $H = \langle x, y, z\rangle$. Since H has also the property (*), $\mathrm{C}_H(x) \trianglelefteq H$ and $x \in \mathrm{Z}(\mathrm{C}_H(x))$. Thus, the normal closure $A = \langle x\rangle^H$ of $\langle x\rangle$ in H is abelian. Similarly, $B = \langle y\rangle^H$ and $C = \langle z\rangle^H$ are abelian normal subgroups of H. We have $H = ABC$ and so by the Fitting's lemma, $\mathrm{cl}(H) \leq 3$.

We use now the following result of Heineken [Hei2]. If $n > 2$ and if every n elements of a group G generate a subgroup of class $\leq n$, then G is of class $\leq n$. Using this for $n = 3$, we get: $\mathrm{cl}(G) \leq 3$. The proof is complete. □

Theorem 27.2. *There exists a 3-group G of class 3 in which $\mathrm{C}_G(x)$ is normal in G for each $x \in G$.*

Proof. We define $G = \langle a, b, c\rangle$ in terms of defining relations as follows:

$$a^3 = b^3 = c^3 = 1, \quad [a,b] = u, \quad [b,c] = v, \quad [c,a] = w,$$
$$u^3 = v^3 = w^3 = 1, \quad [u,v] = [u,w] = [v,w] = 1,$$
$$[u,a] = [u,b] = [v,b] = [v,c] = 1, \quad [u,c] = z, \quad z^3 = 1,$$
$$[v,a] = z, \quad [w,c] = [w,a] = 1, \quad [w,b] = z, \quad [a,z] = [b,z] = [c,z] = 1.$$

Here $G' = \Phi(G) = \langle u, v, w, z\rangle \cong \mathrm{E}_{3^4}$, $G/\Phi(G) \cong \mathrm{E}_{3^3}$, $\mathrm{Z}(G) = \mathrm{K}_3 = [G, G'] = \langle z\rangle \cong \mathrm{C}_3$ so that G is metabelian of order 3^7 and class 3. Next, $\mathrm{cl}(G/\mathrm{K}_3(G)) = 2$.

We want to show that $\mathrm{C}_G(g)$ is normal in G for each $g \in G$. If $g \in G'$, this is clear. Let $g = a^i b^j c^k g'$ be an arbitrary element in $G - G'$, where i, j, k are integers mod 3 not all three equal 0 and $g' \in G'$. We show first that $\mathrm{C}_G(g) = \langle g\rangle \mathrm{C}_{G'}(g)$. Suppose false. Then there is $h \in G - G'$ such that $[g,h] = 1$ and $M = G'\langle g, h\rangle$ is a maximal subgroup of G. Consider $\bar{G} = G/\mathrm{K}_3(G)$ (bar convention). Then $\bar{G}$ is of class 2 with $(\bar{G})' = \overline{G'} \cong \mathrm{E}_{3^3}$, $(\bar{G})' \le \mathrm{Z}(\bar{G})$, and $\bar{M} = (\bar{G})'\overline{\langle g, h\rangle}$ is an abelian maximal subgroup of $\bar{G}$. From Lemma 1.1 follows $3^6 = |\bar{G}| = 3|(\bar{G})'||\mathrm{Z}(\bar{G})|$, which implies $|\mathrm{Z}(\bar{G})| = 3^2$. This is a contradiction since $(\bar{G})' \le \mathrm{Z}(\bar{G})$. We have proved that $\mathrm{C}_G(g) = \langle g\rangle \mathrm{C}_{G'}(g)$.

Since $[\mathrm{C}_{G'}(g), G] \le \mathrm{K}_3(G) = \mathrm{Z}(G) \le \mathrm{C}_{G'}(g)$, we still have to show that $[g,a]$, $[g,b]$, and $[g,c]$ centralize g and then $\langle a, b, c\rangle = G$ would normalize $\mathrm{C}_G(g)$.

Noting that $\mathrm{cl}(G/\mathrm{K}_3(G)) = 2$, we see at once that (for suitable integers s, s', s'')

$$[g,a] = [a^i b^j c^k g', a] = [b,a]^j [c,a]^k z^s,$$
$$[g,b] = [a,b]^i [c,b]^k z^{s'},$$
$$[g,c] = [a,c]^i [b,c]^j z^{s''},$$

and so we still have to show that the following three elements $[b,a]^j [c,a]^k$, $[a,b]^i [c,b]^k$, and $[a,c]^i [b,c]^j$ commute with $g = a^i b^j c^k g'$. Indeed, we get

$$\begin{aligned}
[[b,a]^j [c,a]^k, g] &= [[b,a]^j, g]\,[[c,a]^k, g] \\
&= [[b,a]^j, c^k]\,[[b,a]^j, a^i b^j]\,[[c,a]^k, a^i c^k]\,[[c,a]^k, b^j] \\
&= [[b,a], c]^{jk}\,[[c,a], b]^{jk} = z^{-jk} z^{jk} = 1,
\end{aligned}$$

since $[b,a] \in \mathrm{Z}(\langle a, b\rangle)$ and $[c,a] \in \mathrm{Z}(\langle a, c\rangle)$. Similarly, $[[a,b]^i [c,b]^k, g] = z^{ik} z^{-ik} = 1$ and $[[a,c]^i [b,c]^j, g] = z^{-ij} z^{ij} = 1$.

It remains to be proved that our group G of order 3^7 really exists. We find a faithful transitive permutation representation of degree 81 of $G = \langle a, b, c\rangle$ by setting:

$$\begin{aligned}a =&(2,24,20)(3,11,12)(4,54,55)(5,18,19)(10,58,59)(13,73,66)\\&(14,67,68)(15,74,51)(16,37,38)(17,49,50)(23,46,45)(25,76,77)\\&(26,71,39)(27,33,34)(28,32,35)(29,30,78)(31,81,62)(36,75,80)\\&(40,43,63)(41,42,48)(44,53,47)(52,72,61)(56,79,60)(57,64,65),\end{aligned}$$

$$\begin{aligned}b =&(2,16,25)(3,10,17)(6,69,70)(7,21,22)(11,52,53)(12,40,41)\\&(13,66,73)(14,67,68)(15,61,42)(20,28,29)(23,43,44)(24,80,81)\\&(26,75,30)(27,32,31)(33,37,78)(34,36,77)(35,76,71)(38,62,39)\\&(45,72,50)(46,58,48)(47,51,59)(49,74,63)(56,79,60)(57,65,64),\end{aligned}$$

$$\begin{aligned}c =&(1,2,3)(4,25,10)(5,16,17)(6,24,12)(7,20,11)(8,27,15)\\&(9,26,23)(13,35,47)(14,36,48)(18,75,44)(19,32,42)(21,39,46)\\&(22,34,74)(28,49,67)(29,58,64)(30,43,55)(31,61,54)(33,51,69)\\&(37,41,73)(38,53,68)(40,79,76)(45,70,71)(50,66,80)(52,65,77),\\&(56,78,72)(57,62,63)(59,60,81).\end{aligned}$$

We verify that the even permutations a, b, c satisfy all the above defining relations for G. Since $z = [a, b, c] = (1,8,9)(2,27,36)\cdots \neq 1$, it follows that the obtained representation of G is faithful. Hence, our group G of order 3^7 exists as a subgroup of the alternating group A_{81}. □

§28

p-groups with a uniqueness condition for nonnormal subgroups

Minimal nonabelian p-groups and 2-groups of maximal class have the property that each nonnormal subgroup is contained in exactly one maximal subgroup. It turns out that there are two further infinite families of 2-groups which also have this property. More precisely, we shall prove the following result which gives a complete classification of such p-groups.

Theorem 28.1 ([Jan18]). *Let G be a non-Dedekindian p-group and assume that each nonnormal subgroup of G is contained in exactly one maximal subgroup. Then one of the following holds:*

(a) *G is minimal nonabelian.*

(b) *G is a 2-group of maximal class.*

(c) *$G = \langle a, b\rangle$ is a nonmetacyclic 2-group, where*

$$o(a) = 2^n, \quad n \geq 3, \quad o(b) \in \{2, 4\}, \quad a^b = ak,$$
$$k^2 = a^{-4}, \quad [k, a] = 1, \quad k^b = k^{-1}$$

and we have either:

(c1) *$b^2 \in \langle a^{2^{n-1}}, a^2k\rangle \cong \mathrm{E}_4$, in which case*

$$|G| = 2^{n+2}, \quad \Phi(G) = \langle a^2\rangle \times \langle a^2k\rangle \cong \mathrm{C}_{2^{n-1}} \times \mathrm{C}_2,$$
$$\mathrm{Z}(G) = \langle a^{2^{n-1}}\rangle \times \langle a^2k\rangle \cong \mathrm{E}_4,$$

and $\langle a\rangle \times \langle a^2k\rangle \cong \mathrm{C}_{2^n} \times \mathrm{C}_2$ is a unique abelian maximal subgroup of G, or:

(c2) *$b^2 \notin \langle a^{2^{n-1}}, a^2k\rangle \cong \mathrm{E}_4$, in which case*

$$o(b) = 4, \quad |G| = 2^{n+3},$$
$$\Phi(G) = \langle a^2\rangle \times \langle a^2k\rangle \times \langle b^2\rangle \cong \mathrm{C}_{2^{n-1}} \times \mathrm{C}_2 \times \mathrm{C}_2,$$
$$\mathrm{Z}(G) = \langle a^{2^{n-1}}\rangle \times \langle a^2k\rangle \times \langle b^2\rangle \cong \mathrm{E}_8,$$

and $\langle a\rangle \times \langle a^2k\rangle \times \langle b^2\rangle \cong \mathrm{C}_{2^n} \times \mathrm{C}_2 \times \mathrm{C}_2$ is a unique abelian maximal subgroup of G.

In any case, $G' = \langle k \rangle \cong C_{2^{n-1}}$, *a centralizes* $\Phi(G)$, *and b inverts each element of* $\Phi(G)$, *and so each subgroup of* $\Phi(G)$ *is normal in* G;

(d) $G = \langle a, b \rangle$ *is a splitting metacyclic 2-group, where*

$$a^{2^n} = b^4 = 1, \quad n \geq 3, \quad a^b = a^{-1}z^{\epsilon}, \quad \epsilon = 0, 1, \quad z = a^{2^{n-1}}.$$

Here

$$|G| = 2^{n+2}, \qquad \Phi(G) = \langle a^2 \rangle \times \langle b^2 \rangle \cong C_{2^{n-1}} \times C_2,$$
$$Z(G) = \langle z \rangle \times \langle b^2 \rangle \cong E_4, \qquad G' = \langle a^2 \rangle \cong C_{2^{n-1}},$$

and $\langle a \rangle \times \langle b^2 \rangle \cong C_{2^n} \times C_2$ *is a unique abelian maximal subgroup of* G. *Since a centralizes* $\Phi(G)$ *and b inverts each element of* $\Phi(G)$, *it follows that each subgroup of* $\Phi(G)$ *is normal in* G.

To facilitate the proof of Theorem 28.1, we prove the following

Lemma 28.2 (Berkovich). *Let* G *be a* p*-group,* $p > 2$, *such that all subgroups of* $\Phi(G)$ *are normal in* G. *Then* $\Phi(G) \leq Z(G)$.

Proof. By Lemma 1.18, $\Phi(G)$ is abelian.

Suppose that $\Phi(G)$ is cyclic. Let $U/\Phi(G)$ be a subgroup of order p in $G/\Phi(G)$. Assume that U is nonabelian. Then, by Theorem 1.2, $U \cong M_{p|\Phi(G)|}$ so $U = \Phi(G)\Omega_1(U)$, where $\Omega_1(U)$ is a normal subgroup of type (p, p) in G. In that case, $\Omega_1(U)$ centralizes $\Phi(G)$ so U is abelian, a contradiction. Let $\mathfrak{M} = \{U < G \mid \Phi(G) < U, |U : \Phi(G)| = p\}$. Then $C_G(\Phi(G)) \geq \langle U \mid U \in \mathfrak{M} \rangle = G$ so $\Phi(G) \leq Z(G)$.

Now let $\Phi(G)$ be noncyclic. Then $\Phi(G) = Z_1 \times \cdots \times Z_n$, where $Z_1, \ldots, Z_n$ are cyclic and $n > 1$. By induction, $\Phi(G/Z_i) \leq Z(G/Z_i)$ for all i. Let $f \in \Phi(G)$ and $x \in G$. Then $[f, x] \in \bigcap_{i=1}^{n} Z_i = \{1\}$ so $f \in Z(G)$. It follows that $\Phi(G) \leq Z(G)$. □

Proof of Theorem 28.1. Let G satisfies the hypothesis; then G is nonabelian with $d(G) \geq 2$ so each subgroup of $\Phi(G)$ is G-invariant. By Burnside's argument, $\Phi(G)$ has no nonabelian G-invariant subgroups of order p^3 so, by Theorem 1.20, it is abelian.

Then there is in G a nonnormal cyclic subgroup $\langle a \rangle$ so $a \notin \Phi(G)$ but $a^p \in \Phi(G)$. In that case, $\langle a \rangle\Phi(G)$ must be a unique maximal subgroup of G containing $\langle a \rangle$ so $d(G) = 2$.

If $\Phi(G) \leq Z(G)$, then each maximal subgroup of G is abelian and so G is minimal nonabelian; then G is as in (a).

From now on we assume that $\Phi(G) \nleq Z(G)$. Set $G = \langle a, b \rangle$. Then $[a, b] \neq 1$ and $[a, b] \in \Phi(G)$. Therefore $\langle [a, b] \rangle \lhd G$ and $G/\langle [a, b] \rangle$ is abelian which implies that $G' = \langle [a, b] \rangle \neq \{1\}$. If $|G'| = p$, then the fact $d(G) = 2$ forces that G would be minimal nonabelian (see Lemma 65.2(a)). But then $\Phi(G) \leq Z(G)$, a contradiction.

Hence G' is cyclic of order $\geq p^2$. Set $[a,b] = k$ so that $G' = \langle k \rangle$ and set $Z = \Omega_1(G')$.

(i) First assume that $p > 2$. Then $\Phi(G) \leq Z(G)$ (Lemma 28.2), a contradiction.

(ii) Now assume that $p = 2$. If $\Phi(G)$ is cyclic, then, in view of $\Phi(G) = \mho_1(G)$ and $d(G) = 2$, G has a cyclic subgroup of index 2. But $|G'| \geq 4$ and so G is of maximal class; then G is as in (b). From now on we shall assume that $\Phi(G)$ is not cyclic.

Set $G = \langle a, b \rangle$, $k = [a,b]$, and $\langle z \rangle = \Omega_1(\langle k \rangle)$ so that $G' = \langle k \rangle$, $o(k) \geq 4$, and $\langle z \rangle \leq Z(G)$. Since $\langle a^2 \rangle$ and $\langle b^2 \rangle$ (being contained in $\Phi(G)$) are normal in G, we have $\Phi(G) = \langle a^2 \rangle \langle b^2 \rangle \langle k \rangle$ and so the abelian subgroup $\Phi(G)$ is a product of three cyclic subgroups which implies $d(\Phi(G)) = 2$ or $d(\Phi(G)) = 3$.

From $[a,b] = k$ follows $a^{-1}(b^{-1}ab) = k$ and $b^{-1}(a^{-1}ba) = k^{-1}$ and so

(1) $$a^b = ak,$$

(2) $$b^a = bk^{-1}.$$

From (1) follows

(3) $$(a^2)^b = (a^b)^2 = (ak)^2 = akak = a^2k^ak.$$

From (2) follows

(4) $$(b^2)^a = (b^a)^2 = (bk^{-1})^2 = bk^{-1}bk^{-1} = b^2(k^{-1})^bk^{-1} = b^2(k^bk)^{-1}.$$

We also have

$$a^2 = (a^2)^{b^2} = (a^2k^ak)^b = a^2k^akk^{ab}k^b$$

and so

(5) $$kk^ak^bk^{ab} = 1.$$

Finally, we compute, using (4),

$$(ab)^2 = abab = a^2a^{-1}b^{-1}b^2ab = a^2(a^{-1}b^{-1}ab)(b^2)^{ab} = a^2kb^2(kk^b)^{-1}$$
$$= a^2b^2(k^{-1})^b$$

and so $(ab)^2 = a^2b^2(k^{-1})^b$.

Suppose that $G/\Phi(G)$ acts faithfully on $\langle k \rangle$. In that case $o(k) \geq 2^3$ and we may choose the generators $a, b \in G - \Phi(G)$ so that $k^a = k^{-1}$, $k^b = kz$ (where $\langle z \rangle = \Omega_1(\langle k \rangle)$). Using (3) and (4) we get $(a^2)^b = a^2$ (and so $a^2 \in Z(G)$) and $(b^2)^a = b^2k^{-2}z$. Since $k^a = k^{-1}$, we have $\langle k \rangle \cap \langle a \rangle \leq \langle z \rangle$. The subgroup $\langle b^2 \rangle$ (being contained in $\Phi(G)$) is normal in G and so $k^{-2}z \in \langle b^2 \rangle$ and $k^2 \in \langle b^2 \rangle$ (since $z \in \langle k^2 \rangle$). We have $\langle b \rangle \cap \langle k \rangle = \langle k^2 \rangle$ since $k^b = kz \neq k$ and so $k \notin \langle b \rangle$. If $b^2 \in \langle k^2 \rangle$, then $(b^2)^a = b^{-2}$ since a inverts $\langle k \rangle$ and, on the other hand, $(b^2)^a = b^2k^{-2}z$ and so $b^4 = k^2z$. But $b^2 \in \langle k^2 \rangle$ implies $b^4 \in \langle k^4 \rangle$, a contradiction. Hence $b^2 \notin \langle k^2 \rangle$ and so we can find an element $s \in \langle b^2\rangle - \langle k\rangle$ such that $s^2 = k^{-2}$. Then $(sk)^2 = s^2k^2 = 1$ and so sk is an involution in $\Phi(G)$ which is not contained in $\langle k\rangle$ and therefore $sk \neq z$. But $(sk)^b = sk^b = (sk)z$ and so $\langle sk\rangle$ is not normal in G, a contradiction.

We have proved that $G/\Phi(G)$ does not act faithfully on $\langle k\rangle$. Then we can choose our generator $a \in G - \Phi(G)$ so that $k^a = k$. Using (3) we get $(a^2)^b = a^2k^2$ and so $1 \neq k^2 \in \langle a^2\rangle$ since $\langle a^2\rangle \triangleleft G$. From (5) we get $(k^2)^b = k^{-2}$. Suppose that $\langle k^2\rangle = \langle a^2\rangle$. Then we get $a^{-2} = (a^2)^b = a^2k^2$ and so $k^2 = a^{-4}$, a contradiction. We have obtained:

$$(6)\qquad \begin{aligned} &k^a = k, \quad (a^2)^b = a^2k^2, \quad (k^2)^b = k^{-2},\\ &\{1\} \neq \langle k^2\rangle < \langle a^2\rangle, \quad o(a) = 2^n, \quad n \geq 3. \end{aligned}$$

Suppose that $k^b = kz$. Then (5) and (6) imply $k^4 = 1$ and so $k^b = kz = k^{-1}$. It follows that we have to analyze the following three possibilities for the action of b on $\langle k\rangle$: $k^b = k^{-1}z$ with $o(k) \geq 2^3$, $k^b = k$, and $k^b = k^{-1}$.

(ii1) Suppose that $k^b = k^{-1}z$ with $o(k) \geq 2^3$. Then (4) gives $(b^2)^a = b^2z$ and so $z \in \langle b^2\rangle$ (since $\langle b^2\rangle \triangleleft G$) and $\langle z\rangle < \langle b^2\rangle$ because $b^2 \notin Z(G)$. Since, by (6), $\langle k^2\rangle < \langle a^2\rangle$ and $o(k^2) \geq 4$, it follows that $o(a^2) \geq 2^3$ and $\langle z\rangle = \Omega_1(\langle k\rangle) = \Omega_1(\langle a\rangle) = \Omega_1(\langle b\rangle) \leq Z(G)$. From $o(a^2) \geq 2^3$, $k^2 \in \langle a^4\rangle$, $o(k^2) \geq 4$, and $(k^2)^b = k^{-2}$ follows $(a^2)^b = a^{-2}z^{\epsilon}$ ($\epsilon = 0, 1$) and $C_{\langle a^2\rangle}(b) = \langle z\rangle$ so that $\langle a^2\rangle \cap \langle b^2\rangle = \langle z\rangle$. Let v be an element of order 4 in $\langle a^2\rangle$ so that $v^2 = z$ and $v^b = v^{-1} = vz$. Let s be an element of order 4 in $\langle b^2\rangle$ so that $s^2 = z$. We have $(vs)^2 = v^2s^2 = 1$ and so vs is an involution in $\Phi(G) - \langle a\rangle$ but $(vs)^b = v^{-1}s = (vs)z$, a contradiction.

(ii2) Suppose that $k^b = k$ so that (5) and (6) imply $k^4 = 1$ and $k^2 = z$. Then (4) and (6) imply $(b^2)^a = b^2z$ and $(a^2)^b = a^2z$. Also, $\langle z\rangle < \langle a^2\rangle$ and $\langle z\rangle < \langle b^2\rangle$ since $\langle a^2\rangle$ and $\langle b^2\rangle$ are normal in G, $a^2 \notin Z(G)$ and $b^2 \notin Z(G)$. If $a^2 \in \langle b^2\rangle$, then $a^2 \in Z(G)$ and if $b^2 \in \langle a^2\rangle$, then $b^2 \in Z(G)$. This is a contradiction. Hence $D = \langle a^2\rangle \cap \langle b^2\rangle \geq \langle z\rangle$ and D is a proper subgroup of $\langle a^2\rangle$ and $\langle b^2\rangle$. Because of the symmetry, we may assume $o(a) \geq o(b)$ so that $|\langle a^2\rangle/D| \geq |\langle b^2\rangle/D| = 2^u$, $u \geq 1$. We set $(b^2)^{2^u} = d$ so that $D = \langle d\rangle$. We may choose an element $a' \in \langle a^2\rangle - D$ such that $(a')^{2^u} = d^{-1}$. Then $(a'b^2)^{2^u} = 1$ and $\langle a'b^2\rangle \cong C_{2^u}$ with $\langle a'b^2\rangle \cap D = \{1\}$. On the other hand, $(a'b^2)^a = a'(b^2)^a = (a'b^2)z$, where $z \in D$, a contradiction.

(ii3) Finally, suppose $k^b = k^{-1}$. From (4) follows $(b^2)^a = b^2$ and so $b^2 \in Z(G)$. By (6), $(a^2)^b = a^2k^2$, $\langle k^2\rangle < \langle a^2\rangle$, and so $o(a^2) \geq 4$. Also, $(a^2k)^a = a^2k$, $(a^2k)^b = (a^2k^2)k^{-1} = a^2k$, and so $a^2k \in Z(G)$.

(ii3a) First assume $k \notin \langle a^2\rangle$. We investigate for a moment the special case $o(k) = 4$, where $k^2 = z$, $\langle z\rangle = \Omega_1(\langle k\rangle) = \Omega_1(\langle a\rangle)$ and $(a^2)^b = a^2z$. If $o(a^2) > 4$, then take an element v of order 4 in $\langle a^4\rangle$ so that $v^2 = z$ and $v^b = v$. In that case $(vk)^2 = v^2k^2 = 1$ and so vk is an involution in $\Phi(G) - \langle a^2\rangle$ and $(vk)^b = vk^{-1} = (vk)z$, a contradiction. Hence $o(a^2) = 4$, $a^4 = z$, $k^2 = z = a^{-4}$, $(a^2)^b = a^2z = a^{-2}$, and $\langle a^2, k\rangle$ is an abelian group of type $(4, 2)$ which is inverted by b, and a^2k is a central involution in G. Now suppose $o(k) \geq 8$. In that case $o(k^2) \geq 4$, $k^2 \in \langle a^4 \rangle$, $o(a^2) \geq 8$, and b inverts $\langle k^2 \rangle$, which implies $(a^2)^b = a^{-2}z^\epsilon$, $\epsilon = 0, 1$. On the other hand, $(a^2)^b = a^2k^2$ and so $k^2 = a^{-4}z^\epsilon$. Let v be an element of order 4 in $\langle a^4 \rangle$ so that $v^2 = z$ and $v^b = v^{-1} = vz$. Then we compute:

$$(a^2vk)^2 = a^4zk^2 = z^{\epsilon+1}, \quad (a^2vk)^b = a^2k^2v^{-1}k^{-1} = (a^2vk)z.$$

If $\epsilon = 1$, then a^2vk is an involution in $\Phi(G) - \langle a^2 \rangle$ and $\langle a^2vk \rangle$ is not normal in G. Thus, $\epsilon = 0$, $(a^2)^b = a^{-2}$, $k^2 = a^{-4}$, a^2k is an involution in $\Phi(G) - \langle a^2 \rangle$ and b inverts each element of $\langle a^2, k \rangle = \langle a^2 \rangle \times \langle a^2k \rangle$, where $a^2k \in Z(G)$.

We have proved that in any case $k^2 = a^{-4}$, $o(a^2) \geq 4$, $o(k) \geq 4$, and b inverts each element of the abelian group $\langle a^2, k \rangle = \langle a^2 \rangle \times \langle a^2k \rangle$, where a^2k is an involution contained in $Z(G)$.

It remains to determine $b^2 \in Z(G)$. Suppose $o(b^2) \geq 4$ and let $\langle s \rangle$ be a cyclic subgroup of order 4 in $\langle b^2 \rangle$ so that $s \in Z(G)$. Obviously, $s \notin \langle a^2, k \rangle$ since $Z(G) \cap \langle a^2, k \rangle = \langle z \rangle \times \langle a^2k \rangle \cong E_4$. Let v be an element of order 4 in $\langle a^2 \rangle$ so that $v^2 = z$ and $v^b = v^{-1} = vz$. We have:

$$(vs)^b = v^{-1}s = (vs)z \quad \text{and} \quad (vs)^2 = v^2s^2 = zs^2.$$

If $s^2 = z$, then vs is an involution in $\Phi(G) - \langle a^2, k \rangle$ and $vs \notin Z(G)$, a contradiction. Hence $s^2 \neq z$ so that $\langle v, s \rangle = \langle v \rangle \times \langle s \rangle \cong C_4 \times C_4$. But $(vs)^b = (vs)z$, $(vs)^2 = zs^2 \neq z$, and so $\langle vs \rangle$ is not normal in G, a contradiction. It follows that $o(b^2) \leq 2$. Hence we have either $b^2 \in \langle z, a^2k \rangle$, $\Phi(G) = \langle a^2, k \rangle = \langle a^2 \rangle \times \langle a^2k \rangle$, and we have obtained the possibility (c1) of our theorem or b^2 is an involution in $\Phi(G) - \langle a^2, k \rangle$, $\Phi(G) = \langle a^2 \rangle \times \langle a^2k \rangle \times \langle b^2 \rangle$, and we have obtained the possibility (c2) of our theorem. Note that in both cases a centralizes $\Phi(G)$ and b inverts each element of $\Phi(G)$.

(ii3b) We assume $k \in \langle a^2 \rangle$. Since $o(k) \geq 4$, $k^b = k^{-1}$, $\langle a \rangle$ is normal in G, $o(a) \geq 8$, and b induces on $\langle a \rangle$ an automorphism of order 2, we get $a^b = a^{-1}z^\epsilon$, $\epsilon = 0, 1$, where $\langle z \rangle = \Omega_1(\langle a \rangle) = \Omega_1(\langle k \rangle)$. On the other hand, (1) gives $a^b = ak$ and so $k = a^{-2}z^\epsilon$ which gives $G' = \langle k \rangle = \langle a^2 \rangle \cong C_{2^{n-1}}$, where $o(a) = 2^n$, $n \geq 3$, and $z = a^{2^{n-1}}$.

Since $\Phi(G) = \langle a^2, b^2 \rangle$ and $\Phi(G)$ is noncyclic, we have $b^2 \notin \langle a^2 \rangle$ and we know that $b^2 \in Z(G)$. Suppose $o(b^2) \geq 4$ and let s be an element of order 4 in $\langle b^2 \rangle$. Let v be an element of order 4 in $\langle a^2 \rangle$ so that $v^2 = z$ and $v^b = v^{-1} = vz$. Then

$$(vs)^b = v^{-1}s = (vs)z \quad \text{and} \quad (vs)^2 = v^2s^2 = zs^2.$$

If $s^2 = z$, then vs is an involution in $\Phi(G) - \langle a^2 \rangle$ and $vs \notin Z(G)$, a contradiction. Hence $s^2 \neq z$ so that $\langle v, s \rangle = \langle v \rangle \times \langle s \rangle \cong C_4 \times C_4$. But $\langle vs \rangle$ is not normal in G, a contradiction. Hence b^2 is an involution in $\Phi(G) - \langle a^2 \rangle$ and so $\Phi(G) = \langle a^2 \rangle \times \langle b^2 \rangle \cong C_{2^{n-1}} \times C_2$ and $Z(G) = \langle z \rangle \times \langle b^2 \rangle \cong E_4$. Also note that a centralizes $\Phi(G)$ and b inverts each element of $\Phi(G)$. We have obtained the possibility (d) of our theorem. □

Theorem 28.1 solves Problem 1134 (see 'Research problems and themes II').

It follows from Theorem 28.1 that if $p > 2$ and a p-group G is neither abelian nor minimal nonabelian, then G/H^G is noncyclic for each nonnormal $H < G$.

Exercise. Let G be a 2-group satisfying the hypothesis of Theorem 28.1. Prove that G is isomorphic to a subgroup of a direct product of groups $A_1, \dots, A_k$, where A_i is either abelian or M_{2^n} is of maximal class, $i = 1, \dots, k$. (*Hint.* $\Phi(G)$ is abelian.)

§29

On isoclinism

The starting point of the approach of Philip Hall to the classification of p-groups [Hal3] is the fact that if $x, y \in G$ and $z_1, z_2 \in Z(G)$ then $[xz_1, yz_2] = [x, y]$. In other words, the value of $[x, y]$ depends only on the pair of cosets of $Z(G)$ to which x, y belong. Main results of this section are due to Hall.

Definition 1. Two groups G and H are said to be *isoclinic* (= skew isomorphic) provided there exist two isomorphisms $f : G/Z(G) \cong H/Z(H)$ and $f' : G' \cong H'$ such that if $f(aZ(G)) = a_1Z(H)$ and $f(bZ(G)) = b_1Z(H)$ then $f'([a, b]) = [a_1, b_1]$. In this case, we write $G \sim H$.

By an *isoclinism* between two groups we mean any isomorphism f between their central quotients which induces a corresponding isomorphism between their commutator subgroups. The following easy exercises are designated to help the reader in understanding Definition 1 (see [Suz1, pp. 92–95]).

Exercise 1. Isomorphic groups are isoclinic. Isoclinism is an equivalence relation.

Exercise 2. Any abelian group is isoclinic to the identity group $\{1\}$. Conversely, if $G \sim \{1\}$, then G is abelian. $G \sim G \times A$ if and only if A is abelian.

Exercise 3. If $G \sim H$, then $G^k \sim H^k$ for all $k \in \mathbb{N}$ (here G^k is the direct product of k copies of G).

Exercise 4. Any two 2-groups of maximal class and equal order are isoclinic. For $p > 2$, there exist two p-groups of maximal class and the same order that are not isoclinic.

Exercise 5. If $U \leq G$, then U and $UZ(G)$ are isoclinic. Next, $U \sim G$ if and only if $UZ(G) = G$.

Solution. Note that $(UZ(G))' = U'$ and $Z(UZ(G)) = Z(U)Z(G)$. It follows that

$$UZ(G)/Z(UZ(G)) = UZ(U)Z(G)/Z(U)Z(G) \cong U/(U \cap Z(U)Z(G)) = U/Z(U),$$

and so $U \sim UZ(G)$. Let us prove the second assertion. It suffices to show that if $U \sim G$, then $UZ(G) = G$. Set $UZ(G) = K$. By the second part of Exercise 1 and what has just been proved, $K \sim G$. Therefore, $|K/Z(K)| = |G/Z(G)|$ so that $|Z(G)| \geq |Z(K)|$. Since $Z(G) \leq Z(K)$, we get $Z(G) = Z(K)$.

Exercise 6. (a) If G is nilpotent and $H \sim G$, then H is also nilpotent and $\mathrm{cl}(H) = \mathrm{cl}(G)$. (*Hint.* By definition, $G/\mathrm{Z}(G) \cong H/\mathrm{Z}(H)$.)

(b) If G and H are extraspecial 2-groups of the same order, then $G \sim H$.

(c) If $G_i \sim H_i$, $i = 1, 2$, then $G_1 \times G_2 \sim H_1 \times H_2$.

As Exercise 1 shows the relation of isoclinism is an equivalence. With respect to isoclinism, all groups fall into a number of mutually disjoint *families* — equivalence classes of isoclinism. Any quantity depending on a variable group and which is the same for any two groups of the same family will be called a *family invariant.* We see that the members of the derived series and the central quotient groups are family invariants. It follows that the groups belonging to the same family have the same derived length and class. On the other hand, the commutator quotient group and the center are not family invariants; similarly, the minimal number of generators is not a family invariant (see Exercise 2). If $G \sim H$, then $\mathrm{Inn}(G) \cong \mathrm{Inn}(H)$.

Theorem 29.1 (see [Suz1, Theorem 4.31]). *Let $N \lhd G$. Then $G/N \sim G/(N \cap G')$. Next, $G \sim G/N$ if and only if $N \cap G' = \{1\}$.*

Proof. Let K be the subgroup of G such that $K/N = \mathrm{Z}(G/N)$. Then K is the largest subgroup of G such that $[G, K] \le N$. Since $[G, K] \le G'$, we get $[G, K] \le N \cap G'$ and K is the largest subgroup satisfying this relation. So, $K/(N \cap G') = \mathrm{Z}(G/(N \cap G'))$. Next, $(G/N)' = G'N/N \cong G'/(N \cap G') = (G/(N \cap G'))'$. So, by Definition 1, $G/N \sim G/(N \cap G')$. Let us prove the second assertion. If $N \cap G' = \{1\}$, then $G \sim G/N$ by what has just been proved. Suppose that $G \sim G/N$. Then $|G'| = |(G/N)'|$. Since $|(G/N)'| = |G'N/N| = |G'/(N \cap G')|$, we get $N \cap G' = \{1\}$. (In that case, $[N, G] \le N \cap G' = \{1\}$ so $N \le \mathrm{Z}(G)$.) □

In what follows, we assume that all considered groups are p-groups.

Definition 2. The groups of lowest order in a family are called *stem groups*. If the stem groups in a family Δ are of order p^r, then Δ is of *f-rank r*. The groups of order p^{r+s} in a family Δ of f-rank r are said to form the s-th *branch* of Δ.

The family, containing abelian p-groups, is of f-rank 0 (Exercise 2).

Theorem 29.2. *If G is a nonabelian p-group such that $\mathrm{Z}(G) \le G'$, then G is a stem group. In particular, if the center of a nonabelian p-group is of order p, it is a stem group.*

Proof. Let a p-group H be isoclinic to G. Then $G/\mathrm{Z}(G) \cong H/\mathrm{Z}(H)$ so $G'/\mathrm{Z}(G) = (G/\mathrm{Z}(G))' \cong (H/\mathrm{Z}(H))' = H'\mathrm{Z}(H)/\mathrm{Z}(H)$ and hence

$$\begin{aligned}|G : G'| &= |G/\mathrm{Z}(G) : G'/\mathrm{Z}(G)| = |H/\mathrm{Z}(H) : H'\mathrm{Z}(H)/\mathrm{Z}(H)| \\ &= |H : H'\mathrm{Z}(H)| \le |H : H'|.\end{aligned}$$

Since $G' \cong H'$, we have $|G| = |G : G'||G'| \le |H : H'||H'| = |H|$ so G is a stem group. □

Definition 3. Two p-groups G and H of the same order are in the same *genus*, if there is an isomorphism between the lattices of normal subgroups of G and H such that corresponding normal subgroups belong to the same family. Furthermore, a group and its direct product with a group of order p are in the same genus.

Thus every genus in a branch appears again in the next branch. It is easy to check that two nonisomorphic 2-groups of maximal class and the same order are in the same genus.

Theorem 29.3. *In every nonabelian family (i.e., containing nonabelian groups) there exists a group G such that $Z(G) \le G'$ (so G is a stem group, by Theorem 29.2).*

Corollary 29.4. *If G is a stem group of a family Δ, then $Z(G) \le G'$.*

Proof. Let $H \in \Delta$ be such that $Z(H) \le H'$. By Theorem 29.2, H is a stem group so $|G| = |H|$ and $|G'| = |H'|$. We have, by definition, that $G/Z(G) \cong H/Z(H)$ so that $|Z(G)| = |Z(H)|$. By Theorem 29.1, $G/Z(G) \sim G/(G' \cap Z(G))$ so

$$|G'|/|Z(G)| = |H'/Z(H)| = |(G/Z(G))'| = |G'/(Z(G) \cap G')|.$$

It follows that $|Z(G)| = |G' \cap Z(G)|$ so $Z(G) \le G'$. □

Corollary 29.5. *A group G is a stem group if and only if $Z(G) \le G'$.*

Proposition 29.6. *Let G be a p-group. Then a stem group that is isoclinic to G, has order $|G : Z(G)||G' \cap Z(G)|$.*

Proof. Let H be a stem group isoclinic to G. Then $G/(G' \cap Z(G)) \sim G/Z(G) \cong H/Z(H)$ (Theorem 29.1). By Corollary 29.5,

$$|G'/(G' \cap Z(G))| = |(G/G' \cap Z(G))'| = |H'/Z(H)| = |G'|/|Z(H)|,$$

so $|G' \cap Z(G)| = |Z(H)|$. Thus, we get the equalities: $|H| = |H/Z(H)||Z(H)| = |G/Z(G)||G' \cap Z(G)|$. □

Proposition 29.7. *Let Δ be a family containing a p-group G such that $|G : Z(G)| = p^2$. Then all p-groups with center of index p^2 are members of Δ. The* f-*rank of Δ is three.*

Proof. By Lemma 1.1, $|G'| = p$ and Δ contains nonabelian groups of order p^3 (Proposition 29.6). All nonabelian groups of order p^3 are isoclinic, and Exercise 1 implies the result. □

The family, containing $\{1\}$, we denote by Φ_1 (Φ_1 is the only family containing an abelian member). The family of Proposition 29.7 is denoted by Φ_2.

Theorem 29.8 ([Hal3, HS]). *Let A, B be abelian subgroups of index 2 in nonabelian 2-groups G, H, respectively. Suppose that $A/\mathrm{Z}(G)$ and $B/\mathrm{Z}(H)$ are of types*

$$(2^{\lambda_1},\dots,2^{\lambda_s}),\quad \lambda_1\le\cdots\le\lambda_s,\quad (2^{\mu_1},\dots,2^{\mu_t}),\quad \mu_1\le\cdots\le\mu_t,$$

respectively. Then $G\sim H$ if and only if $s=t$ and $\lambda_i=\mu_i$, all i. In that case, $G'\cong A/\mathrm{Z}(G)$ and the f-rank of G is $\lambda_1+\lambda_2+\cdots+\lambda_s+s+1=\log_2(|G:G'|)+s+1$.

Proof. We have $G=\langle A,x\rangle$ for $x\in G-A$ so $x^2\in A$, $\mathrm{C}_G(x^2)\ge\langle x,A\rangle=G$ whence $x^2\in\mathrm{Z}(G)$. For $y\in A$, we have

$$yx^{-1}yx=x^{-1}yxy=x^{-1}yx^{-1}yx^2=x^{-1}(yx^{-1}yx)x$$

so x and $yx^{-1}yx$ commute. It follows that $yx^{-1}yx\in\mathrm{Z}(G)$, or $x^{-1}yx\equiv y^{-1}$ (mod $\mathrm{Z}(G)$). Since $x^2\in\mathrm{Z}(G)$, it follows that $G/\mathrm{Z}(G)$ is the split extension of the abelian group $A/\mathrm{Z}(G)$ by a group of order 2, and the generator of the last one inverts $A/\mathrm{Z}(G)$. Thus, if $G\sim H$ (in that case, $G/\mathrm{Z}(G)\cong H/\mathrm{Z}(H)$), then $A/\mathrm{Z}(G)\cong B/\mathrm{Z}(H)$ since $G/\mathrm{Z}(H)$ has only one subgroup of index 2 inverted by involution.

Let $A/\mathrm{Z}(G)$ be of type $(2^{\lambda_1},\dots,2^{\lambda_s})$ with $\lambda_1\ge\cdots\ge\lambda_s\ge 1$, $\lambda_1>1$. Let $G^*=\langle u,v_1,\dots,v_s\rangle$ have the following defining relations:

$$u^2=1,\ (v_i)^{2^{\lambda_i+1}}=1,\quad uv_iu=v_i^{-1},\quad [v_i,v_j]=1,\quad i,j=1,\dots,s.$$

Put $z_i=v_i^{2^{\lambda_i}}$. Then $\mathrm{Z}(G^*)=\langle z_1,\dots,z_s\rangle=\Omega_1(\langle v_1,\dots,v_s\rangle)\cong\mathrm{E}_{2^s}$, and if $A^*=\langle v_1,\dots,v_s\rangle=\langle v_1\rangle\times\cdots\times\langle v_s\rangle$, then A^* is abelian of index 2 in G^*, $A^*/\mathrm{Z}(G^*)$ is of type $(2^{\lambda_1},\dots,2^{\lambda_s})$. Since $[v_i,u]=v_i^{-2}$ for $i=1,\dots,s$, it follows that $(G^*)'=\mho_1(A^*)\cong A^*/\mathrm{Z}(G^*)$ and $G/\mathrm{Z}(G)\cong G^*/\mathrm{Z}(G^*)$.

Since A is abelian, the mapping $y\mapsto[y,x]$ ($y\in A$) is a homomorphism of A onto G' with kernel $\mathrm{Z}(G)$. The first and last assertions are obvious. As regards to second one, the image of our homomorphism is $[x,G]$; since $G/[x,G]$ is abelian and $[x,G]\le G'$, we get $[x,G]=G'$. Hence, $G'\cong A/\mathrm{Z}(G)$, and its basis consists of $[y_i,x]=t_i$, $o(t_i)=2^{\lambda_i}$, all i. Write $w_i=[v_i,u]=v_i^{-2}$. Then the above isomorphism of $G/\mathrm{Z}(G)$ and $G^*/\mathrm{Z}(G^*)$ induces the isomorphism of G' and $(G^*)'$, where t_i corresponds to w_i, all i. We see that G and G^* are isoclinic. Since $\mathrm{Z}(G^*)\le(G^*)'$, G^* is a stem group (see Theorem 29.2). Therefore, the f-rank of the family containing G is $\log_2(|G^*|)=\lambda_1+\cdots+\lambda_s+s+1$. We see that $G\sim H$ if and only if $s=t$ and $\lambda_i=\mu_i$, all i. □

267 groups of order 2^6 are divided in 27 families [HS].

Hall also determined all families that contain p-groups of order p^5 [Hal3, pp. 138–140]. The p-groups with derived subgroup of order p are classified in [Blac2].

Mann proved [Man12, Theorem E] the following result, stated in [Hal3] without proof:

Theorem 29.9. *Let G and H be isoclinic and let they have $a_i>0$ and $b_i>0$ irreducible characters of degree p^i, respectively. Then $a_i/b_i=|G|/|H|$, all i.*

The same result is also true for the class sizes.

§30

On p-groups with few nonabelian subgroups of order p^p and exponent p

Let $\Sigma_{p^2} \cong C_p \operatorname{wr} C_p$ be a Sylow p-subgroup of the symmetric group of degree p^2. Let $\mathcal{MA}_k(G)$ denote the set of all $\mathcal{A}_1$-subgroups (= minimal nonabelian subgroups), say L, of G such that $\Omega_k(L) = L$. We denote by $\langle \mathcal{MA}_k(G) \rangle$ the subgroup generated by all members of the set $\mathcal{MA}_k(G)$. Let $e_n(G)$ ($e'_n(G)$) denote the number of all subgroups (all nonabelian subgroups) of G of order p^n and exponent p (in the second case, obviously, $p > 2$). Since a nonabelian p-group G is generated by $\mathcal{A}_1$-subgroups (Theorem 10.28), it is interesting to know whether a certain well described part of that set also generates G. Two results of such kind are contained in our first main theorem.

Theorem 30.1. *Let a nonabelian p-group $G = \Omega_k(G)$, $k > 0$. Then:*

(a) *The set $\mathcal{MA}_k(G)$ is not empty.*

(b) *There are two distinct $U, V \in \Gamma_1$ such that $U = \Omega_k(U)$ and $V = \Omega_k(V)$.*

(c) *Suppose that $\langle \mathcal{MA}_1(G) \rangle < G = \Omega_1(G)$ and, whenever a nonabelian $H \lessdot G$ is such that $\Omega_1(H) = H$, then $\langle \mathcal{MA}_1(H) \rangle = H$. Then $p > 2$ and $G \cong \Sigma_{p^2}$.*

(d) *Suppose that $k = 1$ and G has no subgroups isomorphic to Σ_{p^2}. Then we have $\langle \mathcal{MA}_1(G) \rangle = G$.*

(e) *$G = \langle \mathcal{MA}_{k+1}(G) \rangle$.*

If $G \cong \Sigma_{p^2}$, $p > 2$, then $\langle \mathcal{MA}_1(G) \rangle \in \Gamma_1$ so indeed $\langle \mathcal{MA}_1(G) \rangle < G$. It follows that Σ_{p^2} contains an $\mathcal{A}_1$-subgroup, say B, of exponent p^2 (Theorem 10.28). Let us show that B is of order p^3. Assume that this is false. Let $A \in \Gamma_1$ be elementary abelian. Then $\mathrm{Z}(B) = \Phi(B) \le B \cap A$ since $B \cap A$ is maximal in B so $\mathrm{C}_G(\mathrm{Z}(B)) \ge AB = G$, and we conclude that $|\mathrm{Z}(B)| = p$ since G is of maximal class. Then $|B| = |\mathrm{Z}(B)||B : \mathrm{Z}(B)| = p^3$, as desired. Thus, all $\mathcal{A}_1$-subgroups of Σ_{p^2} have the same order p^3 so Σ_{p^2} is generated by nonabelian subgroups of order p^3.

We study the p-groups G, $p > 2$, satisfying $e'_p(G) \in \{1, \dots, p-1\}$. It appears that if $e'_p(G) = 1$, then $G \cong \Sigma_{p^2}$, unless $|\Omega_1(G)| = p^p$, and, if $1 < e'_p(G) < p$, then G is of maximal class and order p^{p+1}. We also study, in Proposition 30.9, the p-groups G satisfying $e'_p(G) = p$. In conclusion, we prove three new counting theorems for p-groups, $p > 2$. For example, the number of subgroups isomorphic to Σ_{p^2} in a p-group of order $> p^{p+1}$ is a multiple of p. Next, if $3 \le n < m$, where G is of order

p^m, then the number of subgroups of G, which are of maximal class and order p^n and have an abelian subgroup of index p, is a multiple of p.

Remark 1. Let a nonabelian p-group G satisfy the following conditions: (i) $\exp(G) = p^e$, $p > 2$, and (ii) $\exp(\Omega_{e-1}(G)) = p^{e-1}$. We claim that then $\langle \mathcal{MA}_e(G) - \mathcal{MA}_{e-1}(G)\rangle = G$. We proceed by induction on $|G|$. One may assume that G is not an $\mathcal{A}_1$-group. In view of Theorem 10.28, we may assume that $e > 1$. If $K < G$ is of exponent p^e, then $\exp(\Omega_{e-1}(K)) = p^{e-1}$. If $U, V \in \Gamma_1$ are different nonabelian of exponent p^e, then, by induction,

$$\begin{aligned}\langle \mathcal{MA}_e(G) - \mathcal{MA}_{e-1}(G)\rangle &= \langle \mathcal{MA}_e(UV) - \mathcal{MA}_{e-1}(UV)\rangle \\ &\geq \langle \mathcal{MA}_e(U) - \mathcal{MA}_{e-1}(U)\rangle\langle \mathcal{MA}_e(V) - \mathcal{MA}_{e-1}(V)\rangle \\ &= UV = G,\end{aligned}$$

and we are done. Therefore, it suffices to show that the set Γ_1 contains two distinct nonabelian members of exponent p^e. Assume that $A, B \in \Gamma_1$ are distinct of exponent $< p^e$; then $G = AB \leq \Omega_{e-1}(G) < G$, a contradiction. Thus, the set Γ_1 has at most one member of exponent $< p^e$. If the set Γ_1 has no abelian members, it has at least $|\Gamma_1| - 1 \geq (p+1) - 1 = p > 1$ nonabelian members of exponent p^e so our claim is true. Thus, let the set Γ_1 have an abelian member. Let $\mathrm{d}(G) = 2$; then $|\Gamma_1| = p + 1$ and the set Γ_1 has exactly one abelian member since G is not an $\mathcal{A}_1$-group. In that case, the number of nonabelian members of exponent p^e in the set Γ_1 is at least $|\Gamma_1| - 2 = (p+1) - 2 = p - 1 > 1$ since $p > 2$, and our claim follows. Now let $\mathrm{d}(G) > 2$; then $|\Gamma_1| \geq p^2 + p + 1$. Since the set Γ_1 has at most $p + 1$ abelian members (Exercise 1.6(a)) and at most one member of exponent $< p^e$, it has at least $(p^2 + p + 1) - (p + 1) - 1 = p^2 - 1 > 1$ nonabelian members of exponent p^e, and the proof is complete.

Lemma 30.2. *Suppose that $G = \Omega_k(G)$ is a p-group and $E < G$ is such that $\Omega_k(E) = E$, where k is a positive integer. Then:*

(a) There is an element $x \in \mathrm{N}_G(E) - E$ of order $\leq p^k$. It follows that there exists a chain $E = E_1 < E_2 < \cdots < E_n = G$, where $E_{i+1} = \langle x_{i+1}, E_i\rangle$, $o(x_{i+1}) \leq p^k$ and $x_{i+1} \in \mathrm{N}_G(E_i)$ for $i = 1, \dots, n-1$. Moreover, there exists a chain $E = E_1 < E_2 < \cdots < E_n = G$ such that $\Omega_k(E_i) = E_i$ for all i and $|E_2 : E_1| = \cdots = |E_n : E_{n-1}| = p$. In particular, there exists a maximal subgroup U of G such that $E \leq U$ and $\Omega_k(U) = U$.

(b) There are two distinct $U, V \in \Gamma_1$ such that $U = \Omega_k(U)$ and $V = \Omega_k(V)$.

Proof. (a) Assume that such an x does not exist. Then $E = \Omega_k(\mathrm{N}_G(E))$ so E is characteristic in $\mathrm{N}_G(E)$. It follows that $\mathrm{N}_G(E) = G$ so $G - E$ contains an element of order $\leq p^k$, contrary to the assumption. If $o(x)$ is as small as possible and $x^p \in E$, then $|\langle x, E\rangle : E| = p$. Now the second assertion follows easily.

(b) Let $x \in G$ be of order $\leq p^k$. Then, by (a), $x \in U \in \Gamma_1$, where $\Omega_k(U) = U$. If $y \in G - U$ is of order $\leq p^k$, then $y \in V \in \Gamma_1$, where $\Omega_k(V) = V$, by (a). Since $U \neq V$, we are done. □

Lemma 30.3. *Suppose that a 2-group $G = \Omega_1(G)$ is nonabelian. Then G is generated by subgroups isomorphic to D_8. In particular, there are distinct nonabelian $K, L \in \Gamma_1$ such that $\Omega_1(K) = K$ and $\Omega_1(L) = L$.*

Proof. By hypothesis, there are in G two noncommuting involutions, say x and y; then $\langle x, y\rangle$ is dihedral so contains a subgroup $A \cong D_8$ which is an $\mathcal{A}_1$-subgroup and $\Omega_1(A) = A$ so $\mathcal{MA}_1(G) \neq \varnothing$. Assume that $\langle \mathcal{MA}_1(G)\rangle < G$; then $A < G$. By Lemma 30.2(a), $A \leq U \in \Gamma_1$, where $\Omega_1(U) = U$; then U is nonabelian. Take an involution $y \in G - U$. Then $y \in V \in \Gamma_1$, where $\Omega_1(V) = V$ (Lemma 30.2(a) again) and $V \neq U$. Since $U = \Omega_1(U)$ is nonabelian, we have $U = \langle \mathcal{MA}_1(U)\rangle$, by induction. Assume that the lemma is false. Then V must be elementary abelian (otherwise, there is $B \in \mathcal{MA}_1(V)$ such that $B \not\leq U$, and we are done since $G = UB$ and so $\langle \mathcal{MA}_1(UB)\rangle \geq \langle \mathcal{MA}_1(U)\rangle B = UB = G$). We have $V = \langle V - Z(G)\rangle$ so there is (an involution) $z \in V - Z(G)$ such that $z \notin U$. Since z does not centralizes U (otherwise, $z \in Z(G)$), there is an involution $w \in U(= \Omega_1(U))$ such that $zw \neq wz$. Then $L = \langle w, z\rangle$ is dihedral so all its $\mathcal{A}_1$-subgroups are dihedral of order 8 and generate L (Theorem 10.28). Since $L \not\leq U$, there exists an $\mathcal{A}_1$-subgroup $B \leq L$ such that $B \not\leq U$. Since $G = UB$, we get, as above, $\langle \mathcal{MA}_1(G)\rangle \geq \langle \mathcal{MA}_1(U)\rangle B = UB = G$, completing the proof of the first assertion. As to the second assertion, take $U \in \Gamma_1$ as above. By what has just been proved and Theorem 10.28, there is $B \in \mathcal{MA}_1(G)$ such that $B \not\leq U$. Let $B \leq V \in \Gamma_1$ be such that $\Omega_1(V) = V$ (Lemma 30.2(a)). Since the nonabelian $V \neq U$, the proof is completed. □

Lemma 30.4. *Let G be a p-group, $p > 2$.*

(a) If $e'_3(G) = 0$, then $\Omega_1(G)$ is elementary abelian.

(b) Let $e_p(G) > e'_p(G) = 0$. Then $\Omega_1(G)$ is elementary abelian.

(c) If $G = \Omega_1(G)$ is of order $\geq p^p$, then $e_p(G) > 0$.

Proof. (a) Suppose that G is a counterexample of minimal order; then $\Omega_1(G) = G$. Let $R < G$ be a G-invariant elementary abelian subgroup of maximal possible order; then $R < G$ and $|R| > p$. Let $x \in \Omega_1(G) - R$ be of order p and $H = \langle x, R\rangle$; then H is nonabelian and $\Omega_1(H) = H$ so, by induction, $H = G$ and $|G : R| = p$. Let $x \in F \in \Gamma_1$. Then $\Omega_1(F) = F$, by the modular law. By induction, F is abelian so $\mathrm{cl}(G) = \mathrm{cl}(RF) = 2$ (Fitting's lemma); then $\exp(G) = p$ since $p > 2$. In that case, $\mathcal{MA}_1(G) \neq \varnothing$ so $e'_3(G) > 0$ (Exercise 1.8a), contrary to the hypothesis.

(b) Since $p > 2$, this follows from Lemma 30.2(a) and part (a). (c) follows from Lemma 30.2(a) and Theorems 12.1(a) and 7.1(b). □

Proof of Theorem 30.1. We proceed by induction on $|G|$. In view of Lemma 30.3, one may assume that $p^k > 2$. Therefore, if $A \in \mathcal{MA}_k(G)$, then $\exp(A) \le p^k$ (Exercise 1.8a).

(a) In view of Lemma 30.4(a), one may assume that $k > 1$. One may also assume that all proper subgroups of G, generated by elements of orders $\le p^k$, are abelian. By Lemma 30.2(a), there is $E \in \Gamma_1$ such that $\Omega_k(E) = E$; then E is abelian of exponent $\le p^k$. Let $x \in G - E$ be of minimal possible order and $x^p \in E$, then $o(x) \le p^k$, and we have $G = \langle x, E \rangle$. In that case, $Z(G) < E$ and $\langle x, Z(G) \rangle = C_G(x)$ is a maximal abelian subgroup of G hence $\exp(\langle x, Z(G) \rangle) \le p^k$. Let $\langle x, Z(G) \rangle \le T \in \Gamma_1$, where $\Omega_k(T) = T$, by the modular law, and $T \ne E$ so $G = ET$. By assumption, T is abelian; then $E \cap T = Z(G)$ so $G/Z(G) \cong E_{p^2}$ and $cl(G) = 2$. Since $p^k > 2$, we get $\exp(G) \le p^k$ and (a) follows.

(b) follows from Lemma 30.2(b).

(c) Since $\exp(G) > p$ (Theorem 10.28), G is irregular. We have to prove that $G \cong \Sigma_{p^2}$. Assume that this is false. By Lemma 30.3, $p > 2$. By (b), there are distinct $U, V \in \Gamma_1$ such that $\Omega_1(U) = U$ and $\Omega_1(V) = V$. In view of (a) and Lemma 30.2(a), one may assume that U is nonabelian so $U = \langle \mathcal{MA}_1(U) \rangle$, by hypothesis. If V is nonabelian, then, by hypothesis, $V = \langle \mathcal{MA}_1(V) \rangle$ so, since $G = UV$, we get $\langle \mathcal{MA}_1(G) \rangle \ge \langle \mathcal{MA}_1(U) \rangle \langle \mathcal{MA}_1(V) \rangle = UV = G$, a contradiction. Thus, one may assume that V is elementary abelian for every $V \in \Gamma_1 - \{U\}$ with $\Omega_1(V) = V$. Since $\langle V - Z(G) \rangle = V \ne U$, there is $y \in V - Z(G)$ (of order p) not contained in U. Then y does not centralize U (otherwise, $y \in Z(G)$) so there is $x \in U (= \Omega_1(U))$ such that $o(x) = p$ and $xy \ne yx$. Set $K = \langle x, y \rangle$; then $\Omega_1(K) = K$ is nonabelian and $K \not\le U$. Assume that $K < G$. Then $K \le T \in \Gamma_1$, where $\Omega_1(T) = T$ (Lemma 30.2(a)) and $T \ne U$ is nonabelian, a contradiction. Thus, $K = G$, i.e., $G = \langle x, y \rangle$ and $d(G) = 2$. Since $\Omega_1(G) = G$, we get $G' = \Phi(G)$ so $|G : G'| = p^2$. Then $|Z(G)| = p$ (Lemma 1.1) and G is of maximal class. By Exercise 9.13, $G \cong \Sigma_{p^2}$.

(d) follows from (c).

(e) One may assume that G is not an $\mathcal{A}_1$-group. By (b) and (a), there are distinct $U, V \in \Gamma_1$ such that $U = \Omega_k(U)$ and $V = \Omega_k(V)$ and U is nonabelian. By induction, $\langle \mathcal{MA}_{k+1}(U) \rangle = U$, and similarly, for V provided V is nonabelian. In that case, $\langle \mathcal{MA}_{k+1}(G) \rangle \ge UV = G$. Now assume that V is abelian. Then $\exp(V) \le p^k$ so $\exp(G) \le p \cdot \exp(V) \le p^{k+1}$, and the result now follows from Theorem 10.28. □

Recall that $\Omega_k^*(G) = \langle x \in G \mid o(x) = p^k \rangle$. It is natural to believe that under condition $G = \Omega_k^*(G)$, a nonabelian p-group G has an $\mathcal{A}_1$-subgroup, say L, such that $\Omega_k^*(L) = L$. I can prove this only for $p > 2$. In the general case, we prove a weaker result. In the following definitions, k is a positive integer.

Definition 1. A subgroup $H < G$ is said to be k-quasi-maximal in G, if $H \triangleleft G$, G/H is cyclic and $G = \langle x \rangle H$ for some $x \in G - H$ with $o(x) = p^k$.

Definition 2. A p-group G is said to be a $\mathcal{V}_k$-group if it is nonabelian, $\Omega_k^*(G) = G$ and $G/\mathrm{Z}(G)$ is abelian of rank 2 and exponent $\leq p^k$.

Supplement to Theorem 30.1. *Suppose that $k > 1$ and a nonabelian p-group $G = \Omega_k^*(G)$. Then:*

(a) G contains a $\mathcal{V}_k$-subgroup. If, in addition, $p > 2$, then G has an $\mathcal{A}_1$-subgroup, say L, of exponent p^k.

(b) The group G contains two distinct k-quasi-maximal subgroups, say A and B, such that $\Omega_k^*(A) = A$ and $\Omega_k^*(B) = B$.

Proof. Let $u \in G$ be of order p^k and let $u \in X < G$, where $\Omega_k^*(X) = X$ and $|X|$ is as large as possible. Then $X^G < G$ and $\Omega_k^*(X^G) = X^G$, so $X^G = X$, i.e., $X \lhd G$. If $y \in G - X$ is of order p^k, then $\langle y, X\rangle = G$ so X is k-quasi-maximal in G. Thus, for every $u \in G$ of order p^k, there is a k-quasi-maximal subgroup $X < G$ such that $\Omega_k^*(X) = X$ and $u \in X$.

(a) Assume that G is a counterexample of minimal order. Then all k-quasi-maximal subgroups of G, that are generated by elements of order p^k, must be abelian. Let $X < G$ be k-quasi-maximal with $\Omega_k^*(X) = X$; then X is abelian. We have $G = \langle a, X\rangle$, where $o(a) = p^k$. Let $a \in Y$, where Y is a k-quasi-maximal subgroup of G such that $Y = \Omega_k^*(Y)$. By assumption, Y is abelian, and $X \neq Y$. We have $XY \geq \langle X, a\rangle = G$ and so $X \cap Y \leq \mathrm{Z}(G)$. Since G/X and G/Y are cyclic of order $\leq p^k$, it follows that $G/(X \cap Y)$ is a subgroup of the abelian group $(G/X)\times(G/Y)$. Since G is nonabelian, $G/\mathrm{Z}(G)$ is noncyclic so it is abelian of rank 2 and exponent $\leq p^k$. In that case, G is itself a $\mathcal{V}_k$-group. If, in addition, $p > 2$ and a $\mathcal{V}_k$-subgroup $U \leq G$, then U has an $\mathcal{A}_1$-subgroup of exponent p^k (Theorem 10.28).

(b) is obvious. □

Lemma 30.5. *If G is a nonabelian group of order p^m and exponent p, $3 \leq n < m$, then:*

(a) $e_n'(G) \geq p$. If $e_n'(G) = p$, then $n = m - 1$ and $\mathrm{d}(G) = 2$.

(b) If $n < m - 1$, then $e_n'(G) \geq 2p - 1$.

Proof. Since the order of an $\mathcal{A}_1$-subgroup of G equals p^3 (Exercise 1.8a), we get $e_n'(G) > 0$. To prove the first assertion in (a), one may assume that $n = m - 1$. Then $e_n(G) - e_n'(G) \in \{0, 1, p + 1\}$ (Exercise 1.6(a)). Since $|\Gamma_1| \equiv 1 + p \pmod{p^2}$, we are done provided $e_n(G) - e_n'(G) \in \{0, 1\}$. If $e_n(G) - e_n'(G) = p + 1$, then $\mathrm{d}(G) > 2$ (Exercise 1.8a since G is not an $\mathcal{A}_1$-group) so $e_n'(G) \geq (p^2 + p + 1) - (p + 1) = p^2 > p$. Assume that $n < m - 1$. Then $e_{n+1}'(G) \geq p$, by (a). If $A, B < G$ be two distinct nonabelian subgroups of order p^{n+1}, then $e_n'(G) \geq e_n'(A) + e_n'(B) - 1 \geq 2p - 1$, proving (b). Now suppose that $n = m - 1$ and $e_n'(G) = p$. Then, by the above, $e_n(G) - e_n'(G) = 1$ so $|\Gamma_1| = e_n(G) = p + 1$, and hence $\mathrm{d}(G) = 2$. □

Proposition 30.6. *Let G be an irregular p-group with unique nonabelian subgroup R of order p^p and exponent p. Then either $R = \Omega_1(G)$ or else $G \cong \Sigma_{p^2}$.*

Proof. We use induction on $|G|$. Suppose that $R < \Omega_1(G)$. By hypothesis, $p > 2$.

Take $x \in \Omega_1(G) - R$ of order p and set $H = \langle x, R\rangle$; then $|H| = p^{p+1}$. By Lemma 30.5(a), $\exp(H) > p$ so, since $\Omega_1(H) = H$, we conclude that H is irregular hence $\mathrm{cl}(H) = p$, i.e., H is of maximal class. Let $D < R$ be a G-invariant subgroup of index p; then $A = \langle x, D\rangle$ is of order p^p and exponent p. It follows from $A \neq R$ that $A \cong \mathrm{E}_{p^p}$ so $H \cong \Sigma_{p^2}$ (Exercise 9.13). One may assume that $H < G$.

(i) Suppose that $G = \Omega_1(G)$. Let $H < T \leq G$, where $|T : H| = p$. Then T is not of maximal class since $|T| > p^{p+1}$ and $R < T$ (Theorem 9.6(c)). Also, T is irregular and $e_p'(T) = 1$. Therefore, by induction, we must have $T = G$. It follows that $1 < e_p(G) \equiv 1 \pmod p$ (Theorem 13.5) hence $e_p(G) \geq p + 1$. Therefore, there is $y \in \mathrm{N}_G(H) - H$ of order p (here we apply Theorem 13.5 to $\mathrm{N}_G(H)$); then $G = \langle y, H\rangle$ is of order p^{p+2}. Since A is characteristic in H, y normalizes A so $B = \langle y, A\rangle$ is of order p^{p+1}. If B is irregular, then $B \cong \Sigma_{p^2}$ so $e_p'(B) = 1$, a contradiction since $R \not\leq B$ (otherwise, $H = RA \leq B$). Thus, B is regular so $\exp(B) = p$. It follows from $R \not\leq B$ that $e_p'(B) = 0$ so $B \cong \mathrm{E}_{p^{p+1}}$. Since G is not of maximal class, it contains exactly p^2 subgroups of maximal class and index p (Theorem 12.12(c)); let $M \in \Gamma_1$ be one of such subgroups. Since $M \cap B \cong \mathrm{E}_{p^p}$, we conclude that $M \cong \Sigma_{p^2}$. In that case, $e_p'(M) = 1$ so $R < M$. Thus, R is contained in at least p^2 maximal subgroups of G. This is a contradiction since $G/R \cong \mathrm{E}_{p^2}$ contains exactly $p + 1 < p^2$ subgroups of index p (recall that, by assumption, $\Omega_1(G) = G$). Thus, $G = H \cong \Sigma_{p^2}$.

(ii) Now let $\Omega_1(G) < G$. Then, by (i), we get $\Omega_1(G) \cong \Sigma_{p^2}$. In that case, $e_p(G) = e_p(\Omega_1(G)) = 2 \not\equiv 1 \pmod p$ so G is of maximal class (Theorem 13.5). Since G contains a subgroup isomorphic to E_{p^p}, we get $G \cong \Sigma_{p^2}$ (Exercise 9.13) so $G = H$, a final contradiction. □

Theorem 30.7. *Let G be a p-group and $1 < e_p'(G) < p$. Then G is of maximal class and order p^{p+1}.*

Proof. We proceed by induction on $|G|$. By hypothesis, $p > 2$. Then G is irregular (otherwise, applying Lemma 30.5(a) to $\Omega_1(G)$, we get $e_p'(\Omega_1(G)) \geq p$). Therefore, if $|G| = p^{p+1}$, then G is of maximal class. In that case, $G \not\cong \Sigma_{p^2}$. Next we assume that $|G| > p^{p+1}$.

(A) Let $G = \Omega_1(G)$.

Write $t = e_p'(G)$. Let $R_1, \dots, R_t$ be all nonabelian subgroups of order p^p and exponent p in G; then all these subgroups are G-invariant since $t < p$. By Lemma 30.5(a),

(i) G has no nonabelian subgroups of order p^{p+1} and exponent p.

Assume that the theorem is false. Since $R_1 \lhd G$ and $|G| > p^{p+1}$, we get

(ii) G is not of maximal class so $e_p(G) \equiv 1 \pmod{p}$ (Theorem 13.5), and we conclude that $e_p(G) \geq p + 1$.

Since $e_p(G) - t \geq 2$, we obtain

(iii) G contains two distinct subgroups $E, E_1 \cong \mathrm{E}_{p^p}$. Moreover, since $p > 2$ and $e_p(G) - e'_p(G) \equiv p + 1 - t \pmod{p}$ so $e_p(G) - e'_p(G) \not\equiv 0, 1 \pmod{p}$, G has at least $p + 1 - t \geq 2$ normal elementary abelian subgroups of order p^p. Therefore, one can assume that $E, E_1 \lhd G$.

(iv) If $x \in G - R_i$ ($i \leq t$) is of order p, then $L_i = \langle x, R_i \rangle$ is of maximal class of order p^{p+1}. Indeed, the second assertion follows from the first one since $R_i \lhd G$. Since $\Omega_1(L_i) = L_i$ is nonabelian, we get $\exp(L_i) > p$, by (i), so L_i is irregular; then $\mathrm{cl}(L_i) = p$ (Theorem 7.1(b)).

(v) If $i \neq j, i, j \leq t$, then $K = R_i R_j$ is of order p^{p+1} and of maximal class as follows from (i).

(vi) The subgroup $R_i E \cong \Sigma_{p^2}$ ($i \leq t$). Indeed, if $R_i < H_i \leq R_i E$ and $|H_i : R_i| = p$, then $H_i = R_i(H_i \cap E)$, by the modular law. It follows that $\Omega_1(H_i) = H_i$ hence H_i is of maximal class. Thus, all subgroups of $R_i E$ of order p^{p+1}, containing R_i, are of maximal class so $R_i E$ is also of maximal class (Exercise 13.10). Since $E < R_i E$ is of rank p, we get $R_i E \cong \Sigma_{p^2}$ (Exercise 9.13).

(vii) We claim that $|G| = p^{p+2}$. Indeed, let $K_i = R_i E$ ($i \leq t$) and $K_i < M \leq G$, where $|M : K_i| = p$; then $|M| = p^{p+2}$ and M is not of maximal class since $E < M$. Assume that $M < G$. Since $1 \leq e'_p(M) \leq e'_p(G) < p$, we must have, by Proposition 30.6 and induction, that $|M| = p^{p+1}$, a contradiction. Thus, $M = G$.

Let $E, E_1 \lhd G$ be as in (iii). Since $\mathrm{cl}(EE_1) \leq 2$ (Fitting's lemma) and $p > 2$, we get $\exp(EE_1) = p$ so that EE_1 is elementary abelian, by (i). Set $H = R_i R_j$ ($i, j \leq t, i \neq j$). By (v) and (vii), $H \in \Gamma_1$ is of maximal class. The subgroup $EE_1 \cap H \cong \mathrm{E}_{p^p}$ so $H \cong \Sigma_{p^2}$ (Exercise 9.13). But then $e'_p(H) = 1$, a final contradiction: $R_i, R_j < H$ are distinct and nonabelian. Thus, if (A) is satisfied, then G is of maximal class and order p^{p+1}.

(B) Assume that $\Omega_1(G) < G$. By (A), $\Omega_1(G)$ is of maximal class and order p^{p+1} since $e'_p(\Omega_1(G)) = e'_p(G) \in \{2, \ldots, p-1\}$; in particular, $\exp(\Omega_1(G)) = p^2$, and since $\mathrm{d}(\Omega_1(G)) = 2$, we conclude that $e_p(G) = e_p(\Omega_1(G)) < p+1$ (if $e_p(G) = p+1$, then $\exp(\Omega_1(G)) = p$ which is not the case). By hypothesis, $e_p(G) \geq e'_p(G) > 1$ so $e_p(G) \not\equiv 1 \pmod{p}$. It follows that G is of maximal class (Theorem 13.5) and $R_1 \lhd G$, contrary to Theorem 9.6. □

Corollary 30.8. *Let G be a p-group, $p > 2$. If $1 \leq e'_p(G) < p$, then either (a) $|\Omega_1(G)| = p^p$ or (b) G is of maximal class and order p^{p+1}.*

Remark 2. A p-group is said to be an $\mathcal{A}_n$-*group* if it has a nonabelian subgroup of index p^{n-1} but all its subgroups of index p^n are abelian. Suppose that a nonabelian

p-group $G = \Omega_k(G)$, $p > 2$, is not an $\mathcal{A}_1$-group. We claim that then there is in G an $\mathcal{A}_2$-subgroup of exponent $\leq p^{k+1}$. Indeed, by Theorem 30.1(a), G contains an $\mathcal{A}_1$-subgroup A such that $\Omega_k(A) = A$; then $\exp(A) \leq p^k$. Let $A < D \leq G$, where $|D : A| = p$. Then $\exp(D) \leq p \cdot \exp(A) \leq p^{k+1}$. Let $B \leq D$ be an $\mathcal{A}_2$-subgroup; then $\exp(B) \leq \exp(D) \leq p^{k+1}$.

Proposition 30.9. *If G is a p-group such that $e'_p(G) = p$, then one of the following holds:*

(a) G is of maximal class and order $\leq p^{p+2}$. If, in addition, $|G| = p^{p+2}$, then $\Omega_1(G) \in \Gamma_1$ is of maximal class; if $N \in \Gamma_1 - \{\Omega_1(G)\}$, then $|\Omega_1(N)| = p^{p-1}$.

(b) $\Omega_1(G) = H$ is of order p^{p+1} and exponent p, $\mathrm{d}(H) = 2$, H has a subgroup $E \cong \mathrm{E}_{p^p}$, $|\mathrm{Z}(H)| > p$.

(c) $\Omega_1(G) = F$ is of order p^{p+2} and contains exactly p^2 subgroups of maximal class and index p, and exactly p of them are isomorphic to Σ_{p^2}, $e_p(G) = p + 1$.

Proof. It follows from $e'_p(G) > 0$ that $p > 2$. Let $R_1, \dots, R_p$ be all nonabelian subgroups of order p^p and exponent p in G.

Suppose that G is of maximal class; then $e_p(G) = e'_p(G)$ (Exercise 9.13). Then $\mathrm{N}_G(R_1)$ is of maximal class and order p^{p+1} (see Theorem 9.6 or Theorem 13.19). Since $|G : \mathrm{N}_G(R)| \leq e'_p(G) = p$, we get $|G| \leq p^{p+2}$. Assume that $|G| = p^{p+2}$. Let $R_1 < M \in \Gamma_1$; then $M = \mathrm{N}_G(R_1)$ so $e'_p(M) = p = e'_p(G)$. If $N \in \Gamma_1 - \{M\}$, then $N \cap M = \Phi(G)$ is of exponent p^2 so $e_p(N) = 0$ and $\Omega_1(G) = M$. Thus, G is as in (a). In what follows we assume that G is not of maximal class.

(i) Suppose that $H < G$ is nonabelian of order p^{p+1} and exponent p; then $e'_p(H) \geq p = e'_p(G)$ (Lemma 30.5(a)) so $H = \langle R_1, \dots, R_p \rangle$ is characteristic in G.

Let $H < \Omega_1(G)$. Since $e_p(H) \geq p+1$, there is $\mathrm{E}_{p^p} \cong E < H$. Since $e'_p(H) = p$, we get $\mathrm{d}(H) = 2$ (Lemma 30.5(a)) so E is a unique abelian subgroup of index p in H; then E is characteristic in G since H is characteristic in G.

We claim that $E < E_0 \lhd G$, where E_0 is the maximal G-invariant elementary abelian subgroup of G. Indeed, let $x \in G - H$ be of order p and set $E_1 = \langle x, E \rangle$; then $|E_1| = p^{p+1}$ and $\Omega_1(E_1) = E_1$. Since $H \cap E_1 = E$, we get $p = e'_p(G) \geq e'_p(H) + e'_p(E_1) = p + e'_p(E_1)$ so $e'_p(E_1) = 0$, and hence E_1 is elementary abelian (Lemma 30.4(b)). By Theorem 10.1, one may assume that $E_1 \lhd G$. If $\{x \in G \mid x^p = 1\} \neq R_1 \cup \dots \cup R_p \cup E_1$, take $y \in G - (R_1 \cup \dots \cup R_p \cup E_1)$ of order p and set $E_2 = \langle y, E_1 \rangle$. We have $H \cap E_2 = E$ so $e'_p(E_2) = 0$. It follows from $\Omega_1(E_2) = E_2$ that E_2 is elementary abelian (Lemma 30.4(b)). Using Theorem 10.1 again, one may assume that $E_2 \lhd G$. Continuing so, we shall justify our assertion about existence of E_0. In that case, $R_1 \cup \dots \cup R_p \cup E_0 = \{x \in G \mid x^p = 1\}$. We have $|E_0| > p^p$. Let $x \in R_1 - E_0$ and set $F = \langle x, E_0 \rangle$; then F is not of maximal class (Theorem 9.6). Therefore, by Proposition 1.8, we get $|\mathrm{C}_F(x)| > p^2$. Clearly, $\mathrm{C}_F(x)$ is elementary abelian.

Assume that $|C_F(x)| \le p^p$. Let $C_F(x) < T < F$, where $|T| = p^{p+1}$. Since $C_F(x)$ is maximal abelian in F, the subgroup T is nonabelian and we have $|Z(T)| \ge |Z(F)| > p$ so $cl(T) < p$ whence T is regular. By the modular law applied to $\langle x \rangle < T \le F$, we get $T = \langle x \rangle (T \cap E_0)$ so $\Omega_1(T) = T$, $\exp(T) = p$ (Theorem 7.2(b)), and so $T' = \Phi(T)$. By Lemma 30.5(a), $e'_p(T) \ge p$ so $H = \langle R_1, \dots, R_p \rangle = T$. However, by Lemma 1.1,

$$|T : T'| = p|Z(T)| \ge p^3 > p^2 = |H : \Phi(H)| = |H : H'|,$$

contrary to the equality $T = H$.

Now let $|C_F(x)| > p^p$ so $|Z(F)| \ge p^p$. As above, $C_F(x)$ is maximal abelian in F. Let $C_F(x) < T \le F$, where $|T : C_F(x)| = p$; then $Z(T) = Z(F)$. In that case, $|T : Z(T)| = p^2$ and $\Omega_1(T) = T$, by the modular law applied to $\langle x \rangle < T \le F$, so $cl(T) = 2$, and we conclude that $\exp(T) = p$ since $p > 2$. Also we have $|T| > p^{p+1}$ so, by Lemma 30.5(a), $e'_p(T) > p = e'_p(G)$, a contradiction.

We have proved that if G contains a nonabelian subgroup H of order p^{p+1} and exponent p, then $H = \Omega_1(G)$ so G is such as stated in (b).

(ii) Now Let G have no nonabelian subgroups of order p^{p+1} and exponent p. Since $e_p(G) \equiv 1 \pmod p$ (Theorem 13.6), there is in G a normal subgroup $E \cong E_{p^p}$. Let us prove, that $R_i E \cong \Sigma_{p^2}$. By assumption, $\exp(R_i E) > p$. Let $R_i < M \le R_i E$, where $|M : R_i| = p$; then $M = R_i(M \cap E)$ so $\Omega_1(M) = M$. It follows that M is irregular so of maximal class since $|M| = p^{p+1}$ (Theorem 9.5). Thus, all subgroups of $R_i E$ of order p^{p+1} that contain R_i, are of maximal class. Then, by Exercise 13.10, $R_i E$ is also of maximal class so $R_i E \cong \Sigma_{p^2}$ (Exercise 9.13), and this is true for $i = 1, \dots, p$. Since $e'_p(R_i E) = 1$ for all $i \le p$, it follows that $R_1 E, \dots, R_p E$ are pairwise distinct. But $N_G(R_1 E)$ is not of maximal class (Remark 10.5) so $\Omega_1(N_G(R_1 E)) > R_1 E$, by Theorem 13.5 since $e_p(R_i E) = 2 \not\equiv 1 \pmod p$. Therefore, there exists an element $x \in N_G(R_1 E) - R_1 E$ of order p. Set $F = \langle x, R_1 E \rangle$; then $|F| = p^{p+2}$. Therefore, since $R_1 < F$, F is not of maximal class so $e'_p(F) = p$ (Corollary 30.8 and hypothesis). Since R_i and E are characteristic in $R_i E$, we conclude that $R_i, E \lhd F$, $i = 1, \dots, p$. Since, for all i, $R_i \cap E = \Phi(R_i E) = \Phi(F)$ (here Theorem 12.12(c) is used), we get $E \cap R_1 \cap \dots R_p = \Phi(F)$.

Assume that E_0 is another elementary abelian subgroup of order p^p in F. Assume that $E_0 \ntrianglelefteq F$; then $|N_F(E_0)| = p^{p+1}$ so $N_F(E_0)$ is not of maximal class (Remark 10.5) hence regular, and we get $\exp(N_F(E_0)) = p$. Indeed, $N_F(E_0) = N_{E_0 R_1}(E_0)$ (compare orders!) so $\Omega_1(N_F(E_0)) = N_F(E_0)$ so our claim about exponent of $N_F(E_0)$ follows since it is regular. By assumption, $N_F(E_0)$ is elementary abelian. We have $F = R_1 N_F(E_0)$. As above, all subgroups of F of order p^{p+1} that contain R_1, are of maximal class (here we use the modular law and Theorem 7.2(b)) so F is of maximal class (Exercise 13.10), a contradiction. Thus, $E_0 \lhd F$. Set $H = EE_0$. Then $cl(H) \le 2 < p$ (Fitting's lemma) so H is regular hence $\exp(H) = p$. By assumption, H is elementary abelian so H is maximal in F. By what has just been proved, all maximal subgroups of H are normal in F so all subgroups of H are also normal in F. Let $K < H$ be of order p; then $K \le Z(F)$ so $H \le Z(F)$, and we conclude that F is abelian, a contradiction. Thus, E is a unique elementary abelian subgroup of order p^p in F so $e_p(F) = p+1$.

We have $\Omega_1(F) = F$. Assume that $F < \Omega_1(G)$. Let $y \in G - F$ be of order p. Set $E_1 = \langle y, E\rangle$; then $|E_1| = p^{p+1}$ since $E \lhd G$. Since $E_1 \cap F = E$, it follows that $e'_p(E_1) = 0$ so $E_1 \cong \mathrm{E}_{p^{p+1}}$ (Lemma 30.4(b)). One may assume that $E_1 \lhd G$ (Theorem 10.1). Set $F_1 = R_1E_1$; then, by the product formula, $|F_1| = p^{p+2} = |F|$. Since F_1 is not of maximal class, we get $e'_p(F_1) = p$ (Corollary 30.8 and hypothesis). It follows that $R_1, \dots, R_p, E < F_1$ so $F = \langle R_1, \dots, R_p, E\rangle \le F_1$, and we conclude that $F = F_1$, a contradiction since $y \notin F$. Thus, $F = \Omega_1(G)$. Since $F/E \cong \mathrm{E}_{p^2}$, E is contained in exactly $p+1$ maximal subgroups of F and at least one of them is not of maximal class (Exercise 13.10), and $R_1E, \dots, R_pE$ are all subgroups of maximal class and index p that contain E. All remaining assertions follow from Theorem 12.12(c) and Exercise 9.13. Thus, G is such as stated in (c). □

Theorem 30.10. *Let G be a p-group of order $> p^{p+1}$, $p > 2$. Let $\alpha(H)$ be the number of subgroups in H which are isomorphic to Σ_{p^2}. Then p divides $\alpha(G)$.*

Proof. One may assume that $\alpha(G) > 0$; then G is not of maximal class (Exercise 9.13).

(i) Let $|G| = p^{p+2}$ and $\Sigma_{p^2} \cong H < G$; then $H \in \Gamma_1$ and H has a characteristic subgroup $E \cong \mathrm{E}_{p^p}$ so $E \lhd G$. By Theorem 12.12(a), $d(G) = 3$ so $\Phi(G) = \Phi(H)$ and $\Phi(G) = G'$ is of index p in E.

Assume that $E < A \in \Gamma_1$, where A is abelian. Then, by Lemma 1.1, $|Z(G)| = p^2$ so $G = HZ(G)$. Since $\mathrm{cl}(G) = \mathrm{cl}(H)$, all p^2 members of the set Γ_1, that do not contain $Z(G)$, are of maximal class. Let $H_1 \in \Gamma_1$ be of maximal class. If $Z(G) \cong \mathrm{E}_{p^2}$, then $G = H \times C = H_1 \times C$ with $|C| = p$ and so $H_1 \cong H$, and we get $\alpha(G) = p^2$. Suppose that $Z(G)$ is cyclic. Then $E = \Omega_1(A)$. In that case, $H_1 \cap A = E$ so $H_1 \cong \Sigma_{p^2}$ (Exercise 9.13), and $\alpha(G) = p^2$ again. In what follows we assume that E is a maximal abelian subgroup of G.

Let $\mathrm{E}_{p^2} \cong R < H$ be G-invariant. Since H is of maximal class and order $> p^3$, R is uniquely determined so $R \le \Phi(H) = \Phi(G)$. Set $L = C_G(R)$. Then $E < L \in \Gamma_1$. Let $E < H_1 \in \Gamma_1 - \{H, L\}$. Assume that H_1 is not of maximal class. By the previous paragraph, H_1 is nonabelian so E is a maximal abelian subgroup of H_1 whence $Z(H_1) < E$. We have $|Z(H_1)| \ge p^2$ (Lemma 1.1 and Proposition 1.8). Let $R_1 \le Z(H_1)(\lhd G)$ be a G-invariant subgroup of order p^2. Since $R_1 < E < H$, it follows that $R_1 = R$. In that case, $C_G(R) \ge LH_1 = G$ so $R \le Z(H)$, a contradiction since H is of maximal class and $|R| = p^2$. Thus, $p = |Z(H_1)| = \frac{1}{p}|H_1 : H_1'|$ (Lemma 1.1) so $|H : H'| = p^2$, and we conclude that H_1 is of maximal class. Then $H_1 \cong \Sigma_{p^2}$ (Exercise 9.13). Thus, in the case under consideration, E is contained in exactly p subgroups isomorphic to Σ_{p^2}. Let $\mathcal{M}_E$ be the set of all those members of the set Γ_1 that are isomorphic to Σ_{p^2} and contain E. As we have proved, $|\mathcal{M}_E| = p$. Assume that there is in G a subgroup $E_1 \cong \mathrm{E}_{p^p}$ which is different from E. Then, by the above, $|\mathcal{M}_{E_1}|$ is either 0 or p. It follows from the proved property of the set $\mathcal{M}_E$ that $\mathcal{M}_E \cap \mathcal{M}_{E_1} = \varnothing$. Continuing so, we prove that the number of members of the set Γ_1 isomorphic to Σ_{p^2}, is a multiple of p. Thus, the theorem is true for $|G| = p^{p+2}$.

(ii) Now let $|G| > p^{p+2}$. We use induction on $|G|$. By Hall's enumeration principle, we get $\alpha(G) \equiv \sum_{H\in\Gamma_1} \alpha(H) \pmod p$. If $H \in \Gamma_1$ with $\alpha(H) > 0$, then $|H| \geq p^{p+2}$ therefore, by induction, $\alpha(H)$ is a multiple of p so $\alpha(G) \equiv 0 \pmod p$, completing the proof. □

Theorem 30.11. *Let G be a p-group of order p^m and $3 \leq n < m$. Given $H \leq G$, let $\alpha_n(H)$ be the number of subgroups, say L, of maximal class and order p^n in H such that L contains an abelian subgroup of index p. Then $\alpha_n(G) \equiv 0 \pmod p$.*

Proof. One may assume that $\alpha_n(G) > 0$. If $m = 4$, as it is easy to check, $\alpha_3(G) \in \{p, p^2\}$. In what follows we assume that $m > 4$. If $n = 3$, then $\alpha_3(G) \equiv 0 \pmod{p^2}$, by Mann's Theorem 2.3. Next we assume that $n > 3$. If $p = 2$, then every 2-group of maximal class has a cyclic subgroup of index 2 and $\alpha_n(G)$ is even (Theorem 5.4). Now assume that $p > 2$.

(i) Let $n = m - 1$. Take $H \in \Gamma_1$, where H is of maximal class with abelian subgroup A of index p. Since $n > 3$, A is characteristic in H so normal in G.

Let G be of maximal class; then $A = \Phi(G)$. Let $R < A$ be G-invariant of order p^2; then $L = \mathrm{C}_G(R) \in \Gamma_1$. Let $H_1 \in \Gamma_1 - \{L\}$. Assume that H_1 is not of maximal class. Then $|H_1 : H_1'| > p^2$ so $|\mathrm{Z}(H_1)| > p$. Let $R_1 \leq \mathrm{Z}(H_1)$ be G-invariant of order p^2. It follows from $R, R_1 < A < H$ that $R = R_1$. Then $\mathrm{C}_G(R) \geq LH_1 = G$ so $R \leq \mathrm{Z}(G)$, a contradiction since $R \not\leq \mathrm{Z}(H)$. Thus, $\alpha_n(G) = p$.

Next let G be not of maximal class. Since $\alpha_n(G) > 0$, we get $|\mathrm{Z}(G)| = p$, otherwise, p divides $\alpha_n(G)$ (see the proof of Theorem 30.10).

Let $R < A$ be G-invariant of order p^2 and set $U = \mathrm{C}_G(R)$; then $|G : U| = p = |U : A|$ and U is not of maximal class. Using Exercise 13.10, we get $G/A \cong \mathrm{E}_{p^2}$ (otherwise, $\alpha_n(G) = 0$) so A is contained in exactly $p+1$ members of the set Γ_1. Let $A < F \in \Gamma_1 - \{H, U\}$. Assume that F is not of maximal class. Then, by Lemma 1.1, $|\mathrm{Z}(F)| > p$. Since $\mathrm{Z}(F)$ is G-invariant (of order $> p$), it contains a G-invariant subgroup R_1 of order p^2. As in the proof of Theorem 30.10, one may assume that A is a maximal abelian subgroup of G. Then $\mathrm{Z}(F) < A$ so $R_1 < A < H$ and hence $R_1 = R$ since H is of maximal class. In that case, $\mathrm{C}_G(R_1) \geq FU = G$ hence $R \leq \mathrm{Z}(G)$, a contradiction since $R < H$. Thus, the set $\mathcal{M}_A$ of all members of the set Γ_1, that are of maximal class and contain A, is of cardinality p. Now, as in part (i) of the proof of Theorem 30.10, we get $\alpha_n(G) \equiv 0 \pmod p$.

(ii) Now assume that $n < m-1$. Then, as in part (ii) of the proof of Theorem 30.10, we use Hall's enumeration principle to get $\alpha_n(G) \equiv 0 \pmod p$. □

Corollary 30.12. *Let G be a p-group of order p^m and $p+1 \le n < m$. Given $H \le G$, let $\alpha_n(H)$ be the number of metabelian subgroups, say L, of maximal class and order p^n in H. Then p divides $\alpha_n(G)$.*

Proof. In view of Theorem 30.11, one may assume that $n > 3$. Let $n = m - 1$ and let $H \in \Gamma_1$ be metabelian of maximal class. Then $H' = \Phi(H) \le \Phi(G)$ is abelian of index p^3 in G. If $F \in \Gamma_1$, then $\Phi(H)$ is a subgroup of index p^2 in F so F is metabelian. Since the set Γ_1 contains p or p^2 members of maximal class (Theorems 12.12 and 13.6), we are done in this case. Now let $n < m - 1$. Then, using Hall's enumeration principle, we complete the proof. □

Problems

Problem 1. Classify the p-groups $G = \Omega_1(G)$, $p > 2$, such that $\langle \mathcal{MA}_1(G) \rangle < G$.

Problem 2. Study the p-groups $G = \Omega_k^*(G)$, $k > 1$, that are not generated by $\mathcal{V}_k$-subgroups.

Problem 3. Classify the p-groups covered by $\mathcal{A}_1$-subgroups.

Problem 4. Classify the p-groups all of whose $\mathcal{A}_1$-subgroups are isomorphic.

§31

On p-groups with small p'-groups of operators

Throughout this section we assume that p, q, G, Q and W satisfy the following

Hypothesis H1. $p > 2$ and q are distinct primes.

H2. G is a nontrivial p-group.

H3. $Q > \{1\}$ is a q-group of operators of G.

H4. The group $W = Q \cdot G$, the natural semidirect product, is supersolvable.

Lemma 31.1 (see [BZ, Exercise 3.19]). *Let p, q be primes such that q divides $p - 1$. If $W = Q \cdot G$, where $Q \in \mathrm{Syl}_q(W)$ is elementary abelian, $G \in \mathrm{Syl}_p(W)$ is normal in W, then W is supersolvable.*

If p, q, W, G, Q are as in the Hypothesis and Q acts on G nontrivially, then $Q/\mathrm{C}_Q(G)$ is abelian of exponent dividing $p-1$. Hence all results of this section hold if Q is elementary abelian q-group with $q \mid p-1$. Thus, we will assume, in addition, that

H5. $Q/\mathrm{C}_Q(G) > \{1\}$ so q divides $p-1$, $p > 2$.

In what follows we assume that p, q, W, G, Q are as in the Hypothesis and conditions (H1)–(H5) are fulfilled.

Lemma 31.2. *There is a chief series of G with all Q-invariant members.*

This is proved by induction on $|G|$.

Proposition 31.3. *Let the p-group G be regular of exponent p^e.*

(a) G has a Q-invariant subgroup of exponent p of each order $p^n \leq |\Omega_1(G)|$.

(b) G has a Q-invariant cyclic subgroup of order p^n for all $n \leq e$.

(c) If G is abelian, then $G = Z_1 \times \cdots \times Z_d$, where $Z_1, \dots, Z_d$ are Q-invariant cyclic subgroups.

(d) Let $A < B < G$, where A is Q-invariant and $|B : A| = p$. Then G has a Q-invariant subgroup C such that A is of index p in C and $\exp(C) \leq \exp(B)$. *If, in addition, $A \triangleleft G$, then C can be chosen so that it is also normal in G.*

Proof. To prove (a), it suffices to consider a chief series of W containing $\Omega_1(G)$.

(b,c) One may assume that G is noncyclic. We will prove the following stronger result: If $k < e$ and $\exp(\Omega_k(G)) = p^k$, then G has a Q-invariant cyclic subgroup of order p^{k+1} (here we do not assume that G is regular). Indeed, set $\Omega = \Omega_k(G)$. By Lemma 31.2, G/Ω has a Q-invariant subgroup L/Ω of order p. Let $\mathcal{C}$ be the set of all cyclic subgroups of order p^{k+1} in L. Obviously, Q permutes elements of the set $\mathcal{C}$. Set $|\Omega| = p^t$. Then $|\mathcal{C}| = \frac{|L|-|\Omega|}{\varphi(p^{k+1})} = \frac{p^{t+1}-p^t}{(p-1)p^k} = p^{t-k}$. Since $k < t$ and Q is a q-group with $q \neq p$, it fixes a 'point' in $\mathcal{C}$, proving (b). (c) follows from (b) and Corollary 6.2.

(d) We first consider the case where $A \triangleleft G$; then $A \triangleleft W$. Let $\exp(B) = p^n$. Set $D = \Omega_n(G)$; then D is W-invariant of exponent p^n. We have $A < B \leq D$. By Lemma 31.2, D/A has a W-invariant subgroup C/A of order p; then C is the required subgroup. In the general case, apply the above result to $N_G(A)$. □

Lemma 31.4. *Let G be of order p^m.*

(a) *If $A < G$ is abelian of index p, then there is Q-invariant abelian $B < G$ of index p such that* $\exp(B) \leq \exp(A)$.

(b) *If $A \triangleleft G$ is abelian such that G/A is cyclic, then there is a Q-invariant abelian $B < G$ with $|B| = |A|$ and* $\exp(B) \leq \exp(A)$.

(c) *If G is nonabelian, it has a Q-invariant minimal nonabelian subgroup.*

(d) *If G is nonabelian of exponent p, it has a Q-invariant nonabelian subgroup of order p^n for each $n \geq 3$.*

(e) *Suppose that G is not metacyclic. Then G has a Q-invariant minimal nonmetacyclic subgroup, unless $q = 2$, $p = 3$, G is of order* 3^4 *and class* 3 *(such G has exactly two metacyclic Q-invariant maximal subgroups and other two maximal subgroups of G are nonmetacyclic and not Q-invariant).*

(f) *If $A < G$ is Q-invariant, then $A \leq M < G$, where M is Q-invariant and $|G : M| = p$.*

Proof. We use induction on m.

(a) Assume that $A^t \neq A$ for some $t \in Q$. Then $G = AA^t$, $\exp(G) = \exp(A)$ since $\mathrm{cl}(G) = 2$ and $p > 2$, so one may assume that $G' > \{1\}$ (Lemma 31.2). Then $Z(G) = A \cap A^t$ is of index p^2 in G. By Lemma 31.2, $G/Z(G)$ has a Q-invariant subgroup $U/Z(G)$ of order p; then U is the desired subgroup.

(b) In view of (a), we may assume that $|G : A| = p^k > p$. Let $H = A^Q$. Then $H \leq G$ and H/A is cyclic so $H = AA^t$ for some $t \in Q^\#$. In view of Proposition 31.3(c), one may assume that G is nonabelian. By Fitting's lemma and Theorems 7.1 and 7.2, $\mathrm{cl}(H) \leq 2$, H is regular and $\exp(H) = \exp(A)$. It suffices to show that H has a Q-invariant abelian subgroup of order $|A|$. If $H < G$, H has such a subgroup, by induction. Now let $H = G$. Then $A \cap A^t \leq Z(G)$. Next, $Z(G)$ is W-invariant and $G/Z(G)$ is abelian of rank 2, and $\exp(G/Z(G)) \le |G/A| = p^k$. By Proposition 31.3(b), $G/Z(G)$ has a Q-invariant cyclic subgroup $U/Z(G)$ of order $\exp(G/Z(G))$. Then U is abelian and $|G : U| \le p^k$ so $|U| \ge |A|$. Applying Lemma 31.2 to U, we complete the proof.

(c) Assume that G is not minimal nonabelian. By Lemma 31.2, G has a Q-invariant subgroup H of index p. If H is nonabelian, we are done, by induction. Now let H be abelian. Assume that $d(G) = 2$. Then all p maximal subgroups of G which $\neq H$, are nonabelian (otherwise, G is minimal nonabelian), and one of them, say M, is Q-invariant. Then M has a Q-invariant minimal nonabelian subgroup, by induction. In what follows we assume that $d(G) > 2$.

By Lemma 31.2, $Z(G)$ has a Q-invariant subgroup Z of order p; then $Z \triangleleft W$.

(c1) Suppose that G/Z is nonabelian. By induction, G/Z has a Q-invariant minimal nonabelian subgroup A/Z. If $A < G$, we are done, by induction applied to A. Now let $A = G$; then G/Z is minimal nonabelian. It follows from $d(G) = 3$ that $G = Z \times M$, where M is a minimal nonabelian subgroup. In that case, G has exactly p^2 maximal subgroups not containing Z, and Q permutes these (minimal nonabelian) subgroups since Z is Q-invariant. It follows that one of these maximal subgroups is Q-invariant, as required.

(c2) Now let G/Z be abelian for every Q-invariant subgroup Z of order p in $Z(G)$. Then $Z = G'(\le \Phi(G))$ is a unique minimal normal p-subgroup of W. It follows from Maschke's theorem, applied to Q-invariant subgroup $\Omega_1(Z(G))$, that $Z(G)$ is cyclic. By Lemma 1.1, $|G : Z(G)| = p|G'| = p^2$ (recall that $H < G$ is abelian of index p). If $U < G$ is minimal nonabelian, then $UZ(G) = G$ so $d(G) = 3$. Then, since $G/Z(G) \cong E_{p^2}$, there are exactly $(1 + p + p^2) - (1 + p) = p^2$ nonabelian maximal subgroups of G so one of them, say V, is Q-invariant. Now the result follows by induction applied to V.

(d) Let n be as small as possible such that G has no Q-invariant nonabelian subgroups of order p^n. By (c) and Exercise 1.8a, $n > 3$. Let K be a Q-invariant nonabelian subgroup of order p^{n-1} in G. Then $N = N_G(K)$ is Q-invariant. By Lemma 31.2, N/K has a Q-invariant subgroup U/K of order p. Then U is Q-invariant nonabelian of order p^n, contrary to the choice of n.

(e) Assume that the assertion is false. In view of $p > 2$, minimal non-metacyclic p-group is either of order p^3 and exponent p or a 3-group of maximal class and order 3^4 (Theorem 41.1). Suppose that G is a 3-group of maximal class. If $m = 4$, then G is as stated in (e). Let $m > 4$. Then G has exactly three subgroups of maximal class and index 3 so one of them, say F, is Q-invariant so $|F| = 3^4$ and F has exactly two subgroups, say A and B, of order 3^3 and exponent 3. Then $A, B \triangleleft G$, a contradiction. Thus, if G is a 3-group of maximal class, then $m = 4$, and (e) holds. Next we assume that G is not a 3-group of maximal class.

By Lemma 31.2, G has a Q-invariant maximal subgroup H. By induction, H is either metacyclic or a 3-group of maximal class such as in the statement of (e). If

H is metacyclic, then $|\Omega_1(G)| = p^3$ since G is not a 3-group of maximal class (see Theorem 12.1(b)); then $\Omega_1(G)$ is Q-invariant minimal nonmetacyclic, contrary to the assumption. Thus, $p = 3$ and H is of maximal class and order 3^4 such as in the statement of (e). By Theorem 12.12(b), $\mathrm{d}(G) = 3$ and $G/\mathrm{K}_3(G)$ is of order 3^4 and exponent 3. By (a), $G/\mathrm{K}_3(G)$ contains a Q-invariant abelian subgroup $A/\mathrm{K}_3(G)$ of index 3. Then A is regular so $|\Omega_1(A)| = |A : \mho_1(A)| \geq 3^3$ so $\Omega_1(A)$ has a Q-invariant subgroup of order 3^3 which is minimal nonmetacyclic.

(f) Set $N_1 = \mathrm{N}_G(A)$; then N_1 is Q-invariant. If $N_1 < G$, set $N_2 = \mathrm{N}_G(N_1)$. Arguing so, we construct the series $N_1 < N_2 < \cdots < N_k < N_{k+1} = G$ of Q-invariant subgroups. Clearly, $N_k \lhd W$. One may assume that $|G : N_k| > p$. If H/N_k be a Q-invariant maximal subgroup of G/N_k, then H is the desired subgroup. □

Let a group G of maximal class be of order p^p and exponent $p > 3$. Then G has no normal nonabelian subgroups of order p^3. Therefore, it is not true that, in Lemma 31.4(c), G contains a normal W-invariant minimal nonabelian subgroup.

Theorem 31.5. *Suppose that $A < B < G$, where B is abelian of exponent p^n, $|B : A| = p$ and A is Q-invariant. Then G has a Q-invariant abelian subgroup T such that $A < T$, $|T : A| = p$ and $\exp(T) \leq p^n$. If, in addition, $A \lhd G$, then one can choose T so that $T \trianglelefteq G$.*

Proof. Since $\mathrm{C}_G(A)$ is Q-invariant and $B \leq \mathrm{C}_G(A)$, we may assume that $G = \mathrm{C}_G(A)$, i.e., $A \leq \mathrm{Z}(G)$. By Theorem 10.1, we may assume that $B \trianglelefteq G$. Let

$$Q = \{t_1 = 1, t_2, \dots, t_k\}, \quad k = |Q|, \quad B_i = B^{t_i},\ i = 1, \dots, k, \quad H = B_1 \dots B_k.$$

Then $H/A \leq \mathrm{Z}(G/A)$ and H/A is elementary abelian so $\mathrm{cl}(H) \leq 2$ and $\exp(H) \leq p^n$ since $p > 2$. Obviously, H is Q-invariant. By Lemma 31.2, H/A has a Q-invariant subgroup T/A of order p; then T is the desired subgroup. Now let $A \lhd G$. By Theorem 10.1, we may assume that $B \trianglelefteq G$. Defining H and choosing T as above, we see that T is the desired subgroup. □

Now we are ready to prove the main result of this section.

Theorem 31.6. *Suppose that G has a subgroup isomorphic to E_{p^k}, $k = 1, 2, 3, 4$. Then it has a W-invariant subgroup isomorphic to E_{p^k}.*

Proof. Recall that $p > 2$. The group G has a normal subgroup $E \cong \mathrm{E}_{p^k}$, $k \in \{1, 2, 3, 4\}$ (see Theorems 10.1, 10.4 and 10.5). We may assume, in view of Lemma 31.2, that $k > 1$.

(i) Let $k = 2$. Let Z be a W-invariant subgroup of order p in G (Lemma 31.2) and C a subgroup of order p in G, $C \neq Z$ (see Proposition 1.3). Since $|(Z \times C) : Z| = p$, the result follows from Theorem 31.5.

(ii) Let $k = 3$. By (i), G has a W-invariant subgroup $R \cong \mathrm{E}_{p^2}$. Since $\mathrm{C}_{RE}(R) > R$ is elementary abelian, R is contained in a W-invariant subgroup isomorphic to E_{p^3}, by Theorem 31.5.

In the sequel we use the following known fact (see proofs of Theorems 10.4, 10.5 and §33):

(∗) If $T \cong \mathrm{E}_{p^3}$, $p > 2$, then Sylow p-subgroups of $\mathrm{Aut}(T)$ are nonabelian of order p^3 and exponent p.

(iii) Let $k = 4$. Assume that G has no W-invariant subgroup isomorphic to E_{p^4}. By (ii), G has a W-invariant subgroup $K \cong \mathrm{E}_{p^3}$. By Theorem 31.5, $\Omega_1(\mathrm{C}_G(K)) = K$. Hence, by (∗), if M is a subgroup of G of exponent p then $|M| \le p^6$ (since $\exp(\mathrm{C}_{KM}(K)) = p$) and, if $|M| = p^6$, then $K \le M$ and $\mathrm{C}_M(K) = K$. Similarly, $M \not\cong \mathrm{E}_{p^5}$.

As we have noticed, G has a normal subgroup $E \cong \mathrm{E}_{p^4}$. We may assume that there exists in G a normal subgroup $E_1 \cong \mathrm{E}_{p^4}$, which is $\ne E$; set $H = EE_1$. Then $|H| \ge p^5$ and $\exp(H) = p$ since $\mathrm{cl}(H) \le 2$ and $p > 2$. If $|K \cap E| \le p$, then, by (∗), $\mathrm{C}_{KE}(K) > K$ since KE/K is elementary abelian of order $\ge p^3$, a contradiction. Hence, $|K \cap E| = p^2$ for every choice of E. We have $(K \cap E)(K \cap E_1) \le K$. Assume that $K \cap E \ne K \cap E_1$; then $K = (K \cap E)(K \cap E_1) < H$ and $\mathrm{C}_H(K) > K$. Indeed, this is clear if $|\mathrm{Z}(H)| = p^3$. If $|\mathrm{Z}(H)| \le p^2$, then $|H| \ge p^6$ since $E \cap E_1 \le \mathrm{Z}(H)$. In that case, $\mathrm{Z}(H) \le K$ and H/K is elementary abelian of order $\ge p^3$ so $\mathrm{C}_H(K) > K$, by (∗), a contradiction. Thus, $K \cap E = K \cap E_1$ and $K \not\le H$. Therefore, $|H| = p^5$ and H is nonabelian, by (iii) since $K < HE_2$. It follows that any two G-invariant subgroups isomorphic to E_{p^4} have intersection of order p^3; in particular, $\mathrm{Z}(H)$ is of order p^3. Assume that H contains all G-invariant subgroups isomorphic to E_{p^4}. Then H is characteristic in G so Q-invariant. In that case, $\mathrm{Z}(H)$ is Q-invariant. If $U/\mathrm{Z}(H)$ is a W-invariant subgroup of order p in $H/\mathrm{Z}(H)$, then U is the desired subgroup. Thus, we may assume that G has a normal subgroup $\mathrm{E}_{p^4} \cong E_2 \not\le H$. Then $(E \cap E_2)(E_1 \cap E_2) \le E_2 \not\le H$ so $D = E \cap E_2 = E_1 \cap E_2 = E \cap E_1$ has order p^3, by what has just been proved. Hence, $D \le \mathrm{Z}(HE_2)$ and $HE_2/D \cong \mathrm{E}_{p^3}$. Then $|HE_2| = p^6$, $|\mathrm{Z}(HE_2)| \ge p^3$ and $\mathrm{cl}(HE_2) = 2$ so $\exp(HE_2) = p$, contrary to (iii). □

Let $W = Q \cdot G$ be a minimal nonnilpotent group of order $7 \cdot 3^7$, $Q \in \mathrm{Syl}_7(W)$, $G = W' \in \mathrm{Syl}_3(W)$. Then G is extraspecial of exponent 3 and G' is the only Q-invariant abelian subgroup of G. At the same time, G has elementary abelian subgroups of order 3^k, $k = 2, 3, 4$. Therefore, in Theorem 31.6, the condition of supersolvability of W is indispensable.

Exercise 1. Let G be an irregular p-group satisfying the Hypothesis. Then

(a) For $k = 1, \dots, p-1$, G has a W-invariant subgroup of order p^k and exponent p. (*Hint.* Use Theorem 9.8(d).)

(b) If G has no W-invariant subgroups of order p^p and exponent p, it has a Q-invariant cyclic subgroup of order p^2.

Solution. (b) Let $L < G$ be a W-invariant subgroup of order p^{p-1} and exponent p (see (a)). By Lemma 31.2, there is in G/L a W-invariant subgroup H/L of order p. By hypothesis, $\exp(H) = p^2$, and H is regular. Then H has a Q-invariant cyclic subgroup of order p^2 (Proposition 31.3(b)).

Exercise 2. Let all Q-invariant maximal subgroups of a p-group G be abelian. Then G is either abelian or minimal nonabelian. (*Hint.* Use Lemma 31.4(c,f).)

Proposition 31.7. *Suppose that G is neither absolutely regular (see §9) nor of maximal class. Then G has a Q-invariant subgroup of order p^p and exponent p.*

Proof. If G is regular, the result follows from Theorem 7.2 and Proposition 31.3(a). Next let G be irregular. We proceed by induction on $|G|$. By Lemma 31.2, G contains a Q-invariant maximal subgroup H. If H is neither absolutely regular nor irregular of maximal class, the result follows by induction in H.

Now suppose that H is absolutely regular. Then $\Omega_1(G)$ is of order p^p and exponent p (Theorem 12.1(b)), and we are done since $\Omega_1(G)$ is characteristic in G.

Suppose that H is irregular of maximal class. Then we have, by Theorem 12.12(b), $G/\mathrm{K}_p(G)$ is of order p^{p+1} and exponent p, the subgroup $L/\mathrm{K}_3(G) = \mathrm{Z}(G/\mathrm{K}_3(G))$ is of order p^2 and $G/L \cong \mathrm{E}_{p^2}$. Besides, L is characteristic in G. Let T/L be a Q-invariant subgroup of order p in G/L; then T is a Q-invariant maximal subgroup of G. By the choice, T is not of maximal class, and, since $G/\mathrm{K}_p(G)$ is of order p^{p+1} and exponent p, T is not absolutely regular. Then, by induction, T contains a Q-invariant subgroup of order p^p and exponent p. □

Exercise 3. Suppose that G is noncyclic. Then G contains two distinct Q-invariant maximal subgroups. (*Hint.* Use Maschke's theorem.)

Exercise 4. Let $\exp(G) = p$ and $|G| > p^3$. Is it true that if G has a two-generator maximal subgroup, then it has a Q-invariant two-generator maximal subgroup?

Exercise 5. Suppose that G is neither absolutely regular nor of maximal class and $|G| > p^{p+1}$. If G contains an absolutely regular maximal subgroup, then it contains a Q-invariant absolutely regular maximal subgroup. (*Hint.* Use Theorem 12.1(b) and Proposition 12.13.)

Proposition 31.8. *Let G be a metacyclic but noncyclic p-group. Then G has a Q-invariant maximal cyclic subgroup Z such that $G' \le Z$.*

Proof. Let Z be a Q-invariant cyclic subgroup of G such that $G' \le Z$ and assume that $|Z|$ is as large as possible. We will prove that Z is the desired subgroup. If G/Z is cyclic then Z is a maximal cyclic subgroup of G. Indeed, assume that $Z < C$ and C is cyclic. Then $Z \le \Phi(C) \le \Phi(G)$, and so G/Z is not cyclic, a contradiction. In that case, Z is the desired subgroup. Now let G/Z be not cyclic. Assume that $Z < C$, C is cyclic and $|C : Z| = p$. Then G has exactly p cyclic subgroups $C_1 = C, \dots, C_p$ of order $p|Z|$ containing Z (indeed, the abelian group G/Z of rank two has exactly $p+1$ subgroups of order p). Since Z is Q-invariant, Q permutes these p cyclic subgroups. Since $p \neq q$, there exists $i \in \{1, \dots, p\}$ such that C_i is Q-invariant. Then $|C_i| > |Z|$, contrary to the choice of Z. □

Theorem 31.9. *Let G be an extraspecial p-group of order p^{2m+1}, $m > 1$.*

(a) *$G = E_1 * \dots * E_m$, a central product, where $E_1, \dots, E_m$ are Q-invariant nonabelian subgroups of order p^3.*

(b) *$G = A \cdot B$, where A and B are Q-invariant abelian subgroups of order p^{m+1}, $A \cap B = Z(G)$.*

Proof. (a) Note that all minimal nonabelian subgroups of G have the same order p^3. Then, by Lemma 31.4(c), G has a Q-invariant nonabelian subgroup E_1 of order p^3. Since $E_1' = G'$, $E_1 \lhd G$ and $G = E_1 * C_G(E_1)$ (Lemma 4.2; obviously $C_G(E_1)$ is Q-invariant extraspecial of order p^{2m-1}). If $m = 2$, we are done. Now let $m > 2$. Then, by induction on m, $C_G(E_1) = E_2 * \dots * E_m$, where $E_2, \dots, E_m$ are Q-invariant nonabelian subgroups of order p^3. In that case, $G = E_1 * C_G(E_1) = E_1 * \dots * E_m$.

(b) By (a), $G = E_1 * \dots * E_m$, where $E_1, \dots, E_m$ are Q-invariant nonabelian subgroups of order p^3. By Exercise 3, $E_i = A_i B_i$, where $|A_i| = |B_i| = p^2$ and A_i, B_i are Q-invariant for all i. Then $A = A_1 * \dots * A_m$ and $B = B_1 * \dots * B_m$ are desired subgroups. □

Example. Consider $W = E_1 \text{ wr } Q$, where $|Q| = 2$ and E_1 is a nonabelian group of order p^3 and exponent $p > 2$. Set $Q = \langle t \rangle$, $E_1^t = E_2 \cong E_1$; then $G = E_1 \times E_2 \in \text{Syl}_p(W)$. We claim that G has no G-invariant nonabelian subgroups of order p^3 different from E_1 and E_2. Assume that S is such a subgroup. Since $E_1 \cap E_2 = \{1\}$, one of the intersections $E_1 \cap S$, $E_2 \cap S$, say the first one, is equal to $\{1\}$. Assume that $S \cap E_2 > \{1\}$. Then $\{1\} \neq (E_2 \cap S)^t = E_1 \cap S$, contrary to the assumption. Thus, $S \cap E_i = \{1\}$, $i = 1, 2$. In that case, $C_G(S) \geq E_1 E_2 = G$, a contradiction since S is nonabelian.

Problems

Below, G is as in Hypothesis.

Problem 1. Is it true that if $\exp(G) = p^e$, then an irregular p-group G has a Q-invariant cyclic subgroup of order p^e?

Problem 2. Is it true that if G has an abelian subgroup of index p^2, then it has a Q-invariant abelian subgroup of the same index?

Problem 3. Describe the structure of G if it has a normal subgroup of order p^p and exponent p but has no W-invariant subgroups of order p^p and exponent p.

Problem 4. Is it true that if G has an elementary abelian subgroup of order p^5, then G has a W-invariant elementary abelian subgroup of order p^5?

§32

W. Gaschütz's and P. Schmid's theorems on p-automorphisms of p-groups

Gaschütz [Gas2,3] proved the following celebrated theorem: a p-group of order $> p$ admits an outer p-automorphism, i.e., an automorphism whose order is a power of p. Recall that an automorphism α of G is said to be *outer* if $\alpha \in \mathrm{Aut}(G) - \mathrm{Inn}(G)$. In what follows, we write $\mathrm{Out}(G) = \mathrm{Aut}(G)/\mathrm{Inn}(G)$. In this section we will prove the following generalization of Gaschütz's theorem due to P. Schmid:

Theorem 32.1 ([Schm1]). *If G is a nonabelian p-group, then p divides the number $|\mathrm{C}_{\mathrm{Out}(G)}(\mathrm{Z}(G))|$.*

Since every abelian p-group of order $\geq p^2$ has an (outer) p-automorphism, we obtain

Corollary 32.2 ([Gas2,3]). *Every p-group of order at least p^2 has an outer p-automorphism ϕ, i.e., $o(\phi) = p^n$ for some positive integer n.*

Gaschütz's and Schmid's proofs use cohomology theory. The elementary proof of Theorem 32.1, which follows, is due to Webb [Web].

In the sequel the following easy fact is used. Let $N > \{1\}$ be a proper normal subgroup of a p-group G. An automorphism α of G that stabilizes the chain $\{1\} < N < G$, is of order $p^n \leq \exp(N)$.

If $\phi \in \mathrm{Aut}(G)$ and $N \trianglelefteq G$ is ϕ-invariant, then ϕ_N is the restriction of ϕ to N and $\phi_{G/N}$ is the automorphism induced by ϕ on G/N.

In what follows G is a p-group.

Lemma 32.3. *Let N be a maximal subgroup of a p-group G and $g \in G - N$ so $G = \langle g, N \rangle$. Let $\rho : n \mapsto g^{-1}ng$ be the automorphism of N induced by conjugation by g. For $u \in N$, let μ_u denote conjugation in N by u. Then, for $\alpha \in \mathrm{Aut}(N)$, there is a bijection between two sets*

$$A_\alpha = \{\phi \in \mathrm{Aut}(G) \mid \phi_N = \alpha,\ \phi_{G/N} = \mathrm{id}_{G/N}\}\ (\subseteq \mathrm{Aut}(G)),$$
$$N_\alpha = \{u \in N \mid [\rho, \alpha] = \mu_u,\ (g^p)^\alpha = (gu)^p\}\ (\subseteq N),$$

given by $\phi \mapsto g^{-1} \cdot g^\phi$. If α has order a power of p so does any element of A_α.

Proof. Let $\phi, \psi \in A_\alpha$. Suppose that $g^{-1}g^\phi = g^{-1}g^\psi$. Then $g^\phi = g^\psi$. Since $G = \langle g, N\rangle$ and $\phi_N = \alpha = \psi_N$, we get $\phi = \psi$. It follows that the mapping $\phi \mapsto g^{-1}g^\phi$ from A_α into N_α is injective. (Since $\phi_{G/N} = \mathrm{id}_{G/N}$, we get $g^{-1}g^\phi \in N$.)

Given $u \in N_\alpha$, define $\phi : G \to G$ as follows:

$$(g^i n)^\phi = (gu)^i \cdot n^\alpha \quad (i \in \mathbb{N},\ n \in N). \tag{1}$$

In particular, $g^\phi = gu$ and $n^\phi = n^\alpha$ so $\phi_N = \alpha$, $\phi_{G/N} = \mathrm{id}_{G/N}$. Since $(g^p)^\alpha = (gu)^p$ (recall that $g^p \in N$), ϕ is well-defined. We compute

$$n^{\alpha^{-1}\rho\alpha} = (g^{-1}n^{\alpha^{-1}}g)^\phi = n^{g^\phi} = n^{gu} = n^{\rho\mu_u},$$

so $\rho\mu_u = \alpha^{-1}\rho\alpha$ hence $[\rho, \alpha] = \mu_u$. If $i, j \in \{1, \dots, p-1\}$ and $i \neq j$, then $(g^i)^\phi \neq (g^j)^\phi$ so ϕ is bijective. If $n, m \in N$, $i, j \in \mathbb{Z}$, then

$$\begin{aligned}(g^i n)^\phi \cdot (g^j m)^\phi &= (gu)^i \cdot n^\alpha (gu)^j \cdot m^\alpha = (gu)^{i+j}(n^\alpha)^{(gu)^j} \cdot m^\alpha \\ &= (gu)^{i+j} \cdot n^{\alpha(\rho\mu_u)^j} \cdot m^\alpha = (gu)^{i+j} \cdot n^{\alpha\alpha^{-1}\rho^j\alpha} \cdot m^\alpha \\ &= (gu)^{i+j} \cdot n^{\rho^j\alpha} \cdot m^\alpha.\end{aligned}$$

On the other hand,

$$(g^i n \cdot g^j m)^\phi = (g^{i+j}n^{g^j}m)^\phi = (gu)^{i+j} \cdot n^{\rho^j\alpha} \cdot m^\alpha,$$

by (1). Thus, $\phi \in \mathrm{Aut}(G)$, and, by the above, $\phi \in A_\alpha$.

Suppose that $\phi \in A_\alpha$. Then $g^\phi = gu$ for some $u \in N$ since $\phi_{G/N} = \mathrm{id}_{G/N}$, and so $g^{-1} \cdot g^\phi = u \in N$, and $\phi_N = \alpha$ since $n^\phi = n^\alpha$, by definition of A_α. We see that the mapping $\phi \mapsto g^{-1}g^\phi$ from A_α into N_α is bijective. One has

$$\begin{aligned}g^{\phi^{-1}} &= (g^\phi u^{-1})^{\phi^{-1}} = g(u^{-1})^{\phi^{-1}} = g \cdot (u^{-1})^{\alpha^{-1}}, \\ (g^{-1})^\phi &= (gu)^{-1} = u^{-1}g^{-1},\ (g^{-1})^{\phi^{-1}} = u^{\alpha^{-1}} \cdot g^{-1}.\end{aligned}$$

Therefore (we take into account the definition of ρ, $\phi_N = \alpha$ and the above three formulas), we get, for $n \in N$,

$$\begin{aligned}n^{[\rho,\alpha]} &= n^{\rho^{-1}\alpha^{-1}\rho\alpha} = (gng^{-1})^{\phi^{-1}\rho\alpha} \\ &= [g \cdot (u^{-1})^{\alpha^{-1}} \cdot n^{\alpha^{-1}} \cdot u^{\alpha^{-1}} \cdot g^{-1}]^{\rho\alpha} = ((u^{-1}nu)^{\alpha^{-1}\rho^{-1}})^{\rho\alpha} \\ &= u^{-1}nu = n^{\mu_u},\end{aligned}$$

and so $u \in N_\alpha$ since $[\rho, \alpha] = \mu_u$ and $(g^p)^\alpha = (g^p)^\phi = (g^\phi)^p = (gu)^p$. Let $o(\alpha) = p^s$. Then ϕ^{p^s} stabilizes the chain $\{1\} < N < G$, and so $o(\phi^{p^s})$ is a power of p. Then $o(\phi)$ is also a power of p. The proof is complete. □

In what follows we retain the notation introduced in Lemma 32.3. Instead of A_{id_N} and N_{id_N} we write A_1 and N_1, respectively. We use Lemma 32.3 to decide when id_N can be lifted to an outer automorphism of G, that is $A_1 \not\subseteq \mathrm{Inn}(G)$ (that automorphism is a p-automorphism since $o(\mathrm{id}_N) = 1 = p^0$; see Lemma 32.3).

The subgroup $\mathrm{Z}(N)$ is abelian so we have homomorphisms τ and γ of $\mathrm{Z}(N)$ defined by

$$n^\tau = n^{g^{p-1}} \dots n^g n, \quad n^\gamma = [g,n] \quad (n \in \mathrm{Z}(N)).$$

Since $[g,n]^{g^i} = (n^{-g}n)^{g^i} = n^{-g^{i+1}} n^{g^i}$ and $[g,n] \in N$, then

$$\begin{aligned}(n^\gamma)^\tau &= [g,n]^\tau = [g,n]^{g^{p-1}} \dots [g,n]^g [g,n] \\ &= n^{-g^p} n^{g^{p-1}} \cdot n^{-g^{p-1}} n^{g^{p-2}} \dots n^{-g^2} n^g \cdot n^{-g} n = n^{-1} n = 1\end{aligned}$$

so that $\mathrm{im}(\gamma) \le \ker(\tau)$. Next, for $n \in \mathrm{Z}(N)$, we have $(n^\tau)^\gamma = ((n^{\gamma^{-1}})^{\gamma\tau})^\gamma = 1^\gamma = 1$, and we get $\mathrm{im}(\tau) \le \ker(\gamma)$. Now, $n \in \ker(\gamma)$ if and only if $[g,n] = n^\gamma = 1$ so $\ker(\gamma) = \mathrm{Z}(G) \cap \mathrm{Z}(N) = \mathrm{Z}(G) \cap N$.

Corollary 32.4. *Let G be a nonabelian p-group and N a maximal subgroup of G containing $\mathrm{Z}(G)$. Then, in the above notation, G has an outer automorphism of p-power order inducing the identity on G/N and N if and only if* $\mathrm{im}(\tau) \ne \ker(\gamma)$.

Proof. Obviously, A_1 is the stabilizer of the chain $\{1\} < N < G$ so $A_1 \le \mathrm{Aut}(G)$ and the order of each element of A_1 is a power of p (Lemma 32.3). Therefore, G has no outer automorphisms of the required form if and only if A_1 is a subgroup of $\mathrm{Inn}(G)$, which, by Lemma 32.3, happens if and only if each element of N_1 has the form $g^{-1} \cdot g^d$ for some $d \in \mathrm{C}_G(N)$. Indeed, if $\phi \in A_1$, then $g^\phi = g^d$ for $g \in G$ and some $d \in \mathrm{C}_G(N)$ since each element of A_1 is an inner automorphism of G leaving every element of N fixed. Now $\mathrm{Z}(G) \le \mathrm{Z}(N)$, by hypothesis, so $\mathrm{C}_G(N) = \mathrm{Z}(N)$, and this happens if and only if $N_1 \le [g, \mathrm{Z}(N)] = \mathrm{im}(\gamma)$.

In any case, if $n \in N_1$, $\mu_n = [\rho, \mathrm{id}_N] = \mathrm{id}_N$ so $N_1 \subseteq \mathrm{Z}(N)$ and

$$N_1 = \{n \in \mathrm{Z}(N) \mid g^p = (gn)^p\} = \ker(\tau).$$

So $A_1 \le \mathrm{Inn}(G)$ if and only if $\ker(\tau) = N_1 \le \mathrm{im}(\gamma)$; since $\mathrm{im}(\gamma) \le \ker(\tau)$, this happens if and only if $\ker(\tau) = \mathrm{im}(\gamma)$. But then

$$|\mathrm{im}(\tau)| = |\mathrm{Z}(N) : \ker(\tau)| = |\mathrm{Z}(N) : \mathrm{im}(\gamma)| = |\ker(\gamma)|. \qquad \square$$

Now we will prove the special case of Theorem 32.1.

Lemma 32.5. *Suppose that G is a nonabelian p-group such that any maximal subgroup of G containing $\mathrm{Z}(G)$ is abelian. Then p divides $|\mathrm{C}_{\mathrm{Out}(G)}(\mathrm{Z}(G))|$.*

Proof. Since $G/\mathrm{Z}(G)$ is noncyclic, at least two maximal subgroups of G, say M and N, contain $\mathrm{Z}(G)$ so $|G : \mathrm{Z}(G)| = p^2$ and $G/\mathrm{Z}(G) \cong \mathrm{E}_{p^2}$. Let $N = \langle a, \mathrm{Z}(G)\rangle$ and $M = \langle b, \mathrm{Z}(G)\rangle$; then $G = MN$ and $\mathrm{Z}(G) = M \cap N$. By Lemma 1.1, $G' = \langle [a,b]\rangle$ has order p. It is easy to check that if $|G| = p^3$, the result holds (indeed, a Sylow p-subgroup of $\mathrm{Aut}(G)$ is nonabelian of order $p^3 > |\mathrm{Inn}(G)|$ so it contains an outer p-automorphism which centralizes $\mathrm{Z}(G)$). Therefore, one may assume that $|\mathrm{Z}(G)| \geq p^2$.

If α is an outer automorphism of G which centralizes G/N and N, then $o(\alpha)$ is a power of p and α centralizes $\mathrm{Z}(G) < N$ so the conclusion of the lemma holds. Thus, we may assume, in view of Corollary 32.4, that $\mathrm{im}(\tau) = \ker(\gamma) = \mathrm{Z}(G)$.

We have $G = \langle b, N\rangle$. Construct, for b instead of g, the homomorphisms τ and γ. Then (take into account that $\mathrm{cl}(G) = 2$ and $G' < \mathrm{Z}(G)$ so $a^{b^i} = [a[a,b^i] = a[a,b]^i$):

$$a^\tau = a^{b^{p-1}} \dots a^b a = a[a,b]^{p-1} \dots a[a,b]a = a^p [a,b]^{p(p-1)/2}.$$

Now, $N = \langle a, \mathrm{Z}(G)\rangle$ is abelian and $z^\tau = z^p$ for $z \in \mathrm{Z}(G)$ so

$$\mathrm{Z}(G) = \mathrm{im}(\tau) = \langle a^\tau, z^\tau \mid z \in \mathrm{Z}(G)\rangle = \langle a^p[a,b]^{p(p-1)/2}, z^p \mid z \in \mathrm{Z}(G)\rangle.$$

Since $z^p \in \Phi(\mathrm{Z}(G))$ for all $z \in \mathrm{Z}(G)$, we get $\mathrm{Z}(G) = \langle a^p[a,b]^{p(p-1)/2}\rangle$, so $\mathrm{Z}(G)$ is cyclic. Since $|\mathrm{Z}(G)| \geq p^2$ and $o([a,b]) = p$, we get $[a,b] \in \Phi(\mathrm{Z}(G))$ so $\mathrm{Z}(G) = \langle a^p\rangle$.

Thus, $N = \langle a, \mathrm{Z}(G)\rangle = \langle a\rangle$, and we have shown that if h is any element of $G - \mathrm{Z}(G)$, then $\langle h\rangle$ is a maximal subgroup of G (since as a we can take an arbitrary element of $G - \Phi(G)$). It follows from Theorem 1.2 that $G \cong \mathrm{Q}_8$, contrary to $|\mathrm{Z}(G)| \geq p^2$. □

Now we are ready to complete the proof of Theorem 32.1.

Proof of Theorem 32.1. We use induction on $|G|$. By Lemma 32.5 we may assume that G has a nonabelian maximal subgroup N containing $\mathrm{Z}(G)$, and by Corollary 32.4 that $\ker(\gamma) = \mathrm{im}(\tau)$ (where τ and γ are constructed for some $g \in G - N$). Then we may assume, by induction, that p divides $|\mathrm{C}_{\mathrm{Out}(N)}(\mathrm{Z}(N))|$.

For $G = \langle g, N\rangle$, where $g \in G - N$, let ρ denote conjugation in N by g. Then ρ normalizes $\mathrm{C}_{\mathrm{Out}(N)}(\mathrm{Z}(N))$ so there exists (Lemma 32.3) an outer automorphism α of N of p-power order, such that $\alpha \in \mathrm{C}_{\mathrm{Aut}(N)}(\mathrm{Z}(N))$ and $[\rho, \alpha] = \mu_w \in \mathrm{Inn}(N)$, where w is any element of the coset $v\mathrm{Z}(N)$ of $\mathrm{Z}(N)$ in N. Then

$$(\rho^p)^\alpha = (\rho^\alpha)^p = (\rho[\rho,\alpha])^p = (\rho\mu_w)^p, \ \rho^p = \mu_{g^p} \in \mathrm{Inn}(N)$$

so $(g^p)^\alpha \cdot (gw)^{-p} \in \mathrm{Z}(N)$. But $(g^p)^\alpha \cdot (gw)^{-p} = y^\tau$ for some $y \in \mathrm{Z}(N)$, and so $(g^p)^\alpha = (gwy)^p$. As $wy \in v\mathrm{Z}(N)$ and $[\rho, \alpha] = \mu_{wy}$, we get $wy \in N_\alpha$ (see Lemma 32.3).

It follows from Lemma 32.3 that α lifts to an automorphism ϕ of G of p-power order. As $\mathrm{Z}(N) \geq \mathrm{Z}(G)$ and $\phi_N = \alpha$, it follows that ϕ centralizes $\mathrm{Z}(G)$. If ϕ is inner on N, say ϕ is conjugation by some $h \in G$, then h centralizes $Z(N)$. If $h \notin N$, then $G = \langle h, N\rangle$ so $Z(N) = Z(G) = \ker(\gamma) = \mathrm{im}(\tau)$ hence $\ker(\tau) = \{1\}$. But $\{1\} \neq \Omega_1(Z(G)) \leq \ker(\tau)$, and this is a contradiction. Thus, h must lie in N. It follows that ϕ_N is just μ_h so is inner on N, which is impossible since $\phi_N = \alpha \in \mathrm{Aut}(N) - \mathrm{Inn}(N)$. Thus, ϕ is outer, and so p divides the number $|C_{\mathrm{Out}(G)}(Z(G))|$. □

Exercise (P. Schmid). The automorphism group of an abelian p-group G has a nontrivial normal p-subgroup if and only if G is not elementary abelian.

Solution. If G is elementary abelian of order p^n, then $\mathrm{Aut}(G) \cong \mathrm{GL}(n,p)$ has no nontrivial normal p-subgroups. Now suppose that $\exp(G) = p^k > p$. Let $P = \{\phi \in \mathrm{Aut}(G) \mid x^\phi = xf$, some $f \in \Phi(G)$, for all $x \in G\}$. It is easy to check that P is a normal subgroup of $\mathrm{Aut}(G)$. By Hall–Burnside, P is a p-subgroup. It remains to show that $P > \{1\}$. If $G = \langle x\rangle$ is cyclic of order $p^n > 1$, then $\phi : x \to x^{1+p}$ is a nonidentity p-automorphism of G and $x^p \in \Phi(G)$. Now suppose that $G = Z_1 \times \cdots \times Z_k$ is noncyclic with $|Z_1| \geq \cdots \geq |Z_k|$ with $|Z_1| = p^n > p$ and $Z_i = \langle z_i\rangle$. Then $\phi : z_1 \to z_1 \cdot z_1^p$ and $z_i^\phi = z_i$ for $i > 1$ is a nonidentity automorphism of G contained in P.

Proposition 32.6 (P. Schmid). *Suppose that G is an extraspecial p-group. Then we have $O_p(\mathrm{Out}(G)) = \{1\}$ if and only if one of the following holds:*

(a) $p > 2$ *and* $\exp(G) = p$;

(b) $p = 2$ *and* $G \not\cong Q_8$.

Problems

Problem 1. Is it true that a nonabelian p-group G admits an outer automorphism of order p?

Problem 2 (Old problem). Classify the p-groups G such that $|G|$ does not divide $|\mathrm{Aut}(G)|$.

Corollary 32.2 is the first step in solution of Problem 2.

§33

Groups of order p^m with automorphisms of order p^{m-1}, p^{m-2} or p^{m-3}

1°. Let G be a p-group of order p^m, $\phi \in \mathrm{Aut}(G)$. It is clear that $p^m \nmid o(\phi)$ (indeed, ϕ induces a permutation on $G^\#$). On the other hand, if $|G| > p$, then p divides $|\mathrm{Aut}(G)|$. In this section we classify the groups of order p^m admitting an automorphism of order p^{m-1}. We also consider in some detail groups of order p^m with automorphisms of order p^{m-2} and p^{m-3}. (See [Ber0, BM, Mil5].) In what follows G is a p-group of order p^m.

Let $\phi \in \mathrm{Aut}(G)$. If $H < G$ is ϕ-invariant, then ϕ_H is the restriction of ϕ to H. Let us consider the natural semidirect product $W = G(\phi) = \langle\phi\rangle \cdot G$. Obviously, $G(\phi) = \langle x\phi \mid x \in G\rangle$ with composition $x\alpha \cdot u\beta = (xu^{\alpha^{-1}})(\alpha\beta)$, where $\alpha, \beta \in \langle\phi\rangle$. If $x \in G$ and k is the minimal nonnegative integer such that $\phi^k_{\langle x\rangle} = \mathrm{id}_{\langle x\rangle}$, then distinct elements $x, x^\phi, \dots, x^{\phi^{k-1}}$ constitute the ϕ-orbit $\mathrm{O}_\phi(x)$ of x.

Exercise 1. (a) Let $N \trianglelefteq G$ and suppose that $\phi \in \mathrm{Aut}(G)$ induce identity on N and G/N. Prove that $o(\phi)$ divides $\exp(N)$.

(b) If G of order $p^m > p$ has an automorphism of order p^{m-1}, then ϕ is outer.

(c) Let G be a nonabelian group of order p^m, $m > 3$. Suppose that G has an inner automorphism of order p^{m-2}. Then G is a 2-group of maximal class.

(d) Prove that the groups D_{2^m} and Q_{2^m} admit an automorphism α of order 2^{m-1}. In D_{2^m}, α permutes cyclically all noncentral involutions. In Q_{2^m}, α permutes cyclically generators of all nonnormal subgroups of order 4.

Hint. (b) We get $\exp(G/\mathrm{Z}(G)) < p^{m-1}$. (c) $G/\mathrm{Z}(G)$ has a cyclic subgroup of index p and $|\mathrm{Z}(G)| = p$ so $G/\mathrm{Z}(G)$ is nonabelian. If $G/\mathrm{Z}(G) \cong \mathrm{M}_{p^{m-1}}$, then $|\mathrm{Z}(G)| > p$, a contradiction. Thus, $p = 2$ and $G/\mathrm{Z}(G)$ is of maximal class (Theorem 1.2). Since $|\mathrm{Z}(G)| = 2$, we are done. (c) See Theorem 34.8.

Exercise 2 ([BM]). Let $\phi \in \mathrm{Aut}(G)$ be a p-automorphism and let $M, N \in \Gamma_1$ be different ϕ-invariant. If $o(\phi_M), o(\phi_N) \le p^k$, then $o(\phi) \le p^k$. (*Hint.* Set $\psi = \phi^{p^k}$. Then $\psi = \mathrm{id}_G$ since $G = MN$.)

Theorem 33.1 ([Mil5, Ber0]). *If a group G of order p^m admits an automorphism ϕ of order p^{m-1}, then G is one and only one of the following groups: (a) an abelian group,*

$m \le 2$; (b) a cyclic group, $p > 2$, $m > 2$; (c) a noncyclic abelian group of order 8; (d) a dihedral 2-group, $m > 2$; (e) a generalized quaternion group, $m > 2$.

Lemma 33.2. *Let $\phi \in \mathrm{Aut}(G)$ be a p-automorphism and A a ϕ-invariant normal subgroup of G, $\exp(A) \le p^e$. Suppose that $p^{r+1} \nmid \exp(\mathrm{Aut}(G/A))$ and $p^{r+1} \nmid \exp(\mathrm{Aut}(A))$. Then $p^{r+e+1} \nmid o(\phi)$.*

Proof. Set $\psi = \phi^{p^r}$. Since ψ induces the identity on G/A and A, we have $o(\psi)$ divides p^e (Exercise 1(a)); then $o(\phi) \le p^{r+e}$. □

Proof of Theorem 33.1. One may assume that $m > 3$ (check!). Write $W = \langle \phi \rangle \cdot G$. Suppose that G has a W-invariant subgroup A of type (p, p). Since $|G/A| = p^{m-2}$, $p^{m-2} \nmid \exp(\mathrm{Aut}(G/A))$, $p^{m-2} \nmid \exp(\mathrm{Aut}(A))$ (since $m-2 > 1$). Let, in the notation of Lemma 33.2, $r = m-3$ and $e = 1$; then $p^{m-1} = p^{r+e+1} \nmid \exp(\mathrm{Aut}(G))$ (Lemma 33.2), contrary to the hypothesis. Hence A does not exist so G is cyclic or a 2-group of maximal class, by Lemma 1.4. Now an easy check yields the result (see also Lemma 34.8). □

Exercise 3 ([Ber0]). Let G be noncyclic of order $p^m > p^3$, let $\phi \in \mathrm{Aut}(G)$ be of order p^{m-1} and let $x \in G$ be such that $\langle \phi \rangle$-orbit $\mathrm{O}_\phi(x)$ of x has length p^{m-1}.

(a) There is a ϕ-invariant maximal subgroup $H < G$.

(b) $\mathrm{O}_\phi(x^i) = x^i H$ for $i = 1, \dots, p-1$.

(c) Let $x^\phi = xh$ $(h \in H)$. Then $H = \{h, hh^\phi, hh^\phi h^{\phi^2}, \dots, hh^\phi \dots h^{\phi^{p^{m-1}-1}} = 1\}$.

(d) The number of $\langle \phi \rangle$-orbits of length p^{m-1} on G is $p-1$ and these orbits yield a partition of $G - H$.

(e) There is a maximal subgroup $B < G$ such that $\langle \phi \rangle$-orbit of B has length p. One may choose B so that $H \cap B$ is ϕ-invariant (and we do so). In that case, $x(H \cap B) = \mathrm{O}_{\phi^p}(x)$.

2°. Now we consider the groups of order p^m with automorphism of order p^{m-2}. Let $W = P \cdot G$ be the natural semidirect product, where $\phi \in P \in \mathrm{Syl}_p(\mathrm{Aut}(G))$.

Theorem 33.3 ([Mil5]). *If G is a noncyclic group of order $p^m > p^6$, $p > 2$, and $\phi \in \mathrm{Aut}(G)$ is of order p^{m-2}, then G has a cyclic subgroup of index p.*

Proof. By Theorem 1.17(b), there is W-invariant $\mathrm{E}_{p^2} \cong R < G$. Suppose that G/R has no W-invariant subgroups of type (p, p). Then G/R is cyclic. If G has no cyclic subgroups of index p, then $\Omega_1(G)$ is of order p^3 and exponent p. Let $A/\Omega_1(G) < G/\Omega_1(G)$ be of order p; then A is W-invariant of order p^4 and exponent p^2. If G/R has a W-invariant subgroup A/R of type (p, p), then again A is a W-invariant subgroup of order p^4 and exponent at most p^2. Set, as in Lemma 33.2, $r = m - 5$, $e = 2$. Since $|G/A| = p^{m-4} = p^{r+1}$, then $p^{r+1} \nmid \exp(\mathrm{Aut}(G/A))$. Since $r + 1 =$

$m-4 \geq 3$ and $p^3 \nmid \exp(\mathrm{Aut}(A))$ (Theorem 33.1), it follows that $p^{m-2} = p^{r+e+1} \nmid \exp(\mathrm{Aut}(G))$ (Lemma 33.2), a contradiction. □

Lemma 33.4 ([BM]). *If a group G of order 2^m, $m > 5$, admits an automorphism ϕ of order 2^{m-2}, then G' is cyclic.*

Proof. Assume that G' is not cyclic. Then $|G : G'| > 4$, by Taussky's theorem, and G' has a W-invariant subgroup $A \cong \mathrm{E}_4$, by Lemma 1.4. Then G/A is not of maximal class. Setting $r = m-4$ and $e = 1$, we see that $2^{r+1} \nmid \exp(\mathrm{Aut}(G/A))$ (Theorem 33.1) and $\exp(\mathrm{Aut}(A))$ (since $r+1 > 1$). Therefore, by Lemma 33.2, $2^{m-2} = 2^{r+e+1} \nmid \exp(\mathrm{Aut} G))$, a contradiction. □

Lemma 33.5 ([Mil5,BM]). *Let a group G be of order 2^m, $m > 5$, and let $A \cong \mathrm{E}_4$ be W-invariant and such that G/A is neither dihedral nor generalized quaternion. Then G has no automorphisms of order 2^{m-2}.*

Proposition 33.6. *Let G be a group of order p^m, $p > 2$, $m > 7$. If G has an automorphism ϕ of order p^{m-3}, then one of the following holds: (a) G is metacyclic; (b) G has a normal subgroup A of order p^3 and exponent p such that G/A is cyclic.*

Proof. Suppose that G has a W-invariant subgroup A of order p^3 and exponent p. Assume that G/A is not cyclic. Setting $r = m-5$ and $e = 1$, we see that G/A (by Theorem 33.1) and A have no automorphisms of order $p^{r+1} = p^{m-4}$. Therefore, by Lemma 33.2, G has no automorphisms of order $p^{r+e+1} = p^{m-3}$, a contradiction. Thus, if A exists, G/A is cyclic. From now on we suppose that G has no W-invariant subgroups of order p^3 and exponent p. Then by Theorem 13.7 and Theorems 10.4 and 12.1(a), G is either metacyclic or a 3-group of maximal class.

Suppose that G is a 3-group of maximal class. Let $A \triangleleft G$ be of order 3^4 in G; then A is characteristic in G so W-invariant. It is known that A is metacyclic of exponent 3^2 (Theorem 9.6) so A does not admit an automorphism of order 3^3, by Theorem 33.1. Set $r = m-6$ and $e = 2$. Then, by Theorem 33.1, G/A and A has no automorphisms of order $3^{r+1} = 3^{m-5} > 3^2$. In that case, by Lemma 33.2, G has no automorphisms of order $3^{r+e+1} = 3^{m-3}$, a contradiction. Thus, G is metacyclic. □

Exercise 4. If G is a metacyclic group of exponent p^k, then $p^{k+1} \nmid \exp(\mathrm{Aut}(G))$. (*Hint.* Use Lemma 33.2.)

3°. Let G be a group of order p^m, $\mathrm{d}(G) = d$. Then

$$|\mathrm{Aut}(G)| \mid (p^d-1)\dots(p^d-p^{d-1})|\Phi(G)|^d = (p^d-1)\dots(p-1)p^t,$$

where $t = (m-d)d + \frac{1}{2}d(d-1)$. It is easy to see that $t \leq \frac{1}{2}m(m-1)$. Assume that $|\mathrm{Aut}(G)|_p = p^{\frac{1}{2}m(m-1)}$. Then $\frac{1}{2}m(m-1) = (m-d)d + \frac{1}{2}d(d-1)$ or $(m-d)(m-d-1) = 0$ so $m \in \{d, d+1\}$. It follows that either $G \cong \mathrm{E}_{p^m}$ or $|\Phi(G)| = p$. Miller [Mil5] described all such groups with $|\Phi(G)| = p$. In that case, $G = T \times E$, where E is elementary abelian and T is nonabelian of order p^3.

Proposition 33.7. *Let G be a p-group of order p^n, $\mathrm{d}(G) = d$ and let $E = \mathrm{E}_{p^n}$. If $\pi(\mathrm{Aut}(G)) = \pi(\mathrm{Aut}(E))$, then $G \cong E$, unless $G \cong C_{p^2}$ and $p = 2^s - 1$.*

Proof. We have $|\mathrm{Aut}(E)| = p^{\binom{n}{2}}(p-1)(p^2-1)\dots(p^n-1)$, i.e., $\pi(\mathrm{Aut}(E)) = \{p\} \cup \pi(p-1) \cup \pi(p^2-1) \cup \dots \cup \pi(p^n-1)$. By the paragraph preceding the proposition, $\pi(\mathrm{Aut}(G)) \subseteq \{p\} \cup \pi(p-1) \cup \dots \cup \pi(p^d-1)$.

Suppose that $\pi(\mathrm{Aut}(G)) = \pi(\mathrm{Aut}(E))$, however $d < n$. Then, by Zsigmondy's theorem (see [BZ, Chapter 30, Appendix B]) we have $\{p, n\} = \{2, 6\}$ or $\{2^s - 1, 2\}$. In the first case, $\pi(\mathrm{Aut}(E)) = \{2, 3, 5, 7, 31\} = \pi(\mathrm{Aut}(G))$; then $d = 5$. By [HS], the equality $\pi(\mathrm{Aut}(G)) = \{2, 3, 5, 7, 31\}$ does not hold, unless $G \cong E$.

Let $n = 2$ and $p = 2^s - 1$ a Mersenne prime. Assuming that $G \ncong E$, we conclude that $d = 1$ so G is cyclic of order p^2. In that case

$$\pi(\mathrm{Aut}(G)) = \pi(p-1) \cup \{p\} = \pi(2^s - 2) \cup \{p\} = \{2\} \cup \pi(2^{s-1} - 1) \cup \{p\}.$$

Next, $\pi(\mathrm{Aut}(E)) = \pi((p^2-1)(p^2-p)) = \pi(p(p-1)(p+1)) = \{p\} \cup \{2\} \cup \pi(2^{s-1} - 1) = \pi(\mathrm{Aut}(G))$. □

§34

Nilpotent groups of automorphisms

Let $G = G_0 \geq G_1 \geq \cdots \geq G_r = \{1\}$ be a series of subgroups and let A be the set of automorphisms of G such that $[G_{i-1}, A] \leq G_i$ for each $i = 1, \ldots, r$ (this means that $(g_{i-1}G_i)^\alpha = g_{i-1}G_i$ for all $g_{i-1} \in G_{i-1}$ and $\alpha \in A$). All members of our chain are A-invariant and $A \leq \mathrm{Aut}(G)$. A is called the *stability group* of the chain. Hall [Hal4] has proved that A is nilpotent of class at most $\binom{r}{2}$. L. Kaloujnine proved that if all $G_i \trianglelefteq G$, then $\mathrm{cl}(A) \leq r - 1$.

Theorem 34.1. *Let $H, T \leq G$ and let $H = H_0 \geq H_1 \geq \cdots$ be a series of H-invariant subgroups such that $[H_i, T] \leq H_{i+1}$ for each $i = 0, 1, 2, \ldots$. Define $T = T_1$ and let $T_j = \{x \in T \mid [H_i, x] \leq H_{i+j}$ for all $i\}$. Then $[T_j, T_l] \leq T_{j+l}$ for all j, l and $[H_i, \mathrm{K}_j(T)] \leq H_{i+j}$, all i, j.*

Proof. Clearly, $T_j \leq T$, $T \leq \mathrm{N}_G(H)$ and $[H_i, T_j, T_l], [H_i, T_l, T_j] \leq H_{i+j+l}$. By assumption, H_{i+j+l} is normal in TH. By the Three Subgroups Lemma we have $[H_i, [T_j, T_l]] \leq H_{i+j+l}$, and therefore, by definition, $[T_j, T_l] \leq T_{j+l}$. Thus, $T = T_1 \geq T_1 \geq T_2 \geq \cdots$ is a central series of T and $\mathrm{K}_j(T) \leq T_j$ so that $[H_i, \mathrm{K}_j(T)] \leq [H_i, T_j] \leq H_{i+j}$. □

The subgroup T of Theorem 34.1 stabilizes the series $H = H_0 \geq H_1 \geq \cdots$. All parts of Theorem 34.2 were proved at other places.

Theorem 34.2. *For any group G, we have*

(a) $[\mathrm{K}_i(G), \mathrm{K}_j(G)] \leq \mathrm{K}_{i+j}(G)$.

(b) If $i \geq j$, then $[\mathrm{Z}_i(G), \mathrm{K}_j(G)] \leq \mathrm{Z}_{i-j}(G)$ and so $[\mathrm{Z}_i(G), \mathrm{K}_i(G)] = \{1\}$ (the case $i = 2$ of this equality is Grün's lemma).

(c) $G^{(k)} \leq \mathrm{K}_{2^k}(G)$, where $G^{(k)}$ is the k-th derived subgroup of G.

Proof. (a) Take $H_i = \mathrm{K}_{i+1}(G)$ and $T = G$. Then, by Theorem 34.1,

$$[\mathrm{K}_i(G), \mathrm{K}_j(G)] = [H_{i-1}, \mathrm{K}_j(T)] \leq H_{i-1+j} = \mathrm{K}_{i+j}(G).$$

(b) Take $H_j = \mathrm{Z}_{i-j}(G)$ for $j = 0, 1, \ldots, i$ and again $T = G$. We have to show that $[H_0, \mathrm{K}_j(G)] \leq H_j$. This follows from Theorem 34.1.

(c) Since $G' = \mathrm{K}_2(G)$, the claim is true for $k = 1$. So we can use induction on k. Then, by (a), $G^{(k+1)} = [G^{(k)}, G^{(k)}] \leq [\mathrm{K}_{2^k}(G), \mathrm{K}_{2^k}(G)] \leq \mathrm{K}_{2^{k+1}}(G)$. □

If $A \le \mathrm{Aut}(G)$, then $[G, A] = \langle x^{-1}x^{\alpha} \mid x \in G, \alpha \in A\rangle$. We consider G and A as subgroups of the holomorph of G.

Theorem 34.3 (Kaloujnine). *Let $G = G_0 \ge G_1 \ge \cdots \ge G_r = \{1\}$ be a chain of G-invariant subgroups and A ($\le \mathrm{Aut}(G)$) its stability group. Then both A and $[G, A]$ are nilpotent of class $\le r - 1$.*

Proof. We have $[G_{i-1}, A] \le G_i$ for all i. By Theorem 34.1, $[G, \mathrm{K}_r(A)] \le G_r = \{1\}$ so that $\mathrm{K}_r(A) = \{1\}$ since A is faithful on G, and so $\mathrm{cl}(A) < r$. Also $[G_{i-1}, G, A] \le [G_{i-1}, A] \le G_i$ and $[G_{i-1}, A, G] \le [G_i, G] \le G_i$. But $G_i \trianglelefteq \langle A, G\rangle = A \cdot G$. So, by the Three Subgroups Lemma, $[G_{i-1}, [G, A]] \le G_i$ for each i. By Theorem 34.1 again, $[G_1, \mathrm{K}_{r-1}([G, A])] \le G_r = \{1\}$. But $[G, A] \le G_1$ so $\mathrm{K}_{r-1}([G, A]) \le \mathrm{Z}(G_1)$. Hence $\mathrm{K}_r([G, A]) = \{1\}$, i.e., $\mathrm{cl}([G, A]) < r$. □

Exercise 1. Let F be an arbitrary field and $G = \mathrm{UT}(n, F)$ the group of all $n \times n$ upper unitriangular matrices over F. Then G is nilpotent of class $n - 1$.

Solution. (i) Let $V = F^n$. The group G acts on V via matrix multiplication $(v, A) \mapsto vA$ $(v \in V, A \in G)$. Let V_i be the set of vectors whose first $i - 1$ components are all zero. Then $V_1 = V$, $V_{n+1} = \{0\}$, and G stabilizes the normal series of subgroups $V = V_1 > V_2 > \cdots > V_n > V_{n+1} = \{1\}$. By Theorem 34.3, $\mathrm{cl}(G) \le n - 1$.

(ii) To prove that $\mathrm{cl}(G) = n - 1$, we define elements σ_i of G by $\sigma_i(u_i) = u_i + u_{i+1}$, $\sigma_i(u_j) = u_j$ $(j \ne i)$, $i, j = 1, \dots, n$, where u_i is the vector whose k-th component equals δ_{ik} (Kronecker's delta) and u_{n+1} is identified with u_1. Then the element $\xi = [\sigma_1, \dots, \sigma_{n-1}]$ sends u_1 onto $u_1 + u_n$. Thus, $\xi \ne 1_G$ so $\mathrm{cl}(G) \ge n - 1$, and we conclude that $\mathrm{cl}(G) = n - 1$.

Theorem 34.4 ([Hal4]). *Let H and L be subgroups of a group G. Define $H_0 = H$ and inductively $H_{i+1} = [H_i, L]$. Then $H_r = \{1\}$ implies $[H, \mathrm{K}_{1+\binom{r}{2}}(L)] = \{1\}$.*

Proof. This is clear for $r = 0, 1$. Indeed, if $r = 0$, then $H = H_0 = \{1\}$ so $[H, K_1(L)] = [H, L] = \{1\}$. If $r = 1$, then $[H, L] = [H_0, L] = H_1 = \{1\}$ so $[H, \mathrm{K}_{1+\binom{1}{2}}(L)] = [H, L] = H_1 = \{1\}$. Suppose that $r > 1$ and let $M = \langle H, L\rangle$. Then $H_1 = [H, L]$ is normal in $M = \langle H, L\rangle$ so that $H_1 \ge H_2 \ge \cdots \ge H_r = \{1\}$. Let $C = \mathrm{C}_L(H_1)$; then $L/C \cong A$ is the group of automorphisms of H_1 induced by L, by N/C-Theorem. Think of H_1 and A as subgroups of the holomorph of H_1. Then, by the definition of subgroups H_i's, $[H_i, A] = H_{i+1}$ $(i = 1, 2, \dots, r - 1)$. By induction on r, we may suppose that $[H_1, \mathrm{K}_{1+\binom{r-1}{2}}(A)] = \{1\}$ so that $\mathrm{K}_{1+\binom{r-1}{2}}(A) = \{1\}$ (since A acts faithfully on H_1) or, equivalently, $\mathrm{K}_{1+\binom{r-1}{2}}(L) \le C$. Define $C_1 = C$ and inductively $C_{i+1} = [C_i, L]$. It remains only to show that $[H.C_r] = \{1\}$. Since $C = C_1 \le L$, we have $[H, C_1] \le [H, L] = H_1$. Suppose that for some $i < r$ we have proved that $[H, C_i] \le H_i$. Let $y \in L$, $x \in C_i$ and $z \in H$. Then $[y, z^{-1}] \in H_1$ and hence $[y, z^{-1}, x] = 1$ since $C_i \le C$ $(= \mathrm{C}_L(H_1))$. In that case, by the Hall–Witt commutator identity, $[x, y^{-1}, z][z, x^{-1}, y]^{xy^{-1}} = 1$. But

$[z, x^{-1}] \in H_i$, by induction. So $[z, x^{-1}, y]^{xy^{-1}} \in [H_i, L]^{xy^{-1}} = [H_i, L]$ since $xy^{-1} \in L$, in view of $y \in L$ and $z \in C_i \le L$, and $[H_i, L]$ is L-invariant. But $[H_i, L] \le H_{i+1}$. Hence $[x, y^{-1}, z] \in H_{i+1}$. Here $[x, y^{-1}]$ is a typical generator of C_{i+1}. Let $t = u_{j_1}^{r_1} u_{j_2}^{r_2} \dots u_{j_n}^{r_n}$ be any element of C_{i+1}, expressed in terms of such generators $u_1, u_2, \dots$. Hence each $[u_j, z] \in H_{i+1}$ and $z^{-1}tz = \prod_\alpha (u_{j_\alpha}[u_{j_\alpha}, z])^{r_\alpha}$. Each u_j belongs to C and therefore commutes with every element of $H_1 \ge H_{i+1}$. So $z^{-1}tz = t \cdot \prod_\alpha [u_j, z]^{r_\alpha}$ and $[t, z] \in H_{i+1}$. Hence finally $[H, C_{i+1}] \le H_{i+1}$, completing the proof. □

Corollary 34.5. *Let A be the stability group of a chain of subgroups $G = G_0 \ge G_1 \ge \dots \ge G_r = \{1\}$. Then A is nilpotent of class at most $\binom{r}{2}$.*

Proof. Consider G and A as subgroups of the holomorph of G. Apply Theorem 34.4 with $H = G$ and $L = A$ retaining its notation. Then $H_i = G_i$ so $\mathrm{K}_{1+\binom{r}{2}}(A)$ centralizes $H = G$, by Theorem 34.4, what is the same that $\mathrm{K}_{1+\binom{r}{2}}(A) = \{1\}$. It follows that $\mathrm{cl}(A) \le \binom{r}{2}$. □

Exercise 2. Let $X, Y \le G$ be such that $[X, Y, G] = \{1\}$. Then, for $x \in X$, $y \in Y$, the function $[x, y]$ is homomorphic in both arguments.

Hint. If $z \in Y$, then $[x, yz] = [x, z][x, y]^z = [x, y][x, z]$ since $[X, Y] \le \mathrm{Z}(G)$.

Exercise 3. Assume that $\exp(\mathrm{Z}(G)) = m$. Then $\exp(\mathrm{Z}_{r+1}(G)/\mathrm{Z}_r(G))$ divides m for all $r \in \mathbb{N}$. In particular, $\exp(\mathrm{Z}_r(G))$ divides m^r.

Solution. First we show that the exponent of $\mathrm{Z}_2(G)/\mathrm{Z}(G)$ divides m, i.e., that $[x^m, y] = 1$ for all $x \in \mathrm{Z}_2(G)$ and $y \in G$. But $[\mathrm{Z}_2(G), G, G] = \{1\}$, and so $[x^m, y] = [x, y]^m$ by Exercise 2. Since $[x, y] \in \mathrm{Z}(G)$, the result follows. Next, applying this result to the pair $\mathrm{Z}_3(G)/\mathrm{Z}(G) \le G/\mathrm{Z}(G)$, we conclude that $\exp(\mathrm{Z}_3(G)/\mathrm{Z}_2(G))$ divides m. Hence, by induction, all the remaining follows.

Theorem 34.6. *Let I be an ideal in a ring R with unit element 1 and $I^n = \{0\}$. Put $S_j = 1 + I^j$ $(j = 1, 2, \dots)$. Then each S_j is a multiplicative group; $[S_i, S_j] \le S_{i+j}$; $S = S_1$ is nilpotent of class $< n$ and $\mathrm{K}_i(S) \le S_i$.*

Exercise 4. Let G be an arbitrary group and $\phi : x \to x^3$ a homomorphism. If ϕ is an epimorphism or monomorphism, then G is abelian.

Solution. (I am indebted to M. Roitman for solution of this exercise.) By assumption, $x^3y^3 = (xy)^3$. It follows from $y^{-1}x^3y = (y^{-1}xy)^3 = y^{-3}x^3y^3$ that $y^2x^3 = x^3y^2$. Suppose that ϕ is an epimorphism; then $\{x^3 \mid x \in G\} = G$. In that case, $y^2x = xy^2$ for all $x, y \in G$ so $y^2 \in \mathrm{Z}(G)$ for all $y \in G$. Then $xy^2x^2y = x^3y^3 = (xy)^3 = xyxyxy$ so, after canceling, we get $xy = yx$, and G is abelian. Now suppose that ϕ is a monomorphism. It follows from $(xy)^3 = x^3y^3$ that $x^2y^2 = (yx)^2$. Substituting in that formula $x \to x^2$, $y \to y^2$, one obtains $x^4y^4 = (y^2x^2)^2 = ((xy)^2)^2 = (xy)^4$, or $x^3y^3 = (yx)^3$. Thus, $\phi(xy) = (xy)^3 = x^3y^3 = (yx)^3 = \phi(yx)$. Since ϕ is a monomorphism, we get $xy = yx$.

Exercise 5. Let G be an arbitrary group, $n \in \mathbb{N}$, $n > 1$, and the mapping $\phi : x \to x^n$ is a homomorphism. If ϕ is an epimorphism, then $\langle x^{n-1} \mid x \in G\rangle \leq \mathrm{Z}(G)$.

Exercise 6. Suppose that $\alpha \in \mathrm{Aut}_c(G)$, the group of central automorphisms of G. Then $H^\alpha \mathrm{Z}(G) = H\mathrm{Z}(G)$ for each $H \leq G$, and α acts trivially on G'.

Solution. The first part of the conclusion follows from the definition of central automorphisms. For the second part, take $g, h \in G$; then $g^\alpha = gy$, $h^\alpha = hz$ for some $y, z \in \mathrm{Z}(G)$. Then $[g,h]^\alpha = [g^\alpha, h^\alpha] = [gy, hz] = [g,h]$.

Proposition 34.7 ([Gla, Lemma 2.1]). *Suppose that Q is a p-subgroup of a group G. Assume that $\Omega_1(Q) \leq \mathrm{Z}(Q)$ and Q is a direct factor of every p-subgroup of G that contains and normalizes Q. Then every element of G of order p that normalizes Q also centralizes Q.*

Proof. Suppose that $x \in \mathrm{N}_G(Q) - Q$ with $x^p = 1$. Set $Q^* = \langle x, Q\rangle$. By hypothesis, $Q^* = Q \times R$ for some $R \leq Q^*$; obviously, $|R| = p$. Then $x \in \Omega_1(Q^*) = \Omega_1(Q) \times R \leq \mathrm{C}_G(Q)$ in view of $\Omega_1(Q) \leq \mathrm{Z}(Q)$. □

In what follows we use the following fact: Let α be an involutory automorphism of the cyclic 2-group $\langle x\rangle$ of order $2^n > 2$. Then $x^\alpha = x^{-1}$ or, provided $n > 2$, also $x^\alpha = x^{\pm 1 + 2^{n-1}}$. We use this in the following three paragraphs.

Let us find the order of the automorphism group of the dihedral group $G = \langle b, t \mid b^{2^n} = t^2 = 1, b^t = b^{-1}\rangle \cong \mathrm{D}_{2^{n+1}}$. Let $x, y \in G$ with $o(x) = 2^n$, $o(y) = 2$ and $y \notin \langle x\rangle$. Then $\langle x\rangle = \langle b\rangle$ and x, y satisfy the same relations as b, t. It follows that there exists $\alpha \in \mathrm{Aut}(G)$ such that $x^\alpha = b$, $y^\alpha = t$. Since $\mathrm{Aut}(G)$ acts on the set of all such pairs in a fixed-point-free manner, $|\mathrm{Aut}(G)| = \varphi(2^n)(\mathrm{c}_1(G) - 1) = 2^{n-1}2^n = 2^{2n-1}$.

Let us find the order of the automorphism group of the generalized quaternion group $G = \langle b, t \mid b^{2^n} = 1, t^2 = b^{2^{n-1}}, b^t = b^{-1}\rangle \cong \mathrm{Q}_{2^{n+1}}$, $n > 2$. Let $x, y \in G$ with $o(x) = 2^n$, $o(y) = 4$ and $y \notin \langle x\rangle$. Then $\langle x\rangle = \langle b\rangle$ and x, y satisfy the same relations as b, t. It follows that there exists $\alpha \in \mathrm{Aut}(G)$ such that $x^\alpha = b$, $y^\alpha = t$. Therefore, as in the previous paragraph, we have $|\mathrm{Aut}(G)| = \varphi(2^n) \cdot 2(\mathrm{c}_2(G) - 1) = 2^{n-1}2^n = 2^{2n-1}$. If $n = 2$, then x can be chosen in six ways, and, after that choice, y can be chosen in $6 - 2 = 4$ ways. It follows that $|\mathrm{Aut}(G)| = 6 \cdot 4 = 24$; then $\mathrm{Aut}(G) \cong \mathrm{S}_4$.

Let us find the order of the automorphism group of the semidihedral group $G = \langle b, t \mid b^{2^n} = t^2 = 1, b^t = b^{-1+2^{n-1}}\rangle \cong \mathrm{SD}_{2^{n+1}}$. Let $x, y \in G$ with $o(x) = 2^n$, $o(y) = 2$ and $y \notin \langle x\rangle$. Then $\langle x\rangle = \langle b\rangle$ and x, y satisfy the same relations as b, t. It follows that there exists $\alpha \in \mathrm{Aut}(G)$ such that $x^\alpha = b$, $y^\alpha = t$. so $|\mathrm{Aut}(G)| = \varphi(2^n)(\mathrm{c}_1(G) - 1) = 2^{n-1}2^{n-1} = 2^{2n-2}$.

Let G be abelian of type $(2^n, 2)$, $n > 1$. Then it is easily seen that $|\mathrm{Aut}(G)| = (||G| - \Omega_{n-1}(G)|)(\mathrm{c}_1(G) - 1) = 2^n \cdot 2 = 2^{n+1} = |G|$. The same is true for $G \cong \mathrm{M}_{2^{n+1}}$.

Theorem 34.8 (Janko, private communication). *(a) The automorphism group* $\mathrm{Aut}(G)$ *of the dihedral group* $G = \langle b, t \mid b^{2^n} = t^2 = 1, b^t = b^{-1}\rangle \cong D_{2^{n+1}}$ *is generated by the inner automorphism group* $\mathrm{Inn}(G)$ *which is isomorphic to* D_{2^n} *and two outer automorphisms* α *and* β*, where* α *is of order* 2 *and is induced with* $t^\alpha = tb$, $b^\alpha = b^{-1}$ *and* β *is of order* 2^{n-2} *and is induced with* $t^\beta = t$, $b^\beta = b^5$*. We have* $[\alpha, \beta] = i_{b^2}$ *which is the inner automorphism of* G *induced by conjugation with the element* b^2*. Hence the outer automorphism group* $\mathrm{Out}(G) = \mathrm{Aut}(G)/\mathrm{Inn}(G)$ *is abelian of type* $(2, 2^{n-2})$ *so that* $|\mathrm{Aut}(G)| = 2^{2n-1}$*. Furthermore,* $C_G(\beta^{2^j}) \cong D_{2^{j+3}}$ $(0 \le j \le n-2)$*.*

(b) The automorphism group $\mathrm{Aut}(G)$ *of the generalized quaternion group* $G = \langle b, t \mid b^{2^n} = 1, t^2 = b^{2^{n-1}}, b^t = b^{-1}\rangle \cong Q_{2^{n+1}}$, $n \ge 3$*, is generated by the inner automorphism group* $\mathrm{Inn}(G)$ *(which is isomorphic to* D_{2^n}*) and outer automorphisms* α *and* β*, where* α *is of order* 2 *and is induced with* $t^\alpha = tb$, $b^\alpha = b^{-1}$ *and* β *is of order* 2^{n-2} *and is induced with* $t^\beta = t$ *and* $b^\beta = b^5$*. We have* $[\alpha, \beta] = i_{b^2}$ *which is the inner automorphism of* G *induced by conjugation with the element* b^2*. Hence the outer automorphism group* $\mathrm{Out}(G) = \mathrm{Aut}(G)/\mathrm{Inn}(G)$ *is abelian of type* $(2, 2^{n-2})$ *so that* $|\mathrm{Aut}(G)| = 2^{2n-1}$*.*

(c) The automorphism group $\mathrm{Aut}(G)$ *of the semidihedral group* $G = \langle b, t \mid b^{2^n} = 1 = t^2, b^t = b^{-1+2^{n-1}}\rangle \cong SD_{2^{n+1}}$, $n \ge 3$*, is a semidirect product of* $\langle\beta\rangle$ *of order* 2^{n-2} *with* $\mathrm{Inn}(G) \cong D_{2^n}$*, where* β *is an outer automorphism of* G *induced with* $t^\beta = t$, $b^\beta = b^5$*. Hence,* $\mathrm{Out}(G) = \mathrm{Aut}(G)/\mathrm{Inn}(G)$ *is cyclic of order* 2^{n-2} *so that* $|\mathrm{Aut}(G)| = 2^{2n-2}$*.*

(d) The automorphism group $\mathrm{Aut}(G)$ *of the abelian group* $G = \langle b, t \mid b^{2^n} = t^2 = [b, t] = 1\rangle$ *of type* $(2^n, 2)$, $n \ge 2$*, is the central product* $D * B$ *of the dihedral group* $D = \langle\alpha, \beta\rangle \cong D_8$ *with the abelian group* B *of the type* $(2^{n-2}, 2)$ *and* $D \cap B = Z(D)$ *so that* $|\mathrm{Aut}(G)| = 2^{n+1} (= |G|)$*. Here* α *is an involutory automorphism of* G *induced by* $t^\alpha = tz$, $b^\alpha = b$ *with* $z = b^{2^{n-1}}$*;* β *is an involutory automorphism of* G *induced by* $t^\beta = t$, $b^\beta = bt$ *and* B *is the set of all automorphisms of* G *which normalize* $\langle b\rangle$ *and* $\langle t\rangle$ *so that* $B \cong \mathrm{Aut}(\langle b\rangle)$*.*

Proof. (a) The inner automorphisms of G induced with elements contained in $\langle b\rangle$, partition 2^n involutions of $G - Z(G)$ in orbits $O_1 = t\langle b^2\rangle$ and $O_2 = tb\langle b^2\rangle$ of length 2^{n-1} each. The inner automorphism i_t, induced by t, fixes t and inverts $\langle b\rangle$. Consider the automorphism $\alpha \in \mathrm{Aut}(G) - \mathrm{Inn}(G)$ of order 2 given by $t^\alpha = tb$, $b^\alpha = b^{-1}$ so that α fuses O_1 and O_2 since $(tb^2)^\alpha = tbb^{-2} \in tb\langle b^2\rangle = O_2$ and $tb^2 \in O_1$. Let $B \ge \mathrm{Inn}(G)$ be the subgroup of $\mathrm{Aut}(G)$ that fixes O_1 (and so O_2) so that $|\mathrm{Aut}(G) : B| = 2$ and $\mathrm{Aut}(G) = \langle\alpha\rangle \cdot B$, a semidirect product. Let $\gamma \in B - \mathrm{Inn}(G)$. Then multiplying γ with an inner automorphism i induced with an element contained in $\langle b\rangle$, one may assume that $\beta' = \gamma i$ fixes the involution t since the group $\langle b\rangle$ acts transitively on O_1. But β' must act faithfully on the (characteristic) cyclic subgroup $\langle b\rangle$. Multiplying β' with i_t if necessary, one may assume that $\beta_0 = \gamma i i_t^\epsilon$ $(\epsilon = 0, 1)$

fixes t and centralizes a subgroup of order ≥ 4 in $\langle b\rangle$. Then β_0 acts on $\langle b\rangle$ as a power of the automorphism β. Hence, if we consider the outer automorphism β of order 2^{n-2} induced with $t^\beta = t$, $b^\beta = b^5$, we see that B is a semidirect product of $\langle\beta\rangle$ and $\mathrm{Inn}(G)$ and so $\mathrm{Aut}(G) = \langle\alpha,\beta\rangle\mathrm{Inn}(G)$ (as we will see in the last sentence, this product is not direct). Finally, we compute that $[\alpha,\beta] = i_{b^2}$ (in particular, $\alpha\beta \neq \beta\gamma$), and so $\mathrm{Aut}(G)/\mathrm{Inn}(G)$ is abelian of type $(2, 2^{n-2})$. Then it is easy to see that the fixed subgroup of β^{2^j} in G is isomorphic to $\mathrm{D}_{2^{j+3}}$ $(0 \leq j \leq n-2)$.

(b) The proof is identical with the proof of (a). The only difference is that the elements t and tb are of order 4 and not involutions.

(c) The inner automorphisms of G induced by elements in $\langle b\rangle$ fuse all 2^{n-1} involutions from $G - \langle b\rangle$ in a single conjugacy class with the representative t. Other 2^{n-1} elements in $G-\langle b\rangle$ are of order 4 and they also form a single conjugacy class in G. Let $\gamma \in \mathrm{Aut}(G)-\mathrm{Inn}(G)$. Then multiplying γ with an inner automorphism i induced with an element in $\langle b\rangle$, we may assume that $\beta' = \gamma i$ fixes the involution t since $\langle b\rangle$ fuses all involutions in $G - \langle b\rangle$ in a single conjugacy class. But β' must act faithfully on the characteristic subgroup $\langle b\rangle$ of G. Let i_t be the inner automorphism of G induced with t. Then i_t inverts the cyclic subgroup of order 4 in $\langle b\rangle$. Multiplying β' with i_t if necessary, we get that $\beta = \gamma i i_t^\epsilon$ $(\epsilon = 0, 1)$ fixes t and centralizes a subgroup of order ≥ 4 in $\langle b\rangle$. Then β acts on $\langle b\rangle$ as a power of the automorphism β of G of order 2^{n-2} induced by $t^\beta = t$, $b^\beta = b^5$. This gives that $\mathrm{Aut}(G)$ is a semidirect product of $\langle\beta\rangle$ and $\mathrm{Inn}(G) \cong \mathrm{D}_{2^n}$ and so $\mathrm{Aut}(G)/\mathrm{Inn}(G)$ is cyclic of order 2^{n-2}.

(d) Let α and β be defined as in the statement of this part, i.e., $t^\alpha = tz$, $b^\alpha = b$ and $t^\beta = t$, $b^\beta = bt$. Then we have $t^{\alpha\beta} = tz$, $b^{\alpha\beta} = bt$ and so $t^{(\alpha\beta)^2} = t$, $b^{(\alpha\beta)^2} = bz$ and therefore $o(\alpha\beta) = 4$. Thus $\langle\alpha,\beta\rangle = D \cong D_8$. Let $B(\cong \mathrm{Aut}(\langle b\rangle))$ be the set of all automorphisms of G normalizing $\langle b\rangle$ and $\langle t\rangle$. If $\gamma \in B$, then we see that γ commutes with α and β. Also $(\alpha\beta)^2 \in \mathrm{Z}(D)$ and $(\alpha\beta)^2 \in B$ and so we have the central product $D * B$ with $D \cap B = \mathrm{Z}(D)$. Conversely, if $\delta \in \mathrm{Aut}(G)$, then multiplying δ with $\eta \in D$ (if necessary), we may assume that $t^{\delta\eta} = t$ and $\langle b\rangle^{\delta\eta} = \langle b\rangle$ and so $\delta\eta \in B$. Hence $D * B = \mathrm{Aut}(G)$. □

We see that the group $\mathrm{Aut}(G)$, where $G \cong \mathrm{D}_{2^{n+1}}$, acts transitively on the set of 2^n noncentral involutions of G and induces the group $\mathrm{Aut}(\langle b\rangle)$ on $\langle b\rangle$ so again $|\mathrm{Aut}(G)| = 2^n \cdot 2^{n-1} = 2^{2n-1}$. If $G \in \{\mathrm{D}_{2^{n+1}}, \mathrm{Q}_{2^{n+1}}\}$ and $n > 2$, then $\mathrm{Aut}(G)$ is an extension of the group D_{2^n} by the abelian group of type $(2^{n-2}, 2)$. In this case, nonabelian maximal subgroups of G are not characteristic. Comparing the calculations of $\mathrm{Aut}(\mathrm{D}_{2^{n+1}})$ and $\mathrm{Aut}(\mathrm{Q}_{2^{n+1}})$ for $n \geq 3$, it follows at once that $\mathrm{Aut}(\mathrm{D}_{2^{n+1}}) \cong \mathrm{Aut}(\mathrm{Q}_{2^{n+1}})$. If $G \cong \mathrm{SD}_{2^{n+1}}$, then G is a split extension of the group D_{2^n} by the cyclic group of order 2^{n-2}.

Theorem 34.9. *Let $\phi \in \mathrm{Aut}(G)$ and let N be a ϕ-invariant normal subgroup of G. Then $|\mathrm{C}_{G/N}(\phi)| \leq |\mathrm{C}_G(\phi)|$.*

Proof. Let $W = \langle\phi\rangle \cdot G$ be the natural semidirect product; then $N \lhd W$. It suffices to show that if $x \in W$, then, setting $\bar{W} = W/N$, we get $|\mathrm{C}_{\bar{W}}(\bar{x})| \le |\mathrm{C}_W(x)|$. By the Second Orthogonality Relation,

$$|\mathrm{C}_W(x)| = \sum_{\chi\in\mathrm{Irr}(W)} |\chi(x)|^2 \ge \sum_{\chi\in\mathrm{Irr}(W/N)} |\chi(x)|^2 = \sum_{\chi\in\mathrm{Irr}(\bar{W})} |\chi(\bar{x})|^2 = |\mathrm{C}_{\bar{W}}(\bar{x})|. \quad \square$$

Theorem 34.10. *Let N be a normal ϕ-invariant Hall subgroup of a group G, where $\phi \in \mathrm{Aut}(G)$. Then $\mathrm{C}_{G/N}(\phi) = \mathrm{C}_G(\phi)N/N \cong \mathrm{C}_G(\phi)/\mathrm{C}_N(\phi)$.*

Exercise 7. Let $\phi \in \mathrm{Aut}(G)$ with $(o(\phi), |G|) = 1$. If ϕ centralizes all factors of some normal series of G, then $\phi = \mathrm{id}_G$. (*Hint.* One may assume that $o(\phi) = p$. Assume that G is a counterexample of minimal order. Then there exists a minimal nonnilpotent subgroup $H \le W = \langle\phi\rangle \cdot G$ such that $\phi \in H$).

Theorem 34.11. *Let $\phi \in \mathrm{Aut}(G)$ with $(o(\phi), |G|) = 1$. Then (a) $G = \mathrm{C}_G(\phi)[G,\phi]$ and (b) $[G,\phi,\phi] = [G,\phi]$.*

Proof. (a) Since $g^\phi = g[g,\phi]$ for any $g \in G$, ϕ acts trivially on $G/[G,\phi]$. Set $W = \langle\phi\rangle \cdot G$. In that case, $K = \langle\phi\rangle \cdot [G,\phi] \trianglelefteq W$. Then, by Schur–Zassenhaus and Frattini, we get $W = \mathrm{N}_W(\langle\phi\rangle)K = \mathrm{N}_W(\langle\phi\rangle)[G,\phi]$. By the modular law, $G = [G,\phi](\mathrm{N}_W(\langle\phi\rangle) \cap G) = [G,\phi]\mathrm{N}_G(\langle\phi\rangle)$. Since $\mathrm{N}_G(\langle\phi\rangle) = \mathrm{C}_G(\phi)$, we are done.

(b) Applying $[ab,c] = [a,c]^b[b,c]$, we get, by (a), since $[\mathrm{C}_G(\phi),\phi] = \{1\}$,

$$[G,\phi] = [\mathrm{C}_G(\phi)[G,\phi],\phi] = [\mathrm{C}_G(\phi),\phi]^{[G,\phi]}[G,\phi,\phi] = [G,\phi,\phi]. \quad \square$$

Exercise 8 (see Proposition 1.8). Suppose that $\phi \in \mathrm{Aut}(G)$, where $|G| = p^m > p^p$ and $o(\phi) = p$. If ϕ has exactly p fixed points, then G is either absolutely regular or of maximal class. (*Hint.* Use Theorem 9.6.)

Theorem 34.12 (Curran). $|\mathrm{Aut}(\mathrm{M}_{p^m})| = (p-1)p^m$.

Exercise 9. Let G be an arbitrary group. Then $\mathrm{cl}(\mathrm{C}_G(\mathrm{K}_n(G))) \le n$. (*Hint.* Prove that $[c_1,\dots,c_n,c_{n+1}] = \{1\}$ for $c_1,\dots,c_{n+1} \in \mathrm{C}_G(\mathrm{K}_n(G))$.)

It follows from $\mathrm{K}_n(G) \le \mathrm{C}_G(\mathrm{Z}_n(G))$ that $G/\mathrm{C}_G(\mathrm{Z}_n(G))$, as an epimorphic image of $G/\mathrm{K}_n(G)$, has class $\le n-1$.

Exercise 10. If $x \in G$, then $|\{[x,g] \mid g \in G\}| = |G : \mathrm{C}_G(x)|$. (*Hint.* $\{x[x,g] \mid g \in G\}$ is the G-class of x.)

Exercise 11. Let $G = \langle x_1,\dots,x_d\rangle$ and $n = \max\{|G : \mathrm{C}_G(x)| \mid x \in G\}$. Prove that $|G : \mathrm{Z}(G)| \le n^d$. Since $n \le |G'|$, we have $|G : \mathrm{Z}(G)| \le |G'|^d$.

Exercise 12. The subgroup $N = \{x \in G \mid H^x = H \text{ for all } H \le G\}$ is said to be the *norm* of a group G. Prove that $\mathrm{Z}_2(G) \ge N$.

Exercise 13. Let G be a nilpotent group of class c and $H < G$. Define $H_0 = H$ and inductively $H_{i+1} = \mathrm{N}_G(H_i)$. Then $\mathrm{Z}_i(G) \le H_i$ for all i so $H_r = G$ for some $r \le c$.

Example (G. A. Miller). Let $A = \langle x\rangle \times \langle u\rangle \times \langle z\rangle$ be the abelian group of type $(8, 2, 2)$, where $o(x) = 8$, $o(u) = o(z) = 2$. Set $G = \langle y, A\rangle$, where $o(y) = 2$ and $x^y = x^5$, $z^y = z$, $u^y = uz$; then $(x^2)^y = x^2$. We want to prove that $\mathrm{Aut}(G) \cong \mathrm{E}_{2^7}$. We have

$$|G| = 2^6,\ \ Z = \mathrm{Z}(G) = \langle x^2\rangle \times \langle z\rangle \cong \mathrm{C}_4 \times \mathrm{C}_2,\ \ G/Z = \langle yZ\rangle \times \langle xZ\rangle \times \langle uZ\rangle \cong \mathrm{E}_8,$$

so $\mathrm{cl}(G) = 2$. Since $z = u^{-1}u^y \in G' \le \Phi(G)$, we get $G = \langle y, x, u\rangle$ so $\Phi(G) = Z$. Since $|G : Z| = 8$, A is a unique abelian subgroup of index 2 in G so A is characteristic in G. Take $\alpha \in \mathrm{Aut}(G)$; then $A^\alpha = A$. Let us check that α induces the identity on G/Z; then α commutes with all inner automorphisms of G, i.e., α is central. We have $G/Z = \{xZ, xyZ, xuZ, xyuZ, yZ, uZ, yuZ, Z\}$, $A = Z \cup xZ \cup uZ \cup xuZ$, $G - A = yZ \cup xyZ \cup xyuZ \cup yuZ$, so α fixes the sets $\{xZ, uZ, xuZ\}$ and $\{yZ, xyZ, xyuZ, yuZ\}$. Cosets xZ and xuZ contain elements of order 8 only, coset uZ has no elements of order 8. Hence, $(uZ)^\alpha = uZ$. Cosets xyZ and $xyuZ$ contain elements of order 8 only. Coset yZ contains four involutions and four elements of order 4, and coset yuZ contains elements of order 4 only. It follows that $(yZ)^\alpha = yZ$, $(yuZ)^\alpha = yuZ$. Let $z_1 \in Z$. Then

$$(xz_1)^y = x^5z_1 = (xz_1)^5,\ (xuz_1)^y = x^5uzz_1 \ne x^5uz_1 = (xuz_1)^5.$$

It follows that $(xZ)^\alpha \ne xuZ$ and $(xuZ)^\alpha \ne xZ$, and we obtain $(xZ)^\alpha = xZ$, $(xuZ)^\alpha = xuZ$. Next, $(xyZ)^\alpha = (xZ)^\alpha(yZ)^\alpha = xZ \cdot yZ = xyZ$, and so $(xyuZ)^\alpha = xyuZ$. Thus, every automorphism of G is central. Next, all automorphisms of G fix elements of $G' = \langle x^4, z\rangle \cong \mathrm{E}_4$. Every automorphism of G acts in the following way: $x \to xz_1$, where $z_1 \in Z$, $y \to yz_2$, where $z_2 \in \Omega_1(Z) = G'$, $u \to uz_3$, where $z_3 \in \Omega_1(Z) = G'$ since $o(y^\alpha) = o(y) = 2$, $o(u^\alpha) = o(u) = 2$. It follows from the above that $|\mathrm{Aut}(G)| = 8 \cdot 4 \cdot 4 = 2^7$. It remains to check that $\mathrm{Aut}(G)$ is of exponent 2. Let $\alpha \in \mathrm{Aut}(G)^\#$. Clearly, $y^{\alpha^2} = y$ and $u^{\alpha^2} = u$. We have $\alpha : x \to xc$ or $x \to x^3c$ for some $c \in G'$ since G/G' is abelian of type $(4, 2, 2)$. If $x^\alpha = xc$, then $x^{\alpha^2} = (xc)^\alpha = xcc = x$. If $\alpha(x) = x^3c$, then $x^{\alpha^2} = (x^3c)^\alpha = (x^3c)^3c = x$. Thus, $\mathrm{Aut}(G) \cong \mathrm{E}_{2^7}$.

If we set, in the above example, $o(x) = 2^n$, $n > 3$, instead of $o(x) = 8$, we also obtain a group G of order 2^{n+3} with $\mathrm{Aut}(G) \cong \mathrm{C}_{2^{n-2}} \times \mathrm{E}_{2^6}$ (R. R. Struik).

§35

Maximal abelian subgroups of p-groups

In this section we classify the 2-groups in which every two distinct maximal abelian subgroups have cyclic intersection.

Theorem 35.1 (Janko). *Let G be a nonabelian 2-group in which any two distinct maximal abelian subgroups have cyclic intersection. Then $Z(G)$ is cyclic, each abelian subgroup of G is of rank at most 2, the intersection of any two distinct maximal abelian subgroups is equal $Z(G)$, and G has (at least) one abelian subgroup of index 2. Moreover, G is isomorphic to one of the following groups:*

(a) Group of maximal class.

(b) M_{2^n}.

(c) $G = D * C$ (central product), where $D \cong D_{2^n}$, $C \cong C_{2^m}$, $m \geq 2$, is cyclic of order 2^m and $D \cap C = Z(D)$.

(d) The group

$$G = \langle x, t \mid (xt)^2 = a,\ a^{2^m} = t^2 = 1,\ m \geq 2,\ x^2 = ab,\ b^{2^{n-1}} = 1,\ n \geq 3,$$
$$b^t = b^{-1},\ [a, x] = [a, t] = 1,\ t^x = tb,\ a^{2^{m-1}} = b^{2^{n-2}} \rangle,$$

where $|G| = 2^{m+n}$, $m \geq 2$, $n \geq 3$, $Z(G) = \langle a \rangle \cong C_{2^m}$, $G' = \langle b \rangle \cong C_{2^{n-1}}$, and $M = \langle x, a \rangle$ is a unique abelian maximal subgroup of G. We have $C_G(t) = \langle t \rangle \times \langle a \rangle \cong C_2 \times C_{2^m}$ and $\langle b, t \rangle \cong D_{2^n}$.

(e) The group

$$G = \langle g, h \mid g^{2^n} = h^{2^m} = 1,\ m \geq 3,\ n \geq 3,\ g^{2^{n-1}} = h^{2^{m-1}},\ h^g = h^{-1} \rangle,$$

where G is metacyclic, $|G| = 2^{m+n-1}$ since $\langle g \rangle \cap \langle h \rangle = \langle g^{2^{n-1}} \rangle \cong C_2$. Also, $Z(G) = \langle g^2 \rangle \cong C_{2^{n-1}}$, $G' = \langle h^2 \rangle \cong C_{2^{m-1}}$ and $M = \langle h, g^2 \rangle$ is a unique abelian maximal subgroup of G.

The more general problem to determine the structure of a nonabelian p-group G such that $A \cap B = Z(G)$ for any two distinct maximal abelian subgroups A and B is very difficult. First we show that a p-group G has this property if and only if $C_G(x)$ is abelian for each $x \in G - Z(G)$ (Theorem 35.2). Then we show that such a 2-group

G has either an abelian subgroup of index 2 or G is of class 2 and G' is elementary abelian (Theorem 35.3). In Corollary 35.5 we get a new result for an arbitrary 2-group.

We also classify 2-groups G such that $A/Z(G)$ is cyclic for each maximal abelian subgroup A of G (a problem of Heineken–Mann). It is surprising that such groups have the property that $C_G(x)$ is abelian for each element $x \in G - Z(G)$. Then we may use our Theorems 35.2 and 35.3 to classify such groups (Theorem 35.4). In this classification we distinguish the cases, where G has an abelian subgroup of index 2 and the case where $|G : A| > 2$ for each maximal abelian subgroup A of G.

Proof of Theorem 35.1. Let G be a nonabelian 2-group in which any two distinct maximal abelian subgroups have cyclic intersection. Since $Z(G)$ is contained in each maximal abelian subgroup of G, it follows that $Z(G)$ is cyclic.

Suppose that G possesses a subgroup $E \cong E_8$. Let A be a maximal abelian subgroup containing E and set $F = \Omega_1(A)$ so that $E \leq F$. Let $B \leq G$ be such that $A < B$ and $|B : A| = 2$ and let $x \in B - A$. Then $x^2 \in A$ and therefore x induces on F an automorphism of order 2. It follows that $|C_F(x)| \geq 4$ and the abelian subgroup $C_F(x)\langle x\rangle$ is contained in a maximal abelian subgroup C which is distinct from A since $x \notin A$. But $A \cap C \geq C_F(x)$ and so $A \cap C$ is noncyclic, a contradiction. We have proved that each abelian subgroup of G is of rank ≤ 2.

We may assume that G is not of maximal class (case (a) of Theorem 35.1) and so there if $E_4 \cong U \lhd G$. Set $M = C_G(U)$ so that $|G : M| = 2$ since $Z(G)$ is cyclic. Let A be a maximal abelian subgroup of G which contains U so that $A \leq M$. Suppose that $A \neq M$ and let $y \in M - A$ be such that $y^2 \in A$. Let B be a maximal abelian subgroup of G containing the abelian subgroup $U\langle y\rangle$. Then $B \neq A$ (since $y \notin A$) and $A \cap B \geq U$ is noncyclic, a contradiction. Thus, $M = A$. We have proved that whenever U is a normal four-subgroup of G, then $M = C_G(U)$ is an abelian maximal subgroup of G.

If x is any element in $G - M$, then $C_M(x) = Z(G)$ and $Z(G)\langle x\rangle$ is a maximal abelian subgroup of G. Thus, the intersection of any two distinct maximal abelian subgroups of G is equal to $Z(G)$ and this is also true for 2-groups of maximal class.

Suppose that G has two distinct normal four-subgroups. Then, by Theorem 50.2, $G = D * C$ with $D \cong D_8$, $D \cap C = Z(D)$ and C is either cyclic of order ≥ 4 or of maximal class $\not\cong D_8$. Let U be a four-subgroup in D; then $U \lhd G$. By the above, $C_G(U)$ is abelian and so C must be cyclic. We have obtained a group stated in part (c). In the sequel we assume that G has a unique normal four-subgroup U and set $M = C_G(U)$ so that M is an abelian maximal subgroup of rank 2 with $\Omega_1(M) = U$.

(i) First assume $\Omega_2(G) \not\leq M$. Then there is an element $y \in G - M$ of order ≤ 4 so that $y^2 \in U$ (recall that $U = \Omega_1(M)$). We have $U\langle y\rangle \cong D_8$ since y does not centralize U, and so there is an involution $t \in G - M$. Since t does not centralize U and M is abelian of rank two, we get $C_G(t) = \langle t\rangle \times C_M(t)$, where $C_M(t)$ is cyclic of order 2^m, $m \geq 2$. Indeed, if $m = 1$, then G is of maximal class. Also, we have $t \notin \Phi(G)$, G has no elementary abelian subgroups of order 8 and $G \not\cong M_{2^s}$, $s \geq 4$

(since M_{2^s} has only three involutions). We are now in a position to use Theorem 48.1. It follows that G has a subgroup S of index ≤ 2, where $S = AL$, $L \lhd G$,

$$L = \langle b, t \mid b^{2^{n-1}} = t^2 = 1,\ b^t = b^{-1}\rangle \cong D_{2^n}, \quad n \geq 3,$$
$$A = \langle a \rangle \cong C_{2^m}, \quad m \geq 2,$$
$$A \cap L = Z(L) = \langle z \rangle, \quad [a,t] = 1, \quad C_G(t) = \langle t \rangle \times \langle a \rangle,$$
$$\Omega_1(G) = \Omega_1(S) = \Omega_2(A) * L, \quad \Omega_2(A) \cap L = Z(L)$$

and, if $|G : S| = 2$, then there is an element $x \in G - S$ such that $t^x = tb$.

Since $\langle b \rangle$ is a unique cyclic subgroup of index 2 in L, $\langle b \rangle$ is normal in G. Set $B = \Omega_2(A) * L = \Omega_1(G)$, $\Omega_2(A) = \langle l \rangle$, $l^2 = z$, and $\langle v \rangle = \Omega_2(\langle b \rangle)$ so that $\langle v \rangle$ and $\langle l \rangle = Z(B)$ are normal in G. Hence $\langle l, v \rangle \cong C_4 \times C_2$ is normal in G. Set $u = lv$ so that $U = \langle z, u \rangle = \Omega_1(\langle l, v \rangle) \cong E_4$ is a unique normal four-subgroup in G. We know that $M = C_G(U)$ is abelian and $|G : M| = 2$. Note that b centralizes U and $u^t = (lv)^t = lv^{-1} = lvz = uz$. If $v^a = v^{-1} = vz$, then $A = \langle a \rangle > \Omega_2(A) = \langle l \rangle$ and we replace a with $a' = at$. In that case $o(a') = o(a)$, $\Omega_2(\langle a' \rangle) = \langle l \rangle$ and

$$u^{a'} = (lv)^{at} = (avz)^t = av^{-1}z = lv = u,$$

so that $\langle a' \rangle$ centralizes U and $S = \langle a' \rangle L$. Writing again a instead of a', we may assume from the start that $A = \langle a \rangle$ centralizes U. Hence $C_S(U) = \langle a, b \rangle$ is of index 2 in S and therefore $M = C_G(U)$ covers G/S and $G = M\langle t \rangle$. But M is abelian and t centralizes $\langle a \rangle$ and so $\langle a \rangle \leq Z(G)$. On the other hand, $C_G(t) = \langle t \rangle \times \langle a \rangle$ and $C_M(t) = \langle a \rangle$ so that $A = \langle a \rangle = Z(G)$.

If $G = S$, then $G = L * A$, where $L \cong D_{2^n}$, $n \geq 3$, $A \cong C_{2^m}$, $m \geq 2$, and $L \cap A = Z(L)$. We have obtained groups stated in part (c) of Theorem 35.1. In what follows we assume that $|G : S| = 2$ and we know that in that case there is an element $x \in G - S$ such that $t^x = tb$. We may assume that $x \in M - S$. Indeed, if $x = tx'$ with $x' \in M - S$, then $tb = t^x = t^{tx'} = t^{x'}$.

Since M is abelian and $C_M(t) = \langle a \rangle = Z(G)$, it follows that $C_M(xt) = \langle a \rangle$ and so $(xt)^2 \in \langle a \rangle$. Set $(xt)^2 = a'$ and assume that $\langle a' \rangle \neq \langle a \rangle$. This implies that there is an element $a'' \in \langle a \rangle - \langle a' \rangle$ such that $(a'')^2 = (a')^{-1}$. We get $(xt \cdot a'')^2 = (xt)^2(a'')^2 = 1$ and so $x(ta'')$ (with $ta'' \in S$) is an involution in $G - S$, contrary to $\Omega_1(G) = \Omega_1(S)$. It follows that $\langle a' \rangle = \langle a \rangle$ and so replacing a with a' (and writing again a instead of a'), we may assume from the start that $(xt)^2 = a$. From the last relation and $t^x = tb$ we get

$$a = (xt)^2 = xt \cdot xt = x^2(x^{-1}tx)t = x^2tbt = x^2b^{-1},$$

and so $x^2 = ab$. The structure of G is uniquely determined and we have obtained the group stated in part (d) of Theorem 35.1.

(ii) Finally, assume that $\Omega_2(G) \leq M$. Note that M is abelian of rank 2 and so M is metacyclic. Hence G is also metacyclic. If G has a cyclic subgroup of index 2, then

G is either of maximal class or $G \cong \mathrm{M}_{2^n}$, $n \geq 4$, and these are the groups stated in parts (a) and (b) of Theorem 35.1. In what follows we assume that G has no cyclic subgroups of index 2. We have $U = \Omega_1(M) = \Omega_1(G) \cong \mathrm{E}_4$, where $M = \mathrm{C}_G(U)$ is an abelian maximal subgroup of G.

Let H be a normal cyclic subgroup with cyclic G/H so that $|H| \geq 4$ and $|G/H| \geq 4$. We have $U \cap H = \langle z \rangle \cong \mathrm{C}_2$ and $z \in \mathrm{Z}(G)$ so that if $u \in U - \langle z \rangle = U - H$, then $M = \mathrm{C}_G(u)$, where $|G : M| = 2$ and M is abelian. Suppose that u does not centralize H. Then $|H : (H \cap M)| = 2$ and therefore M covers G/H. Let $m \in M$ be such that $\langle m \rangle$ covers $M/M \cap H$ and note that $\mathrm{C}_G(H) = H$ since u does not centralize H. Let $h \in H - M$ so that $H = \langle h \rangle$ and $h^m = hz$. Then $h^{m^2} = (hz)^m = hz \cdot z = h$. This is a contradiction since $|G/H| \geq 4$ and so $m^2 \notin H$.

We have proved that u centralizes H and so $M > H$. Let $g \in G - M$ so that $\langle g \rangle$ covers G/H, $g^2 \in M$ and g^2 centralizes H and therefore g induces on $H = \langle h \rangle$ an involutory automorphism. Also, $u^g = uz$ since $\mathrm{Z}(G)$ is cyclic. If $h^g = hz$, then $G' = \langle z \rangle$ and G is minimal nonabelian. In that case G is splitting metacyclic, i.e., there is $g' \in G - M$ such that $\langle g' \rangle$ covers G/H and $\langle g' \rangle \cap H = \{1\}$. It follows that $\Omega_1(\langle g' \rangle) \leq \mathrm{Z}(G)$ and so $\mathrm{Z}(G) \geq \langle z, \Omega_1(\langle g' \rangle) \rangle \cong \mathrm{E}_4$, a contradiction. We have proved that $h^g = h^{-1} z^{\epsilon}$, $\epsilon = 0, 1$ and $|H| \geq 8$. (Indeed, if $|H| = 4$, then $h^g = h^{-1} = hz$ and we have again $G' = \langle z \rangle$, as above.) In particular, $\mathrm{C}_H(g) = \langle z \rangle$ and so $\langle g \rangle \cap H \leq \langle z \rangle$. However, if $\langle g \rangle \cap H = \{1\}$, then $\mathrm{C}_G(\Omega_1(\langle g \rangle)) \geq \langle M, g \rangle = G$ and so $\mathrm{E}_4 \cong \langle z, \Omega_1(\langle g \rangle) \rangle \leq \mathrm{Z}(G)$, a contradiction.

We have proved that $\langle g \rangle \cap H = \langle z \rangle$ and so $o(g) \geq 8$ and $\mathrm{Z}(G) = \langle g^2 \rangle$. If $h^g = h^{-1} z$, then we replace h with $h' = hu$, where $[h, u] = 1$ and so $o(h') = o(h)$ and

$$(h')^g = (hu)^g = h^{-1} z \cdot uz = h^{-1} u = (hu)^{-1} = (h')^{-1}$$

and $\langle h', g \rangle = \langle hu, g \rangle = \langle h, g \rangle = G$ since $u \in U \leq \Phi(G)$. (Indeed, $\Phi(G) \leq M$ is abelian and so if $\Phi(G) = \mho_1(G)$ were cyclic, then $|G : \Phi(G)| = 4$ implies that G would have a cyclic subgroup of index 2.) Writing h instead of h', we see that we may assume from the start that $h^g = h^{-1}$. We have obtained the group stated in part (e). □

Theorem 35.2. *Let G be a nonabelian p-group. Then $A \cap B = \mathrm{Z}(G)$ for any two distinct maximal abelian subgroups A, B if and only if $\mathrm{C}_G(x)$ is abelian for each $x \in G - \mathrm{Z}(G)$.*

Proof. Let $x \in G - \mathrm{Z}(G)$ and suppose that $\mathrm{C}_G(x)$ is nonabelian. Let A be a maximal abelian subgroup of $\mathrm{C}_G(x)$ so that $A \neq \mathrm{C}_G(x)$ and $A \geq \mathrm{Z}(G)\langle x \rangle > \mathrm{Z}(G)$. Let $b \in \mathrm{C}_G(x) - A$ and let B be a maximal abelian subgroup of $\mathrm{C}_G(x)$ containing $\langle b \rangle$ so that $A \neq B$ and B also contains the abelian subgroup $\mathrm{Z}(G)\langle x \rangle$. Obviously, A and B are also maximal abelian subgroups of G but $A \cap B \geq \mathrm{Z}(G)\langle x \rangle > \mathrm{Z}(G)$.

Conversely, let $C \neq D$ be maximal abelian subgroups of G such that $C \cap D > \mathrm{Z}(G)$. Let $y \in (C \cap D) - \mathrm{Z}(G)$ so that $\mathrm{C}_G(y) \geq \langle C, D \rangle$, where $\langle C, D \rangle$ is nonabelian. □

Theorem 35.3. *Let G be a nonabelian 2-group such that $A \cap B = Z(G)$ for every two distinct maximal abelian subgroups A and B. Then one of the following holds: (a) G has an abelian subgroup of index 2. (b) G is of class 2, G' is elementary abelian and $\Phi(G) \le Z(G)$.*

Proof. Let A be a maximal normal abelian subgroup of G. Then $G/A \neq \{1\}$ acts faithfully on A and $\{1\} \neq Z(G) < A$. Let K be a G-invariant subgroup such that $Z(G) < K \le A$ and $|K : Z(G)| = 2$. Let x be any element in $G - A$. Then $C_A(x) = Z(G)$ and so $\langle x\rangle \cap A \le Z(G)$. Indeed, let B be a maximal abelian subgroup containing the abelian subgroup $C_A(x)\langle x\rangle$. Then $A \neq B$ and $A \cap B \ge C_A(x) = Z(G)$. Let $k \in K - Z(G)$ so that $k^2 \in Z(G)$ and $k^x = kl$ with some $1 \neq l \in Z(G)$. We get $k^2 = (k^2)^x = (k^x)^2 = (kl)^2 = k^2l^2$, and so $l^2 = 1$ and therefore l is an involution in $Z(G)$. This gives $k^{x^2} = (k^x)^x = (kl)^x = k^xl = (kl)l = kl^2 = k$, and so (by the above) $x^2 \in A$ and (since $\langle x\rangle \cap A \le Z(G)$) $x^2 \in Z(G)$. In particular, G/A is elementary abelian. Let $a \in A - Z(G)$ and set $a^x = a' \in A - Z(G)$ so that $(a')^x = (a^x)^x = a^{x^2} = a$ (since $x^2 \in Z(G)$). Therefore, $(aa')^x = a'a = aa'$ which implies that $aa' = z \in Z(G)$ and $a' = a^x = a^{-1}z$ and so x inverts $A/Z(G)$.

Suppose that $|G/A| \ge 4$. Then there are elements $x, y \in G - A$ such that $xy \in G - A$. In this case x and y both invert $A/Z(G)$ and so xy centralizes $A/Z(G)$. But xy also must invert $A/Z(G)$ which implies that $A/Z(G)$ is elementary abelian. Hence $\Phi(G) \le Z(G)$ (noting that for each $x \in G - A$, $x^2 \in Z(G)$) and so G is of class 2. For each $g, h \in G$, $[g,h]^2 = [g^2, h] = 1$ and so G' is elementary abelian. □

Theorem 35.4. *Let G be a nonabelian 2-group such that $A/Z(G)$ is cyclic for each maximal abelian subgroup A of G. Then one of the following holds:*

(a) G has an abelian subgroup M of index 2 and we have either $G = HZ(G)$ with H minimal nonabelian or $G/Z(G) \cong D_{2^n}$, $n \ge 3$, is dihedral of order 2^n with G' cyclic of order ≥ 4, $G' \cap Z(G) \cong C_2$, and if $x \in G - M$, then $x^2 \in Z(G)$ and x inverts G'.

(b) G is of class 2, G' is elementary abelian of order ≥ 8, $\Phi(G) \le Z(G)$ and whenever A is a maximal abelian subgroup of G, then $|A : Z(G)| = 2$.

Proof. Suppose that there is an element $a \in G - Z(G)$ such that $H = C_G(a)$ is nonabelian. Let A be a maximal abelian subgroup of G containing $\langle a\rangle$. Then $Z(G)\langle a\rangle \le A < H < G$. By our assumption, $A/Z(G) \neq \{1\}$ is cyclic. Assume that $H/Z(G)$ contains a subgroup of order 2 distinct from $\Omega_1(A/Z(G))$. In that case there is $x \in H - A$ such that $x^2 \in Z(G)$. Since $[a,x] = 1$, $\langle a,x\rangle$ is abelian but $\langle a,x\rangle Z(G)/Z(G)$ is noncyclic, a contradiction. We have proved that $H/Z(G)$ has only one subgroup of order 2 and so $H/Z(G) \cong Q_{2^n}$, $n \ge 3$, is generalized quaternion of order 2^n. Indeed, if $H/Z(G)$ were cyclic, then H is abelian, a contradiction. Since $H/Z(G) \cong Q_{2^n}$, it follows that $a^2 \in Z(G)$, $Z(H) = Z(G)\langle a\rangle$, $|Z(H) : Z(G)| = 2$ and for each $y \in Z(H) - Z(G)$, $C_G(y) = H = C_G(a)$. Set

$|Z(G)| = 2^m, m \geq 1$, so that $|H| = 2^{m+n}$. Let $A_0/Z(G)$ be a cyclic subgroup of index 2 in $H/Z(G)$ so that A_0 is abelian and $Z(H) < A_0$. Let $A_1/Z(G) = (H/Z(G))'$ so that $Z(H) \leq A_1 < A_0$ and $|H : A_1| = 4$ and therefore $|A_1| = 2^{m+n-2}$. Since $A_1 = H'Z(G)$, H' covers $A_1/Z(G)$ and so $|H'| \geq 2^{n-2} = |A_1/Z(G)|$. By Lemma 1.1, we get $|H| = 2|Z(H)||H'|$ and so $2^{m+n} = 2 \cdot 2^{m+1}|H'|$ and therefore $|H'| = 2^{n-2}$. This gives $H' \cap Z(G) = \{1\}$ and so H' is cyclic with $H' \cap Z(H) = \langle y \rangle \cong C_2$. It follows that $\langle y \rangle$ is characteristic in H and so if T is a subgroup of G such that $H < T \leq G$ and $|T : H| = 2$, then $\langle y \rangle$ is central in T, contrary to the above fact that $C_G(y) = H$, where $y \in Z(H) - Z(G)$. We have proved that for each $a \in G - Z(G)$, $C_G(a)$ is abelian.

By Theorem 35.2, $A \cap B = Z(G)$ for any two distinct maximal abelian subgroups A and B of G. We may use Theorem 35.3 and so either G has an abelian subgroup of index 2 or $G' \leq Z(G)$, $\Phi(G) \leq Z(G)$ and G' is elementary abelian.

(i) First we consider the case, where G has an abelian subgroup M of index 2. Then $Z(G) < M$ and for each $x \in G - M$, $C_M(x) = Z(G)$ and so $x^2 \in Z(G)$. By our assumption, $M/Z(G) \neq \{1\}$ is cyclic. If G has another abelian maximal subgroup N, then $M \cap N = Z(G)$ and this implies that $|M : Z(G)| = 2$. Conversely, suppose that $|M/Z(G)| = 2$. In that case $G/Z(G) \cong E_4$ (because $G/Z(G) \cong C_4$ would imply that G is abelian) and so G has more than one abelian maximal subgroup. We analyze this case further. Let H be a minimal nonabelian subgroup of G. Then $|H : (H \cap Z(G))| = 4$, H covers $G/Z(G)$ and so $G = HZ(G)$ and $G' \cong C_2$. We have obtained the first possibility stated in part (a) of our theorem.

It remains to consider the case, where $M/Z(G) \cong C_{2^n}, n \geq 2$, where M is a unique abelian maximal subgroup of G. We know that for each $x \in G - M$, $x^2 \in Z(G)$. It follows that x inverts the cyclic group $M/Z(G)$ of order ≥ 4 and so $G/Z(G) \cong D_{2^{n+1}}$ is dihedral of order 2^{n+1}. Set $|Z(G)| = 2^m$, $m \geq 1$, and $(G/Z(G))' = L/Z(G)$ so that $G/L \cong E_4$ and $L = G'Z(G)$. Since G has an abelian maximal subgroup, we may use Lemma 1.1 and we get $|G| = 2^{m+n+1} = 2 \cdot 2^m|G'|$ and so $|G'| = 2^n$. Hence $G' \cap Z(G) = \langle z \rangle \cong C_2$, $G'/\langle z \rangle$ is cyclic of order 2^{n-1} and G' is abelian.

Suppose that G' is not cyclic. Then G' splits over $\langle z \rangle = G' \cap Z(G)$. Since $\mho_1(G')$ is normal in G and $\mho_1(G') \cap Z(G) = \{1\}$, it follows that $\mho_1(G') = \{1\}$ and so $G' \cong E_4$ and $G/Z(G) \cong D_8$. Let $a \in M - (Z(G)G')$ so that $\langle a \rangle$ covers $M/Z(G) \cong C_4$ and so $a^2 \notin Z(G)$. For an $x \in G - M$, we have $[a, x] = t \in G' - \langle z \rangle$. Then we get $[a^2, x] = [a, x]^a[a, x] = t^a t = t^2 = 1$. But then $_G(a^2) = \langle M, x \rangle = G$ and so $a^2 \in Z(G)$, a contradiction. We have proved that G' is cyclic of order ≥ 4.

For any $x \in G - M$ and any $m \in M - L$ (where $L = G'Z(G)$), we have $x^2 \in Z(G)$ and $[m, x] = g$ with $\langle g \rangle = G' \cong C_{2^n}$. This gives $m^x = mg$ and so $m = m^{x^2} = (mg)^x = mgg^x$ and this implies $g^x = g^{-1}$ and therefore x inverts G'. We have obtained the second possibility in part (a) of our theorem.

(ii) Now we consider the case, where G has no abelian subgroups of index 2, $G' \leq Z(G)$, $\Phi(G) \leq Z(G)$ and G' is elementary abelian. It follows that $|A : Z(G)| = 2$ for each maximal abelian subgroup A of G. If $|G : Z(G)| = 4$, then G would have an abelian subgroup of index 2, a contradiction. Hence, $G/Z(G) \cong E_{2^m}$, $m \geq 3$, and so there exist elements $g, h, i \in G - Z(G)$ such that $\langle g, h, i\rangle Z(G)/Z(G) \cong E_8$. We have $[g, h] \neq 1$, $[g, i] \neq 1$, and $[h, i] \neq 1$.

Suppose that $|G'| = 2$. Then $[g, h] = [g, i]$ and so $[g, hi] = [g, h][g, i] = 1$ and therefore $\langle g, hi\rangle Z(G)/Z(G) \cong E_4$, a contradiction.

Suppose that $G' \cong E_4$. In that case $[g, h] = t_1$, $[g, i] = t_2$ and $[h, i] = t_3$, where t_1, t_2, t_3 are pairwise distinct involutions in G'. In this case,

$$[gh, gi] = [g, i][h, g][h, i] = t_2 t_1 t_3 = 1,$$

and so $\langle gh, gi\rangle Z(G)/Z(G) \cong E_4$, a contradiction. We have proved that $|G'| \geq 8$. □

Corollary 35.5. *Let G be an arbitrary nonabelian 2-group. Let A and B be any two distinct maximal abelian subgroups in G with intersection $A \cap B$ of maximal possible order. Then the nonabelian subgroup $H = \langle A, B\rangle$ either possesses an abelian subgroup of index 2 or H is of class 2 and H' is elementary abelian.*

Proof. Obviously, $A \cap B = Z(H)$. If C and D are any two distinct maximal abelian subgroups in H, then $C \cap D \geq Z(H)$ and the maximality of $|A \cap B|$ forces $C \cap D = Z(H)$. Then our result follows from Theorem 35.3. □

§36

Short proofs of some basic characterization theorems of finite p-group theory

All proofs in this section are taken from [Ber22].

1°. Blackburn has proved that a p-group G is metacyclic if and only if the quotient group $G/\mathrm{K}_3(G)\Phi(G')$ is metacyclic. This result is a source of some characterizations of metacyclic p-groups. Here we prove this result in slightly another, but the following equivalent form (our proof is shorter).

Theorem 36.1. *The following conditions for a nonabelian p-group G are equivalent: (a) G is metacyclic. (b) G/R is metacyclic for some G-invariant subgroup R of index p in G'. (Since $\mathrm{d}(G) = 2$, R is determined uniquely.)*

Remark 1. Let G be a p-group. If there is a G-invariant subgroup $R < G'$ such that G/R is metacyclic, then G is also metacyclic. Indeed, take $R \le R_1 < G'$, where R_1 is G-invariant of index p in G'; then G/R_1 is metacyclic as an epimorphic image of G/R, whence G is metacyclic (Theorem 36.1).

Theorem 36.1 and Remark 1 imply the original Blackburn's result:

Corollary 36.2 (Blackburn). *If a p-group G is such that $G/\mathrm{K}_3(G)\Phi(G')$ is metacyclic, then G is also metacyclic.*

Lemma 36.3 (= Lemma 65.2(a)). *If a p-group G is such that* $\mathrm{d}(G) = 2$, $G' \le \mathrm{Z}(G)$ *and* $\exp(G') = p$, *then G is minimal nonabelian.*

Lemma 36.4 (= Exercise 1.8a (Redei)). *If G is a nonmetacyclic minimal nonabelian p-group, then*

$$(1)\quad G = \langle a, b \mid a^{p^m} = b^{p^n} = c^p = 1,\ [a,b] = c,\ [a,c] = [b,c] = 1,\ m \ge n\rangle.$$

Here $|G'| = p$, $\mathrm{Z}(G) = \Phi(G)$ has index p^2 in G, $|\Omega_1(G)| = p^3$, G' is a maximal cyclic subgroup of G, $|G/\mho_1(G)| = p^3$ if and only if $p > 2$.

Proof. If $A, B \in \Gamma_1$ are distinct, then $A \cap B = \mathrm{Z}(G)$ and $G/\mathrm{Z}(G)$ is abelian of type (p, p) so $\Phi(G) = \mathrm{Z}(G)$. We have $|G'| = \frac{1}{p}|G : \mathrm{Z}(G)| = p$ (Lemma 1.1). Let $G/G' = (U/G') \times (V/G')$, where both factors are cyclic of orders p^m, p^n, respectively, $m \geq n$; then U and V are noncyclic so $\Omega_1(G) = \Omega_1(U)\Omega_1(V)$ is of order p^3 and exponent p. Assume that $G' < L < G$, where L is cyclic of order p^2. We claim that then G must be metacyclic. This is the case if $m + n = 2$. Suppose that $m + n > 2$. Then $G/G' = (C/G') \times (D/G')$, where $L \leq C$ and C/G' is cyclic (Theorem 6.1). It follows from $G' = \Phi(L) \leq \Phi(C)$ that $1 = \mathrm{d}(C/G') = \mathrm{d}(C)$ so C is cyclic. Thus, G' is a maximal cyclic subgroup of G. All remaining assertions are obvious. □

Lemma 36.5. *(a) If a p-group G is two-generator of class 2, then G' is cyclic.*

(b) (Blackburn; see Lemma 44.1) If G is a nonabelian two-generator p-group, then $G'/\mathrm{K}_3(G)$ is cyclic.

Proof. (a) Since $\mathrm{cl}(G) = 2$, then $[xy, uv] = [x,u][x,v][y,u][y,v]$ for $x, y, u, v \in G$. Let $G = \langle a, b\rangle$ and $w, z \in G$. Expressing w, z in terms of a and b and using the above identity, we see that the commutator $[w, z]$ is a power of $[a, b]$ so $G' = \langle [a, b]\rangle$, and (a) is proven. (b) follows from (a): $G/\mathrm{K}_3(G)$ is two-generator of class 2. □

Remarks. 2. Let G be a 2-group given by (1). We claim that if $G = AB$, where A and B are cyclic, then $n = 1$. Assume that this is false. Set $\bar{G} = G/\langle a^4, b^4\rangle$; then $\bar{G}$ is of order 2^5 and exponent 4 so it is not a product of two cyclic subgroups of order ≤ 4, a contradiction since $\bar{G} = \bar{A}\bar{B}$.

3. Let G be a nonabelian two-generator p-group. It follows from Lemma 36.4 and Theorem 36.1 that if R is a G-invariant subgroup of index p in G', then G is metacyclic if and only if G'/R is not a maximal cyclic subgroup of G/R. Thus, the derived subgroup G' of a 2-group $G = AB$, where A and B are cyclic, is contained properly in a cyclic subgroup of G if and only if G is metacyclic [IO].

4. If G is a nonmetacyclic p-group, then it has a characteristic subgroup R such that either (i) G/R is elementary abelian of order $> p^2$, or (ii) G/R is nonabelian of order p^3 and exponent p, or (iii) G/R is a 2-group, given in (1), with $m = n = 2$, or (iv) G/R is a 2-group, given in (1), with $m = 2$, $n = 1$. Groups (i)–(iii) are not products of two cyclic subgroups. Indeed, If $\mathrm{d}(G) > 2$, we have (i) with $R = \Phi(G)$. Now let $\mathrm{d}(G) = 2$. If $p > 2$, we have (ii) with $R = \mathrm{K}_3(G)\Phi(G')\mho_1(G) = \mathrm{K}_3(G)\mho_1(G)$ (Theorem 36.1 and Lemmas 36.3–5). If $p = 2$, we have (iii) or (iv) with $R = \mathrm{K}_3(G)\Phi(G')\mho_2(G)$ (Corollary 36.2 and Lemma 36.4).

It follows from Remark 4 that, if a 2-group G and all its characteristic maximal subgroups are two-generator, then G is either metacyclic or $G/\mathrm{K}_3(G)\Phi(G')\mho_2(G)$ is a group (iii) of Remark 4. In particular, a 2-group G is metacyclic if and only if G and all its maximal subgroups are two-generator. This also follows from

Corollary 36.6 (Blackburn). *Suppose that a nonabelian p-group G and all its maximal subgroups are two-generator. Then G is either metacyclic or $p > 2$ and $\mathrm{K}_3(G) = \mho_1(G)$ has index p^3 in G (in the last case, $|G : G'| = p^2$).*

Proof. Suppose that G is nonmetacyclic. In cases (iii) and (iv) of Remark 4, G has a maximal subgroup that is not generated by two elements so $p > 2$. By Lemma 36.4, G has no nonmetacyclic epimorphic images which is minimal nonabelian of order $> p^3$. It also has no epimorphic images of order $> p^3$ and exponent p so $|G/\mho_1(G)| = p^3$. Assume that $|G : G'| > p^2$. Let R be a G-invariant subgroup of index p in G'. Then G/R is a nonmetacyclic minimal nonabelian group (Theorem 36.1 and Lemma 36.3) of order $> p^3$, contrary to what has just been said. Thus, $|G : G'| = p^2$. Then $G/\mathrm{K}_3(G)$ is minimal nonabelian; moreover, that quotient group is nonmetacyclic (Corollary 36.2). In that case, by the above, $|G/\mathrm{K}_3(G)| = p^3 = |G/\mho_1(G)|$ so $\mathrm{K}_3(G) = \mho_1(G)$ since $\mho_1(G) \leq \mathrm{K}_3(G)$. □

Corollary 36.7 (= Proposition 1.6 (Taussky)). *Let G be a nonabelian 2-group. If $|G : G'| = 4$, then G is of maximal class.*

Proof. Let R be a G-invariant subgroup of index 2 in G'. Then G/R is nonabelian of order 8 so metacyclic. In that case, G is metacyclic (Theorem 36.1). We have $G' = \Phi(G) = \mho_1(G)$ so G has a cyclic subgroup of index 2 since G' is cyclic. Now the result follows from Theorem 1.2. □

Corollary 36.8 (= Theorem 9.11 (Huppert)). *Let G be a p-group, $p > 2$, and suppose that $|G/\mho_1(G)| \leq p^2$. Then G is metacyclic.*

Proof. Assuming that G is not metacyclic, we must consider cases (i) and (ii) of Remark 4. We have there $|G/\mho_1(G)| > p^2$, a contradiction. □

Supplement 1 to Corollary 36.6. *A p-group G is metacyclic if and only if either (a) $G/\mho^2(G)$ is metacyclic or (b) (see Corollary 44.9) $G/\mho_2(G)$ is metacyclic.*

Proof. Assume that G is nonmetacyclic. Then there is $R \triangleleft G$ such that G/R is one of nonmetacyclic groups (i)–(iv) of Remark 4. Since $\mho_2(G) \leq \mho^2(G) \leq R$, we get a contradiction. □

Supplement 2 to Corollary 36.6. *Suppose that a nonabelian p-group G and all its characteristic subgroups of index $\frac{1}{p^2}|G : G'|$ are two generator. Then either G is metacyclic or $p > 2$ and $G/\mathrm{K}_3(G)$ is of order p^3 and exponent p. If a nonmetacyclic p-group G and all its characteristic subgroups are two-generator, then* $\mathrm{K}_3(G) = \mho_1(G)$.

Proof. A G-invariant subgroup R of index p in G' is characteristic in G. Suppose that G is nonmetacyclic; then G/R is also nonmetacyclic (Theorem 36.1) and minimal nonabelian (Lemma 36.3). Assume that $|G/R| > p^3$. Then $H/R = \Omega_1(G/R)$ is elementary abelian of order p^3 (Lemma 36.4), $\mathrm{d}(H) > 2$, $|G/H| = \frac{1}{p^2}|G/G'|$ and H is characteristic in G, contrary to the hypothesis. Thus, $|G/R| = p^3$ so $|G/G'| = \frac{1}{p}|G/R| = p^2$; then $p > 2$ (Corollary 36.7). It follows that $G/\mathrm{K}_3(G)$ is minimal nonabelian. By Lemma 36.4 again, $|G/\mathrm{K}_3(G)| = p^3$ so $\mathrm{K}_3(G) = R$ and

$\exp(G/R) = p$ since G/R is not metacyclic (Corollary 36.8), completing the proof of the first assertion.

Now suppose that all characteristic subgroups of nonmetacyclic p-group G are two-generator. Set $\bar{G} = G/\mho_1(G)$. Assume that $|\bar{G}| > p^3$. Let $\bar{G}$ be of order p^4; then it contains an abelian subgroup $\bar{A}$ of index p and $\mathrm{d}(A) \geq \mathrm{d}(\bar{A}) \geq 3$ so, by hypothesis, $\bar{A}$ is not characteristic in $\bar{G}$. Then $\bar{G}$ has another abelian maximal subgroup $\bar{B}$. We have $\bar{A} \cap \bar{B} = \mathrm{Z}(\bar{G})$ so $\bar{G}$ is minimal nonabelian. But a minimal nonabelian group of exponent p has order p^3 (Lemma 36.4), a contradiction. Thus, $|\bar{G}| > p^4$. Then $\mathrm{d}(\bar{G}') \leq 2$, by hypothesis, so $|\bar{G}'| = p^2$ since $\exp(G') = p$ (Theorem 44.13). In that case, $|\bar{G}| = |\bar{G} : \bar{G}'||\bar{G}'| \leq p^4$, contrary to the assumption. Thus, $|G/\mho_1(G)| = p^3$ so $\mathrm{K}_3(G) = \mho_1(G)$ since $\mathrm{K}_3(G) \leq \mho_1(G)$ and $G/\mathrm{K}_3(G)$ is minimal nonabelian and nonmetacyclic. □

It follows that if a 2-group G and all its characteristic subgroups of index $\frac{1}{4}|G : G'|$ are two-generator, then G is metacyclic.

Proof of Theorem 36.1. It suffices to show that (b) ⇒ (a). There is a cyclic subgroup $U/R \lhd G/R$ such that G/U is cyclic. Assume that U is noncyclic. Then U has a G-invariant subgroup T such that U/T is abelian of type (p, p). Set $\bar{G} = G/T$. Then $R \not\leq T$ since $\bar{U} = U/T$ cannot be an epimorphic image of the cyclic group U/R; in that case, $G' \not\leq T$ so $\bar{G}$ is nonabelian. Next, $\bar{G}/\bar{G}'$ is noncyclic so $\bar{G}' < \bar{U}$ and $|\bar{G}'| = p$ since $|\bar{U}| = p^2$. Since $\bar{G}' = G'T/T \cong G'/(G' \cap T)$ is of order p, we get $G' \cap T = R$ (Lemma 36.5(b)). Thus, $R = G' \cap T < T$, a contradiction.[1] □

If a p-group G is nonmetacyclic but all its proper epimorphic images are metacyclic, then either G is of order p^3 and exponent p or G is as given in (1) with $p = 2$, $m = 2$ and $n = 1$. Indeed, the result is trivial for abelian G. Now let G be nonabelian and R a G-invariant subgroup of index p in G'; then G/R is not metacyclic (Theorem 36.1) so $R = \{1\}$, and we get $|G'| = p$. By Lemma 36.3, G is minimal nonabelian. Now the assertion follows from Lemma 36.4.

Corollary 36.9. *Suppose that a nonabelian and nonmetacyclic p-group G and all its maximal subgroups are two-generator, $p > 2$ and $|G| = p^m$, $m > 3$. Set $K = \mathrm{K}_4(G)$ and $\bar{G} = G/K$. Then one of the following holds:*

(a) *$|\bar{G}| = p^4$ and $\mathrm{cl}(\bar{G}) = 3$ so, if $p = 3$, then G is of maximal class.*

(b) *$|\bar{G}| = p^5$, all maximal subgroups of $\bar{G}$ are minimal nonabelian.*[2]

Proof. By Corollary 36.6, $\mathrm{K}_3(G) = \mho_1(G)$ has index p^3 in G (so that $\mathrm{cl}(G) > 2$ since $m > 3$) and $|G : G'| = p^2$. Then $\mathrm{Z}(\bar{G}) = \mathrm{K}_3(G)/K$ has index p^3 in $\bar{G}$ since $\mathrm{cl}(\bar{G}) = 3$. Let $\bar{M} < \bar{G}$ be maximal; then $|\bar{M} : \mathrm{Z}(\bar{G})| = \frac{1}{p}|\bar{G} : \mathrm{Z}(\bar{G})| = p^2$ and, since $\mathrm{d}(\bar{M}) = 2$, it follows that $\bar{M}$ is either abelian or minimal nonabelian. In view of

[1] In this proof we use only Lemma 36.5 which is independent of all previously proved results.

[2] See §71 for defining relations of $\bar{G}$.

Lemma 36.4, $\bar{G}$ has a nonabelian maximal subgroup, say $\bar{M}$. By Lemma 1.1, $\bar{G}$ has at most one abelian maximal subgroup.

Suppose that $\bar{G}$ has an abelian maximal subgroup, say $\bar{A}$. Then $|\bar{G}'| \leq p|\bar{M}'\bar{A}'| = p^2$ (Exercise 1.69(a)) so $|\bar{G}| = |\bar{G}'||\bar{G} : \bar{G}'| = p^4$, and $\mathrm{cl}(\bar{G}) = 3$. In particular, if $p = 3$, then G is of maximal class (Theorem 9.7) so it is as in part (a).

Now suppose that all maximal subgroups of $\bar{G}$ are minimal nonabelian; then $|\bar{G}| > p^4$. If $\bar{U}, \bar{V}$ are distinct maximal subgroups of $\bar{G}$, then $|\bar{G}'| \leq p|\bar{U}'\bar{V}'| = p^3$ so $|\bar{G}'| = p^3$ since $p^5 \leq |\bar{G}| = |\bar{G} : \bar{G}'||\bar{G}'|$, and we get $|\bar{G}| = p^5$. □

Corollary 36.10 (Janko). *If every maximal cyclic subgroup of a noncyclic p-group G is contained in a unique maximal subgroup of G, then G is metacyclic.*

Proof. Let $N \triangleleft G$ and let $U/N \leq G/N$ be maximal cyclic. Then $U = AN$ for a cyclic A. Let $B \geq A$ be a maximal cyclic subgroup of G; then $B \cap N = A \cap N$ and $U/N = BN/N$ so $|A| = |B|$ and $A = B$. Assume that $K/N, M/N$ are distinct maximal subgroups of G/N containing U/N. Then $A \leq U \leq K \cap M$, contrary to the hypothesis. Thus, the hypothesis is inherited by epimorphic images.

Let $A < G$ be maximal cyclic. Then $A\Phi(G)/\Phi(G)$ is contained in a unique maximal subgroup of $G/\Phi(G)$ so $\mathrm{d}(G) = 2$ and $A \not\leq \Phi(G)$. Assume that G is nonmetacyclic. Let R be a G-invariant subgroup of index p in G'. Then $\bar{G} = G/R$ is nonmetacyclic (Theorem 36.1) and minimal nonabelian (Lemma 36.3) so $\bar{G}'$ is maximal cyclic in $\bar{G}$ (Lemma 36.4). Since $\bar{G}/\bar{G}'$ is abelian of rank 2, $\bar{G}'$ is contained in $1 + p$ maximal subgroups of $\bar{G}$, contrary to the previous paragraph. □

Remark 5. Obviously, metacyclic p-groups are powerful for $p > 2$. Let us show (this is Janko's result as well) that G of Corollary 36.10 is also powerful for $p = 2$, unless G is of maximal class. Assume that G is not of maximal class. Then $|G/G'| > 4$ (Corollary 36.7) so $W = G/\mho_2(G)$ cannot be nonabelian of order 8. It suffices to show that W is abelian. Assume that this is false. Then $W = \langle a, b \mid a^4 = b^4 = 1, a^b = a^{-1}\rangle$ is a unique nonabelian metacyclic group of order 2^4 and exponent 4 (Corollary 36.10). In that case, $W/\langle b^2\rangle$, being dihedral of order 8, does not satisfy the hypothesis, a contradiction. Thus, G is powerful. Then, by §26, if $X < G$ is maximal cyclic, then $X \not\leq \Phi(G)$ so $X\Phi(G)$ is a unique maximal subgroup of G containing X since $\mathrm{d}(G) = 2$. Thus, G satisfies the hypothesis of Corollary 36.10 if and only if it is powerful and metacyclic.

It follows from Corollary 36.8 that a p-group $G = AB$, where A and B are cyclic, is metacyclic if $p > 2$. This is not true for $p = 2$, however, we have

Corollary 36.11 (Ito–Ohara [IO]). *If a nonmetacyclic 2-group $G = AB$ is a product of two cyclic subgroups A and B, then G/G' is of type $(2^m, 2)$, $m > 1$.*

Proof. Let R be a G-invariant subgroup of index 2 in G'. Then $\bar{G} = G/R$ is nonmetacyclic (Theorem 36.1) and minimal nonabelian (Lemma 36.3) as in (1). Since

$\bar{G} = \bar{A}\bar{B}$, we get $n = 1$ (Remark 2). Next, $m > 1$ (Corollary 36.7). (Supposing that G is a 2-group of Lemma 36.4, given in (1), with $m > 1$ and $n = 1$, we prove that G is a product of two cyclic subgroups. Set $A = \langle a \rangle$. Then $G/\mho_1(A)$ is dihedral of order 8. Let $U/\mho_1(A) < G/\mho_1(G)$ be cyclic of order 4. If B_0 is a cyclic subgroup covering $U/\mho_1(A)$, then $G = AB_0$, by the product formula.) □

Remark 6. Suppose that a nonmetacyclic 2-group $G = AB$ is a product of two cyclic subgroups A and B. Since $A \cap B = \Phi(A) \cap \Phi(B)$, we get $\Phi(G) = \Phi(A)\Phi(B)$, by the product formula, so $\Phi(G)$ is metacyclic (Theorem 44.12). It follows that all subgroups of G are three-generator. By Corollary 36.6, there is $M \in \Gamma_1$ with $\mathrm{d}(M) = 3$. We claim that M is the unique maximal subgroup of G that is not generated by two elements. Indeed, let U, V be maximal subgroups of G, containing A, B, respectively; then $U \neq V$. By the modular law, $U = A(U \cap B)$ and $V = B(V \cap A)$ so $\mathrm{d}(U) = 2 = \mathrm{d}(V)$ since G in nonmetacyclic. Since the set of maximal subgroups of G is $\{M, U, V\}$, our claim follows. In particular, M is characteristic in G. Set $\bar{G} = G/\mho_2(G)$; then $\bar{G} = \bar{A}\bar{B}$ so $|\bar{A}| = 4 = |\bar{B}|$ since $\bar{G}$ is of exponent 4 (in fact, $\bar{G}$ is a group (iv) of Remark 4).

Suppose that X is a 2-group such that $\mathrm{d}(X) = 2$, $\exp(X) > 2$ and $\Phi(X)$ is metacyclic. We claim that $|X/\mho_2(X)| \leq 2^4$. Assume that this is false. To obtain a contradiction, one may assume that $\mho_2(X) = \{1\}$, i.e., $\exp(X) = 4$. Then $2^3 \leq |\Phi(X)| \leq 2^4$ since $\Phi(X)$ is metacyclic of exponent ≤ 4. By Burnside, $\Phi(X)$ cannot be nonabelian of order 8 so it is either abelian of type $(4, 2)$, or abelian of type $(4, 4)$, or $\Phi(X) = \langle a, b \mid a^4 = b^2 = 1, a^b = a^{-1} \rangle$. In any case, every generating system of $\Phi(X)$ must contain an element of order 4. It follows from $\Phi(X) = \mho_1(X)$ that X has an element of order 8, a contradiction since $\exp(X) = 4$.

Let a noncyclic p-group $G = AB$, where A and B are cyclic. We have $U = \langle A, \mho_1(B) \rangle < G$. By the product formula, $|A\mho_1(B)| = \frac{1}{p}|G|$ so $U = A\mho_1(B)$.

Supplement to Corollary 36.11. *Suppose that $G = AB$ is a nonmetacyclic 2-group with cyclic A and B. Set $U = A\mho_1(B)$ and $V = B\mho_1(A)$. Then $\Gamma_1 = \{U, V, M\}$, where $\mathrm{d}(M) = 3$ and U and V are metacyclic.*

Proof. We use the notation introduced above. By Remark 6, $\Phi(G)(= \mho_1(G))$ is metacyclic but not cyclic. Since $\mathrm{d}(G) = 2$ and G is not minimal nonabelian, we get $\mathrm{Z}(G) < \Phi(G)$. By Theorem 12.12(a), U and V are not of maximal class.

Assume that U is nonmetacyclic; then $U/\mho_2(U)$ is nonmetacyclic (Supplement 1 to Corollary 36.6). Since $\mathrm{d}(U) = 2$ and $\Phi(U)$ is metacyclic as a subgroup of $\Phi(G)$, we get $|U/\mho_2(U)| = 2^4$ (see paragraph following Remark 6). We have $\mho_2(U) \lhd G$ and $\mho_2(U) \leq \Phi(\Phi(G)) < \Phi(M)$ (compare indices) so $\mathrm{d}(M/\mho_2(U)) = 3$. To get a contradiction, one may assume that $\mho_2(U) = \{1\}$. In that case, $|G| = 2^5$ and

$$U = \langle x, y \mid x^4 = y^2 = z^2 = 1,\ z = [x, y],\ [x, z] = [y, z] = 1 \rangle$$

is minimal nonabelian. Since U is not metacyclic, it has no normal cyclic subgroups of order 4 (otherwise, if $L \triangleleft U$ is normal cyclic of order 4, then U/L is abelian of type (2.2) so $L = \Phi(U) = \mho_1(U)$, and we get $\exp(U) = 8$). Since $G = AB$ is of order 2^5 and exponent ≤ 8, one of the factors A, B, namely B (since $\exp(U) = 4$) has order 8. It follows from $d(M) = 3$ that $\exp(M) = 4$. Then $B < V$ is cyclic of order 8. It follows from $\Phi(V) = \mho_1(B)$ that $\mho_1(V) \triangleleft G$. Then $U \cap B = \mho_1(V) \triangleleft U$, contrary to what has been said already. Thus, U is metacyclic. Similarly, V is metacyclic. □

Exercise. Let $G = B(p^n, 2)$ be the maximal finite two-generator group of exponent p^n and class 2. Prove that $G/K_3(G) = \langle a, b \mid a^{p^n} = b^{p^n} = c^p = 1, c = [a,b], [a,c] = [b,c] = 1\rangle$. (*Hint.* Use Theorem 36.1 and Exercise 1.8a.)

2°. In this subsection, most proofs are based on counting theorems.

Let G be a p-group of exponent $p^e > p$, $p > 2$, and $1 < k < e$. Suppose that $H < G$ is metacyclic of exponent p^k such that, whenever $H < L \leq G$, then $\exp(L) > p^k$. It is easy to prove that then G is either metacyclic or a 3-group of maximal class. This is also a consequence of the following more general

Theorem 36.12. *Let G be a p-group of exponent $p^e > p^2$ and $1 < k < e$. Let U be a maximal member of the set of subgroups of G with exponent p^k.*

(a) *If U is absolutely regular then G is also absolutely regular, $U = \Omega_k(G)$ and the subgroup U is not of maximal class.*

(b) *If U is irregular of maximal class, then G is also of maximal class.*

Proof. We assume that G is neither absolutely regular nor a 2-group of maximal class.

Let G be of maximal class, $p > 2$ and let U be absolutely regular. Then G is irregular since $e > 2$ (Theorem 9.5). Denote by G_1 the fundamental subgroup of G; then $\exp(G_1) = \exp(G) = p^e > p^k$ (Theorem 9.6). Assume that $U < G_1$. Then $U = \Omega_k(G_1) < G_1$ since $k < e$, hence $U \triangleleft G$. Since $|G : U| > p$, all elements of the set $(G/U) - (G_1/U)$ have the same order p (Theorem 13.19), so there exists $H/U < G/U$ such that $H \not\leq G_1$ and $|H : U| = p$. Then H is of maximal class (Theorem 13.19) so $\exp(H) = \exp(U)$ (Theorem 9.6), contrary to the choice of U. Now suppose that $U \not\leq G_1$. Since $U = \langle x \mid o(x) = p^k\rangle$, we get $k = 2$ (Theorem 13.19) and $|U| \leq p^{2p-2}$. Let $U < L \leq G$, where $|L : U| = p$; then $\exp(L) = p^3$ so L is irregular hence it is of maximal class (Theorem 13.19). Then $|L| \geq p^{2p} > p|U|$, a contradiction. Thus, if G is of maximal class, then U is also of maximal class. Next we assume that G is not of maximal class.

Next we proceed by induction on $|G|$.

(i) Let G be noncyclic and regular; then U is absolutely regular, by hypothesis and Theorem 9.5. In that case, $U = \Omega_k(G)$ (Theorem 7.2) so

$$\Omega_1(G) = \Omega_1(U) \quad \text{and} \quad p^p > |U/\mho_1(U)| = |\Omega_1(U)| = |\Omega_1(G)| = |G/\mho_1(G)|,$$

whence G is absolutely regular. Assume that, in addition, U is of maximal class; then $p > 2$. We have $|U : \Omega_1(U)| = p$ (Theorem 9.5) so $|\Omega_1(G/\Omega_1(G))| = p$. It follows that $G/\Omega_1(G)$ is cyclic (of order $> p$). Let D be a G-invariant subgroup of index p^2 in $\Omega_1(U) = \Omega_1(G)$, and set $C = \mathrm{C}_G(\Omega_1(U)/D)$; then C/D is abelian and $U \le C$ so U/D is abelian of order p^3 so U is not of maximal class, contrary to the assumption. Thus, U is not of maximal class. In what follows we assume that G is irregular.

(ii) Let U be absolutely regular; then $|\Omega_1(U)| = |U/\mho_1(U)| < p^p$. We write $R = \Omega_1(U)$ and $N = \mathrm{N}_G(R)$; then $N < G$. Indeed, assume that $N = G$. Then, by Corollary 13.3, there is in G a normal subgroup S of order $p|\Omega_1(U)|$ and exponent p such that $R < S$. Set $H = US$. Then $H/S \cong U/R$ is of exponent p^{k-1} so, since $U < H$, we get $\exp(H) = p^k$, contrary to the choice of U. Thus, $N < G$. Then N is absolutely regular, by induction. In that case, $U = \Omega_k(N)$ so $R = \Omega_1(N)$ is characteristic in N whence $N = G$, a contradiction. In what follows we assume that U is *irregular* of maximal class.

(iii) Set $V = \Omega_1(\Phi(U))$ and $N = \mathrm{N}_G(V)$. If $N < G$, then, by induction, N is of maximal class so G is also of maximal class (Remark 10.5), contrary to the assumption. Now assume that $N = G$. Then, as in (ii), if G is not of maximal class, it has a normal subgroup R of order p^p and exponent p such that $V < R$. Set $H = UR$; then $H/R \cong U/V$ is of exponent p^{k-1}. This is a contradiction since $\exp(H) = p^k = \exp(U)$ and $U < H$. □

For definition of L_s-groups, see §§17, 18.

Lemma 36.13 (= Lemma 42.1). *Let G be a p-group of order $> p^{p+1}$ with $|\Omega_2(G)| = p^{p+1}$. Then one of the following holds:*

(a) G is absolutely regular;

(b) G is an L_p-group;

(c) $p = 2$ and $G = \langle a, b \mid a^{2^n} = 1, a^{2^{n-1}} = b^4, a^b = a^{-1+2^{n-2}} \rangle$.

It follows from Remark 10.5 that an irregular p-group G has a maximal regular subgroup R of order p^p if and only if it is of maximal class.

Theorem 36.14. *Let G be a p-group and suppose that $H < G$ is a maximal member of the set of subgroups of G of exponent p^2. Suppose, in addition, that $|H| = p^{p+1}$. Then one of the following holds:*

(a) $p = 2$ and G is of maximal class;

(b) $H = \Omega_2(G)$ so G is as in Lemma 36.13.

Proof. By hypothesis, $\exp(H) < \exp(G)$. If G is regular, then $H = \Omega_2(G)$ so G is a group of Lemma 36.13(a,b).

Suppose that G is irregular of maximal class. It follows from Theorems 9.5 and 9.6 that then $p = 2$, and we get case (a). Indeed, assume that $p > 2$. If $H \le G_1$, then

$H = \Omega_2(G_1)$. If $H = G_1$, then $\exp(H) = \exp(G)$, contrary to the choice of H. Thus, $H < G_1$. Let U/H be a subgroup of G/H of order p not contained in G_1/H. Then U is of maximal class and exponent p^2 (Theorem 13.19), contrary to the choice of H. Now let $H \not\leq G_1$; then $\Omega_1(G_1) < H$ and H is of maximal class (Theorem 13.19). Let $H < F \leq G$ with $|F : H| = p$. Then $\exp(F) = \exp(H)$, contrary to the choice of H. The 2-groups of maximal class satisfy the hypothesis.

In what follows we assume that G is neither regular nor of maximal class. Then, in view of Theorem 36.12, one may assume that H is neither absolutely regular nor of maximal class so $\mathrm{cl}(H) < p$ and H is regular (Theorem 7.1) and $\Omega_1(H)$ is of order p^p and exponent p. Set $N = \mathrm{N}_G(\Omega_1(H))$; then $H < N$ since $\Omega_1(H)$ is characteristic in $H < G$. If $H/\Omega_1(H)$ is contained in subgroup $F/\Omega_1(H)$ of type (p, p), then $\exp(F) = p^2$ and $H < F$, contrary to the choice of H. Thus, N/H is either cyclic or generalized quaternion. In that case, $\Omega_1(G) = \Omega_1(H)$, and we conclude that $\Omega_2(G) = H$. □

Let $n \in \mathbb{N}$. A p-group G is said to be a U_n^p-group provided it has a normal subgroup R of order p^n and exponent p such that G/R is irregular of maximal class and, if T/R is absolutely regular of index p in G/R, then $\Omega_1(T) = R$. Let us prove that R contains all G-invariant subgroups of G of exponent p. Assume that $R_1 < G$ is the least G-invariant subgroup of exponent p not contained in R; then $|RR_1 : R| = p$ so $RR_1 < T$ since G/R has only one minimal normal subgroup. This is a contradiction: $RR_1 \leq \Omega_1(T) = R < RR_1$. It follows that R is characteristic in G and $n \geq p - 1$ (Theorems 12.1(a), 9.5 and 9.6). We call R the *kernel* of the U_n^p-group G. It follows from Theorem 12.1(a) that U_{p-1}^p-groups are of maximal class. Note that $\exp(G) = p \cdot \exp(G/R) = \exp(T)$.

Proposition 36.15. *Let G be a p-group and $H < G$ a maximal member of the set of subgroups of G of exponent* $\exp(H)$. *If H is a U_n^p-group, the same is true for G.*

Proof. Let R be the kernel of H, and set $N = \mathrm{N}_G(R)$. If $N < G$, then N is a U_n^p-group, by induction. In that case, R is also kernel of N so characteristic in N hence $N = G$, contrary to the assumption. Thus, $N = G$ so $R \triangleleft G$. Then H/R is a maximal member of the set of subgroups of exponent $\frac{1}{p} \cdot \exp(G)$ in G/R and H/R is irregular of maximal class. Then G/R is of maximal class, by Theorem 36.12. Let us show that G is a U_n^p-group. Let T/R be an absolutely regular subgroup of index p in G/R and set $U/R = (H/R) \cap (T/R)$. Then U/R is an absolutely regular subgroup of index p in H/R so $\Omega_1(U) = R$ since H is a U_n^p-group. Let $F/R < T/R$ be G-invariant of order p. Then

$$F/R < \mathrm{Z}(H/R) < \Phi(H/R) < U/R,$$

and we get $\exp(F) = p^2$ since $F \not\leq R = \Omega_1(U)$. Then $R = \Omega_1(T)$ so G is a U_n^p-group. □

Let G be a p-group and $H < G$ a maximal member of the set of subgroups of G of exponent $\exp(H)$. If H is an L_n-group, then G is an L_n-group. To prove, it suffices to repeat, with small modifications, the proof of Proposition 36.15 and use the following easy fact: If $C < G$ is a cyclic subgroup of order $p^k > p$ which is not contained properly in a subgroup of exponent p^k, then G is cyclic.

Theorem 36.16. *Suppose that a p-group G is such that $G/\mho^2(G)$ is of maximal class. Then G is also of maximal class.*

Proof. One may assume that $\mho^2(G) > \{1\}$. Recall that $\mho^2(G) = \mho_1(\mho_1(G))$.

(a) Suppose that G is regular. Then $|G/\mho^2(G)| \le p^p$ so $|G/\mho_1(G)| = p^k$, where $k < p$ (Theorem 9.5). It follows from Theorem 9.5 that $|G/\mho^2(G)| = p^{k+1}$ so $|\mho_1(G) : \mho_1(\mho_1(G))| = p$, and we conclude that $\mho_1(G)$ is cyclic. Let $|\mho_1(G)| = p^e$; then $\exp(G) = p^{e+1}$. By Theorem 7.2, $|\Omega_1(G)| = p^k$. Since $|G| = p^{k+e}$, it follows that $G/\Omega_1(G)$ is cyclic of order p^e. By hypothesis, $|G : G'| = p^2$ so $e = 1$. In that case, $\mho^2(G) = \{1\}$, a contradiction.

(b) Now suppose that G is irregular. Then $|G| \ge p^{p+1}$ (Theorem 7.1). One may assume that $|G| > p^{p+1}$ (otherwise, in view of Theorem 9.5, there is nothing to prove). By the first Hall's regularity criterion (Theorem 9.8(a)), $|G/\mho_1(G)| \ge p^p$ so $|G/\mho^2(G)| \ge p^{p+1}$. and we conclude that $G/\mho^2(G)$ is irregular (Theorem 9.5). By Theorem 9.5, we get $|G/\mho_1(G)| = p^p$.

Let $H/\mho^2(G)$ be an absolutely regular subgroup of index p in $G/\mho^2(G)$ which exists, by Theorems 9.5 and 9.6. Assume that H is not absolutely regular. Then, by Theorem 9.8(a), we have $|H/\mho_1(H)| \ge p^p$. Clearly, $\mho^2(G) \le \mho_1(H)$ so $H/\mho_1(H)$ of order $\ge p^p$ and exponent p is an epimorphic image of the absolutely regular group $H/\mho^2(G)$, a contradiction. Thus, H is absolutely regular. Assume that G is not of maximal class. Then $G = H\Omega_1(G)$, where $\Omega_1(G)$ is of order p^p and exponent p (Theorem 12.1(b)). We have

$$G/(H \cap \Omega_1(G)) = G/\Omega_1(H) \cong (H/\Omega_1(H)) \times (\Omega_1(G)/\Omega_1(H)).$$

By hypothesis, $|G/G'| = p^2$. Therefore, it follows from the displayed formula that $|H/\Omega_1(H)| = p$ so $|H| = p|\Omega_1(H)| = p^p$ and $|G| = p^{p+1}$. In that case, $\mho^2(G) = \{1\}$, a final contradiction. □

Since $\mho_2(G) \le \mho^2(G)$, we have

Corollary 36.17. *A p-group G is of maximal class if and only if $G/\mho_2(G)$ is.*

Proposition 36.18. *Let a p-group G be irregular. If $H = \Omega_2^*(G)$ is of maximal class, then G is also of maximal class.*

Proof. We use induction on $|G|$. One may assume that $H < G$. We have $H \triangleleft G$ and $c_2(G) = c_2(H)$.

(i) Suppose that H is absolutely regular with $|\Omega_1(H)| = p^k$; then $k < p$. In that case, $c_2(H) = \frac{|\Omega_2(H) - \Omega_1(H)|}{p(p-1)}$ is not divisible by p^{p-1}. Then, by Theorem 13.2(b), G is of maximal class. Next we assume that H is irregular of maximal class.

(ii) Suppose that $|H| = p^{p+1}$; then $\exp(H) = p^2$. Let v be the number of absolutely regular maximal subgroups of H. It follows from $\Omega_2^*(H) = H$ that $v > 1$. Then $c_2(H) = vp^{p-2}$.

Assume that G is not of maximal class. Then, by Theorem 13.2(b), $c_2(G) \equiv 0 \pmod{p^{p-1}}$ so $v = p$.

First suppose that $|G : H| = p$. Let $T_1, \dots, T_{p+1}$ be all members of the set Γ_1 which are not of maximal class; then all T_i are regular (Theorem 12.12(c)). If $\exp(T_i) = p^2$, then $\Omega_2^*(T_i) = T_i \not\leq H$. It follows that $\exp(T_i) = p$ for all i. But $|G : \bigcap_{i=1}^{p+1} T_i| = p^2$ so $\bigcup_{i=1}^{p+1} T_i = G$ and $\exp(G) = p$, a contradiction. In that case, G is of maximal class.

Now suppose that $|G : H| > p$. If $H < F < G$ and $|F : H| = p$, then F is of maximal class, by the previous paragraph. Then G is also of maximal class (Exercise 13.10(a)).

(iii) It remains to consider the case where H is of maximal class and order $> p^{p+1}$. Then, by Theorem 13.2(b), $c_2(G) = c_2(H) \equiv p^{p-2} \pmod{p^{p-1}}$. In that case, by Theorem 13.2(b), G must be of maximal class since G is not absolutely regular since H is irregular. □

Remark 7. If G is a p-group such that $H = \Omega_1(G)$ is of maximal class, then one of the following holds: (a) H is of order $\leq p^p$ and exponent p, (b) G is of maximal class. Indeed, this is the case if G is regular, by Theorem 9.5. Now assume that G is not of maximal class and $|H| > p^p$; then $\exp(H) > p$. Let $E_1, \dots, E_r$ be all subgroups of order p^p and exponent p in G; then $r > 1$ and, by Theorem 13.5, $r \equiv 1 \pmod p$. We have $E_i < H$ for all i so H has a G-invariant subgroup, say E_1, of order p^p and exponent p. It follows that $|H| = p^{p+1}$ (Theorem 9.6). Since $r > 1$ and $d(H) = 2$, we get $\exp(H) = p$, a contradiction.

3°. In conclusion we prove the following

Theorem 36.19. *If a 2-group G and all its nonabelian maximal subgroups are two-generator, then G is either metacyclic or minimal nonabelian.*

Proof. Assume that G is a counterexample of minimal order. Then G is nonabelian and, by Lemma 65.2(a), $|G'| > 2$. Let $R \leq Z(G) \cap G'$ be of order 2. Since $\bar{G} = G/R$ satisfies the hypothesis, it is nonmetacyclic and minimal nonabelian, by induction and Remark 1. Then $\bar{E} = \Omega_1(\bar{G}) \cong E_8$ (Lemma 36.4). Let $\bar{E} \leq \bar{A} \in \Gamma_1$. Since $d(\bar{A}) = 3$, it follows that $d(A) > 2$ so A is abelian, by hypothesis.

Assume that $B \in \Gamma_1 - \{A\}$ is abelian. Then $A \cap B = Z(G)$ so $|G'| = \frac{1}{2}|G : Z(G)| = 2$ so G is minimal nonabelian (Lemma 65.2(a)) and G is not a counterexample. Thus, A is a unique abelian member in the set Γ_1. By the previous paragraph, $\bar{A}$ is a unique maximal subgroup of $\bar{G}$ containing $\bar{E}$. It follows that $\bar{G}/\bar{E}$ is cyclic. By Schreier's theorem, $d(A) = 3$ since $d(A) > 2$ so $E_1 = \Omega_1(A) \cong E_8$.

Let $U < G$ be maximal and $U \neq A$ so U is nonabelian whence $d(U) = 2$. Since $\bar{U}$ is abelian, we get $R = U'$ so U is minimal nonabelian (Lemma 65.2(a)). Thus, all nonabelian maximal subgroups of G are minimal nonabelian so G is an $\mathcal{A}_2$-group (see §65). By Theorem 65.8, G must be metacyclic, a final contradiction. □

Theorem 36.20. *Let G be a nonmetacyclic two-generator* 2*-group. Then the number of two-generator maximal subgroups of G is even.*

For a proof, see Theorem 71.8.

§37

MacWilliams' theorem

Given $g \in G$ and $m \in \mathbb{N}$, we set

$$\vartheta_m(g) = |\{x \in G \mid x^m = g\}|, \quad \chi^{(m)}(g) = \chi(g^m), \quad \nu_m(\chi) = \langle \chi^{(m)}, 1_G \rangle.$$

It is known [BZ, Chapter 4] that $\chi^{(m)}$ is a generalized character of G so $\nu_m(\chi) \in \mathbb{Z}$ and $\vartheta_m = \sum_{\chi \in \mathrm{Irr}(G)} \nu_m(\chi)\chi$, i.e., ϑ_m is also a generalized character. Since $\vartheta_m(1)$ is the number of solutions of $x^m = 1$ in G then, by celebrated Frobenius' theorem (see Introduction, Theorem 8), $\vartheta_m(1)$ is divisible by $(|G|, m)$. Theorem 37.1 was proved in [MacW] for $p = 2$; our proof follows closely the MacWilliams' argument. We also give an alternate proof of Theorem 1.17(a), due to Alperin–Feit–Thompson.

Theorem 37.1. *If G is a p-group such that $|G/\Phi(G)| \geq p^{2k+1}$, then $\vartheta_p(1) \equiv 0 \pmod{p^{k+1}}$ or, what is the same, $c_1(G) \equiv 1 + p + \cdots + p^k \pmod{p^{k+1}}$.*

Proof. We have $\vartheta_p(1) = \sum_{\chi \in \mathrm{Irr}(G)} \nu_p(\chi)\chi(1)$, $\nu_p(\chi) \in \mathbb{Z}$. Set $A = \{\lambda \in \mathrm{Lin}(G) \mid \lambda^p = 1_G\}$; then $|A| = |G/\Phi(G)| \geq p^{2k+1}$ since $G' \leq \Phi(G)$. If $\lambda \in A$ and $g \in G$, then $\lambda(g^p) = \lambda(g)^p = 1$. As we know, A acts on $\mathrm{Irr}(G)$ via multiplication. Let O_χ be the A-orbit of $\chi \in \mathrm{Irr}(G)$. If $\lambda \in A$, then

$$\begin{aligned}\nu_p(\lambda\chi) &= \langle (\lambda\chi)^{(p)}, 1_G \rangle = |G|^{-1} \sum_{g \in G} (\lambda\chi)(g^p) = |G|^{-1} \sum_{g \in G} \lambda(g^p)\chi(g^p) \\ &= |G|^{-1} \sum_{g \in G} \chi(g^p) = |G|^{-1} \sum_{g \in G} \chi^{(p)}(g) = \langle \chi^{(p)}, 1_G \rangle = \nu_p(\chi).\end{aligned}$$

Therefore, if T is a transversal of the set of all A-orbits on the set $\mathrm{Irr}(G)$, then $\vartheta_p(1) = \sum_{\chi \in T} \nu_p(\chi)|O_\chi|\chi(1)|$ so, to prove the theorem, it suffices to show that, for $\chi \in T$, one has $|O_\chi|\chi(1) \equiv 0 \pmod{p^{k+1}}$. We have to consider only those $\chi \in T$ for which $\chi(1) \leq p^k$. Let $\chi(1) = p^{k-e}$ ($e \geq 0$). If A_χ is the A-stabilizer of χ, then $|O_\chi| = |A : A_\chi|$. Let $N = \bigcap_{\lambda \in A_\chi} \ker(\lambda)$; then $|G : N| = |A_\chi|$. If $g \notin N$, there exists $\lambda \in A_\chi$ such that $\lambda(g) \neq 1$. Since $\lambda\chi = \chi$, it follows that $\lambda(g)\chi(g) = \chi(g)$, whence $\chi(g) = 0$, that is, χ vanishes outside N. Since $|G| = |G|\langle \chi, \chi \rangle = \sum_{g \in G} |\chi(g)|^2 = \sum_{g \in N} |\chi(g)|^2 \leq |N|\chi(1)^2$, it follows that $\chi(1)^2 \geq |G : N| = |A_\chi|$. Consequently, $|A_\chi| \leq \chi(1)^2 = (p^{k-e})^2 = p^{2k-2e}$. This yields

$$|O_\chi|\chi(1) = |A : A_\chi|p^{k-e} \geq p^{2k+1-(2k-2e)+k-e} = p^{k+e+1}.$$

Therefore, since $e \geq 0$, we get $\vartheta_p(1) = \sum_{\chi \in T} \nu_p(\chi)|O_\chi|\chi(1) \equiv 0 \pmod{p^{k+1}}$. □

Theorem 1.17(a) does not follow from Theorem 37.1. Below we offer a similar proof of Theorem 1.17(a) (here we follow closely to [Isa1, Theorem 4.9]). Note that, for a 2-group G, $c_1(G) + 1 = \vartheta_2(1)$.

Theorem 37.2 (= Theorem 1.17(a)). *Let G be a 2-group. If $c_1(G) \equiv 1 \pmod 4$ then G is either cyclic or a 2-group of maximal class.*

Proof (Alperin–Feit–Thompson). One may assume that G is not abelian. By Taussky's theorem. it remains to show that $|G : G'| = 4$. Write $t = c_1(G)$. By hypothesis, 4 divides $t - 1$. We use induction on $|G|$.

If $Z(G)$ is not cyclic, choose $K \leq Z(G)$ elementary abelian of order 4. If $x \in G - K$ is an involution, all elements in the coset Kx are involutions. Then the set $\{x \mid x^2 = 1\}$ is a union of cosets of K and hence 4 divides $t + 1$, a contradiction. Therefore, $Z(G)$ is cyclic and G contains only one normal subgroup, say Z, of order 2 so $Z \leq G'$. Also G/Z is not cyclic. By induction, $c_1(G/Z) \not\equiv 1 \pmod 4$ (otherwise, $|G : G'| = |G/Z : G'/Z| = 4$, and we are done, by Taussky's theorem).

Since the function $\nu_2(\chi)$ is inherited by quotient groups, we have:

$$\sum_{\chi \in \mathrm{Irr}(G)} \nu_2(\chi)\chi(1) = t + 1 \equiv 2 \pmod 4, \qquad \sum_{\chi \in \mathrm{Irr}(G/Z)} \nu_2(\chi)\chi(1) \equiv 0 \pmod 4;$$

the second congruence follows, by the previous paragraph. We conclude that

$$(*) \qquad \sum_{\chi \in \mathrm{Irr}(G),\, Z \not\leq \ker(\chi)} \nu_2(\chi)\chi(1) \equiv 2 \pmod 4.$$

Now let $C = \{\lambda \in \mathrm{Lin}(G) \mid \lambda^2 = 1_G\}$. For $\chi \in \mathrm{Irr}(G)$, we have $\lambda\chi \in \mathrm{Irr}(G)$ for every $\lambda \in C$, and since $Z \leq \ker(\lambda)$, we conclude that $Z \not\leq \ker(\chi)$ if and only if $Z \not\leq \ker(\lambda\chi)$. Therefore, C permutes the set $\{\chi \in \mathrm{Irr}(G) \mid Z \not\leq \ker(\chi)\} = \mathrm{Irr}(G \mid Z)$ and partitions this set into orbits $\mathcal{O}_i$, $1 \leq i \leq r$. It is known (see the proof of Theorem 37.1) that $\nu_2(\chi)$ is constant on each C-orbit.

Since $|\mathcal{O}_i|$ is a power of 2, we conclude from (*) that there is $\chi \in \mathrm{Irr}(G)$ such that (i) $Z \not\leq \ker(\chi)$, (ii) $\nu_2(\chi) \neq 0$, (iii) $\chi(1)|\mathcal{O}_\chi| = 2$.

By (i), χ is faithful so $\chi(1) > 1$ since G is nonabelian. It follows from (iii) that $\chi(1) = 2$ and $|\mathcal{O}_\chi| = 1$. Therefore, $\lambda\chi = \chi$ for all $\lambda \in C$.

If $g \in G - \Phi(G)$, then there is $\lambda \in C$ for which $\lambda(g) \neq 1$ and it follows that $\chi(g) = 0$. This means that χ vanishes outside $\Phi(G)$. We have $4 = \chi(1)^2 \geq \langle \chi_{\Phi(G)}, \chi_{\Phi(G)} \rangle = |G : \Phi(G)|$. Since G is not cyclic, $|G : \Phi(G)| > 2$ so we have equality above. This forces $\chi_{\Phi(G)} = 2\mu$, where μ is a faithful linear character of $\Phi(G)$ (recall that χ is faithful). In particular, $\Phi(G)$ is cyclic. Since $\Phi(G) = \mho_1(G)$, G has a cyclic subgroup of index 2, and we are done, by Theorem 1.2. □

Exercise. For a nonabelian 2-group G, the following conditions are equivalent: (a) G is of maximal class, (b) The number of characters of degree 2 in $\mathrm{Irr}(G)$ is odd.

Solution. Assume that G is not of maximal class and has odd number of irreducible characters of degree 2. Then $|G : G'| > 4$ (Taussky's theorem) so we have $|G| = \sum_{\chi \in \mathrm{Irr}(G)} \chi(1)^2 \equiv 4 \pmod 8$, a contradiction. If G is of maximal class and order 2^n, then all its nonlinear irreducible characters have degree 2 (Introduction, Theorem 17) and their number $\frac{2^n-4}{4} = 2^{n-2} - 1$ is odd.

§38

p-groups with exactly two conjugate classes of subgroups of small orders and exponent $p > 2$

Let G be a p-group of order $> p^k$. Let $\mathrm{ce}_k(G)$ be the number of conjugate classes of subgroups of order p^k and exponent p in G. Suppose that $\mathrm{ce}_k(G) = 1, k \le p$, and let $E \triangleleft G$ be of order p^k and exponent p (see Theorem 12.1). Then $\Omega_1(G) = E$. Indeed, assume that there is $x \in G - E$ of order p. There is in E a G-invariant subgroup K of order p^{k-1}. Then $E_1 = \langle x, K\rangle$ is of order p^k and exponent p (Theorem 7.1(b)). Since E_1 and $E(\triangleleft G)$ are not conjugate in G, we get $\mathrm{ce}_k(G) > 1$, contrary to the assumption. If, in addition, $p > 2$ and $\mathrm{ce}_2(G) = 1$, then $\Omega_1(G) = E$ so, by Theorem 13.7, G is either metacyclic or a 3-group of maximal class. The following more interesting case arises when $\mathrm{ce}_2(G) = 2$, and we intend to consider it for $p > 2$.

Assume that a p-group G contains a normal subgroup $N \cong \mathrm{E}_{p^3}$. Let $R < N$ be G-invariant of order p^2. Then N contains $p^2 + p$ subgroups of order p^2 which are $\neq R$, and the set of these subgroups is G-invariant. Since $p^2 + p$ is not a power of p, we get $\mathrm{ce}_2(G) > 2$. Thus, if $\mathrm{ce}_2(G) = 2$, then G has no normal subgroups isomorphic to E_{p^3}; the latter groups are classified for $p > 2$ in Theorem 13.7 (for $p = 2$ this question is considered in §50).

Theorem 38.1. *Suppose that G is a p-group of order $> p^3$, $p > 2$, with $\mathrm{ce}_2(G) = 2$. Then one of the following holds:*

(a) $G = \Omega_1(G)C$, where $\Omega_1(G)$ is nonabelian of order p^3 and exponent p, C is cyclic of order $> p$, $\mathrm{cl}(G) = 3$ (or, what is the same, $[G, C] > \{1\}$, $\mathrm{d}(G) = 2$).

(b) $p = 3$, G is a 3-group of maximal class and exactly one maximal subgroup of G contains a nonabelian subgroup of order 3^3 and exponent 3.

Proof. Let G be a 3-group of maximal class. Then, since $\mathrm{ce}_2(G) = 2$, G contains a normal abelian subgroup R of type $(3, 3)$ and a subgroup E of order 3^3 and exponent 3 such that $R < E$. If E is abelian, then G is isomorphic to a Sylow 3-subgroup of the symmetric group of degree 9 (Exercise 9.13) so $E \triangleleft G$ and $\mathrm{ce}_2(G) > 2$ (see the paragraph preceding the theorem), a contradiction. Thus, E is nonabelian. One may assume that $|G| > 3^4$ (for $|G| = 3^4$ we must have $\Omega_1(G) = E$ and $\mathrm{ce}_2(G) = 2$, and G is as in (b)). Let $E < M < G$, where M is maximal in G. If $R_1 < E$ is abelian of type $(3, 3)$ and $R_1 \neq R$, then $\mathrm{N}_G(R_1) = E$. In that case, $M = \Phi(G) \cup (\bigcup_{x \in G} R_1^x)$ so all non-G-invariant subgroups of type $(3, 3)$ from M are conjugate in G. If $N \in$

$\Gamma_1 - \{M\}$, then $M \cap N = \Phi(G)$ is metacyclic hence M is the unique maximal subgroup containing a subgroup of order 3^3 and exponent 3 since $\mathrm{ce}_2(G) = 2$, and so G is as in (b).

As we know, G has no normal subgroups isomorphic to E_{p^3}. Therefore, in view of Theorem 13.7, it remains to consider the case where $G = \Omega_1(G)C$, where $\Omega_1(G)$ is nonabelian of order p^3 and exponent p, C is cyclic of order $> p$. Then $\Omega_1(G)$ contains exactly $p + 1$ abelian subgroups of type (p, p), and only one of them is G-invariant since $\mathrm{ce}_2(G) = 2$. In that case, $|G'| = p^2$ so $\mathrm{cl}(G) = 3$ and $\mathrm{d}(G) = 2$. □

If $\mathrm{ce}_{p-1}(G) = 1$, then $|\Omega_1(G)| = p^{p-1}$ (see the text preceding Theorem 38.1) and G is either absolutely regular or of maximal class (Theorem 12.1(a)).

Theorem 38.2. *If G is a p-group with $\mathrm{ce}_{p-1}(G) = 2$, then one of the following holds:*

(a) $p > 2$, $\Omega_1(G)$ is nonabelian two-generator subgroup of order p^p and exponent p, exactly one maximal subgroup of $\Omega_1(G)$ is G-invariant.

(b) $p = 2$, $\Omega_1(G) \cong \mathrm{E}_4$, $\mathrm{Z}(G)$ is cyclic (such groups are described in §82).

(c) G is of maximal class and order $> p^{p+1}$ with exactly one maximal subgroup, say M, which contains a subgroup, say E, of order p^p and exponent p; then $\mathrm{d}(E) = 2$. All subgroups of order p^p and exponent p are conjugate in G.

Proof. (i) Suppose that G is of maximal class. If G contains a normal subgroup, say R, of order p^p and exponent p, then $|G| = p^{p+1}$. In view of $\mathrm{ce}_{p-1}(G) = 2$, all subgroups of G of order p^{p-1} and exponent p are contained in R so $\Omega_1(G) = R$. It follows that R contains exactly $|\Gamma_1(R)| - 1$ non-G-invariant members so $|\Gamma_1| = p+1$; then $\mathrm{d}(G) = 2$, and G is as in (a).

Now suppose that G has no normal subgroups of order p^p and exponent p. It follows from $\mathrm{ce}_{p-1}(G) = 2$ that $|\Omega_1(G)| > p^{p-1}$ so G has a subgroup, say R, of order p^p and exponent p (see §7); then $|G : R| > p$ since $R \ntrianglelefteq G$. Let $R < M \in \Gamma_1$; then M is of maximal class (Theorem 9.6). Since M has a G-invariant subgroup of order p^{p-1} and exponent p (this subgroup coincides with $\Omega_1(\Phi(G))$), all subgroups of order p^{p-1} and exponent p are contained in M in view of $\mathrm{ce}_{p-1}(G) = 2$. Therefore, if $N \in \Gamma_1 - \{M\}$, then $M \cap N = \Phi(G)$ is absolutely regular so we must have $|\Omega_1(N)| = p^{p-1}$. It is known that $\mathrm{N}_G(R)$ is of maximal class and order p^{p+1} so there are exactly p^{m-p-1} subgroups conjugate with R in G, where $|G| = p^m$; all these subgroups contain $\Omega_1(\Phi(G))$ (of order p^{p-1}). We have

$$\Big|\Phi(G) \cup \Big(\bigcup_{x \in G} R^x\Big)\Big| = p^{m-2} + (p^p - p^{p-1})p^{m-p-1} = p^{m-1} = |M|.$$

It follows that all subgroups of G of order p^p and exponent p are conjugate with R. (If, in addition, $p = 2$, then G is semidihedral.) Assume that $\mathrm{d}(R) = d > 2$. Then the number of nonnormal subgroups of order p^{p-1} and exponent p in G equals $p^{m-p-1}(p + \cdots + p^{d-1})$ which is not a power of p so these subgroups are not conjugate in G; then $\mathrm{ce}_{p-1}(G) > 2$, a contradiction. Thus, $\mathrm{d}(R) = 2$ and G is as in (c).

In what follows we assume that G is not of maximal class.

(ii) Let $p = 2$ and $\mathrm{E}_4 \cong R \triangleleft G$. Since R contains two non-G-conjugate subgroups of order 2, we get $\Omega_1(G) = R$ so that $\mathrm{Z}(G)$ is cyclic and G is as in (b).

(iii) Let $p > 2$. Then G contains a normal subgroup R of order p^p and exponent p (Theorem 12.1). As above, $R = \Omega_1(G)$ and $\mathrm{d}(R) = 2$ so G is as in (a). □

§39

Alperin's problem on abelian subgroups of small index

1°. We begin with the following known

Remark 1 ([Mil2]). By Exercise 1.6(a), if G is a nonabelian p-group, then the number of abelian members in the set Γ_1 equals one of the numbers $0, 1, p+1$. By Exercise 1.7, if A is an abelian subgroup of index p^2 in G, then G contains a normal abelian subgroup of index p^2. Miller also constructed the p-group containing exactly two abelian subgroups of index p^2.

Theorem 39.1 ([Alp2, Theorem 4]). *If a p-group G, $p > 2$, has an abelian subgroup of index p^3, then it has a normal abelian subgroup of index p^3.*

For the proof of Theorem 39.1 we recommend [Kon1], where the proof of the following more general assertion is presented: Let M be a normal subgroup of a p-group G, $p > 2$ and let M contain an abelian subgroup of index p^2; then M contains a G-invariant abelian subgroup of index p^2.

Alperin [Alp2] has constructed a group of order 2^9 which has exactly two abelian subgroups of index 8, neither of which is a normal subgroup. This group may be constructed as follows: Let H be the group of order 2^8 generated by four involutions a_1, a_2, b_1, b_2 which subject only to the conditions that H be of class two and that $[a_1, a_2] = 1 = [b_1, b_2]$. An automorphism t of order 2 of H can be defined by requiring that it maps a_i to b_i and b_i to a_i for $i = 1, 2$. Then G, as the splitting extension of H by the automorphism t, has order 2^9 and is the required group.

2°. Let us consider the following assertion:

$(\mathfrak{A}_n)$ [Alp2] For $n > 0$ and a sufficiently large prime p, a p-group, containing an abelian subgroup of index p^n, contains a normal abelian subgroup of index p^n.

By Theorem 39.1, $(\mathfrak{A}_3)$ is true for all odd p and fails for $p = 2$ as was shown in the paragraph preceding this subsection. We will show that $(\mathfrak{A}_{\frac{1}{2}(p+3)})$ fails for all $p > 3$; in particular, $(\mathfrak{A}_4)$ is not true for $p = 5$. Alperin and Glauberman have contributed essentially in consideration of $(\mathfrak{A}_n)$.

Let H_1 be the greatest group of class 2, generated by p elements $a_0, a_1, \dots, a_{p-1}$ of order $p > 3$. Then $\exp(H_1) = p$ (see Exercise 1.19). It is known that $|H_1| = p^{\frac{1}{2}p(p+1)}$, $|H_1'| = p^{\frac{1}{2}p(p-1)}$, H_1 is special. Set $t = t_p = \frac{1}{2}(p-3) \in \mathbb{N}$. Let D be the subgroup of H_1, generated by pt linearly independent commutators $c_{i,j} = [a_i, a_j]$, $j > i$, $j - i \in \{1, \dots, t\}$, where $i, j \in \{0, 1, \dots, p-1\}$. Then $|D| = p^{pt}$. Set $H = H_1/D$; then $|H| = p^{\frac{1}{2}p(p+1)-pt} = p^{2p}$ and $|H'| = |H_1'/D| = p^p$. In this section all subscripts under a and c are reduced modulo p. Denote the images $a_i D$ of a_i in H_1/D by a_i again, $i = 0, 1, \dots, p-1$. In H, $c_{i,j} = 1$ if and only if $j - i \in \{1, \dots, t\}$. Let ϕ be an automorphism of H defined on generators $a_0, a_1, \dots, a_{p-1}$ as follows: $a_i^{\phi} = a_{i+1}$, $i = 0, 1, \dots, p-1$ (in particular, $a_{p-1}^{\phi} = a_0$), and extended to G. It is easy to show that such ϕ exists and $o(\phi) = p$. Let $G = \langle \phi \rangle \cdot H$ be the natural semidirect product; then $|G| = p^{2p+1}$. We will show that G does not satisfy $(\mathfrak{A}_{\frac{1}{2}(p+3)})$.

Remark 2. If G is the group constructed above, then (a) $G/H' \cong C_p \operatorname{wr} C_p$ so G is irregular; (b) $|Z(G)| = p$; (c) $G = \langle \phi, a_0 \rangle$ so $d(G) = 2$, $\Omega_1(G) = G$ and $G' = \Phi(G)$; (d) every abelian subgroup of G, not contained in H, has order $\le p^{p+1}$. Indeed, (a) and (c) are obvious. Since $d(G) = 2$, we get $Z(G) < \Phi(G) < H$ so $Z(G) < Z(H)$. Since $\langle \phi, Z(H) \rangle \cong G/H'$, (b) follows. It remains to prove (d). Let $A < G$ be a maximal abelian subgroup of G not contained in H. Set $\bar{G} = G/H'$. It follows from the structure of $\bar{G}$ that $|\bar{A}| \le p^2$. Since $C_G(H') = H$, we get $H' \not\le A$. Therefore, $|A| < |\bar{A}||H'| \le p^{p+2}$.

Take in $H - H'$ two elements

$$x = c_1 \prod_{i=0,\dots,p-1} a_i^{\beta_i}, \quad y = c_2 \prod_{i=0,\dots,p-1} a_i^{\gamma_i}, \quad c_1, c_2 \in H', \quad 0 \le \beta_i, \gamma_j \le p-1.$$

These expressions are determined uniquely. In what follows we assume that x and y have presentations given above. Let $\operatorname{supp}(x) = \{i \mid \beta_i \ne 0\}$ be the *support* of x. For $A \le H$, set $\operatorname{supp}(A) = \bigcup_{x \in A - H'} \operatorname{supp}(x)$.

We have $Z(H) = \bigcap_{i=0}^{p-1} C_H(a_i) = H'$ and

$$C_H(a_i) = \langle H', a_{i-t}, \dots, a_{i-1}, a_i, a_{i+1}, \dots, a_{i+t} \rangle$$

is of order $p^{p+2t+1} = p^{2p-2}$. Since $o(a_i) = p$ for all i, H/H' is elementary abelian so we get the following

Lemma 39.2. *H is special. Set $A_i = \langle H', a_i, a_{i+1}, \dots, a_{i+t} \rangle$, $i = 0, 1, \dots, p-1$. Then A_i is abelian of order p^{p+t+1} so that $|H : A_i| = p^{p-t-1} = p^{(p+1)/2}$.*

Note that $c_{i,j} = [a_i, a_j] \ne 1$ if and only if $j \in \{i+t+1, i+t+2\}$. Next, $[a_i^{\beta_i}, a_j^{\gamma_j}] = c_{i,j}^{\beta_i \gamma_j}$ since $\operatorname{cl}(H) = 2$.

Lemma 39.3. *Let $x, y \in H - H'$ be as above, Then for all $i = 0, 1, \dots, p-1$ and $j \in \{i+t+1, i+t+2\}$:*

(a) $[x, y] = \prod_{i,j} c_{i,j}^{\det \left(\begin{smallmatrix} \beta_i & \beta_j \\ \gamma_i & \gamma_j \end{smallmatrix}\right)}$.

(b) $xy = yx$ *if and only if* $\beta_i \gamma_j - \beta_j \gamma_i \equiv 0 \pmod{p}$.

Lemma 39.4. *A_i is a maximal abelian subgroup of H of order p^{p+t+1}, $i = 0, 1, \ldots, p-1$.*

Proof. Assume that there is $x \in C_H(A) - A$. Since A_i contains all elements whose supports are contained in $\mathcal{S}_i = \{i, i+1, \ldots, i+t\}$, we may assume that $\operatorname{supp}(x) \cap \mathcal{S}_i = \varnothing$. Then x is not permutable with at least one of the following elements of A: $a_i, a_{i+1}, \ldots, a_{i+t}$, a contradiction. □

Theorem 39.5. *Let A be an abelian normal subgroup of the above constructed p-group G, $p \geq 5$. Then $|A| \leq p^{p+1}$ so $|G : A| = p^p$. Hence, since G has an abelian subgroup A_1 of index $p^{p+1-(t+1)} = p^{p-t} = p^{\frac{1}{2}(p+3)} < p^p$, G does not satisfy $\mathfrak{A}_{\frac{1}{2}(p+3)}$.*

Proof. Let $A \lhd G$ be maximal abelian.

Suppose that $A < H$; then $Z(H) < A$ so $|A| \geq p^{p+1}$. Take $x \in A - H'$ and suppose that $\beta_i \neq 0$. Set $y = x^{\phi^{t+1}}$. Then $y \in A$ and $xy = yx$ since $A \lhd G$ is abelian. In that case, $\gamma_i = \beta_{i-t-1}$ and $\gamma_{i+t+1} = \beta_i$. By Lemma 39.3(b), we get

$$\beta_i \gamma_{i+t+1} - \beta_{i+t+1}\gamma_i = \beta_i^2 - \beta_{i+t+1}\beta_{i-t-1} \equiv 0 \pmod{p}.$$

Since $\beta_i \not\equiv 0 \pmod{p}$, we obtain $\beta_{i+t+1} \not\equiv 0 \pmod{p}$. Similarly, $\beta_{i+k(t+1)} \not\equiv 0 \pmod{p}$ for all $k \in \mathbb{N}$. Since p is a prime and $1 < t+1 < p$, we get $\operatorname{supp}(x) = \{0, 1, \ldots, p-1\}$.

Let $y \in A - H'$. Then, since $xy = yx$, we get, by the result of the previous paragraph, $\operatorname{supp}(y) = \operatorname{supp}(x) = \{0, 1, \ldots, p-1\}$ so, by Lemma 39.3(b),

$$\frac{\beta_0}{\gamma_0} = \frac{\beta_{t+1}}{\gamma_{t+1}} = \frac{\beta_{2t+2}}{\gamma_{2t+2}} = \cdots = \frac{\beta_{(p-1)(t+1)}}{\gamma_{(p-1)(t+1)}} = s^{-1},$$

or, what is the same,

$$\frac{\beta_0}{\gamma_0} = \frac{\beta_1}{\gamma_1} = \frac{\beta_2}{\gamma_2} = \cdots = \frac{\beta_{p-1}}{\gamma_{p-1}} = s^{-1},$$

for some $s \in \{1, 2, \ldots, p-1\}$. It follows from the last formula that $y = cx^s$ so that $A = \langle x, H' \rangle$ has order p^{p+1}. Thus, if A is a maximal abelian normal subgroup of G contained in H, then $|A| = p^{p+1}$.

By construction, $G/H' \cong C_p \operatorname{wr} C_p$ so $G/G' \cong E_{p^2}$ since $H' < G'$.

Now suppose that $A \not\leq H$. Then $|A| \leq p^{p+1}$, by Remark 2, so $|G : A| \geq p^p > p^{\frac{1}{2}(p+3)}$.

Thus, G has no abelian normal subgroups of order p^{p+t+1}. Since G has an abelian subgroup of order p^{p+t+1} (and index $p^{p-t} = p^{\frac{1}{2}(p+3)}$), it follows that G does not satisfy $(\mathfrak{A}_{\frac{1}{2}(p+3)})$. □

Lemma 39.6. *If $p > 3$, then normal abelian subgroups of G are contained in H.*

Indeed, if $A \triangleleft G$ is abelian and $A \not\leq H$, then $G = AH$, $G/(A \cap H) \cong (H/(A \cap H)) \times (A/(A \cap H))$ is not two-generator, a contradiction.

It is asserted in [Ber23] that $(\mathfrak{A}_4)$ is not true for $p = 3$. G. Glauberman in his letter at 14/11/07 noticed that the example given in [Ber23], does not disprove $(\mathfrak{A}_4)$ since the 3-group of order 3^{10}, constructed there, has a normal subgroup of index 3^4.

3°. Gillam [Gil1] has proved that, for metabelian p-groups, $(\mathfrak{A}_n)$ is true for all n. Below we present the nice proof of the following stronger result.

Theorem 39.7 ([Man28]). *Let G be a metabelian p-group and let $B < G$ be abelian of maximal order such that, among abelian subgroups of order $|B|$ in B^G, its intersection with G' has maximal order. Then $B \triangleleft G$.*

Lemma 39.8 ([Man28]). *Let $x \in G$ be such that $BB^x = B^x B$. Then $BB^x = B[B, x]$.*

Proof. We have

$$BB^x = \bigcup_{b \in B} Bb^x = \bigcup_{b \in B} Bb^{-1}b^x = \bigcup_{b \in B} B[b, x] \subseteq B[B, x].$$

If $b \in B$, then $[b, x] = b^{-1}b^x \in BB^x$ so $\bigcup_{b \in B} B[b, x] \subseteq BBB^x = BB^x$. □

Proof of Theorem 39.7. Let B be as in the statement and assume that $B \ntriangleleft G$. Choose $x \in \mathrm{N}_G(\mathrm{N}_G(B)) - \mathrm{N}_G(B)$. Then $B^x \neq B$, and B^x, as a subgroup of $\mathrm{N}_G(B)$, normalizes B. Set $A = BB^x$; then $A = B[B, x]$ (Lemma 39.8). Since both B and B^x are maximal abelian subgroups of $A = BB^x$, the equality $\mathrm{Z}(A) = B \cap B^x$ holds since A is nonabelian, by hypothesis. Now set

$$E = \mathrm{Z}(A)((B \cap G')[B, x]);$$

then E is a subgroup of A since $(B \cap G')[B, x]$ is a subgroup of (the abelian subgroup) G'. We have

$$BE = B\mathrm{Z}(A)(B \cap G')[B, x] = \mathrm{Z}(A)B[B, x] = \mathrm{Z}(A)A = A$$

and, since $B \cap E \geq \mathrm{Z}(A) = B \cap B^x$, we get $|E| \geq |B|$, by the product formula (indeed, $|A| = |BB^x| = \frac{|B|^2}{|\mathrm{Z}(A)|}$ and $|A| = |BE| = \frac{|B||E|}{|B \cap E|} \leq \frac{|B||E|}{|\mathrm{Z}(A)|}$). The maximality of $|B|$ implies $|E| = |B|$ since E is abelian as a subgroup of A. But $[B, x] \not\leq B$ since x does not normalize B, so $E \cap G' \geq (B \cap G')[B, x] > B \cap G'$, contrary to the choice of B. □

§40

On breadth and class number of p-groups

If G is a p-group and $x \in G$, then *breadth* $\mathrm{b}_G(x)$ of x (in G) is defined as $p^{\mathrm{b}_G(x)} = |G : \mathrm{C}_G(x)|$, and the breadth $\mathrm{b}_G(G)$ (of G) is $\max\{\mathrm{b}_G(x) \mid x \in G\}$. Clearly, $\mathrm{b}_G(G) = 0$ if and only if G is abelian. We have $p^{\mathrm{b}_G(G)} \le |G'|$.

Theorem 40.1 ([Kno] (see Exercise 2.7)). *If G is a p-group with $\mathrm{b}_G(G) = 1$, then $|G'| = p$.*

Exercise 1. The derived length of a p-group G does not exceed $\mathrm{b}_G(G) + 1$.

Knoche also has proved that if $\mathrm{b}_G(G) = 2$, then $|G'| \le p^3$.
We state some theorems from [GMMPS] and [PS1] without proofs.

Theorem 40.2 ([GMMPS] for $p > 2$, [PS1]). *Let G be a p-group. Then $\mathrm{b}_G(G) = 2$ if and only if (a) $|G'| = p^2$, or (b) $|G : \mathrm{Z}(G)| = p^3 = |G'|$. Moreover, $|A/\mathrm{Z}(G)| = p$ for every maximal abelian normal subgroup A of G.*

We define, for $x \in G$ and an abelian normal subgroup A of G,

$$p^{\mathrm{b}_A(x)} = |A : \mathrm{C}_A(x)|, \quad \mathrm{b}_A(G) = \max\{\mathrm{b}_A(x) \mid x \in G\},$$
$$B_A = \{x \in G \mid \mathrm{b}_A(x) = \mathrm{b}_A(G)\}.$$

Theorem 40.3 ([PS1]). *Let G be a p-group with $\mathrm{b}_G(G) = 3$, and let A be a maximal abelian normal subgroup of G. Then one of the following holds: (a) $|G'| = p^3$. (b) $|G : \mathrm{Z}(G)| = p^4$ and $\mathrm{b}_A(G) \in \{1,2\}$. (c) $|G'| = p^4$, $D = \bigcap_{x \in B_A}[A, x]$ has order p, and $|G/D : \mathrm{Z}(G/D)| = p^3$, $\mathrm{b}_{G/D}(G/D) = 3$; moreover, $\mathrm{b}_A(G) \in \{1,2\}$ and $|[A,G]| \in \{p, p^3\}$. (d) $p = 2$, $|G'| = 2^4$, and $\mathrm{b}_A(G) \in \{1,2\}$.*

Corollary 40.4 ([GMMPS] for $p > 3$, [PS1]). *Let G be a p-group, $p > 2$. Then $\mathrm{b}_G(G) = 3$, if and only if one of the following holds: (a) $|G'| = p^3$ and $|G : \mathrm{Z}(G)| \ge p^4$. (b) $|G : \mathrm{Z}(G)| = p^4$ and $|G'| \ge p^4$. (c) $|G'| = p^4$ and there exists $D \triangleleft G$ with $|D| = p$ and $|G/D : \mathrm{Z}(G/D)| = p^3$.*

The groups of Theorem 40.4(c) are classified in [Wil3].
Recall that $\mathrm{n}(G) = |\mathrm{Irr}_1(G)|$ and $\mathrm{mc}(G) = \frac{\mathrm{k}(G)}{|G|}$. If $N \triangleleft G$, then $\mathrm{mc}(G/N) \ge \mathrm{mc}(G)$ (Gaschütz).

Exercise 2. Let G be a group of order p^m. If $\mathrm{mc}(G) \geq \frac{1}{p}$, then $|G'| \leq p$.

Solution. Put $|G : G'| = p^k$. Since $\mathrm{k}(G) = |G|\mathrm{mc}(G) \geq p^{m-1}$, we have

$$\begin{aligned} p^m &\geq |G : G'| + \mathrm{n}(G)p^2 \geq p^k + (\mathrm{k}(G) - p^k)p^2 \\ &\geq p^k + (p^{m-1} - p^k)p^2 = p^{m+1} - (p^2 - 1)p^k. \end{aligned}$$

It follows that $p^k(p + 1) \geq p^m$ so $k \geq m - 1$; then $|G'| \leq p$.

Exercise 3. Let G be a group of order p^m and $|G'| = p^s$. Suppose that the size of every G-class, contained in $G - G'$, equals p. Then $s \leq 2$.

Solution. Assume that $s \geq 3$. We have $\mathrm{k}(G) > \frac{1}{p}(p^m - p^s) = p^{m-1} - p^{s-1}$ so

$$\begin{aligned} p^m &= |G : G'| + \sum_{\chi \in \mathrm{Irr}_1(G)} \chi(1)^2 \geq p^{m-s} + (\mathrm{k}(G) - p^{m-s})p^2 \\ &> p^{m-s} + (p^{m-1} - p^{s-1} - p^{m-s})p^2 = p^{m+1} - (p^2 - 1)p^{m-s} - p^{s+1}. \end{aligned}$$

It follows that $p^m(p - 1) < (p^2 - 1)p^{m-s} + p^{s+1} \leq (p^2 - 1)p^{m-s} + p^{s+1}(p - 1)$ so

$$p^m < p^{m-s}(p + 1) + p^{s+1} \leq p^{m-3}(p + 1) + p^{m-1} = p^{m-3}(p^2 + p + 1),$$

and we get $p^3 < p^2 + p + 1 = \frac{p^3-1}{p-1} < p^3$, a contradiction. Thus, $s \leq 2$.

Exercise 4. Let G be a group of order p^m. If $p^{m-1} < \mathrm{k}(G) \leq p^{m-1} + p^2$, then $|G'| = p$. Next, $|\mathrm{Z}(G)| \leq p^3$ with equality if $p = 2$. If $|\mathrm{Z}(G)| = p$, then G is extraspecial.

Solution. We have $\mathrm{mc}(G) > \frac{1}{p}$ so, by Exercise 2, $|G'| = p$. Then $|\mathrm{Lin}(G)| = p^{m-1}$ so $\mathrm{n}(G) \leq p^2$. Let $|\mathrm{Z}(G)| = p^z$. In that case, if $\chi \in \mathrm{Irr}_1(G)$, then we have $\chi(1)^2 \leq |G : \mathrm{Z}(G)|$ so

$$n(G) \geq \frac{|G| - |G : G'|}{|G : \mathrm{Z}(G)|} = |\mathrm{Z}(G)| - \frac{|\mathrm{Z}(G)}{|G'|} = p^z - p^{z-1}.$$

Since $\mathrm{n}(G) \leq p^2$, we get $p^2 \geq p^{z-1}(p-1)$ so $z \leq 2$ for $p > 2$ and $z \leq 3$ for $p = 2$. If $z = 1$, G is extraspecial, by Exercise 2.1.

Exercise 5 (Mann and Scoppola). If G is a p-group and $\mathrm{mc}(G) > \frac{1}{p^2}$, then G' is abelian.

Solution. If $N \lhd G$, then $\mathrm{mc}(G/N) \geq \mathrm{mc}(G) > \frac{1}{p^2}$. Let G be a counterexample of minimal order. Then G has a unique minimal normal subgroup, say N, and $N \leq G'$. Set $|G| = p^m$, $|G : G'| = p^k$. Since G' is nonabelian, we get $k \leq m - 4$ (Burnside).

Any faithful irreducible character of G has degree $\geq p^2$ [BZ, Chapter 18]. The number of faithful characters in $\mathrm{Irr}(G)$ is $\leq \frac{1}{p^4}(|G| - |G/N|) = \frac{1}{p^4}(p^m - p^{m-1}) = p^{m-4} - p^{m-5}$. We have

$$\begin{aligned}
\mathrm{k}(G/N) &\leq p^k + \frac{1}{p^2}(p^{m-1} - p^k) = p^{m-3} + p^{k-2}(p^2 - 1) \\
&\leq p^{m-3} + p^{m-6}(p^2 - 1), \\
p^{m-2} &< |G| \cdot \mathrm{mc}(G) = \mathrm{k}(G) = |\mathrm{Irr}(G)| \\
&\leq p^{m-3} + p^{m-4} - p^{m-6} + p^{m-4} - p^{m-5}
\end{aligned}$$

so $p^4 + p + 1 \leq p^3 + 2p^2$, which is impossible. Thus, G' is abelian.

§41

Groups in which every two noncyclic subgroups of the same order have the same rank

Let Θ be a group theoretic property. A group G is said to be a *minimal non-Θ-group* if G is not a Θ-group but all its proper subgroups are Θ-groups. As a rule, we consider only properties inherited by subgroups (otherwise, minimal non-Θ-groups need not necessarily exist). N. Blackburn [Bla5, Theorem 3.2] has proved the following deep and important

Theorem 41.1. *Let G be a minimal nonmetacyclic p-group. Then G is one of the following groups: (a) a group of order p^3 and exponent p; (b) the irregular group of order 3^4 with $\Omega_1(G) \cong E_9$; (c) the direct product $C_2 \times Q_8$; (d) the central product $Q_8 * C_4 \cong D_8 * C_4$ of order 2^4; (e) a special group of order 2^5, $d(G) = 3$, $\Omega_1(G) \cong E_4$.*

For two elementary and easy proofs which are different from the original one, see Theorems 66.1 and 69.1. Theorem 41.1 is crucial in what follows.

Remarks. 1. Let G be a minimal nonmetacyclic p-group. It follows from Theorem 41.1 that (i) if $p = 2$ then $d(G) = 3$ and the exponent of G does not exceed 4, and (ii) if $p > 2$ then $d(G) \leq 3$ and $\exp(G) = p$ or 3^2.

2. It follows from Remark 1 that a p-group G is metacyclic if and only if one of the following holds: (i) $\Omega_2(G)$ is metacyclic, (ii) every subgroup of G, generated by three elements, is metacyclic.

Lemma 41.2. *Let G be a group of order 2^m, $m > 4$. Suppose that all noncyclic subgroups of G of order 2^4 are minimal nonmetacyclic. Then $G = D_8 * Q_8$ (the central product) is extraspecial of order 2^5.*

Proof. If H is minimal nonmetacyclic of order 2^4, then $H \in \{Q_8 \times C_2, Q_8 * C_4\}$. If G has a subgroup $E \cong E_8$ and $E < F < G$ with $|F : E| = 2$, then $F \ncong H$, a contradiction.

(i) Assume that G has a cyclic subgroup $Z \cong C_8$. Note that G is not of maximal class. Let $E_4 \cong R \triangleleft G$ (Lemma 1.4) and let $R_1 \leq R \cap Z(G)$ be of order 2. If $R_1 \leq Z$, then RZ is metacyclic of order 2^4, a contradiction. If $R_1 \nleq Z$, then $R_1 Z = R_1 \times Z$ is metacyclic of order 2^4, a contradiction. Thus, $\exp(G) = 4$.

(ii) Let A be a noncyclic maximal normal abelian subgroup of G; then A is of type $(4, 2)$. By Taussky's theorem, $|G : G'| > 2^2$. Since $C_G(A) = A$, we have $|G : A| \leq |\mathrm{Aut}(A)| = |D_8| = 8$ so $m \leq 6$. Then $|G'| \leq 8$ so G' is abelian. Set $R = \Omega_1(A)$. Since R centralizes $\Phi(G) \geq G'$, RG' is G-invariant noncyclic abelian, and we may assume that $RG' \leq A$. Since $\mathrm{Aut}(A) \cong D_8$ and G/A is abelian, we have $|G : A| = 2^2$, i.e., $m = 5$.

Let $M \in \Gamma_1$; then M is minimal nonmetacyclic. By Theorem 41.1, $|M'| = 2$ and $M/M' \cong E_{2^3}$. Assume that there is $H \in \Gamma_1 - \{M\}$ such that $H' \neq M'$. Then $H' \leq \Phi(H) \leq \Phi(G) < M$ and M/H' is nonabelian. Since $d(M) = 3$, we have $M = H' \times Q$, where $Q \cong Q_8$ (Theorem 41.1). Since $H/H' \cong E_8$, considering the intersection $(M/H') \cap (H/H')$, we get a contradiction: on the one hand, it is the four-group as a subgroup of H/H', and, on the other hand, it is cyclic as a subgroup of $M/H' \cong Q_8$. Thus $H' = M'$ for all $H \in \Gamma_1$. It follows that all maximal subgroups of G/M' are elementary abelian so $G/M' \cong E_{16}$. Then $M' = G' = \Phi(G)$. To prove that G is extraspecial, it suffices to show that $Z(G) = G'$. Assume that this is false. Then $|Z(G)| = 4$. If F is a maximal subgroup of G not containing $Z(G)$ (by assumption, $Z(G) \not\leq G' = \Phi(G)$), then $G = FZ(G)$ so $Z(F) \leq Z(G)$. Since $|Z(F)| = 4 = |Z(G)|$ and $Z(G) \not\leq F$, we get a contradiction. Thus, $Z(G) = G' = \Phi(G)$ is of order 2 so G is extraspecial. Since G has no subgroups isomorphic to E_8, we get $G = D_8 * Q_8$, and this group indeed satisfies the hypothesis. □

Lemma 41.3. *Let G be a group of order 2^m, $m > 5$. If all noncyclic subgroups of G of order 2^5 are minimal nonmetacyclic, then G is special of order 2^6 and $Z(G) = \Omega_1(G) \cong E_4$.*

Proof. If $M < G$, where $|M| = 2^5$ is noncyclic, then, by Theorem 41.1, $\exp(M) = 4$ so $\exp(G) = 4$ and $G' \geq M' \cong E_4$. Let $K \leq Z(G) \cap G'$ be of order 2. Then all subgroups of G/K of order 16 are minimal nonmetacyclic (Theorem 41.1(e)) so, by Lemma 41.2, G/K is extraspecial of order 2^5; then $m = 6$ and $|G'| = 4$. Then $G' = \Phi(G)$ is of order 4. If $M/\Phi(G)$ is maximal in $G/\Phi(G)$, then M is special (Theorem 41.1(e)) and we have $\Phi(G) = Z(M) \cong E_4$ and $\exp(M/\Phi(G)) = 2$ (Theorem 41.1(e)). If $N/\Phi(G)$ is another maximal subgroup of $G/\Phi(G)$, then again $\Phi(G) = Z(N)$. It follows that $C_G(\Phi(G)) \geq MN = G$ so $G' = \Phi(G) = Z(G) \cong E_4$ and G is special. □

Theorem 41.4. *Suppose that G is a p-group such that, whenever $A, B < G$ are noncyclic of the same order, then* $d(A) = d(B)$. *Then one of the following holds:*

(a) all proper subgroups of G are nonmetacyclic;

(b) G is elementary abelian;

(c) $G = \Omega_1(G)C_{p^2}$, where $\Omega_1(G)$ is nonabelian of order p^3 and exponent $p > 2$;

(d) G is the group of Lemma 41.2;

(e) G is the group of Lemma 41.3;

(f) G is a 3-group of maximal class $\not\cong C_3 \mathrm{wr}\, C_3$.

Proof. Groups (a)–(f) satisfy the hypothesis. We may assume that G is not metacyclic and $|G| > p^3$.

(i) Suppose that there is $E_{p^3} \cong E < G$. Assume that G has also a cyclic subgroup Z of order p^2. By Lemma 1.4, G has a normal abelian subgroup R of type (p, p). If $Z \cap R > \{1\}$, then $|RZ| = |E|$ and $\mathrm{d}(RZ) = 2 < 3 = \mathrm{d}(E)$, a contradiction. If $Z \cap R = \{1\}$, take a subgroup R_1 of order p in $R \cap \mathrm{Z}(G)$. Then $|R_1 Z| = |E|$ and $\mathrm{d}(R_1 Z) = 2 < 3 = \mathrm{d}(E)$, a contradiction. Thus Z does not exist so $\exp(G) = p$. It follows that, for $p = 2$, G is elementary abelian. Let $p > 2$. Then G has no nonabelian subgroups of order p^3 and exponent p since $E < G$. Therefore, by Exercise 1.8a, G has no minimal nonabelian subgroups so it is elementary abelian. In what follows we assume that G has no subgroups isomorphic to E_{p^3}; then $\exp(G) > p$.

(ii) Suppose that $p > 2$. By (i) and Theorem 13.7, one of the following holds: (1) G is metacyclic. (2) $G = \Omega_1(G)C$, where $\Omega_1(G)$ is nonabelian of order p^3 and exponent p and C is cyclic with $|C : \mathrm{Z}(G)| \leq p$. (3) G is a 3-group of maximal class $\not\cong C_3 \,\mathrm{wr}\, C_3$. If G is as in (2), then it is easy to check that $|C| = p^2$ (otherwise, G contains two subgroups of order p^4 of ranks 2 and 3, respectively). Groups from (1) and (3) satisfy the hypothesis. The case $p > 2$ is complete.

Now suppose that $p = 2$. Let H be a minimal nonmetacyclic subgroup of the least order in G. We may assume that $H < G$. By (i) and the second paragraph of the proof and Theorem 41.1, $|H| \in \{2^4, 2^5\}$ and all subgroups of G of order $< |H|$ are metacyclic.

(iii) Suppose that $|H| = 2^4$. If K is a noncyclic subgroup of G of order 2^4, then K is minimal nonmetacyclic since $\mathrm{d}(K) = \mathrm{d}(H) = 3$ and all maximal subgroups of K are metacyclic so, by Lemma 41.2, G is as in (d).

(iv) Let $|H| = 2^5$ and let $K < G$ be noncyclic of order 2^5; then $\mathrm{d}(K) = \mathrm{d}(H) = 3$. By the choice of H, every subgroup L of K of order 2^4 is metacyclic, and we conclude that K is minimal nonmetacyclic so, by Lemma 41.3, G is as in (e). □

Suppose that G is a p-group, $p > 2$, such that whenever A, B are two noncyclic subgroups of the same order in G, then $|A : \mho_1(A)| = |B : \mho_1(B)|$. Then one of the following holds: (i) $\exp(G) = p$, (ii) G is metacyclic or minimal nonmetacyclic of order 3^4. Indeed, if G has a subgroup of order p^3 and exponent p, then, as in the proof of Theorem 41.4, $\exp(G) = p$. Otherwise, the result follows from Theorem 13.7. The case $p = 2$ of the above question has considered in Theorem 41.4 since then $\mho_1(G) = \Phi(G)$.

We omit the easy proof of the following

Proposition 41.5. *Let G be a p-group such that whenever $A, B < G$ are noncyclic and $|A| = |B|$, then $c_1(A) = c_1(B)$. Then one of the following holds: (a) $\exp(G) = p$. (b) G is metacyclic not semidihedral. (c) G is a 3-group of maximal class with $c_1(G) = 4$, (d) G is as in Lemma 41.3.*

Exercise 1. Suppose that G is p-group such that, whenever $A < B \le G$, then $|A : A'| < |B : B'|$. Then G is abelian. (*Hint.* A minimal counterexample is minimal nonabelian. Then, if $A \in \Gamma_1$, then $|A : A'| = |G : G'|$, a contradiction).

Exercise 2. Classify the p-groups G such that every proper subgroup of G is either abelian or metacyclic.

Hint. One may assume that G is neither minimal nonabelian nor minimal nonmetacyclic. Then there is a nonmetacyclic $A \in \Gamma_1$ so A is abelian. Since G has a metacyclic maximal subgroup, it follows that $\mathrm{d}(A) = 3$. By hypothesis, all maximal subgroups of G, containing $\Omega_1(A)$, are abelian so one of the following holds: (i) G has at least two maximal subgroups containing $\Omega_1(A)$, (ii) A is a unique maximal subgroup of G containing $\Omega_1(A)$. In case (i), $|G'| = p$ and $\Omega_1(A) \le \mathrm{Z}(G)$; then, if $M \in \Gamma_1$ is nonabelian, then $G = M \times C$, where $|C| = p$ and M is minimal nonabelian. In case (ii), $G/\Omega_1(A)$ is cyclic so either $|G| = p^4$ or G is an L_3-group (see §§17,18).

§42

On intersections of some subgroups

For definition of L_s- and U_2-groups, see §§17, 18.

The following lemma is cited in many places of this book.

Lemma 42.1. *Let G be a group of order p^m, satisfying $|\Omega_2(G)| \le p^{p+1} < |G|$. Then one of the following holds: (a) G is an L_p-group. (b) G is absolutely regular. (c) $p = 2$, G is metacyclic and $G = \langle a, b \mid a^{2^{m-2}} = b^8 = 1, a^b = a^{-1}, a^{2^{m-3}} = b^4, m > 4\rangle$. Here $\mathrm{Z}(G) = \langle b^2\rangle$, $G' = \langle a^2\rangle$ and G/G' is abelian of type $(4,2)$, $\Phi(G) = \langle a^2, b^2\rangle$, $\Omega_2(G) = \langle a^{2^{m-4}}, b^4\rangle$.*

Proof. Since $\Omega_2(G) < G$, then G is not of maximal class and $\exp(G) > p^2$. Suppose that G is not absolutely regular. Then there is $M \lhd G$ of order p^p and exponent p (Theorem 12.1(a)). Then $|\Omega_1(G/M)| = p$ so G/M is either cyclic or generalized quaternion. If G/M is cyclic, then $\Omega_1(G) = M$ so G is an L_p-group.

In what follows we assume that $p = 2$ and G/M is generalized quaternion. Then $\Omega_2(G)$ is abelian of type $(4,2)$ so G is metacyclic (Remark 41.2(i)) and, by Taussky's theorem, G/G' is abelian of type $(4,2)$.

Let $m = 5$. Then G is not minimal nonabelian since $|G'| = 4$ so $\mathrm{C}_G(M) \in \Gamma_1$ is abelian. Then $\mathrm{Z}(G) \cong \mathrm{C}_4$ (Lemma 1.1). Note that $\mathrm{Z}(G) \ne G'$ (otherwise, G is minimal nonabelian so $|G'| = 2$).

Now suppose that $m > 5$. Let $F/M < G/M$ be nonabelian of order 8; then $\mathrm{Z}(F) \cong \mathrm{C}_4$ and $\mathrm{Z}(F) \ne F' < G'$ so $\mathrm{Z}(F')$ is independent of the choice of F since $\mathrm{c}_2(G) = 2$. Since such subgroups, as F, generate G, we conclude that $\mathrm{Z}(F) \le \mathrm{Z}(G)$. Considering $\Omega_2(G)$, we see that $\mathrm{Z}(G)$ is cyclic. Since $\mathrm{d}(G) = 2$, we get $\mathrm{Z}(G) < \Phi(G)$ so $\Phi(G) = G'\mathrm{Z}(G)$ and $|G' \cap \mathrm{Z}(G)| = 2$ hence $|\mathrm{Z}(G)| = 4$, by the product formula, and $\Phi(G)$ is abelian of type $(2^{m-3}, 2)$. We have $\mathrm{C}_G(M) \in \Gamma_1$. Since $|\Omega_2(\mathrm{C}_G(M))| = 8$, it follows from the above that $\mathrm{C}_G(M)$ is abelian of type $(2^{m-2}, 2)$. We conclude that $\mathrm{c}_{m-2}(G) = 2$. Let $A < G$ be a maximal cyclic subgroup containing G'; then $G' < A$ and $G/A \cong \mathrm{C}_4$ since $\Phi(G)$ is noncyclic. Take $bM \in (G/M) - (AM/M)$; then $o(bM) = 4$ (recall that $G/M \cong \mathrm{Q}_{2^{m-2}}$) so $o(b) = 8$. Write $A = \langle a\rangle$. We have $b^4 \in A$ and $b^2 \in \mathrm{C}_G(A)$ so b induces on A an automorphism of order 2.

As $G/\mathrm{Z}(G) \cong \mathrm{D}_{2^{m-2}}$, we get $\mathrm{cl}(G) = m - 2$. We have

$$
\begin{aligned}
G &= G_i \\
&= \langle a, b \mid a^{2^{m-2}} = b^8 = 1,\ a^b = a^{-1+i\cdot 2^{m-3}},\ a^{2^{m-3}} = b^4,\ m > 4,\ i \in \{0,1\}\rangle
\end{aligned}
$$

and $Z(G) = \langle b^2 \rangle$. It remains to show that $G_0 \cong G_1$. This can be seen in the following way. Set $a^{2^{m-3}} = b^4 = z$ and $a^{2^{m-4}} = v$. Then $t = b^2 v$ is an involution which centralizes a. Also, $v^b = v^{-1} = vv^2 = vz$. Thus, $t^b = (b^2 v)^b = b^2 vz = tz$. Now, if $i = 1$, then $a^b = a^{-1}z$. We replace a with $a' = at$ and get $(a')^b = (at)^b = (a^{-1}z)(tz) = a^{-1}t = (at)^{-1} = (a')^{-1}$. Also, note that $o(a') = o(a) = 2^{m-2}$ and $(a')^{2^{m-3}} = b^4$. Thus, replacing a with a', we get the relations for $i = 0$, and so we may assume from the start that $i = 0$, i.e., $a^b = a^{-1}$ and $G_1 \cong G_0$. (The above argument is due to Janko.) □

Definition. Let $n \in \mathbb{N}$, $i \in \mathbb{N} \cup \{0\}$. A p-group G is said to be R_n^i-group if it is not absolutely regular and if, whenever A, B are two nonincident subgroups of indices p^n, p^{n+i} in G, respectively, then $A \cap B$ is absolutely regular.

Obviously, every p-group of order $\leq p^{n+i+p}$ is an R_n^i-group so such group is said to be a trivial R_n^i-group. The R_n^i-groups ($i = 1, 2, 3$) are studied in [Ber3]. By Theorems 9.5 and 9.6, all p-groups of maximal class are R_n^0-groups for all n.

Theorem 42.2. *Let G be a nontrivial R_n^0-group of order p^m. If G is neither of maximal class nor an L_p-group, then $p = 2$, $m = n + 3$, $n > 1$ and G is a group of Lemma 42.1(c).*

Proof. By Theorem 12.1(a), G has a normal subgroup M of order p^p and exponent p. By hypothesis, $|G : M| > p^n$ and G/M has only one subgroup of index p^n. If $m - n - p > 1$, then G/M is cyclic (Proposition 1.3). If G/M is not cyclic, then $m-n-p = 1$ and G/M is generalized quaternion, $p = 2, m = n+3, |G/M| = 2^{n+1}$.

(i) Let G/M be cyclic. We have $|G/M| > p^n > p$. Assume that $M < \Omega_1(G)$; then $|\Omega_1(G)| = p^{p+1}$ and $\exp(\Omega_1(G)) = p$ (Lemma 17.4). Let $H \in \Gamma_1$ be such that $M \nleq H$. Set $M_1 = \Omega_1(G) \cap H$; then M_1 is G-invariant of order p^p and exponent p. Let $A/M_1 < H/M_1$ be of index p^{n-1}; then $|G : A| = p^n$. Let $B/M < G/M$ be of index p^n. Then $A \neq B$ since $M \nleq A$ and $M < B$. It follows from $M_1 \leq A \cap B$ that $A \cap B$ is not absolutely regular, a contradiction. Thus, $M = \Omega_1(G)$ so G is an L_p-group.

(ii) Now let G/M be generalized quaternion and T/M a cyclic subgroup of index 2 in G/M. It is clear that T is an R_{n-1}^0-group. By the result of the previous paragraph, $\Omega_1(T) = M$. It follows that $\Omega_1(G) = M$ and $|\Omega_2(G)| = 2^3$ so G is as in Lemma 42.1(c). □

Exercise 1. Let G be a nontrivial R_n^1-group of order p^m. Then G is an R_{n+1}^0-group.

Exercise 2. If any two distinct maximal subgroups of a nonabelian p-group G have abelian intersection, then $\Phi(G)$ is abelian and either $d(G) = 2$ or $d(G) = 3$ and $\Phi(G) \leq Z(G)$. (*Hint.* Let $H < G$ be minimal nonabelian. Then $H\Phi(G) \in \Gamma_1$ so $d(G) \leq 3$. Let $d(G) = 3$ and $\Phi(G) < T < G$ with $|T : \Phi(G)| = p$; then T must be abelian: T is the intersection of two distinct maximal subgroups of G. Since such T generate G, we get $\Phi(G) \leq Z(G)$.)

Exercise 3. Suppose that G is a nonabelian p-group such that whenever $A, B < G$ are distinct maximal abelian with $|A \cap B|$ maximal possible, then $A/(A \cap B)$ is cyclic. Prove that at least one of indices $|A : (A \cap B)|$, $|B : (A \cap B)|$ equals p. (*Hint.* Set $C = C_G(A \cap B)$. Then $C/(A \cap B)$ has a nontrivial cyclic partition. In that case, either $\exp(C/(A \cap B)) = p$ or $p = 2$ and $C/(A \cap B)$ is dihedral (see §20).)

Exercise 4. Classify the noncyclic p-groups that are not generated by noncyclic subgroups of index p^2.

§43

On 2-groups with few cyclic subgroups of given order

1°. By Theorem 1.17(a), if a 2-group G is neither cyclic nor a 2-group of maximal class, then the number of involutions in G is $\equiv 3 \pmod 4$. In this subsection we study 2-groups with exactly 3 involutions. Such groups we call Ω-*groups* (see also §82 for classification of Ω-groups).

If a nonmetacyclic 2-group G is an Ω-group of order > 4, then $R = \Omega_1(G) \cong \mathrm{E}_4$. Let A be a maximal abelian normal subgroup of exponent ≤ 4 containing R; then $R < A$ and A is either of type $(4,2)$ or $(4,4)$. By Theorem 10.1, we have $\Omega_2(\mathrm{C}_G(A)) = A$.

Theorem 43.1 ([Kon1]). *Suppose that G is a nonmetacyclic Ω-group that has no normal abelian subgroups of type $(4,4)$. Then G contains a normal abelian subgroup A of type $(4,2)$ such that*

(a) $\mathrm{C}_G(A)$ is abelian of type $(2^n, 2)$.

(b) Let $n > 2$. Then $|G/\mathrm{C}_G(A)| \leq 4$ if $\mathrm{C}_G(A) > A$. If $|G/\mathrm{C}_G(A)| = 4$, then $G/\mathrm{C}_G(A) \cong \mathrm{E}_4$.

Proof. As we know, G has a normal abelian subgroup A of type $(4,2)$ such that $\Omega_2(\mathrm{C}_G(A)) = A$ (Theorem 10.1) so $|\Omega_2(\mathrm{C}_G(A))| = 8$. By Lemma 42.1, $\mathrm{C}_G(A)$ is an L_2-group so it is abelian of type $(2^n, 2)$. Next, $G/\mathrm{C}_G(A)$ is isomorphic to a subgroup of $\mathrm{Aut}(A) \cong \mathrm{D}_8$,

Suppose that $n > 2$. Then $\mho_1(\mathrm{C}_G(A))$ is cyclic of order $2^{n-1} \geq 4$. Hence, A has a G-invariant cyclic subgroup Z of order 4. We have $A = Z \times Z_1$ where $|Z_1| = 2$. Then $\mathrm{C}_G(A) = \mathrm{C}_G(Z) \cap \mathrm{C}_G(Z_1)$. As $|G : \mathrm{C}_G(Z)| \leq 2$ and $|G : \mathrm{C}_G(Z_1)| \leq 2$, we get $|G : \mathrm{C}_G(A)| \leq 4$. The last assertion also follows from the proof. □

Lemma 43.2 (= Remark 41.2(i)). *A 2-group G is metacyclic if and only if $\Omega_2(G)$ is metacyclic.*

Theorem 43.3 ([Kon1]). *Let $\mathrm{C}_4 \times \mathrm{C}_4 \cong A \triangleleft G$, where G is an Ω-group.*

(a) If $\Omega_1(A) \leq \mathrm{Z}(G)$, then G has a normal metacyclic subgroup M of index ≤ 4.

(b) If $\Omega_1(A) \not\leq \mathrm{Z}(G)$ and $|\Omega_3(\mathrm{C}_G(A))| = 2^6$, then $|G/\mathrm{C}_G(A)| \leq 8$.

(c) If $\Omega_1(A) \not\leq \mathrm{Z}(G)$ and $|\Omega_3(\mathrm{C}_G(A))| \leq 2^5$, then $\mathrm{C}_G(A) \cong \mathrm{C}_{2^n} \times \mathrm{C}_4$.

(d) If $\Omega_1(A) \not\leq \mathrm{Z}(G)$ and $|G/\mathrm{C}_G(A)| = 2^4$, then $\mathrm{C}_G(A) = A$ and $|G| = 2^8$.

Note that $\Omega_2(\mathrm{C}_G(A)) = A$, by Theorem 10.1. Therefore, by Lemma 43.2, $\mathrm{C}_G(A)$ is metacyclic. It is easy to deduce from hypothesis of (a) that $|G/\mathrm{C}_G(A)| \leq 8$. In general, a desired normal metacyclic subgroup of index 4 does not coincide with $\mathrm{C}_G(A)$. We omit the proof of Theorem 43.3.

2°. In this subsection we classify the 2-groups G with $\mathrm{c}_n(G) = 2$, $n > 1$.

Proposition 43.4. *Suppose that G is a group of order $2^m > 2^4$ and, for some fixed $r \in \{2, \dots, m-2\}$, we have $\mathrm{s}_r(G) = 3$. Then either G is an L_2-group or $r = 2$ and G is a group of Lemma 42.1(c).*

Proof. It is easy to check that G is not of maximal class so it contains a normal subgroup $R \cong \mathrm{E}_4$. If A/R and B/R are two distinct subgroups of order 2^{r-1}, then A and B contain together at least five distinct subgroups of order 2^r, which is not the case. Thus, G/R has only one subgroup of order 2^{r-1}. If $r - 1 > 1$, then G/R is cyclic so G is an L_2-group. Suppose that G/R is noncyclic. Then $p = 2 = r$ and G/R is generalized quaternion. It is easy to see that G is as in Lemma 42.1(c). □

Proposition 43.5. *Suppose that a group G of order p^m, $p > 2$, has exactly one noncyclic subgroup of order p^{r+1} for some fixed r with $1 \leq r < m - 1$. Then one of the following holds:*

(a) $r = 1$, G is either metacyclic or a 3-group of maximal class without subgroups of order 3^3 and exponent 3,

(b) $r > 1$, G is an L_2-group.

Proof. If $r = 1$, G has no subgroups of order p^3 and exponent p; in that case we obtain groups from (a), by Theorem 13.7. Now suppose that $r > 1$. Then G contains a normal subgroup $M \cong \mathrm{E}_{p^2}$. In that case, G/M is cyclic since it has only one subgroup of order p^{r-1}. It is easy to see that then G is as in (b). □

Theorem 43.6. *Suppose that a group G of order $2^m > 2^3$ has exactly two cyclic subgroups U and V of order 4; set $A = \langle U, V \rangle$. Then A is abelian of type $(4, 2)$ and one of the following holds:*

(a) $|\Omega_2(G)| = 8$ (see Lemma 42.1).

(b) $G = \mathrm{D}_{2^{m-1}} \times \mathrm{C}_2$.

(c) $U, V \triangleleft G$, $|G : \mathrm{C}_G(U)| = 2 = |G : \mathrm{C}_G(V)|$ and all elements in the set $I = G - (\mathrm{C}_G(U) \cup \mathrm{C}_G(V))$ are involutions, $\mathrm{c}_1(G) = 2^{m-2} + 3$, $\Omega_2(\mathrm{C}_G(U)) = A = \Omega_2(\mathrm{C}_G(V))$. If $z \in I$, then $\mathrm{C}_G(z)$ is elementary abelian of order 8. Set $D = \langle I \rangle$; then $|G : D| = 2$, $C = D \cap \mathrm{C}_G(U)$ is abelian of type $(2^{m-3}, 2)$, $D - C = I$.

Proof. Clearly, $|G : \mathrm{N}_G(U)| \leq 2$ and $\mathrm{N}_G(U) > V$ (otherwise, U is characteristic in $\mathrm{N}_G(U)$ and then $\mathrm{N}_G(U) = G$ contains V), i.e., V normalizes U, and, by symmetry,

U normalizes V. Then $A = UV \le G$ is of order 8 with $c_2(A) = 2$ so A is abelian of type $(4, 2)$. We have $A \lhd G$.

Suppose that $U \le Z(G)$. If $x \in G - U$ is an involution, then $\langle x, U\rangle = \langle x\rangle \times U = A$ so A contains all elements of G of order ≤ 4. In that case, $\Omega_2(G) = A$ is of order 8 so G is as in (a). Next we assume that $U, V \not\le Z(G)$ so $\exp(Z(G)) = 2$; then $Z(G) < A$.

Set $H = C_G(A)$. Then G/H is isomorphic to a subgroup of $\mathrm{Aut}(A) \cong D_8$. By the above, $\Omega_2(H) = A$ so H is abelian with cyclic subgroup of index 2, by Lemma 42.1. In the sequel we assume that $\Omega_2(G) > A$. Take an involution $y \in G - A$; then $y \in G - H$. (If y does not exist, then $\Omega_2(G) = A$ so G is as in Lemma 42.1.)

Suppose that all elements in the set $G - H$ are involutions. Then $|G : H| = 2$ (Burnside). In that case, $G = D_{2^{m-1}} \times C_2$ is as in (b). Next we assume that there is $x \in G - H$ with $o(x) > 2$; then $o(x) > 4$.

Assume that $G/H \cong D_8$. Let $x^2 \notin H$ and $U < \langle x\rangle$. Then $C_G(U) \ge \langle x, H\rangle$, and the last subgroup is of index 2 in G. Then $|G : C_G(U)| = 2$ since $U \not\le Z(G)$. If U, V are conjugate in G, then also $|G : C_G(V)| = 2$. If U, V are not conjugate in G, then $V \lhd G$, and again $|G : C_G(V)| = 2$. Since in that case $C_G(UV) = H = C_G(U) \cap C_G(V)$, we get $|G : H| \le 4$, a contradiction. Thus, $|G : H| \le 4$.

Assume that $G = \langle x, H\rangle$ and $o(x) > 4$ and, say $U < \langle x\rangle$, then $C_G(U) = G$, $U \le Z(G)$, a contradiction.

It remains to consider the case where $G/H \cong E_4$. Let $x \in G - H$ be of order > 2 and suppose that $U < \langle x\rangle$. Then $|G : C_G(U)| = 2$, and we also have $|G : C_G(V)| = 2$. Since $|G : H| = 4$, we have $H = C_G(U) \cap C_G(V)$ so $C_G(U) \ne C_G(V)$. Set $I = G - (C_G(U) \cup C_G(V))$ and take $z \in I$. Then $C_G(z)$ has no elements of order 4 so it is elementary abelian. Since $\langle z, H\rangle$ is not of maximal class, $|C_H(z)| > 2$ (Proposition 1.8). Since $G = C_G(U)C_G(z)$, we see that $C_G(z) \cap C_G(U) = \Omega_1(C_G(U)) = \Omega_1(A)$ so that $C_G(z) \cong E_8$ whence $I \subset C_G(\Omega_1(A))$. Set $D = \langle I\rangle$. Since $|I| = 2^{m-2}$ and $1 \notin I$, we get $|D| \ge 2^{m-1}$. By the above, $C_G(\Omega_1(A)) \ge \langle I\rangle = D$. It follows from $G > C_G(U) > H = C_G(A)$ and $A = U\Omega_1(A)$ that $\Omega_1(A) \not\le Z(C_G(U))$ so $\Omega_1(A) \not\le Z(G)$. Therefore, $D < G$ so $|G : D| = 2$ and $C_G(\Omega_1(A)) = D$. Set $C = C_G(U) \cap D$; then $|C| = 2^{m-2}$, by the product formula, and $c_1(C) = 3$ since exactly three involutions centralize U. It follows that all elements in $D - C$ are involutions (recall that $I \subset D$ and $I \cap C_G(U) = \varnothing$; next, $|D - C| = 2^{m-2} = |I|$). We deduce from $G = C_G(U) \cup C_G(V) \cup D$ that $c_1(G) = 2^{m-2} + 3$. □

Theorem 43.7. *Let $n > 2$ and let G be a group of order 2^m, $m > n$. If $c_n(G) = 2$, then G is either an L_2- or U_2-group.*

Proof. Since G is not of maximal class, it has a normal subgroup R of type $(2, 2)$ (Lemma 1.4). Let Z_1 be cyclic of order 2^n. Set $H = Z_1R$. If $Z_1 \cap R = \{1\}$, then $c_n(H) = \frac{|H| - |\Omega_{n-1}(H)|}{\varphi(2^n)} = 4 > 2 = c_n(G)$, a contradiction. Thus, H is an L_2-group. Note that H is generated by its cyclic subgroups of order 2^n. Therefore it follows from $c_n(G) = 2 = c_n(H)$ that H is a unique L_2-subgroup of G of order 2^{n+1}, and the result now follows from §18, Theorem $B(p)$ with $p = 2$. □

Theorem 43.8 ([Ber14]). *Suppose that G is a group of order $2^m \geq 2^{n+2}$ such that $\mathrm{sol}_n(G) \not\equiv 0 \pmod{2^{n+2}}$ for some $n > 2$, where $\mathrm{sol}_n(G)$ is the number of solutions of $x^{2^n} = 1$ in G. Then G is either of maximal class or a U_2- or L_2-group.*

This follows from identity $\mathrm{sol}_n(G) = 1 + \sum_{i=1}^{n} 2^{i-1}\mathrm{c}_i(G)$ and Corollary 18.7.

Theorem 43.9. *Suppose that G is a 2-group such that $\mathrm{c}_1(G) = 7$ and $\Omega_1(G)$ is nonabelian. Then $\Omega_1(G) = \mathrm{D}_8 * \mathrm{C}_4$ is of order 2^4 and one of the following holds:*

(a) *$G = D * C$, where $D \cong \mathrm{D}_8$, C is cyclic (of order > 2), $D \cap C = \mathrm{Z}(D)$.*

(b) *$G = DC$, where $D \cong \mathrm{D}_8$ is normal in G, C is nonnormal cyclic of index 4 in G, $D \cap C = \mathrm{Z}(D)$ and $\mathrm{C}_G(D) = \mathrm{Z}(G)$ has index 2 in C, $G/\mathrm{C}_G(D) \cong \mathrm{D}_8$.*

(c) *$G = DQ$, where $D \cong \mathrm{D}_8$ is normal in G, Q is a nonnormal generalized quaternion group of index 4 in G, $D \cap Q = \mathrm{Z}(D)$, $\mathrm{C}_G(D)$ is a cyclic subgroup of index 2 in Q, $G/\mathrm{C}_G(D) \cong \mathrm{D}_8$. If L is a cyclic subgroup of order 4 in Q such that $L \not\leq \mathrm{C}_G(D)$, then $DL \cong \mathrm{SD}_{24}$.*

Proof. Since $\mathrm{c}_1(G) = 7 \equiv 3 \pmod 4$, G is not of maximal class. Since $\Omega_1(G)$ is nonabelian, there exist two noncommuting involutions $u, v \in G$. Set $D = \langle u, v\rangle$; then D is dihedral. Since $\mathrm{c}_1(D) \equiv 1 \pmod 4$, we get $\mathrm{c}_1(D) < 7$ so $D \cong \mathrm{D}_8$. Set $C = \mathrm{C}_G(D)$; then $C \cap D = \mathrm{Z}(D)$ and $C \not\leq D$ (Proposition 10.17). If $x \in C - \mathrm{Z}(D)$ is an involution, then $\langle D, x\rangle = D \times \langle x\rangle$ has $11 > 7$ involutions, a contradiction. Thus, C has only one involution so it is either cyclic or generalized quaternion.

Let C be generalized quaternion and $\mathrm{Q}_8 \cong Q \leq C$. Then $\mathrm{c}_1(D * Q) = 11$ (Appendix 16), a contradiction. Thus, C is cyclic and $|C| > 2$. Next, $\mathrm{c}_1(D * C) = 7 = \mathrm{c}_1(G)$ and $H = \Omega_1(G) = \Omega_1(D * C) = D * \Omega_2(C)$ is characteristic of order 16 in G. Next, $C \triangleleft G$ since $C = \mathrm{Z}(\mathrm{C}_G(D))$. By Appendix 16, H contains exactly one quaternion subgroup and three dihedral subgroups so H has a G-invariant subgroup isomorphic to D_8 which we denote by D again. In that case, G/C contains $DC/C \cong \mathrm{E}_4$, and so $G/C \in \{\mathrm{E}_4, \mathrm{D}_8\}$ since G/C is isomorphic to a subgroup of $\mathrm{Aut}(D) \cong \mathrm{D}_8$.

If $G/C \cong \mathrm{E}_4$, then $G = D * C$. In that case, $\mathrm{c}_1(G) = 7$ so G is as in (a).

Now suppose that $G/C \cong \mathrm{D}_8$. Then $|G : (D * C)| = 2$ and $DC/C \cong \mathrm{E}_4$. Let Q/C be a subgroup of order 2 in G/C such that $Q/C \not\leq DC/C$. Then $G = QD$ and $Q \cap (D * C) = C$ so $\mathrm{Z}(D)$ is a unique subgroup of order 2 in Q since $\Omega_1(G) \leq DC$. Thus, Q is either cyclic or generalized quaternion.

If Q is cyclic, then $G = DQ$, $D \cap Q = \mathrm{Z}(D)$ and $|Q : C| = |Q : \mathrm{C}_G(D)| = 2$ so $\mathrm{C}_G(\mathrm{C}_G(D)) \geq DQ = G$ and $\mathrm{C}_G(D) = \mathrm{Z}(G)$, i.e., G is as in (b).

Let Q be generalized quaternion. Then Q has a cyclic subgroup $L \neq \Omega_2(C)$ of order 4 such that $L \not\leq H (= \Omega_1(G))$. We have $L \cap D = \mathrm{Z}(D)$. Assume that DL is not of maximal class. Then $\mathrm{c}_1(DL) \geq 7$ since $\mathrm{c}_1(DL) \geq 5$ and $\mathrm{c}_1(DL) \equiv 3 \pmod 4$ (Theorem 1.17(a)), so $\Omega_1(DL) = H$, contrary to the choice of L. Thus, DL is of maximal class. Then $\mathrm{c}_1(DL) < 7$ so $\mathrm{c}_1(DL) = 5$ and $DL \cong \mathrm{SD}_{24}$ so G is a group of part (c). □

Exercise 1. Suppose that G is a 2-group with $c_1(G) = 11$. If G has an elementary abelian subgroup E of order 8, then $\Omega_1(G) = D_8 \times C_2$.

Theorem 43.10 (the case $p = 2$ was proved by Janko, independently). *Suppose that G is a p-group with $c_1(G) = 1 + p + 2p^2$. Then $p \leq 3$ and one of the following holds:*

(a) *$p = 2$, $\Omega_1(G) = D \times C$, where $D \cong D_8$ and $C \cong C_2$. Let A be the (characteristic) abelian subgroup of type $(4, 2)$ in $\Omega_1(G)$ and B a maximal abelian G-invariant subgroup of exponent 4 in G containing A. Then $C_G(B)$ is metacyclic. Next, $c_1(C_G(D)) = 3$.*

(b) *$p = 2$, $\Omega_1(G) = D * Q$ is extraspecial of order 2^5, where $D \cong D_8$, $Q \cong Q_8$. Then $C_G(D) = Q$ so that $|G : \Omega_1(G)| \leq 2$.*

(c) *$p = 2$, G has a normal subgroup $D \cong D_{16}$. The subgroup $C = C_G(D)$ is cyclic of order > 2 and $|G : (D * C)| \leq 4$. Next, $\Omega_1(G) = D * \Omega_2(C)$.*

(d) *$p = 3$, G is of maximal class and order 3^4 with exactly two subgroups of order 3^3 and exponent 3.*

Proof. Since $1 + p + 2p^2 \neq \frac{p^n-1}{p-1}$ for all $n \in \mathbb{N}$ and p, we get $\exp(\Omega_1(G)) > p$ so $\Omega_1(G)$ is irregular (Theorem 7.2(b)).

(i) Let $p > 2$, $E_{p^2} \cong R \triangleleft G$ and let $x \in G - R$ be of order p; then $K = \langle x, R\rangle$ is of order p^3 and exponent p (Theorems 7.1 and 7.2). If $\Omega_1(N_G(K)) = K$, then $N_G(K) = G$, a contradiction since $c_1(K) < c_1(G)$. Therefore, there is $y \in N_G(K) - K$ of order p. Set $L = \langle y, K\rangle$. Since $c_1(L) \leq c_1(G)$, we get $\exp(L) = p^2$ so L is irregular of order p^4, $p = 3$ and L is of maximal class (Theorems 7.1(b) and 7.2(b)). Next, $T = \langle y, R\rangle$ is of order 3^3 and exponent 3, $T \neq K$. Since K and T contain together exactly $1 + 3 + 2 \cdot 3^2 = c_1(G)$ distinct subgroups of order 3, $L = KT = \Omega_1(G) \trianglelefteq G$. Next, L has exactly two subgroups of order 3^3 and exponent 3, both these subgroups are normal in G since $L \triangleleft G$. Since $c_1(G) \equiv 1+3+2\cdot3^2 \not\equiv 1+3+3^2 \pmod{3^3}$, G is of maximal class (Theorem 13.2(a)) so, since G has a normal subgroup K of order 3^3 and exponent 3, we get $G = L$, $|G| = 3^4$ (Theorem 9.6) so G is as in (d).

In what follows we assume that $p = 2$.

(ii) Since $c_1(G) = 11 \equiv 3 \pmod 4$, G is not of maximal class (Theorem 1.2) and G has two noncommuting involutions. We have $\sigma(G) = [\frac{11}{4}] = 2$ so $|\Omega_1(G)| \leq 2^{2+1+\sigma(G)} = 2^5$ (see Theorem 64.4). Let x, y be two involutions in G generating a subgroup of maximal possible order. The subgroup $D = \langle x, y\rangle \cong D_{2^n}$; then $11 > c_1(D) = 2^{n-1} + 1$ so $n \leq 4$.

(ii1) Let $D \cong D_8$, i.e., $n = 3$. Since G is not of maximal class, $C = C_G(D) \not\leq D$ (Proposition 10.17(a)). Let $u \in C - D$ be an involution. Then $D \times \langle u\rangle = \Omega_1(G)$ since $c_1(G) = 11 = c_1(D \times \langle u\rangle)$. Let A be *the* abelian subgroup of type $(4, 2)$ in $\Omega_1(G)$; then $A \triangleleft G$ since A is characteristic in $\Omega_1(G)$. Let B be a maximal normal abelian subgroup of exponent 4 in G containing A. Then B is of rank 2 since $B \cap \Omega_1(G) = A$, and $\Omega_2(\mathrm{C}_G(B)) = B$, by Theorem 10.1. Then $\mathrm{C}_G(B)$ is metacyclic (Remark 41.2(i)). Since $C \cap \Omega_1(G) = \mathrm{Z}(\Omega_1(G)) \cong \mathrm{E}_4$, we get $\mathrm{c}_1(C) = 3$ so G is as in (a).

In what follows we assume that $C - D$ has no involutions; then $\mathrm{c}_1(C) = 1$ so C $(= \mathrm{C}_G(D))$ is either cyclic or generalized quaternion.

Suppose that $C = \mathrm{C}_G(D)$ is generalized quaternion. Take in C a subgroup $Q \cong \mathrm{Q}_8$. Then $D * Q$ is extraspecial of order 2^5 with 11 involutions (Appendix 16) so $\Omega_1(G) = D * Q$ and $Q = C \cap \Omega_1(G) \triangleleft C$. If $\mathrm{Q}_8 \cong L < C$ and $L \neq Q$, then again $D * L = \Omega_1(G) = D * Q$, a contradiction. We conclude that $C = Q \cong \mathrm{Q}_8$. Since $\Omega_1(G) = D * C (= D * Q)$ has $\equiv 2 \pmod 4$ dihedral subgroups of order 8 (see §76, Example 1), one may assume that the size of G-orbit of D divides 2. If D is G-invariant, then $G/Q = G/C$ is a subgroup of $\mathrm{Aut}(D) \cong \mathrm{D}_8$ containing $DQ/Q \cong \mathrm{E}_4$; in that case $|G : \Omega_1(G)| \le 2$. Now assume that all dihedral subgroups of $D * C$ are not normal in G. Then, by the choice of D, $N = \mathrm{N}_G(D)$ is of index 2 in G. As above, N/Q is isomorphic to a subgroup of D_8 so $|N : \Omega_1(G)| \le 2$. It follows that $|G : \Omega_1(G)| = |G : N||N : \Omega_1(G)| \le 4$, completing this case.

Now suppose that $C = \mathrm{C}_G(D)$ is cyclic. Let $H = D * \Omega_2(C)$; then $\mathrm{c}_1(H) = 7$. It follows that there exists an involution $z \in \mathrm{N}_G(H) - H$ (otherwise, $\Omega_1(\mathrm{N}_G(H)) = H$ so $\mathrm{N}_G(H) = G$ and again z exists). Since H has exactly three dihedral subgroups of order 8 (Appendix 16), one of them, say D_1, is z-invariant. Set $F = \langle z, D_1 \rangle$; then $|F| = 16 = |H|$ and $F \neq H$. Since $\mathrm{c}_1(F) > \mathrm{c}_1(D_1) = 5$ and $\mathrm{c}_1(F)$ is odd, we get $\mathrm{c}_1(F) \ge 7$ so F is not of maximal class (by the choice of D, $F \not\cong \mathrm{D}_{16}$). It follows that $\mathrm{c}_1(F) \neq 9$ so $\mathrm{c}_1(F) \in \{7, 11\}$. Since $\Omega_1(H) = H \neq F$, we get $\mathrm{c}_1(F) = 7$ so $F = D_1 * Z$, where Z is cyclic of order 4 (Theorem 43.9). We have $F \cap H = D_1$. Since $\Omega_2(C)$ and Z centralize D_1 and $\mathrm{C}_G(D_1)$ is cyclic, we get $\Omega_2(C) = Z$, which is a contradiction since $Z \not\le H$.

(ii2) Now suppose that $D \cong \mathrm{D}_{16}$. We have $\mathrm{c}_1(D) = 9$ so $D < \Omega_1(G)$. We claim that D is a unique subgroup of its structure in G. Indeed, if $G > D_1 \cong D \neq D_1$, then, since every proper subgroup of D contains ≤ 5 involutions, we have $\mathrm{c}_1(G) \ge \mathrm{c}_1(D) + \mathrm{c}_1(D_1) - \mathrm{c}_1(D \cap D_1) \ge 9 + 9 - 5 = 13 > 11$, a contradiction. Set $C = \mathrm{C}_G(D)$. If $x \in C - D$ is an involution, then $\mathrm{c}_1(D \times \langle x \rangle) = 19 > 11$, a contradiction. Thus, $\mathrm{c}_1(C) = 1$ so C is either cyclic or generalized quaternion. Assume that C is generalized quaternion. If D_0 is a nonabelian subgroup of order 8 in D and Q is a nonabelian subgroup of order 8 in C, then $\mathrm{c}_1(D_0 * Q) = 11 = \mathrm{c}_1(G)$. Since $D \not\le D_0 * Q$, we get a contradiction. Thus, C is cyclic. Next, G/C is isomorphic to a subgroup of $\mathrm{Aut}(D)$, containing $\mathrm{Inn}(D) \cong D/\mathrm{Z}(D) \cong \mathrm{D}_8$. Since $|\mathrm{Aut}(D)| = 2^5$ (see the description of $\mathrm{Aut}(\mathrm{D}_{2^n})$ in Theorem 34.8), we get $|G : (D * C)| \le 2^2$.

It remains to find $\Omega_1(G)$. Let $\mathrm{E}_4 \cong R \triangleleft G$; then $R \not\le D$ since $|D| = 16$. Set $H = DR$. In view of $\mathrm{c}_1(H) > 9 = \mathrm{c}_1(D)$, we get $\Omega_1(H) = H = \Omega_1(G)$. Let $\mathrm{D}_8 \cong D_1 < D$. Then R does not centralizes D_1 (otherwise, $\mathrm{c}_1(RD_1) = 11$, which is not the case since $RD_1 < H$ $(= \Omega_1(G))$). It follows that $\mathrm{C}_D(R)$ is cyclic so $\mathrm{C}_H(R)$ is abelian of index 2 in H. Since $|H'| = 4$, we get $|\mathrm{Z}(H)| = 4$ (Lemma 1.1) so

$H = DZ(H)$, where $Z(H) \cong C_4$. Thus $C_G(D) \not\leq D$ and $\Omega_1(G) = D * Z$ with $Z \cong C_4$ and $D \cap Z = Z(D)$. □

3°. In the proof of Theorem 43.6 we considered a 2-group G such that all elements of the set $G - (M \cup N)$ are involutions for some distinct $M, N \in \Gamma_1$. Such G we call I-groups with respect to M, N. I-groups were classified in [BM2] (see Theorem 43.12).

Definition. A nonabelian 2-group H is said to be *quasi-dihedral*, if $H = \langle x, A \rangle$, where A is an abelian maximal subgroup of H, $x^2 = 1$ and $a^x = a^{-1}$ for every $a \in A$. All elements in $H - A$ are involutions.

Exercise 2. A subgroup A of the Definition is uniquely determined as $H_2(H)$. The subgroup A is called *the kernel* of H.

Lemma 43.11. *Suppose that a 2-group G is an I-group with respect to $M, N \in \Gamma_1$. Then:*

(a) *If $g \in G - (M \cup N)$ and $a \in M \cap N$, then $a^g = a^{-1}$.*

(b) *$M \cap N$ is an abelian normal subgroup of index 4 in G.*

(c) *If $g, h \in G - (M \cup N)$ and $a \in M \cap N$, then $a^{hg} = a^{gh}$.*

Proof. Set $A = M \cap N$. Let g, h, a be as in (c); then $1 = (ga)^2 = g^2 a^g a = a^g a$ so $a^g = a^{-1}$, and (a) and (b) are proven. Since $G' \leq A$, we get $a^{[g,h]} = a$, and this is equivalent with $a^{gh} = a^{hg}$. □

Theorem 43.12 ([BM2]). *A 2-group G is an I-group if and only if G contains a maximal subgroup H which is either quasi-dihedral with kernel A or elementary abelian with G-invariant subgroup A of index 2 such that $G/A \cong E_4$.*

Proof. Suppose that G is an I-group with respect to $M, N \in \Gamma_1$. Then, by Lemma 43.11, for $A = M \cap N$ and for any $h \in G - (M \cup N)$, $G/A \cong E_4$ and $H = \langle h, A \rangle$ is either elementary abelian or quasi-dihedral maximal subgroup of G with kernel A.

Conversely, suppose that $H \in \Gamma_1$ is either elementary abelian with G-invariant subgroup A of index 2 or quasi-dihedral with kernel A such that $G/A \cong E_4$. Let M/A and N/A be distinct maximal subgroups of G/A such that $M \neq H \neq N$. Then $H \cup M \cup N = G$ and $G - (M \cup N) = H - A$ is a nonempty set of involutions. □

§44

Some characterizations of metacyclic p-groups

In this section we present some Blackburn's results yielding useful characterizations of metacyclic p-groups.

Lemma 44.1 ([Bla2, Lemma 2.2] = Lemma 36.5). *If G is a nonabelian p-group, $\mathrm{d}(G) = 2$, then $\Phi(G')\mathrm{K}_3(G)$ is the only G-invariant subgroup of index p in G'.*

Theorem 44.2 ([Bla2, Theorem 2.3] = Corollary 36.2). *A p-group G is metacyclic if and only if the factor group $G/\Phi(G')\mathrm{K}_3(G)$ is metacyclic.*

Lemma 44.3 ([Bla2, Lemma 2.5]; see also Theorem 36.1). *If G is a nonmetacyclic p-group with two generators, then $G/\Phi(G')\mathrm{K}_3(G) = \langle a, b \mid [a,b] = c,\ a^{p^m} = b^{p^n} = c^p = 1,\ [a,c] = [b,c] = 1\rangle$ is minimal nonabelian.*

Lemma 44.4 (see Exercise 1.8a). *Suppose that $G = \langle a, b\rangle$, where a, b satisfy the following relations: $a^{p^m} = b^{p^n} = 1$, $[a,b] = c$, $c^p = [a,c] = [b,c] = 1$, $m \geq n$, $m > 1$. Then G is minimal nonabelian and $\langle a^p\rangle \times \langle b\rangle \times \langle c\rangle \in \Gamma_1$ has rank three. If, in addition, $n > 1$, then all maximal subgroups of G are of rank* 3.

Theorem 44.5 (see also Corollary 36.6). *Let G be a normal subgroup of a 2-group W. If G and all its W-invariant subgroups of index 2 are two-generator, then G is metacyclic. In particular (Blackburn), a 2-group G is metacyclic if and only if G and all its maximal subgroups are two-generator.*

Corollary 44.6. *A p-group is metacyclic if and only if one of the following quotient groups is metacyclic: $G/\Phi(G')$, $G/\mathrm{K}_3(G)$, $G/\mho_1(G')$.*

Proposition 44.7. *Let G be a nonabelian and nonmetacyclic p-group, $p > 2$ and $P \in \mathrm{Syl}_p(\mathrm{Aut}(G))$. Suppose that G and all P-invariant maximal subgroups of G have two generators. Then $\mho_1(G) = \mathrm{K}_3(G)$ is of index p^3 in G so $|G : G'| = p^2$.*

Proof. We use induction on $|G|$. As $G/\mho_1(G)$ is of order $> p^2$ (Theorem 9.11), and every P-invariant maximal subgroup of $G/\mho_1(G)$ is two-generator, it follows that $|G/\mho_1(G)| = p^3$ (Theorem 5.8). Set $D = \Phi(G')\mathrm{K}_3(G)$; then D is characteristic in G so P-invariant. Therefore, by Lemma 44.3, $|G/D| = p^3 = |G/\mho_1(G)|$ (otherwise, G/D has a P-invariant maximal subgroup that has no two generators) and G/D is not metacyclic. We conclude that $D = \mho_1(G)$. Assume that $\mathrm{K}_3(G) < D$. To obtain a contradiction, one may assume that $K_3(G) = \{1\}$ and $|D| = p$; then G is minimal nonabelian of order p^4. In that case, $\Omega_1(G) \cong E_{p^3}$ is a P-invariant maximal subgroup of G, contrary to the hypothesis. □

Theorem 44.8. *For $p = 2$, Theorem 44.2 follows from Theorem 44.5.*

Corollary 44.9. *If $G/\mho_2(G)$ is a metacyclic 2-group, then G is also metacyclic.*

Theorem 44.10 (compare with [Bla2, Theorem 4.1(i,ii)]). *Let G be a normal subgroup of a p-group W, $p > 2$, $|G| = p^m$ with $m \geq 6$. Suppose that all W-invariant subgroups of G of order p^r have two generators, where r is fixed and $4 \leq r \leq m-2$. Then G is either metacyclic or a 3-group of maximal class.*

Proof. (a) Suppose that $r = 4$ and G is neither metacyclic nor a 3-group of maximal class. Then, by Theorems 13.7 and 10.4, G has a W-invariant subgroup E of order p^3 and exponent p. Let N be a W-invariant subgroup of order p in E. By hypothesis, G/N has no W-invariant elementary abelian subgroups of order p^3. It follows from Theorems 10.4 and 13.7 that one of the following holds: (i) G/N is metacyclic, (ii) G/N is a 3-group of maximal class, or (iii) $G/N = \Omega_1(G/N)(C/N)$ with $|\Omega_1(G/N)| = p^3$ and C/N is cyclic with $|C/N| \geq p^3$ since $m \geq 6$.

(i) Suppose that G/N is metacyclic. Then G/E is also metacyclic, so its exponent is $> p$ since $m \geq 6$. Then $C_G(E) \nleq E$ since p^2 does not divide $\exp(\mathrm{Aut}(E))$, and $C_G(E)$ is W-invariant since E is W-invariant. If $E \cong E_{p^3}$ and $A/E < EC_G(E)/E$ is W-invariant of order p, then $A < G$ is W-invariant abelian of order $p^4 = p^r$ with $d(A) \geq 3$, contrary to the hypothesis. Now let E be nonabelian. Let $U \leq C_G(E)$ be G-invariant of order p^2 such that $Z(E) < U$. Then $UE < G$ is W-invariant subgroup of order p^4 and rank 3, a contradiction.

(ii) Suppose that G/N is a 3-group of maximal class. Then $|Z(G)| = 3^2$ since G is not of maximal class. It follows from Lemma 1.4 that $Z(G)$ is of type $(3, 3)$ so one may assume that $Z(G) < E$; then $E \cong E_{3^3}$. Since the center of a Sylow 3-subgroup of the holomorph of E is of order 3 and $|\mathrm{Aut}(E)|_3 = 3^3 \leq |G : E|$, it follows that $C_G(E) > E$ is abelian. If $E < A \leq C_G(E)$, where A is W-invariant of order p^4, we get $d(A) > 2$, contrary to the hypothesis.

(iii) Suppose that $G/N = \Omega_1(G/N)(C/N)$, where C/N is cyclic of order $> p^2$ and $|\Omega_1(G/N)| = p^3$. Then G/E is metacyclic so its exponent is greater than p in view of $|C/N| > p^2$. Since a Sylow p-subgroup of $\mathrm{Aut}(E)$ is of order p^3 and exponent p, we get $C_G(E) \nleq E$, and we obtain a contradiction as in (i). Thus the theorem is true for $r = 4$.

(b) For $4 < r \leq m - 2$, we use induction on r. By Theorem 5.8, G is not of exponent p so $\mho_1(G) > \{1\}$. Let N be a W-invariant subgroup of order p in $\mho_1(G)$. By induction, G/N is either metacyclic or a 3-group of maximal class.

(i) If G/N is metacyclic, then G is also metacyclic (Theorem 9.11).

(ii) Suppose that G/N is a 3-group of maximal class. Since $m \geq r+2 \geq 5+2 = 7$, we get $\mathrm{cl}(G) \geq 5$ so $\mathrm{K}_4(G) > \{1\}$. In that case, one may assume that $N \leq \mathrm{K}_4(G)(< \mho_1(G))$ since $|G : \mho_1(G)| = 3^3$). Then $G/\mathrm{K}_4(G)$ is of maximal class as an epimorphic image of G/N so G is also of maximal class (Theorem 9.7). □

It is known that a normal subgroup N of an arbitrary group G contained in $\Phi(G)$ must have very special structure. For example, N cannot be a 2-group of maximal class. Now we consider the case where G is a p-group and $\mathrm{d}(N) = 2$. We need the following

Lemma 44.11 ([Gas1]). *If $N \triangleleft G$ and $N \leq \Phi(G)$, then* $\mathrm{Inn}(N) \leq \Phi(\mathrm{Aut}(N))$.

Proof. Let G act on N by conjugation. Then $g \in G$ induces the automorphism $\tau(g)$ of N; we have $\tau(g) : x \mapsto x^g$ $(x \in N)$. This yields an epimorphism $\tau : G \to \tau(G) = U(\leq \mathrm{Aut}(N))$, namely, $\tau : g \mapsto \tau(g)$. Then N is mapped onto group $\tau(N) = \mathrm{Inn}(N)$. It follows from $N \leq \Phi(G)$ that $\mathrm{Inn}(N) = \tau(N) \leq \tau(\Phi(G)) \leq \Phi(\tau(G)) = \Phi(U)$. On the other hand, $\tau(N) = \mathrm{Inn}(N) \trianglelefteq \mathrm{Aut}(N)$. Since $\mathrm{Inn}(N) \leq \Phi(U)$ and $U \leq \mathrm{Aut}(N)$, we get $\mathrm{Inn}(N) \leq \Phi(\mathrm{Aut}(N))$. □

Theorem 44.12 ([Ber21]). *Suppose that N is a two-generator normal subgroup of a p-group G. If $N \leq \Phi(G)$, then N is metacyclic.*

Proof. Suppose that N is a counterexample of minimal order. Then N is nonabelian. Since $\Phi(N')\mathrm{K}_3(N) \triangleleft G$, $\Phi(N')\mathrm{K}_3(N) < N' \leq \Phi(N) < \Phi(G)$ and $N/\Phi(N')\mathrm{K}_3(N)$ is not metacyclic, by Theorem 44.2, we get $\Phi(N')\mathrm{K}_3(N) = \{1\}$ so $|N'| = p$ since N is minimal nonabelian (Lemma 44.3). Similarly, $\mho_2(N) = \{1\}$ (Corollary 44.9) so $\exp(N) \leq p^2$ and $|N| \leq p^5$.

(a) Let $p > 2$. Then $|N/\mho_1(N)| = p^3$ (Exercise 1.8a). Since $N/\mho_1(N)$ is not metacyclic and $\mho_1(N) \triangleleft G$, we get $\mho_1(N) = \{1\}$ so N is nonabelian of order p^3. Then $\mathrm{Z}(N)$ is cyclic so N is also cyclic (Proposition 1.13), contrary to the assumption.

(b) Let $p = 2$. By the above, Lemma 44.3 and Corollary 44.9,

$$N = \langle a, b \mid c = [a,b],\ a^4 = b^{2^n} = c^2 = [a,c] = [b,c] = 1,\ n \leq 2 \rangle.$$

Assume that $n = 2$. By Lemma 44.4, all maximal subgroups of N are abelian of type $(4,2,2)$. Let K be a G-invariant maximal subgroup of N; then $\Phi(K) \triangleleft G$. Since $K/\Phi(K) \cong \mathrm{E}_8$, then $N/\Phi(K)$ is not metacyclic, so N is not a minimal counterexample. Thus, $n = 1$; then N/N' is abelian of type $(4,2)$ so $\mathrm{Aut}(N/N') \cong \mathrm{D}_8$ and $\mathrm{Aut}(N)$ is a 2-group of order dividing $(2^2-1)(2^2-2)|\Phi(N)|^2 = 3 \cdot 2^5$, i.e., $|\mathrm{Aut}(N)| \mid 2^5$. Then $|\Phi(\mathrm{Aut}(N))| \leq 8$, i.e., $\Phi(\mathrm{Aut}(N))$ is abelian. Since $\Phi(\mathrm{Aut}(N)) = \mho_1(\mathrm{Aut}(N))$ and $\exp(\mathrm{Aut}(N)) \leq 4$ (Theorem 33.1), then $\Phi(\mathrm{Aut}(N))$ is elementary abelian. If $\alpha \in \mathrm{Aut}(N)^{\#}$, then

$$\alpha(a) = a^r b^s c^t, \quad \alpha(b) = a^{2u} b c^v, \quad \alpha(c) = c, \quad r \in \{1,3\}, \quad s,t,u,v \in \{0,1\}.$$

It is easy to check that α^2 is one of the forms: (i) $a \to ac^d$, $b \to b$, (ii) $a \to a^3c^d$, $b \to bc$, (iii) $a \to a^3c^d$, $b \to b$.

It follows that if $\alpha^2 \neq \mathrm{id}_N$, then $\alpha^2 : a \to a^{1+2k}c^x$, $b \to bc^k$ for some $k, x \in \{0, 1\}$. By Lemma 44.11, $\mathrm{Inn}(N) \leq \Phi(\mathcal{A}) = \mho_1(\mathcal{A})$, where $\mathcal{A} = \mathrm{Aut}(N)$, so every inner automorphism β of N is a product of squares of automorphisms of N. It follows that $a^\beta = a^{1+2w}c^z$, $b^\beta = bc^w$ for some $w, z \in \{0, 1\}$. If $\tau(a)$ is the inner automorphism of N induced by the action of a, then $\tau(a) : a \to a$, $b \to bc$ is not contained in $\mho_1(\mathrm{Aut}(G))$. Indeed, it follows from the first equality that $w = 0$ and from the second one that $w = 1$. This contradiction completes the proof. □

Definition. A subgroup H of a group G is said to be *quasinormal* if it is permutable with all subgroups of G. A p-group is said to be *modular* if all its subgroups are quasinormal.

Let $S(p^3)$ be a nonabelian group of order p^3 and exponent p. Let a p-group G, $p > 2$, be modular. Then G is $S(p^3)$-free and D_8-free and $\Omega_1(G)$ is abelian so $\exp(\Omega_n(G)) \leq p^n$. A p-group is said to be *minimal nonmodular* if it is nonmodular, but all its proper subgroups are modular. For example, Q_{16} is minimal nonmodular. If a p-group G has no sections isomorphic to $S(p^3)$ or D_8, it is modular.

Theorem 44.13. *Suppose that G is a minimal nonmodular p-group. Then it contains two non-permutable cyclic subgroups A and B, and we have $G = \langle A, B\rangle$. Set $N = \langle \Phi(A), \Phi(B)\rangle$. Then $N \lhd G$ is metacyclic and $G/N \in \{S(p^3), \mathrm{D}_8\}$.*

Proof. By definition, G contains two non-permutable cyclic subgroups A and B. Then $G = \langle A, B\rangle$ since $\langle A, B\rangle$ is nonmodular. Set $N = \Phi(A)\Phi(B) \leq \Phi(G)$ ($< G$). We have $F = A\Phi(B) < G$ and $H = B\Phi(A) < G$; then

$$FH = A\Phi(B)B\Phi(A) = A\Phi(A)B\Phi(B) = AB \neq G$$

so $FH \neq HF$ since $\langle F, H\rangle = \langle A, B\rangle = G$. We also have $A \cap B = \Phi(A) \cap \Phi(B) = A \cap \Phi(B)$ since A and B are cyclic so

$$\frac{|F|}{|N|} = \frac{|A\Phi(B)|}{|\Phi(A)\Phi(B)|} = \frac{|A||\Phi(B)||\Phi(A) \cap \Phi(B)|}{|A \cap \Phi(B)||\Phi(A)||\Phi(B)|} = \frac{|A|}{|\Phi(A)|} = p.$$

Thus, $N \lhd F$. Similarly, $|H : N| = p$ so $N \lhd H$ whence $N \lhd \langle F, H\rangle = G$. Next, N is metacyclic (Theorem 44.12). Write $\bar{G} = G/N$. Then $\bar{G}$ is generated by two non-permutable subgroups $\bar{F}$ and $\bar{H}$ of order p and $\bar{G}$ is minimal nonmodular.

(i) Let $p = 2$. Then $\bar{G} \cong \mathrm{D}_{2^n}$ for some $n \geq 3$. Since all proper subgroups of $\bar{G}$ are modular, we have $n = 3$ and $\bar{G} = G/N \cong \mathrm{D}_8$.

(ii) Now suppose that $p > 2$. By Lemma 30.2(b), there are $\bar{A} < \bar{U}$, $\bar{B} < \bar{V}$, where $\bar{U}, \bar{V}$ are maximal in $\bar{G}$ and $\Omega_1(\bar{U}) = \bar{U}$, $\Omega_1(\bar{V}) = \bar{V}$. Then $\bar{U}, \bar{V}$ are elementary abelian since they are modular so $\bar{U} \cap \bar{V} = \mathrm{Z}(\bar{G})$. Then $\bar{G}$ is minimal nonabelian since $\mathrm{d}(\bar{G}) = 2$, and we get $\bar{G} \cong S(p^3)$ since $\Omega_1(\bar{G}) = \bar{G}$. □

Exercise 1. Suppose that all proper epimorphic images of a nonmetacyclic p-group G are metacyclic. Then G is either of order p^3 and exponent p or $G = \langle a, b \mid c = [a, b], a^4 = b^2 = c^2 = 1, [a, c] = [b, c] = 1\rangle$ is minimal nonabelian of order 2^4.

Exercise 2. If H is a noncyclic metacyclic maximal subgroup of a p-group G, $p > 2$, then G is nonmetacyclic if and only if $\mho_1(G) = \mho_1(H)$.

Exercise 3. If G is a minimal nonmodular 2-group containing a normal four-subgroup N such that $\Omega_1(G/N) = G/N$, then $N \leq Z(G)$.

Exercise 4. Set $\mho^2(G) = \mho_1(\mho_1(G))$. Prove that a 2-group G is metacyclic if and only if $G/\mho^2(G))$ is metacyclic.

§45

A counting theorem for p-groups of odd order

In this section we prove the following counting

Theorem 45.1. *Let G be a group of order p^m, $p > 2$. Suppose that G has no cyclic subgroups of index p. Let $\mathfrak{M}_n(G)$ be the set of subgroups $H < G$ of order p^n, $m > n \geq 4$, such that H has no cyclic subgroups of index p. Then $|\mathfrak{M}_n(G)| \equiv 1 \pmod{p}$.*

This theorem was proved by Miller for $n = 4$ (see Corollary 45.3).

In our proof of Theorem 45.1 we use the following

Lemma 45.2. *Suppose that $G = AB$ be a p-group, where $p > 2$, $A \lhd G$, $|A| = p^4$, $\exp(A) \leq p^2$ and $\exp(B) \leq p^2$. Then $\exp(G) \leq p^2$.*

Proof. Let $a \in A$, $b \in B$ and H a G-invariant abelian subgroup of index p in A (H exists since the number of abelian subgroups of index p in A is $\equiv 1 \pmod{p}$, by Exercise 1.6(a)). Since $\langle b, A\rangle/H$ is abelian, we get $\langle b, A\rangle' \leq H$. By Theorem 33.1, p^2 does not divide the exponent of $\mathrm{Aut}(H)$ since H is noncyclic and $p > 2$. It follows that b^p centralizes H and $a^p \in H$ so a^p and b^p commute. Set $D = \langle A, b\rangle$. As we have noticed, $D' \leq H$. We claim that $\exp(D) \leq p^2$. By the Hall–Petrescu formula (see Appendix 1), we have

$$(ab)^p = a^p b^p c_2^{\binom{p}{2}} \cdots c_{p-1}^{\binom{p}{p-1}} c_p,$$

where $c_i \leq \mathrm{K}_i(D) \leq H, i = 2, \dots, p$. By the above, a^p, b^p and $c_2, \dots, c_p$ commute pairwise as elements of abelian subgroup H. Since H is noncyclic abelian it contains a G-invariant subgroup R of type (p, p). It is clear that $\mathrm{K}_3(D) \leq R$ so $o(c_i) \leq p$ for $i > 2$ whence $c_i^{\binom{p}{i}} = 1, i = 3, \dots, p-1$, and we have

$$(ab)^p = a^p b^p c_2^{\binom{p}{2}} c_p, \quad \text{and} \quad o(a^p), o(b^p), o(c_2^{\binom{p}{2}}), o(c_p) \leq p.$$

It follows that

$$(ab)^{p^2} = a^{p^2} b^{p^2} c_2^{p\binom{p}{2}} c_p^p = 1.$$

Since $AB = \{ab \mid a \in A, b \in B\}$, the proof is complete. □

Proof of Theorem 45.1. Given $K \le G$, let $\alpha_n(K)$ be the number of members of the set $\mathfrak{M}_n(G)$ contained in K. Let us prove that $\alpha_n(G) > 0$. It suffices to show that $\alpha_4(G) > 0$. By Lemma 1.4, G has a normal subgroup R of type (p, p). If G/R is not cyclic, it has a normal subgroup A/R of type (p, p). In that case, A is a normal subgroup of G of order p^4 and exponent $\le p^2$. Suppose that G/R is cyclic. As G has no cyclic subgroups of index p, $\Omega_1(G)$ is of order p^3 and exponent p (i.e., G is an L_3-group; see §17). Let $A/\Omega_1(G)$ be a subgroup of order p in $G/\Omega_1(G)$; then again A is a normal subgroup of G of order p^4 and exponent p^2. Thus, $\alpha_4(G) > 0$.

(A) Let $n = 4$.

(i) If G is metacyclic, when $\mathfrak{M}_4(G) = \{\Omega_2(G)\}$, and we are done. Now assume that G is nonmetacyclic.

(ii) Let G be a 3-group of maximal class. Then $\Omega_2(G_1)$ is the unique normal member of the set $\mathfrak{M}_4(G)$, and the result holds. Now assume that G is not a 3-group of maximal class.

(iii) If all members of the set Γ_1 have no cyclic subgroups of index p, then $\alpha_4(H) \equiv 1 \pmod{p}$ for all $H \in \Gamma_1$, by induction, and the result follows by Hall's enumeration principle.

(iv) Now suppose that $H \in \Gamma_1$ has a cyclic subgroup of index p. Then $G = H\Omega_1(G)$, where $|\Omega_1(G)| = p^3$, by Theorem 12.1(a) if G is irregular (then $p = 3$) and by Theorem 7.2(b) if G is regular. If $\Omega_1(G) < F \in \Gamma_1$, then $\alpha_4(F) \equiv 1 \pmod{p}$, by induction, and contribution of all such F in $\alpha_4(G)$ is $\equiv 1 \pmod{p}$. Now let $\Omega_1(G) \not\le T \in \Gamma_1$. Then $G = T\Omega_1(G)$ so $\Omega_1(T) \cong \mathrm{E}_{p^2}$ and $T/\Omega_1(T)$ is cyclic as a subgroup of $G/\Omega_1(G)$. In that case, T has a cyclic subgroup of index p so $\alpha_4(T) = 0$. Then, by Hall's enumeration principle, $\alpha_4(G) \equiv 1 \pmod{p}$.

(B) Now let $m > 4$. Since $\alpha_n(G) > 0$, we have $\exp(G) \le p^{m-2}$.

Assume that $\exp(G) < p^{m-2}$. Then, if $H \in \Gamma_1$, then H has no cyclic subgroups of index p so, by induction, $\alpha_n(H) \equiv 1 \pmod{p}$. In that case, $\alpha_n(G) \equiv 1 \pmod{p}$, as in (i).

Next suppose that $\exp(G) = p^{m-2}$. By (A), there is $A \lhd G$ of order p^4 and exponent $\le p^2$; then G/A is cyclic and $\exp(A) = p^2$. Assume that $B < G$ is another subgroup of order p^4 and exponent $\le p^2$. Then $\exp(AB) \le p^2$ (Lemma 45.1) and $AB \lhd G$ since G/A is cyclic, $|AB| \ge p^5$. It follows that AB has no cyclic subgroups of index p^2 so $\exp(G) < p^{m-2}$, contrary to the assumption. Thus, A is the unique subgroup of G of order p^4 and exponent $\le p^2$. Take $H \in \mathfrak{M}_n(G)$. By (A), there is $B < H$ of order p^4 and exponent $\le p^2$ so $B = A$. Since G/A is cyclic, we get $\alpha_n(G) = 1$. □

Corollary 45.3 ([Mil5] for $n = 4$). *Suppose that G is a group of order p^m, $p > 2$, and n is such that $m > n \ge 4$. If G has no cyclic subgroups of index p, then the number of subgroups of order p^n in G having a cyclic subgroup of index p, is divisible by p.*

Corollary 45.3 is a very weak form of Conjectures A(2) from §17.

Appendix 1

The Hall–Petrescu formula

In this section we prove the Hall–Petrescu formula which gives a decomposition of $x^n y^n (xy)^{-n}$ in a product of commutators of increasing weights (reading $x^n y^n (xy)^{-n}$ modulo G', we see that this element is contained in G'). This formula is one of the most important and deep results of general group theory. The original form of this formula (see [Hal1, Theorem 3.1]) is not so exact (but it is sufficient for applications to p-group theory; Hall gave a number of outstanding applications of his formula). Undoubtedly, this formula laid foundations for the commutator calculus. Final version of the formula is due to Petrescu [Pet]. In our exposition we follow closely to the book of Suzuki [Suz1] (see there §3, pages 37–41).

If $x, y \in G$, then the equality $xy = yx[x, y]$ allows us to change the order of elements in a product by adding the commutator of these elements.

Suppose that $x, y \in G$, where G is an arbitrary group. Then $x^2 y^2 = xxyy = xyx[x,y]y = xyxy[x,y][x,y,y] = (xy)^2 c_2 c_3$, where $c_2 = [x,y] \in \mathrm{K}_2(\langle x, y\rangle)$ and $[x,y,y] \in \mathrm{K}_3(\langle x, y\rangle)$. We recommend to the reader to obtain analogous decompositions for $x^3 y^3$ and $x^4 y^4$.

In general, the following *Hall–Petrescu formula* holds:

$$(1) \qquad x^n y^n = (xy)^n c_2^{\binom{n}{2}} c_3^{\binom{n}{3}} \dots c_n^{\binom{n}{n}},$$

where $c_i \in \mathrm{K}_i(\langle x, y\rangle)$, $i = 2, \dots, n$.

Our aim is to prove formula (1). To this end we use the *Hall's commutator collecting process*.

We begin by providing an ordering on the nonempty subsets of the set $\mathbb{N}(n) = \{1, \dots, n\}$ of the first n natural numbers. We will always exhibit $A \subseteq \mathbb{N}(n)$ as $A = \{i_1, \dots, i_r\}$, $i_1 < \dots < i_r$. If $B = \{j_1, \dots, j_k\}$ is another subset of $\mathbb{N}(n)$, we define $A < B$ if either $r < k$, or $r = k$ and the first nonzero difference $j_s - i_s$ is positive. For example, for $n = 4$, we have

$$\{1\} < \{2\} < \{3\} < \{4\} < \{1,2\} < \{1,3\} < \{1,4\} < \{2,3\} < \{2,4\} < \{3,4\}$$
$$< \{1,2,3\} < \{1,2,4\} < \{1,3,4\} < \{2,3,4\} < \{1,2,3,4\} = \mathbb{N}(4).$$

It is clear that the relation $<$ defines a linear order among nonempty subsets of the set $\mathbb{N}(n)$; however, this ordering is not lexicographic.

We now consider the free group F freely generated by $2n$ elements $x_1, \dots, x_n$ and $y_1, \dots, y_n$. For each $i \in \mathbb{N}(n)$, let ϕ_i be the endomorphism of F defined by

$$\phi_i(x_i) = \phi_i(y_i) = 1, \quad \phi_i(x_j) = x_j, \quad \phi_i(y_j) = y_j \quad \text{for all } j \neq i.$$

Obviously, $x_i, y_i \in \ker(\phi_i)$. If $i \neq j$, then $x_i y_j y_i y_j^{-1} \in \ker(\phi_i)$. It is easy to check that $\ker(\phi_i)$ is the normal closure of $\langle x_i, y_i \rangle$ in F.

For any nonempty subset A of $\mathbb{N}(n)$, $|A| = r$, we define

$$F(A) = \langle x_i, y_i \mid i \in A \rangle,$$

$$K(A) = \bigcap_{i \in A} \ker(\phi_i),$$

$$U(A) = \mathrm{K}_r(F) \cap F(A) \cap K(A),$$

where $\mathrm{K}_r(F) = [F, \dots, F]$ (r times).

If $A = \{1\}$, then $F(A) = \langle x_1, y_1 \rangle$, $K(A) = \ker(\phi_1) = F(A)^F = \langle x_1, y_1 \rangle^F$, $U(A) = \langle x_1, y_1 \rangle \cap K(A) = \langle x_1, y_1 \rangle = F(A)$ since $\mathrm{K}_1(F) = F$.

Lemma A.1.1. *There are elements $u_A \in U(A)$ such that*

$$x_1 \dots x_n y_1 \dots y_n = \prod u_A, \tag{2}$$

where the product in the right-hand side ranges over all nonempty subsets A of $\mathbb{N}(n)$ in the increasing order with respect to the relation $<$ defined above.

Proof. Let B be any nonempty subset of $\mathbb{N}(n)$. We will prove the following proposition:

$$(\lambda_B) : x_1 \dots x_n y_1 \dots y_n = \prod_{A<B} u_A z_1 \dots z_m,$$

where $u_A \in U(A)$, $z_i \in U(D_i)$ with $B \leq D_i (\subseteq N(n))$, and the product is taken over all subsets $A < B$ in increasing order (note that $A < B$ does not imply that A is a subset of B).

If B is the first subset $\{1\}$ of $\mathbb{N}(n)$, then the product is taken over the empty set, and $x_i, y_i \in U(\{i\})$ $(B \leq U(\{i\}))$. Thus, λ_B holds for $B = \{1\}$.

We proceed by induction. Suppose that (λ_B) is true for some nonempty subset B of $\mathbb{N}(n)$. We want to collect all z_i which belong to $U(B)$ to the left. Suppose that $z = z_i \in U(B)$ and $u = z_{i-1} \in U(D)$, where $B < D$. Then we have

$$uz = zuw, \quad \text{where } w = [u, z].$$

Since $z \in U(B)$ and $u \in U(D)$, the definitions and the fact that $[\mathrm{K}_i(F), \mathrm{K}_j(F)] \leq \mathrm{K}_{i+j}(F)$, give us $w \in U(B \cup D)$. Since $|B| < |B \cup D|$, we have $B < B \cup D$. Repeating this process for all elements z_i, which are in $U(B)$, we see that one can collect all those elements to the left end, and the added elements (commutators) as well as the remaining elements, belong to $U(D)$ with $B < D$. This proves the proposition $\lambda_{B'}$ for the next (in the above defined ordering) subset B' so for all subsets of the set $\mathbb{N}(n)$. The lemma coincides with proposition $(\lambda_{\mathbb{N}(n)})$. □

Since the word u_A is in $F(A)$, it is written as a word in x_i and y_i with $i \in A$. We will show that u_A is in a sense determined by $|A|$ alone.

Lemma A.1.2. *Let A and B be two nonempty subsets of $\mathbb{N}(n)$ of the same cardinality. Set*

$$A = \{i_1, \dots, i_r\}, \quad B = \{j_1, \dots, j_r\}, \quad |A| = r = |B|.$$

Let E be the free group generated by $2r$ elements $s_1, \dots, s_r; t_1, \dots, t_r$ and let homomorphisms $\alpha, \beta : F \to E$ be defined by

$$\alpha(x_{i_k}) = \beta(x_{j_k}) = s_k, \quad \alpha(y_{i_k}) = \beta(y_{j_k}) = t_k, \quad k = 1, \dots, r,$$
$$\alpha(x_i) = \alpha(y_i) = 1 = \beta(x_j) = \beta(y_j), \qquad (i \notin A,\ j \notin B).$$

Then $\alpha(u_A) = \beta(u_B)$.

Proof. Let θ_A be the endomorphism of F which is defined by

$$\theta_A(x_i) = \theta_A(y_i) = 1 \quad (i \notin A) \quad \text{and} \quad \theta_A(x_j) = x_j,\ \theta_A(y_j) = y_j \quad (j \in A).$$

Then $\mathrm{im}(\theta_A) = F(A)$. Let α' be the homomorphism from $F(A)$ onto E defined by

$$\alpha'(x_{i_k}) = s_k, \quad \alpha'(y_{i_k}) = t_k.$$

Then $\alpha = \theta_A\alpha'$. Apply θ_A to identity (2). The image of the left-hand side is the product $(\prod_{i\in A} x_i)\cdot(\prod_{i\in A} y_i)$. In the right-hand side, if a subset C contains an element, say i, not in A, it follows from $u_C \in K(C)$ that $\theta_A(u_C) = 1$. In fact, $\ker(\phi_i) = \langle x_i, y_i\rangle^F$, the normal closure of $\langle x_i, y_i\rangle$ in F, and this subgroup is contained in $\ker(\theta_A)$. If D is a subset of A, then $\theta_A(u_D) = u_D$ (since u_D is the word in x_i, y_i for $i \in D \subseteq A$ and θ_A induces identity on $F(A)$). Hence, the image of the right-hand side of (2) is $\prod_{D\subseteq A} u_D$.

We prove the lemma by induction on r. Suppose that $r = 1$. In this case, $A = \{i\}$ and $B = \{j\}$. Then, applying θ_A to identity (2), we get $x_i y_i = \prod u_A = u_A$ $(A = \{i\})$. Similarly, we have $x_j y_j = u_B$ and $\alpha(u_A) = s_j t_j = \beta(u_B)$. This proves the lemma for $r = 1$. If $r > 1$, the image by θ_A is $\prod_{i\in A} x_i \cdot \prod_{i\in A} y_i = (\prod_{D<A} u_D)\cdot u_A$. Hence, we get

$$\prod_{k\in A} s_k \cdot \prod_{k\in A} t_k = \Big(\prod_{D<A} \alpha(u_D)\Big)\alpha(u_A).$$

Similarly,

$$\prod_{k\in B} s_k \prod_{k\in B} t_k = \Big(\prod_{E<B} \beta(u_E)\Big)\beta(u_B).$$

Since $|A| = |B|$, there is one-to-one correspondence between the subsets of A and those of B. By the definition of the relation $<$, we can make this correspondence preserve the order relation among the subsets. By induction, each $\alpha(u_D)$ is equal to the corresponding $\beta(u_E)$, and these elements appear in the corresponding position in the products. Cancelling these elements, we get $\alpha(u_A) = \beta(u_B)$. □

Theorem A.1.3. *Formula (1) holds.*

Proof. Let F be the free group of rank $2n$ defined earlier and F_0 the free group freely generated by x and y. Let θ be the homomorphism of F onto F_0 defined by $\theta(x_i) = x$ and $\theta(y_i) = y$ for all $i = 1, \dots, n$. Then $\theta(x_1 \dots x_n \cdot y_1 \dots y_n) = x^n y^n$. We will consider the θ-image of the right-hand side of identity (2). Let A and B be two subsets of $\mathbb{N}(n)$ such that $|A| = |B| = r$. Let E be the free group defined in Lemma A.1.2. Let $\theta'(s_i) = x$ and $\theta'(t_i) = y$ be the homomorphism of E onto F_0, $i = 1, \dots, r$. Let α' be the homomorphism from $F(A)$ to E defined in the proof of Lemma A.1.2. Then the restriction of θ to $F(A)$ is decomposed as $\theta_{F(A)} = \theta'\alpha' : F(A) \to E \to F_0$. Similarly, the restriction of θ on $F(B)$ is a composite mapping of β' and θ'. By Lemma A.1.2, we have $\alpha'(u_A) = \beta'(u_B)$. Hence, $\theta(u_A) = (\theta'\alpha')(u_A) = (\theta'\beta')(u_B) = \theta(u_B)$. Set $c_r = \theta(u_A)$ if $|A| = r$. Since there are $\binom{n}{r}$ distinct subsets of cardinality r in $\mathbb{N}(n)$, the element c_r appears $\binom{n}{r}$ times successively in θ-image of the right-hand side of identity (2). This proves formula (1). □

The above proof shows that the commutators c_r depend only on r and not on n.

Exercise 1. Let $x, y \in F$ and $x^3y^3 = (xy)^3 c_2^3 c_3$, where $c_i \in \mathrm{K}_i(F)$, $i = 2, 3$. Write out c_2 and c_3 in the terms of x, y.

Exercise 2. Using formula (1), prove that if G is a p-group of class $< p$, then $\exp(\Omega_1(G)) = p$.

The following theorem is true for an arbitrary group G however it is most useful in the case where G is a p-group.

Theorem A.1.4 ([HB, Lemma 8.1.1]). *Let G be a p-group and $x, y \in G$.*

(a) $(xy)^p \equiv x^p y^p \pmod{\mho_1(G')\mathrm{K}_p(G)}$.

(b) $[x^p, y] \equiv [x, y]^p \pmod{\mho_1(N')\mathrm{K}_p(N)}$, *where* $N = \langle x, [x, y]\rangle$.

Proof. Set $N = \mho_1(G')\mathrm{K}_p(G)$. By the Hall–Petrescu formula, there exist elements $c_i \in \mathrm{K}_i(G)$, $i = 2, \dots, p$, such that $x^p y^p = (xy)^p c_2^{\binom{p}{2}} \dots c_{p-1}^{\binom{p}{p-1}} c_p$, and (a) follows since $c_2^{\binom{p}{2}}, \dots, c_{p-1}^{\binom{p}{p-1}} \in \mho_1(G')$.

(b) By (a), we get $(x[x, y])^p = x^p[x, y]^p \pmod{\mho_1(N')\mathrm{K}_p(N)}$, where N is given in the part (b) of the theorem. But $(x[x, y])^p = (x^y)^p = (x^p)^y = x^p[x^p, y]$ so $x^p[x, y]^p \equiv x^p[x^p, y] \pmod{\mho_1(N')\mathrm{K}_p(N)}$ and, after cancelling, we obtain the desired result. □

Appendix 2

Mann's proof of monomiality of p-groups

In his letter Mann offered a nice proof of the known fact that p-groups are M-groups (as I know, H. F. Blichfeldt was the first who proved this result). We show that his argument works in a more general situation. A group G is called an M-*group* if, for any $\chi \in \mathrm{Irr}(G)$, there exist $H \le G$ and $\lambda \in \mathrm{Lin}(H)$ such that $\chi = \lambda^G$. Here $\mathrm{Lin}(H)$ is the set of linear characters of H. Abelian groups are M-groups.

Definition. A group G is said to be a C-group if, whenever $H \le G$ is subnormal, there exists a prime $p = p_H$ such that p divides the degrees of all nonlinear irreducible characters of H. We consider abelian groups as C-groups.

Subnormal subgroups of C-groups are also C-groups. Epimorphic images of C-groups are C-groups and p-groups are C-groups.

Theorem A.2.1. *C-groups are M-groups.*

Proof. Let G be a nonabelian C-group. Then p divides $\chi(1)$ for $\chi \in \mathrm{Irr}_1(G)$ and a fixed prime $p = p_G$ so $|G| \equiv |G : G'| \pmod p$ and p divides $|G : G'| = |\mathrm{Lin}(G)|$. Next, $\mathrm{Lin}(G)$ acts on $\mathrm{Irr}(G)$ via multiplication. A linear character λ fixes $\chi \in \mathrm{Irr}_1(G)$ if and only if λ is a constituent of $\chi\overline{\chi}$, and then $\langle \lambda, \chi\overline{\chi}\rangle = 1$. So, if A is a stabilizer of χ in $\mathrm{Lin}(G)$, then the character $\chi\overline{\chi}$ has $|A|$ distinct linear constituents, each with multiplicity 1, and the other irreducible constituents of that character are not linear. Let $\chi\bar{\chi} = \lambda_1 + \cdots + \lambda_{|A|} + a_1\rho_1 + \cdots + a_s\rho_s$, where $\lambda_i \in \mathrm{Lin}(G)$ and $\rho_j \in \mathrm{Irr}_1(G)$ are pairwise distinct. Then $(\chi\bar{\chi})(1) = \chi(1)^2 = |A| + a_1\rho_1(1) + \cdots + a_s\rho_s(1)$ so p divides $|A|$ hence $|A| > 1$. Let $\lambda \in A - \{1_G\}$. Then $\lambda\chi = \chi$, so χ vanishes on $G - \ker(\lambda)$ since $(\lambda - 1_G)\chi = 0$. Let M be a maximal subgroup of G containing $\ker(\lambda)$; then $M \lhd G$ since $G' \le \ker(\lambda) \le M$, and χ vanishes on $G - M$. Let $\chi_M = \mu_1 + \cdots + \mu_t$ be the Clifford decomposition (by Clifford theory, $\mu_1, \dots, \mu_t$ are pairwise distinct since $|G : M| = q$ is a prime). Assume that $t = 1$. Then $\mu_1^G = \chi_1 + \cdots + \chi_q$, where $\chi_1 = \chi, \chi_2, \dots, \chi_q$ are pairwise distinct irreducible characters of G of the same degree $\mu_1(1)$. We have $(\chi_i)_M = \mu_1$ for $i = 1, \dots, q$, by reciprocity. Then, since χ_1 and $\chi_2 + \cdots + \chi_q$ vanish on $G - M$ (recall that μ_1^G vanishes on $G - M$), we get $0 = |G|\langle \chi_1, \chi_2 + \cdots + \chi_q\rangle = |M|(q-1)\langle \mu_1, \mu_1\rangle = |M|(q-1)$, a contradiction. Thus, $t > 1$ (then $t = q = |G : M|$) so $\chi = \mu_1^G$. But M is a C-group so, by induction in M, there exist $H \le M$ and $\tau \in \mathrm{Lin}(H)$ such that $\mu_1 = \tau^M$. We have $\chi = \mu_1^G = (\tau^M)^G = \tau^G$. Since χ is arbitrary, we are done. □

As we have noticed, subnormal subgroups of C-groups are C-groups. However, normal subgroups of M-groups are not necessarily M-groups (S. D. Berman, E. Dade, R. van der Waall).

Appendix 3

Theorems of Isaacs on actions of groups

In this section we, following Isaacs (this material is taken from his papers), study, in particular, the situation where a p-group P acts on an elementary abelian q-group V in such a way that the orbits of the action have pairwise distinct sizes. First we consider the case where p is odd.

Below we use the following fact of elementary number theory. Let p and q be primes such that $p^m = q^n + 1$ for some $m, n \in \mathbb{N}$ one of which is > 1. Then one of the following holds: (i) $m = 1$, $q = 2$ (in that case, p is a Fermat prime), (ii) $p^m = 3^2$, $q^n = 2^3$, (iii) $n = 1$, $p = 2$ (in that case, q is a Mersenne prime).

Theorem A.3.1 (Isaacs; see [BIK, Theorem 3.3]). *Let P be a p-group, $p > 2$, and suppose that P acts on an elementary abelian q-group $V > \{1\}$ in such a way that the P-orbits on V have pairwise distinct sizes. Then $|V| = 1 + p$, and there is a unique nontrivial orbit which has size p; in particular, $q = 2$.*

Proof. It is no loss to assume that the action of P on V is faithful. Since $\{1\}$ is a unique one-element P-orbit, P has no nontrivial fixed points on V, and it follows that $P > \{1\}$, $q \neq p$ and so V is a completely reducible P-module (Maschke). We can therefore decompose V as a direct sum of irreducible P-modules. If this is a nontrivial decomposition, then since the orbit sizes of the action of P on each summand are distinct, we can work by induction on $|V|$ to deduce that there is an orbit of size p on each component. Then the number of P-orbits of size p is > 1, contrary to assumption. It follows that V is an irreducible P-module.

If P is abelian, it must be cyclic, by Schur's lemma, and it acts semiregularly on $V^\#$ since PV is a Frobenius group. In that case, all P-orbits on $V^\#$ have the same size $|P|$ so $V^\#$ itself is a P-orbit of size $|P|$. Thus $|V| = 1 + |P|$, and since $p > 2$, we have $q = 2$ and $|P| = p$. We are done in this case.

Assuming now that P is nonabelian, we work to derive a contradiction. Since $p > 2$ and P is noncyclic, it has a normal subgroup $R \cong \mathrm{E}_{p^2}$ (Lemma 1.4). Since RV is not a Frobenius group, there is $x \in R^\#$ that centralizes an element in $V^\#$. Set $K = \mathrm{C}_{PV}(x)$, $P_1 = K \cap P$, $V_1 = K \cap V$. Then $K = P_1 \cdot V_1$ and $V_1 > \{1\}$ is normal in K since $V \lhd PV$. By assumption, $V_1 < V$, $P_1 < P$ (the first inclusion is proper since P is faithful on V, the second inclusion is proper since V is an irreducible P-module). Since $\mathrm{C}_P(R) = P_1$, we get $|P : P_1| = p$. If $y \in P - P_1$, then $V = V_1 \oplus V_1^y \oplus \cdots \oplus V_1^{y^{p-1}}$ since V is irreducible and the module standing at the right-hand side of the formula, is P-invariant. Since $P_1 \lhd P$, it follows that P_1 stabilizes each of the subspaces $V_1, V_1^y, \ldots, V_1^{y^{p-1}}$. (For example, if $u \in P_1$, then $(V_1^{y^i})^u = (V_1^{y^i u y^{-i}})^{y^i} = V_1^{y^i}$ since $y^i u y^{-i} \in P_1$.) Set $V_i = V_1^{y^{i-1}}$.

Suppose that Y is a P_1-orbit on $V_1^\#$, and note that Y is contained in some P-orbit X on $V^\#$. It is easy to see that $X \cap V_i$ is a single P_1-orbit on V_i and that X is the union of its intersections with the p subgroups V_i. Also, the set of intersections $X \cap V_i$ is transitively permuted by P, and hence each such intersection has the same cardinality $|X \cap V_1| = |Y|$. It follows that $|Y| = \frac{1}{p}|X|$, and since the distinct P-orbits have distinct sizes, we deduce that the distinct P_1-orbits on $V_i^\#$ have distinct sizes. To see that distinct P_1-orbits on V_1 have distinct sizes (so that we can apply the induction to the action of P_1 on V_1), we must show that $|Y| > 1$. We assume, therefore, that $|Y| = 1$, and thus $|X| = p$, and we can write $X = \{v_i \mid 1 \le i \le p\}$ for elements $v_i \in V_i^\#$. Since the sum $\sum_{i=1}^p V_i$ is direct, it follows that in this case that $\sum_{i=1}^p v_i$ is a nonzero P-fixed element of V. But we know that $\mathrm{C}_V(P) = \{1\}$; this contradiction shows that $|Y| > 1$, and thus the P_1-orbits on the elements of V_1 have distinct sizes.

By induction applied to the action of P_1 on V_1, we can now conclude that $q = 2$ and $|V_1| = p+1$. It follows that $|V| = |V_1|^p = (p+1)^p$ and by hypothesis, we know that this number must be a sum of distinct powers of p, and so the base-p expansion of this number can have no digit exceeding 1. If $p > 3$, then $(p+1)^p \equiv 1+p^2+\frac{1}{2}(p-1)\cdot p^3 \pmod{p^4}$, and this is a contradiction since the digit $\frac{1}{2}(p-1)$ exceeds 1. If $p = 3$, however, then $(1+3)^3 = 4^3 = 64 = 1+3^2+2\cdot 3^3$, and this is also a contradiction. □

The situation is more complicated in the case $p = 2$. In that case, as Isaacs has showed, the number of P-orbits on V is not necessarily 2. Therefore, for $p = 2$, we consider only the situation where P acts irreducibly on V, and even then our results are not complete.

Theorem A.3.2 (Isaacs; see [BIK, Theorem 3.4]). *Let a 2-group P act irreducibly on a nontrivial elementary abelian q-group V in such a way that the P-orbits on V have pairwise distinct sizes. Then one of the following occurs.*

(a) P acts transitively on the set $V^\#$. (In this situation, V is either of order q, a Fermat prime, or V is of order 3^2.)

(b) There are P-orbits on $V^\#$ of sizes 2^e and 2^f, where e, f are natural numbers of distinct parity.

(c) $|V| = 3^4$, and there are just two P-orbits on $V^\#$, and these orbits have sizes $16 = 2^4$ and $64 = 2^6$. In this situation, V is the direct product of two subgroups of order 9 that are permuted by P.

Proof. Clearly, $P > \{1\}$. We may assume that P is faithful on V.

(i) First suppose that P has no normal abelian subgroups of type $(2,2)$. Then P is either cyclic or of maximal class (Lemma 1.4) so P has a cyclic subgroup C of index 2. If $B \le C$ is of order 2, then $B \le \mathrm{Z}(P)$, and thus $\mathrm{C}_V(B)$ is P-invariant:

$P \leq C_{PV}(B)$. Since P is faithful on V, we get $C_V(B) = \{1\}$ so $C \cdot V$ is a Frobenius group, and so the C-orbits on the set $V^\#$ all have size $|C|$. Since $|P : C| = 2$, we see that there are just two possible sizes for the P-orbits on $V^\#$: $|C|$ or $2|C|$. Then, as it is easy to check, one of conclusions (a) or (b) holds.

(ii) We can now suppose that P has a normal abelian subgroup of type $(2, 2)$, and hence $V = U \times W$, where P permutes the subspaces U and W (see the proof of the previous theorem). Let N be the P-stabilizer of U, so that $|P : N| = 2$, and observe that N must act irreducibly on U (otherwise, P acts reducibly on V). Arguing as in the proof of Theorem A.3.1, we see that the containment defines an injection from the set of N-orbits on $U^\#$ into the set of P-orbits on $V^\#$, and that there are no one-element N-orbit on $U^\#$. Also, if the non-one-element N-orbit Y is contained in the P-orbit X, we see that $|Y| = \frac{1}{2}|X|$, and it follows that the N-orbits on U have pairwise distinct sizes. The hypotheses of the theorem are thus satisfied for the action of N on U, and working by induction on $|V|$, we see that one of the three conclusions of the theorem must hold for the action of N on U.

If conclusion (b) holds for N acting on U, then it must also hold for P acting on V because we have seen that the double of each orbit size of N on $U^\#$ is an orbit size of P on $V^\#$. If conclusion (c) holds for U, then $|U| = 3^4$, and hence $|V| = 3^8$. But the orbit sizes of P on V correspond to the digits in the binary expansion of $|V|$, and we compute that $3^8 = 2^{12} + 2^{11} + \cdots$. In this case, therefore, conclusion (b) holds for the action of P on V.

Finally, we suppose that N is transitive on $U^\#$, i.e., conclusion (a) holds for N acting on U. Writing $|U| = q^a$, we have $q^a = 1 + 2^e$ for some exponent $e \in \mathbb{N}$. The only possibilities are that $a = 1$ and q is Fermat, in which case e must be a power of 2, or else $q^a = 9$ and $e = 3$. In particular, if $e > 3$, then e must be even. If $e > 1$, we have $|V| = (1 + 2^e)^2 = 1 + 2^{e+1} + 2^{2e}$, and so if e is even we have one even and one odd exponent in the binary expansion of $|V|$, and thus conclusion (b) holds in this situation. The only remaining possibilities are $e = 1$ and $e = 3$. If $e = 1$, then $|V| = 9 = 8 + 1$, and P acts transitively on V^*. In that case P is isomorphic to a subgroup of a Sylow 2-subgroup of $\mathrm{Aut}(P)$ (which is isomorphic to SD_{24}) that has a normal four-subgroup and all P-orbits have distinct sizes. However SD_{24} has no such subgroups. Assuming now that $e = 3$, we have $|V| = |U|^2 = 81 = 1 + 16 + 64$, and conclusion (c) holds. □

Isaacs showed that the conclusion (c) in Theorem A.3.2 really occurs (see [BIK]).

Exercise 1. Let P and Q be nontrivial p-groups. Suppose that P acts on Q faithfully in such a way that all n non-one-element P-orbits have pairwise distinct sizes. Then $p = 2$. Try to describe the structures of P and Q in this case. Study in detail the case $n = 1$.

Using Theorems A.3.1 and A.3.2, Isaacs proved the following

Theorem A.3.3 ([BIK]). *Let N be a nontrivial normal p-subgroup of a solvable group G, let a nonprincipal $\phi \in \mathrm{Irr}(N)$ be G-invariant. Assume that the irreducible con-*

stituents of ϕ^G *have pairwise distinct degrees. Then* $|\mathrm{Irr}(\phi^G)| = 1$ *and* G *is a* p-*group.*

Proposition A.3.4 (Isaacs; see [BIK]). *Let* X, Y *and* Z *be* FG-*modules, where* F *is a field and* G *is an abelian group that acts transitively on the nonzero elements of* Z. *Suppose that there exists an* F-*bilinear map* $B : X \times Y \to Z$ *such that* $B(X, y) = 0$ *only when* $y = 0$, *and such that* $B(xg, yg) = B(x, y)g$ *for all* $x \in X$, $y \in Y$ *and* $g \in G$. *Suppose further, that some subgroup* $H \leq G$ *acts irreducibly on* X *and trivially on* Y. *Then every element of* G *that fixes a nonzero element of* Y, *acts trivially on* Y.

Proof. Let $y \in Y^\# = Y - \{0\}$, and consider the subspace $V \subseteq X$ consisting of all those elements $x \in X$ that $B(x, y) = 0$. If $x \in V$ and $h \in H$, we have $B(xh, y) = B(xh, yh) = B(x, y)h = 0$, and thus $xh \in V$, i.e., V is an FH-submodule of the irreducible FH-module X. Since $y \neq 0$, however, we have $B(X, y) \neq 0$, and thus $V \subset X$. It follows that $V = 0$ since X is an irreducible FH-module, and thus $B(x, y) \neq 0$ for all elements $x \in X^\#$.

Let $a, b \in Y^\#$ be arbitrary, and fix an element $t \in X^\#$. By the previous paragraph, $B(t, a), B(t, b) \in Z^\#$, and hence there exists an element $g \in G$ such that $B(t, a)g = B(t, b)$ since G acts transitively on Z^*. Let $W \subseteq X$ be the subspace consisting of all elements $x \in X$ such that $B(x, a)g = B(x, b)$, and observe that $t \in W$, and hence $W \neq \{0\}$. If $x \in W$ and $h \in H$, we have (G is abelian!)

$$\begin{aligned} B(xh, a)g &= B(xh, ah)g = B(x, a)hg = B(x, a)gh \\ &= B(x, b)h = B(xh, bh) = B(xh, b), \end{aligned}$$

and thus $xh \in W$. Thus W is a nonzero FH-submodule of an irreducible FH-module X, and we conclude that $W = X$ so $B(x, a)g = B(x, b)$ for all $x \in X$.

Let $a, b \in Y^\#$ be arbitrary, and suppose that $k \in G$ is such that $ak = a$. We must show that k also fixes b. Choose g as in the previous paragraph, and let $x \in X$ be arbitrary. We compute that

$$\begin{aligned} B(xk, bk) &= B(x, b)k = B(x, a)gk = B(x, a)kg \\ &= B(xk, ak)g = B(xk, a)g = B(xk, b), \end{aligned}$$

and thus $B(xk, bk - b) = 0$ for all $x \in X$. Thus, $B(X, bk - b) = 0$, and we conclude that $bk - b = 0$. □

Theorem A.3.5 (compare with [Isa14, Lemma 3.1]). *Suppose that* G *acts on some nontrivial* $\pi(G)'$-*group* Q *in such a way that* $\mathrm{C}_G(x) \leq \mathrm{Z}(G) \cap G'$ *for all* $x \in Q^\#$. *Then the action of* G *on* Q *is Frobenius.*

Proof. Let G be a counterexample of minimal order. Then there exists $x \in Q^\#$ such that $\mathrm{C}_G(x) > \{1\}$. Write $Z = \mathrm{C}_G(x)$ and note that $Z \leq G' \cap \mathrm{Z}(G)$, and so $Z \triangleleft G$. Put

$C = \mathrm{C}_Q(Z)$ and note that $C > \{1\}$ and G/Z acts of C (indeed, $\mathrm{C}_{G\cdot Q}(Z) = G \cdot C$ and $C = G \cdot C \cap Q \triangleleft G \cdot C$). Moreover, if Z_1 be a subgroup of prime order, say p, in Z, then G/Z_1 acts on C. If $y \in C^{\#}$, then, since $(|G|, |C|) = 1$, we get

$$\mathrm{C}_{G/Z_1}(y) = \mathrm{C}_G(y)/Z_1 \le (\mathrm{Z}(G) \cap G')/Z_1 \le (\mathrm{Z}(G/Z_1) \cap (G/Z_1)',$$

and so the action of G/Z_1 on C satisfies the hypotheses of the theorem. Since $|G/Z_1| < |G|$, induction tells us that the action of G/Z_1 on C is Frobenius. It follows that $Z_1 = Z$. Then $P/Z \in \mathrm{Syl}_p(G/Z)$ is either cyclic or a generalized quaternion group. Note that then the Schur multiplier of P/Z is trivial. The Schur multiplier $\mathrm{M}(G/Z)$ of G/Z is isomorphic to a subgroup of the direct product of Schur multipliers of Sylow subgroups of G/Z, one for every prime divisor of $|G/Z|$ [BZ, Theorem 6.21(b)]. Since $\mathrm{M}(P/Z) = \{1\}$, we see that p does not divide $|\mathrm{M}(G/Z)|$. This is a contradiction since $Z \le G' \cap \mathrm{Z}(G)$ so p must divide $|\mathrm{M}(G/Z)|$. □

Proposition A.3.6 ([IsN, Lemma 2.2]). *Let a group A act via automorphisms on a group G and suppose that G is the direct product of a collection $\mathcal{X}$ of subgroups, where $\mathcal{X}$ is transitively permuted by A. If $X \in \mathcal{X}$ and B is the A-stabilizer of X, then $\mathrm{C}_G(A) \cong \mathrm{C}_X(B)$.*

Proof. We argue that projection to the direct factor X defines the desired isomorphism from $\mathrm{C}_G(A)$ onto $\mathrm{C}_X(B)$. First, let Y be the product of all members of $\mathcal{X}$ other than X and note that $G = X \times Y$. If $c \in \mathrm{C}_G(A)$, write $c = xy$, with $x \in X$ and $y \in Y$. Since B centralizes c and stabilizes X and Y (indeed, B acts on the set $\mathcal{X}$), we see that B centralizes x since the decomposition $c = xy$ is unique, and so the projection to X does map $\mathrm{C}_G(A)$ into $\mathrm{C}_X(B)$. Furthermore, this map is injective since if c has a trivial projection $x = 1$ to X, then its projection to every member of $\mathcal{X}$ is also trivial because of the transitivity of the action of A on $\mathcal{X}$; it follows that $y = 1$ and $c = xy = 1$.

To see that our map is surjective, take $x \in \mathrm{C}_X(B)$ and note that distinct A-conjugates of x must lie in distinct members of the set $\mathcal{X}$. It follows that the members of the A-orbit containing x commute pairwise, and hence their product is a member of $\mathrm{C}_G(A)$ that projects to x. This proves that $\mathrm{C}_G(A) \cong \mathrm{C}_X(B)$. □

Exercise 2. Let $X > \{1\}$ be a group and $A > \{1\}$ a transitive permutation group. Apply Proposition A.3.6 to $A \cdot G$, where G is the base subgroup of the wreath product $A \operatorname{wr} X$ with active factor A.

Proposition A.3.7 ([IsN, Lemma 3.3]). *Suppose that a group A acts via automorphisms on a group G and that the natural semidirect product $A \cdot G$ acts transitively on some set Ω. If $\alpha \in \Omega$ is fixed by A and $H = G_\alpha$, the G-stabilizer of α, then H is A-invariant and the action of A on the right cosets of H in G is permutation isomorphic to the action of A on Ω.*

Exercise 3 (P. Roquette). Let a p-group G have a normal subgroup D of type (p, p). If $\chi \in \mathrm{Irr}(G)$ is faithful, then $\chi_{\mathrm{C}_G(D)}$ is a sum of p irreducible characters conjugate under G. (*Hint.* If χ is faithful, then $\mathrm{Z}(G)$ is cyclic.)

Exercise 4 ([IsP2, Lemma 5]). Let a 2-group P act faithfully on an elementary abelian q-group Q (q a prime). If P is transitive on $Q^{\#}$, then we have either (a) $P \cong C_{2^n}$, $|Q| = q = 2^n + 1$, or (b) $q = 3$. $|Q| = 9$, $P \in \{C_8, SD_{16}, Q_8\}$.

Solution. Take $x \in Q^{\#}$ and let P_x be the P-stabilizer of x. Then $2^n = |P : P_x| = q^s - 1 = |Q^{\#}|$. In that case, the only solutions are: (i) $s = 1, q = 2^n + 1$ is a Fermat prime, or (ii) $q = 3, s = 2, n = 3$. In the first case, $|Q| = q$ so P is cyclic since P is isomorphic to a subgroup of $\mathrm{Aut}(Q) \cong C_{q-1}$. Hence $P_x = \{1\}$, and (a) follows. In the second case, $|Q| = 9$ and P is a subgroup of SD_{16}, the Sylow 2-subgroup of $\mathrm{Aut}(E_9) \cong \mathrm{GL}(2,3)$. Note that $|P : P_x| = |Q^{\#}| = 8$ so $8|P_x| = |P| \le 16$. If $P_x > \{1\}$, we get $|P| = 16$ and $P \cong SD_{16}$. If $P_x = \{1\}$, then $|P| = 8$ and $P \in \{C_8, Q_8\}$ (in that case, PQ is a Frobenius group).

A group A acts on the set Ω half transitively if all A-orbits have the same size.

Theorem A.3.8 ([IsP2, Theorem 1]). *Let $A > \{1\}$ be a group of automorphisms of a group G, acting half transitively as a permutation group on the set $G^{\#}$. Then either A acts in a fixed-point-free manner on G or G is an elementary abelian q-group for some prime q and A acts irreducibly.*

Proof. Assume that the action of A is not fixed-point-free. Let k denote the common size of all A-orbits on $G^{\#}$. Given $x \in G^{\#}$, set $A_x = \{\alpha \in A \mid \alpha(x) = x\} (= C_A(x))$; then $|A : A_x| = k$. Set $P_x = C_G(A_x)$. If $y \in G$ and $z \in (P_x \cap P_y)^{\#}$, then $A_z \ge \langle A_x, A_y\rangle$ so $A_x = A_z = A_y$ (since $|A : A_z| = |A : A_x|$) and $P_x = P_y$. Since $x \in P_x$, the set $\{P_x\}$ forms a partition of G.

Let $L = A \cdot G$ be the natural semidirect product. We compute the size of x^L for $x \in G^{\#}$. Let x have h conjugates in G. Then for all $\alpha \in A$, $\alpha(x)$ also has h conjugates in G. Hence x^L is a join of G-classes of size h, and therefore h divides $|x^L|$. On the other hand, x^L is the join of A-orbits so k divides $|x^L|$. Now k divides $|G| - 1$ and h divides $|G|$ so $(h,k) = 1$ and therefore hk divides $|x^L|$. This implies that x^L is the join of at least k conjugacy classes of G.

Obviously, A permutes the nonidentity conjugacy classes of G. Let K_x be the G-class containing x and A_{K_x} the subgroup of A fixing K_x as a set. The above argument shows that $|A : A_{K_x}| \ge k$. Now let x, y be nonidentity conjugates in G. Then clearly A_x and A_y are subsets of A_{K_x} so $A_x = A_{K_x} = A_y$, since $|A : A_x| = |A : A_y| = k \le |A : A_{K_x}|$. Hence $P_x = C_G(A_x) = C_G(A_y) = P_y$. Therefore $P_x \lhd G$ for all $x \in G$.

Our partition $\{P_x\}$ is not trivial since the action of $A > \{1\}$ on G is not Frobenius, and G is equally partitioned, and so G is an elementary abelian q-group for some prime q (§68; see also [Isa8]). It is easy to check that the semidirect product $L = A \cdot G$ acts as a permutation group on the elements of the set $G^{\#}$ by $x^{\alpha g} = \alpha(xg)$, and this action is transitive. Indeed, $\{1^{\alpha g} \mid \alpha \in A, g \in G\} = \{\alpha(g) \mid \alpha \in A, g \in G\} = G$. Thus, G is a minimal normal subgroup of L so A acts on G irreducibly. □

Theorem A.3.9 ([Isa14, Theorem A]). *Let N be a nontrivial nilpotent q'-group that acts faithfully on an elementary abelian q-group $H > \{1\}$. Then there exists an element $x \in H$ such that $|\mathrm{C}_N(x)|^2 < |N|$.*

Theorem A.3.9 has the following remarkable consequence.

Theorem A.3.10 (Gagola–Lewis [GL, Theorem A]). *A solvable group G is nilpotent if and only if $\chi(1)^2$ divides $|G/\ker(\chi)|$ for all $\chi \in \mathrm{Irr}(G)$.*

Using the classification of finite simple groups, it is possible to show that Theorem A.3.10 is also true for arbitrary finite groups (see [GL]).

Proof of Theorem A.3.10. If G is nilpotent, then $\chi(1)^2$ divides $|G/\ker(\chi)|$ for all $\chi \in \mathrm{Irr}(G)$. Indeed, in view of the description of irreducible characters of direct products, it suffices to check this assertion for p-groups. However, if $G > \{1\}$ is a p-group and $\chi \in \mathrm{Irr}(G)$, then $\chi(1)^2 < |G|$, and our claim follows since $\chi(1)^2$ is a power of p.

Suppose that $\chi(1)^2$ divides $|G/\ker(\chi)|$ for all $\chi \in \mathrm{Irr}(G)$ and prove that G is nilpotent. Let G be a counterexample of minimal order. Then every proper epimorphic image of G is nilpotent. It follows that $\Phi(G) = \{1\}$, G has only one minimal normal subgroup H that is elementary abelian q-group for some prime q and G/H is nilpotent. Then $G = NH$, where $N < G$ is maximal. Since H is a unique minimal normal subgroup of G, it follows that $N \cap H = \{1\}$ and $\mathrm{C}_N(H) = \{1\}$, i.e., N acts faithfully on H. If $N_q \in \mathrm{Syl}_q(N)$, then $\mathrm{N}_G(N_q) = G$ so $N_q = \{1\}$, i.e., N is a q'-group. By Theorem A.3.9, we may choose $\lambda \in \mathrm{Irr}(H)$ so that the N-orbit of λ has size $> \sqrt{|N|}$. Take $\chi \in \mathrm{Irr}(\lambda^G)$; then, by what has just been said and Clifford's theorem, $\chi(1)^2 > |N|$ and $\chi(1)$ divides $|G : H| = |N|$, by Ito's theorem on degrees (Introduction, Theorem 17). It follows that $\chi(1)^2$ does not divide $|G|$, contrary to the hypothesis. □

Theorem A.3.11. *Suppose that an abelian p-group $P > \{1\}$ acts faithfully on a p'-group Q. Then there is $x \in Q$ such that P-orbit of x is of length $|P|$.*

This follows from

Theorem A.3.12 ([Bro]). *Let $S \in \mathrm{Syl}_p(G)$ be abelian. Then there exists $T \in \mathrm{Syl}_p(G)$ such that $S \cap T = S_G$, the core of S in G.*

We prove the following more general fact which is contained in [Isa5].

Theorem A.3.13. *Let $S \in \mathrm{Syl}_p(G)$. Choose $T \in \mathrm{Syl}_p(G)$ such that $T \cap S$ is minimal by inclusion, and suppose that $K \leq S \cap T$ is normal in S and T. Then $K \leq S_G$.*

Proof. We need to prove that $K \leq P$ for all $P \in \mathrm{Syl}_p(G)$. Fix some $P \in \mathrm{Syl}_p(G)$. Set $M = \mathrm{N}_G(K)$ and note that $S, T \leq M$, and so $S, T \in \mathrm{Syl}_p(M)$ and we have

$(P \cap M)^\mu \le T$ for some $\mu \in M$, by Sylow. Thus $P^\mu \cap M \le T$, and since $S \le M$, we conclude that

$$P^\mu \cap S = P^\mu \cap (M \cap S) = (P^\mu \cap M) \cap S \le T \cap S.$$

By the choice of T, we know that $S \cap T$ is minimal among intersections of two Sylow p-subgroups of G, and thus we cannot have $P^\mu \cap S < T \cap S$. It follows that $P^\mu \cap S = T \cap S \ge K$. Since $\mu \in M = \mathrm{N}_G(K)$, we get $K = K^{\mu^{-1}} \le P$. □

Proof of Theorem A.3.12. Choose $T \in \mathrm{Syl}_p(G)$ so that $S \cap T$ is minimal, and set $N = S \cap T$. By Theorem A.3.13, $N \le S_G$. Since the reverse inclusion is obvious, we get $N = S_G$. □

Proof of Theorem A.3.11. Let $G = P \cdot Q$ be the natural semidirect product. Since P acts faithfully on Q, it follows that $P_G = \{1\}$. By Theorem A.3.12, there exists $x \in Q$ such that $P^x \cap P = \{1\}$. It follows that 1 is the only element of P commuting with x. This means that if u, v are distinct elements of P, then $x^u \ne x^v$ so the P-orbit of x is of size $|P|$. □

Exercise 5. Suppose that a nontrivial abelian p-group P acts faithfully on an abelian q-group Q, $q \ne p$. Then there exists an element $x \in Q$ of order q such that the P-orbit of x is of size $|P|$. (*Hint.* P is faithful on $\Omega_1(Q)$, by the transfer theorem. Use Theorem A.3.11.)

Exercise 6. Suppose that a nontrivial abelian p-group P acts faithfully on a q-group Q such that $\Omega_1(Q)$ is elementary abelian, $q > 2$. Then there exists an element $x \in Q$ of order q such that the P-orbit of x is of size $|P|$.

Hint. Assume that $g \in P^\#$ centralizes $\Omega_1(Q)$. Then the subgroup $\langle g, Q\rangle$ has no minimal nonnilpotent subgroups so it is nilpotent (see Lemma 10.8). In that case, g centralizes Q so the action of P on Q is not faithful. Thus, P acts faithfully on abelian q-group $\Omega_1(Q)$, and the result follows from Theorem A.3.11.

Exercise 7. Suppose that a nontrivial abelian p-group P acts faithfully on a 2-group Q such that $\exp(\Omega_2(Q)) \le 4$. Then there exists an element $x \in Q$ of order ≤ 4 such that the P-orbit of x is of size $|P|$.

Appendix 4

Freiman's number-theoretical theorems

1°. Let $n \in \mathbb{N}$, let p be a prime and r a power of p. In many cases we want to know the standard prime decomposition of the number $|\mathrm{GL}(n,r)| = (r^n - 1)(r^n - r)\dots(r^n - r^{n-1}) = r^{\binom{n}{2}}(r-1)(r^2-1)\dots(r^n-1)$. All number-theoretic results of this section are due to Gregory Freiman (personal communication, July 1980).

Let $a > 1$ and n be natural numbers. Set

$$C(n,a) = (a-1)(a^2-1)\dots(a^n-1).$$

In this section, we present the standard prime decomposition of $C(n,a)$. To do this, it suffices, for every prime q, that does not divide a, to find the greatest power of q, dividing $C(n,a)$.

Suppose that q is a prime that does not divide a, and set $t_q(b) = \log_q(b_q)$, where b_q is the greatest power of q dividing $b \in \mathbb{N}$. Recall that

$$(*) \qquad t_q(n!) = [n/q] + [n/q^2] + \dots < \sum_{i=1}^{\infty} n/q^i = n/(q-1),$$

where $[x]$ is the integer part of a real number x. We want to express $m = t_q(C(n,a))$ in terms of a, n and q (see Theorem A.4.2).

Lemma A.4.1. *Suppose that $B > 1$ and r are natural numbers, q a prime, $B \equiv 1 \pmod{q}$ and $M = 1 + B + \dots + B^{r-1}$. If q is either odd or $q = 2$ and $B \equiv 1 \pmod{4}$, then $t_q(M) = t_q(r)$.*

Proof. One may assume that $q \mid r$ (otherwise, it is nothing to prove since $M \equiv r \pmod{q}$).

(i) For $\theta \in \mathbb{N}$, set

$$M(\theta) = 1 + B^{\theta} + B^{2\theta} + \dots + B^{(q-1)\theta}$$

so that $M(1) = 1 + B + \dots + B^{q-1}$. In that case, we have

$$\begin{aligned} M(\theta) &= q + (B^{\theta} - 1) + (B^{2\theta} - 1) + \dots + (B^{(q-1)\theta} - 1) \\ &= q + (B^{\theta} - 1)[(1 + (B^{\theta} + 1) + (B^{2\theta} + B^{\theta} + 1) \\ &\quad + \dots + (B^{(q-2)\theta} + \dots + B^{\theta} + 1)] \end{aligned}$$

$$= q + (B^\theta - 1)[q - 1 + (q-2)B^\theta + (q-3)B^{2\theta} + \dots + 2B^{(q-3)\theta} + B^{(q-2)\theta}]$$
$$= q + (B^\theta - 1)[(q-1) + (q-2) + \dots + 2 + 1 + (q-2)(B^\theta - 1) + (q-3)(B^{2\theta} - 1) + \dots + 2(B^{(q-3)\theta} - 1) + (B^{(q-2)\theta} - 1)]$$
$$= q + (B^\theta - 1)\left[\frac{1}{2}q(q-1) + (B^\theta - 1)N\right] \equiv q \pmod{q^2}$$

since $B \equiv 1 \pmod q$ and, if $q = 2$, then $B \equiv 1 \pmod 4$; here N is a natural number whose exact value is not important. It follows that $t_q(M(\theta)) = 1$ for all $\theta \in \mathbb{N}$. Thus, we are done if $r = q$.

(ii) Now let $r = q^t v$, where $t = t_q(r)$ so $(v, q) = 1$. Then

$$M = 1 + B + \dots + B^{r-1} = 1 + B + \dots + B^{vq^t - 1}$$
$$= (1 + B + \dots + B^{q-1})(1 + B^q + \dots + B^{q(q-1)})\dots (1 + B^{q^{t-1}} + \dots + B^{q^{t-1}(q-1)})(1 + B^{q^t} + \dots + B^{(v-1)q^t})$$
$$= M(1)M(q)\dots M(q^{t-1})(1 + B^{q^t} + \dots + B^{(v-1)q^t}).$$

We have proved in (i) that $t_q(M(q^i)) = 1$ holds for all nonnegative integers i so $t_q(M(1)M(2)\dots M(q^{t-1})) = t$. Since $B \equiv 1 \pmod q$, we get $1 + B^{q^t} + \dots + B^{(v-1)q^t} \equiv v \not\equiv 0 \pmod q$. Therefore, we get $t_q(M) = t = t_q(r)$, completing the proof. □

In the sequel we set $A = a^\alpha$, where α is the order of $a \pmod q$. Since $A \equiv 1 \pmod q$, Lemma A.4.1 can be applied to A instead of B. Write $m = t_q(C(n,a))$. Let $s = t_q(a^\alpha - 1)$ and $t = \frac{a^\alpha - 1}{q^s}$; then $t_q(A-1) = s$.

Theorem A.4.2. *(a) If $q > 2$ or $q = 2$ and $s > 1$, then*

$$m = t_q(C(n,a)) = s[n/\alpha] + t_q([n/\alpha]!).$$

(b) Let $q = 2$ (then $\alpha = 1$) and $s = 1$ (that is, $a \equiv 3 \pmod 4$). If $s_1 = t_2(a+1)$ (in that case, $s_1 > 1$), then

$$m = t_2(C(n,a)) = n + s_1[n/2] + t_2([n/2]!).$$

Proof. Since q divides $a^k - 1$ if and only if α divides k, we have

$$m = t_q(C(n,a)) = t_q[(A-1)(A^2-1)\dots(A^{[n/\alpha]} - 1)] = t_q(C([n/\alpha], A)),$$

where $A = a^\alpha$. However,

$$(A-1)(A^2-1)\dots(A^{[n/\alpha]}-1) = (A-1)^{[n/\alpha]}(A+1)(A^2+A+1)\dots(A^{[n/\alpha]-1} + \dots + A + 1). \tag{1}$$

We have $t_q((A-1)^{[n/\alpha]}) = s[n/\alpha]$.

(a) Let either $q > 2$ or $q = 2$ and $s > 1$. Then, by Lemma A.4.1,

$$t_q(1 + A) = t_q(2),$$
$$t_q(1 + A + A^2) = t_q(3),$$
$$\vdots$$
$$t_q(1 + A + \cdots + A^{[n/\alpha]-1}) = t_q([n/\alpha]),$$

so that $t_q[(A+1)(A^2+A+1)\dots(A^{[n/\alpha]-1}+\dots+A+1)] = t_q([n/\alpha]!)$, proving (a).

(b) Now let $q = 2$ and $s = 1$, i.e., $a \equiv 3 \pmod 4$. In that case, $\alpha = 1$ so $A = a$. Set $s_1 = t_2(a + 1)$; then $s_1 > 1$ and (1) takes the form

$$\begin{aligned}(a-1)(a^2-1)\dots(a^n-1)\\ = (a-1)^n(1+a)(1+a+a^2)\dots(1+a+\dots+a^{n-1}).\end{aligned} \tag{2}$$

Each factor of the right-hand side product with an odd number of summands is odd since a is. On the other hand, the sum

$$1 + a + \cdots + a^{v2^k-1} = (1 + a)(1 + a^2)\dots(1 + a^{2^{k-1}})(1 + a^{2^k} + \cdots + a^{(v-1)2^k})$$

contains 2 precisely in the power 2^f, where $f = s_1 + k - 1 = s_1 + t_2(2^{k-1}v)$, by Lemma A.4.1 since $t_2(1 + a^{2^i}) = 1$ for $i > 0$. Therefore, the right-hand side of (2) contains 2 precisely in the power $n + \sum_{d=1}^{[n/2]}(s_1 + d) = n + s_1[n/2] + t_2([n/2]!)$, which completes the proof of (b) and thereby the theorem. □

2°. Retaining the above notation, consider the case where the inequality

$$q^m > \frac{1}{2}a^n \tag{3}$$

holds for some prime divisor q of $C(n, a)$. Recall that $q^m = C(n, a)_q$.

(i) Let $a > 2$ and $q > 2$. Then, by Theorem A.4.2, $m = s[n/\alpha] + t_q([n/\alpha]!)$. Set $t = \frac{a^\alpha - 1}{q^s}$, where $q^s = (a^\alpha - 1)_q$. It follows that $q^s = \frac{a^\alpha-1}{t} < \frac{a^\alpha}{t}$ so, by Theorem A.4.2 and (∗), since $\alpha[n/\alpha] \le n$, we have

$$q^m = q^{s[n/\alpha]}q^{t_q([n/\alpha]!)} < (a^\alpha/t)^{[n/\alpha]}q^{[n/\alpha]/(q-1)} \le (q^{1/(q-1)}/t)^{[n/\alpha]} \cdot a^n,$$

and therefore, by (3),

$$((q^{1/(q-1)})/t)^{[n/\alpha]} > 1/2. \tag{4}$$

Assume that $t \ge 2$ (or, what is the same, $a^\alpha - 1 > q^s$).

Suppose that $[n/\alpha] < q$; then $m = s[n/\alpha]$ (Theorem A.4.2(a)), and by (3), $q^{s[n/\alpha]} > \frac{1}{2}a^n$, that is $(\frac{a^\alpha}{t})^{[n/\alpha]} > (\frac{a^\alpha-1}{t})^{[n/\alpha]} = q^{s[n/\alpha]} = q^m > \frac{1}{2}a^n$, and so

$a^n \ge a^{[n/\alpha]\alpha} > \frac{1}{2}t^{[n/\alpha]}a^n$, whence $t^{[n/\alpha]} < 2$, which is a contradiction (by the assumption in this paragraph, $t \ge 2$). Thus, if $n \ge \alpha$, then $[n/\alpha] \ge q$.

Let $n < \alpha$; then $m = 0$ (Theorem A.4.2(a)) and $a^n < 2$ by (3), which is impossible for $a > 1$. Thus, $n \ge \alpha$.

By what has been proved already, $[n/\alpha] \ge q$. By (4), $((q^{\frac{1}{q-1}})/t)^{[n/\alpha]} > 1/2$, and so $t = 2$. Then $2 \cdot q^s = a^\alpha - 1$. It follows that a is odd so α is also odd since $q > 2$. Assume that $\alpha > 1$. Then q is a Zsigmondy prime for the pair $\{a, \alpha\}$ so q does not divide $a - 1$. It follows that $a - 1 = 2$ so $a = 3$. In this case $q \ge 5$ since q does not divide a. If $q = 5$, then $\alpha = 4$ and $t = 16 > 2$, which is a contradiction. It follows that $q > 5$ and hence (4) fails. Thus, $\alpha = 1$ and (4) becomes $(q^{1/(q-1)}/2)^n > 1/2$. Since $q^{\frac{1}{q-1}} \le \sqrt{3}$, we obtain $n \le 4$. But $n \ge q \ge 3$. Let $n = 3$. Then $q = 3$ and $m = 3s + t_3(3!) = 3s + 1$ (Theorem A.4.2(a)), and we get, by (3), $3^{3s+1} > \frac{1}{2}a^3 = \frac{1}{2}(2 \cdot 3^s + 1)^3$, a contradiction. Let $n = 4$. Then $q = 3$, $m = 4s + t_3(4!) = 4s + 1$, $3^{4s+1} > \frac{1}{2}a^4 = \frac{1}{2}(2 \cdot 3^s + 1)^4$, which is impossible.

Thus, we have proved that $t = 1$, and so $q^s = a^\alpha - 1$ and a is even since $q > 2$.

Suppose that $\alpha > 1$. Then $a - 1 = q^\beta$ for some $\beta \in \mathbb{N}$. In that case, $U = a^{\alpha-1} + \cdots + a + 1$ is a power of q. On the other hand, $U \equiv \alpha \pmod{q}$ so $\alpha = q\alpha'$ for some $\alpha' \in \mathbb{N}$. Then $a^q - 1$ is a power of q. However, $a^q - 1 = (1 + q^\beta)^q - 1 = q^\beta + \binom{q}{2}q^{2\beta} + \cdots + \binom{q}{q}q^{q\beta}$ is not a power of q, a contradiction. Thus, $\alpha = 1$ and $a = 1 + q^s$. In this case, $m = sn + t_q(n!)$, and (3) becomes

$$(5) \qquad 2q^{t_q(n!)} > \left(1 + \frac{1}{q^s}\right)^n.$$

This inequality holds if and only if (3) holds. Inequality (5) definitely fails for $s = 1$, large q, and $n = q - 1$ (since then its left-hand side equals 2, while the right-hand side is close to e, the base of natural logarithms).

(ii) Let $q = 2$ and $a \equiv 1 \pmod{4}$, i.e., $s > 1$. In this case $\alpha = 1, m = sn + t_2(n!)$, and $2^s = \frac{a-1}{t}$, where t is odd (Theorem A.4.2(a)).

Let $t \ge 3$. Then (4) becomes $(\frac{2}{t})^n > \frac{1}{2}$, whence $t = 3, n = 1, m = s$ and $\frac{1}{2}a < 2^m = 2^s = \frac{1}{3}(a - 1)$, a contradiction.

Thus, $t = 1$ since $t < 3$ is odd, $a = 1 + 2^s \ge 5$, and $2^m = (2^s)^n 2^{t_2(n!)} = (a-1)^n 2^{t_2(n!)}$. If $n > 3$, then $t_2(n!) > \frac{n}{2}$ and hence $2^m > (a-1)^n 2^{n/2} = (\sqrt{2}(a - 1))^n > \frac{1}{2}a^n$, since $a \ge 5$. Clearly, (3) holds for $n = 1, 2, 3$ (for example, if $n = 3$, then $m = 3s + 1$ and $2^m = 2^{3s+1} = 2(a-1)^3 > \frac{1}{2}a^3$).

(iii) Let $q = 2$ and $a \equiv 3 \pmod{4}$; then $s = 1, \alpha = 1$. Let, as above, $s_1 = t_2(a + 1)$. Set $2^{s_1} = (a + 1)/t_1$; then $(t_1, 2a) = 1$. By Theorem A.4.2(b),

$$m = n + s_1[n/2] + t_2([n/2]!) < n + (s_1 n)/2 + n/2 = 3n/2 + (s_1 n)/2,$$

$$\frac{1}{2}a^n < 2^m < (2^{s_1} \cdot 8)^{n/2} < (8(a + 1)/t_1)^{n/2},$$

and so

$$4^{1/n} \cdot \frac{8(a+1)}{t_1} > a^2. \tag{6}$$

Let $n \geq 2$. Then $\frac{16(a+1)}{t_1} > a^2$, and hence $a \leq 15$.

If $a = 15$, then $s_1 = 4$, $t_1 = 1$ and by (6), $128 \cdot 4^{1/n} > 15^2 = 225$. This is true only if $n = 2$. In this case, by Theorem A.4.2(b), $m = 6$, $2^m = 2^6 < \frac{1}{2} \cdot 15^2$, and (3) fails.

If $a = 11$, then $s_1 = 2$, $t_1 = 3$ and by (6), $32 \cdot 4^{1/n} > 11^2 = 121$, which fails for $n > 1$.

Let $a = 7$. Then $s_1 = 3$, $t_1 = 1$, and

$$2 \cdot 2^m = 2^{n+3[n/2]+t_2([n/2]!)+1} > 7^n \tag{7}$$

must hold. Note that (7) does not hold for some n; it fails, for example, for $n = 1, 3, 5, 7$, and it holds if and only if (3) holds.

Let $a = 3$. Then $s_1 = 2$, $t_1 = 1$, $m = n + 2[n/2] + t_2([n/2]!)$, and the inequality $2^m = 2^{n+2[n/2]+t_2([n/2]!)+1} > 3^n$ must hold. It is not difficult to see that this inequality is always true.

Finally, let $n = 1$. Then $m = 1$, and $2^m = 2 > \frac{1}{2}a$ implies $a = 3$.

(iv) Let $a = 2$ so $q > 2$, and let t be odd. If $t \geq 3$, then $q \geq 5$ (indeed, if $q = 3$, then $t = 1$) and (4) fails. Thus $t = 1$ and $q^s = 2^\alpha - 1$, whence $s = 1$ and $q = 2^\alpha - 1$ is a Mersenne prime. In the case under consideration, $m = [n/\alpha] + t_2([n/\alpha]!)$, by Theorem A.4.2(a). Inequality (3) holds if and only if $q^{[n/\alpha]+t_q([n/\alpha]!)} > 2^{n-1}$.

All possible cases have been considered, and thus the following theorem is proven.

Theorem A.4.3. *If $a > 1$ and n are natural numbers, q a prime which does not divide a, and $q^m > \frac{1}{2}a^n$, where $m = t_q(C(n,a))$, and α, s are defined in the paragraph preceding Theorem A.4.2, then exactly one of the following statements holds:*

(a) $a = 1 + q^s$ *and* $2q^{t_q(n!)} > (1+q^{-s})^n$, $a > 2$, $q > 2$.

(b) $a = 1 + 2^s$, $q = 2$, *n is arbitrary.*

(c) $a = 7$, $q = 2$, $2^{n+2[n/2]+t_2([n/2]!)+1} > 7^n$.

(d) $a = 2$; $q = 2^\alpha - 1$; $q^{[n/\alpha]+t_q([n/\alpha]!)} > 2^{n-1}$.

Corollary A.4.4. *Suppose that p and q are different primes, $n \in \mathbb{N}$ and $m = t_q(C(n,p))$. If $q^m > \frac{1}{2}p^n$, then exactly one of the following statements holds: (a) $q = 2$, p is a Fermat prime, n arbitrary. (b) $q = 2$, $p = 7$, $2^{n+2[n/2]+t_2([n/2]!)+1} > 7^n$. (c) $q = 2^\alpha - 1$ is a Mersenne prime, $p = 2$ and $q^{[n/\alpha]+t_q([n/\alpha]!)} > 2^{n-1}$.*

Corollary A.4.5. *Suppose that p and q are different primes, $n \in \mathbb{N}$ and $m = t_q(C(n,p))$. Then $q^m < p^n$, unless one of the following statements holds: (a) $q = 2$, p is a Fermat prime, n arbitrary. (b) $q = 2$, $p = 7$ (this is true for some n). (c) q is a Mersenne prime, $p = 2$ (this is true for some n).*

Obviously, in case (a) of Corollary A.4.4 we have $q^m > p^n$. If $q = 2$ and $p = 7$, then, for $n = 8$, we have $m = 23$ (see Theorem A.4.2(b)) and $2^{23} > 7^8$. If $q = 7$, $p = 2$ and $n = 21$, then $m = 8$ and $7^8 > 2^{21}$. In particular, if p and q are odd primes, then $q^m < p^n$.

Let G be a p-group such that $\mathrm{d}(G) = n$. It follows from Theorem 1.16 that then $|\mathrm{Aut}(G)|_{p'}$ divides $C(n, p)$. Some applications of Corollary A.4.4 to solvable groups are based on this fact (see [Ber22, §3]).

Corollary A.4.6. *Let P be a p-group and Q a q-group, where p and q are distinct primes and $Q \leq \mathrm{Aut}(P)$. Set $|P : \Phi(P)| = p^n$. Then $|Q| < \frac{1}{2}p^n$, unless the one of the following cases occurs: (a) $q = 2$, p is a Fermat prime, (b) $q = 2$, $p = 7$, (c) q is a Mersenne prime and $p = 2$.*

Proof. Obviously, it suffices to assume that P is elementary abelian [BZ, Lemma 1.21]. Then $\mathrm{Aut}(P) \cong \mathrm{GL}(n, p)$, where $|P| = p^n$, is of order $C(n, p)$. Now the result follows from Corollary A.4.4. □

Proposition A.4.7. *Let $G = PQ > \{1\}$, where $P \in \mathrm{Syl}_p(G)$, $Q \in \mathrm{Syl}_q(G)$, p, q are distinct primes. If $|Q| > \frac{1}{2}|P|$, then $O_q(G) > \{1\}$, unless one of the following cases occurs: (a) $q = 2$, p is a Fermat prime, (b) $q = 2$, $p = 7$, (c) q is a Mersenne prime and $p = 2$.*

Corollary A.4.8 (Burnside). *Let $G = PQ > \{1\}$, where $P \in \mathrm{Syl}_p(G)$, $Q \in \mathrm{Syl}_q(G)$, p, q are distinct primes. If $|Q| > |P|$, then $O_q(G) > \{1\}$, unless one of the following cases occurs: (a) $q = 2$, p is a Fermat prime, (b) $q = 2$, $p = 7$, (c) q is a Mersenne prime and $p = 2$.*

Appendix 5

Another proof of Theorem 5.4

In this section we will give another proof of Theorem 5.4.

Theorem A.5.1 (= Theorem 5.4). *Suppose that a group G of order 2^m is not of maximal class, $3 \leq n < m$. Let $\mu_n(G)$ be the number of subgroups of maximal class and order 2^n in G. Then 4 divides $\mu_n(G)$.*

Proof. We use induction on $|G|$ and assume that $\mu_n(G) > 0$.

(i) Let $n = m - 1$. Then G has a subgroup M of maximal class and order $2^n = 2^{m-1}$.

Let $n = 3$. By Proposition 10.17, $C_G(M) \not\leq M$ so $G = MZ(G)$, $|Z(G)| = 4$ and $d(G) = 3$. Then G has exactly 3 abelian subgroups of order 8 (Exercise 1.6(a)) so $\mu_3(G) = |\Gamma_1| - 3 = 4$.

Let $n > 3$. By Lemma 1.4, G has a normal subgroup R of type $(2, 2)$. Since $n > 3$, we have $R \not\leq M$ so $G = MR$, $M \cap R = Z(M)$, and we have $G/(M \cap R) = (M/(M \cap R)) \times (R/(M \cap R))$. It follows that $d(G) - 3$ and G has exactly 4 maximal subgroups not containing R; let H be one of them. We claim that H is of maximal class. Assume that this is not true. Then H has a G-invariant subgroup D of type $(2, 2)$ (Lemma 1.4), and we have $D \neq R$. Set $S = DR$; then $|S| = 8$. Since $S \cap M$ is a normal subgroup of order 4 in M, it is cyclic of order 4 and hence S is dihedral since $\Omega_1(S) = S$. Set $F = C_G(M \cap S)$; then $F \in \Gamma_1$ is not of maximal class since $|Z(F)| \geq |S \cap M| = 4$. Let D_1 be a G-invariant subgroup of type $(2, 2)$ in F (Lemma 1.4); then $D_1 \neq R, D$ since R and D do not centralize $S \cap M$, and we conclude that $D_1 \not\leq S$. Set $S_1 = D_1R$. As above, $S_1 \cap M$ is cyclic of order 4 so $S_1 \cap M = S \cap M$ since M has a unique normal subgroup of order 4. Therefore, $S_1 = (S_1 \cap M)R = (S \cap M)R = S$. Since $S_1 \neq S$ (indeed, $D_1 \not\leq S$), we get a contradiction. Thus, the theorem is true for $n = m - 1$.

(ii) Let $n < m - 1$. In that case, we obtain the result as in the proof of Theorem 5.4. □

This argument, with some modification, also works in the case $p > 2$ and $n = m-1$, $m > p + 2$ (see Theorem 13.6). Indeed, let G be a group of order p^m, $m > p + 2$. Suppose that G has a subgroup M of maximal class and index p. Let R be a normal subgroup of G of order p^p and exponent p. Since $R \not\leq M$ in view of $|M| > p^{p+1}$, we get $G = MR$, $G/(M \cap R) = (M/(M \cap R)) \times (R/(M \cap R))$ so $d(G) = 3$. We see that G has exactly p^2 maximal subgroups not containing R. Let H be one of them. Let us consider $G/Z(M)$ (note that $Z(M) < \Phi(M) = \Phi(G) < H$). This quotient group has two nonincident subgroups of orders p^p and p^{p-1} and exponent p; one of them is contained in $M/Z(M)$ and another is $R/Z(M)$. It follows from Theorem 12.1(b) that $G/Z(M)$ has no absolutely regular subgroups of index p so $H/Z(M)$ is not absolutely regular. Assume that H is not of maximal class. Then, by Theorem 13.5, H contains a G-invariant subgroup D of order p^p and exponent p. It is possible to choose D so that $H \cap R < D$ (Corollary 13.3). Set $S = DR$; then $|S| = p^{p+1}$, by the product formula (clearly, $D \neq R$ since $R \not\leq H$). Considering $S \cap M$ and taking into account that M has no normal subgroups of order p^p and exponent p, we conclude that $\exp(S \cap M) > p$ so $S \cap M$ is absolutely regular of order p^p and S is irregular of order p^{p+1} since $\Omega_1(S) = S$ and hence S is of maximal class (Theorems 7.2(b) and 7.1(b)). Since the number of normal subgroups of order p^p and exponent p in G is $\equiv 1 \pmod p$ (Theorem 13.5), it follows that G contains a normal subgroup D_1 of order p^p and exponent p that is not contained in S (indeed, the number of such subgroups in S, since $\exp(S) > p$ and $d(S) = 2$, is at most p: S contains exactly $p+1$ subgroups of index p). Set $S_1 = D_1 R$. By the choice of D_1, we get $S_1 \neq S$ and S_1 is of maximal class and order p^{p+1} (considering $R_1 \cap M$, we get $R_1 \cap M = R \cap M$ so $|S_1| = p^{p+1}$, by the product formula). Since M has only one normal subgroup of order p^p, we have $S \cap M = S_1 \cap M$. Since $S = (S \cap M)R = (S_1 \cap M)R = S_1$, we get a contradiction. This proves that all maximal subgroups of G not containing R, are of maximal class. All $p + 1$ maximal subgroups of G containing R, are not of maximal class. Therefore, G has exactly $p^2 = |\Gamma_1| - (p + 1) = p^2$ subgroups of index p that are of maximal class. This is another proof of the case $n = m - 1$ of Theorem 12.12(c) in the considered partial case.

The above argument does not work for $p > 2$ and $|G| = p^{p+2}$.

Appendix 6

On the order of p-groups of given derived length

P. Hall has proved that if the derived length of a p-group G of order p^m is $k+1$, then $m \geq 2^k + k$ (see Theorem A.6.7). In this section we will prove Hall's estimate and state some related results.

Let $G = \mathrm{K}_1(G) \geq \cdots \geq \mathrm{K}_m(G) \geq \cdots$ be the lower central series of a group G. Let $\mathrm{Z}_i(G)$ be the i-th term of the upper central series of G, $\mathrm{Z}_0(G) = \{1\}$. We have $[G, \mathrm{Z}_{i+1}(G)] \leq \mathrm{Z}_i(G)$ and $[G, \mathrm{K}_i(G)] \leq \mathrm{K}_{i+1}(G)$.

Lemma A.6.1 (P. Hall). *If G is an arbitrary group and $j \geq i$ are positive integers, then $[\mathrm{K}_i(G), \mathrm{Z}_j(G)] \leq \mathrm{Z}_{j-i}(G)$. In particular, $[\mathrm{K}_i(G), \mathrm{Z}_i(G)] = \{1\}$.*

Proof. We proceed by induction on i. For $i = 1$, the result has been noticed above. So, assuming $i > 1$ and applying Three Subgroups Lemma (Introduction, Exercise 13(c)), we get

$$\begin{aligned}[\mathrm{K}_i(G), \mathrm{Z}_j(G)] &= [\mathrm{K}_{i-1}(G), G, \mathrm{Z}_j(G)] \\ &\leq [G, \mathrm{Z}_j(G), \mathrm{K}_{i-1}(G)][\mathrm{Z}_j(G), \mathrm{K}_{i-1}(G), G] \\ &\leq [\mathrm{Z}_{j-1}(G), \mathrm{K}_{i-1}(G)] \cdot [\mathrm{Z}_{j-i+1}(G), G] \leq \mathrm{Z}_{j-i}(G).\end{aligned}$$

In particular, $[\mathrm{K}_i(G), \mathrm{Z}_i(G)] \leq \mathrm{Z}_{i-i}(G) = \mathrm{Z}_0(G) = \{1\}$ for all $i \in \mathbb{N}$. □

Lemma A.6.2. *Let G be a p-group, Then $\mathrm{Z}_i(G)$ contains all normal subgroups of G of order $\leq p^i$. In particular, if N is a normal subgroup of G of order $> p^i$, then $|N \cap \mathrm{Z}_i(G)| \geq p^i$.*

Proof. Let $N \trianglelefteq G$ be of order $\leq p^i$. We claim that then $N \leq \mathrm{Z}_i(G)$. It is known that $|N \cap \mathrm{Z}_1(G)| \geq p$ so our claim follows for $i = 1$. Now let $i > 1$. We work by induction on i. Then $N/(N \cap \mathrm{Z}(G)) \leq \mathrm{Z}_{i-1}(G/(N \cap \mathrm{Z}(G))$. Since $\mathrm{Z}_{i-1}(G/(N \cap \mathrm{Z}(G)) \leq \mathrm{Z}_i(G)/(N \cap \mathrm{Z}(G))$, the first assertion is proven. If, in addition, $|N| > p^i$, then N contains a G-invariant subgroup N_i of order p^i and, by what has just been proved, $N_i \leq N \cap \mathrm{Z}_i(G)$ so $|N \cap \mathrm{Z}_i(G)| \geq |N_i| = p^i$. □

Lemma A.6.3 (P. Hall). *Let $N \triangleleft G$ be nonabelian and $N \leq \mathrm{K}_i(G)$. Then $|\mathrm{Z}(N)| \geq p^i$, $|N| \geq p^{i+2}$, $|N : N'| \geq p^{i+1}$.*

Proof. Every G-invariant subgroup of N which has order $\leq p^i$, is contained in $\mathrm{Z}_i(G)$ (Lemma A.6.2) so centralized by $\mathrm{K}_i(G)$; then $|\mathrm{Z}(N)| \geq p^i$. Since N is nonabelian, we get $|N| \geq p^2|\mathrm{Z}(N)| \geq p^{i+2}$.

Let M be a G-invariant subgroup of index p in N'. Since $N/M \leq \mathrm{K}_i(G/M)$ is nonabelian, we get $|N/M| \geq p^{i+2}$ so $|N : N'| = \frac{1}{p}|N : M| \geq p^{i+1}$. □

Lemma A.6.4. *We have* $[\mathrm{K}_i(G), \mathrm{K}_j(G)] \leq \mathrm{K}_{i+j}(G)$ *for all* i, j.

Proof. By definition, $[\mathrm{K}_1(G), \mathrm{K}_j(G)] = [G, \mathrm{K}_j(G)] = \mathrm{K}_{j+1}(G)$. Therefore, assuming $i > 1$ and working by induction on i, we get, by Three Subgroups Lemma,

$$\begin{aligned}[\mathrm{K}_i(G), \mathrm{K}_j(G)] &= [\mathrm{K}_{i-1}(G), G, \mathrm{K}_j(G)] \\ &\leq [G, \mathrm{K}_j(G), \mathrm{K}_{i-1}(G)][\mathrm{K}_j(G), \mathrm{K}_{i-1}(G), G] \\ &\leq [\mathrm{K}_{j+1}(G), \mathrm{K}_{i-1}(G)][\mathrm{K}_{i+j-1}(G), G] \\ &\leq \mathrm{K}_{i+j}(G)\mathrm{K}_{i+j}(G) = \mathrm{K}_{i+j}(G).\end{aligned}$$ □

Corollary A.6.5. $G^{(i)} \leq \mathrm{K}_{2^i}(G)$.

Proof. We have $G^{(1)} = G' = \mathrm{K}_2(G)$, and the corollary holds for $i = 1$. Working by induction on i, we obtain, by Lemma A.6.4,

$$G^{(i+1)} = [G^{(i)}, G^{(i)}] \leq [\mathrm{K}_{2^i}(G), \mathrm{K}_{2^i}(G)] \leq \mathrm{K}_{2^i+2^i}(G) = \mathrm{K}_{2^{i+1}}(G).$$ □

Corollary A.6.6. *If* $G^{(i+1)} > \{1\}$, *then* $|G^{(i)} : G^{(i+1)}| \geq p^{2^i+1}$.

Proof. By hypothesis and Corollary A.6.5, $G^{(i)}$ is a nonabelian subgroup of $\mathrm{K}_{2^i}(G)$. Therefore, by Lemma A.6.3, $|G^{(i)} : G^{(i+1)}| = |G^{(i)} : (G^{(i)})'| \geq p^{2^i+1}$. □

Theorem A.6.7 (P. Hall). *Suppose that a group G has order p^n and derived length $k+1$. Then $n \geq 2^k + k$.*

Proof. We have $|G^{(i)} : G^{(i+1}| \geq 2^i + 1$ for $i = 1, \ldots, k-1$ (Lemma A.6.6) so

$$p^n = |G| = |G : G'||G' : G''| \ldots |G^{(k-1)} : G^{(k)}||G^{(k)}|,$$

so, because of $|G^{(k)}| \geq p$, by hypothesis, we get

$$n \geq 1 + (2^0 + 1) + (2^1 + 1) + \cdots + (2^{k-1} + 1) = 2^k + k.$$ □

It follows from the proof of Theorem A.6.7 that if $n = 2^k + k$, then we must have $|G^{(k)}| = p$ and $|G^{(i)} : G^{(i+1}| = 2^i + 1$ for all $i \in \{1, \ldots, k-1\}$. It is interesting to study the p-groups G satisfying these relations.

In the proof of Theorem A.6.7, we used the inequality $|G : G'| \geq p^{2^0+1} = p^2$ for a nonabelian p-group G. However, if $p = 2$ and G is not of maximal class, then $|G : G'| \geq 2^3$, by Taussky's theorem, and we get $|G| \geq 2^{2^k+k+1}$ (this is true for all 2-groups of derived length > 2.

Mann used the following lemma in the proof of Theorem A.6.9.

Lemma A.6.8. *Let G be an arbitrary group, let $L \leq G'$ be G-invariant and let G'/L be cyclic. Then $G'' = [G', L]$.*

Proof. Obviously, $[G', L] < L$ and $[G', L] \trianglelefteq G$. We have $L/[G', L] \leq Z(G'/[G', L])$ so $G'/[G', L]$ is abelian since G'/L is cyclic. Thus, $G'' \leq [G', L]$. Since $[G', L] \leq [G', G'] = G''$, we are done. □

Theorem A.6.9 (Mann). *Suppose that a group G has order p^n and derived length $k+1$. Then $n \geq 2^k + 2k - 2$.*

This estimate is better than Hall's one for $k > 2$.

Theorem A.6.10 (Mann). *If G is p-group with* $\mathrm{dl}(G) = 4$, *then* $|G| \geq p^{13}$.

This is better than the estimate of Theorem A.6.9.

For $p > 2$, the bound of Theorem A.6.10 is not yet best possible, as it is known that then the order is $\geq p^{14}$ (theses of N. Blackburn and S. Evans-Riley). It is known that, for $p > 3$, the minimal order of a p-group of derived length 4 equals p^{14}.

The following example is due to Hall [Hal3, pp. 54–55]. Let $E_r = \langle a_1, \dots, a_r \rangle \cong \mathrm{E}_{p^r}$, $r > 1$. Let $B_r \in \mathrm{Syl}_p(\mathrm{Aut}(E_r))$. Then $B_r \cong \mathrm{UT}(r, p) \in \mathrm{Syl}_p(\mathrm{GL}(r, p))$ is the group of unimodular upper triangular $r \times r$ matrices over the Galois field $\mathrm{GF}(p)$, $|B_r| = p^{\frac{1}{2}r(r-1)}$. And if we denote by $\tau_{i,j}$ the automorphism of E_r, which replaces a_i by $a_i a_j$ and leaves unchanged all of the a_k with $k \neq i$, then we may suppose that B_r is generated by the $\frac{1}{2}r(r-1)$ elements $\tau_{i,j}$ for which $i < j$. Since $a_i^{\tau_{i,j}^s} = a_i a_j^s$, we get $o(\tau_{i,j}) = p$. Since $(a_i a_j^{-1})^{\tau_{i,j}} = a_i$, we get $a_i^{\tau_{i,j}^{-1}} = a_i a_j^{-1}$. Therefore, if $i \neq \beta$, $j \neq \alpha$, we get $[\tau_{i,j}, \tau_{\alpha,\beta}] = \mathrm{id}$, while $[\tau_{i,j}, \tau_{j,k}] = \tau_{i,k}$. It is easy to verify from these equations, that B_r' is generated by those $\tau_{i,j}$ for which $j \geq i+2$. More generally, $\mathrm{K}_\alpha(B_r)$ is generated by those $\tau_{i,j}$ for which $j \geq i + \alpha$, while $B_r^{(\beta)}$ is generated by those $\tau_{i,j}$ for which $j \geq i + 2^\beta$. Thus, $\mathrm{cl}(B_r) = r-1$ while $B_r^{(\beta)} > \{1\}$ implies $2^\beta < r$. Taking $r = 2^\nu + 1$, we have the group B_r of class 2^ν whose ν-th derived group is greater than $\{1\}$. And in all these cases we have $B_r^{(i)} = \mathrm{K}_{2^i}(B_r)$.

The following theorems contain some additional information on members of the derived series of a p-group G.

Theorem A.6.11. *Suppose that G is a p-group.*

(a) *[Schn2, Theorem 1.1] Let $p > 2$, $|G'/G''| = p^3$ and $G'' > \{1\}$. Then we have $|G'/\mathrm{K}_3(G)| = p$ and $G'' = \mathrm{K}_5(G)$.*

(b) *[Schn2, Lemma 5.2] Let $p > 2$, $|G'/G''| = p^3$ and $G'' > \{1\}$. Then $G'/G'' \cong \mathrm{E}_{p^3}$.*

(c) *[Schn2, Corollary 5.6] If $p > 3$ and $G' \cong S(p^3) \times \mathrm{C}_p$, then $\mho_1(G) \leq Z(G)$. If $p > 2$ and $G' \cong \mathrm{M}_{p^3} \times \mathrm{C}_p$, then $\mho_2(G) \leq Z(G)$. If $\mathrm{cl}(G) < p$ and $\exp(G') = p$, then $\mho_1(G) \leq Z(G)$.*

(d) *[Schn3, Theorem 1.1] If* $p \geq 5$ *and* $G^{(d)} > \{1\}$, *then* $|G'| \geq 2^n$, *where* $n = 2^d + 3d - 6$.

(e) *[Schn3, Theorem 1.2] Let* $d \geq 1$, $G^{(d+1)} > \{1\}$. *If* $|G^{(d)}/G^{(d+1)}| = 2^{2^d+1}$, *then* $|G^{(d)}/[G^{(d)}, G]| = p$.

Exercise 1. Let G be a 2-group with nonabelian derived subgroup G' of order 2^4. Then $G' = A \times C$, where A is nonabelian of order 8.

Hint. We have $Z(G') \cong E_4$ (Proposition 1.13). If G' is metacyclic, it has a characteristic subgroup L such that $G'/L \cong Q_8$, which is impossible (Burnside). Then, by Theorem 44.12, $d(G') = 3$. Let $A < G'$ be minimal nonabelian (by Exercise 1.8a, $d(A) = 2$). If $Z(G') = Z(A) \times L$, then $G' = A \times L$.

Exercise 2. Suppose that G is a 2-group such that $|G'/G''| = 2^3$ and $G'' > \{1\}$. Prove that $G'/G'' \cong E_8$. (*Hint.* Let $L < G''$ be G-invariant of index 2. By Exercise 1, $d(G'/L) = 3$.)

Exercise 3 ([Hall, Theorem 2.47]). If G is a p-group of class $c > 1$ and $i \leq c$, then $K_i(G) > Z_{c-i}(G)$.

Solution. Since $cl(G/K_i(G)) = i - 1$, we have $K_i(G) \leq Z_{c-i+1}(G)$. Assume that $K_i(G) \leq Z_{c-i}(G)$. Then we should have successively

$$K_{i+1}(G) = [K_i(G), G] \leq [Z_{c-i}(G), G] = Z_{c-i-1}(G), \dots, K_c(G) \leq Z_0(G) = \{1\}$$

so $cl(G) < c$, a contradiction. Thus, $K_i(G) > Z_{c-i}(G)$.

Exercise 4 ([Hall, Theorem 2.58]). If $2i \leq c$, the class of a p-group of G, and A is a maximal abelian normal subgroup of G, then $A \not\leq Z_i(G)$ so that $|A| > p^i$.

Solution. Assume that $A \leq Z_i(G)$. Then $[A, K_i(G)] \leq [Z_i(G), K_i(G)] = \{1\}$ (Lemma A.6.1) so $A \leq Z(AK_i(G))$. Since $C_G(A) = A$, we get $K_i(G) \leq A \leq Z_i(G)$. Since $[K_i(G), Z_i(G)] = \{1\}$. it follows that $K_i(G) = A = Z_i(G)$. Then

$$K_i(G) > Z_{c-i}(G) \geq Z_i(G) = K_i(G),$$

by Exercise 3, and this is a contradiction.

In general, if $K_i(G) \leq Z_j(G)$, then the class of G does not exceed $i + j - 1$.

Exercise 5. Suppose that A is a subgroup of a p-group G such that $C_G(A) \leq A$. If $A \leq K_i(G)$, then $Z_i(G) \leq A$. If $A \leq Z_i(G)$, then $K_i(G) \leq A$ and $cl(G) \leq 2i - 1$.

Problem 1. Classify the minimal nonabelian p-groups A such that $A \cong G'$ for some p-group G. (In that case, A is metacyclic, by Theorem 44.12.)

Problem 2. Study the p-groups of derived length k, $k > 1$, which have minimal possible order.

Appendix 7

Relative indices of elements of p-groups

This section was written by M. Roitman.

If G is a group, then several known results indicate that the structure of G is controlled to a large extent by class sizes, that is, by the indices $|G : \mathrm{C}_G(g)|$ for $g \in G$ (see [Bae4]). This topic was extensively studied in the literature.

Here we replace the index $|G : \mathrm{C}_G(g)|$ by $|N : \mathrm{C}_N(g)|$, where N is a normal subgroup of G; thus $|N : \mathrm{C}_N(g)|$ (*the relative index of* g *in* N) is the number of conjugates of g by elements of N. A starting point was the following result [KS, Exercise 5.1.5]: If A is a maximal normal abelian subgroup of a p-group G and $|A : \mathrm{C}_A(x)| \leq p$ for all $x \in G$, then $G' \leq A$. In this section, considering a more general setting, we obtain a stronger conclusion. In particular, it follows from Theorem A.7.5 that in the setting considered in [KS, Exercise 5.1.5], we have $[G, A] \leq \mathrm{Z}(G)$ and $\mathrm{cl}(G) \leq 3$.

The following lemma is well known.

Lemma A.7.1. *A p-group cannot be the union of less then $p + 1$ proper subgroups.*

Proof. Let $|G| = p^k > 1$. Assume that G is a union of n distinct proper subgroups, where $n \leq p$: $G = \bigcup_{i=1}^n H_i$. Since $n \geq 2$ and the sets H_i are not disjoint, we see that $p^k < \sum_{i=1}^n |H_i| \leq n(p^{k-1}) \leq p^k$, a contradiction. □

Let N be a normal subgroup of a group G. We define $G^{[n]}(N)$ for $n \geq 0$ as follows: $G^{[0]}(N) = N$ and

$$G^{[n]}(N) = [G^{[n-1]}(N), G] = [N, \underbrace{G, \dots, G}_{n \text{ times}}] \quad \text{for } n > 0.$$

Clearly $G^{[n]}(N) \trianglelefteq G$. If G is a nilpotent group, then obviously, $G^{[n+1]}(N) \leq G^{[n]}(N)$ with strong inclusion if and only if $G^{[n]}(N) > \{1\}$. We have $G^{[1]}(N) = [G, N]$ so $G^{[0]}(N)/G^{[1]}(N) = N/[G, N] \leq \mathrm{Z}(G/[G, N])$. Similarly,

$$G^{[n]}(N)/G^{[n+1]}(N) \leq \mathrm{Z}(G/G^{[n+1]}(N)),$$

so the series $N = G^{[0]}(N) \geq G^{[1]}(N) \geq \cdots$ is part of a central series of G containing N. Obviously, $G^{[n]}(G) = \mathrm{K}_{n+1}(G)$ is the $(n + 1)$-th term of the lower central series of G.

For the next lemma, compare Lemma A.6.4, above.

Lemma A.7.2. *In our setting, for all n, the group $G^{[n]}(N)$ contains all commutators in $n+1$ elements of G, one of them belonging to N.*

Proof. We proceed by induction on n. For $n = 0$ the assertion is obvious. Let $n > 0$. Assume for all $m < n$ that all commutators in $m+1$ elements, one of them belonging to N, are in $G^{[m]}(N)$. Let u be a commutator of $n+1$ elements of G, one of them belonging to N. Thus $u = [v, w]$, where v and w are commutators so that the sum of their weights is $n+1$, and so that an element of N occurs either in v or in w. Since $u^{-1} = [w, v]$, we obtain by the inductive assumption that $u \in [\mathrm{K}_i(G), G^{[n-i]}(N)]$ for some integer $1 \le i \le n$. We induct also on i. If $i = 1$, then $[\mathrm{K}_1(G), G^{[n-1]}(N)] = [G^{[n-1]}(N), \mathrm{K}_1(G)] = G^{[n]}(N)$. If $i > 1$, by the Three Subgroups Lemma and by the inductive assumptions on $n-1$ and on $i-1$ we obtain:

$$\begin{aligned} u \in {} & [\mathrm{K}_i(G), G^{[n-i]}(N)] \\ & = [\mathrm{K}_{i-1}(G), G, G^{[n-i]}(N)] \\ & \le [[G, G^{[n-i]}(N)], \mathrm{K}_{i-1}(G)][[G^{[n-i]}(N), \mathrm{K}_{i-1}(G)], G] \\ & \le [G^{[1+(n-i)]}(N), \mathrm{K}_{i-1}(G)][G^{[(n-i)+(i-1)]}(N), G] \\ & \le G^{[(1+n-i)+(i-1)]}(N) G^{[n]}(N) = G^{[n]}(N). \end{aligned}$$ □

As usual, if G is a group, $g \in G$ and A is a subset of G, we denote by $[A, g]$ the set of commutators $[a, g]$ for $a \in A$.

We recall from [LGNW] that $\mathrm{C}_G(X : N)$ is defined as $\{g \in G \mid [X, g] \subseteq N\}$, where G is a group, X a subset of G, and N is a normal subgroup of G; thus $\mathrm{C}_G(X : N)$ is a subgroup of G containing N, and $\mathrm{C}_G(X : N)/N$ is the centralizer of $\overline{X}$ in $\overline{G} = G/N$.

For the next lemma, compare [LGNW, Lemma 2.1] and its proof.

Lemma A.7.3. *Let G be a p-group and let N be a normal subgroup of G such that for some fixed integer $0 \le n < p$ we have $|N : \mathrm{C}_N(g)| \le p^n$ for all $g \in G$. Then $G^{[n+1]}(N) = \{1\}$.*

Proof. Given $g \in G$, consider the following sequence of $n+2$ subgroups of N:

$$N = \mathrm{C}_N(g)G^{[0]}(N) \ge \mathrm{C}_N(g)G^{[1]}(N) \ge \cdots \ge \mathrm{C}_N(g)G^{[n+1]}(N).$$

Let $g \in G$. Since $|N : \mathrm{C}_N(g)| \le p^n$, we obtain $\mathrm{C}_N(g)G^{[i]}(N) = \mathrm{C}_N(g)G^{[i+1]}(N)$ for some i with $0 \le i \le n$. For $x \in G$ and $c \in \mathrm{C}_N(g)$ we have

$$[cx, g] = (cx)^{-1}g^{-1}(cx)g = x^{-1}c^{-1}g^{-1}cxg = x^{-1}g^{-1}xg = [x, g].$$

Hence $[\mathrm{C}_N(g)G^{[j]}(N), g] = [G^{[j]}(N), g]$ for all j. It follows that $[G^{[i]}(N), g] = [G^{[i+1]}(N), g] \le G^{[i+2]}(N)$, so $g \in H_i := \mathrm{C}_G([G^{[i]}(N) : G^{[i+2]}(N))$. Hence $G = \bigcup_{i=0}^{n} H_i$. By Lemma A.7.1 we have $G = H_i$ for some i since $n < p$. For this i, we have $[G^{[i]}(N), g] \le G^{[i+2}(N)$ for all $g \in G$ so $G^{[i+1]}(N) = [G^{[i]}(N), G] \le G^{[i+2}(N)$. Hence $G^{[i+1]}(N) = \{1\}$. We conclude that $G^{[n+1]}(N) = \{1\}$. □

Lemma A.7.4. *Let G be a p-group and let N be a normal subgroup of G such that $|N : \mathrm{C}_N(g)| \leq p^p$ for all $g \in G$. Then $G^{[p]}(\Phi(N)) = \{1\}$.*

Proof. Let g be an element of G. We have

$$|\Phi(N) : \mathrm{C}_{\Phi(N)}(g)| = \frac{|\Phi(N)|}{|\Phi(N) \cap \mathrm{C}_N(g)|} = \frac{|\Phi(N)\mathrm{C}_N(g)|}{|\mathrm{C}_N(g)|} \leq |N : \mathrm{C}_N(g)| \leq p^p.$$

If $|\Phi(N) : \mathrm{C}_{\Phi(N)}(g)| = p^p$, then $\Phi(N)\mathrm{C}_N(g) = N$, so $\mathrm{C}_N(g) = N$ and $|\Phi(N) : \mathrm{C}_{\Phi(N)}(g)| = 1$, a contradiction. Hence $|\Phi(N) : \mathrm{C}_{\Phi(N)}(g)| < p^p$ for all $g \in G$. By Lemma A.7.3, we obtain that $G^{[p]}(\Phi(N)) = \{1\}$. □

We put together Lemmas A.7.3 and A.7.4 to obtain:

Theorem A.7.5. *Let G be a p-group and let N be a normal subgroup of G such that for some fixed integer $n \geq 0$ we have $|N : \mathrm{C}_N(g)| \leq p^n$ for all $g \in G$. Then $G^{[n+1]}(N) = \{1\}$ under each of the following conditions:*

(1) $n < p$;

(2) $n = p$ and $[G, N] \leq \Phi(N)$.

Moreover, if $G^{[n+1]}(N) = \{1\}$ and if $\mathrm{C}_G(N) \leq N$, then $\mathrm{cl}(G) \leq 2n + 1$.

Proof. In view of Lemma A.7.3, we assume condition (2). By Lemma A.7.4 we have

$$G^{[p+1]}(N) = G^{[p]}([N, G]) \leq G^{[p]}(\Phi(N)) = \{1\}.$$

Now assume that $\mathrm{C}_G(N) \leq N$. Since by Lemma A.7.2,

$$[N, \mathrm{K}_{n+1}(G)] \leq G^{[n+1]}(N) = \{1\},$$

we have $\mathrm{K}_{n+1}(G) \leq \mathrm{C}_G(N) \leq N$, so

$$\mathrm{K}_{2n+2}(G) = G^{[n+1]}(\mathrm{K}_{n+1}(G)) \leq G^{[n+1]}(N) = \{1\}.$$

Thus $\mathrm{cl}(G) \leq 2n + 1$. □

Since $G' \leq \Phi(G)$, Theorem A.7.5 implies in case $N = G$ that if $n \leq p$, then $\mathrm{K}_{n+2}(G) = G^{[n+1]}(G) = \{1\}$, that is, that $\mathrm{cl}(G) \leq n + 1$ (for this result see [Hup1, Exercise 25, page 310] and [LGNW, Lemma 3.2]).

In case $n = 1$, we may relax the assumptions in Theorem A.7.5 to the effect that $|N : \mathrm{C}_N(x)| \leq p$ just for $x \in G - N$:

Proposition A.7.6. *Let G be a p-group and let $N \trianglelefteq G$. If $|N : \mathrm{C}_N(x)| \leq p$ for all $x \in G - N$, then $|N : \mathrm{C}_N(x)| \leq p$ for all $x \in G$.*

Proof. Assume that $|N : \mathrm{C}_N(x)| \le p$ for all $x \in G - N$, but for some element $h \in N$ we have $|N : \mathrm{C}_N(h)| \ge p^2$. Let x be an element in $G - N$. We have

$$\mathrm{C}_N(x) \cap \mathrm{C}_N(xh) = \mathrm{C}_N(x^{-1}) \cap \mathrm{C}_N(xh) \le \mathrm{C}_N(x^{-1}(xh)) = \mathrm{C}_N(h).$$

Hence

$$p^2 \le |N : \mathrm{C}_N(h)| \le |N : (\mathrm{C}_N(x) \cap \mathrm{C}_N(xh))| \le p^2.$$

It follows that we have equalities everywhere, so $\mathrm{C}_N(h) = \mathrm{C}_N(x) \cap \mathrm{C}_N(xh)$. Thus $\mathrm{C}_N(h) \le \mathrm{C}_N(x)$ for all $x \in G - N$. Since $\langle G - N \rangle = G$, we conclude that $h \in \mathrm{Z}(G)$. It follows that $\mathrm{C}_N(h) = N$, contradicting the choice of h. □

Appendix 8

p-groups with absolutely regular Frattini subgroup

In this section we prove an analog of Theorems 4.4 and 4.5: we will give a certain factorization of p-groups with absolutely regular Frattini subgroup. Recall that a p-group G is said to be *absolutely regular* if $|G : \mho_1(G)| < p^p$ (see §9). Absolutely regular p-groups are regular (Theorem 9.8(a)).

Theorem A.8.1 ([Ber3]). *Suppose that a p-group G is neither absolutely regular nor of maximal class and let $\Phi(G)$ is absolutely regular. Set $\Phi_0 = \Omega_1(\Phi(G))$ and $|\Phi_0| = p^k (< p^p)$. Let $B \leq G$ be such that $\Phi(G) \leq B$, and B has no G-invariant subgroups of order p^{k+1} and exponent p and B is as large as possible; then B is either absolutely regular or of maximal class so $B < G$. Let $G/B = (T_1/B) \times \cdots \times (T_s/B)$, where $|T_i/B| = p$ for all i. Then $T_i = E(T_i)B$, where $E(T_i)$ is a G-invariant subgroup of order p^{k+1} and exponent p. Set $A = E(T_1) \dots E(T_s)$; then*

$$|A| = p^{k+s}, \quad \Phi_0 < E(T_i) \text{ for all } i, \quad \Phi(A) \leq \Phi_0, \quad A/\Phi_0 \leq \Omega_1(Z(G/\Phi_0)).$$

Next, $G = AB$, $A \cap B = \Phi_0$. Let $A_0/\Phi_0 = \langle R/\Phi_0 \mid R/\Phi_0 \lhd G/\Phi_0, |R/\Phi_0| = p, \exp(R) = p\rangle$. Then $|A_0 : A| \leq p$ and

(a) If $k < p - 1$, then G is regular, B is absolutely regular, $A_0 = A = \Omega_1(G)$ and $A \cap B = \Phi_0$.

(b) Let $k = p - 1$. Then B is either absolutely regular or of maximal class, $A \leq A_0$ and A_0 is characteristic in G. If B is absolutely regular, then $|B \cap A_0| \leq p^{2k}$. If B is of maximal class and order $> p^{p+1}$, then $|B \cap A_0| \leq p^p$.

Proof. Let $B < G$ be defined as in the statement of the theorem. Then B is either absolutely regular or of maximal class, by Theorem 13.5, so $B < G$. Let $B < T \leq G$ be such that $|T : B| = p$. Then T has a G-invariant subgroup $E(T)$ of order p^{k+1} and exponent p, by the choice of B. We have $B \cap E(T) = \Phi_0$ since Φ_0 is the unique normal subgroup of order p^k and exponent p in B. Let the elementary abelian p-group $G/B = (T_1/B) \times \cdots \times (T_s/B)$, where $T_1 = T$ and $|T_i/B)| = p$ for all i. Then, by the product formula, $T_i = E(T_i)B$ for all i. Set $A = E(T_1) \dots E(T_s)$. In that case, $AB = (E(T_1)B) \dots (E(T_s)B) = T_1 \dots T_s = G$. Since $A/\Phi_0 \cong \mathrm{E}_{p^s}$ and $|\Phi_0| = p^k$, we have $\mathrm{cl}(A) \leq k + 1$. Since $|A| = p^{k+s}$ and $|G : B| = p^s$, we get $A \cap B = \Phi_0$, by the product formula.

Suppose that $k < p - 1$ (then $p > 2$). In that case, G is regular (Theorem 7.1(c) or Theorem 9.8(c)) so $\exp(A) = p$ since $A \le \Omega_1(G)$ (Theorem 7.2(b)). Since $\exp(\Omega_1(G)) = p$, it follows from $B \cap \Omega_1(G) = \Omega_1(B) = \Phi_0$ that $\Omega_1(G) = A$, by the product formula.

Suppose then $k = p - 1$ (this is the case if G is irregular, by Theorem 9.8(c), however, we do not assume here that G is irregular). Let A_0 be defined as in the statement of the theorem; then $A \le A_0$, $A_0/\Phi_0 \le \Omega_1(Z(G/\Phi_0))$ so A_0/Φ_0 is a characteristic elementary abelian subgroup of G/Φ_0; then A_0 is characteristic in G since Φ_0 is characteristic in G. Set $A_0 \cap B = L$. Then $L/\Phi_0 \le Z(B/\Phi_0)$. If B is irregular of maximal class and order $> p^{p+1}$, then $|L/\Phi_0| \le p$ so $|A_0 : A| \le p$. If B is absolutely regular, then $|L| \le p^{2p-2}$ (otherwise, $\exp(B \cap A_0) > p^2 \ge \exp(A_0)$. which is not the case). If, in addition, $|B \cap A_0| > p^{p-1}$, then $\exp(A_0) \ge \exp(B \cap A_0) > p$ so A_0 is irregular since $\Omega_1(A_0) = A_0$ (Theorem 7.2(b)); then $\mathrm{cl}(A_0) = p$. □

Exercise 1. Let G be an irregular p-group, $p > 2$, and let $\Omega_1(G)$ have order p^p and cyclic center. Prove that G has an absolutely regular maximal subgroup. If, in addition, $|G : \Omega_1(G)| \ge p^p$, then all members of the set Γ_1 not containing $\Omega_1(G)$ are absolutely regular.

Hint. Let $R < G$ be a G-invariant abelian subgroup of type (p, p). Then $C_G(R) = M \in \Gamma_1$ is absolutely regular (Theorem 13.5). To prove the second assertion, use Theorems 13.5 and 9.6. (This is true always.)

Exercise 2. Let G be a p-group such that $\exp(\Omega_1(G)) = p$. Suppose that $\Phi(G) < N \le G$ implies $\Omega_1(N) \not\le \Phi(G)$. Then $\Omega_1(G) = G$ so $G' = \Phi(G)$.

Solution. Set $k = d(G) - 1$. If $N \in \Gamma_k$, then $N = \Omega_1(N)\Phi(G)$. It follows that

$$G = \langle N \mid N \in \Gamma_k \rangle = \langle \Omega_1(N)\Phi(G) \mid N \in \Gamma_k \rangle = \langle \Omega_1(N) \mid N \in \Gamma_k \rangle$$

so $G = \Omega_1(G)$, and we conclude that $\exp(G) = p$ whence $G' = \Phi(G)$.

Remark. Let G be a p-group such that $\exp(\Omega_1(G)) = p$. Suppose that, whenever $\Phi(G) < H \le G$ and $\Omega_1(\Phi(G)) = \Omega_1(H)$, then $H = \Phi(G)$. We claim that then $\exp(G) = p$. Indeed, let $A/\Phi(G) < G/\Phi(G)$ be of order p. Since $A > \Phi(G)$ so, by assumption, $\Omega_1(\Phi(G)) < \Omega_1(A)$, and we have $A = \Omega_1(A)\Phi(G)$ since $\Phi(G)$ is maximal in A. Setting $k = d(G) - 1$, we get

$$G = \langle A \mid A \in \Gamma_k \rangle = \langle \Omega_1(A)\Phi(G) \mid A \in \Gamma_k \rangle = \langle \Omega_1(A) \mid A \in \Gamma_k \rangle \le \Omega_1(G),$$

and the claim follows.

Corollary A.8.2. *Let $G > \{1\}$ be a p-group which is neither absolutely regular nor of maximal class and let $\Phi(G)$ be absolutely regular. Suppose that $\Phi(G)$ is maximal among subgroups N of G such that $\Phi(G) \le N$ and N has no G-invariant subgroups of order $p|\Omega_1(\Phi(G))|$. Then* $\exp(G) = p$.

Proof. Write $\Phi_0 = \Omega_1(\Phi(G))$ and set $|\Phi_0| = p^k$; then $k < p$, by hypothesis. Let, as in Theorem A.8.1, A_0 be a subgroup generated by all G-invariant subgroups of order p^{k+1} and exponent p containing Φ_0. By Theorem A.8.1, $G = \Phi(G)A_0$ so $A_0 = G$. Since $G/\Phi_0 = A_0/\Phi_0$ is elementary abelian, we get $\Phi(G) = \Phi_0$. Let $\Phi(G) < U < G$, where $|U : \Phi(G)| = p$. Then, by hypothesis, $\exp(U) = p$. Since all such U cover G, we get $\exp(G) = p$. □

Problem. Study the 2-groups with metacyclic Frattini subgroup.

Appendix 9

On characteristic subgroups of metacyclic groups

In this section we show that metacyclic groups (not necessarily of prime power order), as a rule, have many characteristic subgroups. Some structure theorems on metacyclic p-groups are also proved.

If $|G| = \prod_{i=1}^{k} p_i^{\alpha_i}$ is a prime decomposition, then set $\lambda(G) = \sum_{i=1}^{k} \alpha_i$ so then $|G|$ is a product of $\lambda(G)$ primes. For example, $\lambda(S_5) = 5$.

Definition. A group $G > \{1\}$ is said to be a $(*)$-group if $G = Z_1 \times Z_2 \times Q$, where either $Q \cong Q_8$ is of odd index in G or $Q = \{1\}$, and Z_1 and Z_2 are isomorphic cyclic groups. (Thus, if $Q > \{1\}$, then $|Z_1| = |Z_2|$ is odd.)

Exercise 1. $(*)$-groups are metacyclic. A $(*)$-group has no characteristic subgroups of prime index.

Exercise 2. Let G be a metacyclic p-group.

(a) If p and $\lambda(G)$ are odd, then for each $n < \lambda(G)$, G has a characteristic subgroup H with $\lambda(H) = n$.

(b) Let G be abelian. If for some $n < \lambda(G)$, G has no characteristic subgroup H with $\lambda(H) = n$, then G is homocyclic so an $(*)$-group.

(c) Let G be nonabelian and $p > 2$. Then for each $n < \lambda(G)$, G has a characteristic subgroup H with $\lambda(H) = n$. (*Hint.* $\Omega_1(G')$ is characteristic in G.)

(d) Let G be a nonabelian and $p = 2$. If for some $n < \lambda(G)$, G has no characteristic subgroup H with $\lambda(H) = n$, then $G \cong Q_8$.

Hint. (d) Assume that $|G'| > 2$. Let $L = \Omega_1 G')$. By induction on $|G|$, $G/L \cong Q_8$. Use Taussky's theorem to get a contradiction. If $|G'| = 2$, then G is minimal nonabelian. Use Exercise 1.8(a) and (b).

Exercise 3. If a nilpotent metacyclic group G is not a $(*)$-group, it has a characteristic subgroup H with $\lambda(H) = n$ for all $n < \lambda(G)$.

Theorem A.9.1. *Let G be a metacyclic group. If, for some $n < \lambda(G)$, G has no characteristic subgroup whose order is a product of n primes, then G is a $(*)$-group.*

Proof. Let G be a counterexample of minimal order; then, by Exercise 3, G is nonnilpotent. Let $R \le G'$ be of prime order, say p; then R is characteristic in G so G/R satisfies the hypothesis, whence, by induction, G/R is a $(*)$-group so nilpotent, and we conclude that $R \not\le Z(G)$; then $p > 2$. Since $G/C_G(R)$ is isomorphic to a proper subgroup of a cyclic group C_{p-1}, it has a characteristic subgroup $H/C_G(R)$ of prime index, say q, where q divides $p-1$. But a $(*)$-group G/R has no characteristic subgroups of prime indices, a final contradiction. □

Corollary A.9.2. *If G is a metacyclic p-group, then $\pi(\mathrm{Aut}(G)) \subseteq \pi(p(p-1))$, unless G is a $(*)$-group.*

Proof. Let $q \ne p$ be a prime divisor of $|\mathrm{Aut}(G)|$ and $\alpha \in \mathrm{Aut}(G)$ with $o(\alpha) = q$. Suppose that G is not a $(*)$-group; then G contains a characteristic subgroup H of index p. We have $G/\Phi(G) = (H/\Phi(G)) \times (L/\Phi(G))$, where L is α-invariant (Maschke's theorem). Then α induces on $G/\Phi(G)$ a nonidentity automorphism, by Burnside–Hall, so, for example, α induces a nonidentity automorphism on $H/\Phi(G)$. Therefore, $o(\alpha) = q$ divides $|\mathrm{Aut}(H/\Phi(G))| = p-1$, i.e., $q \in \pi(p-1)$. □

Lemma A.9.3 ([Tho3, Lemma 5.55]). *Let $G = A \cdot B$, where B is a normal π'-Hall subgroup of G, $A \cap B = \{1\}$. Then $[B, A, A] = [B, A]$.*

Proof. Since $[B, A] \le B$, we get $[B, A, A] \le [B, A]$. Let us prove the reverse inclusion. Choose $a \in A$, $b \in B$, and let $c = [b, a]$, so that $b^a = bc$. For each $n \in \mathbb{N}$, define $x_n \in G$ by $b^{a^n} = bc^n x_n$, so that $x_1 = 1$. Suppose that $x_n \in [B, A, A]$ for some n and prove that $x_{n+1} \in [B, A, A]$. We have

$$bc^{n+1}x_{n+1} = b^{a^{n+1}} = (b^{a^n})^a = (bc^n x_n)^a = bc(c^n)^a x_n^a = bc \cdot c^n[c^n, a]x_n^a,$$

so that $x_{n+1} = [c^n, a]x_n^a \in [B, A, A]$ since $[c^n, a], x_n^a \in [B, A, A]$. Indeed, $c^n = [b, a]^n \in [B, A]$ so $[c^n, a] \in [B, A, A]$. Next,

$$x_n^a \in [B, A, A]^a = [B^a, A^a, A^a] = [B, A, A]$$

since $B \lhd G$. Taking $n = |A|$, we get $b = b^{a^n} = bc^n x_n$ so $c^n x_n = 1$, and hence $c^n = (x_n)^{-1} \in [B, A, A]$. Since c and c^n generate the same cyclic subgroup in view of $(o(c), |A|) \le (|B|, |A|) = 1$, we get $c \in [B, A, A]$, and so $[B, A] \le [B, A, A]$ since elements $c = [b, a]$ generate $[B, A]$. □

Theorem A.9.4 (Huppert; see [BG, Theorem 4.12]). *Suppose that R is a metacyclic p-group and A a p'-group of operators on R. Then $[R, A]$ is abelian, unless R is ordinary quaternion.*

Proof. One may assume that $A/C_A(R) > \{1\}$. Suppose that $R \not\cong Q_8$. Then $p > 2$ (Corollary A.9.2). We use induction on $|R|$. By Lemma A.9.3, $[R, A, A] = [R, A]$. Therefore we may assume that $R = [R, A]$; then we must to prove that R is abelian.

Take a cyclic A-invariant subgroup S of R that is maximal subject to containing R'. Then $S \lhd G = A \cdot R$, the natural semidirect product. Since S is cyclic, $G/\mathrm{C}_G(S)$ is abelian so $G' \leq \mathrm{C}_G(S)$. Since $R = [R, A] \leq G'$, we get $S \leq \mathrm{Z}(R)$ so it suffices to show that R/S is cyclic. Assume that this is false. Since $\Omega_1(R) \cap S = \Omega_1(S)$ is of order p, we get $|\Omega_1(R)S/S| = p$. By Maschke's theorem, there exists an A-invariant complement X/S to $\Omega_1(R)S/S$ in $\Omega_1(R/S)$. Since $\Omega_1(R) \not\leq X$, it follows that $|\Omega_1(X)| = p$ so X is cyclic since $p > 2$. By the maximal choice of S, we must have $X = S$. Thus, $\Omega_1(R/S) = \Omega_1(R)S/S$, and the last group, as we know, is of order p. It follows, that R/S is cyclic since $p > 2$, and so R is abelian since $S \leq \mathrm{Z}(R)$. □

Lemma A.9.5. *Let $G = AR$, where $A < G$ and $R \lhd G$. Let H be a G-invariant subgroup of R such that $(|A|, |H|) = 1$ and $AH \lhd G$. Then $R = H\mathrm{N}_R(A)$. If, in addition, $A \cap R = \{1\}$, then $R = H\mathrm{C}_R(A)$.*

Proof. By Frattini's argument, $G = \mathrm{N}_G(A)AH = \mathrm{N}_G(A)H$ (here we use the Schur–Zassenhaus theorem and the Odd Order theorem) so, by the modular law, we have $R = H(\mathrm{N}_G(A) \cap R) = H\mathrm{N}_R(A)$. If $A \cap R = \{1\}$, then $[\mathrm{N}_R(A), A] \leq R \cap A = \{1\}$ so $\mathrm{N}_R(A) = \mathrm{C}_R(A)$, and we get $R = H\mathrm{C}_R(A)$. □

Corollary A.9.6. *Let $G = A \cdot R$ be a semidirect product with kernel R and complement A such that $(|A|, |R|) = 1$. Then $R = [R, A]\mathrm{C}_R(A)$.*

Theorem A.9.7 (Huppert; see [BG, Theorem 4.12]). *Suppose that R is a metacyclic p-group, $p > 2$, and A is a p'-group of operators on R. Then:*

(a) The subgroup $[R, A]$ is abelian.

(b) $R = [R, A]\mathrm{C}_R(A)$ with $[R, A] \cap \mathrm{C}_R(A) = \{1\}$.

(c) If R is nonabelian and A does not act trivially on R, then $[R, A]$ and $\mathrm{C}_R(A)$ are nonidentity cyclic subgroups and $R' \leq [R, A]$.

Proof. Suppose that A does not act trivially on R (otherwise, there is nothing to prove). Then, by Corollary A.9.2, $p > 2$.

(a) coincides with Theorem A.9.4.

(b) Let $T = [R, A]$; then $T \lhd A \cdot R$. By Lemma A.9.3, $[T, A] = [R, A, A] = [R, A] = T$. By Lemma A.9.6, $R = [R, A]\mathrm{C}_R(A) = T\mathrm{C}_R(A)$. By Fitting's lemma (see §6), $T = [T, A] \times \mathrm{C}_T(A) = T \times \mathrm{C}_T(A)$ since T is abelian so $\{1\} = \mathrm{C}_T(A) = T \cap \mathrm{C}_R(A)$.

(c) Since T (see (b)) is abelian and R is not, we get $\{1\} < T < R = T\mathrm{C}_R(A)$ so $\mathrm{C}_R(A) > \{1\}$. Since $T \cap \mathrm{C}_R(A) = \{1\}$, by (b), it follows that $\Omega_1(R) \not\leq T$, $\Omega_1(R) \not\leq \mathrm{C}_R(A)$, i.e., $|\Omega_1(T)| = p = |\Omega_1(\mathrm{C}_R(A))|$, and so T and $\mathrm{C}_R(A)$ are cyclic since $p > 2$. Since $R/T \cong \mathrm{C}_R(A)$ is cyclic, $R' \leq T$. □

Corollary A.9.8. *Suppose that α is a fixed-point-free π'-automorphism of a metacyclic π-group R. Then R is abelian.*

Proof. Put $A = \langle\alpha\rangle$ and let $G = A \cdot R$ be the natural semidirect product; then $[R, A] = R \le G'$. Since $G/C_G(R')$ is abelian, we get $R \le G' \le C_G(R')$ so R is nilpotent. In that case, one may assume that $\pi = \{p\}$. Then, by Theorem A.9.4, $R = [R, A]$ is abelian. □

Theorem A.9.9 (Blackburn; see [BG]). *Suppose that a nonabelian p-group R, $p > 2$, has no subgroups $\cong E_{p^3}$. If A is a $\{2, p\}'$-group of automorphisms of R such that $[R, A] = R$, then $p > 3$ and $R = S * C$ is a central product, where $S = \Omega_1(R) \cong S(p^3)$ and $C = Z(R)$ is cyclic.*

Proof. By Theorem A.9.4, R is nonmetacyclic. Since A acts nontrivially on $R/\Phi(R)$ and $(|A|, 2p) = 1$, it follows that R is not a 3-group of maximal class (Theorem 1.16). Therefore, by Theorem 13.7, $R = SX$, where $S = \Omega_1(R) \cong S(p^3)$, X is cyclic and $C_X(S)$ is a subgroup of index $\le p$ in X. If A centralizes all elements of S, then $A \cdot R$ has no minimal nonnilpotent subgroups of order multiple p (Theorem 10.8). Then A centralizes R, which is not the case. In view of $(|A|, 2p) = 1$ and $\pi(\mathrm{Aut}(S)) \subseteq \pi(p(p^2 - 1))$, we get $p > 3$ so, by Theorem 12.1(a), R is absolutely regular hence we have $|R/\mho_1(R)| = |\Omega_1(R)| = |S| = p^3$ (Theorem 7.2(d)). Since R has a cyclic subgroup of index p^2, $\mho_1(R)$ is cyclic. It remains to show that $Z(R)$ is cyclic of order $|X|$. One may assume that $|R| > p^3$.

If $d(R) = 3$, then $R = S * C$, where C is cyclic (Lemma 4.2), and we are done.

Now let $d(R) = 2$. Then $R/C_R(S) \cong S$. We also have $|X : C_R(S)| = p$, $R' < S$ and $R' \cong E_{p^2}$ (if $R' \cong C_p$, then $d(R) = 3$). Set $C = C_R(S)$ and $|X| = p^n$; then $\exp(R) = p^n$, $|R : \Omega_{n-1}(R)| = p$ so

$$c_n(R) = \frac{|R - \Omega_{n-1}(R)|}{\varphi(p^n)} = \frac{p^{n+2} - p^{n+1}}{p^{n-1}(p-1)} = p^2.$$

Let $X = \langle x\rangle$. We have $[X, S] \not\le S'$ since R/S' is nonabelian and $R/[X, S]$ is abelian. Taking $y \in S - R'$ and $z \in R' - S'$, we get $S/R' = \langle yR'\rangle$, $R'/S' = \langle zS'\rangle$.

We have $[S, A] \not\le R'$ since S/R' is not a direct factor of $A \cdot (R/R')$ in view of $[R, A] = R$ (Fitting's lemma). Choose $\alpha \in A$ such that $o(\alpha)$ is a power of a prime and $[S, \alpha] \not\le R'$. Since R contains exactly p^2 cyclic subgroups of order $|X|$ and $(p, o(\alpha)) = 1$, one may assume, that $X^\alpha = X$. Then there exist integers i, j and k such that $x^\alpha = x^i$, $y^\alpha \equiv y^j \pmod{R'}$ (since $S/R' = \langle yR'\rangle$ is A-invariant), and $z^\alpha \equiv z^k \pmod{S'}$ (since $R'/S' = \langle zS'\rangle$ is A-invariant). We have $y^{\alpha^2} \equiv y^{j^2} \pmod{R'}$. Since $o(\alpha)$ is odd, α^2 does not act trivially on S/R', and hence $j^2 \not\equiv 1 \pmod{p}$. Since $x^\alpha = x^i$ is a generator of X, $p \nmid i$.

Now $1 \ne [y, z] \in S' < \langle x\rangle$ and $[x, y] \in [X, S] = R'$. Since $[X, S] = R' \not\le S'$, we have $[x, y] \notin S'$ so $[x, y] \in R' - S'$. Since $\mathrm{cl}(S) = 2$, application of Exercise 1.18, in view of $[y, z] \in X$ and $x^i = x^\alpha$, yields

$$[y, z]^i = [y, z]^\alpha = [y^\alpha, z^\alpha] = [y^j, z^k] = [y, z]^{jk},$$

and, in view of $[R, R'] = S'$ and $z^k \equiv z^\alpha \pmod{S'}$, we get, modulo S', $[x, y]^k \equiv [x, y]^\alpha \equiv [x^\alpha, y^\alpha] = [x^i, y^j] = [x, y]^{ij}$. Thus $jk \equiv i \pmod{p}$ and $ij \equiv k \pmod{p}$ so $ij^2 \equiv i \pmod{p}$. Since $p \nmid i$, we get $j^2 \equiv 1 \pmod{p}$, contrary to what has been proved above. □

Corollary A.9.10. *Let R be a p-group of Theorem A.9.9. If R admits a fixed-point-free p'-automorphism α, then $p > 3$ and $R = S * Z$, a central product, where $S = \Omega_1(R)$ is nonabelian of order p^3 and exponent p and $Z = Z(R)$ is cyclic.*

Appendix 10

On minimal characters of p-groups

Let G be a nonabelian group and $d = \min\{\chi(1) \mid \chi \in \mathrm{Irr}_1(G)\}$; irreducible characters of degree d we call, following Mann, *minimal characters* of G. The number d we call the *minimal character degree* of G.

Exercise 1. Let χ be a minimal character of a nonabelian group G. Suppose that $\chi = \mu^G$ for some $H < G$ and $\mu \in \mathrm{Irr}(H)$. Then $G' \le H$.

Solution. We have $|G : H| \le \chi(1)$. Since $(1_H)^G$ is reducible (indeed, by reciprocity, $\langle (1_H)^G, 1_G\rangle = 1$), all its irreducible constituents have degrees $< \chi(1)$ so linear. It follows that

$$G' \le \ker((1_H)^G) = \bigcap_{\tau \in \mathrm{Irr}(1_H^G)} \ker(\tau) = H_G \le H.$$

Exercise 2. Let G be a nonabelian p-group and χ its faithful minimal character. Then all nonlinear irreducible characters of G have the same degree $\chi(1)$.

Hint. We have $\chi = \mu^G$ for some $H < G$ and $\mu \in \mathrm{Lin}(H)$. By Exercise 1, $G' \le H$ so $H \lhd G$. Then $H' \le \ker(\chi) = \{1\}$ so H is abelian. Use Ito's theorem on degrees (Introduction, Theorem 17).

Theorem A.10.1 ([Man12]). *Let G be a nonabelian p-group with minimal character of degree d. Then d is the least index of a subgroup H of G satisfying $H' \ne G'$. The minimal characters of G are those irreducible characters of G that are linearly induced from the above subgroups of minimal index, and they are not linearly induced from other subgroups. If H is such a subgroup of index d, then H has some linear characters which induce to irreducible characters of G. Given a minimal character χ, it is linearly induced from such H (of index d) if and only if $H' \le \ker(\chi)$.*

Proof. The first assertion follows from Exercises 1 and 2. Let $H < G$ be such that $H' < G' \le H$ (in particular, $H \lhd G$). Then there exists $\mu \in \mathrm{Lin}(H)$ such that $G' \not\le \ker(\mu)$. By reciprocity and Clifford, all irreducible constituents of μ^G are nonlinear (of degree at least d). If, in addition, $|G : H| = d$, then induced character μ^G is minimal. Finally, if χ is a minimal character, and H is a subgroup such that $|G : H| = d$ and $H' \le \ker(\chi)$, then all irreducible constituents of χ_H are linear, and $\chi = \lambda^G$, where λ is an arbitrary linear constituent of χ_H, by reciprocity. □

Exercise 3 (Mann). Let d be the minimal character degree of a nonabelian group G (we do not assume that G is nilpotent). Let $H \le G$ satisfy $|G : H| \le d$. Then $G' \le H$. If $|G : H| < d$, then $H' = G'$. Moreover, G contains a subgroup H such that $|G : H| = d$ and $H' < G'$ if and only if G has an irreducible monomial character of degree d, and in that case any such character is linearly induced from some such subgroup, and every such subgroup has a linear character which induces the irreducible character of G.

Solution. For the first assertion, see Exercise 1. Let $|G : H| < d$; then $H \trianglelefteq G$ and G/H' has no nonlinear irreducible characters by Ito's theorem on degrees (Introduction, Theorem 17), so G/H' is abelian and so $H' = G'$. Let $|G : H| = d$ and $H' < G'$. Then there exists $\mu \in \mathrm{Lin}(H)$ such that $G' \nleq \ker(\mu)$. In that case, μ^G has a nonlinear irreducible constituent so μ^G is irreducible since $|G : H| = \mu^G(1) \ge d = |G : H|$; thus μ^G is a minimal character of G. Now let $\chi = \tau^G$ be an irreducible monomial character of G of degree d, where $\tau \in \mathrm{Lin}(H)$ for some $H < G$; then $|G : H| = d$. In that case, clearly, $H' < G'$ (otherwise, $\mathrm{Irr}(\tau^G) \subseteq \mathrm{Lin}(G)$).

Exercise 4 (Mann). Let G and d be as in Theorem A.10.1. Then $D(G)$, the intersection of kernels of all minimal characters of G, is equal to the intersection of the subgroups H' such that $|G : H| = d$ and $H' < G' \le H$. The factor group $G/D(G)$ is metabelian.

Appendix 11

On sums of degrees of irreducible characters

We begin with the following

Definition. Let G be a p-group of maximal class and order p^m, $m \geq 4$, then $G_1 = C_G(K_2(G)/K_4(G))$ is said to be the *fundamental subgroup* of G. If we have $G_1 = C_G(K_i(G)/K_{i+2}(G))$ for $2 \leq i \leq m-2$, then G is said to be *nonexceptional*. Otherwise, G is said to be *exceptional*.

Let G be a p-group of maximal class. If G is nonexceptional and $x \in G - G_1$, then $C_G(x) = \langle x, Z(G)\rangle$. If $|G| > p^{p+1}$ or fundamental subgroup $G_1 < G$ is abelian, then G is nonexceptional. If $|G| < p^5$, then G is nonexceptional.

Suppose that G is a nonabelian p-group. Let $d = p^{\mu(G)} = \min\{\chi(1) \mid \chi \in \mathrm{Irr}_1(G)\}$. In terminology of Appendix 10, $d = p^{\mu(G)}$ is the minimal character degree of G. If G is a p-group of maximal class, then $\mu(G) = 1$ since $G/K_3(G)$ is nonabelian of order p^3. For $H \leq G$, define $\delta(G,H) = T(G) - T(H)$, where $T(G) = \sum_{\chi \in \mathrm{Irr}(G)} \chi(1)$. For example, if $G \cong D_{2^{n+1}}$, $n > 2$, and $H < G$ is cyclic of index 2, then $T(G) = 2^n + 2$, $T(H) = 2^n$, so $\delta(G,H) = 2$. Next, $\delta(S_4, A_4) = 10 - 6 = 4$. If G is a Frobenius group with kernel H, then $\delta(G,H) = T(H) - 1$ (this follows from description of characters of Frobenius groups; see [BZ, Chapter 10]).

Exercise 1. If $H < G$, then $\delta(G,H) > 0$.

Solution. Given $\mu \in \mathrm{Irr}(H)$, there exists $\chi \in \mathrm{Irr}(G)$ such that $\mu \in \mathrm{Irr}(\chi_H)$, by reciprocity, so $T(H) \leq T(G)$. It follows from $H < G$ that $(1_H)^G$ is reducible so there are two distinct $\chi, \tau \in \mathrm{Irr}(G)$ such that $1_H \in \mathrm{Irr}(\chi_H) \cap \mathrm{Irr}(\tau_H)$, and our claim follows.

Exercise 2. Let $H < G$. Then $\delta(G,H) = 1$ if and only if G is a Frobenius group with kernel H of index 2.

Exercise 3. Classify the pairs $H < G$ with $\delta(G,H) = 2$.

Theorem A.11.1 ([BerM]). *Let G be a nonabelian p-group and $\{1\} < H < G$.*

(a) If $|G : H| = p$, then $(p-1)p^{\mu(G)}$ divides $\delta(G,H)$. In particular, $p(p-1)$ divides $\delta(G,H)$.

(b) $\delta(G,H) = p^{\mu(G)}(p-1)$ if and only if $\mu(G) = 1$, G is of maximal class and nonexceptional, $|G : H| = p$, H is abelian or $H = C_G(Z_2(G))$.

Proof. Let $|G : H| = p$. Set

$$\mathcal{L} = \{\phi \in \mathrm{Irr}(H) \mid G' \leq \ker(\phi)\}, \ \mathcal{N} = \{\phi \in \mathrm{Irr}(H) - \mathcal{L} \mid \phi \text{ is } G\text{-invariant}\}.$$

Then $\mathcal{L} \subseteq \mathrm{Lin}(H)$ and $\mathcal{N} \subseteq \mathrm{Irr}_1(H)$ and $\mathcal{L} \neq \varnothing$ since $G' < H$. Characters contained in the set $\mathcal{L}$ are G-invariant. By definition, $|\mathcal{L}| = |H : G'| \geq p$.

Suppose that $H < G$ is abelian (of index p). Then $\mu(G) = 1$ (Ito's theorem on degrees), $|H : G'| = |\mathrm{Z}(G)|$ (Lemma 1.1) and

$$\begin{aligned} T(G) &= |G : G'| + \frac{|G| - |G : G'|}{p^2} \cdot p = p|\mathrm{Z}(G)| + \frac{|G| - p|\mathrm{Z}(G)|}{p} \\ &= |H| + (p-1)|\mathrm{Z}(G)| \end{aligned}$$

so

$$\delta(G, H) = T(G) - |H| = |\mathrm{Z}(G)|(p-1) \equiv 0 \pmod{p(p-1)}.$$

By Brauer's permutation lemma, $\mathcal{L}$ is the set of all G-invariant linear characters of H. In the general case (for nonabelian H), going over to G/H', we see that $\mathcal{L}$ is the set of all G-invariant linear characters of H.

Let $M < G'$ be G-invariant of index p. Then for all $\chi \in \mathrm{Irr}(G/M)$, we have $\chi(1)^2 = |G/M : \mathrm{Z}(G/M)| \leq p^{c(G)}$, where $c(G) = \log_p |G : G'|$. Thus,

(1) $$2\mu(G) \leq c(G) \Rightarrow \mu(G) \leq c(G) - 1.$$

Suppose that $\phi \in \mathcal{N}$. Then $\phi^G = \chi^1 + \cdots + \chi^p$, where $\chi^1, \ldots, \chi^p$ are pairwise distinct irreducible characters of G (and extensions of ϕ to G). Hence, denoting, for $\phi \in \mathrm{Irr}(H)$, $a(\phi) = |\mathrm{Irr}(\phi^G)|$, we get

(2) $$\text{if } \phi \in \mathcal{N}, \text{ then } a(\phi) = p \text{ and } \phi(1) \geq p^{\mu(G)}.$$

Next, if $\phi \in \mathcal{L}$, then $a(\phi) = p$ as well. Now let $\psi \in \mathrm{Irr}(H) - (\mathcal{L} \cup \mathcal{N})$. Then $\psi^G = \chi \in \mathrm{Irr}(G)$ and $\chi_H = \psi_1 + \cdots + \psi_p$, where $\psi_1 = \psi, \ldots, \psi_p$ are distinct of the same degree so $\chi(1) = p\psi(1)$.

(a) Suppose that $|G : H| = p$. Then, by the above,

(3) $$\begin{aligned} \delta(G, H) &= \sum_{\phi \in \mathcal{L}} (a(\phi) - 1) + \sum_{\phi \in \mathcal{N}} (a(\phi) - 1)\phi(1) \\ &= |H : G'|(p-1) + (p-1) \sum_{\phi \in \mathcal{N}} \phi(1). \end{aligned}$$

By (1), $p^{\mu(G)}(p-1)$ divides $|H : G'|(p-1)$. By (2), $p^{\mu(G)}(p-1)$ divides $(p-1) \sum_{\phi \in \mathcal{N}} \phi(1)$. Therefore, by (3), $p^{\mu(G)}(p-1)$ divides $\delta(G, H)$. Since $\mu(G) \geq 1$, the proof of (a) is complete.

(b) Suppose that $\delta(G, H) = p^{\mu(G)}(p-1)$. It is obvious from (a) that H is maximal in G. It follows from (1) and (3) that $|H : G'| = p^{\mu(G)}$, i.e., $\mu(G) = c(G) - 1$, and (1) implies

$$c(G) = 2, \quad \mu(G) = 1, \quad |G : G'| = p^2, \quad \delta(G, H) = p(p-1).$$

It follows from (3) that $\mathcal{N} = \varnothing$. Thus, the set $\mathcal{L}$, which is of cardinality p, coincides with the set of all G-invariant irreducible characters of H. Therefore, there are in H exactly p classes that are also G-classes (Brauer's permutation lemma). This is possible only if $|Z(G) \cap H| = p$. Moreover, $Z(G) \leq \Phi(G) < H$ since $d(G) = 2$. Thus $|Z(G)| = p$. Then also $\delta(G/Z(G), H/Z(G)) \leq \delta(G, H)$, and so $\delta(G/Z(G), H/Z(G)) = p(p-1)$. By induction, $G/Z(G)$ is of maximal class. Since $|Z(G)| = p$, G is also of maximal class.

Take $x \in Z_2(G) - Z(G)$. Since we would like to prove that G is nonexceptional, we may assume that $\log_p |G| \geq 5$. Since $Z_2(G)$ is a unique normal subgroup of order p^2 in G, we get $Z_2(G) < H$, and so $x \in H$. Now $x \cdot Z(G)$ is the conjugacy class of x in G. It cannot be a class of H, for then H would have more than p invariant classes under G. This is possible only if $x \in Z(H)$, so that $H = C_G(Z_2(G))$. By induction, $H = C_G(Z_{i+2}(G)/Z_i(G))$ for all i, so that G is nonexceptional, as was to be shown.

Conversely, if G is of maximal class (then $\mu(G) = 1$) and nonexceptional and $H = C_G(Z_2(G))$, then it is easy to see that G has either only p invariant H-classes (the elements of $Z(G)$), or only p invariant irreducible characters of H; additionally, $\delta(G, H) = p(p-1)$. □

If G is an extraspecial group of order p^5 and H is a nonabelian subgroup of G of order p^3, then

$$\delta(G, H) = p^4 + p^2(p-1) - p^2 - p(p-1) = p^2(p^2-1) + p(p-1)^2$$

is not divisible by $p^{\mu(G)}(p-1) = p^2(p-1)$. Therefore, Theorem A.11.1(a) is not true for $|G : H| > p$.

Exercise 4. Classify the pairs $H < G$ of p-groups with $\delta(G, H) = p^2(p-1)$.

Exercise 5. Let H be a maximal subgroup of an extraspecial group G of order p^{2m+1}, $m > 1$. Show that $\delta(G, H) = p^{2m-1}(p-1)$.

Exercise 6. Let H be an abelian maximal subgroup of a p-group G of maximal class and order p^m. Show that $\delta(G, H) = p(p-1)$. If $m > 3$ and $F \in \Gamma_1$ is nonabelian, then $\delta(G, H) = p^{m-2}(p-1)$.

Appendix 12

2-groups whose maximal cyclic subgroups of order > 2 are self-centralizing

Maximal cyclic subgroups in finite noncyclic p-groups G play an important role. The second author determined all p-groups G with the assumption that each maximal cyclic subgroup of G is contained in exactly one maximal subgroup of G. In §58 all p-groups G such that each maximal cyclic subgroup of order $> p$ is normal in G, are determined. In this section we consider the case, where $p = 2$ and each maximal cyclic subgroup of order > 2 is self-centralizing in G. More precisely, we prove the following result.

Theorem A.12.1 (Janko). *A finite 2-group G of exponent > 2 has the property that each maximal cyclic subgroup of order > 2 is self-centralizing in G if and only if G is one of the following groups:*

(a) G is cyclic of order > 2;

(b) G is of maximal class;

(c) $G = \langle a, t \mid a^8 = t^2 = 1,\ a^t = au,\ u^2 = 1,\ a^u = a^5\rangle$, where $|G| = 2^5$, $Z(G) = \langle a^4\rangle \cong C_2$, $G' = \langle a^4, u\rangle \cong E_4$, $\Phi(G) = \langle a^2, u\rangle \cong C_4 \times C_2$, $\Omega_1(G) = \Omega_2(G) = \langle a^2, t\rangle \times \langle u\rangle \cong D_8 \times C_2$, G has exactly four maximal cyclic subgroups of composite order: $\langle a\rangle$, $\langle au\rangle$, $\langle at\rangle$, $\langle atu\rangle$ and they are all of order 8 (and they are self-centralizing in G). Also, G has exactly two normal elementary abelian subgroups of order 8 and $G = \langle a\rangle\langle at\rangle$ with $\langle a\rangle \cap \langle at\rangle = \langle a^4\rangle = Z(G)$. In fact, G is a subgroup of S_8 generated by the permutations $a = (1,2,3,4,5,6,7,8)$ and $t = (2,6)(3,7)$.

Proof. Let G be a 2-group of exponent > 2 with the property that each maximal cyclic subgroup of order > 2 is self-centralizing. Since cyclic 2-groups and 2-groups of maximal class have this property, we may assume that G is neither cyclic nor of maximal class.

By Lemma 1.4, G has a normal four-subgroup U. Let A be a maximal cyclic subgroup of order 2^n, $n \geq 2$. If $n = 2$, then Proposition 1.8 implies that G is of maximal class, a contradiction. Hence $n \geq 3$. Obviously, $\{1\} \neq Z(G) < A$ and $|Z(G) \cap U| = 2$. Suppose that $Z(G) = \langle r\rangle$ is of order ≥ 4 and take an involution $u \in U - Z(G)$. Let B be a maximal cyclic subgroup of G containing $\langle ur\rangle$, where $o(ur) = o(r) \geq 4$. But then r centralizes B and $r \notin B$, a contradiction. We have proved that $Z(G) = \langle z \rangle$ is of order 2, each cyclic subgroup of composite order contains $\langle z \rangle$ and each maximal cyclic subgroup of composite order in G is of order ≥ 8. In particular, $\exp(G) \geq 8$.

For each $x \in G - \Phi(G)$, $\langle x \rangle$ is a maximal cyclic subgroup of G and so if x centralizes U, then x is an involution. Indeed, if $\langle x \rangle$ is not a maximal cyclic subgroup of G, then x is a square in G and so $x \in \Phi(G)$, a contradiction.

Now we prove that $U \leq \Phi(G)$. Suppose false. Let $M \in \Gamma_1$ be such that $U \not\leq M$ so that $U \cap M = \langle z \rangle = Z(G)$. Set $K = C_G(U)$ so that $|G : K| = 2$ and $G/(M \cap K) \cong E_4$. Each element $x \in K - (M \cap K)$ is contained in $G - \Phi(G)$ and x centralizes U and therefore x is an involution. It follows that all elements in $K - (M \cap K)$ are involutions and so if $u \in U - M$, then u inverts and centralizes each element in $M \cap K$. Thus, $M \cap K$ is elementary abelian and so K is elementary abelian. But then $\exp(G) = 4$, a contradiction.

Now we use the fact that $U \leq \Phi(G)$. Set $K = C_G(U)$ so that $|G : K| = 2$ and $|K/\Phi(G)| \geq 2$. It follows that all elements in $K - \Phi(G)$ are involutions so $H_2(G) \leq \Phi(G)$ so $|K : \Phi(G)| = 2$, and we get $d(G) = 2$.

Now we prove $\Omega_1(G) = U$. Suppose that this is false. Let $E \cong E_8$ be a G-invariant subgroup with $U < E < \Phi(G)$. Let Y be a maximal cyclic subgroup of $\Phi(G)$ of composite order so that $Y \cap E = Y \cap U = \langle z \rangle$. Since Y centralizes $\Phi(G)$, Y is not a maximal cyclic subgroup of G. Let $X = \langle x \rangle$ be a maximal cyclic subgroup of G containing Y so that $|X : Y| = 2$. Then $x \in G - K$ and $x^2 \in \Phi(G)$. Hence x induces an involutory automorphism on E. But then $|C_E(x)| \geq 4$, a contradiction.

Since no element in $U - \langle z \rangle$ is a square in G, it follows that $\Phi(G)$ is abelian of type $(2^m, 2)$, $m \geq 2$, and so taking an involution $t \subset K - \Phi(G)$ and an involution $u \in U - \langle z \rangle$, we see that $\Phi(G) = \langle b \rangle \times \langle u \rangle \cong C_{2^m} \times C_2$, where $o(b) = 2^m$, $\langle b \rangle > \langle z \rangle$, t inverts b and t centralizes u so that $\langle b, t \rangle \cong D_{2^{m+1}}$, and $K = C_G(U) = \langle b, t \rangle \times \langle u \rangle \cong D_{2^{m+1}} \times C_2$. In addition, there is an element $a \in G - K$ such that $a^2 = b$ (since $\langle b \rangle$ cannot be a maximal cyclic subgroup in G) and $u^a = uz$, $C_G(a) = \langle a \rangle \cong C_{2^{m+1}}$, and $\langle a, u \rangle \cong M_{2^{m+2}}$ is another maximal subgroup of G (distinct from K).

We have $at \in G - K$, $(at)^2 \in \Phi(G)$ and $C_{\Phi(G)}(at) = b^{2^{m-2}}u = k$ with $k^2 = b^{2^{m-1}} = z$. Indeed, the assertion about the centralizer of at in $\Phi(G) = \langle b \rangle \times \langle u \rangle$ follows at once because we know the action of a and t on $\Phi(G)$: $b^a = b, u^a = uz, b^t = b^{-1}, u^t = u$ and so (since $k = b^{2^{m-2}}u$ is of order 4 and $k^2 = z$) $k^{at} = (b^{2^{m-2}}u)^{at} = (b^{2^{m-2}}z)(uz) = k$. This gives that $(at)^2 \in \langle k \rangle$. If $(at)^2 = z$, then $o(at) = 4$ and $\langle at \rangle$ would be a self-centralizing maximal cyclic subgroup of order 4 and (by Proposition 1.2) G would be of maximal class, a contradiction. Suppose that $(at)^2 = 1$. Then $a^t = a^{-1}$ so that $\langle a, t \rangle \cong D_{2^{m+2}}$. But in that case $\langle a \rangle$ is normal in G and G is a splitting extension of $\langle a \rangle$ by $\langle t, u \rangle \cong E_4$ and so $\Phi(G) \leq \langle a \rangle$, a contradiction. It follows that $\langle (at)^2 \rangle = \langle k \rangle$ and so replacing u with uz (if necessary), we may assume that $(at)^2 = k = b^{2^{m-2}}u$. This gives $a^t = a^{-1}k$.

Since $\Phi(G)$ is abelian, the centralizer $C_{\Phi(G)}(x)$ of any element $x \in G - K$ is equal to $C_{\Phi(G)}(a) = \langle b \rangle$ or to $C_{\Phi(G)}(at) = \langle k \rangle$, where $b = a^2$. Therefore, if $m > 2$, then the element bu (of order ≥ 8) is not a square of any element in $G - K$ so that $\langle bu \rangle$ (being a maximal cyclic subgroup in $\Phi(G)$) is a maximal cyclic subgroup in G but $C_G(bu) \geq \Phi(G)$, a contradiction.

We have proved that $m = 2$ so that $k = bu = a^2u$ and $a^t = a^{-1}k = au$. In this case $\Phi(G) = \langle b \rangle \times \langle u \rangle = \langle a^2 \rangle \times \langle u \rangle \cong C_4 \times C_2$, $|G| = 2^5$, $G' = \langle u, z \rangle \cong E_4$, $a^u = az = aa^4 = a^5$, and the structure of G is uniquely determined.

The properties of our group G stated in Theorem A.11.1 now follow by direct checking. □

Appendix 13

Normalizers of Sylow p-subgroups of symmetric groups

Let $M = \{1, \dots, n\} \subset \mathbb{N}$ and let $S_n = S_M$ be the symmetric group on the set M. In what follows, a prime p is fixed. Let Σ_n be a Sylow p-subgroup of S_n and N_n its normalizer in S_n. If $n < p$, then $\Sigma_n = \{1\}$ and $N_n = S_n$. For $\pi \in S_n$, let $T(\pi) = \{i \in M \mid \pi(i) \neq i\}$ be the *support* of permutation π. Thus, if $i \in M$, then $\pi(i) \neq i$ if and only if $i \in T(\pi)$. For $H \leq S_n$, we set $T(H) = \bigcup_{\pi \in H} T(\pi)$, the *support* of H. For a group G and $m \in \mathbb{N}$, let G^m be the direct product of m copies of G, $G^0 = \{1\}$.

Let $n \in \mathbb{N}$ be fixed. Denote by $e_{i,j}$ the $n \times n$ matrix $(a_{k,l})$, where $a_{i,j} = \delta_{(i,j),(k,l)}$ (here δ is the Kronecker delta), i.e., $a_{k,l} = 1$ if and only if $k = i$, $l = j$ and all other entries of the matrix $e_{i,j}$ equal 0 (here i, j, k, l are positive integers not exceeding n). Take $\sigma \in S_n$. Denote by M_σ the following $n \times n$-matrix:

$$M_\sigma = e_{1,1^\sigma} + e_{2,2^\sigma} + \dots + e_{n,n^\sigma} = \sum_{i=1}^{n} e_{i,i^\sigma}.$$

In that case, M_σ is said to be the *matrix corresponding to a permutation* σ or, shorter, a *permutation matrix*. If τ is another permutation in S_n, then

$$M_\sigma \cdot M_\tau = \sum_{i=1}^{n} e_{i,i^\sigma} \cdot \sum_{i=1}^{n} e_{i,i^\tau} = \sum_{i=1}^{n} e_{i,i^{\sigma\tau}} = M_{\sigma\tau}.$$

Thus, $\sigma \mapsto M_\sigma$ is a monomorphism of S_n into $\mathrm{GL}(n, \mathbb{R})$.

Let G be a group, B a transitive permutation group of degree n. One may assume that all elements of B are taken in the form of permutation matrices. Then, if $x_1, \dots, x_n \in G$ and $b \in B$, then $\mathrm{diag}(x_1, \dots, x_n) \cdot b$ is called a *monomial matrix* (over G). The set of all such matrices is a group with respect to usual matrix multiplication, which is called the *wreath product* of G by B and denoted by $G \operatorname{wr} B$. The set of all diagonal matrices in $W = G \operatorname{wr} B$ is said to be the *base subgroup* of W (denote this subgroup by $G^B = K$). Then $K \cong G^n$, the direct product of n copies of G, is normal in W and $W = B \cdot K$, a semidirect product with kernel K. Set $G_i = \{\mathrm{diag}\,(1, \dots, 1, x_i, 1, \dots, 1) \mid x_i \in G\}$. Then G_i, a subgroup of K, is called the i-th *coordinate subgroup* of K, $K = G_1 \times \dots \times G_n$ and all G_i are isomorphic to G.

The subgroup $D = \{\mathrm{diag}(x, \dots, x) \mid x \in G\}$ ($\cong G$) is said to be the *diagonal subgroup* of K. Obviously, elements of B induce permutations of the set $\{G_1, \dots, G_n\}$ and D coincides with the centralizer of B in K. We have $|G \mathrm{\,wr\,} B| = |G|^n|B|$. The above wreath product is said to be *standard* if $|B| = n$, i.e., B is regular as a permutation group.

Let $\Sigma_n^{(p)} = \Sigma_n \in \mathrm{Syl}_p(\mathrm{S}_n)$ and $n = a_0+a_1p+\dots+a_tp^t$ the decomposition of $n \in \mathbb{N}$ in the base p, $a_t \neq 0$. Then $|\Sigma_n| = (n!)_p = p^k$, where $k = [n/p] + [n/p^2] + \cdots$, and

$$\Sigma_n = \Sigma_p^{a_1} \times \Sigma_{p^2}^{a_2} \times \cdots \times \Sigma_{p^t}^{a_t}$$

(compare the orders!). In what follows, Σ_{p^m} denotes a p-group always. Next, $\Sigma_p \cong \mathrm{C}_p$ is cyclic of order p and

$$\Sigma_{p^m} = \underbrace{\Sigma_p \mathrm{\,wr\,} \Sigma_p \mathrm{\,wr\,} \dots \mathrm{\,wr\,} \Sigma_p}_{m \text{ times}} = \Sigma_{p^{m-1}} \mathrm{\,wr\,} \Sigma_p = \Sigma_p \mathrm{\,wr\,} \Sigma_{p^{m-1}}$$

(the last wreath product is not standard). It follows that $\Omega_1(\Sigma_{p^m}) = \Sigma_{p^m}$, $\mathrm{d}(\Sigma_{p^m}) = m$. We have $|\Sigma_{p^m}| = p^{j_p(m)}$, where $j_p(m) = 1 + p + \cdots + p^{m-1}$.

A permutation $\pi \in \mathrm{S}_n$ is said to be of type $(b_1, \dots, b_n)$ if, in its decomposition in a product of independent cycles (= standard decomposition), there are exactly b_i cycles of length i, $i = 1, \dots, n$.

Lemma A.13.1. *Let π be a permutation of type $(b_1, b_2, \dots, b_n)$ in $S_M = \mathrm{S}_n$, $|M| = n$. Let M_i be the set of all points which are permuted by i-cycles from the standard decomposition of π; then $|M_i| = ib_i$, $M_1 = M - T(\pi)$. Let Z be the centralizer of π in S_n. For $\sigma \in Z$, we denote by $\sigma_i = \sigma_{M_i}$ the restriction of σ to M_i; then π_i is a product of b_i independent cycles of length i and $\pi = \pi_1 \dots \pi_n$. In that case, $Z = Z_1 \times \cdots \times Z_n$, where $Z_i = \{\sigma_i \mid \sigma \in Z\}$, $T(Z_i) = M_i$, $Z_i = \mathrm{C}_i \mathrm{\,wr\,} \mathrm{S}_{b_i}$, C_i is generated by cycle of length i (if $b_i = 0$, then $Z_i = \{1\}$) and $M = \bigcup_{i=1}^n M_i$ is a partition of M. The base of the wreath product Z_i is generated by the cycles from the standard decomposition of $\pi_i = \pi_{M_i}$, $Z_1 = \mathrm{S}(M_1)$.*

Proof. Let $\sigma \in Z$. Then $\sigma(M_i) = M_i$ for all i so $\sigma = \sigma_1 \dots \sigma_n$, where $\sigma_i \in Z_i$, $T(\sigma_i) \subseteq M_i$ (of course, $T(\sigma_1) = \varnothing$). It follows that $Z = Z_1 \times \cdots \times Z_n$. One may assume that $M = M_i$, $i > 1$; then $\pi = \pi_i$ and all permutations from Z permute b_i independent i-cycles in decomposition of π. Since every permutation of S_n that permutes the above i-cycles, belongs to Z, we see that $Z \cong \mathrm{C}_i \mathrm{\,wr\,} \mathrm{S}_{b_i}$, completing the proof. (Thus, $|Z| = \prod_{i=1}^n i^{b_i} b_i!$.) □

Let $n > 3$. If a maximal subgroup H of S_n has a nontrivial center then either $H \cong \mathrm{S}_2 \times \mathrm{S}_{n-2}$ or $n = 2r$ and $H \cong \mathrm{S}_2 \mathrm{\,wr\,} \mathrm{S}_r$ (Lemma A.13.1). Indeed, since S_n has a trivial center, H is a centralizer of a permutation.

Let us show that $P = \Sigma_{p^m}$ is a transitive subgroup of $\mathrm{S}_{p^m} = G$. Let H be the stabilizer of a point i in G; then $H \cong \mathrm{S}_{p^m-1}$ so $|G : H| = p^m$. By Sylow, one may assume that $P \cap H \in \mathrm{Syl}_p(H)$. Then $PH = G$ so $|P : (P \cap H)| = p^m$. Therefore, since $P \cap H$ is the stabilizer of i in P, we are done.

Let $\delta(G)$ be the minimal degree of a faithful representation of a group G by permutations; then $|G|$ divides $\delta(G)!$ since $\mathrm{S}_{\delta(G)}$ has a subgroup isomorphic to G. We have $\delta(\mathrm{S}_n) = n$ and $\delta(\Sigma_{p^m}) = p^m$. If A and B are nonidentity groups of coprime orders, then $\delta(A \times B) = \delta(A) + \delta(B)$ so $\mathrm{C}_{\mathrm{S}_{p^m}}(\Sigma_{p^m}) = \mathrm{Z}(\Sigma_{p^m})$.

Let $N_n = \mathrm{N}_{\mathrm{S}_n}(\Sigma_n)$ be the normalizer of $\Sigma_n \in \mathrm{Syl}_p(\mathrm{S}_n)$ in S_n.

Theorem A.13.2 ([Ber18]). *Let $n = a_0 + a_1 p + \cdots + a_t p^t$ be the decomposition of $n \in \mathbb{N}$ in the base p, $a_t \neq 0$.*

(a) The normalizer $N_n = N_{a_0} \times N_{a_1 p} \times \cdots \times N_{a_t p^t}$, $|T(N_{a_i p^i})| = a_i p^i$, $T(N_{a_i p^i}) \cap T(N_{a_j p^j}) = \varnothing$ for $i \neq j$, $i, j \in \{1, \dots, t\}$, $n - |T(N_n)| = a_0$.

(b) If $1 \le a < p$, then $N_{ap^m} = N_{p^m} \mathrm{wr}\, \mathrm{S}_a$.

(c) $N_{p^m}/\Sigma_{p^m} \cong \mathrm{C}_{p-1}^m$. In particular, $N_{2^m} = \Sigma_{2^m}$.

(d) $[N_{p^m}, N_{p^m}] = \Sigma_{p^m}$ for $p > 2$. If $p = 2$, then $N_n = \Sigma_n$.

Proof. (i) Assume that $n = p^m$. Then Σ_{p^m} is transitive. The normalizer N_{p^m} has a p'-Hall subgroup K and, by what has been said above about $\delta(A \times B)$ with $(|A|, B|) = 1$, K acts on Σ_{p^m} faithfully. We assume that $p > 2$ (the proof for $p = 2$ is essentially easier and therefore omitted).

Let $m = 1$. Then $\Sigma_p = \langle \pi \rangle$, where $\pi = (1, \dots, p)$ is a p-cycle. For g, a primitive root modulo p, define a permutation σ on $\{1, \dots, p\}$ as follows. We have

$$\pi^g = (1, \overline{1+g}, \overline{1+2g}, \dots, \overline{1+(p-1)g}),$$

where $i_g = \overline{1+(i-1)g}$ is a minimal positive residue of $1 + (i-1)g$ modulo p. Setting $\sigma(i) = i_g$, $i = 1, 2, \dots, p$, we get $\sigma\pi\sigma^{-1} = \pi^g$. If $p = 7$ and $g = 3$, then $\pi^g = \pi^3 = (1, 4, 7, 3, 6, 2, 5)$ so

$$\sigma(1) = 1, \quad \sigma(2) = 4, \quad \sigma(3) = 7, \quad \sigma(4) = 3,$$
$$\sigma(5) = 6, \quad \sigma(6) = 2, \quad \sigma(7) = 5$$

hence $\sigma = (2, 4, 3, 7, 5, 6)$ is a 6-cyclic. We have

$$\langle \pi, \sigma \rangle = \langle \pi, \sigma \mid \pi^p = 1 = \sigma^{p-1}, \sigma\pi\sigma^{-1} = \pi^g \rangle,$$

which is isomorphic to the holomorph of the cyclic group $\mathrm{C}_p = \langle \pi \rangle$. Since there holds $\mathrm{C}_{N_p}(\Sigma_p) = \Sigma_p$, we have

$$N_p = \langle \pi, \sigma \rangle, \quad N_p/\Sigma_p \cong \mathrm{C}_{p-1}, \quad [N_p, N_p] = \Sigma_p,$$

and we are done in case $m = 1$.

Let $m > 1$. In that case we use induction on m. A permutation

$$\pi = (1, \dots, p)(p+1, \dots, 2p) \dots (p^m - p + 1, \dots, p^m)$$

generates $Z(\Sigma_{p^m})$. Let Z be the centralizer of π in S_{p^m}. Then $Z = Z^{(2)} \cdot Z^{(1)}$, a semidirect product with kernel $Z^{(1)}$, generated by independent cycles in the standard decomposition of π so that $Z^{(1)} \cong E_{p^{p^{m-1}}}$, and $Z^{(2)} \cong S_{p^{m-1}}$ (Lemma A.13.1).

Set $U = N_{S_{p^m}}(\langle \pi \rangle)$; then $Z \triangleleft U$ and U/Z is isomorphic to a subgroup of $\mathrm{Aut}(C_p) \cong C_{p-1}$. As before for $m = 1$, one can build the permutation σ of order $p-1$ such that $\sigma\pi\sigma^{-1} = \sigma^g$, where g is a primitive root modulo p; then $\sigma \in U$ and $\langle \sigma \rangle \cap Z = \{1\}$. Hence, $U = \langle \sigma \rangle \cdot Z$ is a semidirect product of Z and $\langle \sigma \rangle$. Since $\sigma\tau\sigma^{-1} = \tau^g$ for all $\tau \in Z^{(1)}$ (this is true if τ is a p-cycle in the standard decomposition of π, and so for $Z^{(1)}$ which is the direct product of cyclic groups generated by such cycles), we see that $Z^{(1)} \triangleleft U = \langle Z, \sigma \rangle$. Since $\langle \pi \rangle = Z(\Sigma_{p^m})$ is characteristic in Σ_{p^m}, we get $N_{p^m} \le U$. Since $Z/Z^{(1)} \cong Z^{(2)} \cong S_{p^{m-1}}$ and $\Sigma_{p^m}/Z^{(1)} \cong \Sigma_{p^{m-1}}$, it follows that the normalizer of $\Sigma_{p^m}/Z^{(1)}$ in $Z/Z^{(1)}$ is isomorphic with $N_{p^{m-1}}$, and the structure of latter is known, by the inductive hypothesis. Since the group $Z/Z^{(1)} \cong S_{p^{m-1}}$ is complete, by Hölder's theorem, we get $U/Z^{(1)} \cong (Z/Z^{(1)}) \times C_{p-1}$. So the normalizer of $\Sigma_{p^m}/Z^{(1)}$ in $U/Z^{(1)}$ is isomorphic, on the one hand, to $N_{p^{m-1}} \times C_{p-1}$, and on the other hand, to $N_{p^m}/Z^{(1)}$. Thus, we obtain

$$N_{p^m}/\Sigma_{p^m} \cong C_{p-1} \times (N_{p^{m-1}}/\Sigma_{p^{m-1}}) \cong C_{p-1} \times C_{p-1}^{m-1} \cong C_{p-1}^m,$$

and (c) is proved for $n = p^m$.

Now we prove that $[N_{p^m}, N_{p^m}] = \Sigma_{p^m}$ provided $p > 2$ (if $p = 2$, then, as we have noticed, $N_{2^m} = \Sigma_{2^m}$ so $[N_{2^m}, N_{2^m}] < \Sigma_{2^m}$). Since N_{p^m}/Σ_{p^m} is abelian, by (c), we get $[N_{p^m}, N_{p^m}] \le \Sigma_{p^m}$; so it is enough to prove the reverse inclusion. We use induction on m. Suppose that we have proved already that $\Sigma_{p^{m-1}} = [N_{p^{m-1}}, N_{p^{m-1}}]$. By the previous paragraph, $N_{p^m}/Z^{(1)} \cong N_{p^{m-1}} \times C_{p-1}$, and by induction we get $[N_{p^m}/Z^{(1)}, N_{p^m}/Z^{(1)}] = \Sigma_{p^m}/Z^{(1)}$. The above-constructed permutation σ satisfies the equality $\sigma\tau\sigma^{-1} = \tau^g$ for all $\tau \in Z^{(1)}$ (here g is a primitive root modulo p). Hence,

$$[\langle Z^{(1)}, \sigma \rangle, \langle Z^{(1)}, \sigma \rangle] = Z^{(1)}, \ Z^{(1)} \le [N_{p^m}, N_{p^m}].$$

Therefore, $\Sigma_{p^m} \le [N_{p^m}, N_{p^m}]$, completing the proof of (d) for $n = p^m$.

Next we assume that n is not a prime power.

(ii) Now suppose that $n = ap^m$, where $1 < a < p$ so $p > 2$. Let us partition the set $M = \{1, \dots, ap^m\}$ into a subsets R_i of equal cardinality p^m:

$$R_i = \{(i-1)p^m + 1, (i-1)p^m + 2, \dots, ip^m\}, \quad i = 1, \dots, a.$$

On each of these sets R_i we construct the symmetric group S_{R_i}, and take there a Sylow p-subgroup $\Sigma_{p^m}^{(i)}$. Then, comparing the orders, we get $\Sigma_{ap^m} = \Sigma_{p^m}^{(1)} \times \dots \times \Sigma_{p^m}^{(a)}$. The permutation

$$\pi_i = ((i-1)p^m + 1, \dots, (i-1)p^m + p) \dots (ip^m - p + 1, \dots, ip^m)$$

generates the center of $\Sigma_{p^m}^{(i)}$. Let $\pi = \pi_1 \dots \pi_a$; then $Z(\Sigma_{ap^m}) = \langle \pi \rangle$. Let $\sigma = \sigma_1 \dots \sigma_a \in S_{ap^n}$, where $\sigma_i \in S_{R_i}$ is constructed as in (i) (with the same g for all i). We claim that σ normalizes $\langle \pi \rangle$. It is enough to prove that $\sigma \langle \pi_i \rangle \sigma^{-1} = \langle \pi_j \rangle$ for $i \in \{1, \dots, a\}$ and suitable $j = j(i) \in \{1, \dots, a\}$. Set $L = \langle \pi_1 \rangle \cup \langle \pi_2 \rangle \cup \dots \cup \langle \pi_a \rangle$. Then any nonidentity permutation from L is a product of p^{m-1} pairwise independent p-cycles, and every permutation from $C - L$, where $C = \langle \pi_1 \rangle \times \langle \pi_2 \rangle \times \dots \times \langle \pi_a \rangle$, has no such form. Since $C = Z(\Sigma_n)$, then $\sigma C \sigma^{-1} = C$ so $\sigma \langle \pi_i \rangle \sigma^{-1} = \sigma \langle \pi_j \rangle$ for some $j = j(i) \in \{1, \dots, a\}$. It follows from the obtained equality that $\sigma(R_i) = R_j$ for the same i, j. Thus, σ can be considered as a permutation of a set system $\{R_i\}_1^a$. Since S_n contains the permutations which produce all possible permutations of this system of sets, we get

$$N_{ap^m} = N_n = N_{S(R_1)}(\Sigma_{p^m}^{(1)}) \operatorname{wr} S_a \cong N_{p^m} \operatorname{wr} S_a \quad \text{and} \quad T(N_n) = M,$$

completing the proof of (b).

(iii) Now let $n = a_0 + a_1 p + \dots + a_t p^t$ be the decomposition of n in the base p, $a_t \neq 0$, and let at least two coefficients in this decomposition be $\neq 0$. The set M can be partitioned into $t + 1$ subsets $Q_0, Q_1, \dots, Q_t$ with $|Q_i| = a_i p^i$, $i = 0, 1, \dots, t$. Let $H = S(Q_0) \times S(Q_1) \times \dots \times S(Q_t)$. Then, since $a_0 < p$, we have

$$\Sigma_n = \Sigma_{a_1 p} \times \dots \times \Sigma_{a_t p^t} \cong \Sigma_p^{a_1} \times \dots \times \Sigma_{p^t}^{a_t}.$$

Suppose that $\Sigma_{a_i p^i} \in \mathrm{Syl}_p(S(Q_i))$, $i = 0, 1, \dots, t$. Let F_i be the center of $\Sigma_{a_i p^i}$, $i = 0, 1, \dots, t$ and $F = F_1 \times \dots \times F_t$. Then F is the center of Σ_n. If $\sigma \in N_n$, then $\sigma F \sigma^{-1} = F$. Since the permutations of F_i are of definite type among the elements of F, we get $\sigma F_i \sigma^{-1} = F_i$ for all i. This means that $\sigma(Q_i) = Q_i$ for all $i > 0$, and now it is clear that $\sigma(Q_0) = Q_0$. Thus, $\sigma \in S(Q_0) \times S(Q_1) \times \dots \times S(Q_t) = H$, i.e.,

$$N_n = N_H(\Sigma_n) = N_{a_0} \times N_{a_1 p} \times \dots \times N_{a_t p^t}, \quad \text{where } N_{a_0} = S(Q_0) \cong S_{a_0}.$$

The statement about $T(N_{a_i p^i})$ now is obvious.

(iv) Let us show that $[N_n, N_n] \geq \Sigma_n$ for $p > 2$. The obtained decomposition of N_n in (a) and the first assertion of (d) imply that it suffices to consider the case where $n = ap^m$, $1 < a < p$. We have

$$\Sigma_{ap^m} (\cong \Sigma_{p^m}^a) \leq N_{p^m}^a \leq N_{ap^m} \quad \text{and} \quad \Sigma_{ap^m} = [N_{p^m}^a, N_{p^m}^a] \leq [N_{ap^m}, N_{ap^m}],$$

by the first assertion of (d). Since the reverse inclusion follows from the above, the proof is completed. □

Corollary A.13.3 (P. Hall, 1956). *$N_n^{(p)} = \Sigma_n^{(p)}$ if and only if $p = 2$.*

Exercise 1. If a p-group P is lattice isomorphic with Σ_{p^m}, then $P \cong \Sigma_{p^m}$.

Exercise 2. Prove that any normal abelian subgroup of $G = \Sigma_{p^m} = \Sigma_{p^{m-1}} \operatorname{wr} C_p$, $m > 1$, is contained in the base K of G, unless $p = 2$ and $m = 2$.

Solution. Let $A < G$ be a maximal normal abelian subgroup and $A \not\leq K = G_1 \times \cdots \times G_p$, where $G_i \cong \Sigma_{p^{m-1}}$. Take $x \in A - K$; then x induces the p-cycle on the set $\{G_1, \dots, G_p\}$. It follows that $A \cap G_i = (A \cap K) \cap G_i = \{1\}$ for all i. Since $A \cap K, G_i \triangleleft K$, $i = 1, \dots, p$, we get $C_G(A \cap K) \geq AG_1 \dots G_p = AK = G$ so $A \cap K \leq Z(G)$. However, $|Z(G)| = p$ so $|A| = p^2$. We have $C_G(A) = A$ so $|G| = p^3$, $p = 2$ and $m = 2$.

N. Ito asked me, in his letter at Feb. 21, 2006, to prove the following fact. For any maximal subgroup H of $W = \Sigma_{p^{n+1}}$, $p > 2$, distinct from K, the base of W, we have $Z(H) = Z(W)$. Below we prove that, indeed, this is true for all $p^{n+1} > 4$ (here $p = 2$ is admissible). Assume, by the way of contradiction, that $Z(H) > Z(W)$. Let A be a G-invariant subgroup of order p^2 in $Z(H)$. By Exercise 2, $A < K = U_1 \times \cdots \times U_p$, the base of W, where $U_1, \dots, U_p$ are coordinate subgroups of the wreath product $W = \Sigma_{p^n} \operatorname{wr} \Sigma_p$, $U_i \cong \Sigma_{p^n}$, all i. As above, $A \cap U_i = \{1\}$, all i. It follows that $C_W(A) \geq H(U_1 \times \cdots \times U_p) = HK = W$, a contradiction since $A \not\leq Z(W)$.

Exercise 3. Prove that $W = \Sigma_{p^m}$, where $p^m > 2^2$, has no normal cyclic subgroups of order p^2.

Solution. Assume that $p^m > 2^2$ and $C_{p^2} \cong A \triangleleft W = \Sigma_{p^m}$. Let $K = G_1 \times \cdots \times G_p$ be the base of W, $G_i \cong \Sigma_{p^{m-1}}$. Then $A < K$, by Exercise 2. Since $\Omega_1(A) = Z(W)$, we get $A \cap G_i = \{1\}$ for all i. It follows that $C_G(A) = K$, i.e., $A \leq Z(K)$, a contradiction since $Z(K) \cong E_{p^p}$ is of exponent p.

Exercise 4. Describe the Sylow 2-subgroups of A_n.

Exercise 5. Prove that if $n > 5$ and $P \in \mathrm{Syl}_2(A_n)$, then $N_{A_n}(P) = P$.

Problem. Let $G = \Sigma_{2^n}$. Classify all $H < G$ such that $d(H) = 2^{n-1}$.

Appendix 14

2-groups with an involution contained in only one subgroup of order 4

N. Blackburn proposed to classify 2-groups G which possess an involution contained in only one subgroup of G of order 4 (see [Bla13]). We have classified here all such 2-groups.

Theorem A.14.1 (Bozikov–Janko). *Let G be a 2-group which has an involution z contained in at most one subgroup of G of order 4. Then one of the following holds:*

(a) G is cyclic.

(b) $C_G(z) = \langle z \rangle \times C_{2^n}$, $n \geq 1$, and such groups are classified in §48.

(c) $C_G(z) = \langle z \rangle \times Q_{2^n}$, $n \geq 3$, and such groups are classified in §49.

(d) $G = \langle a, b \mid a^{2^m} = b^4 = 1,\ m \geq 2,\ u = a^{2^{m-1}},\ b^2 = uz,\ z^2 = 1,\ a^b = a^{-1}z^{\epsilon},\ \epsilon = 0, 1 \rangle$.

Here (in (d)) $Z(G) = \langle z, u \rangle \cong E_4$, $G/\langle z \rangle \cong Q_{2^{m+1}}$, $G' = \langle a^2 z^{\epsilon} \rangle$, and G is metacyclic if and only if $\epsilon = 0$. Also, G is a U_2-group (as defined in §18 or §64).

Proof. Let z be an involution in a 2-group G which is contained in only one subgroup of G of order 4. Set $Z = \langle z \rangle$ and consider the subgroup $H = C_G(Z)$. Our assumption implies that H/Z is either cyclic or generalized quaternion.

Suppose that $Z \not\leq \Phi(H)$. Then $H = Z \times H_0$, where H_0 is either cyclic or generalized quaternion. If $|H_0| > 1$, then we have obtained the cases (b) and (c) of our theorem. If $|H_0| = 1$, then $Z = H = G$ and G is cyclic (case (a) of the theorem).

Suppose that $Z \leq \Phi(H)$. If H/Z is cyclic, then H is also cyclic. In that case $N_G(H)$ centralizes Z and so $H = G$ is cyclic. This is the case (a) of our theorem.

It remains to consider the case $Z \leq \Phi(H)$ and $H/Z \cong Q_{2^n}$, $n \geq 3$. Let T/Z be a cyclic subgroup of index 2 in H/Z so that T is abelian. If T is cyclic, then H must be of maximal class. But in that case H has no a proper homomorphic images which are isomorphic to a generalized quaternion group. Thus T is noncyclic and so we may set $T = \langle z \rangle \times \langle a \rangle$, where the element a is of order 2^m, $m \geq 2$.

If $|H'| \geq \langle z \rangle$, then $|H : H'| = 4$ and so, by Taussky's theorem, H is of maximal class, a contradiction. Hence $H' = \langle a^2 z^{\epsilon} \rangle$ with $\epsilon = 0, 1$ and setting $u = a^{2^{m-1}}$, we get $Z(H) = \langle z, u \rangle$ and $\Phi(T) = \langle a^2 \rangle \geq \langle u \rangle$. Since $\Phi(H) \geq \langle z \rangle$, there is $b \in H - T$ such that $b^2 \in T - \langle z \rangle$ and so $b^2 = uz$ and $a^b = a^{-1}z^{\epsilon}$. If $\epsilon = 0$, then $\langle a \rangle$ is normal in H and H is metacyclic (since $H = \langle a\rangle\langle b\rangle$). Since $H/\langle z\rangle \cong Q_{2^{m+1}}$, there is no $x \in H - T$ such that $x^2 = z$. We have $Z(H) = \langle z, u\rangle$ and both u and uz are squares in H whereas z is not a square in H. Thus, $\langle z\rangle$ is characteristic in H which implies that $N_G(H)$ centralizes $\langle z\rangle$. This forces $H = G$ and so the structure of G is completely determined as stated in the case (d) of our theorem. □

Theorem A.14.2. *Let G be a 2-group having an involution z which is contained in exactly one four-subgroup V. Then one of the following holds:*

(a) G is dihedral or semidihedral.

(b) $C_G(z) = \langle z\rangle \times Q$, where Q is either cyclic or generalized quaternion.

(c) V is normal in G.

Proof. If G is of maximal class, then it is dihedral or semidihedral. Next, we assume that G is not of maximal class. Then G has a normal subgroup R of type $(2, 2)$ (Lemma 1.4). Set $Z = \langle z\rangle$ and assume that $V \neq R$. Then $H = RZ$ is dihedral of order 8 so $Z \not\leq T = C_G(R)$. It follows that $|G : T| = 2$ so $G = TZ$. By the modular law, $C_G(Z) = Z \times Q$, where $Q = C_G(T)$. By the hypothesis, Q has no abelian subgroups of type $(2, 2)$ so Q is either cyclic or generalized quaternion. □

Another version of N. Blackburn's problem (see 'Research problems and themes II', #1204) is to classify the 2-groups G containing an involution z such that z is contained in exactly one four-subgroup V of G. According to Theorem A.14.2, it suffices to consider the case where V is normal in G. This problem is surprisingly complicated. Indeed, all 2-groups with exactly three involutions satisfy this hypothesis; the classification of the last groups is one of the outstanding problems of p-group theory. As the following theorem shows, a solution of this extended Blackburn's problem follows from the classification of 2-groups with exactly three involutions.

Theorem A.14.3 (see Theorem A.14.2). *Let G be a finite 2-group having an involution z which is contained in exactly one four-subgroup V. Suppose that $V \trianglelefteq G$. Then one of the following holds:*

(a) G has exactly three involutions.

(b) G has an involution t such that $C_G(t) = \langle t\rangle \times Q$, where Q is either cyclic or generalized quaternion. (Such groups are classified in §§48,49).

Proof. Set $T = C_G(V)$. Then T contains exactly three involutions and all of them lie in V; in particular, V is unique abelian subgroup of type $(2, 2)$ in T. Assume that there is an involution $t \in G - T$. By the modular law, $C_G(t) = \langle t\rangle \times Q$, where $Q = C_T(t)$. By hypothesis, $V \not\leq Q$. It follows that Q has no abelian subgroups of type $(2, 2)$ and so Q is either cyclic or generalized quaternion. □

Appendix 15

A criterion for a group to be nilpotent

1°. For a finite group G define the *class frequency function* $w_G : \mathbb{N} \to \mathbb{Z}$ by

$$w_G(n) = \frac{1}{n} \cdot |\{g \in G \mid |G : \mathrm{C}_G(g)| = n\}|,$$

which is the number of conjugacy classes of G of size n. We prove the following

Theorem A.15.1 ([CHM]). *If G is a nilpotent group and H is a group with $w_H = w_G$, then H is nilpotent.*

Let $\mathcal{S}_p(G)$ be the union of conjugacy classes of G whose size is a power of p. Then

$$(1) \qquad |\mathcal{S}_p(G)| = \sum_{\alpha \geq 0} p^\alpha w_G(p^\alpha).$$

Let $\mathrm{H}(G)$ be the last member of the upper central series of G (the hypercenter of G) and $\mathrm{O}^p(G) = \langle x \in G \mid p \notin \pi(x) \rangle$.

If $G = P \times L$ is nilpotent, $P \in \mathrm{Syl}_p(G)$, then $\mathcal{S}_p(G) = P \times \mathrm{Z}(L) = P\mathrm{Z}(G)$.

Theorem A.15.2. *If G is an (arbitrary) group, then, in the above notation, we have $|\mathrm{H}(G)|_p = |\mathcal{S}_p(G)|_p$.*

This theorem implies that the order of $\mathrm{H}(G)$ is determined by function w_G. Therefore, Theorem A.15.1 follows from Proposition A.15.2.

Lemma A.15.3. *Let G be a group.*

(a) Let $R = \mathrm{O}^p(G)$. Then $\mathrm{C}_G(R)$ is nilpotent and its Sylow p-subgroup is contained in $\mathrm{H}(G)$.

(b) A p-element $x \in G$ belongs to $\mathrm{H}(G)$ if and only if it commutes with each p'-element of G.

Proof. (a) We have $\mathrm{C}_G(R) \trianglelefteq G$ and $\mathrm{C}_G(R) \cap R = \mathrm{Z}(R)$. Next, $\mathrm{Z}(R)$ contains a p'-Hall subgroup T of $\mathrm{C}_G(R)$ since

$$\mathrm{C}_G(R)R/R \cong \mathrm{C}_G(R)/(\mathrm{C}_G(R) \cap R) = \mathrm{C}_G(R)/\mathrm{Z}(R)$$

is a p-group as a subgroup of G/R. Therefore, if $P \in \mathrm{Syl}_p(\mathrm{C}_G(R))$, we get $\mathrm{C}_G(R) = PT = P \times T$, by Burnside's normal p-complement theorem. It remains to show that

$P \le \mathrm{H}(G)$. Let $P \le S \in \mathrm{Syl}_p(G)$. Then $P \cap \mathrm{Z}(S) > \{1\}$ since P is G-invariant. We have $\mathrm{C}_G(P \cap \mathrm{Z}(S)) \ge SR = G$ so $P \cap \mathrm{Z}(S) \le \mathrm{Z}(G)$. Now the result follows by induction applied to $P/(P \cap \mathrm{Z}(S)) \le G/(P \cap \mathrm{Z}(S))$.

(b) It is enough to show the sufficiency part. Set $R = \mathrm{O}^p(G)$. Then $x \in P \in \mathrm{Syl}_p(\mathrm{C}_G(R))$. Since $P \le \mathrm{H}(G)$, by (a), the proof is complete. □

Proof of Theorem A.15.2. We assert that

(2) If N is a G-invariant p-subgroup of $\mathrm{H}(G)$, then $\mathcal{S}_p(G)$ is a union of cosets of N and $\mathcal{S}_p(G/N) = \{gN \mid g \in \mathcal{S}_p(G)\}$.

Indeed, let $x \in \mathcal{S}_p(G)$ and $q \in \pi(G) - \{p\}$. Then x is centralized by some Sylow q-subgroup Q of G, and, by Lemma A.15.3, Q centralizes $\langle N, x\rangle$ so, if $y \in xN$, then $q \nmid |G : \mathrm{C}_G(y)|$, and the first part of (2) follows. Taking for N the Sylow p-subgroup of $\mathrm{H}(G)$, we get

(3) $|\mathrm{H}(G)|_p$ divides $|\mathcal{S}_p(G)|$.

To justify the second assertion in (2), let $xN \in \mathcal{S}_p(G/N)$ and $q \in \pi(G) - \{p\}$. Some Sylow q-subgroup of G/N, necessarily of the form QN/N with $Q \in \mathrm{Syl}_q(G)$, centralizes xN, and so $(xN)^g = xN$ for all $g \in Q$. Since $|xN|$ is a power of p, the q-group Q, acting on xN by conjugation, has an orbit of size 1; but then, as above, Q centralizes xN, and it follows that $xN \subseteq \mathcal{S}_p(G)$, as desired.

Let $|\mathcal{S}_p(G)|_p = p^r$. We will now argue by induction on r that $|\mathrm{H}_p(G)|_p = p^r$. If $r = 0$, by (3) we have $|\mathrm{H}(G)|_p = 1$, and the result is true. Suppose that $r \ge 1$. Then follows from (1) that p divides $w_G(p^0)$. Since $w_G(p^0) = |\mathrm{Z}(G)|$, the group G has a nonidentity normal subgroup, N, say, contained in $\mathrm{Z}(G)$. From (2), we have $|\mathcal{S}_p(G/N)|_p = |\mathcal{S}_p(G)|_p/|N|$. Moreover, since $\mathrm{H}(G/N) = \mathrm{H}(G)/N$, we get $|\mathrm{H}(G/N)|_p = |\mathrm{H}(G)|_p/|N|$. The inductive hypothesis yields $|\mathrm{H}(G/N)|_p = |\mathcal{S}_p(G/N)|_p$, and the last three equalities now give the conclusion. □

Corollary A.15.4. *A group* $G = \mathrm{O}_p(G) \times \mathrm{O}_{p'}(G)$ *if and only if* $|G|_p$ *divides the number* $\sum_{\alpha \ge 0} p^\alpha w_G(p^\alpha)$.

Note that w_G determines the orders of the first and the last terms of the upper central series of G but does not determines the orders of the intermediate terms as the following example shows. Let M be an elementary abelian maximal subgroup of a p-group G so we can regard it as a vector space over $\mathrm{GF}(p)$. The element $x \in G - M$ induces a linear transformation on this space, and if t is the number of blocks of the Jordan form of this transformation, then $|\mathrm{Z}(G)| = p^t$. The class of G, on the other hand, is the maximal size of these blocks. Now each Jordan block has size at most p. Let, in addition, $p > 2$ and $\dim(M) = 4$. Let the linear transformation of x have two blocks of size 2 while x_1 has a block of size 1 and a block of size 3. Then the class of $G = \langle x\rangle \cdot M$ equals 2 and the class of $G_1 = \langle x_1\rangle \cdot M$ equals 3, however, $w_G = w_{G_1}$. It is not known whether a knowledge of $\mathrm{h}(G)$, the conjugate type vector of G, is sufficient to establish the p-nilpotence of G.

2°. Let $X_1(G)$ be the first column of the character table of G (this column consists of the degrees of irreducible characters of G). Isaacs [Isa4] has proved the following result whose proof is presented below: If $X_1(G) = X_1(G_1)$ and G_1 is p-nilpotent, then G is p-nilpotent, too.

Recall that if $\phi \in \mathrm{Irr}(N)$ then $o(\phi)$ is the order of the linear character $\det(\phi)$ in the group $\mathrm{Lin}(N)$. (If T is a representation of G affording ϕ, then $\det(\phi)(g) = \det(T(g))$ for $g \in G$.)

Lemma A.15.5. *Let p be a prime, $N = \mathrm{O}^p(G)$,*

$$\begin{aligned} Y(G) &= \{\chi \in \mathrm{Irr}(G) \mid p \nmid \chi(1)\}, \\ Y_1(N) &= \{\phi \in Y(N) \mid \phi \text{ is invariant with respect to } G\}, \\ u(G) &= \sum_{\chi \in Y(G)} \chi(1)^2, \ u_1(N) = \sum_{\phi \in Y_1(N)} \phi(1)^2. \end{aligned}$$

Then

$$u(G) = |G : G'|u_1(N), \quad u_1(N) \equiv |N| \pmod{p}.$$

In particular, $p \nmid |N|$ if and only if $p \nmid u(G)\,|G : G'|^{-1}$.

Proof. If $\phi \in Y_1(N)$, then $p \nmid \phi(1) \cdot o(\phi)$, since $p \nmid |N : N'|$. Therefore, by Gallagher's extension theorem (see [BZ, Chapter 7]), there exists $\chi \in \mathrm{Irr}(G)$ such that $\chi_N = \phi$. By [BZ, Theorem 14.18], there exist exactly $|G : G'|$ such χ's. If $\chi \in Y(G)$, then, by Clifford theory, $\chi_N \in Y_1(N)$ since $|G : N|$ is a power of p and $p \nmid \chi(1)$. Therefore,

$$u(G) = \sum_{\chi \in Y(G)} \chi(1)^2 = |G : G'| \sum_{\phi \in Y_1(N)} \phi(1)^2 = |G : G'|u_1(N).$$

Furthermore,

$$|N| = \sum_{\phi \in \mathrm{Irr}(N)} \phi(1)^2 \equiv \sum_{\phi \in Y_1(N)} \phi(1)^2 \equiv u_1(N) \pmod{p}.$$

Indeed, if $\psi \in \mathrm{Irr}(N) - Y_1(N)$, then either p divides $\psi(1)$ or $|G : \mathrm{I}_G(\psi)| = p^s$ for some $s > 0$ (here $\mathrm{I}_G(\phi)$ is the inertia subgroup of ϕ in G). Thus,

$$p \nmid u(G)|G : G'|^{-1} \iff p \nmid u_1(N) \iff p \nmid |N|.$$

(It follows that G is p-nilpotent if and only if $p \nmid u(G)|G : G'|$.) □

Corollary A.15.6 ([Isa4]). *Let $X_1(G) = X_1(G_1)$ and let G_1 be p-nilpotent. Then G is also p-nilpotent.*

Proof. Set $N_1 = \mathrm{O}^p(G_1)$. By assumption, $p \nmid |N_1|$. Therefore, by Lemma A.15.5,

$$p \nmid u_1(N_1) = \frac{u(G_1)}{|G_1 : G_1'|} = \frac{u(G)}{|G : G'|} = u_1(N),$$

where $N = \mathrm{O}^p(G)$. Since $u_1(N) \equiv |N| \pmod p$, it follows that $p \nmid |N|$, and so N is the normal p-complement in G. □

By Molien's theorem on the structure of complex group algebras (see [BZ, Chapter 2]), $\mathrm{X}_1(G) = \mathrm{X}_1(G_1)$ if and only if $\mathbb{C}G \cong \mathbb{C}G_1$. Therefore, if $\mathbb{C}G \cong \mathbb{C}G_1$, then G and G_1 are either both p-nilpotent or both not p-nilpotent.

Theorem A.15.7. *A group G is nilpotent if and only if, for every $\chi \in \mathrm{Irr}(G)$, we have $\pi(\chi(1)) = \pi(G/\mathrm{Z}(\chi))$.*

Proof. Obviously, it is enough to prove that our condition is sufficient. Suppose that G is a counterexample of minimal order. All proper epimorphic images of G satisfy the hypothesis so nilpotent, by induction. So G is a monolith; let N be the unique minimal normal subgroup of G. If $N = G$, G is simple so it is of prime order, by [Isa1, Corollary 12.2]. Now we assume that $N < G$. By induction, G/N is nilpotent. By Wielandt's theorem, $N \not\leq \Phi(G)$.

Suppose that G is solvable. Then N is a p-subgroup for some prime $p \in \pi(G)$ and $G = M \cdot N$, a semidirect product, where $M < G$ is maximal. Since G is a monolith and M is nilpotent, we see that $p \nmid |M|$. Since $\bigcap_{\chi \in \mathrm{Irr}(G)} \ker(\chi) = \{1\}$, there exists a faithful irreducible character χ of G. Clearly, $\mathrm{Z}(\chi) = \{1\}$. By Ito's theorem on degrees (see Introduction, Theorem 17), $p \nmid \chi(1)$, contrary to the hypothesis since, in our situation, we must have, by hypothesis, $\pi(\chi(1)) = \pi(G/\mathrm{Z}(\chi)) = \pi(G)$.

Assume that G is nonsolvable; then N is nonsolvable. Let p be a prime divisor of $|N|$ and $P \in \mathrm{Syl}_p(G)$. By Tate's theorem (see [BZ, Exercise 7.23]), $P \cap N \not\leq \Phi(P)$; then $P \cap N \not\leq P'$. Therefore, there exists $\lambda \in \mathrm{Lin}(P)$ such that $P \cap N \not\leq \ker(\lambda)$. Since $p \nmid \lambda^G(1) = |G : P|$, there exists $\chi \in \mathrm{Irr}(\lambda^G)$ such that $p \nmid \chi(1)$. Since N is unique, we get $N \leq G'$ so χ is nonlinear, by reciprocity. Since $\mathrm{Z}(\chi) = \{1\}$, we get $\pi(\chi(1)) \neq \pi(G) = \pi(G/\mathrm{Z}(\chi))$, contrary to the hypothesis. □

Research problems and themes I

A tremendous effort has been made by mathematicians for more than a century to clear up the chaos in group theory. Still, we cannot answer some of the simplest questions.

Richard Brauer (Quoted in Michael Artin, Algebra, Prentice Hall, Englewood Cliffs, 1991.)

To ask the right question is harder than to answer it.

Georg Cantor (Quoted in Arnold's Problems, Springer-Phasis)

You are never sure whether or not a problem is good unless you actually solve it.

Mikhail Gromov (Quoted in Arnold's Problems, Springer-Phasis)

As long as a branch of science offers an abundance of problems, so long it is alive; a lack of problems foreshadows extinction or the cessation of independent development.

D. Hilbert, Mathematical Problems, Bull. Amer. Math. Soc. **8** (1902), 437–479.

Almost all problems in this list were posed by the author and only few ones by other mathematicians. About all problems arose in the time of writing this book. I am indebted to Avinoam Mann, Zvonimir Janko and Lev Kazarin for numerous discussions and constant comments of this list. Zvonimir Janko also solved a lot of problems from this and previous versions of the list, and this inspired a number of new problems. Few problems from the list were solved by other mathematicians. I did not make any attempts to sort problems according to their themes. Only some problems are commented. A lot of comments are due to Mann (not all of them are presented below). The list contains information about solved problems. Notice that we use notation G_1 only in the case where G is of maximal class and order $> p^3$; then $G_1 = C_G(K_2(G)/K_4(G))$ is its fundamental subgroup. Below G is a p-group.

1. (Zhmud) Let G be such that $|G/\ker(\chi)| = p\chi(1)^2$ for every nonprincipal $\chi \in \mathrm{Irr}(G)$. Is the derived length of G bounded? Moreover, is the derived length of G at most 2 if $p = 2$?

2. A p-group G is said to be a CM$_k$-*group* if, for every normal subgroup N of G, the quotient group G/N has at most k faithful irreducible characters. Let $\gamma(G)$ be the least $n \in \mathbb{N}$ such that G is a CM$_n$-group. Mann (personal communication) showed that there exists a 3-group G such that $\gamma(G) \neq \varphi(3^a)$ for all $a \in \mathbb{N}$. Describe the range of the function γ : finite p-groups $\to \mathbb{N}$.

3. Let a 2-group G be a product of n pairwise permutable cyclic subgroups. Find the number $d_n = d_n(G)$ such that every subgroup of G is generated by d_n elements. It is known that $d_2 = 3$. In the similar situation, when p is odd, the answer is $d_n = n$, as follows from results of §26 due to Lubotzky and Mann. Indeed, let the p-group $G = C_1 \dots C_n$, where $C_1, \dots, C_n$ are cyclic and pairwise permutable, $p > 2$. The group $G/\mho_1(G)$ is a group of exponent p that is a product of pairwise permutable cyclic subgroups, and so it is elementary abelian. It follows that G is powerful (see §26). But if a powerful p-group ($p > 2$) is generated by n elements, then all its subgroups are also generated by n elements (§26), and we are done. Note that a powerful p-group, $p > 2$, of rank d is a product of d cyclic subgroups not necessarily pairwise permutable; converse is also true. However, a p-group which is a product of not necessarily pairwise permutable cyclic subgroups may be non-powerful (such, for example, is a nonabelian group of exponent p).[1]

[1]Below we present the letter of Mann which solves Problem 3. In that letter he uses the following above noticed property of powerful p-groups (see §26): If a powerful p-group G is generated by n elements, then this is true for all subgroups of G. "6/08/03, Jerusalem. I thought a little about that problem, what is the [maximal] rank [of subgroups] of a product of n cyclic 2-groups? You denote it by d_n. I can now show that $d_n \leq 2n - 1$. I don't know if that is best possible for $n > 2$. There are groups which are the product of two cyclic 2-groups and have subgroups generated by 3 elements, and by taking direct powers of these groups we find for each k a group which is the product of $2k$ cyclic 2-groups and has subgroups requiring $3k$ generators. Here are the proofs. I let $G = C_1C_2 \dots C_n$, where each C_i is a cyclic 2-group, and $C_iC_j = C_jC_i$. (i) If $C \leq H$, C cyclic, and $|H| \leq |C|^2$, then C contains a non-trivial normal subgroup of H. This was proved by Herzog–Kaplan [HK] for all finite groups. For p-groups a simple proof is as follows: If $D = C^x$, then $H \neq CD$, therefore $C \cap D \neq \{1\}$, so the subgroup E of order p in C is contained in D. Thus E is contained in all conjugates of C, so the intersection of these conjugates is not trivial. (ii) Now let G be as above, assume that $C_1, \dots, C_k$ are the factors of largest order, and let this order be $|C_1| = 2^{e+1}$. By (i), the subgroup E_1 of order 2 in C_1 is normalized by each C_j, $j \leq n$, so E_1 is normal in G, hence central. If E_i is the subgroup of order 2 in C_i, $i \leq k$, we see that $E = E_1 \dots E_k$ is an elementary abelian central subgroup of G. G/E is a product of n cyclic subgroups of smaller order than C_1. Now easy induction shows that G has a central series of length $e+1$ with elementary abelian factors, and thus $\exp(G) = 2^{e+1}$, $\mathrm{cl}(G) \leq e+1$, and $\mho_r(G) = \prod_{i=1}^n \mho_r(C_i)$. One may assume that $e > 0$. Let $e = 1$; then $\Phi(G) = \mho_1(G) = \prod_{i=1}^n \mho_1(C_i)$ is generated by n elements. Considering $G/\mho_2(G)$, we conclude, in view of $\mho_2(G) < \Phi(\Phi(G))$, that $\mathrm{d}(\Phi(G)) \leq n$. (iii) Let $x \in \mho_{e-1}(G)$ and $y \in G$. Then, working in $\bar{G} = G/\mho_e(G)$ and taking into account that $\bar{x} \in \mathrm{Z}(\bar{G})$, we get $x^y = xz$, with $z \in \mho_e(G)$, so $z \in \mathrm{Z}(G)$ and z has order 2. It follows that $x^{y^2} = x$, i.e. $\mho_{e-1}(G) \leq \mathrm{Z}(\mho_1(G))$. In particular, if $e = 2$, then $N = \mho_1(G)$ is abelian. (iv) Apply (iii) to the group $G/\mho_3(G)$. This shows that $\mho_1(G)/\mho_3(G)$ is abelian. This means that $N = \mho_1(G)$ is powerful since $\mho_3(G) \leq \mho_2(N)$. (v) Let $H \leq G$. We may assume that $H \neq G$. Then $HN \neq G$ since $N = \Phi(G)$, so $H/(H \cap N) \cong HN/N$ can be generated by $n-1$ elements as a proper subgroup of $G/N \leq \mathrm{E}_{2^n}$, while $H \cap N$ can be generated by n elements, because N is powerful. Thus, $\mathrm{d}(H) \leq (n-1) + n = 2n - 1$, as claimed." We see that the exponent of any product of pairwise permutable cyclic p-groups equals to

4. Let $n > 1$. (i) Classify the p-groups G such that $c_n(G) = p^p$, $p > 2$. (ii) Study the p-groups G such that p^p does not divide $c_n(G)$.

5. Study the p-groups G containing a subgroup H of maximal class such that $C_G(H) < H$. Consider the case $|H| = p^4$ in detail.

6. Classify the p-groups G of exponent p^e such that the set Γ_1 has only one element of exponent p^e.

7. Classify the p-groups G such that $\text{cl}(N_G(H)) = \text{cl}(H)$ $(\text{cl}(N_G(H)) \leq \text{cl}(H) + 1)$ for all nonnormal nonabelian $H < G$.

8. Describe the irregular p-groups all of whose subgroups of exponent $\leq p^2$ have orders $\leq p^{2p}$.

9. Study the p-groups G in which $C_G(A) = Z(A)$ for all nonabelian $A \leq G$.

10. Study the 2-groups G with $|G : \Phi(G)| = 2^{2k+1}$ and $|\Omega_1(G)| = 2^{k+1}$. (See §37. As Mann noticed, this is impossible for p-groups with $p > 2$, by [Laf1].)

11. (Isaacs) Let X be a set of powers of p, $1 \in X$, $|X| > 1$. Isaacs [Isa4] has showed that there exists a p-group G of class 2 with $\text{cd}(G) = \{\chi(1) \mid \chi \in \text{Irr}(G)\} = X$ (see also Appendix 21.) Does there exist a p-group G of given class $c > 2$ with $\text{cd}(G) = X$ if X is not class bounding? (See [Sla]. Recall that X is class bounding if there exists $n \in \mathbb{N}$ such that for every p-group G with $\text{cd}(G) = X$ we have $\text{cl}(G) \leq n$.)

12. (i) Classify the p-groups G in which every maximal cyclic subgroup of composite order is contained in a unique maximal subgroup of G. (ii) Study the p-groups G all of whose cyclic subgroups of maximal order are contained in $\Phi(G)$ (in $\mho_1(G)$).

13. Study the p-groups G of order p^m such that $k_m = k(G)$ is as small as possible for given m. Estimate k_m in terms of m and p.

14. Does there exist a nonmetacyclic two-generator 2-group containing exactly one two-generator maximal subgroup? (The answer is 'no'; see Theorem 71.6.)

15. (Mann and Scoppola) Classify the 2-groups G with $\text{mc}(G) \geq 1/4$, where $\text{mc}(G) = k(G)/|G|$, the *measure of commutativity* of G. (For $p > 2$, see [GMMPS].)

16. Let $H \in \Gamma_1$ and $k(G) = pk(H) - (p^2 - 1)$. Describe the structures of H and G (see §2).

17. Let H be a maximal member of the set of subgroups of exponent p^2 in a p-group G. Study the structure of G provided H is extraspecial.

18. Study the p-groups G such that (a) every irreducible character of G assumes at most p distinct values, (b) every nonlinear irreducible character of G assumes at most four distinct values.

---

the maximal order of a factor since above argument is also working for $p > 2$.

19. If G is nonabelian, then $\mathrm{T}(G)^2 < |G|\mathrm{k}(G)$, where $\mathrm{T}(G) = \sum_{\chi \in \mathrm{Irr}(G)} \chi(1)$ [BZ, Chapter 11]. Classify the p-groups G such that $(\mathrm{T}(G)+1)^2 \geq |G|\mathrm{k}(G)$ (this question is also interesting for arbitrary groups). All 2-groups of maximal class satisfy this condition. (According to the report of Mann, for all such groups we have $\mathrm{mc}(G) = \frac{\mathrm{k}(G)}{|G|} \geq \frac{1}{9}$. For $p > 2$, the last groups are determined in [GMMPS].)

20. (Kazarin) Let $G = \mathrm{GF}(p^m) \times \mathrm{GF}(p^m)$, where $p > 2$ and $m > 1$, and let θ be an automorphism of $\mathrm{GF}(p^m)$ of order $k > 1$. Let us define the multiplication in G as follows: $(a,b)(c,d) = (a+c, b+d+a\theta(c))$; then G is a p-group of order p^{2m}. (See §25.) Describe the structure of $\mathrm{Aut}(G)$.

21. (Isaacs–Passman) Classify the p-groups G with $\mathrm{cd}(G) = \{1, p^k\}$, $k > 1$.

22. Let $\mathrm{cd}(G) = \{d_0, d_1, \dots, d_s\}$, $s > 1$, $1 = d_0 < d_1 < \cdots < d_s$. Let a_i be the number of characters of degree d_i in $\mathrm{Irr}(G)$ $(i \in \{0, 1, \dots, s\})$. Note that $p-1$ divides a_i for $i = 1, \dots, s$ (Mann). Study the structure of a p-group G with $a_2 = p-1$ (see [BZ, Chapters 3, 31]).

23. Let $\mathrm{Kern}(G) = \{\ker(\chi) \mid \chi \in \mathrm{Irr}(G)\}$. Classify the p-groups G such that $|\mathrm{Kern}(G)| \leq 3$. The same question for quasikernels instead of kernels.

24. Describe the p-groups G for which $\mathrm{Kern}(G)$ (see #23) is a chain with respect to inclusion. Consider the analogous question for quasikernels instead of kernels.

25. Let $H \in \Gamma_1$. Suppose that all abelian subgroups of G not contained in H, have order $\leq p^3$. Study the structure of G. (All p-groups of maximal class satisfy the above condition.)

26. Construct a p-group G that has a normal subgroup H such that $\mathrm{f}(G/H) < \mathrm{f}(G)$, where $\mathrm{f}(G) = \mathrm{T}(G)/|G|$ (see #19).

27. Classify the 2-groups with > 1 distinct metacyclic maximal subgroups.

28. Classify the 2-groups G with $\mathrm{n}(G) = |\mathrm{Irr}_1(G)| \leq 16$.

29. Characterize the p-groups with faithful irreducible character of degree p^n, $n \in \{2, 3\}$ (see [BZ, Chapter 18]).

30. Study the p-groups G such that $\chi(1)^2 = |G/\mathrm{Z}(\chi)|$ for all $\chi \in \mathrm{Irr}(G)$. (Here $\mathrm{Z}(\chi) = \langle x \in G \mid |\chi(x)| = \chi(1) \rangle$ is the quasikernel of χ.)

31. Given $\mathrm{X}_1(G)$, the first column of the character table of a p-group G ($\mathrm{X}_1(G)$ consists of degrees of irreducible characters of G), is there a way to obtain some information on $|\mathrm{Z}(G)|$? Let $\mathrm{X}_1(G) = \mathrm{X}_1(H)$ and let the derived length of G equals 2. Is it possible to estimate the derived length of H in terms of $\mathrm{X}_1(G)$?

32. Let $|G| = p^{2n+e}$, where $e \in \{1, 2\}$. Suppose that $\mathrm{cd}(G) = \{1, p, \dots, p^n\}$. Study the structure of G.

33. Classify the p-groups G such that $\mathrm{mc}(G)-\mathrm{f}(G)^2 < \frac{1}{2p^2}$ (see §22 or [BZ, Chapter 11]). Classify the p-groups G such that (i) $\mathrm{f}(G) > \frac{1}{p+1}$, (ii) $p > 2$ and $\mathrm{mc}(G) > \frac{1}{p^2+1}$. (iii) Does there exist estimate of $|G'|$ in terms of $\mathrm{mc}(G)$?

34. Study the irregular p-groups G, $p > 2$, such that $|G/\mathrm{K}_p(G)| = p^p$ (see Theorem 9.7).

35. Study the p-groups G with $|G : \Phi(G)| = p^d$, satisfying one of conditions:

(a) $|\mathrm{Aut}(G)| = (p^d-1)\dots(p^d-p^{d+1})|\Phi(G)|^d$,

(b) $|\mathrm{Aut}(G)| = (p^d-1)\dots(p^d-p^{d-1})$,

(c) $|\mathrm{Aut}(G)| = p^{\binom{d}{2}}|\Phi(G)|^d$.

Note that D_8 satisfies (c). If G satisfies (a), then all maximal subgroups of G are isomorphic (Mann).

36. (Ito) (i) Classify the p-groups G such that $|G : \mathrm{C}_G(x)| = p^n$ for all $x \in G-\mathrm{Z}(G)$ and a fixed $n \in \mathbb{N}$. (ii) Study the p-groups G such that $|G : \mathrm{C}_G(x)| \in \{p^m, p^n\}$ for all $x \in G-\mathrm{Z}(G)$, $m < n$. Is it possible to estimate the derived length (or nilpotence class) of G in terms of p, m, n? (Commentary of Mann at 1/03/08 to (ii): One cannot bound the nilpotency class, because there exist p-groups with an abelian maximal subgroup of any given class. For the derived length, all I know that for $p = 2$ the groups are metabelian, and for $p = 3$ the derived length is at most 4. I don't even know if 4 is best possible in the last result.)

37. Classify the p-groups G such that $|\{x \in G \mid x^p = 1\}| \geq \frac{1}{p+1} \cdot |G|$.

38. Classify the 2-groups G such that all members of the set Γ_1 but one are metacyclic. (For a solution, see §87.)

39. Classify the p-groups G all of whose proper subgroups H satisfy $|H'| \leq p$.

40. Study the irregular p-groups G containing a maximal regular subgroup R of order p^{p+1}. (See Lemma 17.2. For $p = 2$, see §§51, 77.)

41. Let H be a subgroup of exponent 4 in a 2-group G such that every subgroup of G properly containing H, has exponent > 4. Study the structure of G provided $|H| \leq 2^5$.

42. Classify the p-groups G all of whose maximal subgroups have cyclic Frattini subgroup but $\Phi(G)$ is noncyclic.

43. Study the p-groups G such that every maximal subgroup of G contains an absolutely regular subgroup of index p.

44. Let G be a group of order p^m and exponent p, $2 < n < m-1$. Suppose that the number of two-generator subgroups of G of order p^n is not divisible by p^2. Study the structure of G. (See Theorem 5.8.)

45. (Old problem) Give a good upper estimates of the order, the class, the derived length of maximal (finite) two-generator group $B(p,2)$ of exponent p. (It is known that $|B(5,2)| = 5^{34}$.)

46. Classify the p-groups all of whose proper subgroups are either abelian or generated by two elements. (This was solved in [XZA].)

47. Classify the 2-groups G with elementary abelian $\Phi(G)$ and exactly $\mathrm{d}(\Phi(G))$ nonidentity squares.

48. Let G be the standard wreath product of two p-groups A and B. Find $\Phi(G)$, G', $|\Omega_1(G)|$, $|G/\mho_1(G)|$, $\mathrm{k}(G)$, $|\mathrm{cd}(G)|$, $\mathrm{c}_k(G)$ in terms of A and B.

49. Study the p-groups G such that for every nonabelian $H < G$ we have $|H : \mathrm{Z}(H)| \le p^3$.

50. Let $G = \Sigma_{p^n} \in \mathrm{Syl}_p(\mathrm{S}_{p^n})$. Find $\mathrm{c}_i(G)$ for all i and the number of solutions of $x^{p^k} = 1$ in G. The same questions for $\mathrm{UT}(n,p) \in \mathrm{Syl}_p(\mathrm{GL}(n,p))$.

51. Find $\mathrm{c}_k(G)$, where G is the standard wreath product of two 2-groups of maximal class. Describe all members of lower and upper central series of that group.

52. (Old problem) Study the 2-groups $G = \langle x, y\rangle$, where $o(x) = 2$, $o(y) = 4$.

53. (S.D. Berman) Let $P \in \mathrm{Syl}_p(\mathrm{AGL}(n,p))$. Find all normal subgroups N of P such that $\mathrm{Z}(P/N)$ is cyclic.

54. Let H be a metacyclic subgroup of exponent $2^k > 2$ of a 2-group G. Study the structure of G provided every subgroup of G containing H properly, has exponent $> 2^k$.

55. Study the structure (i) of the standard wreath product of two cyclic (elementary abelian, homocyclic) p-groups, (ii) $\mathrm{Aut}(G)$, where G is the standard wreath product of two cyclic p-groups.

56. Study the p-groups G such that $\mathrm{C}_G(H) = \mathrm{Z}(H)$ for all minimal nonabelian $H < G$.

57. Find the character degree vector of $G = A \,\mathrm{wr}\, B$, the standard wreath product of abelian p-groups A and B.

58. Let G be a p-group, $\exp(G) = p^n > p$. Study the structure of G if all its elements of order p^n are contained in G'. See [Macd3]. Does there exist a p-group G such that all elements of G of maximal order are contained in $\Phi(\Phi(G))$ (in $\mho_2(G)$)?

59. Study the p-groups all of whose maximal abelian subgroups are isomorphic.

60. Study the p-groups in which $|\mathrm{C}_G(x)\Phi(G) : \Phi(G)| = p$ for all $x \in G - \Phi(G)$.

61. Study the p-groups all of whose subgroups of index p (of index p^2) are characteristic.

62. Study the p-groups all of whose proper characteristic subgroups are abelian.

63. Classify the nonabelian p-groups with only one nontrivial characteristic subgroup.

64. (Mann) Classify the irregular p-groups, $p > 2$, all of whose proper sections are regular. (See §11.)

65. Study the generalized homocyclic p-groups G with $|\Omega_1(G)| = p^3$ (see §8).

66. Study the irregular p-groups G, all of whose characteristic subgroups of exponent p are of order $\leq p^{p-1}$ (see Theorem 9.8(d)).

67. Classify the p-groups G such that every maximal subgroup of G contains an abelian (metacyclic) subgroup of index p.

68. (Mann) Study the p-groups, $p > 2$, all of whose (a) subgroups of index p (of index p^2) are powerful (for definition, see §26), (b) subgroups of indices p and p^2 are powerful.

69. (Freiman) If $M \subseteq G$, then $M^2 = \{xy \mid x, y \in M\}$ is the square of the set M. Let n be a fixed natural number, $1 < n \leq p + 1$. Classify the p-groups G such that for every n-element subset M of G one has $|M^2| < n^2$. (See [BFP].)

70. Study the p-groups without sections of order p^4 and exponent p.

71. A subset $M \subset G$ is *product free* if $M^2 \cap M = \varnothing$. Write $pf(G) = \max\{|M| \mid M \subset G$ is product free $\}$. Study the p-groups with "small" $pf(G)$.

72. Study the structure of the group of automorphisms of G fixing all noncyclic subgroups of G.

73. Let $|G| > p$. Is it true that $\mathrm{Aut}(G) - \mathrm{Inn}(G)$ contains an element of order p?

74. (Janko). Suppose that a two-generator 2-group G has exactly one nonmetacyclic maximal subgroup. Is it true that then $G = AB$ with cyclic A and B? (See §89, where more general problem is solved.)

75. Given an abelian p-group G, study the group $A = \langle \alpha \in \mathrm{Aut}(G) \mid \alpha_{G/\Phi(G)} = \mathrm{id}_{G/\Phi(G)} \rangle$.

76. Study the p-groups G such that $\exp(\mathrm{Aut}(G))_p < \exp(G)$.

77. Let G be a group. Two subsets X, Y of G are said to be *automorphic* if $\phi(X) = Y$ for some $\phi \in \mathrm{Aut}(G)$. Classify the p-groups G (a) with ≤ 2 classes of automorphic subgroups of order p, (b) all of whose cyclic subgroups of order p^2 are automorphic, (c) with exactly $k \in \{1, 2\}$ classes of automorphic subgroups of index p^2, (d) all of whose maximal subgroups are automorphic. (e) Given $n \in \mathbb{N}$, find the minimal

(maximal) number of classes of automorphic subsets of cardinality ≤ 2 in groups of order p^n?

78. Does there exist, for each $n > 2$, a p-group G such that $|\mathrm{cd}(G)| - 1 = n = \mathrm{dl}(G)$, where $\mathrm{dl}(G)$ is the derived length of G? (See [Sla].) If the answer is 'yes', study the structure of such groups.

79. Study the nonabelian p-groups all of whose maximal subgroups of class two have the same order.

80. (Old problem) Let $\alpha(G) = \max\{\mathrm{d}(H) \mid H \leq G\}$. Let $G = AB$, where A, B are p-groups. Find the dependence between $\alpha(G)$, $\alpha(A)$ and $\alpha(B)$. (Kazarin conjectured that if $r = \max\{\alpha(A), \alpha(B)\}$, then there exists a constant $C \leq 5$ such that $\alpha(G) \leq Cr$. He also noticed that in all examples $C \leq 3$.) The same question for $\alpha_a(G) = \max\{\mathrm{d}(H) \mid H$ is an abelian subgroup of $G\}$.

81. Given an abelian p-group G of exponent $p^e > p$, study the structure of the stability groups of the chain $G > \mho_1(G) > \cdots > \mho_e(G)$.

82. (Old problem) Study the p-groups G with multiplier of order $\leq p$.

83. Classify the p-groups G such that $|\mathrm{Aut}(G)| \leq |G|$.

84. Let $A < B < G$, where A and B are abelian, $|B : A| = p^2$ and $\exp(B) \leq p^n$, $p^n > 2$. Is it true that the number of abelian subgroups of G of order $p^2|A|$ containing A and having exponent at most p^n, is $\equiv 1 \pmod{p}$?

85. Describe all representation groups of $\Sigma_{p^n} \in \mathrm{Syl}_p(\mathrm{S}_{p^n})$.

86. Let $P = \mathrm{UT}(n, p) \in \mathrm{Syl}_p(\mathrm{GL}(n, p))$. Find the multiplier $\mathrm{M}(P)$ and describe all representation groups of P.

87. (Old problem) Study the p-groups of class 3 all of whose proper subgroups are of class ≤ 2.

88. (Old problem) Study the non-metabelian p-groups all of whose proper subgroups are metabelian.

89. Study the p-groups G of exponent $> p > 2$, all of whose noncyclic abelian subgroups are elementary abelian.

90. (Mann) Classify the p-groups G of order 2^{2n+e}, $e \in \{0, 1\}$, satisfying $\mathrm{k}(G) = p^e + (p^2 - 1)n + (p^2 - 1)(p - 1)$ (see §2).

91. Classify the 2-groups all of whose proper two-generator subgroups have cyclic derived subgroups.

92. Study the regular p-groups G such that $G \times M$ is regular for all metacyclic p-groups M. Is it true that, for each such G, the direct product $G \times R$ is regular for all regular p-groups R?

93. Study the p-groups all of whose 2-generator subgroups are of exponent p or metacyclic.

94. Study the p-groups all of whose non-faithful nonlinear irreducible characters are of degree p.

95. Study the p-groups all of whose nonabelian subgroups of index p^2 are isomorphic (have the same rank).

96. Study the p-groups G such that $C_G(H)$ is abelian for all nonabelian $H < G$.

97. Suppose that, for all abelian $A < G$, one has $|AZ(G)/Z(G)| \le p^2$. Study the structure of $G/Z(G)$. (See §20.)

98. A subgroup $H \le G$ is called an NR-*subgroup* if $K \triangleleft H$ implies $H \cap K^G = K$. Classify the p-groups all of whose nonnormal subgroups are NR-subgroups.

99. Let $K < G$ be generated by all minimal irregular subgroups of an irregular p-group G. Study the structure of G/K.

100. Study the nonabelian p-groups G such that whenever H has the same order as G, then (i) $T(G) \le T(H)$, (ii) $T(G) \ge T(H)$, where $T(G) = \sum_{\chi \in \mathrm{Irr}(G)} \chi(1)$.

101. Let G be an abelian p-group G. Study the structure of the group $\langle \alpha \in \mathrm{Aut}(G) \mid \alpha_{\Omega_1(G)} = \mathrm{id}_{\Omega_1(G)} \rangle$.

102. Find $\mathrm{Aut}_c(P)$, the group of central automorphisms of P, where P is a Sylow p-subgroup of S_{p^n} or $\mathrm{GL}(n, p)$.

103. Classify the special p-groups whose representation groups are special.

104. Classify the nonabelian 2-groups G such that $G/\mho_2(G)$ is a nonmetacyclic nonabelian of order 16.

105. Study the p-groups G such that $H' \triangleleft G$ for all $H < G$.

106. Classify the p-groups G such that, whenever $A < G$ is an $\mathcal{A}_1$-subgroup, then $A < B < G$, where B is an $\mathcal{A}_2$-subgroup (see §72).

107. For a p-group G, define its *lower* $\mho$-*series* as follows:

$$\mho^0(G) = G, \quad \mho^1(G) = \mho_1(G), \quad \mho^2(G) = \mho_1(\mho^1(G)), \ldots,$$
$$\mho^{i+1}(G) = \mho_1(\mho^i(G)), \ldots.$$

Since $\exp(G/\mho^n(G)) \le p^n$, we get $\mho^n(G) \ge \mho_n(G)$, and the strong inequality is possible. Let $\mathfrak{E}_n$ be the set of all groups of exponent $p^n > p$. Prove that, for every $m \in \mathbb{N}$, the set $\mathfrak{E}_n$ contains a member G such that the length of the lower $\mho$-series of G exceeds m. (Recently Wilkens [Wil2] showed that, for $p > 2$, the answer is 'yes'.)

108. Classify the p-groups G such that $N_G(H)$ and H^G are incident for all $H < G$.

109. Classify the 2-groups G such that (i) $\Omega_2(G) \cong B(4,2)$, (ii) $\Omega_2^*(G) \cong B(4,2)$. (Here $B(4,2)$ is the restricted Burnside group.)

110. (i) (Old problem) Classify the irregular p-groups of order p^{p+1}. (ii) Classify the irregular L_p-groups (see §17,18).

111. Let H be a p-group. (i) Is it true that there exists a p-group G containing H and such that G has a characteristic subgroup of order p^n for all $p^n \leq |G|$. (ii) Try to construct a p-group $G > H$ such that H is characteristic in G. (According to Mann's report, B. Wilkens has proved that for every p-group H there exists a p-group $M \geq H$ such that M is not characteristic in every p-group containing M properly.)

112. Let M be a metacyclic p-group and let G be a p-group such that $\mathrm{c}_n(G) = \mathrm{c}_n(M)$ for all $n > 1$. Study the structure of G.

113. Study the p-groups G such that, for every $\chi \in \mathrm{Irr}_1(G)$, there is $H \leq G$ and $\mu \in \mathrm{Lin}(H)$ with $\chi = \mu^G$ and $|H : \ker(\mu)| = p$. (All nonabelian groups of exponent p satisfy the above property.)

114. Study the p-groups G of order at least p^{p+3} such that $|\mathrm{Z}_{p+1}(G)| = p^{p+1}$. Are these groups irregular?

115. Study the p-groups all of whose nonabelian subgroups are generated by elements of order p. Is the derived length of such groups bounded?[2] Since nonabelian p-groups G are generated by minimal nonabelian subgroups, it follows that if all minimal nonabelian subgroups of G are generated by elements of order p, then the same holds for all nonabelian subgroups of G. If $p = 2$, then the (abelian) H_2-subgroup has index 2 in G, and this solves problem in this case; see Theorem 10.32 and [Jan19].

116. (i) Classify the p-groups G such that $|\mathrm{N}_G(H) : H| = p$ for all nonnormal $H < G$. (Commentary of Mann at 1/03/08: Partial answer. First, assume that x is a non-central element of order p. Take $H = \langle x \rangle$. Then $\mathrm{N}_G(H)$ is a self-centralizing subgroup of order p^2, hence G is of maximal class. Now assume that all elements of order p are central. If p is odd, such groups are called *p-central*. We may assume that G is not a Dedekind group, and take H to be a non-normal cyclic subgroup. Then $\mathrm{Z}(G) \leq \mathrm{N}_G(H)$, therefore $|\mathrm{Z}(G) : (\mathrm{Z}(G) \cap H)| \leq p$, and $\mathrm{Z}(G)$ has a cyclic maximal subgroup, and has two generators. By properties of p-central groups, all subgroups can be generated by two elements (Thompson; see Theorem 15.1). Such groups are

[2] Commentary of Mann (in fact this is the solution of the problem for $p > 2$): Suppose that G is a p-group in which all non-abelian subgroups can be generated by elements of order p, $p > 2$, and suppose that G is not of exponent p. Let x have order $> p$, and let y be a non-central element in $\mathrm{Z}_2(G)$. Then $\langle x, y \rangle$ is of class 2 at most. If it is non-abelian, it is generated by elements of order p, and therefore has exponent p since $p > 2$. This is a contradiction. Therefore x centralizes y, and thus x centralizes $\mathrm{Z}_2(G)$. Thus all elements of order $> p$ generate a proper subgroup H (the Hughes subgroup). Applying the same argument to H shows that H is abelian. Suppose that $|G : H| > p$. Let $H < K \leq G$ be such that $|K : H| = p^2$. Then K is metabelian. But it is known that in a metabelian group the index of the Hughes subgroup is at most p, a contradiction. Thus $|G : H| = p$, and G is metabelian.

known (see Theorem 13.7). For $p = 2$, 2-central means that all elements of order 2 and 4 are central. A similar argument shows that either G is 2-central, and then again all subgroups are two-generator (see also Theorems 15.3 and 15.4), or there exists a self-centralizing subgroup of order 8.) (ii) Classify the p-groups G such that $C_G(H)/H$ is cyclic for all noncentral cyclic $H < G$.

117. Study the p-groups all of whose cyclic subgroups (elementary abelian subgroups) have at most p conjugates. (Commentary of Mann: in the first case, A. Shalev has proved, that the size of every conjugacy class of G is p^3 at most.)

118. (i) Does there exist a nonabelian group of exponent $p > 2$ admitting a nontrivial partition with all nonabelian components? (ii) Is it true that every nontrivial partition of a p-group has an abelian component? (iii) Does there exist a p-group ($p > 2$) of composite exponent which admits a nontrivial partition Σ such that all components of Σ of exponent p are nonabelian.

119. Classify the p-groups in which the centralizer of every noncentral element is metacyclic.

120. Study the p-groups all of whose nonnormal subgroups are of the same exponent. (See §63.)

121. Study the structure of G/E, where E is generated by all minimal non-metacyclic subgroups of a p-group G.

122. Classify the 2-groups all of whose maximal subgroups are of the form $H = MZ(H)$, where M is of maximal class (metacyclic).

123. Let G be a nonabelian group of order p^n, $P \in \mathrm{Syl}_p(\mathrm{Aut}(G))$, $W = P \cdot G$ the natural semidirect product. Study the structure of G in the case where $G = Z_m(W)$ for some $m \in \mathbb{N}$ (here $Z_m(W)$ is the m-th term of the upper central series of W).

124. Let $n \in \mathbb{N}$ be fixed. Study the p-groups all of whose subgroups of index p^n have the same rank.

125. Classify the p-groups all of whose nonnormal nonabelian subgroups are extraspecial (special).

126. Study the p-groups all of whose proper sections are pyramidal (see §8).

127. Is it true that if every two elements of equal order of a p-group G generate the regular subgroup, then G is regular? (For $p = 2$, the answer is 'yes': G has no minimal nonabelian subgroups.)

128. Study the 2-groups G in which $\langle x, y\rangle$ is minimal nonabelian or metacyclic for all $x, y \in G$.

129. Let G be a 2-group such that the number of metacyclic subgroups of index 2 in G is odd. Describe the structure of G.

130. Let $H < G = \Omega_1(G)$ be of order p^p and exponent p. Suppose that for every $x \in \mathrm{N}_G(H) - H$ with $o(x) = p$, the subgroup $\langle x, H\rangle$ is of maximal class. Study the structure of G.

131. Study the p-groups that are products of normal cyclic (abelian) subgroups.

132. Study the p-groups G such that p^2 does not divide $|\mathrm{Aut}(G) : \mathrm{Inn}(G)|$.

133. Describe the structure of a normal subgroup H of a 2-group G such that H has no G-invariant subgroups $\cong \mathrm{E}_8$. Is it true that $\mathrm{d}(H)$ is bounded? (See §50.)

134. Classify the p-groups G such that there is only one maximal chain connecting G with each its nonnormal subgroup.

135. Classify the groups G of exponent 4 without subgroups $\mathcal{H}_2 \cong \langle a, b \mid a^4 = b^4 = 1, a^b = a^{-1}\rangle$.

136. Study the irregular p-groups G with cyclic $\mho_1(G)$ and $|G/\mho_1(G)| = p^p$.

137. Let $\mathfrak{N} = \mathfrak{N}_d$ be the set of all representation groups of E_{p^d}. Assuming $p = 2$, find the number of irreducible characters of $G \in \mathfrak{N}$ of degree p^n for all $n \in \mathbb{N}$. (For $p > 2$, see [IM].)

138. Study the 2-groups in which the number of subgroups $\cong \mathrm{E}_8$ is even.

139. Study the p-groups all of whose nonabelian subgroups of equal order are isomorphic.

140. Classify the irregular p-groups G of order p^{p+1} with $|\Omega_1(G)| = p^p$. (According to Mann, the set of such G is nonempty.)

141. Give examples of p-groups G such that $\exp(\mho_k(G)) = \exp(G)$. (According to Mann's letter, such groups were constructed by Caranti and Scoppola for $k = 1$.)

142. Let $H \in \Gamma_1$. Describe the structure of H if $\mathrm{C}_H(x) \le \mathrm{Z}(H)$ for all $x \in G - H$.

143. Describe all automorphisms of order p of an arbitrary abelian p-group.

144. Classify the metacyclic p-groups containing maximal cyclic subgroup of order p.

145. Study the p-groups G containing exactly one maximal subgroup, say M, with $\exp(M) = \exp(G)$.

146. Let H be a p-group which is $\cong \Phi(G)$ for some p-group G. Is it true that there exists a p-group W such that (i) $H \cong \mho_1(W)$ ($p > 2$), (ii) $H \cong W'$?

147. Does there exist $n \in \mathbb{N}$ not depending on p such that every p-group G is embedded isomorphically into a suitable p-group generated by n elements of order p?

148. Study the p-groups G such that, for all nonnormal $H < G$, we have (i) $\exp(H^G) = \exp(H)$, (ii) $|H^G : H| = p$, (iii) $H^G = HG'$.

149. Let $\mathbb{F} = \mathrm{GF}(p^m)$, where $m > 2$, and let θ be a nonidentity automorphism of $\mathbb{F}$. Let $\mathrm{A}_p(m, \theta)$ be the set of all pairs (x, y), where $x, y \in \mathbb{F}$. Define the multiplication in $\mathrm{A}_p(m, \theta)$ as follows: $(x, y)(u, v) = (x + u, y + v + x\theta(u))$ (see §46). Find: (a) $|\mathrm{Aut}_c(\mathrm{A}_p(m, \theta))|$, where $\mathrm{Aut}_c(G)$ is the group of central automorphisms of G, (b) $|\mathrm{Aut}(\mathrm{A}_p(m, \theta))|$.

150. Let $P \in \{\Sigma_{p^n}(\in \mathrm{Syl}_p(\mathrm{S}_{p^n})), \mathrm{UT}(n, p)(\in \mathrm{Syl}_p(\mathrm{GL}(n, p))\}$. Find (a) $|\mathrm{Aut}_c(P)|$; (b) $|\mathrm{Aut}(P)|$; (c) $\mathrm{k}(P)$; (d) $|P/\mho_1(P)|$; (e) $\mathrm{T}(P)$ (see #19).

151. Classify the p-groups all of whose nonnormal abelian subgroups (i) are cyclic, (ii) have the same order.

152. Let G be representation group of E_{p^n}. Find all epimorphic images of G that are special.

153. Study the set $\mathscr{G}(m, p, 2)$ of all groups of order p^m, $m > 3$, and class $m - 2$. Find the minimal $m = m(p)$ such that all groups in $\mathscr{G}(m, p, 2)$ $(p > 2)$ are irregular.

154. Classify the 2-groups in which only one maximal subgroup (nonabelian maximal subgroup) is not generated by involutions.

155. Study the p-groups G with $|AG' : G'| \leq p$ for all abelian $A < G$.

156. Study the p-groups G such that, for all nonnormal subgroups H of G, the factor group $\mathrm{N}_G(H)/H$ is cyclic.

157. Classify the 2-groups G such that $\Omega_2(G)$ is extraspecial. (This problem was solved in Theorem 83.1.)

158. Study the p-groups G such that $\mathrm{Aut}(G)$ is (i) metabelian, (ii) minimal nonabelian.

159. Classify the p-groups G of normalized degree $\Delta(G) = \frac{p+1}{p^3}|G|$ (here $\Delta(G) = \frac{\delta(G)}{|G|}$, where $\delta(G)$ is the minimal degree of representations of G by permutations).

160. Find $\mathrm{k}(G)$ and $\mathrm{T}(G)$, where $G = A * B$, in terms of A, B and $A \cap B$ (see #19).

161. Study the p-groups G such that (i) whenever $A < B \leq G$, then $\mathrm{d}(A) \leq \mathrm{d}(B)$ (for $p > 2$, such groups G determined by Mann [Man11] in the case $\exp(G) \neq p^2$) (ii) the ranks of all maximal subgroups of G are equal to $\mathrm{d}(G) - 1$.

162. Let G be a p-group, $P \in \mathrm{Syl}_p(\mathrm{Aut}(G))$. Estimate $\mathrm{d}(P)$ in terms of G.

163. Study the p-groups G with (i) $p^2 \nmid \exp(\mathrm{Aut}(G))$, (ii) $\exp(G) \nmid \exp(\mathrm{Aut}(G))$.

164. (G. A. Miller) Classify the p-groups G with abelian $\mathrm{Aut}(G)$.

165. Study the p-groups G such that $\mho_1(G)$ is cyclic and $\exp(\Omega_1(G)) = p$.

166. Study the structure of G/G' if, for every abelian subgroup A of G, the quotient group AG'/G' is cyclic.

167. Classify the p-groups in which every two isomorphic nonnormal abelian subgroups are conjugate.

168. (Mann) Study the powerful p-groups all of whose proper subgroups are regular.

169. (Heineken–Mann) Study the p-groups G such that, for every abelian subgroup A of G, the quotient group $AZ(G)/Z(G)$ is cyclic. (For $p = 2$, see §91.)

170. Suppose that H is a proper subgroup of a p-group G such that every nonlinear irreducible character of G is induced from H. Study the structure of G.

171. Classify the p-groups in which every two distinct subgroups of the same order have abelian intersection.

172. Study the p-groups G all of whose noncyclic subgroups are quasinormal.

173. Classify the p-groups all of whose cyclic subgroups of order $> p$ are quasinormal.

174. Suppose that G is a p-group such that $|G : C_G(x)| \leq p$ for all elements x of G of order p. Study the structure of $\Omega_1(G)$ and its embedding in G.

175. Study the p-groups G such that, for each nonnormal subgroup H of G, the quotient group $N_G(H)/C_G(H)$ is isomorphic to a Sylow p-subgroup of $\mathrm{Aut}(H)$.

176. Give a method for calculating of $|\mathrm{Aut}(A \times B)|$, where A, B are p-groups.

177. Suppose that $E_{p^2} \cong A < G$ and $C_G(A)$ is abelian of type (p^n, p). Study the structure of G.

178. Study the two-generator p-groups G all of whose maximal subgroups are not two-generator.

179. Classify the p-groups G such that every abelian subgroup of G is contained in an $\mathcal{A}_1$-subgroup of G. (For a dual result, see Theorem 92.2.)

180. Write $c_m = \max\{|\mathrm{cd}(G)| \mid G$ is a p-group of maximal class and $|G| = p^m\}$. Estimate c_m in terms of m.

181. Study the irregular p-groups G such that H^G is regular, where H runs over all abelian subgroups of G.

182. Study the p-groups G such that $|\langle x\rangle^G| \leq p^{1+(p-1)k}$ for every element x of order p^k in G. (See §23.)

183. Study the p-groups G such that, for every $H < G$ and $|G : H| \leq p^k$ (k is fixed), one has $\Phi(H) = \Phi(G)$.

184. Classify the p-groups all of whose nonabelian maximal subgroups are of the form $M \times C$, where M is minimal nonabelian (extraspecial) and C is cyclic.

185. Describe the structure of $\mathrm{Aut}(G \times C_{p^n})$ in terms of G.

186. (i) Describe the structure of $\mathrm{Aut}(G)$, where G is the standard wreath products $A \operatorname{wr} C_p$ and $C_p \operatorname{wr} A$, in terms of A, (ii) Describe the structures of $W = C_{p^n} \operatorname{wr} C_{p^m}$ and $\mathrm{Aut}(W)$.

187. Study the two-generator p-groups G of class $k+1$, where $p^k = |G'|$.

188. Classify the 2-groups G such that $c_2(G) = 4$. (This problem was solved in §53.)

189. Find the least upper bound of the class of p-groups G which are $\mathcal{A}_n$-groups with $d(G) = n + 1$.

190. Study the 2-groups G such that $\langle x, y\rangle'$ is cyclic for all $x, y \in G$.

191. Study the irregular p-groups G such that $\langle x, y\rangle$ is regular for all $x, y \in G$ with $o(x)o(y) > p^2$. (Mann has showed that the set of such G is nonempty.)

192. (Mann) Classify the p-groups G such that $\mathrm{Aut}(G) \cong G$ (D_8 is the unique known such a group).

193. Study the irregular p-groups G such that $\mathrm{cl}(\langle x, y, z\rangle) \leq p$ for all $x, y, z \in G$.

194. Study the pyramidal p-groups G (see §8) such that $\exp(\Omega_i(G)) = p^i$ for all $p^i \leq \exp(G)$.

195. Study the p-groups all of whose subgroups of the same order have the same class.

196. Study the p-groups G with $|\langle x, y\rangle| \leq o(x)o(y)$ for all $x, y \in G$.

197. Let G be an irregular p-group, $p > 2$, $H < G$. Study the structure of H and its embedding in G if $\langle h, x\rangle$ is regular for all $h \in H$ and $x \in G - H$.

198. Let $H < G$. Study the structure of H and its embedding in G if $\langle h, x\rangle$ is powerful for all $h \in H$ and $x \in G$ (or $x \in G - H$).

199. Study the p-groups G such that $G/Z(G)$ is regular.

200. Study the p-groups all of whose maximal subgroups are decomposable in nontrivial direct (central) products. (Some partial cases of this very difficult problem have appeared in this list; see also Appendix 17.)

201. Classify the p-groups G which are extensions of a homocyclic group by a cyclic group.

202. Classify the p-groups all of whose nonnormal cyclic subgroups have the same order.

203. Characterize the powerful p-groups, $p > 2$, that are regular.

204. Classify the 2-groups all of whose subgroups of index 4 are Dedekindian.

205. Study the p-groups G such that $d(H) \geq d(G) > 2$ for all $H \in \Gamma_1$.

206. Describe the p-groups G of maximal class such that $\mathrm{Aut}(G)$ is not a p-group.

207. Classify the groups of exponent p, all of whose nonabelian subgroups of order p^4 are isomorphic.

208. Describe the structure of $\mathrm{Aut}(G)$, where G is

(a) abelian of type $(p, p^2, \dots, p^{n-1}, p^n)$,

(b) homocyclic of exponent $p^e > p$.

209. Classify the 2-groups G such that $\Omega_k(G) \cong D * C$ $(k \leq 2)$, where D is of maximal class, C is cyclic of order 4 and $D \cap C = \mathrm{Z}(D)$.

210. Study the p-groups G such that, for all $n \in N$, $\mathrm{c}_n(G) = \mathrm{c}_n(M \times C_p)$ for some irregular p-group of maximal class M.

211. Does there exist p-groups $H < G$ such that $\mathrm{Aut}(G) \cong \mathrm{Aut}(H)$?

212. Study the p-groups G having the lower central series $G > \mathrm{K}_2(G) > \mathrm{K}_3(G) > \mathrm{K}_{c+1}(G) = \{1\}$ such that

(i) $\mathrm{K}_i(G)/\mathrm{K}_{i+1}(G)$ is cyclic for $i = 2, 3, \dots, c$,

(ii) $\mathrm{K}_i(G)/\mathrm{K}_{i+2}(G)$ is metacyclic for $i = 2, 3, \dots, c-1$.

213. (Isaacs–Passman) Study the p-groups G such that $\mathrm{cd}(G) = \{1, p, p^2\}$.

214. Suppose that an irregular p-group G has no subgroups of order p^{p+1} and exponent p. Study the structure of G if it has at most p subgroups $\cong \mathrm{E}_{p^p}$.

215. Study the p-groups G such that $\langle H^\phi \mid \phi \in \mathrm{Aut}(G)\rangle = G$ for a fixed $\mathcal{A}_1$-subgroup $H < G$.

216. Study the p-groups G such that $2|G : G'| \geq \mathrm{T}(G)$ (see #19).

217. (Berkovich–Mann) Classify the pairs of p-groups $H < G$ such that $p^2 \nmid \delta(G, H) = \mathrm{T}(G) - \mathrm{T}(H)$ (see Theorem A.11.1).

218. Study the p-groups such that, whenever A and B are two nonincident nonabelian subgroups of G, then $A \cap B$ is maximal either in A or in B.

219. Study the p-groups G such that $G/\Phi(G')$ $(G/\Phi(\Phi(G)))$ is special.

220. Classify the 2-groups G such that, for every involution $t \in G - \mathrm{Z}(G)$, $\mathrm{C}_G(t)$ is elementary abelian.

221. Classify the p-groups G all of whose $\mathcal{A}_1$-subgroups are $\mathrm{Aut}(G)$-conjugate.

222. Find (mod p), the number of abelian subgroups of type (p^2, p^2) in a p-group.

223. Let $\delta(G)$ is the minimal degree of a faithful representation of G by permutations. Classify the p-groups G such that (i) $\delta(G) = \exp(G)$, (ii) $\delta(G) = \delta(A)$ for some abelian $A < G$.

224. (i) Study the p-groups G satisfying $\delta(H) = \delta(G)$ for all nonabelian $H \in \Gamma_1$ (see #223). (ii) Find $\delta(H)$ for all $H \in \Gamma_1$, where $G = \Sigma_{p^n}$.

225. Let $\delta_t(G)$ be the minimal degree of faithful transitive representations of a group G by permutations. Find $\delta_t(H)$ for all $H \in \Gamma_1$, where $G = \Sigma_{p^n}$.

226. Study the nonabelian p-groups G such that (i) $\delta_t(H) = \exp(H)$ for all nonabelian $H \leq G$. (ii) $\delta(H) = \delta_t(H)$ for all nonabelian $H \leq G$.

227. Study the nonabelian p-groups G such that (i) $\delta_t(H) = \exp(H)$ for all nonabelian $H \leq G$. (ii) $\delta(H) = \delta_t(H)$ for all nonabelian $H \leq G$.

228. Study the p-groups G such that, whenever $Z < G$ is cyclic of order $> p$, then $Z_G = \{1\}$.

229. Let $H < G$. Describe the embedding of H in G if $|\langle H, x\rangle| \leq |H|o(x)$ for all $x \in G - H$.

230. Does there exist a p-group $G = AB$ of exponent p with $A_G B_G = \{1\}$?

231. Classify the p-groups with exactly $p + 1$ nonabelian subgroups of order p^p and exponent p. (See §13.)

232. Let Σ be a nontrivial partition of a nonabelian group G of exponent p and $T \in \Sigma$. Suppose that all members of the set $\Sigma - \{T\}$ have the same order p^k. Is it true that then $k = 1$?

233. (Isaacs) Suppose that a p-group G is equally partitioned (see §68). Is it true that then all components of Σ are abelian?

234. Study the p-groups G such that $U' \cap V' = \{1\}$ for all distinct $U, V \in \Gamma_1$.

235. Let $N \lhd G$. Suppose that $C_G(x) \leq N$ for all $x \in N - Z(G)$. Study the structure of N and its embedding in G.

236. Study the p-groups G such that, whenever $C < G$ is cyclic, then either $C \leq Z(G)$ or $C \cap Z(G) = \{1\}$.

237. Classify the p-groups G such that, whenever $H \leq G$ with $d(H) = 2$, then $|H| \leq p^2 \exp(H)$.

238. Does there exist a p-group G such that $H_G = \{1\}$ for all minimal nonabelian $H < G$?

239. Classify the p-groups containing an $\mathcal{A}_1$-subgroup of index p.

240. Classify the p-groups of maximal class, $p > 2$, with trivial Schur multiplier.

241. Let $n > 2$ and $p > 2$. Classify the p-groups G such that $\Omega_n^*(G) = \langle x \in G \mid o(x) = p^n\rangle$ is absolutely regular.

242. (Mann) Find the p-groups G, $p > 2$, in which $\mathrm{cl}(G) > \mathrm{b}(G) + 1$, where $\mathrm{b}(G)$ is the breadth of G (see §40). For $p = 2$ there are infinitely many examples.

243. (Mann–Vaughan-Lee) Vaughan-Lee and Wiegold showed that if a p-group G is generated by elements of breadth n at most, then $\mathrm{cl}(G) \le n^2$. In fact, as Mann showed, in that case $\mathrm{cl}(G) \le n^2 - n + 1$. Improve this bound if possible. Is there a linear estimate of $\mathrm{cl}(G)$?

244. (I. D. Macdonald and [LMM]) Does there exist a p-group, $p > 2$, that contain only $p - 1$ conjugate classes of elements of maximal breadth. (According to Mann's report, M. Newman, E. O'Brien and A. Jaikin-Zapirain constructed infinitely many 3-groups of such type.) If so, classify such groups.

245. Suppose that p-groups G and G_0 are lattice isomorphic p-groups. Does there exist an estimate of $|\mathrm{Z}(G)|$ ($|G'|$) is terms of G_0?

246. (Isaacs) Let X be a set of powers of p such that $1 \in X$. The set X is of *bounded type* if $\mathrm{cl}(G)$ is bounded for all p-groups G such that $\mathrm{cd}(G) = X$. Classify all sets of bounded type.

247. (Ito; ##248–253 concerning Hadamard groups, were taken or inspired by papers of Ito and his associates.) Let e^* be a central involution of a group G, $|G| = 8n$. The group G is said to be *Hadamard* with respect to e^* if, for some transversal D of $\langle e^* \rangle$ in G and all $a \in G - \langle e^* \rangle$, $|D \cap Da| = 2n$. The set of Hadamard groups is very large (for example, every 2-group G is a subgroup of an appropriate Hadamard 2-group; Ito, unpublished). In the time of this writing only one nonsolvable Hadamard group was known: $\mathrm{SL}(2,5)$. Classify (i) abelian Hadamard groups. (ii) Hadamard groups containing a normal abelian subgroup of prime index. (iii) Suppose that G is Hadamard; is $P \in \mathrm{Syl}_2(G)$ Hadamard?

248. Classify the groups H such that $\mathrm{C}_2 \times H$ is Hadamard.

249. Classify the groups H such that $\mathrm{C}_4 \times H$ is Hadamard.

250. Classify the metacyclic groups (in particular, metacyclic 2-groups) that are Hadamard.

251. Describe all 2-groups of order 2^n, $n \le 7$, that are Hadamard.

252. Classify all the minimal nonabelian 2-groups that are Hadamard.

253. Classify all the special 2-groups that are Hadamard.

254. Study the p-groups G such that, whenever $H \le G$, then $|H| \le \exp(H)^{\mathrm{d}(H)}$.

255. Study the p-groups G such that whenever $A, B \le G$ then $\log_p(|\langle A, B\rangle|) \le \log_p |A| + \log_p |B| - \log_p |A \cap B| + 1$.

256. Study the p-groups G with normal centralizers of noncyclic subgroups.

257. Study the p-groups with normal normalizers of cyclic subgroups.

258. Study the nonabelian p-groups G such that $A \cap B = \mathrm{Z}(G)$ for distinct maximal abelian $A, B < G$.

259. Let $H_1, \dots, H_{p+2}$ be such subgroups of a p-group G that $G = \bigcup_{i=1}^{p+2} H_i$ but G is not covered by every $p + 1$ members of the set $\{H_1, \dots, H_{p+2}\}$. What can be said about $H_1, \dots, H_{p+2}$ and G?

260. Classify the p-groups G such that all $H < G$ with $|H'| = p$ are normal.

261. (Old problem) Study the p-groups G with nilpotent $\mathrm{Aut}(G)$.

262. Study the p-groups without normal subgroups of order $p^{2+(p-1)k}$ and exponent $\leq p^k$, $k > 1$ is fixed (see §24).

263. Let $\Sigma = \{H_\lambda\}_{\lambda \in \Lambda}$ be a nontrivial partition of a group X. Suppose that $\pi(H_\lambda)$ is the same for all $\lambda \in \Lambda$. Is X a prime power group? Study the groups for which $\{\pi(H_\lambda) \mid \lambda \in \Lambda\}$ is a chain with respect to inclusion. (See §§20, 68.)

264. Given an abelian p-group G, study the structures of

$$\mathrm{O}_p(\mathrm{Aut}(G)) \quad \text{and} \quad \mathrm{Aut}(G)/\mathrm{O}_p(\mathrm{Aut}(G)).$$

265. (Isaacs) Let a p-group P act on a q-group V (where p and q are primes not necessarily distinct) in such a way that the sizes of different non-one-element P-orbits are distinct. Study the structure of P and V.

266. (Isaacs) Let $\{1\} < A < G$. Study the structure of G provided all conjugacy classes of G that are not contained in A, have the same size.

267. Classify the 2-groups G with $\mathrm{c}_n(G) = 4$ $(n > 2)$. (For a solution, see §53.)

268. (i) Classify the groups of exponent p all of whose maximal subgroups are isomorphic. (ii) Study the p-groups G all of whose maximal subgroups are isomorphic (such groups were treated by P. Hermann and Mann).

269. Study the p-groups G such that, whenever $A, B < G$ are distinct minimal nonabelian subgroups, then $A \cap B = A' \cap B'$.

270. (Isaacs) Let R be a finite algebra over $\mathrm{GF}(q)$, where q is a power of a prime p, and let $J = \mathrm{J}(G)$ be the Jacobson radical of R. Then $G = 1 + J$ is a p-group. Prove that $|G : G'|$ is a power of q.

271. Classify the p-subgroups G of S_{p^n} such that $\mathrm{d}(G) = p^{n-1}$. Does there exist, for $p = 2$, such subgroups having class exceeding 2?

272. Classify the representation groups of minimal nonabelian p-groups.

273. A subgroup H of a group X is said to be a CR-subgroup if every irreducible character of H is the restriction of an irreducible character of X. Study the p-groups G all of whose subgroups not contained in G' are CR-subgroups.

274. Classify the p-groups containing extraspecial maximal subgroup (two distinct extraspecial maximal subgroups).

275. Classify the p-groups G such that H/H_G is cyclic for all $H \leq G$.

276. Given a p-group G, set $\mho^2(G) = \mho_1(\mho_1(G))$. Study the p-groups G such that $G/\mho^2(G)$ is abelian.

277. Study the p-groups without normal subgroups of class 2.

278. Study the p-groups all of whose epimorphic images of equal order are isomorphic.

279. Study the p-groups without isomorphic maximal subgroups.

280. Find the exponent of the automorphism group of an abelian p-group.

281. Describe a Sylow p-subgroup of the automorphism group of $E * \mathrm{C}_{p^n}$, where E is an extraspecial p-group and $E \cap \mathrm{C}_{p^n} = \mathrm{Z}(E)$.

282. It is known (see §37) that, if $|G : \Phi(G)| \geq p^{2k+1}$, then $\mathrm{c}_1(G) \equiv 1 + p + \cdots + p^k \pmod{p^{k+1}}$. Give a character free proof of this result.

283. Study the p-groups G such that $\mathrm{N}_G(H) = H\mathrm{Z}(G)$ for all nonnormal nonabelian subgroups H of G.

284. Study the p-groups G satisfying one of the following conditions: (a) $\mathrm{C}_G(x) \leq \Phi(G)$ for all $x \in \Phi(G) - \mathrm{Z}(G)$, (b) $\mathrm{C}_G(x) \leq G'$ for all $x \in G' - \mathrm{Z}(G)$. (As E. Khukhro noticed, the free 2-generator p-group of class 3 and exponent $p > 3$, satisfies (a) and (b); there $G' = \Phi(G)$.)

285. Study the p-groups G such that $\exp(\langle x \rangle^G) = o(x)$ for all $x \in G$.

286. Study the irregular p-groups without sections isomorphic to $\mathrm{C}_p \operatorname{wr} \mathrm{C}_p$. (For $p = 2$, see Theorem 44.18.)

287. (Old problem) Study the normal and power structure of irregular p-groups of class p.

288. Study the p-groups G such that if $A < B$ are consecutive members of the upper or lower central series of G with $B < G$, then B/A is cyclic. (Commentary of Mann: Such groups were discussed in the unpublished Ph.D. thesis of Arye Juhasz.)

289. Is it true that, for each minimal nonabelian p-group A, there exists a powerful p-group G that contains a subgroup $\cong A$? (L. Wilson has showed that there are many groups that are not contained in any powerful p-group.)

290. Let G be a p-group such that $\Omega_1(G) = G$ and the product of every two elements of G of order p is of order $\leq p^2$. Is it true that $\exp(G)$ bounded?

291. Classify the p-groups with two conjugate classes of nonnormal cyclic subgroups.

292. The intersection $\mathcal{N}(G)$ of normalizers of all subgroups of G is said to be the *norm* of G (Baer). It is known that $\mathcal{N}(G) \le \mathrm{Z}_2(G)$. Study the p-groups G such that $G/\mathcal{N}(G)$ has a cyclic subgroup of index p.

293. Study the p-groups G such that $|\mathrm{Z}(G/\ker(\chi))| = p$ for all nonprincipal $\chi \in \mathrm{Irr}(G)$.

294. Mann termed a p-group G to be *adequate* if the class of G is less than the class of its representation group. All noncyclic abelian p-groups are adequate. Study the p-groups of class 2 that are adequate.

295. Let R be a regular p-group. Is it true that there exists a regular p-group G such that $R \cong R_0 \le \Phi(G)$?

296. (Mann) If a p-group G of order p^n has derived length $k+1$, $k > 1$, then $n \ge 2^k + 2k - 2$ (see Appendix 6). Improve this estimate if possible. (C. Schneider improved this estimate.)

297. Study the p-groups G such that, whenever $x \in G - M$, where $M \in \Gamma_1$ is fixed, then $|G : \mathrm{C}_G(x)| \le p^2$. (According to Mann [Man20, §4], the breadth of such G is at most three.)

298. Study the nonabelian p-groups G such that $H \cap \mathrm{Z}(G) > \{1\}$ (i) for every $H \le G$ of composite order, (ii) for every noncyclic $H \le G$.

299. Classify the p-groups such that, whenever $G = AB$, then one of the factors A, B is normal in G (see [BlaDM]).

300. Let M be a regular subgroup of maximal class in a p-group G, and let G be not of maximal class. Is it true that the number of subgroups $L < G$ of maximal class and order $p^k|M|$ such that $M < L$, is a multiple of p for a fixed k? (See §13.)

301. Let G be a group of exponent p^n, $n > 2$. Suppose that every two distinct cyclic subgroups of G of order p^n have intersection of order at least p^{n-2}. Study the structure of G.

302. Classify the p-groups that are not generated by their noncyclic abelian subgroups. (The semidihedral groups satisfy the above condition.)

303. (M. Roitman) Let N be a normal p-subgroup of an (arbitrary) group X and $N \le \mathrm{H}(X)$, the hypercenter of X. Suppose that $|N : \mathrm{C}_X(x)| \le p^n$ for all $x \in X$. Is it true that $[N, X, \dots, X] \le \mathrm{Z}(X)$ (X appears n times)? (See Appendix 7).

304. Construct, for each $n \in \mathbb{N}$, a p-group G with $|\mathrm{Z}_n(G)| = p^n$ that is not of maximal class.

305. Find the number of families (see §29), containing irregular subgroups of order p^{p+1}.

306. Describe all members of the family of p-groups containing (a) a Sylow p-subgroup of the symmetric group of degree p^n; (b) an irregular group of maximal class; (c) a metacyclic group.

307. Study the groups isoclinic to absolutely regular p-groups.

308. Study a family of p-groups containing a nonabelian group of order p^3 and exponent p.

309. Study the irregular p-groups G, $p > 2$, with $\exp(G/\mathrm{K}_{p+1}(G)) = p$.

310. Let $e(H)$ be the maximal order of subgroups of exponent p in a p-group H. Let a p-group $G = AB$, where subgroups A and B are regular. Is it possible to estimate $e(G)$ in terms of $e(A)$ and $e(B)$?

311. Classify the p-groups all of whose normal (nonnormal) two-generator subgroups are abelian.

312. Study the p-groups G such that $\Phi(G)$ is abelian of type (p, p^m).

313. Classify the 2-groups G containing an element y of order 4 such that $\mathrm{C}_G(y)$ is abelian of type $(4, 2)$. Moreover, classify the 2-groups containing an abelian subgroup of type $(4, 2)$, coinciding with its centralizer. (§77 contains a lot of information on such groups.)

314. Let $H \triangleleft G$ be p-groups and let G/H be noncyclic of order $> p$. Suppose that there is only one subgroup of G, containing H as a subgroup of index p, which is not of maximal class. Study the structure of G.

315. Classify the p-groups G all of whose (a) maximal cyclic subgroups are complemented, (b) subgroups not contained in $\Phi(G)$, are complemented.

316. Classify the irregular p-groups G such that, whenever R is a maximal absolutely regular subgroups of G, then R is a maximal regular subgroup of G.

317. Study the p-groups all of whose $\mathcal{A}_1$-subgroups are complemented.

318. Does there exist, for every $n \in \mathbb{N}$, an interval $]u, v[$ such that $v - u \geq n$ and there is no p-group G with $\alpha_1(G) \in]u, v[$ (see §76).

319. Classify the p-groups without three nonnormal subgroups of distinct orders.

320. Classify the p-groups G such that, whenever $A, B < G$ are non-incident, then $A \cap B \triangleleft G$.

321. Study the p-groups G such that two irreducible characters of G have the same kernel if and only if they have the same degree.

322. Describe the automorphism groups of all metacyclic p-groups.

323. Study the p-groups G, $p > 2$, such that $\Omega_2(G) = E \times M$, where E is elementary abelian and M is of maximal class. (For $p = 2$, see §75.)

324. Study the p-groups all of whose maximal elementary abelian subgroups are maximal abelian.

325. Study the p-groups such that for any $\chi \in \mathrm{Irr}_1(G)$, there is an $\mathcal{A}_1$-subgroup $A < G$ and $\lambda \in \mathrm{Lin}(A)$ such that $\chi = \lambda^G$.

326. Classify the 2-groups G containing a metacyclic subgroup H of index 2.

327. Let Z be a cyclic subgroup of order 8 of a 2-group G. Suppose that $C_G(Z) = Z$. Study the structure of G.

328. Suppose that, whenever $M \in \Gamma_1$, then all maximal abelian subgroups of M are also maximal abelian subgroups of G. Study the structure of G.

329. Describe the p-groups G such that, whenever Z is a maximal cyclic subgroup of G, $|Z| = p^n$, where $n > 1$ for $p > 2$ and $n > 2$ for $p = 2$, then $N_G(Z) \cong M_{p^{n+1}}$.

330. Study the irregular p-groups (i) in which the normalizers of all nonnormal regular subgroups are regular, (ii) all of whose maximal regular subgroups are normal.

331. Suppose that $G = H_1 H_2$, where H_1, H_2 are homocyclic p-groups and $H_1 \cap H_2 = \{1\}$. Study the power structure of G.

332. Suppose that $G = Z_1 Z_2 \dots Z_k$, where $Z_1, Z_2, \dots, Z_k$ are pairwise permutable cyclic 2-groups. Study the power structure of G. (See #3.)

333. Study a p-group $G = E_1 E_2$, where E_1 and E_2 are elementary abelian.

334. Study a p-group $G = S_1 S_2$, where S_1, S_2 are extraspecial.

335. (Thompson) Let q be a power of p, $G = \mathrm{UT}(n, p)$. $G_0 = \mathrm{UT}(n, q)$. Find the connection between $\mathrm{cd}(G)$ and $\mathrm{cd}(G_0)$.

336. Classify the minimal nonmodular p-groups. (For solution, see §78.)

337. Study maximal abelian p-subgroups of $\mathrm{Aut}(A)$, where A is a homocyclic p-group.

338. Let G be an extraspecial group of order p^{1+2m}. Find the number of abelian subgroups of order p^k, $k \le m + 1$, in G.

339. Let A be a metacyclic p-group, $|A'| > p > 2$. Describe all (regular) p-groups B such that $A \times B$ is regular.

340. Study the p-groups G such that, whenever $\psi, \chi \in \mathrm{Irr}_1(G)$ with $\psi(1) < \chi(1)$, then $\ker(\psi) > \ker(\chi)$.

341. Let $G \in \mathrm{Syl}_p(\mathrm{GL}(n, p))$. Find in G the orders of (i) maximal elementary abelian subgroups, (ii) maximal normal elementary abelian subgroups, (iii) maximal subgroups of exponent p, (iv) maximal normal subgroups of exponent p.

342. Study the p-groups all of whose maximal regular subgroups are (i) abelian, (ii) of class ≤ 2.

343. Study the p-groups with abelian H_p-subgroup.

344. Study the p-groups G with $C \leq \mathrm{Z}(\mathrm{N}_G(C))$ for all cyclic C.

345. Find all m such that there is a group of maximal class and order p^{2m+1} with $p^m \in \mathrm{cd}(G)$.

346. Study the p-groups G such that whenever A, B are distinct maximal abelian subgroups of G, then $A/(A \cap B)$ is cyclic.

347. Classify the 2-groups all of whose epimorphic images are not $\mathcal{A}_2$-groups.

348. Study the nonabelian p-groups G with $\mathrm{Z}(N) \leq \mathrm{Z}(G)$ for all nonabelian $N \leq G$.

349. Study the p-groups all of whose normal abelian subgroups have exponent p. ($\Sigma_{p^n} \in \mathrm{Syl}_p(\mathrm{S}_{p^n})$, $p^n > 4$, have the above property.)

350. Study the p-groups satisfying $\Omega_1(G) = \mho_1(G) \leq \mathrm{Z}(G)$.

351. Study the p-groups G such that, whenever $A, B < G$ are abelian of type (p, p) (cyclic of order p^2), there exists $\phi \in \mathrm{Aut}(G)$ such that $B = A^{\phi}$.

352. Classify the p-groups with exactly one maximal subgroup H such that $|\mathrm{Z}(H)| > p$.

353. Study the p-groups such that, whenever $x, y \in G$, there exists $z \in \mathrm{Z}(\langle x, y\rangle)$ such that $(xy)^p = x^p y^p z$.

354. Study the $\mathcal{A}_m$-groups G, $m > 2$ such that, whenever $n < m - 1$ and $F < G$ is an $\mathcal{A}_n$-subgroup, then $F < H$, where $H < G$ is an $\mathcal{A}_{n+1}$-subgroup.

355. Classify the irregular 3-groups all of whose nonabelian regular subgroups are minimal nonabelian.

356. Let $P < G$. Study the structure of G if P is the unique subgroup of G of its order and exponent.

357. Classify the p-groups G such that $A < B \leq G$ implies $|A : A'| \leq |B : B'|$.

358. Study the p-groups G containing an element x such that $\mathrm{C}_G(x)$ is special (metacyclic).

359. Classify the p-groups G such that, whenever A, B are nonabelian subgroups of G of the same order, then $|A : A'| = |B : B'|$.

360. Classify the 2-groups G of exponent $2^e > 2$ such that $\Omega_e^*(G) = \langle x \mid o(x) = 2^e\rangle = Z \times M$, where Z is cyclic and M is of maximal class.

361. Study the p-groups all of whose $\mathcal{A}_1$-subgroups are metacyclic.

362. Let $A < G$ be abelian. Suppose that all subgroups of G of order $p|A|$, are minimal nonabelian. Study the structure of G.

363. Classify the groups of order p^m admitting an automorphism of order p^{m-3}. (See §33.)

364. Let G be a p-group and let $A < \Phi(G)$ be G-invariant. Does there exist a p-group W such that $A = \Phi(W)$? The same problem for derived subgroups instead of Φ-subgroups.

365. Classify the p-groups with exactly p nonnormal subgroups of given order p^n.

366. Let $N \lhd G$ be of order p^2 and suppose that $\mathrm{k}(G) - \mathrm{k}(G/N) = p^2 - 1$. Is it true that then the derived length of G is bounded?

367. Classify the p-groups which are lattice isomorphic to $\mathcal{A}_n$-groups, $n = 1, 2$.

368. Study the p-groups containing exactly two conjugate classes of subgroups of order p^p and exponent p.

369. Suppose that p-groups G and H are lattice isomorphic via ϕ. Is it true that there exists a minimal normal subgroup N of G such that $N^\phi \lhd H$?

370. (i) Classify the p-groups with exactly one minimal nonmetacyclic subgroup. (ii) Study the p-groups all of whose minimal nonmetacyclic subgroups are isomorphic.

371. Classify the 2-groups G containing a cyclic subgroup Z of order 4 such that $C_G(Z) = Z * M$, where M is of maximal class.

372. Study the p-groups all of whose subgroups of class 2 are minimal nonabelian.

373. Let $p > 2$ and $H \in \Gamma_1$. Suppose that all regular subgroups of G not contained in H, are abelian. Study the structure of G.

374. Study the 2-groups all of whose noncentral involutions are not squares.

375. Study a p-group G such that $\mathrm{Z}(\Gamma) \not\cong \mathrm{M}(G)$, where Γ is a representation group of G and $\mathrm{M}(G)$ the Schur multiplier of G.

376. A group G is said to be *capable* if there exists a group H such that $H/\mathrm{Z}(H) \cong G$. If G is finite then it is possible to take H to be finite (Isaacs; see §21). Suppose that $|H|$ is as small as possible. Estimate $|\mathrm{Z}(H)|$ in terms of G.

377. Study the p-groups of class 2 all of whose epimorphic images are not $\mathcal{A}_1$-groups.

378. Study the irregular p-groups G satisfying $|\Omega_2(G)| = p^{p+2} < |G|$. (See §52, where this problem is solved for $p = 2$.)

379. Classify the groups G of order 2^n with exactly $2^n - 1$ maximal chains of subgroups.

380. Find the number of maximal chains of subgroups of (a) an extraspecial group of order p^{2m+1}, (b) an abelian p-group of given type.

381. Let H be a p-group and $n \in \mathbb{N}$. Is it true that there exist a p-group G such that G contains a characteristic subgroup $H_1 \cong H$ of index p^n?

382. Is it true that, for every abelian p-group A, there exists a *nonabelian* p-group G of the same order whose Schur multiplier is isomorphic with A?

383. Find as many as possible properties of p-groups H such that there exists a p-group G satisfying one of the following conditions: (a) $H \cong \Phi(G)$; (b) $H \cong G'$; (c) $G/\mathrm{Z}(G) \cong H$; (d) $G/\mathrm{Z}_2(G) \cong H$.

384. Is it possible to estimate $|\mathrm{M}(G)|$ in terms of $|\mathrm{M}(H)|$ for all $H \in \Gamma_1$?

385. Does there exist a p-group G such that $\mathrm{C}_G(\ker(\chi)) \leq \ker(\chi)$ for all $\chi \in \mathrm{Irr}(G)$?

386. Classify the nonabelian p-groups with the unique nontrivial characteristic subgroup (such groups are special).

387. Study the p-groups in which the centralizers of noncentral elements of order p are elementary abelian.

388. Study the p-groups G having an automorphism α of order p such that $\mathrm{C}_G(\alpha)$ is abelian of type (p, p). (For $p = 2$, see §51).

389. Let $L < G$ be of order p^2 such that there exists only one maximal chain connecting L and G. Study the structure of G.

390. Is it true that every p-group of class 2 is isomorphic to a maximal subgroup of the Frattini subgroup (the derived subgroup) of an appropriate p-group?

391. Classify the special p-groups that are not isomorphic to Frattini subgroups (derived subgroups) of all p-groups.

392. Classify the groups H of order p^k for $k \leq 6$ that are isomorphic to Frattini subgroups (derived subgroups) of some p-group. (For $p^k = 2^4$, see §85.)

393. Study the p-groups G such that $\mathrm{cl}(G) = \mathrm{cl}(\Phi(G))$.

394. Is it true that for every nonabelian p-group A there exists a p-group G such that $\Phi(G) \cong A \times A$?

395. Classify the special p-groups G such that G/N is extraspecial for all maximal subgroups N of $\mathrm{Z}(G)$. (Nonabelian normal subgroups of minimal nonnilpotent groups satisfy the above property.)

396. Classify the special p-groups G such that G/K is special for all $K < \mathrm{Z}(G)$.

397. Study the p-groups G such that whenever $\{1\} < A < B \leq G$ and A is characteristic in a nonabelian subgroup B, then $\mathrm{N}_G(A) = \mathrm{N}_G(B)$.

398. Study the p-groups without nonabelian metacyclic subgroups.

399. Classify the p-groups G such that (i) $\mathrm{Aut}(G) \cong \mathrm{Aut}(A)$ for some abelian group A, (ii) $|\mathrm{Aut}(G)| = |\mathrm{Aut}(A)|$ for some abelian group of order $|G|$.

400. Study the p-groups G of order p^n with $|\pi(\mathrm{Aut}(\mathrm{E}_{p^n})) - \pi(\mathrm{Aut}(G))| \le 1$.

401. Let $H \in \Gamma_1$, where G is a nonabelian p-group. Study the structure of G if, whenever $A < G$ is abelian, then either $A \le H$ or $|A| \le p^3$. Is it true that the coclass of G is bounded?

402. Find all i such that, whenever a p-group G of maximal class and large order has a subgroup $E \cong \mathrm{E}_{p^i}$, then G has a normal subgroup $\cong E$.

403. Study the p-groups G in which any maximal abelian subgroup is equal to the centralizer of an element $x \in G$.

404. Study the p-groups G such that (i) $\mathrm{H}_p(G) \le \Phi(G)$, (ii) $\mathrm{H}_p(G) \le \mho_1(G)$.

405. Classify the 2-groups G containing a cyclic subgroup Z of order 4 such that $\mathrm{N}_G(Z) = Z \times M$, where M is of maximal class.

406. Study the p-groups G with extraspecial (special) $\mathrm{N}_G(B)$ for some $B < G$.

407. Study the p-groups G such that (i) $\mathrm{C}_G(A)$ is metacyclic (minimal nonabelian) for some $A < G$ with $A \cong \mathrm{E}_{p^2}$, (ii) $\mathrm{N}_G(H)$ is metacyclic for some $H < G$.

408. Let P be a p-group and $n > 2$. Does there exist a nonabelian p-group $G > P$ of order $|P|p^n$ such that G has the unique normal subgroup of index p^n and that subgroup is isomorphic with P? (According to a recent paper of B. Wilkens, Isr. J. Math., to appear, the answer is 'no'.)

409. Let G be a special p-group of exponent p^2, $p > 2$. Describe all possible structures of $G/\mho_1(G)$.

410. Classify the p-groups in which the centralizer of any $x \in G - \mathrm{Z}(G)$ has order $\le p^2 \cdot o(x)$.

411. Let G be the maximal n-generator group of class 2 and exponent p^e. Find $\mathrm{cd}(G)$ and the number of irreducible characters of G of given degree. (For $e = 1$ and $p > 2$, see [IM].)

412. Study the p-groups G with $|A\mathrm{Z}(G) : \mathrm{Z}(G)| \le p^2$ for all abelian $A < G$.

413. (Isaacs) Let G be a group of order p^m and class c. The number $m - c$ is said to be the *coclass* of G. Suppose that $x \in G$ with $|\mathrm{C}_G(x)| = p^r$. Find all r such that the coclass of G is bounded in terms of r? (As E. Khukhro noticed, in general, the coclass of G cannot be bounded in terms of r.)

414. Denote by $a_n(G)$ the number of characters of degree p^n in $\mathrm{Irr}(G)$. (According to Mann, if $n \ge 1$ then $p - 1 \mid a_n(G)$.) Find all m such that there exists a p-group G with $a_1(G) = (p - 1) \cdot m$? (As Isaacs noticed, there is m for which relevant groups G do not exist.)

415. Let G be an irregular p-group with $|G/\mho_1(G)| = p^w$. Does there exist a function f such that $|H/\mho_1(H)| \leq f(w)$ for all $H < G$? (The answer is no; see [Man24].)

416. Study the p-groups G such that, for all $M \in \Gamma_1$, we have $|G/\mho_1(G)| = |M/\mho_1(M)|$.

417. Classify the p-groups G of maximal class, $p > 3$, such that there is a nonabelian $H \in \Gamma_1$ with $\mathrm{d}(H) = p - 1$.

418. Does there exist a 3-group G, containing an abelian subgroup of index 3^4 but not containing a normal abelian subgroup of the same index? (See §39.)

419. Classify the 2-groups with unique normal subgroup of index 8.

420. Classify the p-groups G such that, whenever noncyclic (normal noncyclic) $A, B < G$ with $|A| = |B|$, then (i) $|A : \mho_1(A)| = |B : \mho_1(B)|$, (ii) $|\Omega_1(A)| = |\Omega_1(B)|$, (iii) $|A : A'| = |B : B'|$; (iv) $|\Omega_2(A)| = |\Omega_2(B)|$.

421. Given $n > 1$, classify the p-groups G such that $\Omega_n(G)$ is an $\mathcal{A}_1$-subgroup.

422. Study the p-groups G such that for every two nonmetacyclic subgroups A, B of G of the same order we have $\mathrm{d}(A) = \mathrm{d}(B)$.

423. (Old problem) Classify characteristic subgroups of abelian p-groups G. Is it possible to describe these subgroups in terms of $\Omega_m(G)$ and $\mho_n(G)$ only ($m, n \in \mathbb{N}$)?

424. Study the p-groups $G = AB > \{1\}$ with $A_G B_G = \{1\}$.

425. (i) Classify the 2-groups G with $\mathrm{c}_2(G) = 6$. (For solution, see §89.) (ii) Classify the 2-groups G with $\mathrm{c}_2(G) \equiv 2 \pmod 4$ (see Corollary 18.7).

426. Classify the groups G of order 2^m with $\mathrm{c}_1(G) > 2^{m-2}$.

427. Study the 2-groups G containing two distinct maximal subgroups M and N such that all elements of the set $G - (M \cup N)$ are involutions. (This problem was solved by Baginsky–Malinowska in [BM2].)

428. Study the p-groups all of whose characteristic subgroups are two-generator.

429. Study the p-groups G such that $\mathrm{c}_n(G) = \mathrm{c}_n(A)$ for all $n \in \mathbb{N}$ and some abelian p-group A.

430. Let G be an extraspecial group of order p^{1+2m} and exponent p^2, $p > 2$. Classify all groups H that are lattice isomorphic to G.

431. Classify the 2-groups with a unique subgroup of order 2^6 and exponent ≤ 8.

432. Study the p-groups that are lattice isomorphic to special p-groups.

433. Study the p-groups G having only one normal subgroup of each order $\leq |\Phi(G)|$. (If $p = 2$, then $\Phi(G)$ is cyclic; see Proposition 4.9.)

434. Classify the irregular p-groups all of whose absolutely regular subgroups have orders $\leq p^{p+1}$.

435. Let M be a nonabelian maximal subgroup of a p-group G. Suppose that all maximal abelian (cyclic) subgroups of M are also maximal abelian (cyclic) subgroups of G. Study the structure of G.

436. Classify the p-groups all of whose nonnormal subgroups are (a) cyclic (for solution, see Theorem 16.2), (b) metacyclic, (c) absolutely regular.

437. Classify the 2-groups with unique subgroup of order 2^5 and exponent ≤ 8. (For solution, see [BozJ1].)

438. Classify the p-groups G in which the normalizer of each nonnormal cyclic subgroup of order $> p$ is maximal in G (see §58).

439. Study the p-groups G such that, whenever H is a nonnormal subgroup of G, then $N_G(H) \leq H^G$.

440. Study the p-groups G such that, whenever A is an $\mathcal{A}_1$-subgroup of G, there exists a unique maximal chain connecting A and G.

441. Classify the groups of exponent p which have no characteristic maximal subgroups.

442. Classify the p-groups all of whose nonnormal cyclic subgroups generate a metacyclic subgroup.

443. Study the p-groups G such that whenever $H < G$ is nonabelian (minimal nonabelian), then $H \cap G' = H'$.

444. Classify the groups G of order 2^m, $m > 5$ such that for some fixed r with $4 \leq r < m-1$, all G-invariant subgroups of orders 2^r and 2^{r+1} are two-generator.

445. Classify the p-groups that are not generated by $\alpha_1(G) - 1$ minimal nonabelian subgroups.

446. Classify the p-groups all of whose nonnormal cyclic subgroups are maximal cyclic.

447. Study the non-Dedekindian p-groups G such that, for every nonnormal cyclic subgroup Z of composite order, G/Z^G is cyclic.

448. Study the p-groups G such that $N_G(A) \in \Gamma_1$ for every noncyclic $A \ntrianglelefteq G$.

449. Classify the absolutely regular p-groups H of order $> p^p$, $p > 2$, which are the fundamental subgroups of p-groups of maximal class and order $> p^{p+1}$.

450. Does there exist a group H of maximal class of order p^p and exponent p such that, whenever G is an irregular p-group of maximal class, then (i) G has no subgroups isomorphic to H? (ii) $G/\mho_1(G) \ncong H$?

451. Classify the p-groups G containing a normal subgroup R of order p such that G/R is metacyclic.

452. Study the p-groups G such that, whenever A is a maximal abelian subgroup of G and $A < H$ with $|H : A| = p$, then $\mathrm{cl}(H) = 2$.

453. Classify the p-groups of maximal class of order p^p and exponent p.

454. Let G be a p-group of maximal class and order $> p^{p+1}$ such that all elements in $G - G_1$ have order p. Is it true that the Schur multiplier of G is nontrivial? (If $p = 2$, the answer is 'yes'.)

455. Classify the p-groups that are generated by any distinct $p+1$ their $\mathcal{A}_1$-subgroups.

456. Study the structure and embedding of the intersection of all normalizers of maximal cyclic subgroups of a p-group.

457. Given a p-group G, let $\mathrm{b}(G) = \max\{\chi(1) \mid \chi \in \mathrm{Irr}(G)\}$. Estimate $|G : A|$, where $A < G$ is of class 2 of maximal order, in terms of $\mathrm{b}(G)$.

458. Let $M \in \Gamma_1$. Study the structure of G if $\langle x, y\rangle$ is of class ≤ 2 for all $x \in M$ and $y \in G - M$.

459. For an irregular p-group G, describe the set of all elements $x \in G$ such that $\langle x, y\rangle$ is regular for all $y \in G$.

460. For a p-group G, describe the set of all elements $x \in G$ such that $\langle x, y\rangle$ is metacyclic for all $y \in G$.

461. Let $k \in \mathbb{N}$ and $x \in G$ with $o(x) = p^k$. Study the set of all elements $y \in G$ such that $|\langle x, y\rangle| \leq p^{k+2}$.

462. Does there exist a p-group G of exponent p^{e+1}, $p > 2$, such that $\Omega_e(G) = \Omega_{e-1}(G)$ is of index p in G?

463. Gaschütz [Gas4] has proved that, for every finite group G there exists a finite group W such that, if $\Psi(W)$ is generated by all subgroups of prime orders in W, then $W/\Psi(W) \cong G$. Is it true that if G is a p-group, then there exists a p-group W with the above property?

464. Classify the p-groups G such that, for each nonabelian subgroup $H \not\leq \Phi(G)$, there is only one maximal chain connecting H with G.

465. Study the p-groups G of class $c > p$ such that $|\Omega_1(\mathrm{Z}_i(G))| = p^i$ for all $i \leq c$.

466. Find the maximum of orders of elementary abelian p-subgroups in $\mathrm{Aut}(A)$, where A is a homocyclic p-group (an abelian p-group).

467. Give a good estimate of orders of maximal abelian subgroups in groups of order p^m and exponent p.

468. Classify the non-Dedekindian p-groups G such that $|G/H^G| \leq p^2$ for all nonnormal $H < G$.

469. Classify the pairs $N \triangleleft G$ of p-groups such that $N < G'$, $|N| = p^2$ and $\mathrm{n}(G) - \mathrm{n}(G/N) = p^2 - 1$. (See §2.)

470. Classify the p-groups with ≤ 3 conjugate classes of subgroups of fixed order $p^k > p$.

471. Classify the p-groups all of whose nonnormal abelian subgroups are either cyclic or of exponent p.

472. Let a 2-group G admit an involutory automorphism α such that $C_G(\alpha)$ is cyclic or generalized quaternion. Study the structure of G in detail. (See §§48, 49.)

473. Classify the p-groups G containing a subgroup $E \cong E_{p^3}$ coinciding with its centralizer in G. (For $p = 2$, see §51 about such groups.)

474. Study the p-groups all of whose nonlinear irreducible characters are induced from (i) abelian subgroups, (ii) $\mathcal{A}_1$-subgroups.

475. Classify the p-groups of exponent $> p$ in which the normalizer of every nonnormal cyclic subgroup of order $> p$ is abelian.

476. (Ito) Let G be a p-group with k conjugacy class sizes. Is $\mathrm{dl}(G)$ bounded?

477. Study the p-groups without normal absolutely regular subgroups of order p^p.

478. Does there exist a p-group all of whose maximal subgroups are isomorphic to nontrivial standard wreath products.

479. Let $L < G$, where G is abelian. Set $\mathrm{Aut}_L(G) = \{\phi \in \mathrm{Aut}(G) \mid L^\phi = L\}$. Find $|\mathrm{Aut}_L(G)|$ in terms of invariants of G, L and G/L.

480. Let M be a normal subgroup of a 2-group G such that M has no G-invariant subgroups $\cong E_8$. Is it true that (i) there exists $n \in \mathbb{N}$ such that $\mathrm{d}(A) \leq n$ for all $A \leq M$? (ii) the derived length of M is bounded?

481. Find the number of elements of order p in a Sylow p-subgroup of $\mathrm{GL}(n, p^m)$.

482. Study the p-groups G such that, whenever $A, B \triangleleft G$ are of equal order, then $G/A \cong G/B$.

483. Describe the normal and power structures of Sylow p-subgroups of $\mathrm{Aut}(H)$, where H is a homocyclic p-group.

484. Study the p-groups G such that, whenever $M, N \in \Gamma_1$, then $M/\mathrm{Z}(M) \cong N/\mathrm{Z}(N)$.

485. Study the irregular p-groups G with $|G/\mho_1(G)| = p^{p+1}$ but $|M/\mho_1(M)| < p^{p+1}$ for all $M < G$.

486. Study the p-groups without normal subgroup of order p^{p+1} and exponent p. (For $p = 2$, see §50.)

487. Let G be abelian and let P be a Sylow p-subgroup of the holomorph of G. Express $\mathrm{d}(P)$ and $|P/\Omega_1(P)|$ in terms of G.

488. Study the p-groups G containing an element t of order p such that $\mathrm{C}_G(t) = \langle t \rangle \times M$, where M is (i) absolutely regular, (ii) of maximal class. (For $p = 2$, see §§48, 49, 51.)

489. Study the groups G of order 2^m all of whose nonabelian subgroups of order 2^r are two-generator for $r \in \{4, m-1\}$ (see §70).

490. Classify the p-groups G such that, whenever Z is a maximal cyclic subgroup of G, then $\mathrm{N}_G(Z)/Z$ is cyclic or generalized quaternion.

491. Does there exist, for each $n \in \mathbb{N}$, a p-group G such that $\mathrm{c}_1(G) = 1 + p + \cdots + p^{n-1}$ and G has exactly n conjugate classes of subgroups of order p.

492. (i) Classify the 2-groups containing exactly one subgroup of order 2^4 and exponent 4, (ii) in particular. classify the 2-groups G with $|\Omega_2(G)| = 2^4$. (Now both these problems are solved; see §52.)

493. Classify the p-groups all of whose noncyclic normal subgroups of the same order have the same minimal number of generators.

494. (Mann) Suppose that $T = T_{n_1,\dots,n_s}$ is the subgroup of G generated by G-classes of sizes $n_1, \dots, n_s$. Does there exist a function $f : \mathbb{N} \to \mathbb{N}$ such that T has derived length at most $f(s)$?

495. (Mann) Let $M < G$ be generated by all minimal classes of G. Does there exist a function $g : \mathbb{N} \to \mathbb{N}$ such that $\mathrm{cl}(M) \leq g(p)$? (Mann [Man28] has proved that $g(2) = 2$. It is proved in [Man32] that $g(p) = 3$ for $p > 2$, and this estimate is best possible.)

496. Given a p-group G, the *Mann subgroup* $\mathcal{M}(G)$ is defined as follows: $M = \langle x \in G \mid \mathrm{C}_G(x^p) = \mathrm{C}_G(x) \rangle$; then $\mathcal{M}(G) = M$ is abelian (Mann). Classify the p-groups G for which $\mathcal{M}(G)$ is cyclic.

497. Study the p-groups G such that whenever $Z < G$ is cyclic, then either Z and G' are incident or $Z \cap G' = \{1\}$.

498. Study the p-groups G with $|\mathrm{C}_G(x^p) : \mathrm{C}_G(x)| \leq p$ for all $x \in G$ of order $> p$.

499. Study the p-groups G satisfying one of the following conditions:

(i) $\mathrm{N}_G(\mho_1(H)) = \mathrm{N}_G(H)$ for all $H \leq G$ of composite exponent,

(ii) $\mathrm{N}_G(\Phi(H)) = \mathrm{N}_G(H)$ for all $H \leq G$ that are not elementary abelian,

(iii) $\mathrm{N}_G(H') = \mathrm{N}_G(H)$ for all nonabelian $H \leq G$,

(iv) $\mathrm{N}_G(\mathrm{Z}(H)) = \mathrm{N}_G(H)$ for all $H \leq G$.

500. Classify the p-groups G with $|N_G(F) : N_G(H)| \leq p$ for each characteristic $F > \{1\}$ in $H < G$.

501. Construct a p-group, $p > 2$, containing exactly one proper irregular subgroup. (For $p = 2$ this is impossible; see §76.)

502. Classify the p-groups G such that a Sylow p-subgroup of $\mathrm{Aut}(G)$ is regular.

503. Classify the p-groups all of whose subgroups of index p^2 are modular.

504. Let G be a 2-group such that $|G : H_2(G)| = 2$ (such G is said to be quasidihedral). Describe $\mathrm{Aut}(G)$.

505. Let G be an irregular p-group of exponent p^e. Study the quotient group G/T, where T is generated by all cyclic subgroups of G of order p^e.

506. Classify the irregular p-groups H of maximal class such that H is not isomorphic to a subgroup of every p-group of maximal class and order $p|H|$ (see Theorem 9.18).

507. Classify the p-groups of exponent p all of whose representation groups have exponent p.

508. Study the p-groups in which normalizers of all subgroups are normal.

509. Study the p-groups G such that, whenever $A, B < G$ are two nonnormal subgroups with $A_G = B_G$, then (i) A and B are conjugate in G, (ii) $|A| = |B|$.

510. Construct a p-group G of exponent p containing an elementary abelian subgroup E of order p^p but not containing a normal subgroup isomorphic to E.

511. Study the p-groups of maximal class all of whose maximal subgroups are characteristic (pairwise nonisomorphic).

512. Study the p-groups without nonabelian subgroups with cyclic subgroup of index p.

513. Classify the p-groups in which the centralizer of every maximal cyclic subgroup is abelian.

514. Study the p-groups G such that (i) $C_G(H) < H$ for all nonabelian $H \leq G$, (ii) the intersection of all nonabelian subgroups of G is $> \{1\}$.

515. Classify the 2-groups with $|\Omega_2(G)| = 2^5$.

516. Given an irregular p-group H, does there exist a p-group G with $H \in \Gamma_1$ and such that H is the unique irregular maximal subgroup of G?

517. Classify the irregular p-groups, all of whose regular subgroups are either absolutely regular or of exponent p.

518. Study the p-groups which are M_{p^n}-free for all $n \in \mathbb{N}$.

519. Study the non-Dedekindian p-groups G such that $HG' = H^G$ for all nonnormal $H < G$.

520. (Old problem) Estimate the rank of a 2-group without normal subgroup $\cong E_{2^n}$.

521. Classify the 2-groups in which every two distinct maximal abelian subgroups have cyclic intersection. (This is solved, for $p = 2$, in §35.)

522. Classify the p-groups G containing a self normalizing abelian subgroup A of type (p^2, p). (See §77 for $p = 2$.)

523. Study the powerful p-groups all of whose proper subgroups are powerful.

524. Suppose that $\exp(G/\mathcal{M}(G)) = p^k$ (see #496). Then G has at least $k + 1$ class sizes (see paragraph following Exercise 7.20). Consider the case where G has exactly $k + 1$ class sizes.

525. Find all $n \in \mathbb{N}$ such that every group of exponent p containing an abelian subgroup of order (index) p^n, contains a normal abelian subgroup of the same order (index).

526. Classify the 2-groups G satisfying (i) $|\Omega_3(G)| = 2^6$, (ii) $|\Omega_4(G)| \leq 2^8$.

527. Does there exist a p-group G such that, whenever $A < G$ is minimal nonabelian, then $A \cap Z(G) = \{1\}$.

528. Classify the p-groups G such that $\Omega_2(G) = E \times C$, where E is extraspecial and $C \cong C_{p^2}$.

529. It follows from Theorem 13.2(a) that an irregular p-group G has at least $p - 1$ conjugate classes of subgroups of order p and, if that number equals $p - 1$, then G is of maximal class and $|\Omega_1(G)| = p^{p-1}$. Study the irregular p-groups with exactly p conjugate classes of subgroups of order p.

530. Classify the p-groups G such that $|N_G(L) : L| \leq p$ for all maximal cyclic subgroups $L < G$.

531. Suppose that $G = M \times C_{2^k}$, where M is a 2-group of maximal class and order 2^n, $n > 3$. Describe the structure of $\mathrm{Aut}(G)$.

532. (Old problem) Classify the p-groups all of whose normal subgroups are characteristic.

533. Study the special p-groups E such that $E = \Phi(G)$ for some p-group G.

534. Study the p-groups all of whose two-generator subgroups are either absolutely regular or minimal irregular (here, in contrast to §7, a p-group is said to be minimal irregular if it is irregular but all its proper subgroups are regular).

535. Classify the p-groups G such that, whenever $R \leq G' \cap Z(G)$ is of order p, then G/R is an $\mathcal{A}_1$-group.

536. Does there exist a 2-group G with $\Omega_1(G) \cong E_8$ and $d(G) > 6$?

537. Let G be a p-group. Suppose that $M \in \Gamma_1$ is of maximal class and such that $C_G(x) \nleq M$ for all $x \in M$. Is it true that then $C_G(M) \nleq M$?

538. Study the p-groups all of whose minimal nonabelian subgroups are normal. (Then all nonabelian subgroups are also normal.)

539. Let G be a p-group such that $c_n(G) = c_n(A)$ for some abelian p-group A and all $n \in \mathbb{N}$. Is it true that then $\exp(\Omega_k(G)) \leq p^k$ for all $k \in \mathbb{N}$?

540. Study the regular p-groups G such that, whenever $H < G$ is an $\mathcal{A}_2$-subgroup, then $\exp(H) < \exp(G)$. (There is, in nonabelian regular p-group G, an $\mathcal{A}_1$-subgroup H with $\exp(H) = \exp(G)$, by Theorems 10.28 and 7.2.)

541. Study the p-groups G such that $|G : H^G| \leq p^2$ for all nonnormal $H < G$.

542. Classify the p-groups all of whose nonnormal cyclic subgroups of the same order are conjugate.

543. Study the p-groups G, $p > 2$, all of whose epimorphic images of order $\frac{1}{p}|G|$ are of maximal class.

544. Construct a p-group G such that, whenever $H \ntrianglelefteq G$, then $|G : H^G| > p$.

545. Classify the p-groups all of whose nonnormal subgroups of the same order are conjugate.

546. Study the p-groups G with unique extraspecial subgroup of given order.

547. Study the 2-groups G such that all subgroups of $\Phi(G)$ are normal in G and $\Phi(G) \nleq Z(G)$.

548. Classify the 2-groups in which the centralizer of every noncentral element is Dedekindian.

549. Study the p-groups all of whose powerful subgroups are abelian.

550. Construct a p-group in which all classes of maximal size are contained in its Φ-subgroup.

551. Classify the 2-groups G with $C_G(t) \cong C_{2^n} \times E_4$, $n > 1$, for an involution $t \in G$.

552. Find Schur multipliers of all $\mathcal{A}_2$-groups (see §71).

553. Study the p-groups G such that $G/\mathcal{M}(G)$ (see #496) is (i) of order p^2, (ii) of exponent p, (iii) extraspecial.

554. The number $r_n(G) = \max\{\log_p(|E|) \mid E \trianglelefteq G \text{ is elementary abelian}\}$ is said to be the *normal rank* of G. Let a p-group $G = AB$, where $A, B \triangleleft G$. Is it possible to estimate $r_n(G)$ in terms of A and B?

555. Suppose that a 2-group $G = ML$, where $M \triangleleft G$ is of maximal class and $L \triangleleft G$ is cyclic and $M \cap L = Z(M)$. Find (i) $c_n(G)$ for all n, (ii) $C_G(M)$.

556. Let a 2-group $G = M \times L$, where M is of maximal class and L is cyclic. (i) Describe all two-generator subgroups of G. (ii) Find $\alpha_1(G)$, the number of $\mathcal{A}_1$-subgroups in G, (iii) Find the orders of $\mathcal{A}_1$-subgroups of G.

557. Let a 2-group $G = \langle t \rangle \cdot M$, where $o(t) = 2$. Describe the structure of M in the following cases: (i) $C_M(t) \cong E_4$, (ii) $C_M(t)$ is of maximal class. (See §§48, 49, 51).

558. Classify the 2-groups G with (i) $|\Omega_2^*(G)| = 16$ (For a solution, see §55.), (ii) $|\Omega_2^*(G)| = 2^5$.

559. Study the p-groups G that contain a proper special subgroup E such that $C_G(x) < E$ for all $x \in E - Z(G)$.

560. Let G be a generalized homocyclic p-group (see §8). Is it true that $G/K_i(G)$ generalized homocyclic?

561. Classify the p-groups G such that (i) $\Omega_1(G) \cong \Sigma_{p^n}$, (ii) $\Omega_1(G) \cong E_{p^k} \times \Sigma_{p^n}$, $k \in \mathbb{N}$, (iii) $\Omega_n(G) \cong E_{p^k} \times \Sigma_{p^n}$, $n > 2$, where $\Sigma_{p^n} \in \mathrm{Syl}_p(S_{p^n})$.

562. Classify the 2-groups G such that $\Omega_1(G) = D_{2^n} * C_4$ is of order 2^{n+1}.

563. Classify the 2-groups G such that $\Omega_2^*(G) \cong Q_{2^n} \times E_{2^m}$. (This problem was solved in §75 for $m = 1$.)

564. Study the irregular p-groups all of whose sections of order p^{p+1} are regular.

565. Study the 2-groups without abelian sections of type (4, 4).

566. Classify the 2-groups G with $c_1(G) = 11$ (see Theorem 43.10).

567. Classify the 2-groups G with $\Omega_2(G) = M \times N$, where M, N are of maximal class.

568. Classify the 2-groups G with $\Omega_1(G) = D_1 \times D_2$, where D_1, D_2 are dihedral.

569. Classify the nonabelian p-groups all of whose nonnormal nonabelian subgroups have the same order.

570. Let $n > 2$. Classify the 2-groups G such that $\Omega_n^*(G)$ is metacyclic.

571. Study the 2-groups G such that there are exactly two Aut(G)-classes of involutions in G.

572. (Glauberman) Suppose that a p-group G has an (elementary) abelian subgroup of order p^n; does it have one that is normalized by all its G-conjugates? (This is true for $p > 3$, and D_{32} is a counterexample for elementary abelian case with $p = 2$. The other cases are open.)

573. Study the p-groups G such that $c_2(G)$ is not divisible by p^p. (For $c_n(G)$, $n > 3$, see Corollary 18.7.)

574. Study the 2-groups with metacyclic Frattini subgroup.

575. For $n > 1$, estimate $a_n = \min\{\alpha_1(G) \mid G \in \mathcal{A}_n\}$ and $b_n = \max\{\alpha_1(G) \mid G \in \mathcal{A}_n\}$ (see §72).

576. Study the p-groups G such that G' is the unique G-invariant subgroup of its order.

577. Classify the p-groups G with $\alpha_1(G) \leq p^2 + p + 1$. Is it true that such groups are $\mathcal{A}_n$-groups with $n < 4$. (The second problem is solved in §76.)

578. Let G is a p-group and $M \in \Gamma_1$. Suppose that, for any $H \ntrianglelefteq M$, we have $N_G(H) < M$. Study the structures of M and G.

579. Study the p-groups all of whose nonnormal subgroups are either abelian or $\mathcal{A}_1$-subgroups.

580. Classify the p-groups of exponent $p^e > p$ covered by cyclic subgroups of order p^e.

581. Give the best possible upper estimate for the class c_k of a group of order p^{p+k}, $k \leq p$, and exponent p. (We have $c_1 = p - 1$, by Theorem 9.5.)

582. Study the irregular p-groups G with absolutely regular G'.

583. Classify the p-groups G such that, whenever $H < G$ is minimal nonabelian, then G/H^G is cyclic.

584. Study the p-groups G containing a nonabelian subgroup B of order p^3 such that $C_G(B) \cong B$ (so that $BC_G(B)$ is extraspecial of order p^5).

585. Classify the p-groups, $p > 2$, with maximal elementary abelian subgroup of order p^2.

586. Let H be a p-group. Does there exist a p-group G such that $H \cong H_1 \leq \Phi(G)$ and $\exp(G) = \exp(H)$?

587. Let H be a p-group of exponent $p > 2$. Study the structure of H if there exists an extension G of C_p by H such that $\exp(G) = p^2$.

588. Let $n, k \in \mathbb{N}$ and $k \equiv 1 + p \pmod{p^2}$ be fixed. Does there exist a p-group G such that $s_n(G) = k$?

589. Let $E < G$ be extraspecial and, for every nonnormal $H < E$, we have $N_G(H) < E$. Study the structure of G.

590. Study the 2-groups G admitting a four-group V of automorphisms such that $C_G(V)$ is of order 2. (See §51.)

591. Let $n > 1$. Let G be a p-group of minimal order such that it contains a copy of every group of order $\leq p^n$. Estimate $|G|$.

592. Classify the p-groups containing a nonnormal subgroup $H \cong M_{p^n}$ such that $|N_G(H) : H| = p$.

593. Classify the p-groups G containing a subgroup H of maximal class such that $|\mathrm{N}_G(H) : H| = p$.

594. Study the irregular p-groups G with cyclic $\mho_1(G)$ and $\mathrm{cl}(G) = p$.

595. Classify the 2-groups G containing a subgroup H such that $\mathrm{N}_G(H)$ is isomorphic to a Sylow 2-subgroup of Suzuki simple group $\mathrm{Sz}(2^{2m+1})$.

596. Study the p-groups G such that for all $x, y \in G$, there exists $z \in \langle x, y\rangle'$ such that $x^{p^2} y^{p^2} = (xy)^{p^2} z^{p^2}$.

597. (N. Ito in his letter at 28/01/05) Classify the groups G of exponent p with faithful irreducible character of degree p^2 and $\mathrm{dl}(G) = 3$. (The set of such G is not empty.) Is it true that, for $n > 2$, the set of groups G of exponent p with faithful irreducible character of degree p^n and $\mathrm{dl}(G) = n + 1$ is not empty?

598. Let $H \triangleleft G$ and suppose that all elements in $G - H$ have the same order $p^k > p$. Describe all possible structures of G/H.

599. Classify the 2-groups all of whose proper characteristic subgroups have order ≤ 4.

600. Find all possible values of $\mathrm{c}_n(G)$, where G is a product of two cyclic 2-groups, $n \in \mathbb{N}$.

601. Classify the p-groups all of whose subgroups (cyclic subgroups) of order $> p$ are quasinormal.

602. Is it true that if G is a p-group such that $|\mho_1(G)| = p$, then $|G : \Omega_1(G)| \leq p^n$ for some $n \in \mathbb{N}$ depending only on p?

603. Study the p-groups G such that the normalizer of each nonnormal $\mathcal{P}_i$-subgroup $H < G$ is also a $\mathcal{P}_i$-group (three questions, for i equals 1, 2 and 3 separately; for definition of $\mathcal{P}_i$-groups, see §11).

604. Classify the p-groups G such that whenever $H < G$ is nonnormal abelian, then $\mathrm{N}_G(H)$ is an $\mathcal{A}_1$-subgroup.

605. Let G be a p-group of maximal class and let A be its abelian (elementary abelian) subgroup. Is it true that G contains a normal abelian (elementary abelian) subgroup of order $|A|$?

606. Let G be an irregular p-group of order $> p^{p+2}$. Describe the structure of G if $|\Omega_2^*(G)| \leq p^{p+2}$. (For $p = 2$, see §55.)

607. Study the p-groups, $p > 2$, all of whose $\mathcal{A}_1$-subgroups are $\cong \mathrm{M}_{p^3}$. (See #115.)

608. Study the p-groups with exactly one $\mathcal{A}_1$-subgroup of fixed order.

609. Classify the 2-groups G with $|\Omega_2(M)| \leq 2^4$ ($|\Omega_2^*(M)| \leq 2^4$) for all $M \in \Gamma_1$.

610. Classify the 2-groups containing exactly two L_2-subgroups of fixed order.

611. Is it true that the number of members of the set Γ_i, $i > 2$, that are of maximal class, is a multiple of p (see §12).

612. Given a group M of maximal class, order p^p and exponent $p > 2$, does there exist a p-group G of maximal class and order p^{p+1} such that $G/\mho_1(G) \cong M$?

613. Let $1 < s < k < m - 1$. Suppose that a group G of order p^m is such that, whenever S is a subgroup of G of order p^s, then the number of subgroups of G of order p^k containing S is at least $\varphi_{m-s,k-s}$. Study the structure of G. (See Theorem 5.17.)

614. Study the p-groups G such that, whenever $x \in A \leq G$, where A is minimal nonabelian, then $|G : C_G(x)| \leq p$.

615. Classify the 2-groups with G/E is of maximal class for $E_{2^n} \cong E \triangleleft G, n \leq 3$.

616. Set $\Omega_n^*(G) = \langle x \in G \mid o(x) = 2^n \rangle$, where G is a 2-group. Classify all possible structures of $\Omega_2^*(G), n > 1$, in the case where $c_2(G) \equiv 2 \pmod 4$.

617. Study the 2-groups G generated by three involutions.

618. Study the p-groups G such that a Sylow p-subgroup of Aut(G) is isomorphic to a Sylow p-subgroup of Aut(A) for some abelian p-group A.

619. Let $N \triangleleft G$. Describe the structure of a p-group G provided all cyclic subgroups of G that are not contained in N, are G-invariant.

620. Suppose that a p-group $A * B$ is regular. Is it true that $A \times B$ regular?

621. Study the p-groups H such that there exists a p-group G satisfying the following conditions: (i) G contains H as a subgroup of index p, (ii) $\Omega_1(G) \leq H$.

622. Classify the p-groups G possessing a subgroup H of order p such that H is contained in exactly one noncyclic subgroup of G of order p^2.

623. Give sufficient conditions for a p-group H to satisfy the following condition: $H \cong \Omega_1(G) < G$ ($H \cong \Omega_2(G) < G$) for some p-group G.

624. Give conditions sufficient for a p-group G to be such that, for every p-automorphism α of G, the semidirect product $\langle \alpha \rangle \cdot G$ has the same class as G.

625. Let a nonabelian $H = H_0 \triangleleft G$. Set $H_k = [H, G, \ldots, G]$ (k times). A normal subgroup H of a group G is *densely embedded* in a p-group G if $|H_i : H_{i+1}| = p$ for $i = 0, 1, \ldots$. Are there any restrictions on the structure of G/H, if a subgroup H of large order is densely embedded in a p-group G?

626. Let H be a group of maximal class and order p^{p+1} containing p subgroups of order p^p and exponent p. Does there exist a p-group of maximal class G such that G contains a maximal subgroup isomorphic to H?

627. Classify the 2-groups G with normal metacyclic subgroup M such that $G/M \cong E_4$.

628. Classify the p-groups all of whose nonnormal regular subgroups are absolutely regular.

629. Study the p-groups all of whose normal abelian subgroups have exponent p. (This is a partial case of #1251.)

630. Let G be 2-group and $M < G$, where M is of maximal class, $|M| = 16$. Describe the structure of G if $C_G(M) = Z(M)$.

631. Classify the p-groups G such that the normal closure C^G is metacyclic (minimal nonabelian) for every cyclic subgroup C of G.

632. Let $\mathcal{C}^p$ be the set of p-groups such that all indices of their characteristic series equal p. Is it true that for any p-group G there is $W \in \mathcal{C}^p$ such that W contains a section isomorphic with G?

633. Let G be a p-group of exponent $> p$ in which the intersection of all cyclic subgroups of order $> p$ is nontrivial. Is it true that $\mho_1(G)$ is cyclic?

634. Study the p-groups G such that $\exp(H/H_G) \leq p$ for all $H < G$.

635. Let G be a p-group. Study the set of all elements $x \in G$ such that $\langle x, y \rangle$ is p-abelian for all $y \in G$.

636. Let a p-group G be irregular. Is it true that the number of members of the set Γ_i, $i < d(G)$, that are absolutely regular, is a multiple of p?

637. Study the p-groups G such that $\Omega_1(G)$ is extraspecial (special).

638. Classify the p-groups all of whose $\mathcal{A}_2$-subgroups are metacyclic.

639. Given a p-group G of exponent $> p$, let H be a p-group with $c_k(H) = c_k(G)$, all k; then $|H| = |G|$. Estimate $|H/\mho_1(H)|$, $|\Omega_1(H)|$, $|Z(H)|$ in terms of G.

640. Suppose that p-groups G and H have the same irreducible character degrees vectors. Is it possible to estimate $|Z(H)|$ in terms of G.

641. Study the p-groups G of order $> p^{p+1}$ such that, whenever $R < G$ is of order p^p and exponent p, then $R < U < G$, where U is irregular of order p^{p+1}.

642. Classify the 2-groups G such that $G/Z(G)$ is of maximal class.

643. Classify the p-groups G satisfying $c_1(G) = 1 + p + \cdots + p^p$ and $\exp(\Omega_1(G)) > p$. (For $p = 2$, see Theorems 43.9 and 64.17.)

644. Study the irregular p-groups G such that, whenever H is a regular subgroup of G, then $H/\Omega_1(H)$ is cyclic.

645. Let G be a p-group of maximal class and order $> p^{p+1}$. Estimate the number of p-groups of maximal class with fundamental subgroup $\cong G_1$.

646. Study the p-groups all of whose maximal abelian subgroups are complemented.

647. Study the irregular p-groups G, $p > 3$, all of whose subgroups of order p^p are 2-generator.

648. Study the p-groups in which the elements of order p that are p-th powers, generate a subgroup of order p^2.

649. Study the p-groups G containing a subgroup H such that $N_G(H)$ is minimal nonabelian.

650. (Old problem) Classify, using elementary methods, the p-groups with abelian subgroup of index p.

651. Let $G = M * N$, where M and N are p-groups of maximal class. Study the p-groups which are lattice isomorphic with G.

652. Classify the 2-groups containing exactly one subgroup of order 2^5 and exponent ≤ 8. (This question is solved; see Theorem 52.13.)

653. Let G be the abelian group of type $(\alpha_1 \cdot p^{e_1}, \dots, \alpha_n \cdot p^{e_n})$. Find $d(P)$, where $P \in \mathrm{Syl}_p(\mathrm{Aut}(G))$.

654. Study the p-groups $G > \Omega_1(G)$ provided $\Omega_1(G)$ is the unique normal subgroup of its order in G.

655. Suppose that $C < G$ is self centralizing cyclic of order p^n. Is it true that $d(G)$ is bounded? If not, find all n for which the answer is 'yes'.

656. Describe all possible structures of $\Omega_1(G)$ in the case $\Omega_1(G) < G$, where G runs over all irregular p-groups of maximal class, $p > 5$.

657. Study the p-groups all of whose $\mathcal{A}_1$-subgroups of minimal order are conjugate.

658. Study the irregular p-groups G such that $|\Omega_1(G)| > p^p$, and, whenever $R < G$ is maximal regular, then $|\Omega_1(R)| \leq p^p$.

659. Does there exist a group G of order p^n and exponent p such that $\exp(\mathrm{Aut}(G)) > \exp(\mathrm{UT}(n, p))$?

660. (Inspired by papers of Zhmud) Study the p-groups G such that, for every $x, y \in G$ with $\langle x \rangle^G = \langle y \rangle^G$, we have $o(x) = o(y)$ and $|G : C_G(x)| = |G : C_G(y)|$. (Compare with #1.)

661. A p-group G is said to be k-stepped if $\Omega_k^*(G) = G$. Classify the k-stepped p-groups of exponent $> p^k$ with exactly one nonabelian k-stepped maximal subgroup.

662. Let $fc(G)$ be the number of faithful characters in $\mathrm{Irr}(G)$. Let a p-group $G = A * B$. Express $fc(G)$ in terms A, B and $A \cap B$.

663. Let $G = \mathrm{UT}(n, p) \in \mathrm{Syl}_p(\mathrm{GL}(n, p))$. Find the maximal order of subgroups of exponent p in G. Find $\max\{d(H) \mid H \leq G\}$.

664. Study the p-groups G with $\exp(\Omega_1(G)) > p$ and $\exp(\Omega_1(H)) = p$ for all $H < G$.

665. Classify the p-groups such that, whenever $H < G$ is an $\mathcal{A}_1$-subgroup, then $N_G(H)$ is an $\mathcal{A}_2$-subgroup.

666. Let G be a p-group with $c_1(G) = 1 + p + \cdots + p^n$, $n > p$. Is it possible to estimate $e_p(G)$, the number of subgroups of G of order p^p and exponent p?

667. Classify the 2-groups G with $d(G) = 3$ all of whose maximal subgroups are two-generator (see §70).

668. Given a p-group H and $n > 1$, does there exist a p-group G of order $p^n|H|$ containing a nonnormal subgroup $\cong H$?

669. Study the 2-groups G such that $N_G(H)/C_G(H) \cong P \in \mathrm{Syl}_p(\mathrm{Aut}(H))$ for all $H < G$.

670. Study the 2-groups without elementary abelian sections of order 16.

671. Study the p-groups G with special $\Phi(G)$ and G'.

672. Let $H < G$. Suppose that all abelian subgroups of G not contained in H, are cyclic. Study the structure of G.

673. Study the p-groups G such that, whenever $F < G$ is an $\mathcal{A}_1$-subgroup, there exists an $\mathcal{A}_2$-subgroup $H < G$ containing F.

674. Study the p-groups G such that $|H'| \le p$ for all $H < G$ with $|G : H| = p^2$.

675. Study the p-groups without sections of maximal class.

676. Let G be a p-group. Set $\mho^1(G) = \mho_1(G)$ and $\mho^{(i+1)}(G) = \mho_1(\mho^i(G))$ (see #107). Find the length of the $\mho^i$-series $G > \mho_1(G) > \mho^2(G) > \cdots$ for $G \in \{\Sigma_{p^n}, \mathrm{UT}(n, p)\}$.

677. Suppose that G is a group of order p^m, $m > n \ge 3$. Find all possible numbers of normal subgroups N of G such that G/N is of maximal class and order p^n.

678. Study the irregular p-groups without characteristic subgroup of order $> p^{p-1}$ and exponent p.

679. Classify the p-groups covered by subgroups of maximal class.

680. Study the p-groups G, $p > 2$, such that $H_p(G) = E \times M$, where E is elementary abelian and M is of maximal class.

681. Study the p-groups G such that every $H < G$ is a nontrivial direct product, unless H is cyclic or minimal nonabelian.

682. Suppose that G and G_0 are p-groups of the same order containing, for each n, the same number of conjugacy classes of size n. Is there connection between (i) $\mathrm{cl}(G)$ and $\mathrm{cl}(G_0)$, (ii) $|G : G'|$ and $|G_0 : G_0'|$.

683. Classify the p-groups G such that $\mathrm{Aut}(G) \cong \mathrm{GL}(n, p)$ for some $n \in \mathbb{N}$.

684. Study the p-groups G such that $\mathrm{Aut}(G)$ is a nontrivial direct product.

685. Classify the 2-groups G such that $\mathrm{Aut}(G) \cong \mathrm{Aut}(M)$, where M is a 2-group of maximal class.

686. Study the p-groups all of whose $\mathcal{A}_2$-subgroups are two-generator.

687. Let a p-group $G = \langle x \rangle \cdot M$, where $M \in \Gamma_1$. Estimate $\mathrm{cl}(G)$ in terms of the structure of M only. Find all M such that $\mathrm{cl}(G) < \mathrm{cl}(M \operatorname{wr} \mathrm{C}_p)$ for all extensions G?

688. Study the p-groups all of whose two noncyclic subgroups are permutable. In particular, study the p-groups all of whose noncyclic subgroups are quasinormal.

689. Classify the p-groups G, $p > 2$, satisfying $G > \mathrm{H}_p(G) \cong \Sigma_{p^n} \in \mathrm{Syl}_p(\mathrm{S}_{p^n})$.

690. Study the p-groups G such that $\Phi(H) = H'$ for all nonabelian $H \leq G$.

691. Study the p-groups G such that $H\mathrm{Z}(G)/\mathrm{Z}(G)$ is metacyclic for all $H \in \Gamma_1$.

692. (L. K. Hua and H. F. Tuan; reported by M. Y. Xu) Let G be a group of order p^m and $n \leq m$. Is it true that, for $p > 2$, a residue of $\mathrm{s}_n(G) \pmod{p^3}$ is one of the following numbers: $1, 1+p, 1+p+p^2, 1+p+2p^2$? ('Yes, for $\exp(G) = p$; see Theorem 5.9.)

693. Let G and G_0 be p-groups such that for each n, $\mathrm{Irr}(G)$ and $\mathrm{Irr}(G_0)$ contain the same number of characters of degree n. (i) Is it true that G_0 is of maximal class if G is? (ii) Is there connection between $\mathrm{cl}(G)$ and $\mathrm{cl}(G_0)$?

694. Is it true that if B is a regular p-group, $p > 2$, such that $A \times B$ is regular for all regular p-groups A, then $\exp(B') \leq p$? (The converse is true, according to O. Grün.)

695. Let G be a p-group of maximal class and order $> p^{p+1}$, $p > 2$, with $\mathrm{c}_1(G) = 1 + p + \cdots + p^{p-2} + kp^p$ for some $k \geq 0$ (see §12). Find all possible values of k.

696. Classify the p-groups G, $p > 2$, with $\mathrm{c}_2(G) = p^p$ for a fixed $n > 1$. (For $p = n = 2$, see §53.)

697. Suppose that a group G of order p^m has an automorphism of order p^{m-k}, $k \in \mathbb{N}$. Is it true that if m is large, then G has a cyclic subgroup of index p^k?

698. For a subset M of a group G, denote $\mathrm{k}_G(M)$ the number of G-classes having nonempty intersection with M. Let H be a Sylow p-subgroup of the holomorph of G. Classify all p-groups G such that $\mathrm{k}_H(G) = \mathrm{k}_G(G) (= \mathrm{k}(G))$.

699. Study the p-groups, $p > 2$, all of whose nonabelian two-generator subgroups are either minimal nonabelian or of maximal class.

700. Describe automorphism groups of minimal nonabelian p-groups.

Bibliography

[Alp1] J. L. Alperin, On a special class of regular p-groups, *Trans. Amer. Math. Soc.* **106** (1963), 77–99.

[Alp2] J. L. Alperin, Large abelian subgroups of p-groups, *Trans. Amer. Math. Soc.* **117** (1965), 10–20.

[Alp3] J. L. Alperin, Centralizers of abelian normal subgroups of p-groups, *J. Algebra* **1** (1964), 110–113.

[AlpG] J. L. Alperin and G. Glauberman, Limits of abelian subgroups of finite p-groups, *J. Algebra* **203** (1998), 533–566.

[AlpK] J. L. Alperin and Kuo Tzee-Nan, The exponent and the projective representations of a finite group, *Illinois J. Math.* **11** (1967), 410–414.

[AK1] B. Amberg and L. Kazarin, On the rank of a finite product of two p-groups, in: *Groups – Korea 94*, pp. 1–8, W. de Gruyter, Berlin, 1995.

[AK2] B. Amberg and L. Kazarin, On the rank of a product of two finite p-groups and nilpotent p-algebras, *Comm. Algebra* **27(8)** (1999), 3895–3907.

[Arg] D. E. Arganbright, The power-commutator structure of finite p-groups, *Pacific J. Math.* **29** (1969), 11–17.

[Bae1] R. Baer, Groups with abelian central quotient groups, *Trans. Amer. Math. Soc.* **44** (1938), 357–386.

[Bae2] R. Baer, Partitionen endlicher Gruppen, *Math. Z.* **75** (1961), 333-372.

[Bae3] R. Baer, Gruppen mit Hamiltonschem Kern, *Comp. Math.* **2** (1935), 241–246.

[Bae4] R. Baer, Group elements of prime power index, *Trans. Amer. Math. Soc.* **75** (1953), 20–47.

[BM1] C. Baginski and I. Malinowska, On groups of order p^n with automorphism of order p^{n-2}, *Demonstr. Math.* **23** (1996), 565–575.

[BM2] C. Baginski and I. Malinowska, On finite 2-groups with many involutions, *Arch. Math.* **81** (2003), 241–244.

[Bal] F. Balogh, Finite groups in which different conjugacy classes have different cardinalities, *J. Algebra* **181** (1996), 286–287.

[Ban] W. Bannusher, Über Gruppen mit genau zwei irreduziblen Charaktergraden I, II, *Math. Nachr.* **154** (1991), 253–563.

[BarI] Y. Barnea and I. M. Isaacs, Lie algebras with few centralizer dimensions, *J. Algebra* **259** (2003), 284–299.

[Bec] H. Bechtell, Frattini subgroups and Φ-central groups, *Pacific J. Math.* **18** (1966), 15–23.

[Bei1] B. Beisiegel, Semi-extraspezielle p-Gruppen, *Math. Z.* **156** (1976), 247–254.

[Bei2] B. Beisiegel, Die Automorphismengruppen homozyklischer p-Gruppen, *Arch. Math.* **29**, 4 (1977), 363–366.

[Ben1] H. A. Bender, A determination of the groups of order p^5, *Ann. Math.* **(2) 29** (1927), 61–72.

[Ben2] H. A. Bender, Determination of all prime power groups containing only one invariant subgroup of every index which exceeds this prime number, *Trans. Amer. Math. Soc.* **26**, 4 (1924), 427–434.

[BenG] H. Bender and G. Glauberman, *Local Analysis for the Odd Order Theorem*, London Math. Soc. Lect. Note Series 188, Cambridge Univ. Press, Cambridge, 1994.

[BKN] T. R. Berger, L. G. Kovacs and M. F. Newman, Groups of prime power order with cyclic Frattini subgroup, *Nederl. Akad. Wetensch. Indag. Math.* **42** (1980), no. 1, 13–18.

[Ber0] V. G. Berkovich, Groups of order p^n possessing an automorphism of order p^{n-1}, *Algebra i Logika* **9** (1970), no. 1, 4–8 (in Russian).

[Ber1] Y. Berkovich, On p-groups of finite order, *Siberian Math. Zh.* **9** (1968), 1284–1306 (in Russian).

[Ber2] Y. Berkovich, Subgroups, normal divisors and epimorphic images of a finite p-group, *Soviet Math. Dokl.* **10** (1969), 878–881.

[Ber3] Y. Berkovich, A generalization of theorems of Ph.Hall and Blackburn and an application to non-regular p-groups, *Math. USSR Izv.* **35** (1971), 815–844.

[Ber4] Y. Berkovich. Some consequences of Maschke's theorem, *Algebra Coll.* **5, 2** (1998), 143–158.

[Ber5] Y. Berkovich, Alternate proofs of some basic theorems of finite group theory, *Glas. Mat.* **40** (2005), no. 2, 207–233.

[Ber6] Y. Berkovich, Finite metacyclic groups, *Soobsch. Akad. Nauk Gruzin. SSR* **68** (1972), 539–542 (in Russian).

[Ber7] Y. Berkovich, On finite metacyclic groups, in: *Structural Properties of Algebraic Systems*, pp. 12–19, Nalchik, 1985 (in Russian).

[Ber8] Y. Berkovich, Relations between some invariants of finite solvable groups, *Soobsch. Akad. Nauk Gruzin. SSR* **123** (1986), no. 3, 469–472 (in Russian).

[Ber9] Y. Berkovich, On subgroups of finite p-groups, *J. Algebra* **224** (2000), 198-240.

[Ber10] Y. Berkovich, Alternate proofs of two theorems of Philip Hall on finite p-groups, and related results, *J. Algebra* **294** (2005), no- 2, 463–477.

[Ber11] Y. Berkovich, On abelian subgroups of p-groups, *J. Algebra* **199** (1998), 262–280.

[Ber12] Y. Berkovich, On the order of the commutator subgroup and the Schur multiplier of a finite p-group, *J. Algebra* **144** (1991), no. 2, 269–272.

[Ber13] Y. Berkovich, On the number of subgroups of given order in a finite p-group of exponent p, *Proc. Amer. Math. Soc.* **109** (1990), no. 4, 875–879.

[Ber14] Y. Berkovich, On the number of elements of given order in a finite p-group, *Israel. J. Math.* **73** (1991), 107–112.

[Ber15] Y. Berkovich, On the number of subgroups of given order and exponent p in a finite irregular p-group, *Bull. London Math. Soc.* **24** (1992), 259–266.

[Ber16] Y. Berkovich, Counting theorems for finite p-groups, *Arch. Math.* **59** (1992), 215–222.

[Ber17] Y. Berkovich, On the number of solutions of equation $x^{p^k} = a$ in a finite p-group, *Proc. Amer. Math. Soc.* **116** (1992), no. 3, 585–590.

[Ber18] Y. Berkovich, On p-subgroups of finite symmetric and alternating groups, Contemporary Mathematics **93** (1989), 67–76.

[Ber19] Y. Berkovich, On the number of solutions of equation $x^{p^k} = 1$ in a finite group, Rendiconti Lincei, *Matematica e applicazioni Ser. 9* **6** (1995), 5–12.

[Ber20] Y. Berkovich, On the number of subgroups of a given structure in a finite p-group, *Arch. Math.* **63** (1994), 111–118.

[Ber21] Y. Berkovich, Normal subgroups in a finite group, *Soviet Math. Dokl.* **9** (1968), 1117–1120.

[Ber22] Y. Berkovich, Short proofs of some basic characterization theorems of finite p-group theory, *Glas. Mat.* **41**(61) (2006), 239–258.

[Ber23] Y. Berkovich, On an irregular p-group, *Siberian J. Math.* **12** (1971), no. 4, 907–911.

[Ber24] Y. Berkovich, On subgroups and epimorphic images of finite p-groups, *J. Algebra* **248** (2002), 472–553.

[Ber25] Y. Berkovich, Selected Topics of Finite Group Theory, Parts I, II, in preparation.

[Ber26] Y. Berkovich, Nonnormal and minimal nonabelian subgroups of a finite group, *Glas. Mat.*, to appear.

[Ber27] Y. Berkovich, Hall chains in finite p-groups, *Israel J. Math.*, to appear.

[Ber28] Y. Berkovich, Alternate proofs of characterization theorems of Miller and Zvonimir Janko on p-groups, and some related results, *Glas. Math.* **43**(62) (2007), 319–343.

[Ber29] Y. Berkovich, On the metacyclic epimorphic images of a finite p-group, *Glas. Mat.* **41**(61) (2007), 259–269.

[Ber30] Y. Berkovich, Finite p-groups with few minimal nonabelian subgroups. With an appendix by Z. Janko, *J. Algebra* **297** (2006), no. 1, 62–100.

[Ber31] Y. Berkovich, Alternate proofs of some basic theorems of finite group theory, *Glas. Mat.* **40** (2005), no. 2, 207–233.

[Ber32] Y. Berkovich, A property of p-groups of order $p^{p(e+1)}$ and exponent p^e, *Glas. Mat.* **40**(60) (2005), 51–58.

[Ber33] Y. Berkovich, p-groups in which some subgroups are generated by elements of order p, submitted.

[BFP] Y. Berkovich, G. Freiman and C. E. Praeger, Small squaring and cubing properties for finite groups, *Bull. Aust. Math. Soc.* **44** (1991), no. 3, 429–450.

[BIK] Y. Berkovich, I. M. Isaacs and L. S. Kazarin, Distinct monolithic character degrees, *J. Algebra* **216** (1999), 448–480.

[BJ1] Y. Berkovich and Z. Janko, Structure of finite p-groups with given subgroups, *Contemporary Mathematics* **402** (2006), 13–93.

[BJ2] Y. Berkovich and Z. Janko, *Groups of Prime Power Order*, Volume 2, Walter de Gruyter, Berlin, 2008.

[BJ3] Y. Berkovich and Z. Janko, On subgroups of finite p-groups, *Israel J. Math.*, to appear.

[BerM] Y. Berkovich and A. Mann, On sums of degrees of irreducible characters, *J. Algebra* **199** (1998), 646–665.

[BZ] Y. Berkovich and E. M. Zhmud, *Characters of Finite Groups*, Parts 1, 2, Translations of Mathematical Monographs 172, 181, American Mathematical Society, Providence, RI, 1998, 1999.

[Bert] E. A. Bertram, Large centralizers in finite solvable groups, *Israel J. Math.* **47** (1984), 335–344.

[BEOB1] H. U. Besche, B. Eick and E. A. O'Brien, The groups of order at most 2000, *Electronic Research Announcements of AMS* **7** (2001), 1–4.

[BEOB2] H. U. Besche, B. Eick and E. A. O'Brien, A millennium project: Constructing small groups, Internat. *J. Algebra and Comp.* **12** (2002), 623–644.

[Bey] F. R. Beyl, The Schur multiplicator of metacyclic groups, *Proc. Amer. Math Soc.* **40** (1973), 413–418.

[BeyT] F. R. Beyl and J. Tappe, *Group Extensions, Representations and the Schur Multiplicator*, Lect. Notes in Math. 958, Springer, Berlin, 1982.

[Bla1] N. Blackburn, On prime-power groups in which the derived group has two generators, *Proc. Cambridge Phil. Soc.* **53** (1957), 19–27.

[Bla2] N. Blackburn, On prime power groups with two generators, *Proc. Cambridge Phil. Soc.* **54** (1958), 327–337.

[Bla3] N. Blackburn, On a special class of p-groups, *Acta Math.* **100** (1958), 45–92.

[Bla4] N. Blackburn, Über das Produkt von zwei zyklischen Gruppen, *Math. Z.* **68** (1958), 422–427.

[Bla5] N. Blackburn, Generalizations of certain elementary theorems on p-groups, *Proc. London Math. Soc.* **11** (1961), 1–22.

[Bla6] N. Blackburn, Automorphisms of finite p-groups, *J. Algebra* **3** (1966), 28–29.

[Bla7] N. Blackburn, Finite groups in which the nonnormal subgroups have nontrivial intersection, *J. Algebra* **3** (1966), 30–37.

[Bla8] N. Blackburn, Note on a paper of Berkovich, *J. Algebra* **24** (1973), 323–334.

[Bla9] N. Blackburn, Some homology groups of wreath products, *Illinois J. Math.* **16** (1972), 116–129.

[Bla10] N. Blackburn, Über Involutionen in 2-Gruppen, *Arch. Math.* **35** (1980), 75–78.

[Bla11] N. Blackburn, The derived group of a 2-group, *Math. Proc. Camb. Phil. Soc.* **101** (1987), 193–196.

[Bla12] N. Blackburn, On centralizers in p-groups, *J. London Math. Soc. (2)* **9**, (1975), 478–482.

[Bla13] N. Blackburn, Groups of prime-power order having an abelian centralizer of type $(r, 1)$, *Mh. Math.* **99** (1985), 1–18.

[Bla14] N. Blackburn, Conjugacy in nilpotent groups, *Proc. Amer. Math. Soc.* **16** (1965), 143–148.

[Bla15] N. Blackburn, Nilpotent groups in which the derived group has two generators, *J. London Math. Soc.* **35** (1960), 33–35.

[BlaDM] N. Blackburn, M. Deaconescu and A. Mann, Equilibrated groups, *Proc. Cambridge Philos. Soc.* **120** (1996), no. 2, 579–588.

[BlaEs] N. Blackburn, A. Espuelas, The power structure of metabelian p-groups, *Proc. Amer. Math. Soc.* **92** (1984), 478–484.

[BlaEv] N. Blackburn and L. Evens, Schur multipliers of p-groups, *J. reine angew. Math.* **309** (1979), 100–113.

[Blac1] S. R. Blackburn, Enumeration within isoclinism classes of groups of prime power order, *J. London Math. Soc.* **50** (1994), 293–304.

[Blac2] S. R. Blackburn, Groups of prime power order with derived subgroup of prime order, *J. Algebra* **219** (1999), 625–657.

[BDM] H. F. Blichfeldt, L. E. Dickson and G. A. Miller, *Theory and Applications of Finite Groups*, New York, Stechert, 1938.

[BosI] N. Boston and I. M. Isaacs, Class numbers of p-groups of given order, *J. Algebra* **279** (2004), 810–819.

[BosW] N. Boston and J. L. Walker, 2-groups with few conjugacy classes, *Proc. Edinburgh Math. Soc. (2)* **43** (2000), no. 1, 211–217.

[BozJ1] Z. Bozikov and Z. Janko, Finite 2-groups G with $|\Omega_3^*(G)| = 2^5$, *J. Group Theory* **7** (2004), 65–73.

[BozJ2] Z. Bozikov and Z. Janko, On a question of N. Blackburn about finite 2-groups, Israel J. Math. 147 (2005), 329–331.

[BozJ3] Z. Bozikov and Z. Janko, On finite p-groups in which the centralizer of each element is a normal subgroup, manuscript.

[BozJ4] Z. Bozikov and Z. Janko, Finite p-groups all of whose nonmetacyclic subgroups are generated by involutions, *Arch. Math.* **90** (2008), 14–17.

[BozJ5] Z. Bozikov and Z. Janko, A complete classification of finite p-groups all of whose noncyclic subgroups are normal, *Math. Z.*, to appear.

[Bran] A. Brandis, Beweis einer Satzes von Alperin und Kuo Tzee-Nan, *Illinois J. Math.* **13** (1969), 275.

[Bro] J. Brodkey, A note on finite groups with an abelian Sylow group, *Proc. Amer. Math. Soc.* **14** (1963), 132–133.

[Bur1] W. Burnside, *The Theory of Groups of Finite Order*, Dover. Publ., N.Y., 1955.

[Bur2] W. Burnside, On some properties of groups whose orders are powers of primes I, *Proc. London. Math. Soc. (2)* **11** (1912), 225–245; II. ibid **13** (1913), 6–12.

[Bur3] W. Burnside, On the outer automorphisms of a group, *Proc. London Math. Soc. (2)* **11** (1913), 40–42.

[Bur4] W. Burnside, On an unsettled question in the theory of discontinuous groups, *Quarterly J. Math.* **33** (1902), 230–238.

[Bus] K. Buzasi, On the structure of the wreath product of a finite number of cyclic groups of prime order, *Publ. Math. Debrecen* **15** (1968), 107–129.

[Ca] A. Caranti, Projectivity of p-groups of maximal class, *Rend. Sem. Mat. Padova* **61** (1979), 393–404 (in Italian).

[Cha] E. I. Chankov, p-groups with five nonlinear irreducible characters, manuscript.

[Che] Y. Cheng, On finite p-groups with cyclic commutator subgroups, *Arch. Math.* **39** (1982), 295–298.

[CHe] D. Chillag and M. Herzog, Finite groups with almost distinct character degrees, to appear.

[CI] M. D. E. Conder and I. M. Isaacs, Derived subgroups of products of an abelian and a cyclic subgroup, *J. London Math. Soc.* **69** (2004), no. 2, 333–348.

[Con] S. B. Conlon, p-groups with an abelian maximal subgroup and cyclic centre, *J. Aust. Math. Soc. Ser. A* **22** (1976), no. 2, 221–233.

[Cor] G. Corsi Tani, Automorphisms fixing every normal subgroup of a p-group, *Bull. Un. Mat. Ital. B (6)* 4 (1985), 245–252.

[CHa] J. Cossey and T. Hawkes, Sets of p-powers as conjugacy classes sizes, *Proc. Amer. Math. Soc.* **128** (2000), 49–51.

[CHM] J. Cossey, T. Hawkes and A. Mann, A criterion for a group to be nilpotent, *Bull. London Math. Soc.* **24** (1992), 267–270.

[Cut] G. Cutolo, On a question about automorphisms of finite p-groups, *J. Group Theory* **9** (2006), 231–250.

[Dad] E. C. Dade, Products of orders of centralizers, *Math. Z.* **96** (1967), 223–225.

[DS] R. Dark and C. Scoppola, On Camina groups of prime power order, *J. Algebra* **181** (1996), 787–802.

[Dav1] R. M. Davitt, The automorphism group of finite p-abelian p-groups, *Illinois J. Math.* **16** (1972), 76–85.

[Dav2] R. M. Davitt, The automorphism group of a finite metacyclic p-group, *Proc. Amer. Math. Soc.* **25** (1970), 876 879.

[Dav3] R. M. Davitt, On the automorphism group of a finite p-group with a small central quotient, *Can. J. Math.* **32** (1980), 1168–1176.

[DO] R. M. Davitt and A. D. Otto, On the automorphism group of a finite p-group with the central quotient metacyclic, *Proc. Amer. Math. Soc.* **30** (1971), 467–472.

[DO2] R. M. Davitt and A. D. Otto, On the automorphism group of a finite modular p-group, *Proc. Amer. Math. Soc.* **35** (1972), 399–404.

[Ded] R. Dedekind, Über Gruppen, deren sämtliche Teiler Normalteiler sind, *Math. Ann.* **48** (1897), 548–561.

[Del] P. Deligne, Congruences sur le nombre de sous-groupes d'ordre p^k dans un groupe fini, *Bull. Soc. Math. Belg.* **18** (1966), 129–132.

[Die] J. Dietz, Automorphisms of p-groups given as cyclic-by-elementary abelian extensions, *J. Algebra* **242**, (2001), 417–432.

[DdSMS] J. Dixon, M. P. F. du Sautoy, A. Mann and D. Segal, *Analytic Pro-p-Groups*, London Math. Soc. Lecture Notes Series 157, Cambridge University Press, 1991.

[Dol] S. Dolfi, Arithmetical conditions on the length of the conjugacy classes of a finite group, *J. Algebra* **174** (1995), 753–771.

[Dra] S. V. Draganyuk, On the structure of finite primary groups all 2-maximal subgroups of which are abelian, in: *Complex Analysis, Algebra and Topology*, pp. 42–51, Kiev, 1990.

[Eas] T. E. Easterfield, The orders of products and commutators in prime-power groups, *Proc. Cambridge Philos. Soc.* **36** (1940), 14–26.

[ENOB] B. Eick, M. F. Newman, and E. A. O'Brien. The class-breadth conjecture revisited. *J. Algebra* **300** (2006), 384–393.

[ELGOB] B. Eick, C. R. Leedham-Green and E. A. O'Brien, Constructing automorphism groups of p-groups, *Comm. Algebra* **30** (2002), no. 5, 2271–2295.

[Fal] K. Faltings, Automorphismengruppen endlicher abelscher p-Gruppen, in: *Studies on Abelian Groups* (Symposium, Montpellier, 1967), pp. 101–119, Springer, Berlin, 1968.

[Fan] Y. Fan, A characterization of elementary abelian p-groups by counting subgroups, *Math. Practice Theory* **1** (1988), 63–65 (in Chinese); MR 89h: 20030.

[Fei] W. Feit, Theory of finite groups in the twentieth century, *Amer. Math. Heritage: Algebra and Applied Math.* **13** (1881), 37–60.

[FT] W. Feit and J. G. Thompson, Solvability of groups of odd order, *Pacific J. Math.* **13** (1963), 775–1029.

[Fer] G. A. Fernandez-Alcober, An introduction to finite p-groups: regular p-groups and groups of maximal class, *Math. Contemp.* **20** (2001), 155-226.

[F-AM] G. A. Fernandez-Alcober and A. Moreto, Groups with two extreme character degrees and their normal subgroups, *Trans. Amer. Math. Soc.* **353** (2001), 2271–2292.

[Fitt] H. Fitting, Die Gruppe der zentralen Automorphismen einer Gruppe mit Hauptreihe, *Math. Ann.* **114** (1937), 355–372.

[FMHOBS] J. Flynn, D. MacHale, E. A. O'Brien and R. Sheely, Finite groups whose automorphism groups are 2-groups, *Proc. Roy. Ir. Acad.* **94A (2)** (1994), 137–145.

[Fom] A. N. Fomin, Finite 2-groups in which the centralizer of a certain involution is of order 8, *Ural Gos. Univ. Mat. Zap.* **8** 1972), no. 3, 122–132 (in Russian).

[For] C. E. Ford, Characters of p-groups, *Proc. Amer. Math. Soc.* **101** (1987), 595–601.

[FrT] J. S. Frame und O. Tamaschke, Über die Ordnungen der Zentralizatoren der Elemente in endlichen Gruppen, *Math. Z.* **83** (1964), 41–45.

[FS] G. Frobenius and L. Stickelberger, Über Gruppen von vertauschbaren Elementen, *J. reine angew. Math.* **86** (1879), 217–262.

[Gag] S. M. Gagola, Jr., A character theoretic condition for $\mathrm{F}(G) > 1$, *Comm. Algebra* **33** (2005), 1369–1382.

[GL] S. M. Gagola and M. L. Lewis, A character theoretic condition characterizing nilpotent groups, *Comm. Algebra* **27 (3)** (1999), 1053–1056.

[Gal] J. A. Gallian, Finite p-groups with homocyclic central factors, *Can. J. Math.* **26** (1974), 636–643.

[Gas1] W. Gaschütz, Über die Φ-Untergruppe endlicher Gruppen, *Math. Z.* **58** (1953), 160–170.

[Gas2] W. Gaschütz, Kohomologische Trivialitäten und äußere Automorphismen von p-Gruppen, *Math. Z.* **88** (1965), 432–433.

[Gas3] W. Gaschütz, Nichtabelsche p-Gruppen besitzen äußere p-Automorphismen, *J. Algebra* **4** (1966), 1–2.

[GNY] W. Gaschütz, J. Neubüser and Ti Yen, Über den Multiplikator von p-Gruppen, *Math. Z.* **100** (1970), 93–96.

[GMMPS] N. Gavioli, A. Mann, V. Monti, A. Previtali and C. Scoppola, Groups of prime power order with many conjugacy classes, *J. Algebra* **202** (1998), 129–141.

[GMS] N. Gavioli, A. Mann, and C. Scoppola, Two applications of the Hudges subgroup of finite group, in: *Ischia Group Theory 2006*, pp. 138–146, World Scientific Books, Singapore, 2007.

[Gil] J. D. Gillam, A note on finite metabelian p-groups, *Proc. Amer. Math. Soc.* **25** (1970), 189–190.

[Gla1] G. Glauberman, Large abelian subgroups of finite p-groups, *J. Algebra* **196** (1997), 301–338.

[Gla2] G. Glauberman, Large abelian subgroups of groups of prime exponent, *J. Algebra* **237** (2001), 735–768.

[Gla3] G. Glauberman, On Burnside's other $p^a g^b$-theorem, *Pacific J. Math.* **56**, (1975), 469–476.

[Gla4] G. Glauberman, Isomorphic subgroups of finite p-groups, I, II, *Canad. J. Math.* 23 (1971), 983–1022, 1023–1039.

[Gla5] G. Glauberman, Large subgroups of small class in finite p-groups, *J. Algebra* **272** (2004), 128–153.

[Gla6] G. Glauberman, Centrally large subgroups of finite p-groups, *J. Algebra* **300** (2006), 480–508.

[Gla7] G. Glauberman, Existence of normal subgroups in finite p-groups, *J. Algebra* **319** (2008), 800–805.

[Gol] Y. A. Golfand, On groups all of whose subgroups are nilpotent, *Dokl. Akad. Nauk SSSR* **125** (1948), 1313–1315.

[Gor1] D. Gorenstein, *Finite Groups*, Harper and Row, N.Y., 1968.

[Gor2] D. Gorenstein, On a theorem of Philip Hall, *Pacific J. Math.* **19** (1966), 77–80.

[Gor3] D. Gorenstein (Editor), *Reviews on Finite Groups*, Amer. Math. Soc., Providence, RI, 1974.

[GLS] D. Gorenstein, R. Lyons and R. Solomon, *The Classification of the Finite Simple Groups*, Part 1, Chapter G: General Group Theory, AMS, Providence, RI, 1995.

[Gre] J. A. Green, On the number of automorphisms of a finite group, *Proc. Roy. Soc. London A* **237** (1956), 574–580.

[Gro] F. Gross, 2-automorphic 2-groups, *J. Algebra* **40** (1976), 348–353.

[Gro2] F. Gross, Automorphisms of permutational wreath products, *J. Algebra* **117** (1988), 472–493.

[Grov] L. C. Grove, *Groups and Characters*, Pure and Applied Mathematics, John Wiley and Sons, New York, 1997.

[Grov1] J. R. J. Groves, On minimal irregular p-groups, *J. Aust. Math. Soc.* **16** (1973), 78–89.

[Grov2] J. R. J. Groves, On direct products of regular p-groups, *Proc. Amer. Math. Soc.* **37** (1973), 377–379.

[Grov3] J. R. J. Groves, Some criteria for the regularity of a direct product of regular p-groups, *J. Aust. Math. Soc. Ser. A* **24** (1977), 35–49.

[Gr1] O. Grün, Beiträge zur Gruppentheorie. V, Über endliche p-Gruppen, *Osaka Math. J.* **5** (1953), 117–146.

[Gr2] O. Grün, Über das direkte Produkt regulärer p-Gruppen, *Arch. Math.* **5** (1954), 241–243.

[Gr3] O. Grün, Eine obere Grenze für die Klasse einer h-stufigen p-Gruppe, *Abh. Math. Sem. Univ. Hamburg* **21** (1957), 90–91.

[Gr4] O. Grün, Einige Sätze über Automorphismen abelscher p-Gruppen, *Abh. Math. Sem. Univ. Hamburg* **24** (1960), 54–58.

[HalM] M. Hall, *The Theory of Groups*, Macmillan, New York, 1959.

[HS] M. Hall and J. K. Senior, *On Groups of Order 2^n, ($n \leq 6$)*, Macmillan, New York, 1964.

[Hal1] P. Hall, A contribution to the theory of groups of prime power order, *Proc. London Math. Soc.* **36** (1933), 29–95.

[Hal2] P. Hall, On a theorem of Frobenius, *Proc. London Math. Soc.* **40** (1936), 468–501.

[Hal3] P. Hall, The classification of prime power groups, *J. reine angew. Math.* **182** (1940), 130–141.

[Hal4] P. Hall, *Nilpotent Groups*, Can. Math. Congr., Alberta, 1957.

[Hal5] P. Hall, On groups of automorphisms, *J. Math.* **182** (1940), 194–204.

[Hal6] P. Hall, Some sufficient conditions for a group to be nilpotent, *Illinois J. Math.* **2** (1958), 787–801.

[Hal7] P. Hall, Verbal and marginal subgroups, *J. reine angew. Math.* **182** (1940), 156–167.

[HH] P. Hall and G. Higman, On the p-length of p-solvable groups and the reduction theorems for Burnside's problem, *Proc. London Math. Soc.* **(3)** (1956), 1–42.

[Han] A. Hanaki, A condition on lengths of conjugacy classes and character degrees, *Osaka J. Math.* **33** (1996), 207–216.

[Har] K. Harada, On some 2-groups of normal 2-rank 2, *J. Algebra* **20** (1972), no. 1, 90–93.

[Harr] M. E. Harris, On decomposing an abelian p-group under a p'-operator group, *Algebra Colloquium* **7** (2000), 291–294.

[Haw] T. O. Hawkes, On the automorphism group of a 2-group, *Proc. London Math. Soc.* **26** (1973), 207–225.

[HMH] P. Hegarty and D. MacHale, Two-groups in which an automorphism inverts precisely half of elements, *Bull. London Math. Soc.* **30** (1998), 129–135.

[Hei1] H. Heineken, Gruppen mit kleinen abelschen Untergruppen, *Arch. Math.* **29** (1977), 20–31.

[Hei2] H. Heineken, Über ein Levisches Nilpotenzkriterium, *Arch. Math.* **12** (1961), 176–178.

[Hei3] H. Heineken, Nilpotente Gruppen, deren sämtliche Normalteiler charakteristisch sind, *Arch. Math.* **33** (1979/80), 497–503.

[Hei4] H. Heineken, Bounds for the nilpotency class of a group, *J. London Math. Soc.* **37** (1962), 456–458.

[HL1] H. Heineken and H. Liebeck, On p-groups with odd order automorphism groups, *Arch. Math.* **24** (1973), 465–471.

[Hel1] G. T. Helleloid, A survey of automorphism groups of finite p-groups, arXiv.mathGR/0610294 v2 25 Oct 2006, 1–20.

[Hel2] G. T. Helleloid, *Automorphism groups of finite p-groups: structure and applications*, PhD Thesis, Stanford Univ., 2007.

[HelM] G. T. Helleloid and U. Martin, The automorphism group of a finite p-group is almost always is a p-group, *J. Algebra* **312** (2007), 294–329 (see also arXiv.mathGR/0602039 v5 Oct 2006, 1–38).

[Herz] M. Herzog, Counting group elements of order p modulo p^2, *Proc. Amer. Math. Soc.* **66** (1977), 247–250.

[HK] M. Herzog and G. Kaplan, Large cyclic subgroups contain non-trivial normal subgroups, *J. Group Theory* **4** (2001), 247–253.

[HKL] M. Herzog, G. Kaplan and A. Lev, On the commutator and the center of finite groups, *J. Algebra* **278** (2004), 494–501.

[HLMM] M. Herzog, P. Longobardi, M. Maj and A. Mann, On generalized Dedekind groups and Tarski super monsters, *J. Algebra* **226** (2000), 690–613.

[Het] L. Hethelyi, On powerful normal subgroups of a p-group, *Monatsh. Math.* **130** (2000), 201–209.

[HL] L. Hethelyi and L. Levai, On elements of order p in powerful p-groups, *J. Algebra* **270** (2003), 1–6.

[Hig1] G. Higman, Suzuki 2-groups, *Illinois J. Math.* **7** (1963), 79–96.

[Hig2] G. Higman, Enumerating p-groups, I, inequalities, *Proc. London Math. Soc.* **10** (1960), 24–30.

[Hig3] G. Higman, Enumerating p-groups, II, Problems whose solution is PORC, *Proc. London Math. Soc.* **10** (1960), 566–582.

[Hob1] C. Hobby, The Frattini subgroup of a p-group, *Pacific J. Math.* **10** (1960), 209–211.

[Hob2] C. Hobby, A characteristic subgroup of a p-group, *Pacific J. Math.* **10** (1960), 853–858.

[Hob3] C. Hobby, Generalizations of a theorem of N. Blackburn on p-groups, *Illinois J. Math.* **5** (1961), 225–227.

[Hob4] C. Hobby, The derived series of a finite p-group, *Illinois J. Math.* **5** (1961), 228-233.

[Hob5] C. Hobby, Nearly regular p-groups, *Can. J. Math.* **19** (1967), 520–522.

[HW] C. Hobby and C. R. B. Wright, A generalization of a theorem of N. Ito on p-groups, *Proc. Amer. Math. Soc.* **1** (1960), 707–709.

[HogK] G. T. Hogan and W. P. Kappe, On the H_p-problem for finite p-groups, *Proc. Amer. Math. Soc.* **20** (1969), 450–454.

[Hop1] C. Hopkins, Metabelian groups of order p^m, $p > 2$, *Trans. Amer. Math. Soc.* **37** (1935), 161–195.

[Hop2] C. Hopkins, Non-abelian groups whose groups of automorphisms are abelian, *Ann. Math.* **29** (1927/28), 508–520.

[Hua] L. K. Hua, Some "Anzahl" theorems for groups of prime power order, *Sci. Rep. Nat. Tsing Hua Univ.* **4** (1947), 313–327.

[HT1] L. K. Hua and H. F. Tuan, Determination of the groups of odd-prime-power order p^n which contain a cyclic subgroup of index p^2, *Sci. Rep. Nat. Tsing. Hua Univ. A* **4** (1940), 145–154.

[HT2] L. K. Hua and H. F. Tuan, Some "Anzahl" theorems for groups of prime-power orders, *J. Chinese Math.* **2** (1940), 313–319.

[HT] D. R. Hughes and J. G. Thompson, The H_p-problem and the structure of H_p-groups, *Pacif. J. Math.* **9** (1959), 1097–1101.

[Hug] N. J. S. Hughes, The structure and order of the group of central automorphisms of a finite group, *Proc. London Math. Soc.* **53** (1951), 377–385.

[Hum] K. Hummel, The order of the automorphism group of a central product, *Proc. Amer. Math. Soc.* **47** (1975), 37–40.

[Hup1] B. Huppert, *Endliche Gruppen*, Band 1, Springer, Berlin, 1967.

[Hup2] B. Huppert, Über das Produkt von paarweise vertauschbaren zyklischen Gruppen, *Math. Z.* **58** (1953), 243–264.

[HupB] B. Huppert and N. Blackburn, *Finite Groups II*, Springer, Berlin, 1982.

[HupM] B. Huppert and O. Manz, Orbit sizes of p-groups, *Arch. Math.* **54** (1990), 105–110.

[Isa1] I. M. Isaacs, *Character Theory of Finite Groups*, Acad. Press, N.Y., 1976.

[Isa2] I. M. Isaacs, An alternate proof of the Thompson replacement theorem, *J. Algebra* **15** (1970), 149–150.

[Isa3] I. M. Isaacs, The number of generators of a linear p-group, *Can. J. Math.* **24** (1972), 852–858.

[Isa4] I. M. Isaacs, Sets of p-powers as irreducible character degrees, *Proc. Amer. Math. Soc.* **96** (1986), 551–552.

[Isa5] I. M. Isaacs, *Algebra: A Graduate Course*, Brooks/Cole, 1994.

[Isa6] I. M. Isaacs, Commutators and commutator subgroup, *Amer. Math. Monthly* **84** (1977), 720–722.

[Isa7] I. M. Isaacs, Automorphisms fixing elements of prime order in finite groups, *Arch. Math.* **68** (1997), 359–366.

[Isa8] I. M. Isaacs, Equally partitioned groups, *Pacific J. Math.* **49** (1973), 109–116.

[Isa9] I. M. Isaacs, Normal subgroups and nonabelian quotients in p-groups, *J. Algebra* **247** (2002), 231–243.

[Isa10] I. M. Isaacs, Recovering information about a group from its complex group algebra, *Arch. Math.* **47** (1986), 293–295.

[Isa11] I. M. Isaacs, Characters of groups associated with finite algebras, *J. Algebra* **177** (1995), 708–730.

[Isa12] I. M. Isaacs, Groups with many equal classes, *Duke Math. J.* **37** (1970), 501–506.

[Isa13] I. M. Isaacs, Coprime group actions fixing all nonlinear irreducible characters, *Can. J. Math.* **41, 1** (1989), 68–82.

[Isa14] I. M. Isaacs, Large orbits in nilpotent action, *Proc. Amer. Math. Soc.* **127** (1999), 45–50.

[IsM] I. M. Isaacs and A. Moreto, The character degrees and nilpotence class of a p-group, *J. Algebra* **238** (2001), 827–842.

[INW] I. M. Isaacs, G. Navarro, T. R. Wolf, Finite group elements where no irreducible character vanishes, *J. Algebra* **222** (1999), 413–423.

[IsP1] I. M. Isaacs and D. S. Passman, A characterization of groups in terms of the degrees of their characters I, *Pacific J. Math.* **15** (1965), 877–903; II, ibid. **24** (1968), 467–510.

[IsP2] I. M. Isaacs and D. S. Passman, Half-transitive automorphism groups, *Can. J. Math.* **18** (1966), 1243–1250.

[IsR] I. M. Isaacs and G. R. Robinson, On a theorem of Frobenius: solutions of $x^n = 1$ in finite groups, *Amer. Math. Monthly* **99** (1992), 352–354.

[IsS] I. M. Isaacs and M. C. Slattery, Character degree sets that do not bound the class of a p-group, *Proc. Amer. Math. Soc.* **129** (2002), 119–123.

[Ish] K. Ishikawa, On finite p-groups which have only two conjugacy lengths, *Isr. J. Math.* **129** (2002), 119–123.

[Ito1] N. Ito, On the degrees of irreducible representations of a finite group, *Nagoya Math. J.* **3** (1951), 5–6.

[Ito2] N. Ito, On finite groups with given conjugate types, I, *Nagoya Math. J.* **6** (1953), 17–28.

[Ito3] N. Ito, *Lectures on Frobenius and Zassenhaus Groups*, Chicago, 1969.

[Ito4] N. Ito, On a theorem of L. Redei and J. Szep concerning p-groups, *Acta Sci. Math Szeged* **14** (1952), 186–187.

[Ito5] N. Ito, Note on p-groups, *Nagoya Math. J.* **1** (1950), 113-116.

[Ito6] N. Ito, Über das Produkt von zwei zyklischen 2-Gruppen, *Publ. Math. Debrecen* **4** (1956), 517–520.

[Ito7] N. Ito, Über das Produkt von zwei abelschen Gruppen, *Math. Z.* **62** (1955), 400–401.

[Ito8] N. Ito, A conjecture on p-groups, manuscript.

[Ito9] N. Ito, On Hadamard 2-groups, manuscript.

[IM] N. Ito and A. Mann, Counting classes and characters of groups of prime exponent, *Israel J. Math.* **156** (2006), 205–220.

[IO] N. Ito and A. Ohara, Sur les groupes factorisables par deux 2-groupes cycliques, I, II, *Proc. Japan Acad.* **32** (1956), 736–743.

[Iwa] K. Iwasawa, Über die endlichen Gruppen und die Verbände ihrer Untergruppen, *J. Univ. Tokyo* **4** (1941), 171–199.

[Jai] A. Jaikin-Zapirain, On almost regular automorphisms of finite p-groups, *Adv. Math.* **153** (2000), 391–402.

[JNOB] R. James, M. F. Newman and E. A. O'Brien, The groups of order 128, *J. Algebra* **129** (1990), 136–158.

[Jam] R. James, 2-groups of almost maximal class, *J. Austral. Math. Soc. (Ser. A)* **19** (1975), 343–357; corrigendum, ibid **35** (1983), 307.

[Jan1] Z. Janko, Finite 2-groups with small centralizer of an involution, *J. Algebra* **241** (2001), 818–826.

[Jan2] Z. Janko, Finite 2-groups with small centralizer of an involution, 2, *J. Algebra* **245** (2001), 413–429.

[Jan3] Z. Janko, Bemerkung über eine Arbeit von N. Ito, *Glasnik Mat.-Fiz. Astronom. Drustvo Mat. Fiz. Hroatske Ser. II* **11** (1961), 75-77.

[Jan4] Z. Janko, A theorem on nilpotent groups, *Glasnik Mat.-Fiz. Astronom. Drustvo Mat. Fiz. Hroatske Ser. II* **115** (1960), 247-249.

[Jan5] Z. Janko, Finite 2-groups with no normal elementary abelian subgroups of order 8, *J. Algebra* **246** (2001), 951–961.

[Jan6] Z. Janko, Finite 2-groups with a self centralizing elementary abelian subgroup of order 8, *J. Algebra* **269** (2003), 189–214.

[Jan7] Z. Janko, Finite 2-groups G with $|\Omega_2(G)| = 16$, *Glas. Mat.* **40**(60) (2005), 71–86.

[Jan8] Z. Janko, Finite 2-groups with exactly four cyclic subgroups of order 2^n, *J. reine angew. Math.* **566** (2004), 135–181.

[Jan9] Z. Janko, Minimal nonmodular p-groups, *Glas. Mat.* **39** (2004), 221–233.

[Jan10] Z. Janko, 2-groups with self-centralizing subgroup of type $(4, 2)$, *Glas. Mat.* **39** (2004), 235–243.

[Jan11] Z. Janko, Elements of order at most 4 in finite 2-groups, *J. Group Theory* **7** (2004), 431–436.

[Jan12] Z. Janko, The structure of the Burnside group of order 2^{12}, manuscript.

[Jan13] Z. Janko, Nonmodular quaternion-free 2-groups, *Israel J. Math.* **154** (2006), 157–184.

[Jan14] Z. Janko, On maximal cyclic subgroups in finite p-groups, *Math. Z.* **254** (2006), 29–31.

[Jan15] Z. Janko, Minimal non-quaternion-free finite 2-groups, *Israel J. Math.* **154** (2006), 185–189.

[Jan16] Z. Janko, A classification of finite 2-groups with exactly three involutions, *J. Algebra* **291** (2005), 505–533.

[Jan17] Z. Janko, Elements of order at most 4 in finite 2-groups 2, *J. Group Theory* **8** (2005), 683–686.

[Jan18] Z. Janko, Finite p-groups with a uniqueness condition for non-normal subgroups, *Glas. Mat.* **40**(60) (2005), 235–240.

[Jan19] Z. Janko, Finite 2-groups all of whose nonabelian subgroups are generated by involutions, *Math. Z.* **252** (2006), 419–420.

[Jan20] Z. Janko, On finite 2-groups generated with three involutions, manuscript.

[Jan21] Z. Janko, Finite p-groups with $\Omega_2^*(G)$ is metacyclic, *Glas. Mat.* **41**(61) (2006), 71–76.

[Jan22] Z. Janko, New results in the theory of finite p-groups, *Cont. Math.* **402**, 193–195.

[Jan23] Z. Janko, Nonabelian 2-groups in which any two noncommuting elements generate a group of maximal class, *Glas. Mat.* **41**(61) (2006), 271–274.

[Jan24] Z. Janko, On maximal abelian subgroups in finite p-groups, *Math. Z.* **258** (2008), 629–635.

[Jan25] Z. Janko, Finite nonabelian 2-groups all of whose minimal nonabelian subgroups are of exponent 4, *J. Algebra* **315** (2007), 801–808.

[Jan26] Z. Janko, Finite 2-groups with exactly one nonmetacyclic maximal subgroup, *Israel J. Math.* (2008), to appear.

[Jan27] Z. Janko, Finite 2-groups all of whose maximal cyclic subgroups of composite order are self-centralizing, *J. Group Theory* **10** (2007), 1–4.

[Jan28] Z. Janko, Cyclic subgroups of order 4 in finite 2-groups, *Glas. Mat.* **46**(62) (2007), 345–355.

[Jan29] Z. Janko, Some peculiar minimal situations by finite p-groups, *Glas. Mat.* (2008), to appear.

[Jan30] Z. Janko, On minimal nonabelian subgroups of p-groups, *J. Group Theory* (2008), to appear.

[Jan31] Z. Janko, Some exceptional minimal situations by finite p-groups, in: *Ischia Group Theory 2008*, to appear.

[Joh] D. L. Johnson, A property of finite p-groups with trivial multiplicator, *Amer. J. Math.* **98** (1976), 105–108.

[JonK1] D. Jonah and M. W. Konvisser, Abelian subgroups of p-groups, an algebraic approach, *J. Algebra* **34** (1975), 386–402.

[JKon2] D. Jonah and M. W. Konvisser, Some nonabelian p-groups with abelian automorphism groups, *Arch. Math.* **26** (1975), 131–133.

[Jon1] M. R. Jones, Multiplicators of p-groups, *Math. Z.* **127** (1972), 165–166.

[Jon2] M. R. Jones, A property of finite p-groups with trivial multiplicators, *Trans. Amer. Math. Soc.* **210** (1975), 179–183.

[Kal] L. Kaloujnine, La structure des p-groupes de Sylow des groupes symetriques finis, *Ann. Sci Ecole Norm. Super.* **65** (1968), 239–276.

[Kal2] L. Kaloujnine, Zum Problem der Klassifikation der endlichen metabelschen p-Gruppen, *Wiss. Z. Humboldt-Univ. Berlin, Math.-Nat. Reihe* **4** (1955), 1–7.

[Kar] G. Karpilovsky, *Group Representations*, vol. 2, North-Holland, Amsterdam, 1993.

[Kaz1] L. S. Kazarin, Groups with certain conditions for normalizers of subgroups, *Uchen. zapiski Perm Univ.* **218** (1969), 268–279 (in Russian).

[Kaz2] L. S. Kazarin, On some classes of finite groups, *Soviet Math. Dokl.* **12** (1971), no. 2, 549–553 (in Russian).

[Kaz3] L. S. Kazarin, Groups with restrictions on normalizers of subgroups, *Izv. vuzov (mathematics)*, **2** (1973), 41–50 (in Russian).

[Kaz4] L. S. Kazarin, On product of two nilpotent groups, *Questions of group theory and homological algebra* (1981), 62–66; II, ibid (1982), 47–49 (in Russian).

[Kaz5] L. S. Kazarin, On groups with factorization, *Soviet Math. Dokl.* **23** (1981), no. 1, 19–22 (in Russian).

[Keg] O. H. Kegel, Die Nilpotenz der H_p-Gruppen, *Math. Z.* **75** (1960), 373–376.

[Khu] E. I. Khukhro, *Nilpotent Groups and their Automorphisms*, Walter de Gruyter, Berlin, 1993.

[Kim1] I. Kiming, Some remarks on a certain class of finite p-groups, *Math. Scand.* **76** (1995), 35–49.

[Kim] I. Kiming, Structure and derived length of finite p-groups possessing an automorphism of p-power order having exactly p fixed points, *Math. Scand.* **62** (1988), 153–172.

[Kin1] B. W. King, Normal subgroups of groups of prime-power order, in: *Proc. 2nd Int. Con. Theory Groups*, pp. 401–408, Lecture Notes in Math. 372, Springer, Berlin, 1973.

[Kin2] B. W. King, Normal structure of p-groups, *Bull. Aust. Math. Soc.* **10** (1974), 317–318.

[Klu] F. L. Kluempen, The power structure of 2-generator 2-groups of class two, *Algebra Colloq.* **9**, 3 (2002), 287–302.

[Kno] H. G. Knoche, Über den Frobeniusschen Klassenbegriff in nilpotenten Gruppen, I,II, *Math. Z.* **55** (1951), 71–83; ibid **59** (1953), 8–16.

[Kon1] M. W. Konvisser, Embedding of abelian subgroups in p-groups, *Trans. Amer. Math. Soc.* **153** (1971), 469–481.

[Kon2] M. W. Konvisser, 2-groups which contain exactly three involutions, *Math. Z.* **130** (1973), 19–30.

[Kon3] M. W. Konvisser, Metabelian p-groups which contain a self-centralizing element, *Illinois J. Math.* **14** (1970), 650–657.

[KonJ] M. W. Konvisser and D. Jonah, Counting abelian subgroups of p-groups. A projective approach, *J. Algebra* **34** (1975), 309–330.

[KovN] L. G. Kovacs and M. F. Newman, Direct complementation in groups with operators, *Arch. Math.* **13** (1962), 427–433.

[KLG] L. G. Kovacs and C. R. Leedham-Green, Some normally monomial p-groups of maximal class and large derived length, *Quart. J. Math. (2)* **37** (1986), 49–54.

[KM] J. Krempa and I. Malinowska, Groups of p-automorphisms for finite p-groups, *Publ. Math. Debrecen* **61** (2002), no. 3–4, 495–509.

[Kul] A. Kulakoff, Über die Anzahl der eigentlichen Untergruppen und der Elemente von gegebener Ordnung in p-Gruppen, *Math. Ann.* **104** (1931), 779–793.

[KS] H. Kurzweil und B. Stellmacher, *Theorie der endlichen Gruppen. Eine Einführung*, Springer, 1998.

[Laf1] T. J. Laffey, The minimum number of generators of a finite p-group, *Bull. London Math. Soc.* **5** (1973), 288–290.

[Laf2] T. J. Laffey, Bounding the order of a finite p-group, *Proc. R. Ir. Acad. 80a* **2** (1980), 131–134.

[Laf3] T. J. Laffey, A lemma on finite p-groups and some consequences, *Proc. Camb. Philos. Soc.* **75** (1974), 133–137.

[Laf4] T. J. Laffey, Centralizers of elementary abelian subgroups in finite p-groups, *J. Algebra* **51** (1978), 88–96.

[Laf5] T. J. Laffey, The number of solutions of $x^3 = 1$ in a 3-group, *Math. Z.* **149** (1976), no. 1, 43–45.

[Lam1] T.-Y. Lam, Artin exponent of finite groups, *J. Algebra* **9** (1968), 94–119.

[Lam2] T.-Y. Lam, On the number of solutions of $x^{p^k} = a$ in a p-group, *Illinois J. Math.* **32** (1988), 575–583.

[Lan] G. L. Lange, Two-generator Frattini subgroups of finite groups, *Israel J. Math.* **29** (1978), 357–360.

[LGMK] C. R. Leedham-Green and S. McKay, *The Structure of Groups of Prime Power Order*, London Mathematical Monographs, New Series, Oxford Science Publications, Oxford Univ. Press, Oxford, 2002.

[LGNW] C. R. Leedham-Green, P. M. Neumann and J. Wiegold, The breadth and the class of a finite p-group, *J. London Math. Soc. (2)* **1** (1969), 409–420.

[Leo] A. Leone, Finite minimal non-KC-groups, Matematiche **38** (1987), 191–200.

[Leong1] Y. K. Leong, Finite 2-groups of class two with cyclic centre, *J. Aust. Math. Soc. Ser. A* **27** (1979), 125–140.

[Leong2] Y. K. Leong, Odd order nilpotent groups of class two with cyclic centre, *J. Aust. Math. Soc.* **17** (1974), 142–153.

[Lev] F. W. Levi, Groups in which the commutator operations satisfy certain algebraic conditions, J. *Indian Math. Soc.* **6** (1942), 87–97.

[Li1] Li Shirong, The structure of NC-groups, *J. Algebra* **241** (2001), 611–619.

[Li2] Li Shirong, Finite 2-groups with large centralizers of abelian subgroups, *Math. Proc. Roy. Irish Acad.* **A104** (2004), no. 2, 191–197.

[Li3] Li Shirong, The number of conjugacy classes of nonnormal cyclic subgroups in nilpotent groups of odd order, *J. Group Theory* **1** (1998), 165–171.

[Lie1] H. Liebeck, A note an prime-power groups with symmetrical generating relations, *Proc. Cambridge Philos. Soc.* **51** (1955), 594–595.

[Lie2] H. Liebeck, The automorphism group of a finite p-group, *J. Algebra* **4** (1966), 426–432.

[Lie3] H. Liebeck, Outer automorphisms in nilpotent groups of class 2, *J. London Math. Soc.* **40** (1965), 268–275.

[LM] P. Longobardi and M. Maj, On p-groups of breadth two, *Algebra Colloq.* **6** (1999), 121–124.

[LMM] P. Longobardi, M. Maj and A. Mann, Minimal classes and maximal class in p-groups, *Israel J. Math.* **110** (1999), 93–102.

[LubM] A. Lubotzky and A. Mann, Powerful p-groups 1, *J. Algebra* **105** (1987), 484–505.

[Macd1] I. D. Macdonald, Generalizations of a classical theorem on nilpotent groups, *Illinois J. Math.* **8** (1964), 556–570.

[Macd2] I. D. Macdonald, A question of C. R. Hobby on regular p-groups, *Proc. Edin. Math. Soc. (2)* **18** (1973), 207–208.

[Macd3] I. D. Macdonald, Commutators and their products, *Amer. Math. Monthly* **93** (1986), 440–444.

[Macd4] I. D. Macdonald, Finite p-groups with unique maximal classes, *Proc. Edinburgh Math. Soc.* **26** (1983), 233–239.

[Macd5] I. D. Macdonald, The breadth of finite p-groups. I, *Proc. Roy. Soc. Edinburgh* **A78** (1978), 1–39.

[Macd6] I. D. Macdonald, Groups of breadth four have class five, *Glasgow Math. J.* **19** (1978), 141–148.

[Macd7] I. D. Macdonald, Computer results on Burnside groups, *Bull. Aust. Math. Soc.* **9** (1973), 433–438.

[Macd8] I. D. Macdonald, Solution of the Hudges problem for finite groups of class $2p-2$, *Proc. Amer. Math. Soc.* **27** (1971), 39–42.

[Macd9] I. D. Macdonald, Some examples in the theory of groups, in: *Mathematical Essays dedicated to A. J. Macintyre*, pp. 263–269, Ohio Univ. Press, Athens, Ohio, 1970.

[Macd10] I. D. Macdonald, On cyclic commutator subgroups, *J. London Math. Soc.* **38** (1963), 419–422.

[Macd12] I. D. Macdonald, On central series, *Proc Edinburgh Math. Soc. (2)* **3** (1962/63), 175–178.

[MacW] A. R. MacWilliams, On 2-groups with no normal abelian subgroup of rank 3 and their occurrence as Sylow 2-subgroups of finite simple groups, *Trans. Amer. Math. Soc.* **150** (1970), 345–408.

[Mal1] I. Malinowska, Finite p-groups with few automorphisms, *J. Group Theory* **4** (2001), 395–400.

[Mal2] I. Malinowska, p-automorphisms of finite p-groups: problems and questions, in: *Advances in Group Theory (2002)*, pp. 111–127, Napoli, Ilaly, 2002.

[Mal3] I. Malinowska, On quasi-inner automorphisms of a finite p-group, *Publ. Math. Debrecen* **41** (1992), 73–77.

[Mal4] I. Malinowska, On automorphism groups of finite p-groups, *Rend. Sem. Mat. Univ. Padova* **91** (1994), 265–271.

[Man1] A. Mann, Generators of 2-groups, *Israel J. Math.* **10** (1971), 158–159.

[Man2] A. Mann, Regular p-groups, I, *Israel J. Math.* **10** (1971), 471–477.

[Man3] A. Mann, Regular p-groups, II, *Israel J. Math.* **14** (1973), 294–303.

[Man4] A. Mann, Regular p-groups, III, *J. Algebra* **70** (1981), 89–101.

[Man5] A. Mann, The power structure of p-groups I, *J. Algebra* **42** (1976), 121–135; II, ibid **318** (2007), 953–956.

[Man6] A. Mann, Regular p-groups and groups of maximal class, *J. Algebra* **42** (1976), 136–141.

[Man7] A. Mann, Conjugacy classes in finite groups, *Israel J. Math.* **31** (1978), 78–84.

[Man8] A. Mann, Groups with small abelian subgroups, *Arch. Math.* **50** (1988), 210–213.

[Man9] A. Mann, Extreme elements of finite p-groups, *Rend. Sem. Mat. Univ. Padova.* **83** (1990), 45–54.

[Man10] A. Mann, On p-groups whose maximal subgroups are isomorphic, *J. Aust. Math. Soc. A* **59** (1995), 143–147.

[Man11] A. Mann, The number of generators of finite p-groups, *J. Group Theory* **8** (2005), 317–337.

[Man12] A. Mann, Minimal characters of p-groups, *J. Group Theory* **2** (1999), 225–250.

[Man13] A. Mann, On the splitting of extensions by a group of prime order, *Arch. Math.* **56** (1991), 105–106.

[Man14] A. Mann, Some finite groups with large conjugacy classes, *Israel J. Math.* **71** (1990), 55–63.

[Man15] A. Mann, Generators of p-groups, in: *Proceedings of Groups – St. Andrews 1985*, pp. 273–281, Cambridge, 1986.

[Man16] A. Mann, Some applications of powerful p-groups, *Proceedings of Groups – St. Andrews 1989*, pp. 370–385, Cambridge, 1991.

[Man17] A. Mann, A transfer result for powerful Sylow subgroups, *J. Algebra* **178** (1995), 299–301.

[Man18] A. Mann, Finite groups with maximal normalizers, *Illinois J. Math.* **12** (1968), 67–75.

[Man19] A. Mann, Enumerating finite groups and their defining relations, *J. Group Theory* **1** (1998), 59–64.

[Man20] A. Mann, Some questions about p-groups, *J. Aust. Math. Soc. (Series A)* **67** (1999), 356–379.

[Man21] A. Mann, The derived length of p-groups, *J. Algebra* **224** (2000), 263–267.

[Man22] A. Mann, Finite p-Groups, in preparation.

[Man23] A. Mann, Groups generated by elements of small breadth, *J. Group Theory* **4** (2001), 241–246.

[Man24] A. Mann, On the power structure of some p-groups, *Circ. Mat. Palermo II* **23** (1990), 227-235.

[Man25] A. Mann, Groups with few class sizes and the centralizer equality subgroup, *Isr. J. Math.* **142** (2004), 367–380.

[Man26] A. Mann, Philip Hall's 'rather curious' formula for abelian p-groups, *Israel J. Math.* **96B** (1996), 445–448.

[Man27] A. Mann, An inequality for group presentations, *Bull. Aust. Math. Soc.* **62** (2000), 467–469.

[Man28] A. Mann, A remark on class sizes in 2-groups, manuscript.

[Man29] A. Mann, On characters-classes duality and orders of centralizers, *Cont. Math.* **402** (2006), 215–217.

[Man30] A. Mann, Normally monomial p-groups, *J. Algebra* **300** (2006), 2–9.

[Man31] A. Mann, On the exponent of the product of two groups, *Rend. Sem. Mat. Univ. Padova* **115** (2006), 205–207.

[Man32] A. Mann, Elements of minimal breadth in finite p-groups and Lie algebras, *J. Aust. Math. Soc.* **81** (2006), 209–214.

[MM] A. Mann and C. Martinez, The exponent of finite groups, *Arch. Math.* **67** (1996), 8–10.

[MS] A. Mann and C. Scoppola, On p-groups of Frobenius type, *Arch. Math.* **56** (1991), 320–332.

[Mat] S. Mattarei, An example of p-groups with identical character tables and different derived lengths, *Arch. Math.* **62** (1994), 12–20.

[Maz] V. D. Mazurov, 2-groups with an automorphism of odd order fixing all involutions, *Algebra and Logika* **8** (1969), no. 6, 874–885 (in Russian).

[McK] S. McKay, *Finite p-Groups*, Queen Mary Math. Notes 18, London, 2000.

[McKel] A. M. McKelven, Groups of order 2^m that contain cyclic subgroups of order 2^{m-3}, *Amer. Math. Monthly* **13** (1906), 121–136.

[Men] F. Menegazzo, Automorphisms of p-groups with cyclic commutator subgroup, *Rend. Sem. Mat. Padova* **90** (1993), 81–101.

[Mie1] R. J. Miech, Metabelian p-groups of maximal class, *Trans. Amer. Math. Soc.* **152** (1970), 331–373.

[Mie2] R. J. Miech, On p-groups with a cyclic commutator subgroup, *J. Aust. Math. Soc.* **20** (1975), 178–198.

[Mie3] R. J. Miech, The metabelian p-groups of maximal class, *Trans. Amer. Math. Soc.* **236** (1978), 93–119.

[Mie4] R. J. Miech, The metabelian p-groups of maximal class, II, *Trans. Amer. Math. Soc.* **272** (1982), 465–484.

[Mil1] G. A. Miller, An extension of Sylow's theorem, *Proc. London Math. Soc. (2)* **2** (1904), 142–143.

[Mil2] G. A. Miller, Number of abelian subgroups in every prime power group, *Amer. J. Math.* **51** (1929), 31–34.

[Mil3] G. A. Miller, A nonabelian group whose group of isomorphisms is abelian, *Messenger Math.* **43** (1913), 124–125 (or G. A. Miller, *Collected Works*, vol. 5, 415–417).

[Mil4] G. A. Miller, On the groups of order p^m which contain operators of order p^{m-2}, *Trans. Amer. Math. Soc.* **26** (1902), 383–387.

[Mil5] G. A. Miller, Isomorphisms of a group whose order is a power of a prime, *Trans. Amer. Math. Soc.* **12** (1911), 387–402.

[Mil6] G. A. Miller, The groups of order p^m which contain exactly p cyclic subgroups of order p^α, *Trans. Amer. Math. Soc.* **7** (1906), 228–232.

[Mil7] G. A. Miller, Determination of all the groups of order 2^m which contain an odd number of cyclic subgroups of composite order, *Trans. Amer. Math. Soc.* **6** (1905), 58–62.

[Mil8] G. A. Miller, On the holomorph of the cyclic group of order p^m, *Trans. Amer. Math. Soc.* **9** (1908), 232–236.

[Mil9] G. A. Miller, The groups in which every subgroup is either abelian or Hamiltonian, *Trans. Amer. Math. Soc.* **8** (1907), 25–29.

[MilM] G. A. Miller and H. Moreno, Non-abelian groups in which every subgroup is abelian, *Trans. Amer. Math. Soc.* **4** (1903), 398–404.

[Mill] W. H. Mills, The automorphisms of the holomorph of a finite abelian group, *Trans. Amer. Math. Soc.* **85** (1956), 1–34.

[MLC] E. Morgado Morales and M. Lazo Cortis, On the Sylow p-groups of the automorphism group of a finite homocyclic p-group, *Rev. Cienc. Mat.* **6** (1985), 35–44 (in Spanish).

[Mori1] M. Morigi, A note on factorized (finite) p-groups, *Rend. Sem. Math. Univ. Padova* **98** (1997), 101–105.

[Mori2] M. Morigi, Power automorphisms of finite p-groups, *Comm. Algebra* **70** (1999), 4853–4877.

[Mori3] M. Morigi, On the minimal number of generators of finite non-abelian p-groups having an abelian automorphism group, *Comm. Algebra* **23** (1995), 2045–2064.

[Mori4] M. Morigi, On p-groups with abelian automorphism group, *Rend. Sem. Mat. Univ. Padova* **92** (1994), 47–58.

[Nak1] K. Nakamura, Über den Quasinormalteiler der regulären p-Gruppe von der Klasse 2, *Nagoya Math. J.* **26** (1966), 61–67.

[Nak2] K. Nakamura, Über einige Beispiele der Quasinormalteiler einer p-Gruppe, *Nagoya Math. J.* **31** (1968), 97–103.

[Nap1] F. Napolitani, Sui p-gruppi modulari finiti, *Rend. Sem. Mat. Univ. Padova* **39** (1967), 296–303.

[Nap2] F. Napolitani, Gruppi finite minimal non-modulari, *Rend. Sem. Mat. Univ. Padova* **45** (1971), 229–248.

[Nei] L. I. Neikirk, Groups of order p^m which contain cyclic subgroups of order p^{m-3}, *Trans. Amer. Math. Soc.* **6** (1905), 316–325.

[Nek1] K. G. Nekrasov, On finite 2-groups with small Frattini subgroup, in: *Logical-Algebraic Constructions*, pp. 75–82, Tver, 1992 (in Russian).

[Nek2] K. G. Nekrasov, On some 2-groups with a small noncyclic Frattini subgroup, in: *Algebraic and Logical Constructions*, pp. 53–65, Tver, 1994 (in Russian).

[NekB] K. G. Nekrasov and Y. Berkovich, Necessary and sufficient condition for cyclicity of the Frattini subgroup of a finite p-group, in: *Questions of Group Theory and Homological Algebra*, pp. 35–37, Yaroslavl, 1985 (in Russian).

[Neu] B. H. Neumann, On some finite groups with trivial multiplicator, *Publ Math. Debrecen* **4** (1955), 190–194.

[NO] M. F. Newman and E. A. O'Brien, Classifying 2-groups by coclass, *Trans. Amer. Math. Soc.* **351** (1999), 131–169.

[Nin] Y. Ninomiya, Finite p-groups with cyclic subgroups of index p^2, *Math. J. Okayama Univ.* **36** (1994), 1–21.

[Ols] A. Y. Olshanski, The number of generators and orders of abelian subgroups of finite p-groups, *Math. Notes* **23** (1978), 183–185

[Ott] A. D. Otto, Central automorphisms of a finite p-group, *Trans. Amer. Math. Soc.* **125** (1966), 280–287.

[PS] P. P. Palfy and M. Szalay, The distribution of the character degrees of the symmetric p-groups, *Acta Math. Hung.* **41** (1983), 137–150.

[PR] C. Parker and P. Rowley, *Symplectic Amalgams*, Springer, Berlin, 2002.

[PS1] G. Parmeggiani and B. Stellmacher, p-groups of small breadth, *J. Algebra* **213** (1999), 52–68.

[Pas] D. S. Passman, Nonnormal subgroups of p-groups, *J. Algebra* **15** (1970), no. 3, 352–370.

[Pat1] A. R. Patterson (=MacWilliams), On Sylow 2-subgroups with no normal Abelian subgroups of rank 3, in finite fusion-simple groups, *Trans. Amer. Math. Soc.* **187** (1974), 1–67.

[Pat2] A. R. Patterson, The minimal number of generators for p-subgroups of $\mathrm{GL}(n, p)$, *J. Algebra* **32** (1974), 132–140.

[Paz] G. Pazdersky, Prime power groups which are cyclic extensions of elementary Abelian groups, *Math. Nachr.* **97** (1980), 57–68.

[Pet] J. Petrescu, Sur les commutateurs, *Math. Z.* **61** (1954), 348–356.

[Pol] J. Poland, Two problems on finite groups with k conjugate classes, *J. Aust. Math. Soc.* **8** (1968), 49–55.

[Red1] L. Redei, Das schiefe Produkt in der Gruppentheorie, *Comment. Math. Helvet.* **20** (1947), 225–267.

[Red2] L. Redei, Die endlichen einstufig nichtnilpotenten Gruppen, *Publ. Math. Debrecen* **4** (1956), 303–324.

[Rie] J. M. Riedl, Character degrees, class sizes and normal subgroups of a certain class of p-groups, *J. Algebra* **218** (1999), 190–215.

[Rocc] N. R. Rocco, On week commutativity between finite p-groups, *J. Algebra* **76** (1982), 471–488.

[Rock] D. M. Rocke, p-groups with abelian centralizers, *Proc. London Math. Soc. (3)* **30** (1975), 55–75.

[Rod] E. Rodemich, The groups of order 128, *J. Algebra* **67** (1980), 129–142.

[Ron] C. Ronse, On centralizers of involutions in 2-groups, *Math. Ser. Cambridge Philos. Soc.* **86** (1979), 199–204.

[Roi] M. Roitman, Relative indices of elements of finite p-groups, manuscript.

[Roq] P. Roquette, Realisierungen von Darstellungen endlicher nilpotenter Gruppen, Arch. Math. **9** (1958), 241–250.

[Rus] D. J. Rusin, What is the probability that two elements of a finite group commute, *Pacific J. Math.* **82** (1979), 237–247.

[Sag1] I. A. Sagirov, Degrees of irreducible characters of 2-groups of Suzuki, *Math. Notes* **66** (1999), 258–263.

[Sag2] I. A. Sagirov, Degrees of irreducible characters of p-groups of Suzuki $A_p(m, \theta)$, $p > 2$, to appear.

[Sag3] I. A. Sagirov, Finite groups having exactly two degrees of monolithic characters, in: *Questions of Group Theory and Homological Algebra*, pp. 1–8, Univ. Yaroslavl, Yaroslavl, 1998.

[Sak] A. I. Saksonov, Answer on a Brauer question, *Izv. Akad. Nauk BSSR, fiz.-mat. nauki* **1** (1967), 129–130.

[San] P. J. Sanders, The coexponent of a regular p-group, *Comm. Algebra* **28** (2000), 1309–1333.

[SanW] P. J. Sanders and T. S. Wilde, The class and coexponent of a finite p-group, manuscript.

[Sand] P. R. Sanders, The central automorphisms of a finite group, *J. London Math. Soc.* **44** (1969), 225–228.

[Sano] I. N. Sanov, Solution of Burnside's problem for exponent four, *Leningrad State Univ. Ann. Math. Ser.* **10** (1940), 166–170.

[Schm1] P. Schmid, Normal p-subgroups in the group of outer automorphisms of a finite p-group, *Math. Z.* **147** (1976), 271–277.

[Schm2] P. Schmid, Frattinian p-groups, *Geom. Dedicata* **6**, (1990), 359–364.

[Schm3] P. Schmid, On the automorphism group of extraspecial 2-groups, *J. Algebra* **234** (2000), 492–506.

[Sch1] O. Y. Schmidt, A new proof of the theorem of A. Kulakoff in group theory, *Mat. Sb.* **39** (1932), 66–71 (in Russian).

[Sch2] O. Y. Schmidt, Groups all whose subgroups are nilpotent, *Mat. Sb.* **31** (1924), 366–372.

[Scm3] O. Y. Schmidt, Groups having only one class of nonnormal subgroups (Russian), *Mat. Sb.* **33** (1926), 161–172.

[Sch4] O. Y. Schmidt, Groups with two classes of nonnormal subgroups (Russian), *Proc. Seminar on Group Theory* (1938), 7–26.

[Schn1] C. Schneider, On the derived subgroup of a finite p-group, *Austral. Math. Soc. Gaz.* **26** (1999), 232–237.

[Schn2] C. Schneider, Groups of prime-power order with a small second derived quotient, *J. Algebra* **286** (2003), 539–551.

[Schr] O. Schreier, Über die Erweiterung von Gruppen, I, *Monatsh. Math, Physik* **34** (1926), 165–180; II, *Abh. Math. Sem. Univ. Hamburg* **4** (1926), 321–346.

[Sco] C. M. Scoppola, Groups of prime power order as Frobenius-Wielandt complements, *Trans. Amer. Math. Soc.* **325** (1991), 855–874.

[Scot] W. R. Scott, *Group Theory*, Prentice Hall, 1964.

[Sei1] G. Seitz, Finite groups having only one irreducible representation of degree greater than one, *Proc. Amer. Math. Soc.* **19** (1968), 459–461.

[Sha1] A. Shalev, The structure of finite p-groups: effective proof of coclass conjectures, *Invent. Math.* **115** (1994), 315–345.

[Sha2] A. Shalev, Finite p-groups, in: *Finite and locally finite groups*, pp. 401–450, Kluwer Acad. Publ., Dodrecht, 1995.

[She] V. A. Sheriev, A description of the class of finite p-groups whose 2-maximal subgroups are abelian, in: *Proc. Sem. Algebraic Systems* **2**, pp. 25–76, Krasnoyarsk, 1970 (in Russian).

[Shu] P. Shumyatsky, Involutory automorphisms of finite groups and their centralizers, *Arch. Math.* **71** (1998), 425–432.

[Sim] C. C. Sims, Enumerating p-groups, *Proc. London Math. Soc.* **15** (1965), 151–166.

[Sla1] M. C. Slattery, Character degrees of finite p-groups, in: *The Arcata Conf. on Representations of Finite Groups*, pp. 89–92, Proc. Symp. Pure Math. 47, Part 2, Amer. Math. Soc., Providence, RI, 1987.

[Sla2] M. C. Slattery, Character degrees and nilpotent class in p-groups, *J. Aust. Math. Soc. (Series A)* **57** (1994), 76–80.

[Sla3] M. C. Slattery, Character degrees and derived length in p-groups, *Glasgow. Math. J.* **30** (1988), 221–230.

[Sla4] M. C. Slattery, Computing character degrees in p-groups, *J. Symb. Comput.* **2** (1986), 51–58

[Spe] W. Specht, Isomorphic subgroups of finite p-groups revisited, *Canad. J. Math.* **26** (1974), 574–579.

[SS] E. G. Straus and G. Szekeres, On a problem of D. R. Hughes, *Proc. Amer. Math. Soc.* **9** (1958), 157–158.

[Str] R. R. Struik, Some nonabelian 2-groups with abelian automorphism groups, *Arch. Math.* **39** (1982), 299–302.

[Suz1] M. Suzuki, *Group Theory I, II*, Springer, Berlin, 1982, 1986.

[Suz2] M. Suzuki, *Structure of a Group and the Structure of its Lattice of Subgroups*, Ergebnisse der Mathematik und ihrer Grenzgebiete 10, Springer, Berlin, 1956.

[Tau] O. Taussky, Remark on the class field tower, *J. London Math. Soc.* **12** (1937), 82–85.

[Tes] L. Teschke, Über die Normalteiler der p-Sylowgruppe der symmetrischen Gruppe vom Grade p^m, *Math. Nachr.* **87** (1979), 197–212.

[Tho1] J. G. Thompson, A replacement theorem for p-groups and a conjecture, *J. Algebra* **13** (1969), 149–151.

[Tho2] J. G. Thompson, Finite groups with fixed-point-free automorphisms of prime order, *Proc. Nat. Acad. Sci. USA* **45** (1959), 578–581.

[Tho3] J. G. Thompson, Nonsolvable finite groups all of whose local subgroups are solvable, 1, *Bull. Amer. Math. Soc.* **74** (1968), 383–437.

[Tho4] J. G. Thompson, Fixed points of p-groups acting on p-groups, *Math. Z.* **86** (1964), 12–13.

[Tho5] J. G. Thompson, Centralizers of elements in p-groups, *Math. Z.* **96** (1967), 292–293.

[Tow] M. J. Towers, *Modular Representations of p-Groups*, PhD Thesis, Hertford College, University of Oxford, 2005.

[Tua1] H. F. Tuan, A theorem about p-groups with abelian subgroup of index p, *Acad. Sinica Science Record* **3** (1950), 17–23.

[Tua2] H. F. Tuan, An Anzahl theorem of Kulakoff's type for p-groups, *Sci. Rep. Nat. Tsing-Hua Univ. A* **5** (1948), 182–189.

[Ust] A. D. Ustjuzaninov, Finite 2-groups in which the set of self-centralizing abelian normal subgroups of rank ≥ 3 is empty ($SCN_3(2) = \varnothing$), *Izv. Akad. Nauk SSSR* **37** (1973), 251–283 (in Russian).

[Ust2] A. D. Ustjuzaninov, Finite 2-groups with three involutions, *Sibirsk. Mat. Z.* **13** (1972), 182–197.

[VL] M. R. Vaughan-Lee, Breadth and commutator subgroups of p-groups, *J. Algebra* **32** (1976), 278–285 (in Russian).

[VLW1] M. R. Vaughan-Lee and J. Wiegold, Breadth, class and commutator subgroups of p-groups, *J. Algebra* **32** (1974), 268–277.

[Ver] L. Verardi, A class of finite groups of exponent p in which every normal subgroup is characteristic, *Boll. Un. Mat. Ital. B (6)* **4** (1988), 307–317.

[Waa] R. W. van der Waall, On finite p-groups whose commutator subgroups are cyclic, *Indag. Math.* **35** (1973), 342–345.

[Wal] G. E. Wall, On Hughes' H_p-problem, in: *Proc. Internat. Conf. Theory of Groups (Canberra, 1965)*, pp. 357–362, Gordon and Breach, New York 1967.

[Wal2] G. E. Wall, Finite groups with class-preserving outer automorphisms, *J. London Math. Soc.* **22** (1947), 315–320.

[Wal3] G. E. Wall, Secretive prime-power groups of large rank, *Bull. Austral. Math. Soc.* **12** (1975), 963–969.

[War] H. N. Ward, Automorphisms of quaternion-free 2-groups, *Math. Z.* **112** (1969), 52–58.

[Web1] U. H. M. Webb, An elementary proof of Gaschütz theorem, *Arch. Math.* **35** (1980), 23–26.

[Web2] U. H. M. Webb, The number of stem covers of an elementary abelian p-group, *Math. Z.* **182** (1983), no. 3, 327–337.

[Web3] U. H. M. Webb, On the rank of a p-group of class 2, *Canad. Math. Bull.* **26** (1983), 101–105.

[Web4] U. H. M. Webb, The Schur multiplier of a nilpotent group, *Trans. Amer. Math. Soc.* **291** (1985), 755–763.

[Wei] P. M. Weichsel, On isoclinism, *J. London Math. Soc.* **38** (1963), 63–65.

[Weir1] A. Weir, Sylow p-subgroups of the classical groups over finite fields with characteristic prime to p, *Proc. Amer. Math. Soc.* **6** (1955), 529–533.

[Weir] A. Weir, The Sylow subgroups of the symmetric groups, *Proc. Amer. Math. Soc.* **6** (1955), 534–541.

[Wie1] J. Wiegold, Multiplicators and groups with finite central factor-groups, *Math. Z.* **89** (1965), 245–247.

[Wie2] J. Wiegold, The Schur multiplier: an elementary approach, in: Groups – St. Andrews, 1981, pp. 137–154, London Math. Soc. Lect. Notes 71, Cambridge, 1982.

[Wie3] J. Wiegold, Commutator subgroups of finite p-groups, *J. Aust. Math. Soc.* **10** (1969), 480–484.

[Wil1] B. Wilkens, On quaternion-free 2-groups, *J. Algebra* **258** (2002), 477–492.

[Wil2] B. Wilkens, On the upper exponent of a finite p-group, *J. Algebra* **277** (2004), 249–263.

[Wil3] B. Wilkens, 2-groups of breadth 3, *J. Algebra* **318** (2007), 202–224.

[Wilk] D. F. Wilkinson, The groups of order p^7 (p any prime), *J. Algebra* **118** (1988), 109–119.

[Wils] L. Wilson, On the power structure of powerful p-groups, *J. Group Theory* **5** (2002), no. 2, 129–144.

[Xu] M. Y. Xu, Regular p-groups and their generalizations, manuscript.

[XZA] M. Y. Xu, Q. Zhang and L.-J. An, Finite p-groups all of whose nonabelian subgroups are generated by two elements, in preparation.

[Zap] G. Zappa, Finite groups in which all nonnormal subgroups have the same order II (Italian), *Atti Accad. Naz. Lincei Cl. Sci. Fiz. Mat. Natur. Rend. Lincei (9) Mat. Appl.* **14** (2003), no. 1, 13–21.

[ZAX] Q. Zhang, L.-J. An and M. Y. Xu, Finite p-groups all of whose non-abelian proper subgroups are metacyclic, *Arch. Math.* **87** (2006), 1–5.

[Zha] J. P. Zhang, Finite groups with many conjugate elements, *J. Algebra* **170** (1994), 608–624.

[Zhm1] E. M. Zhmud, Finite groups with uniquely generated normal subgroups, *Mat. Sb.* **72 (114)** (1967), 135–147 (in Russian).

[Zhm2] E. M. Zhmud, On the multiplier of a finite group with nontrivial center, *Ukrainian Math. J.* **47** (1995), 546–550 (in Russian).

[Zhm3] E. M. Zhmud, Symplectic geometries and projective representations of finite abelian groups, *Mat. Sb.* **87 (129)** (1971) 3–17 (in Russian).

[Zhm4] E. M. Zhmud, Symplectic geometries on finite abelian groups, *Math. Sb.* **86** (1972), 9–33 (in Russian).

[Zhm5] E. M. Zhmud, The isomorphisms of the lattice of normal subgroups of a finite nilpotent group, *Vestnik Kharkov Univ.* **26** (1967), 3–17 (in Russian).

[Zho] X. Zhou, On the order of Schur multipliers of finite p-groups, *Comm. Algebra* **22** (1994), 1–8.

Author index

A

Alperin, J. L., §§1, 10, 37, 39

B

Baer, R., §§20, 21
Baginski, C., §§33, 43
Bannusher, W., §22
Bender, H., Introduction, §10
Berkovich, V., §33
Berkovich, Y., §§1, 2, 4–6, 8, 10, 12–14, 17–19, 21–24, 31, 33, 36, 38, 39, 41–45, Appendices 5, 8, 9, 11, 13
Berman, S. D., Appendix 2
Blackburn, N., §§1, 7, 9–13, 15, 21, 23, 36, 41, 42, 44, Appendices 6, 9, 13
Bozikov, Z., §§10, 16, 27, Appendix 14
Brodkey, J. S., Appendix 3

D

Dade, E. C., Appendix 2
Dedekind, R., §1
Dolfi, S., Introduction

E

Evens, L., §21

F

Fan, Y., §5
Feit, W., §§1, 10, 37
Fitting, G., Introduction, §6
Frame, J. S., §22
Freiman, G., Introduction, Appendix 4
Frobenius, G., Introduction

G

Gagola, S., Appendix 3
Gaschütz, W., §§1, 32
Gillam, J. D., §39
Glauberman, G., §§10, 34, 39, Problems
Golfand, Y. A., §10, Appendix 22
Gorenstein, D., §4, 9, 10
Grün, O., §1

H

Hall, M., §29
Hall, P., Introduction, §§1–7, 9, 23, 24, 26, 29, 34, Appendices 1, 6, 13, 15
Harris, M., §6
Heineken, H., §20
Hering, C., §4
Herzog, M., §1
Hethelyi, L., §26
Hobby, C., §1
Hua, L. K., §5
Huppert, B., §§9, 36, 44, Appendix 9

I

Isaacs, I. M., Introduction, §§1, 2, 10, 20–22, Problems
Ito, N., Introduction, §§1, 4, 22, 36, Appendices 3, 10

J

Janko, Z., §§1, 10, 16, 22, 23, 27, 28, 34-36, Appendices 12, 14, Problems
Johnson, D. L., §21
Jonah, D., §10

K

Kaloujnine, L., Introduction, §34
Karpilovsky, G., §§21
Kazarin, L. S., §§1, 2, 4, 22, Problems
Khukhro, E., Problems
King, M., §1
Knoche, H. G., §2
Konvisser, M., §§10,39,43
Kronecker, L., Introduction
Kulakoff, A., §§1, 5

L

Levai, L., §26
Lewis, M., Appendix 3
Longobardi, P., §3
Lubotzky, A., §26
Lyons, R., §§9, 10

M

MacWilliams, A., §§5, 37
Maj, M., §3
Malinowska, I., §§33, 43
Mann, A., §§1–3, 5, 7, 9–11, 13, 15, 20–24, 26, 39, 40, Appendices 2, 6, 10, Problems
Maschke, H., §6, 8
Matsuyama, H., Introduction
Miller, G. A., Introduction, §§1, 13, 33, 34, 45
Moreto, A., §2
Morigi, M., §1

N

Navarro, G., Appendix 3
Nekrasov, K. G., §§4, 18
Neumann, B. H., Introduction, §21

O

Ohara, A., §36

P

Parker, C., §1
Passman, D. S., §§1, 16, 21, 22
Petrescu, J., Appendix 1
Poland, J., §2

R

Redei, L., §1
Robinson, G. R., Introduction
Rocke, D. M., §27
Roitman, M., §34, Appendix 7
Roquette, P., §1
Rowley, P., §1

S

Sanders, P. J., §12
Schmid, P., §32
Schmidt, O. Y., §10
Schneider, C., §1, Appendix 6
Schur, I., §21
Senior, J., §29
Solomon, R., §§9, 10
Stickelberger, L., Introduction
Struik, R. R., §34
Suzuki, M., §§1
Sylow, L., Introduction, §1
Szep, J., Introduction

T

Taunt, D., §6
Taussky, O., §§1, 36
Thompson, J. G., §§1, 4, 14, 15, 37, Appendix 9
Tobin, S., §1
Tuan, H. F., §§1, 5, Problems

W

van der Waall, R., Appendix 2
Webb, U. H. M., §32
Weichsel, P., §29
Wiegold, J., §21
Wielandt, H., Introduction, §7
Wilde, T. S., §12
Wilkens, B., §26, Problems
Wilson, L., §26
Witt, E., Introduction

Z

Zassenhaus, H., §9
Zhmud, E. M., Introduction, §22

Subject index

A

abelian group, Introduction, §6, Appendix 19
abelian subgroups, §35
abelian subgroups of extraspecial p-groups, §4
abelian subgroups, the number of, §§4, 10, 20, 39
absolutely regular maximal subgroup, regularity of §§9, 12
absolutely regular p-group, §§7, 9
Alperin's conjecture, §39
action of a group, §§1, 8, 10
$\mho_n(G)$, Introduction, §7
$\mho^n(G)$, §§23,24, 36
$\Omega_n(G)$, Introduction
automorphisms groups of 2-groups with cyclic subgroup of index 2, §34
automorphism group, order of, §§1, 6, 33, 34
automorphism groups of abelian p-groups, order of, §6
automorphisms of large orders, §33

B

basic theorem on abelian groups, Introduction
branch, sth branch, §29
breadth of an element, of a group, §40, Appendix 18
broad subgroup, §15

C

capable group, §21
$c_n(G)$, the number of cyclic subgroups of order p^n, §§1, 5, 13
center of a group, §20
centralizers of elements, §§2, 10
central product, §4
character, Introduction, §§2, 22, 37
character degrees, Introduction
characteristic subgroups, §§4, 9, 14
characteristic subgroups of abelian p-groups, §6
characteristic subgroups of exponent p in irregular p-groups, the order of, §9
characterization of extraspecial p-groups in the terms of class numbers, §2
characterization of metacyclic p-groups, §9, 36, 43, 44
characterizations of p-groups of maximal class, §§1, 2, 5, 9, 10, 12, 13, 36
characterization of Σ_{p^n} as group with maximal possible subgroups of given order $> p$, §5
class of a p-group, §§1, 7, 9, 34
class number, Introduction, §2
Clifford's decomposition, Introduction
Clifford's theorem, Introduction
coexponent of a group, §12
coexponent and class, §12
collecting process, Appendix 1
commutator identities, Introduction
commutator identities for p-groups of class 2, Introduction
commutator (= derived) subgroup, §§1, 4, 7, Appendix 6
components of a group, Introduction
conjecture $A(s)$, §17
conjecture $B(s)$, §18
conjugacy class, §2
counting theorems, §§1, 5, 10, 12, 13, 17–19, 45
counting theorems for groups of exponent p, §5

counting theorems for p-groups of maximal class, §12
covering group (= representation group), §21
critical subgroup of solvable group, of p-group, §14
cyclic Frattini subgroup, §§1, 4
cyclic maximal subgroup, §1
cyclic subgroups, the number of §§1, 4, 5, 12, 13, 18, 19

D

D_{2^n}, the dihedral group of order 2^n, §1
decomposition of abelian p-group under p'-group of operators, §6
Dedekindian groups, structure of, §1
derived length, §22
derived subgroup of order p^4, Appendix 6
dihedral group, §1
direct decomposition of p-groups, §§6, 8

E

elementary abelian subgroups, the number of, characterizations of, §§1, 5, 10, 13
enumeration principles for subgroups, §5
enumeration principle for normal subgroups, §12
$\eta(G)$, §§1, 9
existence of subgroups of given exponent, §23
exponent of a group, §§6, 7–9, 13, 23, 24, 26
extraspecial p-groups, decomposition in a central product, abelian subgroups of, §§1, 4, §22

F

factorized groups, §1
faithful irreducible character, §22
family of isoclinic groups, family rank, §29
Fan's characterization of elementary abelian p-groups, a new proof of, §5
Fitting subgroup, Introduction
Fitting's lemma on action, §6
Fitting's lemma on the class of product of two normal nilpotent subgroups, §1
fixed-point-free (= regular) automorphism, §10
Frattini subgroup, §§1, 4, Appendix 8
Freiman's number-theoretic theorems, Appendix 4
Frobenius theorem on the number of solutions to $x^n = 1$ in a group, Isaacs–Robinson's proof of, Introduction
Frobenius reciprocity, Introduction
$\varphi_2(G)$, §2
fundamental subgroup of a p-group of maximal class, §§9, 12

G

Gaschütz' theorem on outer automorphisms, §32
generalized homocyclic group, §8
generalized quaternion group, §1
generators of p-groups, the number of, §§1, 11, 15, 26
groups in which the intersection of all nonnormal subgroups is nontrivial, §1
groups of order p^4, §10
groups with large measure of commutativity, §38
Grün's lemma, §1

H

Hall chains, §24
Hall–Petrescu identity, §7, Appendix 1
Hall–Witt commutator identity, Introduction
Hall's enumeration principle, §5
Hall's regularity criteria, §9
Hall's theorem on the class number, Mann's proof of, §2
Hall's theorems on regular p-groups, §7
Hall's theorem on the stability group of a chain of subgroups, §34
homocyclic group, §6

I

image and kernel of homomorphism, §1
induced character, Introduction

inertia subgroup of a character, Introduction
intersection of nonnormal subgroups, §1
intersections of subgroups, §42
irreducible character, Introduction, §2
irreducible representation, Introduction
irregularity of p-groups of maximal class, §9
Isaacs–Moreto's theorems, §2
Isaacs' theorems on actions, Appendix 3
isoclinic family, §29
isoclinic groups, isoclinism, §29
Ito–Ohara's theorem on product of two cyclic 2-subgroups, §36, Appendix 13
Ito's theorem on degrees, Introduction
Ito's theorem on product of two abelian groups, §1
Iwasawa'a theorem on modular p-groups, §44

K

Kaloujnine's theorem on the stability group of a normal chain, §34
kernel of a character, of a representation, Introduction
kernel of a homomorphism, §1
King's theorems, §1
Knoche's theorems, §40
Kulakoff's theorem, new proof of, §§1, 5

L

lattice isomorphism, projectivity, §25
lattice isomorphisms of p-groups, solvable groups, §25
lower central series, Introduction
lower pyramidal p-groups, §8

M

M_{p^n}, §1
MacWilliams' theorems, §§5, 37
major subgroup, §5
Mann's counting theorems, §§2, 5
Mann's regularity criterion, §11
Mann's theorems, §§1, 2, 5, 7, 9, 11
Mann's proof of monomiality of p-groups, Appendix 2
Mann subgroup, §7
Maschke's theorem, for abelian p-groups, for arbitrary groups, for pyramidal p-groups, generalizations, §§6, 8
maximal chains, the number of, §5
maximal class quotient groups, §§9, 12, 13
maximal subgroups, Introduction, §5
maximal subgroups of p-groups of maximal class, §9
measure of commutativity mc(M), §2
metacyclic group, §§9, 10, 43, 44
metacyclic subgroups, the number of, §10
$\mathcal{M}(G)$, §1
$\mho^n(G)$, §§23, 24
Miller's example of 2-groups G with abelian Aut(G), §34
Miller's theorems, §§1, 10, 45
minimal characters, the number of, §22
minimal classes, §3
minimal irregular p-groups, §§7, 11
minimal nonabelian p-group (subgroup), §§1, 10, 35
minimal nonmetacyclic p-groups, §41
minimal nonmodular p-groups, §44
minimal nonnilpotent groups, §10
minimal non-$\mathcal{P}_i$-groups, §11
minimal number of generators, §§1, 14
minimal order of p-groups with given derived length, Appendix 6
modular law, Introduction
modular p-groups, classification of, §44
monomiality of p-groups, Mann's proof, Appendix 2
multiplier (= Schur multiplier), the order of, §21

N

N/C-theorem, Introduction
nilpotence class, Introduction, §§1, 12
nilpotent groups, Introduction
nonlinear irreducible characters, the number of, §2, 22
nonnormal subgroups are cyclic, §16
normalizer of a Sylow p-subgroup in the symmetric group, Appendix 13
normal subgroups, §§1, 3, 5, 6, 9, 10, 24
number of abelian subgroups, §10

number of absolutely regular subgroups of given order, §13
number of commuting pairs of elements, §2
number of cyclic subgroups, of order p, $p^n > p$, §§1, 5, 12, 13, 18
number of elementary abelian subgroups of orders p^2, p^3 and p^4, §§1, 10
number of generators, §1, 14
number of L_s-subgroups, §§17, 18
number of thin members in the set Γ_2, §18
number of members of maximal class in the set Γ_1, §13
number of metacyclic subgroups of given order, §10
number of minimal characters, characterization of extraspecial groups in terms of, Appendix 10
number of nonabelian subgroups of order p^3, §2
number of subgroups of given order in a group of exponent p, §5
number of two-generator subgroups of given order in a group of exponent p, §5
number of subgroups of maximal class of given order, §§1, 13, Appendix 5
number of subgroups of maximal class containing a given irregular p-group of maximal class, §13
number of subgroups of order p^p and exponent p, §13

O

$\Omega_n(G)$, Introduction
$\Omega_n^*(G)$, §43
one-stepped p-groups, §1
ordinary quaternion group, §1
orthogonality relations, Introduction
outer p-automorphisms, Gaschütz's theorem, §32

P

partitions of p-groups, §20
$\mathcal{P}$- and $\mathcal{P}_i$-groups, $i = 1, 2, 3$, §§11, 24
p'-automorphisms acting identically on $G/\Phi(G)$, §1
p-groups all of whose characteristic abelian subgroups are cyclic, §4
p-groups all of whose characteristic subgroups are two-generator, §44
p-groups all of whose nonnormal subgroups have order p, §1
p-groups all of whose nonnormal subgroups are cyclic, §16
p-groups all of whose subgroups of composite exponent are normal, §1
p-groups all of whose noncyclic subgroups of the same order have the same rank, §41
p-groups all of whose nonnormal subgroups have order p, §1
p-groups G in which $\Omega_2^*(G)$ is absolutely regular, §13
p-groups of maximal class, characterizations of, §§1, 2, 5, 9, 10, 12, 13, 16, 18, 19, 23, 36
p-groups of maximal class, structure of, §§1, 9
p-groups of odd order without normal subgroups of order p^p and exponent p, §12
p-groups of odd order without normal elementary abelian subgroup of order p^3, §13
p-group with G abelian Aut(G), §34
p-groups with abelian subgroup of index p, §29
p-groups with absolutely regular maximal subgroup, §12
p-groups G with $c_1(G) = 1 + p + \cdots + p^{p-2} \pmod{p^p}$, §13
p-groups G with $c_1(G) = 1 + p + \cdots + p^p$, §13
p-groups G with $p^{p-1} \nmid c_k(G)$ $(k > 1)$. §13
p-groups with cyclic subgroup of index p, §1
p-groups with cyclic Frattini subgroup, §4
p-groups with derived subgroup of order p, §4

p-groups with exactly k class sizes, §7
p-groups with exactly $p+1$ subgroups of order p^p and exponent p, §13
p-groups with few cyclic subgroups of order p^2, §19
p-groups with few nonabelian subgroups of order p^p and exponent p, Appendix 13
p-groups with few nonlinear irreducible characters, §22
p-groups G with $G/\mho_2(G)$ of maximal class, §36
p-groups with large derived subgroup, §21
p-groups G with large exp(Aut(G)), §33
p-groups with large Schur multiplier, §21
p-groups with nonabelian G', $|G'/G''| = p^3$, Appendix 6
p-group with nonabelian self centralizer of order p^3, §10
p-groups with $\Omega_2^*(G)$ absolutely regular, §13
p-groups G with $\Omega_1(G)$ ($\Omega_1^*(G)$) of maximal class, §13
p-groups with $\Omega_2^*(G)$ of maximal class, §13
p-groups with only one abelian subgroup of order p^3, §10
p-groups with small abelian subgroups, §20
p-groups with small operator p'-groups, §31
p-groups G with subgroup H such that $G' = H'\mathrm{K}_3(G)$, §1
p-groups with subgroup of maximal class and index p, §12
p-groups without normal elementary abelian subgroup of order p^2, §1
p-groups without normal subgroup of order p^p and exponent p, §12
p-groups with trivial Schur multiplier, §21
p-subgroups of symmetric groups, Appendix 13
Passman's characterization of Dedekindian p-groups, §1
Passman's theorems on groups close to Dedekindian, §1
power closed p-group, §26
power structure of p-groups, §§7, 8, 9, 11, 26
powerful groups, numbers of generators of subgroups of, §26
powerfully embedded subgroups, §26
principal character, Introduction
product formula, Introduction
products of cyclic p-subgroups, §26
products of subgroups, §1
pyramidal p-groups, §8

Q

Q-group, §1
Q_{2^n}, the generalized quaternion group of order 2^n, subgroup structure of, §1
quasikernel of a character, Introduction

R

regular character, decomposition of, Introduction
regular p-groups, properties of, §§7–9, 11, 17
regularity criteria, §§7, 9
relative index of an element, Appendix 7
representation groups of metacyclic p-groups, §47
representation of a group, Introduction
Roitman's theorems, Appendix 7

S

Schneider theorem, §1
SD_{2^n}, the semidihedral group of order 2^n, subgroup structure of, §1
Schmid's theorem on outer p-automorphism, Webb's proof of, §32
Schur multiplier, the order of, §21
section of a group, §§7, 11
sectional rank, §26
semidihedral group, §1
semidirect product, §1
short enumeration identity, §5
sizes of classes, §2
small self centralizers, §§1, 10
socle of a group, §13
special p-groups, §14, 26
stem, §29

structure of a p-group with absolutely regular subgroup of index p, §12
structure of a p-group with subgroup of maximal class and index p, §12
subgroups, the number of, §§1, 5, 9, 10, 12, 13, 17–19, Appendix 2
subgroups of given exponent, the order of, §23
subgroups of maximal class, the properties of, the number of, §§5, 9, 12, 13
Suzuki's theorem on small centralizer, §1

T

Taussky's theorem, §§1, 36
2-groups of maximal class, §1
Thompson's replacement theorem, Isaacs' proof of, §3
Thompson's theorems, §§3, 10, 14, 15
Three Subgroups Lemma, Introduction
two-generator G-invariant subgroups of $\Phi(G)$, §44
2-groups G with abelian $\mathrm{Aut}(G)$, Miller example, §44
2-groups with an involution contained in only one subgroup of order 4, Appendix 14
2-groups G with nonabelian G' of order 16, Appendix 6
2-groups G with $c_n(G) = 2, n > 1$, §44
2-groups with 7 and 11 involutions, §43
2-groups with small abelian subgroups, §§20, 35
2-groups with three involutions, §43

U

upper central series, Introduction, §1
upper pyramidal p-groups, §8

W

Wielandt's example, §7

Z

Zassenhaus' identity, §9

Bei Fragen zur Produktsicherheit wenden Sie sich bitte an:
If you have any questions regarding product safety,
please contact:

Walter de Gruyter GmbH
Genthiner Straße 13
10785 Berlin
productsafety@degruyterbrill.com